QUANTUM FIELDS

Quantum Fields

From the Hubble to the Planck Scale

MICHAEL KACHELRIESS

Department of Physics
The Norwegian University of Science and Technology, Trondheim

OXFORD
UNIVERSITY PRESS

OXFORD
UNIVERSITY PRESS

Great Clarendon Street, Oxford, OX2 6DP,
United Kingdom

Oxford University Press is a department of the University of Oxford.
It furthers the University's objective of excellence in research, scholarship,
and education by publishing worldwide. Oxford is a registered trade mark of
Oxford University Press in the UK and in certain other countries

First Edition published in 2018
Impression: 2

Published in the United States of America by Oxford University Press
198 Madison Avenue, New York, NY 10016, United States of America

British Library Cataloguing in Publication Data
Data available

Library of Congress Control Number: 2017943891

ISBN 978–0–19–880287–7

DOI 10.1093/oso/9780198802877.001.0001

Printed and bound by
CPI Group (UK) Ltd, Croydon, CR0 4YY

Preface

Why this book? The number of excellent introductory books on quantum field theory and on cosmology has grown much in the last years. Teaching a one-semester course on Gravitation and Cosmology and a one-year course on Quantum Field Theory (QFT) since 2009, I profited enormously from these textbooks. Working out my own lectures, I tried however to teach the two courses in a more unified manner than is usually done. One motivation for doing so was the belief that studying a subject in depth is only half the premise; the remaining—and not least—struggle is to put the pieces into a comprehensible picture. This is particularly true for students who aim to work at the interface between theoretical particle physics, cosmology and astroparticle physics. Thus I tried to stress the basic principles and methods with which rather dispersed phenomena in these fields can be analysed. Moreover, this approach saves also time and makes it thus possible to discuss additional applications within the restricted time for lectures.

This book reflects this approach and aims to introduce QFT together with its most important applications to processes in our universe in a coherent framework. As in many modern textbooks, the more universal path-integral approach is used right from the beginning. Massless spin one and two fields are introduced on an equal footing, and gravity is presented as a gauge theory in close analogy with the Yang–Mills case. Concepts relevant to modern research as helicity methods, effective theories, decoupling, or the stability of the electroweak vacuum are introduced. Various applications as topological defects, dark matter, baryogenesis, processes in external gravitational fields, inflation and black holes help students to bridge the gap between undergraduate courses and the research literature.

How to use this book. I tried to present all derivations in such detail that the book can be used for self-studies. It should be accessible for students with a solid knowledge of calculus, classical mechanics, electrodynamics including special relativity and quantum mechanics. As always, it is indispensable to work through the text and the exercises to get a grip on the material. Although the book is written with the intention to be read from cover to cover, time constraints and special interests will typically push students to omit several topics in a first round. A chart showing the interdependence of the chapters is shown below.

Additionally to being suitable for self-study, the book may serve as basis for a course in quantum field theory or an advanced course in astroparticle physics and cosmology. For a standard two-semester course on QFT, one can use chapters 2–12 plus, depending on preferences and the time budget, material from chapters 13–18. For an advanced course in astroparticle physics and cosmology, one may select suitable chapters from the second half of the book. The order of some of the topics in the book may be reshuffled:

- Section 4.3 introduces some basic tools needed to perform loop calculations and applies them to three examples. If one prefers a more systematic approach, this section could be shifted to the end of section 11.4.

- Section 5.3 discusses symmetries on the quantum level. It could be postponed and used as introductory section to chapter 17.

- Chapter 8 and 10 on fermions and on gauge theories could be omitted in a first round, restricting the discussion of renormalisation to the scalar case.

- Chapter 9 on scattering is rather independent of the main text. While the prediction of scattering cross-sections is the main occupation of most theorists working in particle physics, it will be needed only rarely in the latter parts of the text. Section 9.1 introduces the optical theorem which will be applied in chapter 14 and 21. Section 9.4 is useful as preparation for chapter 18, and explains why we consider only fields with spin $s \leq 2$.

A minimal path through the QFT oriented chapters is shown graphically in the first two rows, with round boxes denoting material that could be omitted in a first round and shifted to latter places. The two lines at the bottom show a similar path collecting the chapters discussing gravitation and cosmology.

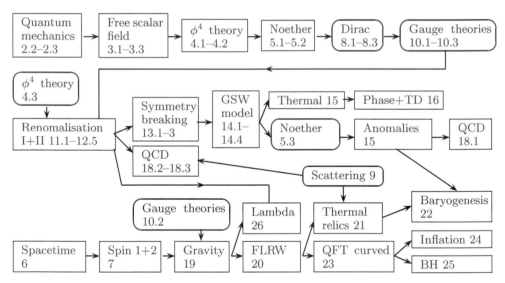

Some will miss important topics in this chart. For instance, grand unified theories or supersymmetry are two aspects of "beyond the standard model physics" (BSM) which are not only very attractive from a theoretical point of view but have also often been invoked to explain dark matter or baryogenesis. Having digested the material presented in this book, students may consult e.g. Dine (2016) as an entrée into the world of BSM. Moreover, I adapted from the field of astroparticle physics and cosmology only few topics directly relevant to the main theme of this book. Thus all more phenomenological aspects like, for example, neutrino oscillations or cosmic ray physics are omitted.

As this book is intended as an introduction, I suggest mainly review articles and textbooks in the "Further reading" sections at the end of each chapter. References to the original research literature are almost absent—I have to apologise to all those whose papers have been only indirectly referenced via these reviews. Moreover, even a minimal account of the historical development of this field is missing in this book. To compensate this deficit, I recommend the reader to consult some references specialised on the history of physics: Schweber (1994) gives a very readable account how QED was created, while O'Raifeartaigh (1997) reviews the development of the gauge principle. The story how the hot big bang model became the leading cosmological theory is told by Kragh (2013) and Peebles *et al.* (2009).

Website. A list of corrections, updates, solutions to more than 100 exercises as well as some software is available on the website of the book, `www.oup.co.uk/companion/` `quantumfields2018`. Comments and corrections are welcome and can be submitted via this website.

Acknowledgements. First of all, I would like to thank the students of my courses which had to digest various test versions of these lectures. I am grateful to Peder Galteland, Jonas Glesaaen and William Naylor for working out a first LaTeX version of the lecture notes for FY3466, and to all the students of the following semesters who spotted errors and pointed out obscure passages in the draft versions of this book. Jens Andersen, Eugeny Babichev, Sergey Ostapchenko and Pasquale Serpico read parts of the book and made valuable comments which helped to improve the text. Last but not least I would like to thank all my collaborators for sharing their insights with me.

Acknowledgements for the figures.
- Figure 12.3 is courtesy of D. Kazakov (hep-ph/0012288) who adapted it from Fig. 1 in U. Amaldi, W. de Boer, W. and H. Fürstenau, "Comparison of grand unified theories with electroweak and strong coupling constants measured at LEP", Phys. Lett. **B260**, 447 (1991). It has been reproduced with permission of Elsevier.
- Figure 14.1 has been adapted from Fig. 5 in G. Degrassi *et al.*, JHEP **08**, 098 (2012), published under the Creative Commons Noncommercial License.
- Figure 24.2 has been adapted from Fig. 5.3 in V. Mukhanov, "Physical Foundations of Cosmology" (2005) with permission of Cambridge University Press.
- Figure 24.5 has been reproduced from Fig. 2 in A. Challinor, "Cosmic Microwave Background Polarization Analysis", in "Data Analysis in Cosmology", Lecture Notes in Physics 665 (2008) with permission of Springer.

Notation and conventions

We use natural units with $\hbar = c = 1$, but mostly keep Newton's gravitational constant $G_N \neq 1$. Then all units can be expressed as powers of a basic unit which we choose as mass or energy. Instead of G_N, we use also $\kappa = 8\pi G_N$, the Planck mass $M_{Pl} = 1/\sqrt{G_N}$ or the reduced Planck mass $\widetilde{M}_{Pl} = 1/\sqrt{8\pi G_N}$. Maxwell's equations are written in the Lorentz–Heaviside version of the cgs system. Thus there is a factor 4π in the Coulomb law, but not in Maxwell's equations. Sommerfeld's fine-structure constant is $\alpha = e^2/(4\pi) \simeq 1/137$.

We choose as signature of the metric -2, thence the metric tensor in Minkowski space is $\eta_{\mu\nu} = \eta^{\mu\nu} = \mathrm{diag}(1, -1, -1, -1)$. If not otherwise specified, Einstein's summation convention is implied.

The d'Alembert or wave operator is $\Box \equiv \partial_\mu \partial^\mu = \frac{\partial^2}{\partial t^2} - \Delta$, while the four-dimensional nabla operator has the components $\partial_\mu \equiv \frac{\partial}{\partial x^\mu} = \left(\frac{\partial}{\partial t}, \frac{\partial}{\partial x}, \frac{\partial}{\partial y}, \frac{\partial}{\partial z} \right)$.

A boldface italic letter denotes the components of a three-vector $\boldsymbol{V} = \{V_x, V_y, V_z\} = \{V_i, i = 1, 2, 3\}$ or the three-dimensional part of a contravariant vector with components $V^\mu = \{V^0, V^1, V^2, V^3\} = \{V^0, \boldsymbol{V}\}$; a covariant vector has in Minkowski space the components $V_\mu = (V_0, -\boldsymbol{V})$. Scalar products of four-vectors are also denoted by $p_\mu q^\mu = p \cdot q$, of three-vectors by $\boldsymbol{p} \cdot \boldsymbol{q} = p_i q^i$. If there is no danger of confusion, the dot is omitted. Vectors and tensors in index free notation are denoted by boldface Roman letters, $\mathbf{V} = V^\mu \partial_\mu$ or $\mathbf{g} = g_{\mu\nu} \mathrm{d}x^\mu \otimes \mathrm{d}x^\nu$.

Greek indices $\alpha, \beta, \ldots$ encompass the range $\alpha = \{0, 1, 2, \ldots d - 1\}$, Latin indices $i, j, k, \ldots$ the range $i = \{1, 2, \ldots d - 1\}$, where d denotes the dimension of the space-time. In chapter 19, Latin indices $a, b, c, \ldots$ denote tensor components with respect to the vielbein field e^a_μ.

Our convention for the Fourier transformation is asymmetric, putting the factor $1/(2\pi)^n$ into

$$f(x) = \int \frac{\mathrm{d}^4 k}{(2\pi)^4} f(k) \mathrm{e}^{-ikx} \quad \text{and} \quad f(\boldsymbol{x}) = \int \frac{\mathrm{d}^3 k}{(2\pi)^3} f(\boldsymbol{k}) \mathrm{e}^{i\boldsymbol{k}\boldsymbol{x}}.$$

If no borders are specified in definite integrals, integration from $-\infty$ to ∞ is assumed.

Our nomenclature for disconnected, connected and one-particle irreducible (1PI) n-point Green functions and their corresponding generating functionals is as follows:

	Green function	generating functional
(dis-) connected	$\mathcal{G}(x_1, \ldots, x_n)$	$Z[J, \ldots]$
connected	$G(x_1, \ldots, x_n)$	$W[J, \ldots]$
1PI	$\Gamma(x_1, \ldots, x_n)$	$\Gamma[\phi, \ldots]$

Dirac spinors are normalised as $\bar{u}(p,s)u(p,s) = 2m$.

We use as covariant derivative $D_\mu = \partial_\mu + igA_\mu^a T^a$ with coupling $g > 0$, field strength $F_{\mu\nu}^a = \partial_\mu A_\nu^a - \partial_\nu A_\mu^a - gf^{abc}A_\mu^b A_\nu^c$ and generators T^a satisfying $[T^a, T^b] = if^{abc}T^c$ for all gauge groups. Special cases used in the SM are the groups $U_{em}(1)$, $U_Y(1)$, $SU_L(2)$ and $SU(3)$ with $g = \{q, g', g, g_s\}$ and $T^a = \{1, 1, \tau^a/2, \lambda^a/2\}$ in the fundamental representation. In particular, the electric charge of the positron is $q = e > 0$.

Employing dimensional regularisation (DR), we change the dimension of loop integrals from $d = 4$ to $d = 2\omega = 4 - 2\varepsilon$.

The results of problems marked by ♣ are used later in the text, those marked by ♥ require more efforts and time than average ones. Solutions to selected problems can be found on the webpage of this book. Commonly used symbols are

a	scale factor in FLRW metric
$\delta(x)$	Dirac's delta function, $\int dx\, f(x)\delta(x) = f(0)$
ε	infinitesimal quantity, slow-roll parameter
$\varepsilon_\mu, \varepsilon_{\mu\nu}$	polarisation vector and tensor for spin $s = 1, 2$
η	boost parameter, conformal time, slow-roll parameter
$g = \det(g_{\mu\nu})$	determinant of the metric tensor $g_{\mu\nu}$
$g_*, g_{*,S}$	relativistic degrees of freedom entering ρ, S
$H(q,p),\ \mathscr{H}(\phi,\pi)$	Hamiltonian, Hamiltonian density
$H^\dagger$	Hermitian conjugate (h.c.) with $M^\dagger = M^{*T}$ or $M_{ij}^\dagger = M_{ji}^*$
$H = \dot{a}/a,\ \mathscr{H} = a'/a$	Hubble parameter
$L(q,\dot{q}),\ \mathscr{L}(\phi,\partial_\mu\phi)$	Lagrangian, Lagrangian density
$\Omega_i = \rho_i/\rho_{cr}$	fraction of critical energy density in component i
$p^\mu; P; P_{ij}$	four momentum $p^\mu = (E, \boldsymbol{p})$, pressure
$\bar{\psi}$	adjoint spinor with $\bar{\psi} = \psi^\dagger\gamma^0$
$R^\alpha{}_{\beta\rho\sigma} = [\partial_\rho\Gamma^\alpha{}_{\beta\sigma} - \ldots$	Riemann or curvature tensor
$R_{\alpha\beta} = R^\rho{}_{\alpha\rho\beta}$	Ricci tensor
$R = g_{\mu\nu}R^{\mu\nu}$	curvature scalar R
$S[\phi],\ S[\phi,\partial_\mu\phi]$	action functional
$T_{\mu\nu}$	(energy–momentum) stress tensor
$\vartheta(x)$	Heaviside step function, $\vartheta(x) = 1$ for $x > 0$, 0 for $x < 0$.
tr	trace of a matrix $\mathrm{tr}(A) = \sum_i A_{ii}$, of a tensor $\mathrm{tr}(T) = T^\mu{}_\mu$
Tr	sum/integration over a complete set of quantum numbers
u, v	solutions of Dirac equation, light-cone coordinates $t \pm x$
$w = P/\rho$	equation of state (EoS) parameter
$Y = n_X/s$	abundance of particle type X relative to entropy density
$X = n_X/n_B$	abundance of particle type X relative to baryon density

Contents

1
Classical mechanics

To begin, in this chapter we review those concepts of classical mechanics which are essential for progressing towards quantum theory. First we recall briefly the Lagrangian and Hamiltonian formulation of classical mechanics and their derivation from an action principle. We also illustrate the Green function method using as example the driven harmonic oscillator and recall the action of a relativistic point particle.

1.1 Action principle

Variational principles. Fundamental laws of nature as Newton's axioms or Maxwell's equations were discovered in the form of differential equations. Starting from Leibniz and Euler, it was realised that one can re-express differential equations in the form of variational principles. In this approach, the evolution of a physical system is described by the extremum of an appropriately chosen functional. Various versions of such variational principles exist, but they have in common that the functionals used have the dimension of "energy × time"; that is, the functionals have the same dimension as Planck's constant $\hbar$. A quantity with this dimension is called action S. An advantage of using the action as main tool to describe dynamical systems is that this allows us to implement easily both spacetime and internal symmetries. For instance, choosing as ingredients of the action local functions that transform as scalars under Lorentz transformations leads automatically to relativistically invariant field equations. Moreover, the action S economically summarises the information contained typically in a set of various coupled differential equations.

If the variational principle is formulated as an integral principle, then the functional S will depend on the whole path $q(t)$ described by the system between the considered initial and final time. In the formulation of quantum theory we will pursue, we will look for a direct connection from the classical action $S[q]$ of the path $[q(t): q'(t')]$ to the transition amplitude $\langle q', t'|q, t\rangle$. Thus the use of the action principle will not only simplify the discussion of symmetries of a physical system but it also lies at the heart of the approach to quantum theory we will follow.

1.1.1 Hamilton's principle and Lagrange's equations

A functional $F[f(x)]$ is a map from a certain space of functions $f(x)$ into the real or complex numbers. We will consider mainly functionals from the space of (at least) twice differentiable functions between fixed points a and b. More specifically, Hamilton's principle uses as functional the action S defined by

Quantum Fields–From the Hubble to the Planck Scale. Michael Kachelriess. © Michael Kachelriess 2018.
Published in 2018 by Oxford University Press. DOI 10.1093/oso/9780198802877.001.0001

$$S[q^i] = \int_a^b \mathrm{d}t \, L(q^i, \dot{q}^i, t), \tag{1.1}$$

where L is a function of the $2n$ independent functions q^i and $\dot{q}^i = \mathrm{d}q^i/\mathrm{d}t$ as well as of the parameter t. In classical mechanics, we call L the Lagrange function of the system, q^i are its n generalised coordinates, $\dot{q}^i$ the corresponding velocities and t is the time. The extrema of this action give those paths $q(t)$ from a to b which are solutions of the equations of motion for the system described by L.

How do we find those paths that extremize the action S? First of all, we have to prescribe which variables are kept constant, which are varied and which constraints the variations have to obey. Depending on the variation principle we choose, these conditions and the functional form of the action will differ. Hamilton's principle corresponds to a smooth variation of the path,

$$q^i(t, \varepsilon) = q^i(t, 0) + \varepsilon \eta^i(t),$$

that keeps the endpoints fixed, $\eta^i(a) = \eta^i(b) = 0$ but is otherwise arbitrary. The scale factor ε determines the magnitude of the variation for the one-parameter family of paths $\varepsilon \eta^i(t)$. The notation $S[q^i]$ stresses that we consider the action as a functional only of the coordinates q^i. The velocities $\dot{q}^i$ are not varied independently because ε is time-independent. Since the time t is not varied in Hamilton's principle, varying the path $q^i(t, \varepsilon)$ requires only to calculate the resulting change of the Lagrangian L. Following this prescription, the action has an extremum if

$$0 = \left. \frac{\partial S[q^i(t, \varepsilon)]}{\partial \varepsilon} \right|_{\varepsilon = 0} = \int_a^b \mathrm{d}t \left(\frac{\partial L}{\partial q^i} \frac{\partial q^i}{\partial \varepsilon} + \frac{\partial L}{\partial \dot{q}^i} \frac{\partial \dot{q}^i}{\partial \varepsilon} \right) = \int_a^b \mathrm{d}t \left(\frac{\partial L}{\partial q^i} \eta^i + \frac{\partial L}{\partial \dot{q}^i} \dot{\eta}^i \right). \tag{1.2}$$

Here we applied—as always in the following—Einstein's convention to sum over a repeated index pair. Thus, for example, the first term in the bracket equals

$$\frac{\partial L}{\partial q^i} \eta^i \equiv \sum_{i=1}^n \frac{\partial L}{\partial q^i} \eta^i$$

for a system described by n generalised coordinates. We can eliminate $\dot{\eta}^i$ in favour of η^i, integrating the second term by parts, arriving at

$$\left. \frac{\partial S[q^i(t, \varepsilon)]}{\partial \varepsilon} \right|_{\varepsilon = 0} = \int_a^b \mathrm{d}t \left[\frac{\partial L}{\partial q^i} - \frac{\mathrm{d}}{\mathrm{d}t} \left(\frac{\partial L}{\partial \dot{q}^i} \right) \right] \eta^i + \left[\frac{\partial L}{\partial \dot{q}^i} \eta^i \right]_a^b. \tag{1.3}$$

The boundary term $[\ldots]_a^b$ vanishes because we required that the functions η^i are zero at the endpoints a and b. Since these functions are otherwise arbitrary, each individual term in the first bracket has to vanish for an extremal curve. The n equations resulting from the condition $\partial S[q^i(t, \varepsilon)]/\partial \varepsilon = 0$ are called the (Euler–) Lagrange equations of the action S,

$$\frac{\partial L}{\partial q^i} - \frac{\mathrm{d}}{\mathrm{d}t} \frac{\partial L}{\partial \dot{q}^i} = 0 \tag{1.4}$$

and give the equations of motion for the system specified by L. In the future, we will use a more concise notation, calling

$$\delta q^i \equiv \lim_{\varepsilon \to 0} \frac{q^i(t,\varepsilon) - q^i(t,0)}{\varepsilon} = \left. \frac{\partial q^i(t,\varepsilon)}{\partial \varepsilon} \right|_{\varepsilon=0} \tag{1.5}$$

the variation of q^i, and similarly for functions and functionals of q^i. Thus we can rewrite, for example, Eq. (1.2) in a more evident form as

$$0 = \delta S[q^i] = \int_a^b \mathrm{d}t\, \delta L(q^i, \dot{q}^i, t) = \int_a^b \mathrm{d}t\, \left(\frac{\partial L}{\partial q^i} \delta q^i + \frac{\partial L}{\partial \dot{q}^i} \delta \dot{q}^i \right). \tag{1.6}$$

We close this paragraph with three remarks. First, we note that Hamilton's principle is often called the principle of least action. This name is somewhat misleading, since the extremum of the action can be also a maximum or a saddle-point. Second, observe that the Lagrangian L is not uniquely fixed. Adding a total time derivative, $L \to L' = L + \mathrm{d}f(q,t)/\mathrm{d}t$, does not change the resulting Lagrange equations,

$$S' = S + \int_a^b \mathrm{d}t\, \frac{\mathrm{d}f}{\mathrm{d}t} = S + f(q(b), t_b) - f(q(a), t_a), \tag{1.7}$$

since the last two terms vanish varying the action with the restriction of fixed endpoints a and b. Finally, note that we used a Lagrangian that depends only on the coordinates and their *first* derivatives. Such a Lagrangian leads to second-order equations of motion and thus to a mechanical system specified by the $2n$ pieces of information $\{q_i, \dot{q}_i\}$. Ostrogradsky showed 1850 that a stable ground-state is impossible, if the Lagrangian contains higher derivatives $\ddot{q}, q^{(3)}, \ldots$, cf. problem 1.3. Therefore such theories contradict our experience that the vacuum is stable. Constructing Lagrangians for the fundamental theories describing Nature, we should restrict ourselves thus to Lagrangians that lead to second-order equations of motion.

Lagrange function. We illustrate now how one can use symmetries to constrain the possible form of a Lagrangian L. As example, we consider the case of a free non-relativistic particle with mass m subject to the Galilean principle of relativity. More precisely, we use that the homogeneity of space and time forbids that L depends on $\boldsymbol{x}$ and t, while the isotropy of space implies that L depends only on the norm of the velocity vector $\boldsymbol{v}$, but not on its direction. Thus the Lagrange function of a free particle can be only a function of v^2, $L = L(v^2)$.

Let us consider two inertial frames moving with the infinitesimal velocity $\boldsymbol{\varepsilon}$ relative to each other. (Recall that an inertial frame is defined as a coordinate system where a force-free particle moves along a straight line.) Then a Galilean transformation connects the velocities measured in the two frames as $\boldsymbol{v}' = \boldsymbol{v} + \boldsymbol{\varepsilon}$. The Galilean principle of relativity requires that the laws of motion have the same form in both frames, and thus the Lagrangians can differ only by a total time derivative. Expanding the difference δL in ε gives with $\delta v^2 = 2\boldsymbol{v} \cdot \boldsymbol{\varepsilon} + \mathcal{O}(\varepsilon^2)$

$$\delta L = \frac{\partial L}{\partial v^2}\, \delta v^2 = 2\boldsymbol{v} \cdot \boldsymbol{\varepsilon} \frac{\partial L}{\partial v^2}. \tag{1.8}$$

Since $v^i = \mathrm{d}x^i/\mathrm{d}t$, the term $\partial L/\partial v^2$ has to be independent of v such that the difference δL is a total time derivative. Hence, the Lagrangian of a free particle has the form

$L = av^2 + b$. The constant b drops out of the equations of motion, and we can set it therefore to zero. To be consistent with usual notation, we call the proportionality constant $m/2$, and the total expression kinetic energy T,

$$L = T = \frac{1}{2}mv^2. \tag{1.9}$$

For a system of non-interacting particles, the Lagrange function L is additive, $L = \sum_a \frac{1}{2}m_a v_a^2$. If there are interactions (assumed for simplicity to depend only on the coordinates), then we subtract a function $V(\boldsymbol{x}_1, \boldsymbol{x}_2, \dots)$ called potential energy. One confirms readily that this choice for L reproduces Newton's law of motion.

Energy. The Lagrangian of a closed system does not depend on time because of the homogeneity of time. Its total time derivative is

$$\frac{\mathrm{d}L}{\mathrm{d}t} = \frac{\partial L}{\partial q^i}\,\dot{q}^i + \frac{\partial L}{\partial \dot{q}^i}\,\ddot{q}^i. \tag{1.10}$$

Using the equations of motion and replacing $\partial L/\partial q^i$ by $(\mathrm{d}/\mathrm{d}t)\partial L/\partial \dot{q}^i$, it follows

$$\frac{\mathrm{d}L}{\mathrm{d}t} = \dot{q}^i\,\frac{\mathrm{d}}{\mathrm{d}t}\frac{\partial L}{\partial \dot{q}^i} + \frac{\partial L}{\partial \dot{q}^i}\,\ddot{q}^i = \frac{\mathrm{d}}{\mathrm{d}t}\left(\dot{q}^i\,\frac{\partial L}{\partial \dot{q}^i}\right). \tag{1.11}$$

Hence the quantity

$$E \equiv \dot{q}^i\,\frac{\partial L}{\partial \dot{q}^i} - L \tag{1.12}$$

remains constant during the evolution of a closed system. This holds also more generally, for example in the presence of static external fields, as long as the Lagrangian is not time-dependent.

We have still to show that E coincides indeed with the usual definition of energy. Using as Lagrange function $L = T(q, \dot{q}) - V(q)$, where the kinetic energy T is quadratic in the velocities, we have

$$\dot{q}^i\,\frac{\partial L}{\partial \dot{q}^i} = \dot{q}^i\,\frac{\partial T}{\partial \dot{q}^i} = 2T \tag{1.13}$$

and thus $E = 2T - L = T + V$.

Conservation laws. In a general way, we can derive the connection between a symmetry of the Lagrangian and a corresponding conservation law as follows. Let us assume that under a change of coordinates $q^i \to q^i + \delta q^i$, the Lagrangian changes at most by a total time derivative,

$$L \to L + \delta L = L + \frac{\mathrm{d}\delta F}{\mathrm{d}t}. \tag{1.14}$$

In this case, the equations of motion are unchanged and the coordinate change $q^i \to q^i + \delta q^i$ is a symmetry of the Lagrangian. The change $\mathrm{d}\delta F/\mathrm{d}t$ has to equal δL induced by the variation δq^i,

$$\frac{\partial L}{\partial q^i}\delta q^i + \frac{\partial L}{\partial \dot{q}^i}\delta \dot{q}^i - \frac{\mathrm{d}\delta F}{\mathrm{d}t} = 0. \tag{1.15}$$

Replacing again $\partial L / \partial q^i$ by $(\mathrm{d}/\mathrm{d}t)\partial L / \partial \dot{q}^i$ and applying the product rule gives as conserved quantity

$$Q = \frac{\partial L}{\partial \dot{q}^i}\delta q^i - \delta F. \tag{1.16}$$

Thus any continuous symmetry of a Lagrangian system results in a conserved quantity. In particular, energy conservation follows for a system invariant under time translations with $\delta q^i = \dot{q}^i \delta t$. Other conservation laws are discussed in problem 1.7.

1.1.2 Palatini's principle and Hamilton's equations

Legendre transformation and the Hamilton function. In the Lagrange formalism, we describe a system specifying its generalised coordinates and velocities using the Lagrangian, $L = L(q^i, \dot{q}^i, t)$. An alternative is to use generalised coordinates and their canonically conjugated momenta p_i defined as

$$p_i = \frac{\partial L}{\partial \dot{q}^i}. \tag{1.17}$$

The passage from $\{q^i, \dot{q}^i\}$ to $\{q^i, p_i\}$ is a special case of a Legendre transformation:[1] Starting from the Lagrangian L we define a new function $H(q^i, p_i, t)$ called Hamiltonian or Hamilton function via

$$H(q^i, p_i, t) = \frac{\partial L}{\partial \dot{q}^i}\,\dot{q}^i - L(q^i, \dot{q}^i, t) = p_i\dot{q}^i - L(q^i, \dot{q}^i, t). \tag{1.18}$$

Here we assume that we can invert the definition (1.17) and are thus able to substitute velocities $\dot{q}^i$ by momenta p_i in the Lagrangian L.

The physical meaning of the Hamiltonian H follows immediately comparing its defining equation with the one for the energy E. Thus the numerical value of the Hamiltonian equals the energy of a dynamical system; we insist, however, that H is expressed as function of coordinates and their conjugated momenta. A coordinate q_i that does not appear explicitly in L is called cyclic. The Lagrange equations imply then $\partial L / \partial \dot{q}_i = $ const., so that the corresponding canonically conjugated momentum $p_i = \partial L / \partial \dot{q}^i$ is conserved.

Palatini's formalism and Hamilton's equations. Previously, we considered the action S as a functional only of q^i. Then the variation of the velocities $\dot{q}^i$ is not independent and we arrive at n second–order differential equations for the coordinates q^i. An alternative approach is to allow independent variations of the coordinates q^i and of the velocities $\dot{q}^i$. We trade the latter against the momenta $p_i = \partial L / \partial \dot{q}^i$ and rewrite the action as

$$S[q^i, p_i] = \int_a^b \mathrm{d}t \, \left[p_i \dot{q}^i - H(q^i, p_i, t) \right]. \tag{1.19}$$

The independent variation of coordinates q^i and momenta p_i gives

---

[1] The concept of a Legendre transformation may be familiar from thermodynamics, where it is used to change between extensive variables (e.g. the entropy S) and their conjugate intensive variables (e.g. the temperature T).

$$\delta S[q^i, p_i] = \int_a^b \mathrm{d}t \, \left[p_i \delta \dot{q}^i + \dot{q}^i \delta p_i - \frac{\partial H}{\partial q^i} \delta q^i - \frac{\partial H}{\partial p_i} \delta p_i \right]. \tag{1.20}$$

The first term can be integrated by parts, and the resulting boundary terms vanishes by assumption. Collecting then the δq^i and δp_i terms and requiring that the variation is zero, we obtain

$$0 = \delta S[q^i, p_i] = \int_a^b \mathrm{d}t \, \left[-\left(\dot{p}_i + \frac{\partial H}{\partial q^i} \right) \delta q^i + \left(\dot{q}^i - \frac{\partial H}{\partial p_i} \right) \delta p_i \right]. \tag{1.21}$$

As the variations δq^i and δp_i are independent, their coefficients in the round brackets have to vanish separately. Thus we obtain in this formalism directly Hamilton's equations,

$$\dot{q}^i = \frac{\partial H}{\partial p_i}, \qquad \text{and} \qquad \dot{p}_i = -\frac{\partial H}{\partial q^i}. \tag{1.22}$$

Consider now an observable $O = O(q^i, p_i, t)$. Its time dependence is given by

$$\frac{\mathrm{d}O}{\mathrm{d}t} = \frac{\partial O}{\partial q^i} \dot{q}^i + \frac{\partial O}{\partial p_i} \dot{p}_i + \frac{\partial O}{\partial t} = \frac{\partial O}{\partial q^i} \frac{\partial H}{\partial p_i} - \frac{\partial O}{\partial p_i} \frac{\partial H}{\partial q^i} + \frac{\partial O}{\partial t}, \tag{1.23}$$

where we used Hamilton's equations. If we define the Poisson brackets $\{A, B\}$ between two observables A and B as

$$\{A, B\} = \frac{\partial A}{\partial q^i} \frac{\partial B}{\partial p_i} - \frac{\partial A}{\partial p_i} \frac{\partial B}{\partial q^i}, \tag{1.24}$$

then we can rewrite Eq. (1.23) as

$$\frac{\mathrm{d}O}{\mathrm{d}t} = \{O, H\} + \frac{\partial O}{\partial t}. \tag{1.25}$$

This equations gives us a formal correspondence between classical and quantum mechanics. The time evolution of an operator O in the Heisenberg picture is given by the same equation as in classical mechanics, if the Poisson bracket is changed to a commutator. Since the Poisson bracket is antisymmetric, we find

$$\frac{\mathrm{d}H}{\mathrm{d}t} = \frac{\partial H}{\partial t}. \tag{1.26}$$

Hence the Hamiltonian H is a conserved quantity, if and only if H is time-independent.

1.2 Green functions and the response method

We can test the internal properties of a physical system, if we impose an external force $J(t)$ on it and compare its measured to its calculated response. If the system is described by linear differential equations, then the superposition principle is valid. We can reconstruct the solution $x(t)$ for an arbitrary applied external force $J(t)$, if we know the response to a normalised delta function-like kick $J(t) = \delta(t - t')$. Mathematically, this corresponds to the knowledge of the Green function $G(t - t')$ for the differential equation $D(t)x(t) = J(t)$ describing the system. Even if the system is described by a non-linear differential equation, we can often use a linear approximation in case of a sufficiently small external force $J(t)$. Therefore the Green function method is extremely useful and we will apply it extensively in discussing quantum field theories.

We illustrate this method with the example of the harmonic oscillator which is the prototype for a quadratic, and thus exactly solvable, action. In classical physics, causality implies that the knowledge of the external force $J(t')$ at times $t' < t$ is sufficient to determine the solution $x(t)$ at time t. We define therefore two Green functions $\widetilde{G}$ and G_R by

$$x(t) = \int_{-\infty}^{t} \mathrm{d}t' \, \widetilde{G}(t - t')J(t') = \int_{-\infty}^{\infty} \mathrm{d}t' G_R(t - t')J(t'), \tag{1.27}$$

where the retarded Green function G_R satisfies $G_R(t - t') = \widetilde{G}(t - t')\vartheta(t - t')$. The definition (1.27) is motivated by the trivial relation $J(t) = \int \mathrm{d}t' \, \delta(t - t')J(t')$: an arbitrary force $J(t)$ can be seen as a superposition of delta functions $\delta(t - t')$ with weight $J(t')$. If the Green function $G_R(t-t')$ determines the response of the system to a delta function-like force, then we should obtain the solution $x(t)$ integrating $G_R(t-t')$ with the weight $J(t')$.

We convert the equation of motion $m\ddot{x} + m\omega^2 x = J$ of a forced harmonic oscillator into the form $D(t)x(t) = J(t)$ by writing

$$D(t)x(t) \equiv m\left(\frac{\mathrm{d}^2}{\mathrm{d}t^2} + \omega^2\right)x(t) = J(t). \tag{1.28}$$

Inserting (1.27) into (1.28) gives

$$\int_{-\infty}^{\infty} \mathrm{d}t' \, D(t)G_R(t - t')J(t') = J(t). \tag{1.29}$$

For an arbitrary external force $J(t)$, this relation can be only valid if

$$D(t)G_R(t - t') = \delta(t - t'). \tag{1.30}$$

Thus a Green function $G(t - t')$ is the inverse of its defining differential operator $D(t)$. As we will see, Eq. (1.30) does not specify uniquely the Green function, and thus we will omit the index "R" for the moment. Performing a Fourier transformation,

$$G(t - t') = \int \frac{\mathrm{d}\Omega}{2\pi} G(\Omega)\mathrm{e}^{-\mathrm{i}\Omega(t-t')} \quad \text{and} \quad \delta(t - t') = \int \frac{\mathrm{d}\Omega}{2\pi} \mathrm{e}^{-\mathrm{i}\Omega(t-t')}, \tag{1.31}$$

we obtain

$$\int \frac{\mathrm{d}\Omega}{2\pi} G(\Omega)D(t)\mathrm{e}^{-\mathrm{i}\Omega(t-t')} = \int \frac{\mathrm{d}\Omega}{2\pi} \mathrm{e}^{-\mathrm{i}\Omega(t-t')}. \tag{1.32}$$

The action of $D(t)$ on the plane waves $\mathrm{e}^{-\mathrm{i}\Omega(t-t')}$ can be evaluated easily, since the differentiation has become equivalent with multiplication, $\mathrm{d}/\mathrm{d}t \to -\mathrm{i}\Omega$. Comparing

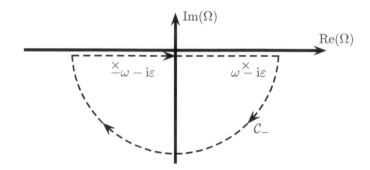

Fig. 1.1 Poles and contour in the complex Ω plane used for the integration of the retarded Green function.

then the coefficients of the plane waves on both sides of this equation, we have to invert only an algebraic equation, arriving at

$$G(\Omega) = \frac{1}{m}\frac{1}{\omega^2 - \Omega^2}. \tag{1.33}$$

For the back-transformation with $\tau = t - t'$,

$$G(\tau) = \int \frac{d\Omega}{2\pi m}\frac{e^{-i\Omega\tau}}{\omega^2 - \Omega^2}, \tag{1.34}$$

we have to specify how the poles at $\Omega^2 = \omega^2$ are avoided. It is this choice by which we select the appropriate Green function. In classical physics, we implement causality ("cause always precedes its effect") selecting the retarded Green function.

We will use Cauchy's residue theorem, $\oint dz\, f(z) = 2\pi i \sum \mathrm{res}_{z_0} f(z)$, to calculate the integral. Its application requires to close the integration contour adding a path which gives a vanishing contribution to the integral. This is achieved, when the integrand $G(\Omega)e^{-i\Omega\tau}$ vanishes fast enough along the added path. Thus we have to choose for positive τ the contour $\mathcal{C}_-$ in the lower plane, $e^{-i\Omega\tau} = e^{-|\Im(\Omega)|\tau} \to 0$ for $\Im(\Omega) \to -\infty$, while we have to close the contour in the upper plane for negative τ. If we want to obtain the retarded Green function $G_R(\tau)$ which vanishes for $\tau < 0$, we therefore have to shift the poles $\Omega_{1/2} = \pm\omega$ into the lower plane as shown in Fig. 1.1 by adding a small negative imaginary part, $\Omega_{1/2} \to \Omega_{1/2} = \pm\omega - i\varepsilon$, or

$$G_R(\tau) = -\frac{1}{2\pi m}\int d\Omega \frac{e^{-i\Omega\tau}}{(\Omega - \omega + i\varepsilon)(\Omega + \omega + i\varepsilon)}. \tag{1.35}$$

The residue $\mathrm{res}_{z_0} f(z)$ of a function f with a single pole at z_0 is given by

$$\mathrm{res}_{z_0} f(z) = \lim_{z \to z_0} (z - z_0)f(z). \tag{1.36}$$

Thus we pick up at $\Omega_1 = -\omega - i\varepsilon$ the contribution $2\pi i\, e^{+i\omega\tau}/(-2\omega)$, while we obtain $2\pi i\, e^{-i\omega\tau}/(2\omega)$ from $\Omega_2 = \omega - i\varepsilon$. Combining both contributions and adding a minus sign because the contour is clockwise, we arrive at

$$G_R(\tau) = \frac{\mathrm{i}}{2m\omega} \left[\mathrm{e}^{-\mathrm{i}\omega\tau} - \mathrm{e}^{\mathrm{i}\omega\tau} \right] \vartheta(\tau) = \frac{1}{m} \frac{\sin(\omega\tau)}{\omega} \vartheta(\tau) \tag{1.37}$$

as result for the retarded Green function of the forced harmonic oscillator.

We can now obtain a particular solution solving (1.27). For instance, choosing $J(t') = \delta(t - t')$, results in

$$x(t) = \frac{1}{m} \frac{\sin(\omega t)}{\omega} \vartheta(t). \tag{1.38}$$

Thus the oscillator was at rest for $t < 0$, got a kick at $t = 0$, and oscillates according $x(t)$ afterwards. Note the following two points: first, the fact that the kick proceeds the movement is the result of our choice of the retarded (or causal) Green function. Second, the particular solution (1.38) for an oscillator initially at rest can be generalised by adding the solution to the homogeneous equation $\ddot{x} + \omega^2 x = 0$.

1.3 Relativistic particle

In special relativity, we replace the Galilean transformations as symmetry group of space and time by Lorentz transformations. The latter are all those coordinate transformations $x^\mu \to \tilde{x}^\mu = \Lambda^\mu{}_\nu x^\nu$ that keep the squared distance

$$s_{12}^2 \equiv (t_1 - t_2)^2 - (x_1 - x_2)^2 - (y_1 - y_2)^2 - (z_1 - z_2)^2 \tag{1.39}$$

between two spacetime events x_1^μ and x_2^μ invariant. The distance of two infinitesimally close spacetime events is called the line element $\mathrm{d}s$ of the spacetime. In Minkowski space, it is given by

$$\mathrm{d}s^2 = \mathrm{d}t^2 - \mathrm{d}x^2 - \mathrm{d}y^2 - \mathrm{d}z^2 \tag{1.40}$$

using a Cartesian inertial frame. We can interpret the line element $\mathrm{d}s^2$ as a scalar product, if we introduce the metric tensor $\eta_{\mu\nu}$ with elements

$$\eta_{\mu\nu} = \begin{pmatrix} 1 & 0 & 0 & 0 \\ 0 & -1 & 0 & 0 \\ 0 & 0 & -1 & 0 \\ 0 & 0 & 0 & -1 \end{pmatrix} \tag{1.41}$$

and a scalar product of two vectors as

$$a \cdot b \equiv \eta_{\mu\nu} a^\mu b^\nu = a_\mu b^\mu = a^\mu b_\mu . \tag{1.42}$$

In Minkowski space, we call a four-vector any four-tuple V^μ that transforms as $\tilde{V}^\mu = \Lambda^\mu{}_\nu V^\nu$. By convention, we associate three-vectors with the spatial part of four-vectors with upper indices, for example we set $x^\mu = \{t, x, y, z\}$ or $A^\mu = \{\phi, \boldsymbol{A}\}$. Lowering then the index by contraction with the metric tensor result in a minus sign of the spatial components of a four-vector, $x_\mu = \eta_{\mu\nu} x^\mu = \{t, -x, -y, -z\}$ or $A_\mu = \{\phi, -\boldsymbol{A}\}$. Summing over an index pair, typically one index occurs in an upper and one in a lower position. Note that in the denominator, an upper index counts as a lower index and vice versa; cf. for example with Eqs. (1.18) and (1.17). Additionally to four-vectors, we will meet tensors $T^{\mu_1 \cdots \mu_n}$ of rank n which transform as $\tilde{T}^{\mu_1 \cdots \mu_n} = \Lambda^{\mu_1}{}_{\nu_1} \cdots \Lambda^{\mu_n}{}_{\nu_n} T^{\nu_1 \cdots \nu_n}$.

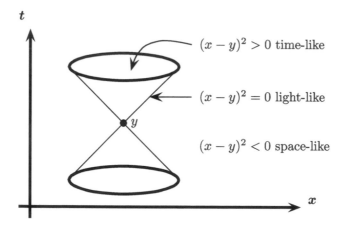

Fig. 1.2 Light-cone at the point $P(y^\mu)$ generated by light-like vectors. Contained in the light-cone are the time-like vectors, outside the space-like ones.

Since the metric $\eta_{\mu\nu}$ is indefinite, the norm of a vector a^μ can be

$$a_\mu a^\mu > 0, \quad \text{time-like,} \tag{1.43a}$$

$$a_\mu a^\mu = 0, \quad \text{light-like or null-vector,} \tag{1.43b}$$

$$a_\mu a^\mu < 0, \quad \text{space-like.} \tag{1.43c}$$

The cone of all light-like vectors starting from a point P is called light-cone, cf. Fig. 1.2. The time-like region inside the light-cone consists of two parts, past and future. Only events inside the past light-cone can influence the physics at point P, while P can influence only the interior of its future light-cone. The proper-time τ is the time displayed by a clock moving with the observer. With our conventions—negative signature of the metric and $c = 1$—the proper-time elapsed between two spacetime events equals the integrated line element between them,

$$\tau_{12} = \int_1^2 \mathrm{d}s = \int_1^2 [\eta_{\mu\nu}\mathrm{d}x^\mu\mathrm{d}x^\nu]^{1/2} = \int_1^2 \mathrm{d}t[1 - v^2]^{1/2} < t_2 - t_1. \tag{1.44}$$

The last part of this equation, where we introduced the three-velocity $v^i = \mathrm{d}x^i/\mathrm{d}t$ of the clock, shows explicitly the relativistic effect of time dilation, as well as the connection between coordinate time t and the proper-time τ of a moving clock, $\mathrm{d}\tau = (1 - v^2)^{1/2}\mathrm{d}t \equiv \mathrm{d}t/\gamma$. The line describing the position of an observer is called world-line. Parameterising the world-line by the parameter σ, $x = x(\sigma)$, the proper-time is given by

$$\tau = \int \mathrm{d}\sigma \left[\eta_{\mu\nu}\frac{\mathrm{d}x^\mu}{\mathrm{d}\sigma}\frac{\mathrm{d}x^\nu}{\mathrm{d}\sigma}\right]^{1/2}. \tag{1.45}$$

Note that τ is invariant under a reparameterisation $\tilde\sigma = f(\sigma)$.

The only invariant differential we have at our disposal to form an action for a free point-like particle is the line element, or equivalently the proper-time,

$$S_0 = \alpha \int_a^b \mathrm{d}s = \alpha \int_a^b \mathrm{d}\sigma \, \frac{\mathrm{d}s}{\mathrm{d}\sigma} \tag{1.46}$$

with $L = \alpha \mathrm{d}s/\mathrm{d}\sigma = \alpha \mathrm{d}\tau/\mathrm{d}\sigma$. We check now if this choice which implies the Lagrangian

$$L = \alpha \frac{\mathrm{d}\tau}{\mathrm{d}\sigma} = \alpha \left[\eta_{\mu\nu} \frac{\mathrm{d}x^\mu}{\mathrm{d}\sigma} \frac{\mathrm{d}x^\nu}{\mathrm{d}\sigma} \right]^{1/2} \tag{1.47}$$

for a free particle is sensible. The action has the correct non-relativistic limit,

$$S_0 = \alpha \int_a^b \mathrm{d}s = \alpha \int_a^b \mathrm{d}t \, \sqrt{1 - v^2} = \int_a^b \mathrm{d}t \left(-m + \frac{1}{2}mv^2 + \mathcal{O}(v^4) \right), \tag{1.48}$$

if we set $\alpha = -m$. The mass m corresponds to a potential energy in the non-relativistic limit and has therefore a negative sign in the Lagrangian. Moreover, a constant drops out of the equations of motion, and thus the term $-m$ can be omitted in the non-relativistic limit. The time t enters the relativistic Lagrangian in a Lorentz invariant way as one of the dynamical variables, $x^\mu = (t, \boldsymbol{x})$, while σ assumes now t's purpose to parameterise the trajectory, $x^\mu(\sigma)$. Since a moving clock goes slower than a clock at rest, solutions of this Lagrangian maximise the action.

> **Example 1.1:** Relativistic dispersion relation. We extend the non-relativistic definition of the momentum, $p_i = \partial L/\partial \dot{x}^i$, to four dimensions setting $p_\alpha = -\partial L/\partial \dot{x}^\alpha$. Note the minus sign that reflects the minus in the spatial components of a covariant vector, $p_\alpha = (E, -\boldsymbol{p})$. Then
>
> $$p_\alpha = -\frac{\partial L}{\partial \dot{x}^\alpha} = m \frac{\mathrm{d}x_\alpha/\mathrm{d}\sigma}{\mathrm{d}\tau/\mathrm{d}\sigma} - m \frac{\mathrm{d}x_\alpha}{\mathrm{d}\tau} \equiv m u_\alpha. \tag{1.49}$$
>
> In the last step, we defined the four-velocity $u^\alpha = \mathrm{d}x^\alpha/\mathrm{d}\tau$. Using $\mathrm{d}t = \gamma \mathrm{d}\tau$, it follows $u_\alpha u^\alpha = 1$ and $p_\alpha p^\alpha = m^2$. The last relation expresses the relativistic dispersion relation $E^2 = m^2 + \boldsymbol{p}^2$.

The Lagrange equations are

$$\frac{\mathrm{d}}{\mathrm{d}\sigma} \frac{\partial L}{\partial (\mathrm{d}x^\alpha/\mathrm{d}\sigma)} = \frac{\partial L}{\partial x^\alpha}. \tag{1.50}$$

Consider, for example, the x^1 component, then

$$\frac{\mathrm{d}}{\mathrm{d}\sigma} \frac{\partial L}{\partial (\mathrm{d}x^1/\mathrm{d}\sigma)} = \frac{\mathrm{d}}{\mathrm{d}\sigma} \left(\frac{1}{L} \frac{\mathrm{d}x^1}{\mathrm{d}\sigma} \right) = 0. \tag{1.51}$$

Since $L = -m \mathrm{d}\tau/\mathrm{d}\sigma$, Newton's law follows for the x^1 coordinate after multiplication with $\mathrm{d}\sigma/\mathrm{d}\tau$,

$$\frac{d^2 x^1}{d\tau^2} = 0, \tag{1.52}$$

and similar for the other coordinates.

An equivalent but often more convenient form for the Lagrangian of a free particle is

$$L = -m\eta_{\mu\nu}\dot{x}^\mu \dot{x}^\nu, \tag{1.53}$$

where we set $\dot{x}^\mu = \mathrm{d}x^\mu/\mathrm{d}\tau$. If there are no interactions (except gravity), we can neglect the mass m of the particle and one often sets $m \to -1$.

Next we want to add an interaction term S_{em} between a particle with charge q and an electromagnetic field. The simplest possible action is to integrate the potential A_μ along the world-line $x^\mu(\sigma)$ of the particle,

$$S_{\mathrm{em}} = -q \int \mathrm{d}x^\mu \, A_\mu(x) = -q \int \mathrm{d}\sigma \, \frac{\mathrm{d}x^\mu}{\mathrm{d}\sigma} \, A_\mu(x). \tag{1.54}$$

Using the choice $\sigma = \tau$, we can view $q\dot{x}^\mu$ as the current j^μ induced by the particle and thus the interaction has the form $L_{\mathrm{em}} = -j^\mu A_\mu$. Any candidate for S_{em} should be invariant under a gauge transformation of the potential,

$$A_\mu(x) \to A_\mu(x) - \partial_\mu \Lambda(x). \tag{1.55}$$

This is the case, since the induced change in the action,

$$\delta_\Lambda S_{\mathrm{em}} = q \int_1^2 \mathrm{d}\sigma \, \frac{\mathrm{d}x^\mu}{\mathrm{d}\sigma} \frac{\partial \Lambda(x)}{\partial x^\mu} = q \int_1^2 \mathrm{d}\Lambda = q[\Lambda(2) - \Lambda(1)], \tag{1.56}$$

depends only on the endpoints. Thus $\delta_\Lambda S_{\mathrm{em}}$ vanishes keeping the endpoints fixed. Assuming that the Lagrangian is additive,

$$L = L_0 + L_{\mathrm{em}} = -m \left[\eta_{\mu\nu} \frac{\mathrm{d}x^\mu}{\mathrm{d}\sigma} \frac{\mathrm{d}x^\nu}{\mathrm{d}\sigma} \right]^{1/2} - q \frac{\mathrm{d}x^\mu}{\mathrm{d}\sigma} A_\mu(x) \tag{1.57}$$

the Lagrange equations give now

$$\frac{\mathrm{d}}{\mathrm{d}\sigma} \left[\frac{m\mathrm{d}x_\alpha/\mathrm{d}\sigma}{[\eta_{\mu\nu}\mathrm{d}x^\mu/\mathrm{d}\sigma \, \mathrm{d}x^\nu/\mathrm{d}\sigma]^{1/2}} + qA_\alpha \right] = q \, \frac{\mathrm{d}x^\lambda}{\mathrm{d}\sigma} \frac{\partial A_\lambda(x)}{\partial x^\alpha}. \tag{1.58}$$

Performing then the differentiation of $A(x(\sigma))$ with respect to σ and moving it to the RHS, we find

$$m\frac{\mathrm{d}}{\mathrm{d}\sigma} \left[\frac{\mathrm{d}x_\alpha/\mathrm{d}\sigma}{\mathrm{d}\tau/\mathrm{d}\sigma} \right] = q \left(\frac{\mathrm{d}x^\lambda}{\mathrm{d}\sigma} \frac{\partial A_\lambda}{\partial x^\alpha} - \frac{\mathrm{d}x^\lambda}{\mathrm{d}\sigma} \frac{\partial A_\alpha}{\partial x^\lambda} \right) = q \, \frac{\mathrm{d}x^\lambda}{\mathrm{d}\sigma} F_{\alpha\lambda}, \tag{1.59}$$

where we introduced the electromagnetic field-strength tensor $F_{\mu\nu} = \partial_\mu A_\nu - \partial_\nu A_\mu$. Choosing $\sigma = \tau$ we obtain the covariant version of the Lorentz equation,

$$m\frac{\mathrm{d}^2 x^\alpha}{\mathrm{d}\tau^2} = q \, F^\alpha{}_\lambda u^\lambda. \tag{1.60}$$

You should work through problem 1.9, if this equation and the covariant formulation of the Maxwell equations are not familiar to you.

Summary

The Lagrange and Hamilton function are connected by a Legendre transformation, $L(q^i, \dot{q}^i, t) = p_i \dot{q}^i - H(q^i, p_i, t)$. Lagrange's and Hamilton's equations follow extremizing the action $S[q^i] = \int_a^b dt\, L(q^i, \dot{q}^i, t)$ and $S[q^i, p_i] = \int_a^b dt\, [p_i \dot{q}^i - H(q^i, p_i, t)]$, respectively, keeping the endpoints a and b in coordinate space fixed. Knowing the Green function $G(t - t')$ of a linear system, we can find the solution $x(t)$ for an arbitrary external force $J(t)$ by integrating $G(t - t')$ with the weight $J(t)$.

Further reading. The series of Landau and Lifshitz on theoretical physics is a timeless resource for everybody studying and working in this field; its Volume 1 (Landau and Lifshitz, 1976) presents a succinct treatment of classical mechanics.

Problems

1.1 Units. ♣ a.) The four fundamental constants $\hbar$ (Planck's constant), c (velocity of light), G_N (gravitational constant) and k (Boltzmann constant) can be combined to obtain the dimension of a length, time, mass, energy and temperature. Find the relations and calculate the numerical values of two of them. What is the physical meaning of these "Planck units"? b.) Find the connection between a cross-section σ expressed in units of cm^2, mbarn and GeV^{-2}.

1.2 $d\delta = \delta d$. Use the definition (1.5) to show that variation and differentiation commute, i.e. that "$d\delta = \delta d$".

1.3 Higher derivatives. a.) Find the Lagrange equation for a Lagrangian containing higher derivatives, $L = L(q, \dot{q}, \ddot{q}, \ldots)$. b.) Consider $L = L(q, \dot{q}, \ddot{q})$ choosing as canonical variables $Q_1 = q$, $Q_2 = \dot{q}$, $P_1 = \frac{\partial L}{\partial \dot{q}} - \frac{d}{dt}\frac{\partial L}{\partial \ddot{q}}$ and $P_2 = \frac{\partial L}{\partial \ddot{q}}$ and defining as Hamiltonian $H(Q_1, Q_2, P_1, P_2) = \sum_{i=1}^{2} P_i q^{(i)} - L$. Show that the resulting Hamilton equations give the correct time evolution and that H corresponds to the energy. Why does H describe a unstable system?

1.4 Oscillator with friction. Consider a one-dimensional system described by the Lagrangian $L = \exp(2\alpha t) L_0$ and $L_0 = \frac{1}{2} m \dot{q}^2 - V(q)$. a.) Show that the equation of motion corresponds to an oscillator with friction term. b). Derive the energy lost per time dE/dt of the oscillator, with $E = \frac{1}{2} m \dot{q}^2 + V(q)$. c.) Show that the result in b.) agrees with the one obtained from the Lagrange equations of the first kind, $\frac{d}{dt}\frac{\partial L}{\partial \dot{q}} - \frac{\partial L}{\partial q} = Q$, where the generalised force Q perform the work $\delta A = Q\delta q$.

1.5 Classical driven oscillator. Consider an harmonic oscillator satisfying $\ddot{q}(t) - \Omega^2 q(t) = 0$ for $0 < t < T$ and $\ddot{q}(t) + \omega^2 q(t) = 0$ otherwise, with ω and Ω as real constants. a.) Show that for $q(t) = A_1 \sin(\omega t)$ for $t < 0$ and $\Omega T \gg 1$, the solution $q(t) = A_2 \sin(\omega_0 t + \alpha)$ with $\alpha = $ const. satisfies $A_2 \approx \frac{1}{2}(1 + \omega^2/\Omega^2)^{1/2} \exp(\Omega T)$. b.) If the oscillator was in the ground-state at $t < 0$, how many quanta are created?

1.6 Functional derivative. ♣ We define the derivative of a functional $F[\phi]$ by

$$\int dx\, \eta(x) \frac{\delta F[\phi]}{\delta \phi(x)} = \lim_{\varepsilon \to 0} \frac{1}{\varepsilon} \left\{ F[\phi + \varepsilon\eta] - F[\phi] \right\}.$$

a.) Find the functional derivative of $F[\phi] = \int dx\, \phi(x)$ and show thereby that $\delta\phi(x)/\delta\phi(x') = \delta(x - x')$. b.) Re-derive the Lagrange equations.

1.7 Conservation laws. Discuss the symmetries of the Galilean transformations and the resulting conservation laws, following the example of time-translation invariance and energy conservation.

1.8 Step function. Heaviside's step function $\vartheta(\tau)$ is defined by $\vartheta(\tau) = 0$ for $\tau < 0$ and $\vartheta(\tau) = 1$ for $\tau > 0$. a.) Use Chauchy's residuum theorem to show that the integral representation

$$\vartheta(\tau) = -\frac{1}{2\pi \mathrm{i}} \lim_{\varepsilon \to 0} \int_{-\infty}^{\infty} \mathrm{d}\omega \, \frac{\mathrm{e}^{-\mathrm{i}\omega\tau}}{\omega + \mathrm{i}\varepsilon}$$

is valid. b.) Show that $\mathrm{d}\vartheta(\tau)/\mathrm{d}\tau = \delta(\tau)$.

1.9 Electrodynamics. Compare Eq. (1.60) to the three-dimensional version of the Lorentz force and derive thereby the elements of the field-strength tensor $F_{\mu\nu}$. Find the Lorentz invariants that can be formed out of $F_{\mu\nu}$ and express them through $\boldsymbol{E}$ and $\boldsymbol{B}$. What is the meaning of the zero component of the Lorentz force?

1.10 Transformation between inertial frames. ♣ Consider two inertial frames K and K' with parallel axes at $t = t' = 0$ that are moving with the relative velocity v in the x direction. a.) Show that the linear transformation between the coordinates in K and K' can be written as $t' = At + Bx$, $x' = A(x - vt)$, $y' = y$, and $z' = z$. b.) Show that requiring (1.39) leads to Lorentz transformations. c.) What is the condition leading to Galilean transformations?

1.11 Relativity of simultaneity. ♣ Draw a spacetime diagram (in $d = 2$) for two inertial frames connected by a boost with velocity β. What are the angles between the axes t and t', x and x'? Draw lines of constant t and t' and convince yourself that the time order of two space-like events is not invariant.

1.12 Wave equation for a string. Consider a string of length L, mass density ρ and tension κ in one spatial dimension. Denoting its deviation from its equilibrium position x_0 with $\phi(x,t) \equiv x(t) - x_0$, write down its kinetic and potential energy (density) and the corresponding action. Derive its equation of motion. [Note: $\phi(x,t)$ depends on t and x, and the Lagrange equation for $L(\phi, \partial_t\phi, \partial_x\phi)$ will contain $\mathrm{d}/\mathrm{d}t$ and $\mathrm{d}/\mathrm{d}x$ terms.]

2
Quantum mechanics

The main purpose of this chapter is to introduce Feynman's path integral as an alternative to the standard operator approach to quantum mechanics. Most of our discussion of quantum fields will be based on this approach and thus becoming familiar with this technique using the simpler case of quantum mechanics is of central importance. Instead of employing the path integral directly, we will use as a basic tool the vacuum persistence amplitude $\langle 0, \infty | 0, -\infty \rangle_J$. This quantity is the probability amplitude that a system under the influence of an external force J stays in its ground-state. Since we can apply an arbitrary force J, the amplitude $\langle 0, \infty | 0, -\infty \rangle_J$ contains all information about the system. Moreover, it serves as a convenient tool to calculate Green functions which will become our main target studying quantum field theories.

2.1   Reminder of the operator approach

A classical system described by a Hamiltonian $H(q^i, p_i, t)$ can be quantised promoting q^i and p_i to operators[1] $\hat{q}^i$ and $\hat{p}_i$ which satisfy the canonical commutation relations $[\hat{q}^i, \hat{p}_j] = \mathrm{i}\delta^i_j$. The latter are the formal expression of Heisenberg's uncertainty relation. Apart from ordering ambiguities, the Hamilton operator $H(\hat{q}^i, \hat{p}_i, t)$ can be directly read from the Hamiltonian $H(q^i, p_i, t)$. The basic features of any quantum theory can be synthesised into a few principles.

General principles.   A physical system in a pure state is fully described by a probability amplitude

$$\psi(a, t) = \langle a | \psi(t) \rangle \in \mathbb{C}, \tag{2.1}$$

where $\{a\}$ is a set of quantum numbers specifying the system and the states $|\psi(t)\rangle$ form a complex Hilbert space. The probability p to find the specific values a_* in a measurement is given by $p(a_*) = |\psi(a_*, t)|^2$. The possible values a_* are the eigenvalues of Hermitian operators $\hat{A}$ whose eigenvectors $|a\rangle$ form an orthogonal, complete basis. In Dirac's bra-ket notation, we can express these statements by

$$\hat{A}|a\rangle = a|a\rangle, \qquad \langle a|a'\rangle = \delta(a - a'), \qquad \int \mathrm{d}a\, |a\rangle\langle a| = 1, \tag{2.2}$$

In general, operators do not commute. Their commutation relations can be obtained by the replacement $\{A, B\} \to \mathrm{i}[\hat{A}, \hat{B}]$ in the definition (1.24) of the Poisson brackets.

---

[1] When there is the danger of an ambiguity, operators will be written with a "hat"; otherwise we drop it.

Quantum Fields–From the Hubble to the Planck Scale. Michael Kachelriess. © Michael Kachelriess 2018.
Published in 2018 by Oxford University Press. DOI 10.1093/oso/9780198802877.001.0001

The state of a particle moving in one dimension in a potential $V(q)$ can be described either by the eigenstates of the position operator $\hat{q}$ or of the momentum operator $\hat{p}$. Both eigenstates form a complete, orthonormal basis, and they are connected by a Fourier transformation which we choose to be asymmetric,

$$\psi(q) = \langle q | \psi \rangle = \int \frac{\mathrm{d}p}{2\pi} \, \mathrm{e}^{\mathrm{i}px} \, \psi(p) = \int \frac{\mathrm{d}p}{2\pi} \, \langle q | p \rangle \, \langle p | \psi \rangle \tag{2.3a}$$

$$\psi(p) = \langle p | \psi \rangle = \int \mathrm{d}q \, \mathrm{e}^{-\mathrm{i}kx} \, \psi(x) = \int \mathrm{d}q \, \langle p | q \rangle \, \langle q | \psi \rangle. \tag{2.3b}$$

Choosing this normalisation has the advantage that the factor $1/(2\pi)$ in the Fourier integral over momenta is the same as in the density of free states, $L \, \mathrm{d}p/(2\pi)$, which will enter quantities like decay rates or cross-sections. From Eq. (2.3), it follows that the asymmetry in the Fourier transformation is reflected in the completeness relation of the states,

$$\int \mathrm{d}q \, |q\rangle \, \langle q| = 1 \quad \text{and} \quad \int \frac{\mathrm{d}p}{2\pi} \, |p\rangle \, \langle p| = 1. \tag{2.4}$$

Time evolution.   Since the states $|\psi(t)\rangle$ form a complex Hilbert space, the superposition principle is valid: If ψ_1 and ψ_2 are possible states of the system, then also

$$\psi(t) = c_1\psi_1(t) + c_2\psi_2(t), \qquad c_i \in \mathbb{C}. \tag{2.5}$$

In quantum mechanics, a stronger version of this principle holds which states that if $\psi_1(t)$ and $\psi_2(t)$ describe the possible time evolution of the system, then so does also the superposed state $\psi(t)$. This implies that the time evolution is described by a linear, homogeneous differential equation. Choosing it as first order in time, we can write the evolution equation as

$$\mathrm{i}\partial_t |\psi(t)\rangle = D|\psi(t)\rangle, \tag{2.6}$$

where the differential operator D on the RHS has to be still determined.

We call the operator describing the evolution of a state from $\psi(t)$ to $\psi(t')$ the time-evolution operator $U(t', t)$. This operator is unitary, $U^{-1} = U^{\dagger}$, in order to conserve probability and forms a group, $U(t_3, t_1) = U(t_3, t_2)U(t_2, t_1)$ with $U(t, t) = 1$. For an infinitesimal time step δt,

$$|\psi(t + \delta t)\rangle = U(t + \delta t, t) \, |\psi(t)\rangle, \tag{2.7}$$

we can set with $U(t, t) = 1$

$$U(t + \delta t, t) = 1 - \mathrm{i}H\delta t. \tag{2.8}$$

Here we introduced the generator of infinitesimal time-translations H. The analogy to classical mechanics suggests that H is the operator version of the classical Hamilton function $H(q, p)$. Inserting Eq. (2.8) into (2.7) results in

$$\frac{|\psi(t + \delta t)\rangle - |\psi(t)\rangle}{\delta t} = -\mathrm{i}H \, |\psi(t)\rangle. \tag{2.9}$$

Comparing then Eqs. (2.6) and (2.9) reveals that the operator D on the RHS of Eq. (2.6) coincides with the Hamiltonian H. We call a time-evolution equation of this type for arbitrary H Schrödinger equation.

Next we want to determine the connection between H and U. Plugging $\psi(t) = U(t,0)\psi(0)$ in the Schrödinger equation gives

$$\left[i\frac{\partial U(t,0)}{\partial t} - HU(t,0)\right]\psi(0) = 0. \tag{2.10}$$

Since this equation is valid for an arbitrary state $\psi(0)$, we can rewrite it as an operator equation,

$$i\partial_{t'}U(t',t) = HU(t',t). \tag{2.11}$$

Integrating it, we find as formal solution

$$U(t',t) = 1 - i\int_t^{t'} dt''\,H(t'')U(t'',t) \tag{2.12}$$

or, if H is time-independent,

$$U(t,t') = \exp(-iH(t-t')). \tag{2.13}$$

Up to now, we have considered the Schrödinger picture where operators are constant and the time evolution is given by the change in the state vectors $|\psi(t)\rangle$. In the Heisenberg picture, the time evolution is driven completely by the one of the operators. States and operators in the two pictures are connected by

$$O_S(t) = U(t,t_0)O_H(t)U^\dagger(t,t_0), \tag{2.14a}$$
$$|\psi_S(t)\rangle = U(t,t_0)|\psi_H\rangle, \tag{2.14b}$$

if they agree at the time t_0.

Propagator.   We insert the solution of U for a time-independent H into $|\psi(t')\rangle = U(t',t)|\psi(t)\rangle$ and multiply from the left with $\langle q'|$,

$$\psi(q',t') = \langle q'|\psi(t')\rangle = \langle q'|\exp[-iH(t'-t)]|\psi(t)\rangle. \tag{2.15}$$

Then we insert $1 = \int d^3q|q\rangle\langle q|$,

$$\psi(q',t') = \int d^3q\,\langle q'|\exp[-iH(t'-t)]|q\rangle\langle q|\psi(t)\rangle = \int d^3q\,K(q',t';q,t)\psi(q,t). \tag{2.16}$$

In the last step we introduced the propagator or Green function K in its coordinate representation,

$$K(q',t';q,t) = \langle q'|\exp[-iH(t'-t)]|q\rangle. \tag{2.17}$$

The Green function K equals the probability amplitude for the propagation between two spacetime points; $K(q',t';q,t)$ is therefore also called more specifically two-point

Green function. We can express the propagator K by the solutions of the Schrödinger equation, $\psi_n(q,t) = \langle q|n(t)\rangle = \langle q|n\rangle \exp(-iE_n t)$ as

$$K(q',t';q,t) = \sum_{n,n'} \langle q'|n\rangle \underbrace{\langle n|\exp(-iH(t'-t))|n'\rangle}_{\delta_{n,n'}\exp(-iE_n(t'-t))} \langle n'|q\rangle$$

$$= \sum_n \psi_n(q')\psi_n^*(q)\exp(-iE_n(t'-t)),$$

(2.18)

where n represents the complete set of quantum numbers specifying the energy eigenvalues of the system. Note that this result is very general and holds for any time-independent Hamiltonian.

Let us compute the propagator of a free particle in one dimension, described by the Hamiltonian $H = p^2/2m$. We write with $\tau = t' - t$

$$K(q',t';q,t) = \langle q'|e^{-iH\tau}|q\rangle = \langle q'|e^{-i\tau\hat{p}^2/2m} \int \frac{dp}{2\pi}|p\rangle\langle p|q\rangle$$

$$= \int \frac{dp}{2\pi} e^{-i\tau p^2/2m} \langle q'|p\rangle\langle p|q\rangle = \int \frac{dp}{2\pi} e^{-i\tau p^2/2m+i(q'-q)p},$$

(2.19)

where we used $\langle q'|p\rangle = \exp(iq'p)$ in the last step. The integral is Gaussian if we add an infinitesimal factor $\exp(-\varepsilon p^2)$ to the integrand in order to ensure the convergence of the integral. Thus the physical value of the energy $E = p^2/(2m)$ seen as a complex variable is approached from the negative imaginary plane, $E \to E - i\varepsilon$. Taking afterwards the limit $\varepsilon \to 0$, we obtain

$$K(q',t';q,t) = \left(\frac{m}{2\pi i\tau}\right)^{1/2} e^{im(q'-q)^2/2\tau}.$$

(2.20)

Knowing the propagator, we can calculate the solution $\psi(t')$ at any time t' for a given initial state $\psi(t)$ via Eq. (2.16).

Example 2.1: Calculate the integrals $A = \int dx \exp(-x^2/2)$, $B = \int dx \exp(-ax^2/2 + bx)$, and $C = \int dx \cdots dx_n \exp(-x^T Ax/2 + J^T x)$ for a symmetric $n \times n$ matrix A.
a.) We square the integral and calculate then A^2 introducing polar coordinates, $r^2 = x^2 + y^2$,

$$A^2 = \int_{-\infty}^{\infty} dx \int_{-\infty}^{\infty} dy \exp(-(x^2+y^2)/2) = 2\pi \int_0^{\infty} dr\, re^{-r^2/2} = 2\pi \int_0^{\infty} dt\, e^{-t} = 2\pi,$$

where we substituted $t = r^2/2$. Thus the result for the basic Gaussian integral is $A = \sqrt{2\pi}$. All other solvable variants of Gaussian integrals can be reduced to this result.
b.) We complete the square in the exponent,

$$-\frac{a}{2}\left(x^2 - \frac{2b}{a}x\right) = -\frac{a}{2}\left(x - \frac{b}{a}\right)^2 + \frac{b^2}{2a},$$

and shift then the integration variable to $x' = x - b/a$. The result is

$$B = \int_{-\infty}^{\infty} \mathrm{d}x \exp(-ax^2/2 + bx) = e^{b^2/2a} \int_{-\infty}^{\infty} \mathrm{d}x' \exp(-ax'^2/2) = \sqrt{\frac{2\pi}{a}} \, e^{b^2/2a} . \quad (2.21)$$

c.) We should complete again the square and try $x' = x - A^{-1}J$. With

$$(x - A^{-1}J)^T A(x - A^{-1}J) = x^T Ax - x^T AA^{-1}J - J^T A^{-1}Ax + J^T A^{-1}AA^{-1}J$$
$$= x^T Ax - 2J^T x + J^T A^{-1}J,$$

we obtain after shifting the integration vector,

$$C = \exp\left(J^T A^{-1}J/2\right) \int \mathrm{d}x_1' \cdots \mathrm{d}x_n' \exp\left(-x'^T Ax'/2\right) . \quad (2.22)$$

Since the matrix A is symmetric, we can diagonalise A via an orthogonal transformation $D = OAO^T$. This corresponds to a rotation of the integration variables, $y = Ox'$. The Jacobian of this transformation is one, and thus the result is

$$C = \exp(J^T A^{-1}J/2) \prod_{i=1}^{n} \int \mathrm{d}y_i \exp(-a_i y_i^2/2) = \sqrt{\frac{(2\pi)^n}{\det A}} \, \exp\left(\frac{1}{2}J^T A^{-1}J\right) . \quad (2.23)$$

In the last step we expressed the product of eigenvalues a_i as the determinant of A.

2.2 Path integrals in quantum mechanics

In problem 2.1 you are asked to calculate the classical action of a free particle and of a harmonic oscillator and to compare them to the corresponding propagators found in quantum mechanics. Surprisingly, you will find that in both cases the propagator can be written as $K(q', t'; q, t) = N \exp(iS)$ where S is the classical action along the path $[q(t) : q'(t')]$ and N a normalisation constant. This suggests that we can reformulate quantum mechanics, replacing the standard operator formalism used to evaluate the propagator (2.17) "somehow" by the classical action.

To get an idea how to proceed, we look at the famous double-slit experiment sketched in Fig. 2.1: According to the superposition principle, the amplitude A for a particle to move from the source at q_1 to the detector at q_2 is the sum of the amplitudes A_i for the two possible paths,

$$A = K(q_2, t_2; q_1, t_1) = \sum_{\text{paths}} A_i. \quad (2.24)$$

Clearly, we could add in a gedankenexperiment more and more screens between q_1 and q_2, increasing at the same time the number of holes. Although in this way we replace continuous spacetime by a discrete lattice, the differences between these two descriptions should vanish for sufficiently small spacing τ. Moreover, for $\tau \to 0$, we can expand $U(\tau) = \exp(-iH\tau) \simeq 1 - iH\tau$. Applying then $H = \hat{p}^2/(2m) + V(\hat{q})$ to eigenfunctions $|q\rangle$ of $V(\hat{q})$ and $|p\rangle$ of $\hat{p}^2$, we can replace the operator H by its eigenvalues. In this way, we hope to express the propagator as a sum over paths, where the individual amplitudes A_i contain only classical quantities.

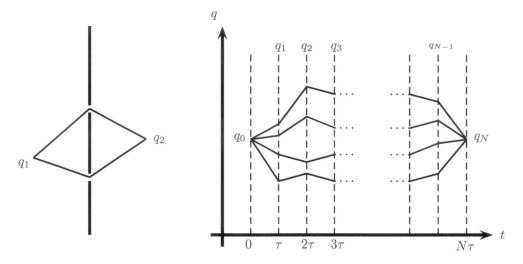

Fig. 2.1 Left: The double slit experiment. Right: The propagator $K(q_N, \tau; q_0, 0)$ expressed as a sum over all N-legged continuous paths.

We apply now this idea to a particle moving in one dimension in a potential $V(q)$. The transition amplitude A for the evolution from the state $|q, 0\rangle$ to the state $|q', t'\rangle$ is

$$A \equiv K(q', t'; q, 0) = \langle q'| e^{-iHt'} |q\rangle. \tag{2.25}$$

This amplitude equals the matrix element of the propagator K for the evolution from the initial point $q(0)$ to the final point $q'(t')$. Let us split the time evolution into two smaller steps, writing $e^{-iHt'} = e^{-iH(t'-t_1)} e^{-iHt_1}$. Inserting also $\int dq_1 |q_1\rangle \langle q_1| = 1$, the amplitude becomes

$$A = \int dq_1 \langle q'| e^{-iH(t'-t_1)} |q_1\rangle \langle q_1| e^{-iHt_1} |q\rangle = \int dq_1 \, K(q', t'; q_1, t_1) K(q_1, t_1; q, 0). \tag{2.26}$$

This formula expresses simply the group property, $U(t', 0) = U(t', t_1)U(t_1, 0)$, of the time evolution operator U evaluated in the basis of the continuous variable q. More physically, we can view this equation as an expression of the quantum mechanical rule for combining amplitudes. If the same initial and final states can be connected by various ways, the amplitudes for each of these processes should be added. A particle propagating from q to q' must be somewhere at the intermediate time t_1. Labelling this intermediate position as q_1, we compute the amplitude for propagation via the point q_1 as the product of the two propagators in Eq. (2.26) and integrate over all possible intermediate positions q_1.

We continue to divide the time interval t' into a large number N of time intervals of duration $\tau = t'/N$. Then the propagator becomes

$$A = \langle q'| \underbrace{e^{-iH\tau} e^{-iH\tau} \cdots e^{-iH\tau}}_{N \text{ times}} |q\rangle. \tag{2.27}$$

We insert again a complete set of states $|q_i\rangle$ between each exponential, obtaining

$$A = \int dq_1 \cdots dq_{N-1} \langle q'| e^{-iH\tau} |q_{N-1}\rangle \langle q_{N-1}| e^{-iH\tau} |q_{N-2}\rangle \cdots \langle q_1| e^{-iH\tau} |q\rangle$$

$$\equiv \int dq_1 \cdots dq_{N-1} K_{q_N,q_{N-1}} K_{q_{N-1},q_{N-2}} \cdots K_{q_2,q_1} K_{q_1,q_0}, \qquad (2.28)$$

where we have defined $q_0 = q$ and $q_N = q'$. Note that these initial and final positions are fixed and therefore are not integrated. Figure 2.1 illustrates that we can view the amplitude A as the integral over the partial amplitudes A_{path} of the individual N-legged continuous paths.

We ignore the problem of defining properly the limit $N \to \infty$, keeping N large but finite. We rewrite the amplitude as sum over the amplitudes for all possible paths, $A = \sum_{\text{paths}} A_{\text{path}}$, with

$$\sum_{\text{paths}} = \int dq_1 \cdots dq_{N-1}, \qquad A_{\text{path}} = K_{q_N,q_{N-1}} K_{q_{N-1},q_{N-2}} \cdots K_{q_2,q_1} K_{q_1,q_0}.$$

Let us look at the last expression in detail. We can expand the exponential in each propagator $K_{q_{j+1},q_j} = \langle q_{j+1}| e^{-iH\tau} |q_j\rangle$ for a single sub-interval, because τ is small,

$$K_{q_{j+1},q_j} = \langle q_{j+1}| \left(1 - iH\tau - \frac{1}{2}H^2\tau^2 + \cdots \right) |q_j\rangle$$

$$= \langle q_{j+1}| q_j\rangle - i\tau \langle q_{j+1}| H |q_j\rangle + \mathcal{O}(\tau^2). \qquad (2.29)$$

In the second term of (2.29), we insert a complete set of momentum eigenstates between H and $|q_j\rangle$. This gives

$$-i\tau \langle q_{j+1}| \left(\frac{\hat{p}^2}{2m} + V(\hat{q})\right) \int \frac{dp_j}{2\pi} |p_j\rangle \langle p_j| q_j\rangle$$

$$= -i\tau \int \frac{dp_j}{2\pi} \left(\frac{p_j^2}{2m} + V(q_{j+1})\right) \langle q_{j+1}| p_j\rangle \langle p_j| q_j\rangle \qquad (2.30)$$

$$= -i\tau \int \frac{dp_j}{2\pi} \left(\frac{p_j^2}{2m} + V(q_{j+1})\right) e^{ip_j(q_{j+1}-q_j)}.$$

The expression (2.30) is not symmetric in q_j and q_{j+1}. The reason for this asymmetry is that we could have inserted the factor 1 either to the right or to the left of the Hamiltonian H. In the latter case, we would have obtained p_{j+1} and $V(q_j)$ in (2.30). Since the difference $[V(q_{j+1}) - V(q_j)]\tau \simeq V'(q_j)(q_{j+1} - q_j)\tau \simeq V'(q_j)\dot{q}_j\tau^2$ is of order τ^2, the ordering problem should not matter in the continuum limit which we will take eventually; we set therefore $V(q_{j+1}) \simeq V(q_j)$.

The first term of (2.29) gives a delta function, which we can express as

$$\langle q_{j+1}| q_j\rangle = \delta(q_{j+1} - q_j) = \int \frac{dp_j}{2\pi} e^{ip_j(q_{j+1}-q_j)}. \qquad (2.31)$$

Now we can combine the two terms, obtaining as propagator for the step $q_j \to q_{j+1}$

$$K_{q_{j+1},q_j} = \int \frac{dp_j}{2\pi} \, e^{ip_j(q_{j+1}-q_j)} \left[1 - i\tau \left(\frac{p_j{}^2}{2m} + V(q_j) \right) + \mathcal{O}(\tau^2) \right]. \qquad (2.32)$$

Since we work at $\mathcal{O}(\tau)$, we can exponentiate the factor in the square bracket,

$$1 - i\tau \, H(p_j, q_j) + \mathcal{O}(\tau^2) = e^{-i\tau H(p_j, q_j)}. \qquad (2.33)$$

Next we rewrite the exponent in the first factor of Eq. (2.32) using $\dot{q}_j = (q_{j+1} - q_j)/\tau$, such that we can factor out the time-interval τ. The amplitude A_{path} consists of N such factors. Combining them, we obtain

$$A_{\mathrm{path}} = \left(\prod_{j=0}^{N-1} \int \frac{dp_j}{2\pi} \right) \exp i\tau \sum_{j=0}^{N-1} [p_j \dot{q}_j - H(p_j, q_j)]. \qquad (2.34)$$

We recognise the argument of the exponential as the discrete approximation of the action $S[q, p]$ in the Palatini form of a path passing through the points $q_0 = q, q_1, \cdots, q_{N-1}, q_N = q'$. The propagator $K = \int dq_1 \cdots dq_{N-1} A_{\mathrm{path}}$ becomes then

$$K = \left(\prod_{j=1}^{N-1} \int dq_j \right) \left(\prod_{j=0}^{N-1} \int \frac{dp_j}{2\pi} \right) \exp i\tau \sum_{j=0}^{N-1} [p_j \dot{q}_j - H(p_j, q_j)]. \qquad (2.35)$$

For $N \to \infty$, this expression approximates an integral over all functions $p(t)$, $q(t)$ consistent with the boundary conditions $q(0) = q$, $q(t') = q'$. We adopt the notation $\mathcal{D}p\mathcal{D}q$ for the *functional* or *path integral* over all functions $p(t)$ and $q(t)$,

$$K \equiv \int \mathcal{D}p(t)\mathcal{D}q(t) e^{iS[q,p]} = \int \mathcal{D}p(t)\mathcal{D}q(t) \exp \left(i \int_0^{t'} dt \, (p\dot{q} - H(p, q)) \right). \qquad (2.36)$$

This result expresses the propagator as a path integral in phase space. It allows us to obtain for any classical system which can be described by a Hamiltonian the corresponding quantum dynamics.

If the Hamiltonian is of the form $H = p^2/2m + V$, as we have assumed[2] in our derivation, we can carry out the quadratic momentum integrals in (2.35). We can rewrite this expression as

$$K = \left(\prod_{j=1}^{N-1} \int dq_j \right) \exp -i\tau \sum_{j=0}^{N-1} V(q_j) \left(\prod_{j=0}^{N-1} \int \frac{dp_j}{2\pi} \right) \exp i\tau \sum_{j=0}^{N-1} (p_j \dot{q}_j - p_j^2/2m). $$
$$(2.37)$$

The p integrals are all uncoupled Gaussians. One such integral gives

$$\int \frac{dp}{2\pi} \, e^{i\tau(p\dot{q}-p^2/2m)} = \sqrt{\frac{m}{2\pi i\tau}} \, e^{i\tau m\dot{q}^2/2}, \qquad (2.38)$$

---

[2]Since we evaluated $\exp(-iH\tau)$ for infinitesimal τ, the result (2.36) holds also for a time-dependent potential $V(q, t)$.

where we added again an infinitesimal factor $\exp(-\varepsilon p^2)$ to the integrand. Thence the propagator becomes

$$K = \left(\frac{m}{2\pi i\tau}\right)^{N/2} \left(\prod_{j=1}^{N-1} \int dq_j\right) \exp i\tau \sum_{j=0}^{N-1} \left(\frac{m\dot{q}_j^2}{2} - V(q_j)\right). \tag{2.39}$$

The argument of the exponential is again a discrete approximation of the action $S[q]$ of a path passing through the points $q_0 = q, q_1, \cdots, q_{N-1}, q_N = q'$, but now seen as functional of only the coordinate q. As earlier, we can write this in a more compact form as

$$K = \langle q_f, t_f | q_i, t_i \rangle = \int \mathcal{D}q(t) e^{iS[q]} = \int \mathcal{D}q(t) \exp\left(i \int_{t_i}^{t_f} dt\, L(q, \dot{q})\right), \tag{2.40}$$

where the integration includes all paths satisfying the boundary condition $q(t_i) = q_i$ and $q(t_f) = q_f$. This is the main result of this section, and is known as the *path integral in configuration space*. It will serve us as starting point discussing quantum field theories of bosonic fields.

Knowing the path integral and thus the propagator is sufficient to solve scattering problems in quantum mechanics. In a relativistic theory, the particle number during the course of a scattering process is however not fixed, since energy can be converted into matter. In order to prepare us for such more complex problems, in the next section we will generalise the path integral to a *generating functional* for *n-point Green functions*. In this formalism, the usual propagator giving the probability amplitude that a single particle moves from $q_i(t_i)$ to $q_f(t_f)$ becomes the special case of a two-point Green function, while Green functions with $n > 2$ describe processes involving more points. For instance, the four-point Green function will be the essential ingredient to calculate $2 \to 2$ scattering processes in a quantum field theory (QFT). The corresponding generating functional is the quantity which n.th derivative returns the n-point Green functions.

2.3   Generating functional for Green functions

Having re-expressed the transition amplitude $\langle q_f, t_f | q_i, t_i \rangle$ of a quantum mechanical system as a path integral, we first want to generalise this result to the matrix elements of an arbitrary potential $V(q)$ between the states $|q_i, t_i\rangle$ and $|q_f, t_f\rangle$. For all practical purposes, we can assume that we can expand $V(q)$ as a power series in q; thus it is sufficient to consider the matrix elements $\langle q_f, t_f | q^m | q_i, t_i \rangle$. In a QFT, the initial and final states are generally free particles which are described mathematically as harmonic oscillators. In this case, we are able to reconstruct all excited states $|n\rangle$ from the ground-state,

$$|n\rangle = \frac{1}{\sqrt{n!}} (a^\dagger)^n |0\rangle.$$

Therefore it will be sufficient to study matrix elements between the ground-state $|0\rangle$. With this choice, we are able to extend the integration limit in the path integral (2.40) to $t = \pm\infty$. This will not only simplify its evaluation but also avoid the need to choose a specific inertial frame. As a result, the generating functional will have an obviously Lorentz invariant form in a relativistic theory.

Time-ordered products of operators and the path integral.   In a first step, we try to include the operator q^m into the transition amplitude $\langle q_f, t_f | q_i, t_i \rangle$. We can reinterpret our result for the path integral as follows,

$$\langle q_f, t_f | \mathbf{1} | q_i, t_i \rangle = \int \mathcal{D}q(t)\, 1\, \mathrm{e}^{\mathrm{i}S[q]}. \tag{2.41}$$

Thus we can see the LHS as matrix element of the unit operator $\mathbf{1}$, while the RHS corresponds to the path-integral average of the classical function $f(q, \dot{q}) = 1$. Now we want to generalise this rather trivial statement to two operators $\hat{A}(t_a)$ and $\hat{B}(t_b)$ given in the Heisenberg picture. In evaluating the unknown function f on the RHS of

$$\int \mathcal{D}q(t)\, A(t_a)B(t_b)\, \mathrm{e}^{\mathrm{i}S[q(t)]} = \langle q_f, t_f | f\{A(t_a)B(t_b)\} | q_i, t_i \rangle, \tag{2.42}$$

we go back to Eq. (2.28) and insert $\hat{A}(t_a)$ and $\hat{B}(t_b)$ at the correct intermediate times,

$$= \begin{cases} \int \mathrm{d}q_1 \cdots \mathrm{d}q_{N-1} \cdots \langle q_{a+1}, t_{a+1} | \hat{A} | q_a, t_a \rangle \cdots \langle q_{b+1}, t_{b+1} | \hat{B} | q_b, t_b \rangle \cdots & \text{for } t_a > t_b, \\ \int \mathrm{d}q_1 \cdots \mathrm{d}q_{N-1} \cdots \langle q_{b+1}, t_{b+1} | \hat{B} | q_b, t_b \rangle \cdots \langle q_{a+1}, t_{a+1} | \hat{A} | q_a, t_a \rangle \cdots & \text{for } t_a < t_b. \end{cases} \tag{2.43}$$

Since the time along a classical path increases, the matrix elements of the operators $\hat{A}(t_a)$ and $\hat{B}(t_b)$ are also ordered with time increasing from the right to the left. If we define the *time-ordered product* of two operators as

$$T\{\hat{A}(t_a)\hat{B}(t_b)\} = \hat{A}(t_a)\hat{B}(t_b)\vartheta(t_a - t_b) + \hat{B}(t_b)\hat{A}(t_a)\vartheta(t_b - t_a), \tag{2.44}$$

then the path-integral average of the classical quantities $A(t_a)$ and $B(t_b)$ corresponds to the matrix element of the time-ordered product of these two operators,

$$\langle q_f, t_f | T\{\hat{A}(t_a)\hat{B}(t_b)\} | q_i, t_i \rangle = \int \mathcal{D}q(t)\, A(t_a)B(t_b)\, \mathrm{e}^{\mathrm{i}S[q(t)]}, \tag{2.45}$$

and similar for more than two operators.

External sources.   Next in our formalism, we want to include the possibility that we can change the state of our system by applying an external driving force or source term $J(t)$. In quantum mechanics, we could imagine, for example, a harmonic oscillator in the ground-state $|0\rangle$, making a transition under the influence of an external force J to the state $|n\rangle$ at the time t and back to the ground-state $|0\rangle$ at the time $t' > t$. Including such transitions, we can mimic the relativistic process of particle creation and annihilation as follows. We identify the vacuum (i.e. the state containing zero real particles) with the ground-state of the quantum mechanical system, and the creation and annihilation of particles with the (de-) excitation of states that have a higher energy than the ground-state by an external source J. Schwinger realised that adding a linear coupling to an external source,

$$L \to L + J(t)q(t), \tag{2.46}$$

also leads to an efficient way to calculate the matrix elements of an arbitrary polynomial of operators $q(t_n) \cdots q(t_1)$. If the source $J(t)$ would be a simple number instead of a time-dependent function in the augmented path integral

$$\langle q_f, t_f | q_i, t_i \rangle_J \equiv \int \mathcal{D}q(t) e^{i \int_{t_i}^{t_f} dt(L+Jq)}, \tag{2.47}$$

then we could obtain $\langle q_f, t_f | q^m | q_i, t_i \rangle_J$ simply by differentiating $\langle q_f, t_f | q_i, t_i \rangle_J$ m-times with respect to J. However, the LHS is a functional of $J(t)$ and thus we need to perform instead functional derivatives with respect to $J(t)$. By analogy with the rules for the differentiation of functions, e.g. $\partial x^l / \partial x^k = \delta^l_k$, we define a functional derivative as

$$\frac{\delta}{\delta J(x)} 1 = 0 \quad \text{and} \quad \frac{\delta J(x)}{\delta J(x')} = \delta(x - x'). \tag{2.48}$$

Thus we replace for a continuous index the Kronecker delta by a delta function. Moreover, we assume that the Leibniz and the chain rule holds for sufficiently nice functions $J(x)$. As the notation suggest, the variation of a functional defined in Eq. (1.5) is the special case of a *directional* functional derivative, cf. problem 1.6

Now we are able to differentiate $\langle q_f, t_f | q_i, t_i \rangle_J$ with respect to the source J. Starting from

$$\frac{\delta}{\delta J(t_1)} \int \mathcal{D}q(t)\, e^{i \int_{t_i}^{t_f} dt J(t) q(t)} = i \int \mathcal{D}q(t)\, q(t_1) e^{i \int_{t_i}^{t_f} dt J(t) q(t)}, \tag{2.49}$$

we obtain

$$\langle q_f, t_f | T\{\hat{q}(t_1) \cdots \hat{q}(t_n)\} | q_i, t_i \rangle = (-i)^n \frac{\delta^n}{\delta J(t_1) \cdots \delta J(t_n)} \langle q_f, t_f | q_i, t_i \rangle_J \bigg|_{J=0}. \tag{2.50}$$

Thus the source $J(t)$ is a convenient tool to obtain the functions $q(t_1) \cdots q(t_n)$ in front of $\exp(iS)$. Having performed the functional derivatives, we set the source $J(t)$ to zero, coming back to the usual path integral. Physically, the expression (2.50) corresponds to the probability amplitude that a particle moves from $q_i(t_i)$ to $q_f(t_f)$, having the intermediate positions $q(t_1), \ldots, q(t_n)$.

Vacuum persistence amplitude. As a last step, we want to eliminate the initial and final states $|q_i, t_i\rangle_J$ and $|q_f, t_f\rangle$ in favour of the ground-state or vacuum, $|0\rangle$. In this way, we convert the transition amplitude $\langle q_f, t_f | q_i, t_i \rangle_J$ into the probability amplitude that a system which was in the ground-state $|0\rangle$ at $t_i \to -\infty$ remains in this state at $t_f \to \infty$ despite the action of the source $J(t)$. Inserting a complete set of energy eigenstates, $1 = \sum_n |n\rangle\langle n|$, into the propagator, we obtain

$$\langle q', t' | q, t \rangle = \sum_n \psi_n(q') \psi_n^*(q) \exp(-iE_n(t' - t)). \tag{2.51}$$

We can isolate the ground-state $n = 0$ by adding either to the energies E_n or to the time difference $\tau = (t' - t)$ a small negative imaginary part. In this case, all terms are exponentially damped in the limit $\tau \to \infty$, and the ground-state as the state with the smallest energy dominates more and more the sum. Alternatively, we can add a term $+i\varepsilon q^2$ to the Lagrangian.

Remark 2.1: Wick rotation and Euclidean action. Instead of adding the infinitesimally small term $i\varepsilon q^2$ to the Lagrangian, we can do a more drastic change, rotating in the action the time axis clockwise by 90° in the complex plane. Inserting $t_E = \mathrm{i}t$ into $x_\mu x^\mu$, we see that this procedure called Wick rotation corresponds to the transition from Minkowski to Euclidean space,

$$x^2 = t^2 - \boldsymbol{x}^2 = (-\mathrm{i}t_E)^2 - \boldsymbol{x}^2 = -[t_E^2 + \boldsymbol{x}^2] = -x_E^2 \,.$$

Performing the changes $t = -\mathrm{i}t_E$ and $\mathrm{d}t = -\mathrm{i}\mathrm{d}t_E$ in the action of a particle moving in an one-dimensional potential gives

$$S = -\mathrm{i} \int \mathrm{d}t_E \left(-\frac{1}{2}m\dot{q}_E^2 - V(q) \right) \equiv \mathrm{i}S_E \,. \tag{2.52}$$

Note that the Euclidean action $S_E = T + V$ is bounded from below. The phase factor in the path integral transforms as $\mathrm{e}^{\mathrm{i}S} = \mathrm{e}^{-S_E}$, and thus contributions with large S_E are exponentially damped in the Euclidean path integral.

Finally, we have only to connect the results we obtained so far. Adding a coupling to an external source $J(t)$ and a damping factor $+\mathrm{i}\varepsilon q^2$ to the Lagrangian gives us the ground-state or vacuum-persistence amplitude

$$Z[J] \equiv \langle 0, \infty | 0, -\infty \rangle_J = \int \mathcal{D}q(t)\, \mathrm{e}^{\mathrm{i} \int_{-\infty}^{\infty} \mathrm{d}t(L + Jq + \mathrm{i}\varepsilon q^2)} \tag{2.53}$$

in the presence of a classical source J. This amplitude is a functional of J which we denote by $Z[J]$. Taking derivatives w.r.t. the external sources J, and setting them afterwards to zero, we obtain

$$\left. \frac{\delta^n Z[J]}{\delta J(t_1) \cdots \delta J(t_n)} \right|_{J=0} = \mathrm{i}^n \int \mathcal{D}q(t)\, q(t_1) \cdots q(t_n) \mathrm{e}^{\mathrm{i} \int_{-\infty}^{\infty} \mathrm{d}t(L + \mathrm{i}\varepsilon q^2)} \,. \tag{2.54}$$

The RHS corresponds to the path integral in Eqs. (2.45), augmented by the factor $\mathrm{i}\varepsilon q^2$. This factor damps in the limit of large t everything except the ground-state. Thus we found that $Z[J]$ is the generating functional for the vacuum expectation value of the time-ordered product of operators $\hat{q}(t_i)$,

$$(-\mathrm{i})^n \left. \frac{\delta^n Z[J]}{\delta J(t_1) \cdots \delta J(t_n)} \right|_{J=0} = \langle 0, \infty | T\{\hat{q}(t_1) \cdots \hat{q}(t_n)\} | 0, -\infty \rangle = G(t_1, \dots, t_n). \tag{2.55}$$

In the last step, we defined also the n-point Green function $G(t_1, \dots, t_n)$. These functions will be the main building block we will use to perform calculations in quantum field theory, and the formula corresponding to Eq. (2.55) will be our master formula in field theory. For the special case $n = 2$, we will see that the n-point Green function coincides up to a phase with the Feynman propagator $K_F(t', t)$ for the system described by L: The $\mathrm{i}\varepsilon q^2$ prescription selects from the set of possible propagators (retarded, advanced, ...) the one suggested by Feynman.

2.4   Oscillator as a one-dimensional field theory

Canonical quantisation.   A one-dimensional harmonic oscillator can be viewed as a free quantum field theory in one time and zero space dimensions. In order to exhibit this equivalence clearer, we rescale the usual Lagrangian

$$L(x, \dot{x}) = \frac{1}{2}m\dot{x}^2 - \frac{1}{2}m\omega^2 x^2, \tag{2.56}$$

where m is the mass of the oscillator and ω its frequency as

$$\phi(t) \equiv \sqrt{m}x(t). \tag{2.57}$$

We call the variable $\phi(t)$ a "scalar field", and the Lagrangian now reads

$$L(\phi, \dot{\phi}) = \frac{1}{2}\dot{\phi}^2 - \frac{1}{2}\omega^2\phi^2. \tag{2.58}$$

After the rescaling, the kinetic term $\dot{\phi}^2$ has the dimensionless coefficient $1/2$. This choice is standard in field theory and therefore such a field is called "canonically normalised".

We derive the corresponding Hamiltonian, determining first the conjugate momentum π as $\pi(t) = \partial L/\partial\dot{\phi} = \dot{\phi}(t)$. Thus the classical Hamiltonian follows as

$$H(\phi, \pi) = \frac{1}{2}\pi^2 + \frac{1}{2}\omega^2\phi^2. \tag{2.59}$$

The transition to quantum mechanics is performed by promoting ϕ and π to operators which satisfy the canonical commutation relations $[\phi, \pi] = \mathrm{i}$. The harmonic oscillator is solved most efficiently introducing creation and annihilation operators, $a^\dagger$ and a. They are defined by

$$\phi = \frac{1}{\sqrt{2\omega}}\left(a^\dagger + a\right) \quad \text{and} \quad \pi = \mathrm{i}\sqrt{\frac{\omega}{2}}\left(a^\dagger - a\right), \tag{2.60}$$

and satisfy $\left[a, a^\dagger\right] = 1$. The Hamiltonian follows as

$$H = \frac{\omega}{2}\left(aa^\dagger + a^\dagger a\right) = \left(a^\dagger a + \frac{1}{2}\right)\omega. \tag{2.61}$$

We interpret $N \equiv a^\dagger a$ as the number operator, counting the number n of quanta with energy ω in the state $|n\rangle$.

We now work in the Heisenberg picture where operators are time-dependent. The time evolution of the operator $a(t)$ can be found from the Heisenberg equation,

$$\mathrm{i}\frac{\mathrm{d}a}{\mathrm{d}t} = [a, H] = \omega a, \tag{2.62}$$

from which we deduce that

$$a(t) = a(0)\mathrm{e}^{-\mathrm{i}\omega t} = a_0\mathrm{e}^{-\mathrm{i}\omega t}. \tag{2.63}$$

As a consequence, the field operator $\phi(t)$ can be expressed in terms of the creation and annihilation operators as

$$\phi(t) = \frac{1}{\sqrt{2\omega}}\left(a_0\mathrm{e}^{-\mathrm{i}\omega t} + a_0^\dagger\mathrm{e}^{\mathrm{i}\omega t}\right). \tag{2.64}$$

If we look at $\phi(t)$ as a classical variable, then a_0 and $a_0^\dagger$ have to satisfy $a_0 = a_0^\dagger \equiv a_0^*$ in order to make ϕ real. Thus they are simply the Fourier coefficients of the single eigenmode $\sin(\omega t)$. This suggests that we can shortcut the quantisation procedure as follows. We write down the field as sum over its eigenmodes $i = 1, \ldots, k$. Then we reinterpret the Fourier coefficients as creation and annihilation operators, requiring $[a_i, a_j^\dagger] = \delta_{ij}$.

Path-integral approach.   We solve now the same problem, the rescaled Lagrangian (2.58), in the path-integral approach. Using this method, we have argued that it is convenient to include a coupling to an external force J. Let us define therefore the effective action S_{eff} as the sum of the classical action S, the coupling to the external force J and a small imaginary part $i\varepsilon\phi^2$ to make the path integral well-defined,

$$S_{\text{eff}} = S + \int_{-\infty}^{\infty} dt \left(J\phi + i\varepsilon\phi^2 \right) = \int_{-\infty}^{\infty} dt \left[\frac{1}{2}\dot{\phi}^2 - \frac{1}{2}\omega^2\phi^2 + J\phi + i\varepsilon\phi^2 \right]. \quad (2.65)$$

The function $e^{iS_{\text{eff}}}$ is the integrand of the path integral. We start our work by massaging S_{eff} into a form such that the path integral can be easily performed. The first two terms in S_{eff} can be viewed as the action of a differential operator $D(t)$ on $\phi(t)$, writing

$$\frac{1}{2}\left(\dot{\phi}^2 - \omega^2\phi^2 \right) = -\frac{1}{2}\phi(t)\left(\frac{d^2}{dt^2} + \omega^2 \right)\phi(t) = \frac{1}{2}\phi(t)D(t)\phi(t). \quad (2.66)$$

Here we performed a partial integration and dropped the boundary term. This is admissible, because boundary terms vanish varying the action.

We can evaluate this operator going to Fourier space,

$$\phi(t) = \int \frac{dE}{2\pi} e^{-iEt} \phi(E) \quad \text{and} \quad J(t) = \int \frac{dE}{2\pi} e^{-iEt} J(E). \quad (2.67)$$

To keep the action real, we have to write all bilinear quantities as $\phi(E)\phi^*(E') = \phi(E)\phi(-E')$, etc. Since only the phases depend on time, the time integration gives a factor $2\pi\delta(E - E')$, expressing energy conservation,

$$S_{\text{eff}} = \frac{1}{2}\int \frac{dE}{2\pi} \left[\phi(E)(E^2 - \omega^2 + i\varepsilon)\phi(-E) + J(E)\phi(-E) + J(-E)\phi(E) \right]. \quad (2.68)$$

In the path integral, this expression corresponds to a Gaussian integral of the type (2.21), where we should "complete the square". Shifting the integration variable to

$$\tilde{\phi}(E) = \phi(E) + \frac{J(E)}{E^2 - \omega^2 + i\varepsilon},$$

we obtain

$$S_{\text{eff}} = \frac{1}{2}\int \frac{dE}{2\pi} \left[\tilde{\phi}(E)(E^2 - \omega^2 + i\varepsilon)\tilde{\phi}(-E) - J(E)\frac{1}{E^2 - \omega^2 + i\varepsilon}J(-E) \right]. \quad (2.69)$$

Here we see that the "damping rule" for the path integral makes also the integral over the energy denominator well-defined. The physical interpretation of this way of shifting the poles—which differs from our treatment of the retarded Green function in the classical case—will be postponed to the next chapter, where we will discuss this issue in detail.

We are now in the position to evaluate the generating functional $Z[J]$. The path-integral measure is invariant under a simple shift of the integration variable, $\mathcal{D}\tilde{\phi} = \mathcal{D}\phi$, and we omit the tilde from now on. Furthermore, the second term in S_{eff} does not depend on ϕ and can be factored out,

$$
\begin{aligned}
Z[J] = \exp\left(-\frac{\mathrm{i}}{2}\int\frac{\mathrm{d}E}{2\pi}J(E)\frac{1}{E^2 - \omega^2 + \mathrm{i}\varepsilon}J(-E)\right) \\
\times \int \mathcal{D}\phi\, \exp\frac{\mathrm{i}}{2}\int\frac{\mathrm{d}E}{2\pi}\left[\phi(E)(E^2 - \omega^2 + \mathrm{i}\varepsilon)\phi(-E)\right].
\end{aligned}
\tag{2.70}
$$

Setting the external force to zero, $J = 0$, the first factor becomes one and the generating functional $Z[0]$ becomes equal to the path integral in the second line. For $J = 0$, however, the oscillator remains in the ground-state and thus $Z[0] = \langle 0, \infty | 0, -\infty \rangle = 1$. Therefore

$$
Z[J] = \exp\left(-\frac{\mathrm{i}}{2}\int\frac{\mathrm{d}E}{2\pi}J(E)\frac{1}{E^2 - \omega^2 + \mathrm{i}\varepsilon}J(-E)\right).
\tag{2.71}
$$

Inserting the Fourier transformed quantities, we arrive at

$$
Z[J] = \exp\left(-\frac{\mathrm{i}}{2}\int\mathrm{d}t'\,\mathrm{d}t\, J(t')K_F(t' - t)J(t)\right),
\tag{2.72}
$$

where we introduced also the Feynman propagator

$$
K_F(t - t') = \int\frac{\mathrm{d}E}{2\pi}\,\mathrm{e}^{-\mathrm{i}E(t-t')}\frac{1}{E^2 - \omega^2 + \mathrm{i}\varepsilon}.
\tag{2.73}
$$

This Green function differs from the retarded propagator G_R defined in Eq. (1.35) by the position of its poles.

The generating functional $Z[J]$ given by (2.72) is in the form most suitable for deriving arbitrary n-point Green functions using our master formula (2.55). Thus finding $Z[J]$ for a general quadratic action requires only to determine the inverse of the differential operator $D(t)$, accounting for the right boundary conditions induced by the $\mathrm{i}\varepsilon\phi^2$ term. This inverse is the Feynman propagator or two-point function $K_F(t'-t)$ which we can determine directly solving

$$
D(t)K_F(t' - t) = \delta(t' - t).
\tag{2.74}
$$

Going to Fourier space, we find immediately

$$
K_F(E) = \frac{1}{E^2 - \omega^2 + \mathrm{i}\varepsilon}.
\tag{2.75}
$$

Hence we can short-cut the calculation of $Z[J]$ by determining the Feynman propagator and using then directly Eqs. (2.71) or (2.72).

These results allow us also to calculate arbitrary matrix elements between oscillator states. For instance, we obtain the expectation value $\langle 0 | \phi^2 | 0 \rangle$ from

$$\langle 0 | T\{\phi(t')\phi(t)\} | 0 \rangle = (-\mathrm{i})^2 \frac{\delta^2 Z[J]}{\delta J(t')\delta J(t)}\bigg|_{J=0} = \mathrm{i}K_F(t'-t) = \frac{1}{2\omega}\,\mathrm{e}^{\mathrm{i}\omega|t-t'|}. \qquad (2.76)$$

Here, we used in the last step the explicit expression for K_F which you should check in problem 2.6. Taking the limit $t' \searrow t$ and replacing $\phi^2 \to mx^2$, we reproduce the standard result $\langle 0 | x^2 | 0 \rangle = 1/(2m\omega)$. Matrix elements between excited states $|n\rangle = (n!)^{-1/2}(a^\dagger)^n | 0 \rangle$ are obtained by expressing the creation operator $a^\dagger$ using $\pi(t) = \dot\phi(t)$ as

$$a^\dagger = \sqrt{\frac{\omega}{2}} \left(1 - \frac{\mathrm{i}}{\omega}\frac{\mathrm{d}}{\mathrm{d}t} \right) \phi(t). \qquad (2.77)$$

2.5 The need for quantum fields

We have already argued that any relativistic quantum theory has to be a many-particle theory. Such a theory has to include infinitely many degrees of freedom—as field theories like electrodynamics do. Before we move on to introduce the simplest quantum field theory in the next chapter, we present an argument that relativity and the single particle picture are incompatible.

In classical mechanics, the principle of relativity implies that all trajectories of massive particles are time-like, while massless particles move along light-like trajectories. This implements causality, that is, the requirement that no signal can be transmitted faster than light. How should we translate this principle into a quantum theory? Causality would be clearly satisfied, if the relativistic propagator $K(\boldsymbol{x}', t'; \boldsymbol{x}, t)$ vanishes for space-like distances. Another, less restrictive translation of the principle of relativity would be to ask that measurements performed at space-like separated points do not influence each other. This is achieved if all observables $O(x)$ commute for space-like distances,

$$[\hat{O}(\boldsymbol{x}, t), \hat{O}(\boldsymbol{x}', t')] = 0 \qquad \text{for} \quad (t - t')^2 < (\boldsymbol{x} - \boldsymbol{x}')^2. \qquad (2.78)$$

In quantum mechanics, the Heisenberg operators $\hat{\boldsymbol{x}}(t)$ and $\hat{\boldsymbol{p}}(t)$ depend, however, only on time. Therefore we cannot implement the condition (2.78) in such a framework.

The only rescue for causality in relativistic quantum mechanics is therefore the vanishing of the propagator $K(t', \boldsymbol{x}'; t, \boldsymbol{x})$ outside the light-cone. We evaluate the propagator as in the non-relativistic case,

$$K(\boldsymbol{x}', t'; \boldsymbol{x}, t) = \langle \boldsymbol{x}' | \mathrm{e}^{-\mathrm{i}H(t'-t)} | \boldsymbol{x} \rangle = \int \frac{\mathrm{d}^3 p}{(2\pi)^3} \langle \boldsymbol{x}' | \mathrm{e}^{-\mathrm{i}E_p(t'-t)} | \boldsymbol{p} \rangle \langle \boldsymbol{p} | \boldsymbol{x} \rangle \qquad (2.79)$$

adapting, however, the relativistic dispersion relation, $E_{\boldsymbol{p}} = \sqrt{m^2 + \boldsymbol{p}^2}$. Next we use that the momentum operator $\hat{\boldsymbol{p}}$ generates space translations, $\exp(-\mathrm{i}\hat{\boldsymbol{p}}\boldsymbol{x})|0\rangle = |\boldsymbol{x}\rangle$, to obtain

$$K(\boldsymbol{x}', t'; \boldsymbol{x}, t) = K(x' - x) = \int \frac{\mathrm{d}^3 p}{(2\pi)^3} |\langle \boldsymbol{0} | \boldsymbol{p} \rangle|^2\, \mathrm{e}^{-\mathrm{i}p(x'-x)}. \qquad (2.80)$$

Here we introduced also the four-vector $p^\mu = (E_p, \boldsymbol{p})$, rewriting the plane wave thereby in a Lorentz-invariant way. In order that the complete propagator is invariant, we have to choose as integration measure $\propto \mathrm{d}^3p/E_{\boldsymbol{p}}$, cf. problem 2.8, and we set therefore $|\langle 0|\boldsymbol{p}\rangle|^2 = 1/(2E_{\boldsymbol{p}})$. Knowing its explicit expression, it is a straight-forward exercise to show that the propagator does not vanish outside the light-cone, but goes only exponentially to zero, $K(\boldsymbol{x}',0;\boldsymbol{x},0) \propto \exp(-m|\boldsymbol{x}'-\boldsymbol{x}|)$. Thus we failed to implement both versions of causality into relativistic quantum mechanics. Instead, we will develop quantum field theory with the aim to implement causality via the condition (2.78).

Before starting this endeavour, we can draw still some important conclusion from Eq. (2.80). For space-like distances, $(x-x')^2 < 0$, a Lorentz boost can change the time order of two spacetime events, cf. problem 1.11. Consistency requires thus to include both time-orderings: if a particle is created at t and absorbed at $t' > t$, then it can be created necessarily also at t' and absorbed at $t > t'$. We extend therefore the propagator as

$$ K(x'-x) = \int \frac{\mathrm{d}^3p}{(2\pi)^3 2E_{\boldsymbol{p}}} \left[\vartheta(t'-t)\mathrm{e}^{-\mathrm{i}p(x'-x)} + \vartheta(t-t')\mathrm{e}^{\mathrm{i}p(x'-x)} \right], \qquad (2.81) $$

where we chose the opposite sign for the plane wave in the second factor. In this way, the phase of the plane waves observed in both frames agree, $-E_{\boldsymbol{p}}\tau\vartheta(\tau) < 0$ and $+E_{\boldsymbol{p}}\tau\vartheta(-\tau) < 0$, and similarly for the momenta. If we imagine that the propagating particle carries a conserved charge, then we can associate the positive frequencies to the propagation of a particle (with charge q) and the negative frequencies to the propagation of an antiparticle (with charge $-q$). Then the resulting current is frame-independent, if the antiparticle has the same mass but the opposite additive charges. This prediction of relativistic quantum field theory is experimentally confirmed with extreme accuracy. For instance, the limits on the mass and charge difference of electrons and positrons are smaller than 8×10^{-9} and 4×10^{-8}, respectively. The best experimental limit is currently the relative mass difference of the K^0 and $\bar{K}^0$ mesons, which is bounded by 10^{-18} (Olive *et al.*, 2014).

Finally, we should mention an alternative way to implement causality. Instead of defining quantum fields $\hat{\phi}(x^\mu)$ on classical spacetime, we could promote time t to an operator, parameterising the world-line $\hat{x}^\mu(\tau)$ of a particle, for example, by its proper-time τ. Considering then the surface $\hat{x}^\mu(\tau,\sigma)$ generated by a set of world-lines is the starting point of string theory.

Summary

Using Feynman's path-integral approach, we can express a transition amplitude as a sum over all paths weighted by a phase which is determined by the classical action, $\langle q_f, t_f | q_i, t_i \rangle = \int \mathcal{D}q(t)\exp(\mathrm{i}S[q])$. Adding a linear coupling to an external source J and a damping term to the Lagrangian, we obtain the ground-state persistence amplitude $\langle 0, \infty | 0, -\infty \rangle_J$. This quantity serves as the generating functional $Z[J]$ for n-point Green functions $G(t_1, \ldots, t_n)$ which are the time-ordered vacuum expectation values of the operators $\hat{q}(t_1), \ldots, \hat{q}(t_n)$.

Further reading.   For additional examples for the use of the path integral and Green functions in quantum mechanics see, for example, MacKenzie (2000) or Das (2006). Schweber (2005) sketches the historical development that lead to Schwinger's Green functions, including his quantum action principle.

Problems

2.1 Classical action. Calculate the classical action $S[q]$ for a free particle and an harmonic oscillator. Compare the results with the expression for the propagator $K = \langle x', t' | x, t \rangle = N \exp(\mathrm{i}\phi)$ of the corresponding quantum mechanical system and express both ϕ and N through the action S.

2.2 Propagator as Green function. Show that the Green function or propagator $K(x', t'; x, t) = \langle x' | \exp[-\mathrm{i}H(t' - t)] | x \rangle$ of the Schrödinger equation is the inverse of the differential operator $(\mathrm{i}\partial_t - H)$.

2.3 Classical limit. Sketch (without detailed calculation) why in the path integral the allowed paths dominate in the classical limit.

2.4 Commutation relations. Show that the commutation relations for the field, $[\phi, \pi] = \mathrm{i}$, imply $[a, a^\dagger] = 1$. What happens, if we change the normalisation (2.57)?

2.5 Mode functions. Consider the generalisation of (2.64) to $\phi(t) = u a_0 e^{-\mathrm{i}\omega t} + u^* a_0^\dagger e^{\mathrm{i}\omega t}$, where the functions $u(t)$ are called mode functions. a.) Show that the usual commutation relations are valid, if $\Im(u\dot{u}) = 1$. b.) Show that the standard choice $u = 1/\sqrt{2\omega}$ minimises the energy of the groundstate.

2.6 Feynman propagator. Find the explicit expression for the Feynman propagator used in (2.76) from a.) its definition as time-ordered product of fields ϕ, and b.) evaluating (2.73) using Cauchy's theorem.

2.7 Matrix elements from $Z[J]$. Evaluate the matrix element $\langle 0 | \phi^2 | 1 \rangle$ of an harmonic oscillator from $Z[J]$.

2.8 Lorentz invariant integration measure. ♣ Show that $\mathrm{d}^3 k/(2\omega_k)$ is a Lorentz invariant integration measure by a) calculating the Jacobian of a Lorentz transformation, b) showing that

$$\int \mathrm{d}^4 k \, \delta(k^2 - m^2) \vartheta(k^0) f(k^0, \boldsymbol{k}) = \int \frac{\mathrm{d}^3 k}{2\omega_k} f(\omega_k, \boldsymbol{k}) \tag{2.82}$$

holds for any function f.

2.9 Propagator at large $|\boldsymbol{x}|$. Show that the propagator $K(\boldsymbol{x}, 0; 0, 0)$ defined in Eq. (2.81) decays exponentially outside the light-cone for $m > 0$. Find the propagator for $m = 0$.

2.10 Statistical mechanics. Derive the connection between the partition function $Z = \mathrm{tr} \, e^{-\beta H}$ of statistical mechanics and the path integral of quantum mechanics in Euclidean time $t_E = -\mathrm{i}t$. (Hint: compare to remark 2.1.)

2.11 Scattering at short-range potential. ♣ Consider in $d = 1$ the scattering of modes with large wavelengths λ on a short-range potential, $V(x) = 0$ for $|x| > a$ and $\lambda \gg a$. i) Show that the potential V can be approximated by $V(x) = c_0 \delta(x) + c_1 \delta'(x) + \mathcal{O}(Va^2/\lambda^2)$. ii) Find the transmission and reflection coefficients setting $V(x) = c_0 \delta(x)$. Argue that $T \simeq \mathrm{i}p/c_0$ holds for any short-range potential in the limit $p \ll 1/a$. iii) Show that no consistent solution exists setting $c_0 = 0$.

3
Free scalar field

In this chapter we extend the path-integral approach from quantum mechanics to the simplest field theory, containing a single real scalar field $\phi(x)$. Such a field may either represent an elementary particle like the Higgs boson, a bound-state like a scalar meson, or a scalar parameter describing a specific property of a more complex theory. Proceeding similar to our approach in quantum mechanics, we will introduce the generating functional $Z[J] = \langle 0 + |0-\rangle_J$ of n-point Green functions as our main tool to calculate the time-ordered vacuum expectation value of a product of fields $\phi(x_1)\cdots\phi(x_n)$. Calculating the vacuum energy of the scalar field, we will encounter for the first time the fact that many calculations in quantum field theories return a formally infinite result. In order to extract sensible predictions, we therefore have to introduce the concepts of regularisation and renormalisation.

3.1 Lagrange formalism and path integrals for fields

A field is a map which associates to each spacetime point x a k-tuple of values $\phi_a(x)$, $a = 1, \ldots, k$. We require that the fields $\phi_a(x)$ transform under a definite representation of the Poincaré group which is the group combining Lorentz transformations and translations[1]. For massive particles, these representations are labelled by the mass m and the spin s of one-particle states. Thus this condition guarantees that observers in all inertial frames agree what, for example, a spin $1/2$ particle with mass m is. Additionally, particles can be characterised by their transformation properties under internal symmetry groups. These internal symmetries may lead to conserved quantum numbers like the electric charge q, by which we can distinguish further various particles types.

Except for a real scalar field ϕ, a field has several components. Thus we have to generalise Hamilton's principle to a collection of fields $\phi_a(x)$, where the index a includes all internal as well as Lorentz indices. Moreover, the Lagrangian for a field $\phi_a(x)$ will contain not only time but also space derivatives. To ensure Lorentz invariance, we consider a scalar Lagrange density $\mathscr{L}(x)$ that depends as a local function on the fields $\phi_a(x)$ and their derivatives $\partial_\mu\phi_a(x)$. By analogy to $L(q, \dot{q})$, we restrict ourselves to first derivatives. We include no explicit time-dependence, since "everything" should be explained by the fields and their interactions. The Lagrangian $L(\phi_a, \partial_\mu\phi_a)$ is obtained by integrating the density $\mathscr{L}$ over a given space volume V. The action S is thus the four-dimensional integral

---

[1]The basic properties of these two groups are collected in the appendices B.3 and B.4.

Quantum Fields–From the Hubble to the Planck Scale. Michael Kachelriess. © Michael Kachelriess 2018.
Published in 2018 by Oxford University Press. DOI 10.1093/oso/9780198802877.001.0001

$$S[\phi_a] = \int_{t_a}^{t_b} \mathrm{d}t \, L(\phi_a, \partial_\mu \phi_a) = \int_\Omega \mathrm{d}^4 x \, \mathscr{L}(\phi_a, \partial_\mu \phi_a) \tag{3.1}$$

with $\Omega = V \times [t_a : t_b]$. A variation $\delta\phi_a(x)$ of the fields leads to a variation of the action,

$$\delta S = \int_\Omega \mathrm{d}^4 x \left[\frac{\partial \mathscr{L}}{\partial \phi_a} \delta\phi_a + \frac{\partial \mathscr{L}}{\partial(\partial_\mu \phi_a)} \delta(\partial_\mu \phi_a) \right], \tag{3.2}$$

where we have to sum over field components $(a = 1, \ldots, k)$ and the Lorentz index $\mu = 0, \ldots, 3$. The correspondence $q(t) \to \phi(x^\mu)$ implies that the scale factor ε parameterising the variations $\phi_a(x^\mu, \varepsilon)$ depends not on x^μ. We can therefore eliminate again the variation of the field gradients $\partial_\mu \phi_a$ by a partial integration using Gauss' theorem,

$$\delta S = \int_\Omega \mathrm{d}^4 x \left[\frac{\partial \mathscr{L}}{\partial \phi_a} - \partial_\mu \left(\frac{\partial \mathscr{L}}{\partial(\partial_\mu \phi_a)} \right) \right] \delta\phi_a = 0. \tag{3.3}$$

The surface term vanishes, since we require that the variation is zero on the boundary $\partial\Omega$. Thus the Lagrange equations for the fields ϕ_a are

$$\frac{\partial \mathscr{L}}{\partial \phi_a} - \partial_\mu \left(\frac{\partial \mathscr{L}}{\partial(\partial_\mu \phi_a)} \right) = 0. \tag{3.4}$$

If the Lagrange density $\mathscr{L}$ is changed by a four-dimensional divergence, $\delta\mathscr{L} = \partial_\mu K^\mu$, and surface terms can be dropped, the same equations of motion result. Note also that it is often more efficient to perform directly the variation $\delta\phi_a$ in the action $S[\phi_a]$ than to use the Lagrange equations.

The path integral in configuration space becomes now a functional integral over the k fields ϕ_a,

$$K = \int \mathcal{D}\phi_1 \cdots \mathcal{D}\phi_k \, \mathrm{e}^{\mathrm{i}S[\phi_a]} = \int \mathcal{D}\phi_1 \cdots \mathcal{D}\phi_k \, \mathrm{e}^{\mathrm{i} \int_\Omega \mathrm{d}^4 x \, \mathscr{L}(\phi_a, \partial_\mu \phi_a)}. \tag{3.5}$$

A major problem we have to address later is that the k fields ϕ_a are often not independent. For instance, in electrodynamics all potentials A^μ connected by a gauge transformation describe the same physics. This redundancy makes the path integral (3.5) ill-defined. We therefore start with the simplest case of a single, real scalar field ϕ where such problems are absent. Moreover, we restrict ourselves in this chapter to a free field without interactions.

3.2   Generating functional for a scalar field

Lagrangian.   The (free) Schrödinger equation $\mathrm{i}\partial_t\psi = H_0\psi$ can be obtained substituting $\omega \to \mathrm{i}\partial_t$ and $\boldsymbol{k} \to -\mathrm{i}\boldsymbol{\nabla}_x$ into the non-relativistic energy–momentum relation $\omega = \boldsymbol{k}^2/(2m)$. With the same replacements, the relativistic $\omega^2 = m^2 + \boldsymbol{k}^2$ becomes the Klein–Gordon equation

$$(\Box + m^2)\phi = 0 \qquad \text{with} \quad \Box = \eta_{\mu\nu}\partial^\mu\partial^\nu = \partial_\mu\partial^\mu. \tag{3.6}$$

The relativistic energy–momentum relation implies that the solutions to the free Klein–Gordon equation consist of plane waves with positive and negative energies

$\pm\sqrt{k^2 + m^2}$. For the stability of a quantum system it is, however, essential that its energy eigenvalues are bounded from below. Otherwise, we could generate, for example, in a scattering process $\phi + \phi \to n\phi$, an arbitrarily high number of ϕ particles with sufficiently low energy, and no stable form of matter could exist. Interpreting the Klein–Gordon equation as a relativistic wave equation for a single particle cannot therefore be fully satisfactory, since the energy of its solutions is not bounded from below.

How do we guess the correct Lagrange density $\mathscr{L}$? Plane waves can be seen as a collection of coupled harmonic oscillators at each spacetime point. The correspondence $\dot{q} \to \partial_\mu\phi$ means that the kinetic field energy is quadratic in the field derivatives. Relativistic invariance implies that the Lagrange density is a scalar, leaving as the only two possible terms containing derivatives

$$\eta_{\mu\nu}(\partial^\mu\phi)(\partial^\nu\phi) \quad \text{and} \quad \phi\Box\phi.$$

Using the action principle to derive the equation of motion, we can however drop boundary terms performing partial integrations. Thus these two terms are equivalent, up to a minus sign. The Klein–Gordon equation $\Box\phi = -m^2\phi$ suggests that the mass term is also quadratic in the field ϕ. Therefore we try as Lagrange density

$$\mathscr{L} = \frac{1}{2}\eta_{\mu\nu}\left(\partial^\mu\phi\right)\left(\partial^\nu\phi\right) - \frac{1}{2}m^2\phi^2 \equiv \frac{1}{2}\eta_{\mu\nu}\partial^\mu\phi\partial^\nu\phi - \frac{1}{2}m^2\phi^2. \tag{3.7}$$

From now on, we will drop the parenthesis around $\partial^\mu\phi$ and it should be understood from the context that the derivative ∂^μ acts only on the first field ϕ. Even shorter alternative notations are $(\partial_\mu\phi)^2$ and the concise $(\partial\phi)^2$. Swapping the indices in the Lagrangian (3.7), we obtain for the second part of the Lagrange equation

$$\frac{\partial}{\partial(\partial_\alpha\phi)}\left(\eta^{\mu\nu}\partial_\mu\phi\partial_\nu\phi\right) = \eta^{\mu\nu}\left(\delta^\alpha_\mu\partial_\nu\phi + \delta^\alpha_\nu\partial_\mu\phi\right) = \eta^{\alpha\nu}\partial_\nu\phi + \eta^{\mu\alpha}\partial_\mu\phi = 2\partial^\alpha\phi. \tag{3.8}$$

Hence the Lagrange equation becomes

$$\frac{\partial\mathscr{L}}{\partial\phi} - \partial_\alpha\left(\frac{\partial\mathscr{L}}{\partial(\partial_\alpha\phi)}\right) = -m^2\phi - \partial_\alpha\partial^\alpha\phi = 0, \tag{3.9}$$

and the Lagrange density (3.7) leads indeed to the Klein–Gordon equation. We can understand the relative sign in the Lagrangian splitting the relativistic kinetic energy into the "proper" kinetic energy $(\partial_t\phi)^2/2$ and the gradient energy density $(\nabla\phi)^2/2$,

$$\mathscr{L} = \frac{1}{2}\dot{\phi}^2 - \frac{1}{2}(\nabla\phi)^2 - \frac{1}{2}m^2\phi^2. \tag{3.10}$$

The last two terms correspond to a potential energy and therefore carry the opposite sign of the first one.

Instead of guessing, we can derive the correct Lagrangian $\mathscr{L}$ as follows: we multiply the free field equation for ϕ by a variation $\delta\phi$ that vanishes on $\partial\Omega$. Then we integrate over Ω, perform a partial integration of the kinetic term, use $\partial_\mu \delta = \delta \partial_\mu$, the Leibniz rule and ask that the variation vanishes,

$$A \int_\Omega \mathrm{d}^4 x \, \delta\phi \, (\Box + m^2)\phi = A \int_\Omega \mathrm{d}^4 x \, \left[-\delta(\partial_\mu \phi)\partial^\mu \phi + \delta\phi\phi m^2 \right] = \tag{3.11a}$$

$$= A \int_\Omega \mathrm{d}^4 x \, \delta \left[-\frac{1}{2}(\partial_\mu \phi)^2 + \frac{1}{2}\phi^2 m^2 \right] = 0. \tag{3.11b}$$

The term in the square brackets agrees with our guess (3.7), taking into account that the source-free field equation fixes the Lagrangian only up to the overall factor A. In analogy with a quantum mechanical oscillator, we want the coefficients of the two terms to be $\pm 1/2$ and thus we set $|A| = 1$.

We can determine the correct overall sign of $\mathscr{L}$ by calculating the energy density ρ of the scalar field and requiring that it is bounded from below and stable against small perturbations. We identify the energy density ρ of the scalar field with its Hamiltonian density $\mathscr{H}$, and use the connection between the Lagrangian and the Hamiltonian known from classical mechanics. The transition from a system with a finite number of degrees of freedom to one with an infinite number of degrees of freedom proceeds as follows,

$$p_i = \frac{\partial L}{\partial \dot{q}^i} \quad \Rightarrow \quad \pi_a = \frac{\partial \mathscr{L}}{\partial \dot{\phi}_a}, \tag{3.12a}$$

$$H = p_i \dot{q}^i - L \quad \Rightarrow \quad \mathscr{H} = \sum_a \pi_a \dot{\phi}_a - \mathscr{L}. \tag{3.12b}$$

The canonically conjugated momentum π of a real scalar field is

$$\pi = \frac{\partial \mathscr{L}}{\partial \dot{\phi}} = \dot{\phi}. \tag{3.13}$$

Thus the Hamilton density is

$$\mathscr{H} = \pi\dot{\phi} - \mathscr{L} = \pi^2 - \mathscr{L} = \frac{1}{2}\dot{\phi}^2 + \frac{1}{2}(\boldsymbol{\nabla}\phi)^2 + \frac{1}{2}m^2\phi^2 \geq 0 \tag{3.14}$$

and thus obviously positive definite. Moreover, generating fluctuations $\delta\phi$ costs energy and thus the system is stable against small perturbations. Hence the transition from a single-particle interpretation of the Klein–Gordon equation to a field theory has been sufficient to cure the problem of the negative energy solutions.

Note that we could subtract a constant ρ_0 from the Lagrangian which would drop out of the equation of motion. From Eq. (3.14) we see that such a constant corresponds to a uniform energy density of empty space. Such a term would act as an additional source of the gravitational field, but would be otherwise unobservable. Next we generalise the Lagrangian by subtracting a polynomial in the fields, $V(\phi)$, subject to the stability constraint discussed above. Hence the potential V should be bounded from below, and we can expand it around its minimum at $\phi \equiv v$,

$$\left. \frac{\mathrm{d}V}{\mathrm{d}\phi} \right|_{\phi=v} = 0, \qquad \left. \frac{\mathrm{d}^2 V}{\mathrm{d}\phi^2} \right|_{\phi=v} \equiv m^2 > 0. \tag{3.15}$$

The term $V''(v)$ acts as mass term for the field ϕ. We will see soon that terms ϕ^n with $n \geq 3$ generate interactions between n particles, as expected from the analogy of a quantum field to coupled quantum mechanical oscillators. The field ϕ has the non-zero value $\phi = v$ everywhere, if the minimum v of $V(\phi)$ is not at zero, $v \neq 0$. If the value of $V(\phi)$ at the minimum v is not zero, $V(v) \neq 0$, then the non-zero potential implies a non-zero uniform energy density $\rho = V(v)$.

Generating functional. Now we move on to the quantum theory of a scalar field, which we define by the path-integral over $\exp \mathrm{i} S[\phi]$. Since the Hamiltonian (3.14) is quadratic in the momentum, we can start directly from the path integral in configuration space. Then the Green functions which encode all information about this theory can be obtained from the generating functional

$$Z[J] = \langle 0 + | 0 - \rangle_J = \mathcal{N} \int \mathcal{D}\phi \exp \mathrm{i} \int \mathrm{d}^4 x \left(\frac{1}{2}\partial_\mu \phi \partial^\mu \phi - \frac{1}{2}m^2\phi^2 + J\phi \right), \qquad (3.16)$$

where we appended to the action a linear coupling between the field and an external source. To ensure the convergence of the integral, we included an infinitesimal small imaginary part into the squared mass of the particle, $m^2 \to m^2 - \mathrm{i}\varepsilon$. Moreover, we added a normalisation factor $\mathcal{N}$ which we will have to determine. We start performing an integration by part of the first term, where we exploit the fact that the boundary term vanishes,

$$Z[J] = \mathcal{N} \int \mathcal{D}\phi \exp \mathrm{i} \int \mathrm{d}^4 x \left(-\frac{1}{2}\phi(\Box + m^2)\phi + J\phi \right). \qquad (3.17)$$

The first two terms, $\phi A \phi = -\phi(\Box + m^2)\phi$, are quadratic and symmetric in the field ϕ,

$$-\frac{1}{2}\int \mathrm{d}^4 x \, \phi(x)(\Box_x + m^2)\phi(x) = \frac{1}{2}\int \mathrm{d}^4 x \mathrm{d}^4 x' \, \phi(x)A(x,x')\delta(x - x')\phi(x'). \qquad (3.18)$$

Note that the operator A is local, $A(x) \propto A(x,x')\delta(x - x')$. Since special relativity forbids action at a distance, non-local terms like $\phi(x')A(x,x')\phi(x)$ should not appear in a relativistic Lagrangian.

The expression on the RHS of Eq. (3.18) is the continuous version of the matrix equation $\phi_i A_{ij} \phi_j$. If we discretise continuous spacetime x^μ into a lattice, we can use Eq. (2.23) to perform the path integral,

$$Z[J] = \mathcal{N} \left(\frac{(2\pi\mathrm{i})^N}{\det[A]} \right)^{1/2} \exp \left(-\frac{1}{2}\mathrm{i} \, J A^{-1} J \right) \equiv \mathcal{N} Z[0] \exp(\mathrm{i} W[J]). \qquad (3.19)$$

The pre-factor of the exponential function does not depend on J and is thus given by $\mathcal{N} Z[0] = \langle 0 + | 0 - \rangle$. The vacuum should be stable and normalised to one in the absence of sources, $\langle 0 + | 0 - \rangle = 1$. Therefore the proper normalisation of $Z[J]$ implies that $\mathcal{N}^{-1} = Z[0]$. Thus we can omit the normalisation factor, if we normalise the path-integral measure $\mathcal{D}\phi$ such that the Gaussian integral over a free field is one. In the last step of Eq. (3.19), we defined a new functional $\mathrm{i} W[J] \equiv \ln(Z[J])$ that depends only quadratically on the source J; therefore it should be easier to handle than $Z[J]$.

Going for $N \to \infty$ back to continuous spacetime, the matrix multiplications become integrations,

$$Z[J] = \exp(\mathrm{i}W[J]) = \exp\left(-\frac{\mathrm{i}}{2}\int \mathrm{d}^4x \mathrm{d}^4x' J(x) A^{-1}(x,x') J(x')\right) \qquad (3.20)$$

and

$$W[J] = -\frac{1}{2}\int \mathrm{d}^4x \mathrm{d}^4x' J(x) A^{-1}(x,x') J(x'). \qquad (3.21)$$

Propagator.   In order to evaluate the functional $W[J]$ we have to find the inverse $\Delta(x,x') \equiv A^{-1}(x,x')$ of the differential operator A, defined by

$$-(\Box + m^2)\Delta(x,x') = \delta(x-x'). \qquad (3.22)$$

Because of translation invariance, the Green function $\Delta(x,x')$ can depend only on the difference $x-x'$. Therefore it is advantageous to perform a Fourier transformation and to go to momentum space,

$$-\int \frac{\mathrm{d}^4k}{(2\pi)^4}(\Box + m^2)\Delta(k)\mathrm{e}^{-\mathrm{i}k(x-x')} = \int \frac{\mathrm{d}^4k}{(2\pi)^4}\mathrm{e}^{-\mathrm{i}k(x-x')}, \qquad (3.23)$$

or

$$\Delta_F(k) = \frac{1}{k^2 - m^2 + \mathrm{i}\varepsilon}, \qquad (3.24)$$

where the pole at $k^2 = m^2$ is avoided by the $\mathrm{i}\varepsilon$. Thus the $m^2 \to m^2 - \mathrm{i}\varepsilon$ prescription introduced to ensure the convergence of the path integral tells us also how to handle the poles of the Green function. The index F specifies that the propagator Δ_F is the Green function obtained with the $m^2 - \mathrm{i}\varepsilon$ prescription proposed by Feynman. (Some authors use instead D_F for the propagator of massive bosons and Δ_F for the propagator of massless bosons.)

Note that the four momentum components k^μ are independent. Therefore $\Delta_F(k)$ describes the propagation of a *virtual* particle that has—in contrast to a real or external particle—not to be on "mass-shell": in general

$$k_0 \neq \pm\omega_k \equiv \pm\sqrt{\boldsymbol{k}^2 + m^2}.$$

We can evaluate the k_0 integral in the coordinate representation of $\Delta_F(x-x')$ explicitly,

$$\Delta_F(x-x') = \int \frac{\mathrm{d}^4k}{(2\pi)^4}\frac{e^{-\mathrm{i}k(x-x')}}{k_0^2 - \boldsymbol{k}^2 - m^2 + \mathrm{i}\varepsilon} \qquad (3.25\mathrm{a})$$

$$= \int \frac{\mathrm{d}^3k}{(2\pi)^3}\int \frac{\mathrm{d}k_0}{2\pi}\frac{e^{-\mathrm{i}k_0(t-t')}\mathrm{e}^{\mathrm{i}\boldsymbol{k}(\boldsymbol{x}-\boldsymbol{x}')}}{(k_0 - \omega_k + \mathrm{i}\varepsilon)(k_0 + \omega_k - \mathrm{i}\varepsilon)}, \qquad (3.25\mathrm{b})$$

using Cauchy's theorem.[2] The integrand has two simple poles at $+\omega_k - \mathrm{i}\varepsilon$ and $-\omega_k + \mathrm{i}\varepsilon$, cf. Fig. 3.1. For negative $\tau = t - t'$, we can close the integration contour $\mathcal{C}_+$ on the

---

[2]Since ε is infinitesimal and $\omega_k > 0$, we can set $2\mathrm{i}\omega_k\varepsilon + \varepsilon^2 \to \mathrm{i}\varepsilon$.

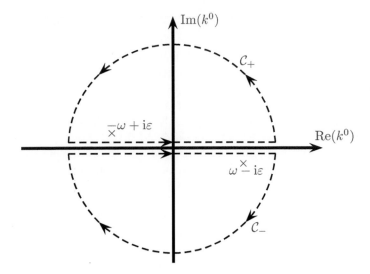

Fig. 3.1 Poles and contours in the complex k^0 plane used for the integration of the Feynman propagator.

upper half-plane, including the pole at $-\omega_k$,

$$\int \mathrm{d}k_0 \frac{\mathrm{e}^{-\mathrm{i}k_0\tau}}{(k_0 - \omega_k + \mathrm{i}\varepsilon)(k_0 + \omega_k - \mathrm{i}\varepsilon)} = 2\pi\mathrm{i} \, \mathrm{res}_{-\omega_k} = 2\pi\mathrm{i} \, \frac{\mathrm{e}^{\mathrm{i}\omega_k\tau}}{-2\omega_k} \qquad \text{for} \quad \tau < 0. \quad (3.26)$$

For positive τ, we have to choose the contour $\mathcal{C}_-$ in the lower plane, picking up $2\pi\mathrm{i} \, \mathrm{e}^{-\mathrm{i}\omega_k\tau}/(2\omega_k)$ and an additional minus sign since the contour is clockwise. Combining both results, we obtain

$$\mathrm{i}\Delta_F(x) = \int \frac{\mathrm{d}^3k}{(2\pi)^3 2\omega_k} \left[\mathrm{e}^{-\mathrm{i}\omega_k t}\vartheta(t) + \mathrm{e}^{\mathrm{i}\omega_k t}\vartheta(-t) \right] \mathrm{e}^{\mathrm{i}\boldsymbol{k}\boldsymbol{x}}, \qquad (3.27)$$

or after shifting the integration variable $\boldsymbol{k} \to -\boldsymbol{k}$ in the second term

$$\mathrm{i}\Delta_F(x) = \int \frac{\mathrm{d}^3k}{(2\pi)^3 2\omega_k} \left[\mathrm{e}^{-\mathrm{i}(\omega_k t - \boldsymbol{k}\boldsymbol{x})}\vartheta(t) + \mathrm{e}^{\mathrm{i}(\omega_k t - \boldsymbol{k}\boldsymbol{x})}\vartheta(-t) \right]. \qquad (3.28)$$

Comparing this expression to our guess (2.81) at the end of the last chapter, we see that our intuitive arguments about the structure of a Lorentz-invariant propagator in a quantum theory were correct. We stress once again the salient features of the Feynman propagator: first, the propagator contains positive and negative frequencies, as expected from the existence of solutions to the Klein–Gordon equation with positive and negative energies. Second, positive frequencies propagate forward in time, while negative frequencies propagate backward. This implies the existence of antiparticles. Third, the relativistic normalisation of (on-shell) plane waves includes a factor $1/\sqrt{2\omega_k}$, or

$$\langle k|k'\rangle = 2\omega_k(2\pi)^3\delta(\boldsymbol{k} - \boldsymbol{k}'), \qquad (3.29)$$

while the non-relativistic normalisation uses $\langle \boldsymbol{k}|\boldsymbol{k}'\rangle = \delta(\boldsymbol{k} - \boldsymbol{k}')$.

Remark 3.1: Other Green functions are obtained, if we choose different prescriptions for the handling of the poles. For positive $\tau = t - t'$, we have to close the circle on the lower half-plane. Shifting both poles to the lower half-plane, $\pm\omega_k - i\varepsilon$, gives thus the retarded propagator $\Delta_{\mathrm{ret}}(x)$ vanishing for all $\tau < 0$. In the opposite case, we shift both poles to the upper half-plane, $\pm\omega_k + i\varepsilon$, and obtain the advanced propagator $\Delta_{\mathrm{adv}}(x)$. Both propagators are real-valued, propagating a real solution of the wave equation into another real one at a different time, as required in classical physics. Moreover, both Green functions have support inside the light-cone, the retarded in the forward and the advanced in the backward part of the light-cone. This behaviour should be contrasted with the Feynman propagator Δ_F which is complex-valued and non-zero in $\mathbb{R}(1,3)$. Another way to handle the singularities is to use Cauchy's Principal value prescription, obtaining $\Delta(x) = \frac{1}{2}[\Delta_{\mathrm{adv}}(x) + \Delta_{\mathrm{ret}}(x)]$. This choice corresponds to an action-at-distance which seems to have no relevance in physics. Finally, we can shift one pole up and the other one down. The choice $\pm(\omega_k - i\varepsilon)$ used in the Feynman propagator allows us to rotate the integration contour anticlockwise to $-i\infty : +i\infty$ avoiding both poles in the complex k_0 plane. Since $k_0 = i\partial_t$, this transformation is consistent with the clockwise rotation in coordinate space required to obtain an Euclidean action bounded from below. Thus the Feynman prescription is the only one in which the physics in Minkowski and Euclidean space are analytically connected, cf. with remark 2.1.

We are now in the position to evaluate the generating functional

$$W[J] = -\frac{1}{2} \int \mathrm{d}^4x \mathrm{d}^4x' J(x) \Delta_F(x - x') J(x') \,. \tag{3.30}$$

in Fourier space. Inserting the Fourier transformations for the propagator as well as for the sources J gives

$$W[J] = -\frac{1}{2} \int \mathrm{d}^4x \mathrm{d}^4x' \int \frac{\mathrm{d}^4k}{(2\pi)^4} \frac{\mathrm{d}^4k'}{(2\pi)^4} \frac{\mathrm{d}^4\tilde{k}}{(2\pi)^4} J(k)^* \mathrm{e}^{ikx} \frac{\mathrm{e}^{-i\tilde{k}(x-x')}}{\tilde{k}^2 - m^2 + i\varepsilon} J(k')\mathrm{e}^{-ik'x'} \,. \tag{3.31}$$

Exchanging the integration order and performing the spacetime integrations leads to the conservation of the four-momenta entering and leaving the two interaction points, $(2\pi)^8\delta(k - \tilde{k})\delta(\tilde{k} - k')$. The source $J(k)^*$ produce a scalar particle with momentum k, and thus only the Fourier component k of the scalar propagator contributes. This is a very general behaviour, based solely on the translation invariance of the free particle states we are using. In the final step, we cancel two of the three momentum integrations with the two momentum delta functions and are left with

$$W[J] = -\frac{1}{2} \int \frac{\mathrm{d}^4k}{(2\pi)^4} J(k)^* \frac{1}{k^2 - m^2 + i\varepsilon} J(k) \,. \tag{3.32}$$

The functional $W[J]$ has the same structure as the one for the harmonic oscillator found in the last chapter. We will see that, as in the one-dimensional case, it contains all information about a free scalar field, not only about its ground-state. Note that the contribution of Fourier modes to $W[J]$ increases, the closer they are on-shell, $k^2 \to m^2$. For $k^2 = m^2$, the propagator diverges finally. In order to interpret this unphysical result, we can compare it to a classical harmonic oscillator. If an external source is applied in resonance with the eigenfrequency ω of the oscillator, the oscillation amplitude will increase until friction cannot be longer neglected. In our case at hand, the non-zero life time of unstable particles plays the role of friction. Including in our formalism the fact that the exchanged particle is unstable, the infinitesimal ε would be replaced by half its decay-width, $\varepsilon \to i\Gamma/2$.

Attractive Yukawa potential by scalar exchange.   From our macroscopic experience, we know the two cases of electromagnetism, where equal electric charges repel each other, and of gravity where two masses attract each other. The first physics question we want to answer with our newly developed formalism is if the scalar field falls into the category of a fundamentally attractive or repulsive interaction.

In order to address this question, we consider two static point charges as external sources, $J = J_1(x_1) + J_2(x_2)$ with $J_i = \delta(\boldsymbol{x} - \boldsymbol{x}_i)$, in $W[J]$. Multiplying out the terms in $J(x)\Delta_F(x - x')J(x')$ gives four contributions, $W_{ij} \propto J_i J_j$. The terms $W_{11}[J]$ and $W_{22}[J]$ correspond to the emission and re-absorption of the particle by the same source J_i. They are examples of self-interactions that we will neglect for the moment. The interaction between two different charges is given by

$$W_{12}[J] = W_{21}[J] = -\frac{1}{2} \int \mathrm{d}^4 x \mathrm{d}^4 x' \int \frac{\mathrm{d}^4 k}{(2\pi)^4} J_1(x) \frac{\mathrm{e}^{-ik(x-x')}}{k^2 - m^2 + i\varepsilon} J_2(x') \tag{3.33a}$$

$$= -\frac{1}{2} \int \mathrm{d}t \mathrm{d}t' \int \frac{\mathrm{d}^4 k}{(2\pi)^4} \frac{\mathrm{e}^{-ik^0(t-t')} \mathrm{e}^{i\boldsymbol{k}(\boldsymbol{x}_1-\boldsymbol{x}_2)}}{k^2 - m^2 + i\varepsilon}. \tag{3.33b}$$

Performing one of the two time integrals, for example the one over t', gives $2\pi\delta(k_0)$. Hence our assumption of static sources implies that the virtual particle carries zero energy and is space-like, $k^2 = -\boldsymbol{k}^2 < 0$. Eliminating then the k_0 integral with the help of the delta function, we obtain next

$$W_{12}[J] = \frac{1}{2} \int \mathrm{d}t \int \frac{\mathrm{d}^3 k}{(2\pi)^3} \frac{\mathrm{e}^{i\boldsymbol{k}\boldsymbol{r}}}{\boldsymbol{k}^2 + m^2} \tag{3.34}$$

with $\boldsymbol{r} = \boldsymbol{x}_1 - \boldsymbol{x}_2$. The denominator is always positive, and we can therefore omit the $i\varepsilon$. Before we can go on, we have to make sense out of the infinite time integral: Looking at

$$Z[J] = \langle 0| \exp(-iH[J]\tau)|0\rangle = \exp(iW[J]), \tag{3.35}$$

we see that $W[J] = -E\tau$ with $\tau = t - t'$ as the considered time interval. Hence the potential energy V of two static point charges separated by the distance $\boldsymbol{r}$ is

$$V = -(W_{12} + W_{21})/\tau = -\int \frac{\mathrm{d}^3 k}{(2\pi)^3} \frac{\mathrm{e}^{i\boldsymbol{k}\boldsymbol{r}}}{\boldsymbol{k}^2 + m^2} = -\frac{\mathrm{e}^{-mr}}{4\pi r} < 0. \tag{3.36}$$

Thus the potential energy of two equal charges is reduced by the exchange of a scalar particle, which means that the scalar force between them is attractive. If the exchanged particle is massive, the range of the force is of order $1/m$. These two observations were the basic motivation for Yukawa to suggest in 1935 the exchange of scalar particles as model for the nuclear force. Note also that we obtain in the limit $m \to 0$ a $1/r$ potential as in Newton's and Coulomb's law. Thus we learnt the important fact that the only two known forces of infinite range, the electromagnetic and the gravitational force, are transmitted by massless particles, the photon and the graviton, respectively. The result $V \propto 1/r$ for $m = 0$ and $n = 4$ spacetime dimensions, or more generally $V \propto 1/r^{n-3}$ for $n \geq 4$, follows from simple dimensional analysis. For $m = 0$, the only remaining dimensionful parameter after the integration over $\boldsymbol{k}$ is the distance r. From Eq. (3.36), we read off that the potential energy V has the dimension $[V] = k^{n-3}$. Thus the potential energy due to the exchange of a massless particle scales as r^{-n+3}. Finally, note that the amplitude $W_{12} + W_{21}$ or $J(x)\Delta_F(x - x')J(x') = J(x')\Delta_F(x' - x)J(x)$ is symmetric against the exchange $1 \leftrightarrow 2$ or $x_1 \leftrightarrow x_2$, reflecting the fact that the scalar propagator is an even function. Thus scalar particles are bosons and follow Bose–Einstein statistics.

3.3 Green functions for a free scalar field

In the last section, we obtained the scalar Feynman propagator or two-point Green function as the inverse of the Klein–Gordon operator. As next step, we want to derive n-point Green functions from their generating functional. Consider the expansion of the exponential in the functional

$$Z[J] = \mathrm{e}^{\mathrm{i}W[J]} = \sum_{n=0}^{\infty} \frac{\mathrm{i}^n}{n!} W^n = \sum_{n=0}^{\infty} \frac{\mathrm{i}^n}{n!} \int \mathrm{d}^4 x_1 \cdots \mathrm{d}^4 x_n \mathcal{G}_n(x_1, \cdots, x_n) J(x_1) \cdots J(x_n),$$
$$(3.37)$$

where we assume that $Z[J]$ is normalised so that $Z[0] = 1$. The RHS serves as definition of the n-point Green functions $\mathcal{G}(x_1, \cdots, x_n)$. They can be calculated as the functional derivatives of $Z[J]$,

$$\mathcal{G}(x_1, \ldots, x_n) = \frac{1}{\mathrm{i}^n} \frac{\delta^n}{\delta J(x_1) \cdots \delta J(x_n)} Z[J] \Big|_{J=0}. \tag{3.38}$$

For $n = 2$ we should re-derive the Feynman propagator. Starting from

$$\frac{1}{\mathrm{i}} \frac{\delta Z[J]}{\delta J(x)} = \frac{1}{\mathrm{i}} \frac{\delta}{\delta J(x)} \exp\left(-\frac{\mathrm{i}}{2} \int \mathrm{d}^4 x_1 \mathrm{d}^4 x_2 J(x_1) \Delta_F(x_1 - x_2) J(x_2)\right)$$
$$= -\int \mathrm{d}^4 x_1 \Delta_F(x - x_1) J(x_1) \exp(\mathrm{i}W[J]), \tag{3.39}$$

we obtain

$$\frac{1}{\mathrm{i}} \frac{\delta}{\delta J(y)} \frac{1}{\mathrm{i}} \frac{\delta}{\delta J(x)} Z[J] = \mathrm{i}\Delta_F(x - y) \exp(\mathrm{i}W[J])$$
$$+ \left(\int \mathrm{d}^4 x_1 \Delta_F(x - x_1) J(x_1)\right) \left(\int \mathrm{d}^4 x_1 \Delta_F(y - x_1) J(x_1)\right) \exp(\mathrm{i}W[J]). \tag{3.40}$$

Setting $J = 0$ gives the desired result for the two-point function,

$$\mathcal{G}(x, y) = \mathrm{i}\Delta_F(x - y). \tag{3.41}$$

It is straightforward to continue. Another functional derivative gives the three-point function,

$$
\frac{\delta}{i\delta J(x_1)} \frac{\delta}{i\delta J(x_2)} \frac{\delta}{i\delta J(x_3)} Z[J] = - \left(\int d^4x \Delta_F(x_1 - x)J(x) \right)
$$
$$
\times \left(\int d^4x \Delta_F(x_2 - x)J(x) \right) \left(\int d^4x \Delta_F(x_3 - x)J(x) \right) \exp(iW[J])
$$
$$
-i\Delta_F(x_2 - x_3) \int d^4x \Delta_F(x_1 - x)J(x) \exp(iW[J])
$$
$$
-i\Delta_F(x_2 - x_1) \int d^4x \Delta_F(x_3 - x)J(x) \exp(iW[J])
$$
$$
-i\Delta_F(x_3 - x_1) \int d^4x \Delta_F(x_2 - x)J(x) \exp(iW[J]) . \tag{3.42}
$$

For odd n, we obtain always a source J in the pre-factor because $W[J]$ is an even polynomial in J. Hence all odd n-point functions are zero. We continue with the four-point function. Taking another derivative and setting $J = 0$, only terms linear in J in Eq. (3.42) contribute and thus we obtain

$$
\begin{aligned}
\mathcal{G}(x_1, x_2, x_3, x_4) = - \big[& \Delta_F(x_1 - x_2)\Delta_F(x_3 - x_4) \\
& + \Delta_F(x_1 - x_3)\Delta_F(x_2 - x_4) \\
& + \Delta_F(x_1 - x_4)\Delta_F(x_2 - x_3) \big].
\end{aligned} \tag{3.43}
$$

We see that the four-point function is the sum of all permutations of products of two two-point functions. For instance, the first term $\Delta_F(x_1 - x_2)\Delta_F(x_3 - x_4)$ in the four-point function describes the independent propagation of a scalar particle from x_1 to x_2 and of another one from x_3 to x_4. Thus our approach leads indeed to a many-particle theory. Since we did not include interactions, particles are propagating independently and the n-point function factorises into products of two-point functions. Thus the functional $Z[J]$ generates Green functions which can be connected or disconnected. The statement that the n-point function is the sum of all permutations of the product of the $n/2$ two-point functions holds for all n and is called "Wick's theorem".

We introduce next the connected n-point functions $G(x_1, \ldots, x_n)$. Their generating functional is $W[J]$,

$$
G(x_1, \ldots, x_n) = \frac{1}{i^n} \frac{\delta^n}{\delta J(x_1) \cdots \delta J(x_n)} iW[J] \bigg|_{J=0} . \tag{3.44}
$$

For a free theory, W is quadratic in the sources J. Hence, all connected n-point functions $G(x_1, \ldots, x_n)$ with $n > 2$ vanish and the only non-zero one is the two-point function with

$$
G(x, y) = i\Delta_F(x - y) = \mathcal{G}(x, y). \tag{3.45}
$$

To summarise, there exists only one non-zero connected n-point function in a free theory which is determined by the Feynman propagator, $G(x, y) = i\Delta_F(x - y)$. All non-zero disconnected n-point functions can be obtained by permuting the product of $n/2$

two-point functions ("Wick's theorem"). Hence any higher-order Green function can be constructed out of a single building block, the Feynman propagator. In perturbation theory, we will recast the interacting theory—loosely speaking—in "interaction vertices times free propagators". This enables us to derive simple Feynman rules that tell us how one constructs an arbitrary Green function out of vertices and propagators.

Remark 3.2: Let us assume that the source in $Z[J]$ consists of two terms, $J(x, y) = J_1(x) + J_2(y)$, localised at large space-like distances. Then the requirement that the experiment $J_1(x)$ performed on the Moon should not influence $J_2(y)$ on Earth means that their probability amplitude factorises, $Z[J] = Z[J_1]Z[J_2]$. Sometimes this property called the "cluster decomposition theorem" is included instead of locality in the axioms of QFT, e.g. by Weinberg (2005). Assuming now in (3.43) that the coordinates $x = (x_1, x_2)$ and $y = (x_3, x_4)$ are separated by a large space-like distance, and neglecting thus the exponentially suppressed propagators, the disconnected Green function becomes $\mathcal{G}(x, y) = -\Delta_F(x_1 - x_2)\Delta_F(x_3 - x_4) = G(x)G(y)$. This factorisation reflects the one for the generating functional, $Z[J] = Z[J_1]Z[J_2]$. In contrast, the generating functional $W[J]$ has to satisfy $W[J] = W[J_1] + W[J_2]$. It is this additivity property which excludes that disconnected Green functions contribute to $W[J]$. Since we can construct disconnected Green functions by simple multiplication, $\mathcal{G}(x, y, \ldots) = G(x)G(y)\ldots$, it is sufficient to study only $W[J]$ and connected Green functions.

Causality and the Feynman propagator. We already discussed in section 2.5 that any valid relativistic theory should implement the requirement of causality. No signal using ϕ particles as carrier should travel with a speed larger than the one of light. We also saw that the Feynman prescription leads to a relativistic consistent interpretation of the propagator, although the propagator does not vanish outside the light-cone but goes only exponentially to zero, cf. problem 2.9. One may therefore wonder, if this means that the uncertainty principle makes the light-cone "fuzzy" and thus the axiom of special relativity that no signal can be transmitted with $v > c$ is violated on scales smaller $\lesssim 1/m$.

We can address this question considering the field $\phi(x)$ as operator $\hat{\phi}(x)$ and asking then when a measurement of $\hat{\phi}(x)$ influences $\hat{\phi}(x')$. Recall first that the Feynman propagator equals the two-point Green function which in turn corresponds to the vacuum expectation value of the time-ordered product of field operators,

$$\mathcal{G}(x_1, x_2) = \langle 0|T\{\hat{\phi}(x_1)\hat{\phi}(x_2)\}|0\rangle = \mathrm{i}\Delta_F(x_1 - x_2). \tag{3.46}$$

The property $\Delta_F(x_1 - x_2) = \Delta_F(x_2 - x_1)$ implies that the field operators $\hat{\phi}(x_1)$ and $\hat{\phi}(x_2)$ commute,

$$\begin{aligned}
\langle 0|T\{\hat{\phi}(x_1)\hat{\phi}(x_2)\}|0\rangle &= \langle 0|\hat{\phi}(x_1)\hat{\phi}(x_2)|0\rangle \vartheta(t_1 - t_2) \\
&+ \langle 0|\hat{\phi}(x_2)\hat{\phi}(x_1)|0\rangle \vartheta(t_2 - t_1).
\end{aligned} \tag{3.47}$$

Using the analogy of a free quantum field to an infinite set of oscillators, we try to express the field operator $\hat{\phi}(x)$ through annihilation and creation operators. Comparing to the expansion (2.64) of an oscillator in $d = 1$, i.e. $\phi(t) = (2\omega)^{-1/2}(ae^{-i\omega t} + a^\dagger e^{i\omega t})$, suggests the ansatz

$$\hat{\phi}(x) = \int \frac{\mathrm{d}^3 k}{\sqrt{(2\pi)^3 2\omega_k}} \left[a(\boldsymbol{k})e^{-ikx} + a^\dagger(\boldsymbol{k})e^{+ikx} \right], \qquad (3.48)$$

with $k^0 = \omega_k$, $a(\boldsymbol{k})$ and $a^\dagger(\boldsymbol{k})$ as annihilation and creation operators that satisfy

$$a(\boldsymbol{k})\,|0\rangle = 0\,, \qquad a^\dagger(\boldsymbol{k})\,|0\rangle = |\boldsymbol{k}\rangle \qquad \text{and} \qquad [a(\boldsymbol{k}), a^\dagger(\boldsymbol{k}')] = \delta(\boldsymbol{k} - \boldsymbol{k}'). \qquad (3.49)$$

Hence the vacuum state $|0\rangle$ is defined by $a(\boldsymbol{k})\,|0\rangle = 0$ for all $\boldsymbol{k}$.

If this ansatz is correct, then we should be able to reproduce the known form of the Feynman propagator. Inserting our ansatz for the field into $\langle 0|\hat{\phi}(\boldsymbol{x}, t)\hat{\phi}(0)|0\rangle$ for $t > 0$, we obtain four terms containing the products aa, $aa^\dagger$, $a^\dagger a$, $a^\dagger a^\dagger$. Only $aa^\dagger$ survives, resulting in

$$\langle 0|\int \frac{\mathrm{d}^3 k\, \mathrm{d}^3 k'}{(2\pi)^3 \sqrt{2\omega_k 2\omega_{k'}}} a(\boldsymbol{k})e^{-ikx} a^\dagger(\boldsymbol{k}')\vartheta(t)|0\rangle = \int \frac{\mathrm{d}^3 k}{(2\pi)^3 2\omega_k} e^{-ikx}\vartheta(t). \qquad (3.50)$$

In the second step, we used the commutation rule $[a(\boldsymbol{k}), a^\dagger(\boldsymbol{k}')] = \delta(\boldsymbol{k} - \boldsymbol{k}')$. Performing the same exercise for $t < 0$, we see that in this case too we reproduce the corresponding term of the Feynman propagator. Thus we conclude that our ansatz for the field and commutation rules for the annihilation and creation operators are consistent. Note that we could create in Eq. (3.50) alternatively one-particle states, $\langle 0|a(\boldsymbol{k})a^\dagger(\boldsymbol{k}')|0\rangle = \langle \boldsymbol{k}|\boldsymbol{k}'\rangle$, and consistency requires thus that the states $|\boldsymbol{k}\rangle$ are non-relativistically normalised, $\langle \boldsymbol{k}|\boldsymbol{k}'\rangle = \delta(\boldsymbol{k} - \boldsymbol{k}')$. This should come as no surprise, since we started from the analogy to the non-relativistic oscillator. If one prefers states satisfying the relativistic normalisation, one can rescale the creation and annihilation operators such that $[\tilde{a}(k), \tilde{a}^\dagger(k')] = (2\pi)^3 2\omega_k\delta(\boldsymbol{k} - \boldsymbol{k}')$ and the field-operator becomes

$$\hat{\phi}(x) = \int \frac{\mathrm{d}^3 k}{(2\pi)^3 2\omega_k} \left[\tilde{a}(k)e^{-ikx} + \tilde{a}^\dagger(k)e^{+ikx} \right]. \qquad (3.51)$$

Both normalisations lead to canonical commutation relations between the field $\hat{\phi}$ and its canonically conjugated momentum density $\hat{\pi} = \dot{\hat{\phi}}$ at equal times,

$$[\hat{\phi}(\boldsymbol{x}, t), \hat{\pi}(\boldsymbol{x}', t)] = i\delta(\boldsymbol{x} - \boldsymbol{x}'). \qquad (3.52a)$$

$$[\hat{\phi}(\boldsymbol{x}, t), \hat{\phi}(\boldsymbol{x}', t)] = [\hat{\pi}(\boldsymbol{x}, t), \hat{\pi}(\boldsymbol{x}', t)] = 0. \qquad (3.52b)$$

We come back to the question regarding what happens if the commutator of two fields vanishes for space-like separation. We evaluate first

$$[\hat{\phi}(x), \hat{\phi}(x')] = \int \frac{\mathrm{d}^3 k\, \mathrm{d}^3 k'}{(2\pi)^3 \sqrt{2\omega_k 2\omega_{k'}}} \left[a(\boldsymbol{k})e^{-ikx} + a^\dagger(\boldsymbol{k})e^{+ikx}, a(\boldsymbol{k}')e^{-ik'x'} + a^\dagger(\boldsymbol{k}')e^{+ik'x'} \right]$$

$$= \int \frac{\mathrm{d}^3 k}{(2\pi)^3 2\omega_k} \left(e^{-ik(x-x')} - e^{+ik(x-x')} \right) \equiv D(x - x'). \qquad (3.53)$$

The commutator $D(x - x')$ is the sum of two Lorentz invariant expressions, and thus its value has to be the same in all inertial frames. For equal times, $t = t'$, exchanging the dummy variable $\boldsymbol{k} \to -\boldsymbol{k}$ in the second term shows that the contribution from positive and negative energies cancel. Hence the equal-time commutator of two fields is zero, as claimed in (3.52b). For space-like distances, $(x-x')^2 < 0$, we can find a Lorentz boost which changes the ordering of the spacetime events, $x-x' \to -(x-x')$. But the commutator is antisymmetric, $D(x) = -D(-x)$, and therefore $D(x)$ has to vanish, if x is outside the light-cone of x' and vice versa. Hence we have shown that the commutator of two space-like separated fields is zero,

$$[\hat{\phi}(x), \hat{\phi}(x')] = 0 \qquad \text{for} \quad (x-x')^2 < 0, \tag{3.54}$$

which is the condition for causality. The transmission of a signal corresponds not only to the propagation of a virtual particle but it also includes its measurement. Thus the fact that the Feynman propagator does not vanish outside the light-cone does not contradict causality by itself. However, it excludes the interpretation of $\phi(x)$ as a wave-function, i.e. as the probability amplitude to find the particle ϕ at x, because expressions like (2.16) violate causality. In our approach, ϕ denotes either a classical field or a quantum field as in Eq. (3.48).

There are two main differences between the Feynman propagator and the commutator of two fields. First, $[\hat{\phi}(x), \hat{\phi}(x')]$ is an operator, while $i\Delta(x_1 - x_2)$ is a vacuum expectation value. The quantum vacuum fluctuates, and these fluctuations are correlated also on space-like distances, similar to the ERP correlations in quantum mechanics. The Feynman propagator $i\Delta_F(x_1 - x_2)$ is designed to describe not only the propagation of time-like particles, but includes also the space-like propagation of virtual particles: the most "extreme" case is the *instantaneous* exchange of particles transmitting the Coulomb or Yukawa force between static sources, cf. Eq. (3.34). Second, in $[\hat{\phi}(x), \hat{\phi}(x')]$ we subtract the contribution of positive and negative frequencies, while we add them in the Feynman propagator. As a result, the contributions from a particle travelling the distance $\boldsymbol{x}$ and from an antiparticle travelling the distance $-\boldsymbol{x}$ cancel in the commutator, while they add up in the Feynman propagator. Since causality relies on the cancellation between positive and negative energy modes in $[\hat{\phi}(x), \hat{\phi}(x')]$, we conclude that a relativistic quantum theory has to incorporate antiparticles.

3.4 Vacuum energy and the Casimir effect

Vacuum energy. We now seek to calculate the energy of the vacuum state of a free scalar quantum field. The energy density ρ of the quantum field ϕ (omitting from now on the hat) is given by the vacuum expectation value of its Hamiltonian density $\mathscr{H}$,

$$\rho = \langle 0|\mathscr{H}|0\rangle = \rho_0 + \frac{1}{2}\langle 0|\pi^2 + (\boldsymbol{\nabla}\phi)^2 + m^2\phi^2|0\rangle = \rho_0 + \rho_1 . \tag{3.55}$$

Here we added the constant energy density ρ_0 to (3.14) and used that the vacuum is normalised, $\langle 0|0\rangle = 1$. For the calculation of ρ_1, we can recycle our result for the propagator of a scalar field by considering $\phi^2(x)$ as the limit of two fields at nearby points,

$$\langle 0|\phi(x')\phi(x)|0\rangle_{x' \searrow x} = \int \frac{\mathrm{d}^3k}{(2\pi)^3 2\omega_k} \mathrm{e}^{-ik(x'-x)}\bigg|_{x' \searrow x} = \int \frac{\mathrm{d}^3k}{(2\pi)^3 2\omega_k} . \tag{3.56}$$

We perform first the differentiation in $\langle \pi^2 \rangle = \langle \dot{\phi}^2 \rangle$ and $\langle (\boldsymbol{\nabla}\phi)^2 \rangle$ and send then $x' \searrow x$. Thus π^2 and $(\boldsymbol{\nabla}\phi)^2$ add a ω_k^2 and $\boldsymbol{k}^2$ term, respectively,

$$\rho = \langle 0 | \mathcal{H} | 0 \rangle = \rho_0 + \int \frac{\mathrm{d}^3 k}{(2\pi)^3 2\omega_k} \left[\frac{1}{2}(\omega_k^2 + \boldsymbol{k}^2 + m^2) \right] = \rho_0 + \int \frac{\mathrm{d}^3 k}{(2\pi)^3} \frac{1}{2} \omega_k \,. \quad (3.57)$$

If we insert $\hbar$ and c into this expression, we see that ρ_0 as a classical contribution to the energy density of the vacuum is $\propto \hbar^0$, while the second term $\rho_1 \propto \hbar\omega_k/V$ as a quantum correction is linear in $\hbar$. The total energy density ρ of the vacuum state of a free scalar field has a very intuitive interpretation. Additionally to the classical energy density ρ_0, it sums up the zero-point energies of all individual modes $\boldsymbol{k}$ of a free field. Despite its simplicity, we cannot make sense out of this result. Since both the density of modes and their energy increases with $|\boldsymbol{k}|$, the integral diverges. This is the first example that momentum integrals in quantum field theories are often ill-defined and require some care. One calls momentum integrals which are divergent for $k \to 0$ infrared- (IR) divergent, while one calls integrals which diverge for $k \to \infty$ ultraviolet- (UV) divergent.

Let us now consider the case that the Hamiltonian (3.14) describes physics correctly only up to the energy scale Λ, while the modes with $|\boldsymbol{k}| \gtrsim \Lambda$ do not contribute to ρ_1. Such a possibility exists, for example, in supersymmetric theories where the contributions of different particle types cancel each other above the scale Λ_{SUSY} where supersymmetry is broken. Integrating the contribution to the vacuum energy density by field modes up to the cut-off scale Λ, we find

$$\rho_1 = \int_0^\Lambda \frac{\mathrm{d}k \, k^2}{2\pi^2} \frac{1}{2} \omega_k \sim \Lambda^4 \quad (3.58)$$

in the limit $\Lambda \gg m$. Since only the total energy density ρ is observable, the unknown ρ_0 can be always chosen such that $\rho_0 + \rho_1$ agrees with observations, even if $|\rho_0|, |\rho_1| \gg |\rho|$. Nevertheless, the strong sensitivity of ρ_1 on the value of the cut-off scale Λ is puzzling for two reasons: first, cosmological observations determine the total vacuum energy density ρ_Λ to which all types of fields contribute as $\rho_\Lambda \sim (\mathrm{meV})^4$. On the other hand, accelerator experiments give no indications that a cancellation mechanism as supersymmetry works at energy below few TeV. Thus we expect naively at least $\rho_\Lambda \sim (\Lambda_{\mathrm{SUSY}})^4 \gtrsim (\mathrm{fewTeV})^4$, which is 60 orders of magnitude larger than observed, if no strong cancellation of the various contributions to ρ_Λ takes place. Second, the behaviour $\rho_1 \sim \Lambda^4$ implies that all scalar particles with mass $m \lesssim \Lambda$ contribute equally to ρ. This poses the question of whether we have to know the physics at energy scales much larger than those we probe experimentally in order to make predictions using QFT. Such a behaviour would contradict the successful development of chemistry, atomic or nuclear physics using only the experimental data and models of the corresponding relevant energy scale E. Something similar should happen in QFT too and we will study later the conditions that heavy particles with mass m "decouple" at energies $E \ll m$. In this case, their effects are either suppressed by factors E/m, or are hidden in unobservable quantities like ρ_1.

Casimir effect. Although we cannot unambiguously calculate the vacuum energy, we can determine the energy difference of different vacua. As a concrete example, we consider the suggestion by Casimir that the vacuum between two conducting plates is disturbed. As a result, the vacuum energy density between the plates becomes a function of their distance d. The difference of the vacuum energy density inside and outside the plates is finite and leads to a measurable force between them.

Let us consider two parallel, uncharged, perfectly conducting plates at distance d. Standing waves between them have the form $\sin(n\pi x/d)$ with discrete energies $\omega_n = n\pi/d$. The vacuum fluctuations of a photon have the same form as the one of massless scalar field, except that there is an additional factor two due to its two spin degrees of freedom. Thus the vacuum energy inside the box of volume $L_y L_z d$ per single polarisation mode is given by

$$E = L_y L_z \sum_{n=1}^{\infty} \int \frac{\mathrm{d}k_y \mathrm{d}k_z}{(2\pi)^2} \frac{1}{2} \sqrt{\left(\frac{n\pi}{d}\right)^2 + k_y^2 + k_z^2}. \tag{3.59}$$

To simplify the calculations, we consider a $1+1$ dimensional system of two plates separated by the distance d. Then the energy density $\rho = E/d$ of a massless field per polarisation mode inside the plates is

$$\rho(d) = \frac{\pi}{2d^2} \sum_{n=1}^{\infty} n. \tag{3.60}$$

Next we introduce a cut-off function $f(a) = \exp(-an\pi/d)$ which suppresses the high-energy modes,

$$\rho(d) \to \rho(a,d) = \frac{\pi}{2d^2} \sum_{n=1}^{\infty} n\mathrm{e}^{-an\pi/d}. \tag{3.61}$$

This procedure is called *regularisation*: for $a > 0$, we obtain a well-defined mathematical sum which we can manipulate following the usual rules of analysis, while we recover for $a \to 0$ the original divergent sum. We have chosen as argument of the exponential $an\pi/d$, because the physically relevant quantities are the energy levels $\omega_n = n\pi/d$ of the system. Note that we would obtain the same cut-off function, if we would keep the two fields in Eq. (3.56) the distance $\delta t = -ia$ separated. Thus UV divergences are caused, if fields or propagators coincide at the same spacetime point.

Now we can evaluate the regularised sum, rewriting it as a geometrical sum,

$$\rho(a,d) = \frac{\pi}{2d^2} \sum_{n=1}^{\infty} n\mathrm{e}^{-an\pi/d} = -\frac{1}{2d} \frac{\partial}{\partial a} \sum_{n=0}^{\infty} \mathrm{e}^{-an\pi/d} \tag{3.62a}$$

$$= -\frac{1}{2d} \frac{\partial}{\partial a} \frac{1}{1 - \mathrm{e}^{-a\pi/d}} = \frac{\pi}{2d^2} \frac{\mathrm{e}^{-a\pi/d}}{(1 - \mathrm{e}^{-a\pi/d})^2}. \tag{3.62b}$$

Then we use $\mathrm{e}^x (1 - \mathrm{e}^{-x})^2 = 4 \sinh^2(x/2)$ and expand $\rho(a,d)$ for small a in a Laurent series,

$$\rho(a,d) = \frac{\pi}{8d^2} \frac{1}{\sinh^2(a\pi/2d)} = \frac{1}{2\pi a^2} - \frac{\pi}{24d^2} + \mathcal{O}(a^2 d^{-4}). \tag{3.63}$$

Note that we isolated thereby the divergence into a term which does not depend on the distance d of the plates. Thus the divergence cancels in the difference of the vacuum energy with and without plates,

$$\rho_{\text{Cas}}(d) \equiv \lim_{a \to 0} \left[\rho(a, d) - \rho(a, d \to \infty) \right] = -\frac{\pi}{24d^2}. \tag{3.64}$$

This final step in order to obtain a finite result is called *renormalisation*. One can verify that the result is not only independent of the cut-off parameter a, but also of the shape of a reasonable[3] cut-off function $f(a)$. In contrast, the a dependent terms in Eq. (3.63) may depend on the form of $f(a)$. The quantity measured in actual experiments is the force F with which the plates attract (or repel) each other. This force is given by

$$-F = \frac{\partial E}{\partial d} = \frac{\partial(d\rho_{\text{Cas}})}{\partial d} = \frac{\pi}{24d^2}. \tag{3.65}$$

Thus two parallel plates attract each other. The experimentally relevant case of electromagnetic waves between two parallel plates in $3 + 1$ dimensions can be calculated analogously, cf. problem 4.5. The experimental confirmation of the Casimir effect has been achieved only in the 1990s, with a precision on the 1% level.

How can we understand that the Casimir force is independent of the details of the regularisation procedure? Let us compare the impact of the two plates on modes with different wave-number $k = 2\pi/\lambda$. In a typical experimental set-up, the plates are separated by a distance of the order $d \sim 1\,\text{mm}$ and thus $k_0 \equiv 2\pi/d \sim \text{meV}$. The plates eliminate all low-energy modes with $k < k_0$ between them, while the modes with $k > k_0$ attain a discrete spectrum. However, for $k_n \gg k_0$, the spacing between the modes becomes negligible and experimentally one cannot distinguish the discrete spectrum from a continuous one. In particular, we can approximate the sum over the discrete energies by an integral and the contributions of modes with $k \gg k_0$ with and without plates cancel calculating the energy difference. Since the main contribution to the Casimir energy comes from cutting off modes with $k \lesssim 2\pi/d \sim \text{meV}$, we conclude that the Casimir energy is an IR effect. Therefore the details of the UV regularisation should not influence the result and any reasonable cut-off function that makes the mathematical manipulations (3.62a–3.63) well-defined should lead to the same result.

Summary

The exchange of space-like scalar quanta with zero energy between two static sources leads to the Yukawa potential. The corresponding force mediated by a scalar field is attractive. The Feynman propagator obtained by the $m^2 - i\varepsilon$ prescription is the unique Green function which can be analytically continued to an Euclidean Green function. It propagates particles (with positive frequencies) forward in time, while antiparticles (with negative frequencies) propagate backward in time. While these two contributions add up in the scalar Feynman propagator, they cancel in the commutator of field-operators at space-like distances, as required by causality. Disconnected n-point Green functions are generated by the functional $Z[J]$, while $iW[J] = \ln Z[J]$ generates connected Green functions. Wick's theorem says that an n-point function can be obtained as the permutation

---

[3]Reasonable means that $f(a)$ is normalised, $f(0) = 1$, and that all its derivatives vanish for large a, $\lim_{a \to \infty} f^{(n)}(a) = 0$.

of products of two-point functions. The Casimir effect shows that the zero-point energies of quantum fields have real, measurable consequences.

Further reading. The quantisation of free fields using both canonical quantisation and the path-integral approach is discussed extensively by Greiner and Reinhardt (2008).

Problems

3.1 Complex scalar field. Derive the Lagrangian and the Hamiltonian for a complex scalar field, considering $\phi = (\phi_1 + i\phi_2)/\sqrt{2}$ and $\phi^* = (\phi_1 - i\phi_2)/\sqrt{2}$ as the dynamical variables. Find the conserved current of this complex field. [Hint: Proceed similar as in the case of the Schrödinger equation.]

3.2 Maxwell Lagrangian. a.) Derive the Lagrangian for the photon field A_μ from the source-free Maxwell equation $\partial_\mu F^{\mu\nu} = 0$ following the steps from (3.11a) to (3.11b) in the scalar case. b.) What is the meaning of the unused set of Maxwell equations? c.) Determine the canonically conjugated momenta π_μ of A^μ.

3.3 Dimension of ϕ. a.) Determine the mass dimension of a scalar field ϕ in d spacetime dimensions. b.) For which d has $\mathscr{L} = \lambda\phi^3$ ($\mathscr{L} = \lambda\phi^4$) a dimensionless coupling constant? c.) Has the numerical value of λ for a $\lambda\phi^4$ theory classically a physical meaning? In quantum theory? [Hint: Can you rescale fields such that $\mathscr{L}(\lambda) = \lambda\mathscr{L}(1)$?]

3.4 Generating functional. Follow the approach of section 2.4 to evaluate the free functional $Z[J]$.

3.5 Yukawa potential. Show that the Yukawa potential $V(r) = e^{-mr}/(4\pi r)$ is the Fourier transform of $(\mathbf{k}^2 + m^2)^{-1}$, cf. Eq. (3.36).

3.6 Vacuum energy. Re-derive (3.57) expressing ϕ by annihilation and creation operators. Show that rewriting all creation operators on the left of the annihilation operators results in $\rho_1 = 0$. (This prescription is called "normal ordering".)

3.7 Canonical commutation relations. Derive (3.52a) assuming the validity of (3.49).

3.8 Green functions. Derive the conditions that the connected and the unconnected n-point Green functions are identical for $n = 2$, $G(x_1, x_2) = \mathcal{G}(x_1, x_2)$. Show that they differ in general for $n \geq 3$.

3.9 ζ function regularisation. a.) The function $f(t) = t/(e^t - 1)$ is the generating function for the Bernoulli numbers B_n, i.e. $f(t) = \frac{t}{e^t - 1} = \sum_{n=0}^{\infty} \frac{B_n}{n!} t^n$. Calculate the first Bernoulli numbers up to B_3. b.) Connect $\sum_{n=1}^{\infty} n e^{-an}$ to $f(t)$: split the sum into a divergent and a finite part for $a \to 0$ and compare to our old result. c.) The Riemann ζ function can be defined as $\zeta(s) = \sum_{n=1}^{\infty} n^{-s}$ for $s > 1$ and then analytically continued into the complex s plane. The Bernoulli numbers are connected to the Riemann ζ function with negative odd argument as $\zeta(1-2n) = -B_{2n}/(2n)$. Find the Casimir energy using the Riemann ζ function.

4
Scalar field with $\lambda\phi^4$ interaction

We know from quantum mechanics that adding an anharmonic term to an oscillator forces us to use either perturbative or numerical methods. The same happens in field theory. No analytic solution for a realistic interacting theory is at present known in $n = 4$ spacetime dimensions. Therefore, in this chapter we develop a perturbative method to evaluate the generating functionals $Z[J]$ and $W[J]$. We continue to work with the simplest case of a single real scalar field and choose as interaction a $\lambda\phi^4$ term. Then the coupling constant λ is dimensionless in the, for us, interesting case $n = 4$. If λ is small enough, we may hope that a perturbative series expansion in λ provides a useful approximation scheme. As motivation, we note that a scalar field with $\lambda\phi^4$ interaction cannot only model a wide range of phenomena in statistical physics but also describes the Higgs field of the SM and its self-interactions.

4.1 Perturbation theory for interacting fields

General formalism. The Lagrange density $\mathscr{L}$ in the functional $Z[J]$ for the scalar field considered up to now was at most quadratic in the fields and its derivatives. On the one hand, this allowed us to evaluate the path integral, while on the other hand this means that the field has no interactions. Two wave packets described by the free propagator just pass each other without interaction, as the superposition principle prescribes. As a next step we therefore add an interaction term $\mathscr{L}_I$ to the free Lagrangian $\mathscr{L}_0$, that is we set $\mathscr{L} = \mathscr{L}_0 + \mathscr{L}_I$. Then the generating functional $Z[J]$ for an interacting real scalar field ϕ becomes

$$Z[J] = \int \mathcal{D}\phi \exp i \int \mathrm{d}^4 x \left(\mathscr{L}_0 + \mathscr{L}_I + J\phi \right), \tag{4.1}$$

while from now on we denote the free functionals we have considered so far as $Z_0[J]$ and $W_0[J]$.

Starting from the full generating functional $Z[J]$ we can define *exact* Green functions which we denote by boldface letters. For instance, the exact two-point function or propagator is given analogous to Eq. (3.44) by

$$\boldsymbol{G}(x_1, x_2) = \frac{1}{\mathrm{i}^2} \frac{\delta^2 Z[J]}{\delta J(x_1)\delta J(x_2)} \bigg|_{J=0} = \int \mathcal{D}\phi \, \phi(x_1)\phi(x_2)\mathrm{e}^{\mathrm{i}\int \mathrm{d}^4 x \, [\mathscr{L}_0 + \mathscr{L}_I]}. \tag{4.2}$$

In general, we are not able to calculate the exact Green functions, and we will apply therefore perturbation theory. We assume that the interaction term $\mathscr{L}_I$ is a polynomial

Quantum Fields–From the Hubble to the Planck Scale. Michael Kachelriess. © Michael Kachelriess 2018.
Published in 2018 by Oxford University Press. DOI 10.1093/oso/9780198802877.001.0001

$\mathcal{P}(\phi)$ of degree ≥ 3 in the field ϕ and contains an expansion parameter λ which is small in the considered kinematic regime, $\mathscr{L}_I = \lambda\mathcal{P}(\phi)$ with $\lambda \ll 1$. This suggests expanding the interaction term,

$$\exp i \int d^4x\,\mathscr{L}_I(\phi) = 1 + i\lambda \int d^4x\,\mathcal{P}(\phi(x)) + \frac{(i\lambda)^2}{2!} \int d^4x_1 d^4x_2 \mathcal{P}(\phi(x_1))\mathcal{P}(\phi(x_2)) + \dots \tag{4.3}$$

Since

$$i\phi(x)e^{i\int d^4x'(\mathscr{L}_0+J\phi)} = \frac{\delta}{\delta J(x)}\,e^{i\int d^4x'(\mathscr{L}_0+J\phi)}, \tag{4.4}$$

we can perform the replacement

$$\mathscr{L}_I(\phi(x)) \to \mathscr{L}_I\left(\frac{1}{i}\frac{\delta}{\delta J(x)}\right). \tag{4.5}$$

Then the interaction $\mathscr{L}_I$ does not depend longer on ϕ and can be pulled out of the functional integral,

$$Z[J] = \exp\left[i\int d^4x\,\mathscr{L}_I\left(\frac{1}{i}\frac{\delta}{\delta J(x)}\right)\right] \int \mathcal{D}\phi \exp i \int d^4x\,(\mathscr{L}_0 + J\phi) \tag{4.6a}$$

$$= \exp\left[i\int d^4x\,\mathscr{L}_I\left(\frac{1}{i}\frac{\delta}{\delta J(x)}\right)\right] Z_0[J]. \tag{4.6b}$$

The solution of the free functionals $Z_0[J]$ and $iW_0 = \ln Z_0[J]$ was given in Eq. (3.19) as

$$Z_0[J] = Z_0[0]\exp\left(-\frac{i}{2}\int d^4x d^4x'\,J(x)\Delta_F(x-x')J(x')\right) = Z_0[0]\sum_{n=0}^{\infty}\frac{i^n}{n!}W_0^n. \tag{4.7}$$

Perturbation theory consists in a double expansion of the two exponentials in $Z[J]$: one in the coupling constant λ and one in the number of external sources J. The latter is fixed by the number of external particles in a scattering process, while the former is chosen according to the desired precision of the calculation.

Choosing the interaction term. Let us recall from our discussion of the free Lagrangian the physical requirements we should impose on the Lagrangian: each term should be a Lorentz scalar which is local in the fields. The corresponding Hamiltonian has to be bounded from below, stable against small perturbations and real. These conditions assure that the vacuum in the absence of external sources is stable.

Additional restrictions follow from a surprisingly simple argument employing dimensional analysis: using natural units, $\hbar = c = 1$, the dimension of all physical quantities can be expressed as powers of one basic unit which we choose as mass m. Then we use that the action has dimension zero, $[S] = m^0$, and thus the Lagrangian $[\mathscr{L}] = m^4$ in four spacetime dimensions. We consider next the free Lagrangian. From the kinetic term, we conclude that the dimension of a scalar field is $[\phi] = m^1$. Thus simple dimensional analysis shows that the term m^2 in front of ϕ^2 has the interpretation of a mass squared. Furthermore, we can order possible self-couplings of a scalar field according to their dimension as

$$\mathscr{L}_I = g_3 M \phi^3 + g_4 \phi^4 + \frac{g_5}{M}\phi^5 + \dots, \tag{4.8}$$

where the coupling constants g_i are dimensionless and we introduced the mass scale M to ensure $[\mathscr{L}] = m^4$. We call ϕ^d an operator of dimension d. Similarly as in the case of the Fermi constant, $G_F = \sqrt{2}g^2/(8m_W^2)$, the scale M could be connected to the exchange of a heavy particle.

Let us now estimate by dimensional analysis which energy scaling of the interaction probability we expect for the different coupling terms in $\mathscr{L}_I$. At lowest-order perturbation theory, the interaction probability is $\mathrm{d}W \propto |\mathscr{L}_I|^2$. Hence the interaction probability scales as $\propto (g_d/M^{d-4})^2$. Now we consider the ultra-relativistic limit, so that we can neglect the mass m of the scalar particle compared to the centre-of-mass (cms) energy $\sqrt{s}$. A probability has to be dimensionless, and for $s \gg m^2$ the only remaining dimensionful variable that can enter the total interaction probability W is s. Thus W has to scale as $g_d^2(s/M^2)^{d-4}$ in the limit $s \gg m^2$. Let us now distinguish the two ranges $m^2 \ll s \ll M^2$ and $s \gg M^2$. In the latter case, the interaction terms with $d > 4$ contain the large factors $(\sqrt{s}/M))^{d-4} \gg 1$ and perturbation theory becomes thus unreliable. In contrast, these terms are smaller than one below the scale M and thus suppressed relative to the operators with dimension $d \leq 4$. Therefore, in a first approach, we neglect all operators with dimension $d \geq 5$. Simplifying further $\mathscr{L}_I$, we want to include only one interaction term. In this case, a ϕ^3 term would lead to an unstable vacuum. Therefore our choice for the scalar self-interaction is $\mathscr{L}_I = -\lambda\phi^4/4!$, where the factor $1/4!$ was added for later convenience. If this choice of interaction is realised in nature for a specific particle has to be decided by experiment.

4.2 Green functions for the $\lambda\phi^4$ theory

We start now with the perturbative evaluation of Eq. (4.6a) for a $\lambda\phi^4/4!$ interaction. From

$$Z[J] = \left(1 - \frac{\mathrm{i}\lambda}{4!}\int \mathrm{d}^4x \frac{\delta^4}{\mathrm{i}^4\delta J(x)^4} + \dots\right) Z_0[J] = Z_0[J] - \frac{\mathrm{i}\lambda}{4!}\int \mathrm{d}^4x \frac{\delta^4 Z_0[J]}{\delta J(x)^4} + \dots =$$
$$= Z_0[J] + Z_1[J] + \dots \tag{4.9}$$

we see that we will generate a series of the type free Green functions plus higher-order corrections in λ. The calculation of the first-order correction is very similar to the calculation of the free four-point function, with the difference that now the four sources sit at the same point. You should find in problem 4.1 as the result

$$\left(\frac{\delta}{\mathrm{i}\delta J(x)}\right)^4 \exp(\mathrm{i}W_0[J]) = \left[3(\mathrm{i}\Delta_F(0))^2 + 6\mathrm{i}\Delta_F(0)\left(\int \mathrm{d}^4y\Delta_F(x-y)J(y)\right)^2\right.$$
$$\left. + \left(\int \mathrm{d}^4y\Delta_F(x-y)J(y)\right)^4\right]\exp(\mathrm{i}W_0[J]). \tag{4.10}$$

Next we introduce a graphical representation for the various terms in Eq. (4.10). Each Feynman propagator $\Delta_F(x - y)$ is represented by a line,

$$i\Delta_F(x - y) = \quad \text{———} \quad , \tag{4.11}$$

a source term $J(x)$ by an open dot,

$$i \int d^4x \, J(x) = \quad \circ\!\!-\!\!- \quad , \tag{4.12}$$

and an interaction vertex by a filled dot,

$$-i\lambda \int d^4x = \bullet \, . \tag{4.13}$$

Each source and vertex has its own coordinates and an integration over all coordinates is implied. In the case of the ϕ^4 interaction, a vertex connects four lines. Using this notation,[1] we can express the order $\mathcal{O}(\lambda)$ contribution to $Z[J]$ as

$$Z_1[J] = \frac{1}{4!}\left(3 \, \infty + 6 \, \underset{\bullet}{\circ\!-\!\!\bigcirc\!\!-\!\circ} + \times \right) \exp\left(\frac{1}{2}\circ\!\!-\!\!\circ\right). \tag{4.14}$$

A graph consists of lines and dots, where the latter may be vertices or sources. We distinguish internal and external lines: a line which ends on both sides at a dot with at least two lines attached is called internal, otherwise it is an external line. The three graphs contained in $Z_1[J]$ differ by the number of loops, that is, by the number L of closed lines. A graph with loop number $L = 0$ (as the third one in $Z_1[J]$) is called a tree graph, otherwise it is a loop graph. An inspection of the three graphs shows their loop number L is connected to the number n of lines and d of dots as $L = n - d + 1$. Expressing L via the number of vertices and sources, $d = V + S$, and the number of internal and external lines, $n = I + E$, we have $L = I + E - V - S + 1$. Since each external line comes with one source, we can express therefore the loop number also as $L = I - V + 1$, a formula which is valid for all types of interactions. Note also that the first and the second graph contained in $Z_1[J]$ can be obtained from the third one by joining two and one lines, respectively. There are six ways to join one line, and three ways to join two lines. Thus the pre-factors of the various graphs can be derived by simple symmetry arguments.

Knowing $Z_1[J]$, we can derive disconnected Green functions valid at $\mathcal{O}(\lambda)$ by performing functional derivatives,

$$\mathcal{G}^{(n)}(x_1, \dots, x_n) = \frac{1}{i^n}\frac{\delta^n}{\delta J(x_1) \cdots \delta J(x_n)}\left(Z_0[J] + Z_1[J]\right)\Big|_{J=0}. \tag{4.15}$$

In the graphical notation, differentiating with respect to $J(x)$ amounts to replace the open dot denoting the source $i \int d^4y J(y)$ by its position x,

$$\frac{1}{i}\frac{\delta}{\delta J(x)}\circ\!\!-\!\!- = \quad x\ \text{———} \tag{4.16}$$

---

[1] This graphical notation first introduced by Stückelberg was made popular by Feynman. The graphs are therefore often called Feynman diagrams or Feynman graphs.

Vacuum diagrams. We call terms in the perturbative evaluation of $Z[J]$ which contain no source vacuum diagrams. Since setting $J = 0$ eliminates all graphs containing at least one source, the vacuum diagrams correspond to loops without external lines. The corresponding Green functions are the "zero-point" Green functions $\mathcal{G}^{(0)}$.

Let us assume that the path-integral measure $\mathcal{D}\phi$ is chosen such that the free vacuum is normalised, $Z_0[0] = 1$. Switching on interactions will change the free vacuum into the true vacuum of the interacting theory. Therefore the true vacuum and thus $Z[J]$ are not normalised. As example, we obtain setting $J = 0$ in our result (4.14) for $Z[J]$ at lowest order perturbation theory,

$$\mathcal{G}^{(0)} \equiv Z[0] = 1 - \frac{i\lambda}{8} \int d^4x (i\Delta_F(0))^2 \neq 1. \tag{4.17}$$

Because of $\mathcal{N} = \exp\ln(\mathcal{N})$, a normalisation different from 1 is equivalent to adding a constant term to the Lagrangian,

$$\mathscr{L} \to \mathscr{L} + \ln(\mathcal{N})/(VT) = \mathscr{L} - \rho, \tag{4.18}$$

where VT is the four-dimensional integration volume in the action. Since vacuum diagrams only change the vacuum energy density ρ but do not contribute to scattering processes, one often prefers to eliminate these diagrams multiplying $Z[J]$ with the normalisation constant

$$\mathcal{N}^{-1} = Z[0] = \int \mathcal{D}\phi\, e^{iS}. \tag{4.19}$$

Thus one uses the normalised generating functional $\widetilde{Z}[J] \equiv \mathcal{N}Z[J] = Z[J]/Z[0]$ which corresponds to a vacuum with zero energy density ρ. We now show that this normalisation eliminates all vacuum graphs. Expanding the numerator and denominator of $\widetilde{Z}[J]$ up to $\mathcal{O}(\lambda)$, we have at lowest order perturbation theory

$$\widetilde{Z}[J] = \frac{Z[J]}{Z[0]} = \frac{Z_0[J] + Z_1[J] + \mathcal{O}(\lambda^2)}{1 + Z_1[0] + \mathcal{O}(\lambda^2)} \tag{4.20a}$$

$$= Z_0[J] + (Z_1[J] - Z_1[0]) + \mathcal{O}(\lambda^2). \tag{4.20b}$$

Thus dividing $Z[J]$ by the source-free functional subtracts the vacuum graph at $\mathcal{O}(\lambda)$. It becomes obvious that this procedure works at any order perturbation theory, if we look at the generating functional for connected graphs, $W[J]$. As dividing $Z[J]$ by the source-free functional $Z[0]$ corresponds to

$$i\widetilde{W}[J] = \ln \widetilde{Z}[J] = \ln Z[J] - \ln Z[0], \tag{4.21}$$

it is clear that this procedure eliminates indeed all vacuum graphs.

Two-point functions. We start by taking one derivative of the normalised generating functional,

$$
\frac{1}{i}\frac{\delta}{\delta J(x_1)}\left[1+\frac{1}{4!}\left(6\ \text{⬡}\ +\ \text{✕}\ \right)\right]\exp\left(\frac{1}{2}\ \text{⊶}\ \right)= \tag{4.22a}
$$

$$
=\left[\frac{1}{4!}\left(6\times 2\ \text{⬡}\ +4\ \text{✕}\ \right)+\left(1+\frac{1}{4!}\left(6\ \text{⬡}\ +\ \text{✕}\ \right)\right)\text{⊸}\right]
$$

$$
\times\exp\left(\frac{1}{2}\ \text{⊶}\ \right)= \tag{4.22b}
$$

$$
=\left[\ \text{⊸}\ +\frac{1}{4!}\left(12\ \text{⬡}\ +4\ \text{✕}\ +6\ \text{⬡}\ \text{⊸}\right.\right.
$$

$$
\left.\left.+\ \text{✕}\ \text{⊸}\right)\right]\exp\left(\frac{1}{2}\ \text{⊶}\ \right). \tag{4.22c}
$$

Every term in this expression contains at least one source J, and the one-point function $\mathcal{G}^{(1)}(x)$ vanishes therefore. If we proceed to the two-point function $\mathcal{G}^{(2)}(x_1,x_2)$, we have to differentiate only those terms with one source,

$$
\frac{1}{i^2}\frac{\delta^2}{\delta J(x_1)\delta J(x_2)}\widetilde{Z}[J]=
$$

$$
=\frac{1}{i}\frac{\delta}{\delta J(x_2)}\left[\ \text{⊸}\ +\frac{1}{4!}\left(12\ \text{⬡}\ +\ \begin{array}{c}\text{vanishing terms}\\\text{for }J=0\end{array}\right)\right]\exp\left(\frac{1}{2}\ \text{⊶}\ \right)
$$

$$
=\left(\ \text{—}\ +\frac{1}{2}\ \text{⬡}\ \right)\exp\left(\frac{1}{2}\ \text{⊶}\ \right). \tag{4.23}
$$

Setting then the sources J to zero, the exponential factor becomes 1. Converting the graphical formula back into standard notation, we find the two-point function $\mathcal{G}^{(2)}(x_1,x_2)$ at order $\mathcal{O}(\lambda)$ as the sum of the free two-point function $\mathcal{G}_0^{(2)}(x_1,x_2)$ and a correction term,

$$
\mathcal{G}^{(2)}(x_1,x_2)=\mathcal{G}_0^{(2)}(x_1,x_2)-\frac{i\lambda}{2}\int d^4x\,i\Delta_F(x_1-x)i\Delta_F(x-x)i\Delta_F(x-x_2). \tag{4.24}
$$

This correction is called the self-energy $\Sigma(x_1,x_2)$ of the scalar particle. Note that the pre-factors combine to $6\times 2/4!=1/2$, so that there appears an extra factor $1/2$. Such factors are called symmetry factors. They appear because we included a factor $1/4!$ in $\mathscr{L}_I$ to compensate for the $4!$ permutations of four sources. This cancellation works in most diagrams however only partially and a pre-factor different from 1 is left over.

Example 4.1: Let us illustrate how one can determine the symmetry factor of more complicated diagrams. As first step, we express the Green function that corresponds to a given Feynman diagram as the time-ordered product of fields. Consider, for example, the so-called sunrise diagram, which is a second-order diagram, corresponding to the term

$$x_1 \; \begin{array}{c} y_1 \; y_2 \end{array} \; x_2 \;\; = \frac{1}{2!} \left(\frac{-i\lambda}{4!} \right)^2 \int d^4 y_1 d^4 y_2 \langle 0 | T[\phi(x_1)\phi(x_2)\phi^4(y_1)\phi^4(y_2)] | 0 \rangle + (y_1 \leftrightarrow y_2)$$

in the perturbative expansion. The exchange graph $y_1 \leftrightarrow y_2$ is identical to the original one, cancelling the factor $1/2!$ from the Taylor expansion. This cancellation takes place in general: the $1/n!$ factor from the Taylor expansion of an nth order contains n interaction points which leads to $n!$ permutations.

Next consider how the fields ϕ are combined in $T[\cdots]$: the internal points y_1 and y_2 denote interaction points, which have four fields attached. In contrast, the external points x_1 and x_2 carry each only one field. We have to count the number of possible ways to combine the fields in the time-ordered product into the five propagators of the graph. As shorthand notation, we mark a possible combination as $\phi(x_1)\phi(y_1)$. We have four possibilities to combine $\phi(x_1)$ with $\phi(y_1)$, $\phi(x_1)\phi(y_1)$. Similarly, there are four possibilities for $\phi(x_2)\phi(y_2)$. The remaining six fields can be combined in $3!$ ways into pairs, as, for example, in $\phi(y_1)\phi(y_1)\phi(y_1)\phi(y_2)\phi(y_2)\phi(y_2)$. Thus the symmetry factor of this diagram is given by

$$S = \left(\frac{1}{2!} \times 2! \right) \left(\frac{1}{4!} \right)^2 (4 \times 4 \times 3!) = \frac{1}{3!} \, .$$

Four-point functions. The disconnected four-point function $\mathcal{G}^{(4)}(x_1, x_2, x_3, x_4)$ is shown graphically in Fig. 4.1. The first three graphs correspond to the free four-point function from (3.43), the next six graphs are the corresponding $\mathcal{O}(\lambda)$ corrections. Finally, the last diagram corresponds to the connected four-point function $G^{(4)}(x_1, x_2, x_3, x_4)$. Next we want to derive $G^{(4)}(x_1, x_2, x_3, x_4)$ from its generating functional $\widetilde{W}[J]$. We insert $1 + \widetilde{Z}_1[J]$ into

$$i\widetilde{W}[J] = \ln \exp \left(\frac{1}{2} \; \circ\!\!-\!\!\circ \; \right) + \ln \left[1 + \frac{1}{4!} \left(6 \; \bigcirc \; + \; \times \; \right) \right] + \mathcal{O}(\lambda^2)$$

$$= \frac{1}{2} \; \circ\!\!-\!\!\circ \; + \frac{1}{4!} \left(6 \; \bigcirc \; + \; \times \; \right) + \mathcal{O}(\lambda^2) \, , \qquad (4.25)$$

where we expanded the logarithm, $\ln(1+x) \simeq x$. Taking four derivatives with respect to J, only the last term survives and we obtain as connected Green function

$$G^{(4)}(x_1, x_2, x_3, x_4) = -i\lambda \int d^4 x \, i\Delta_F(x_1 - x) i\Delta_F(x_2 - x) i\Delta_F(x_3 - x) i\Delta_F(x_4 - x) \, . \tag{4.26}$$

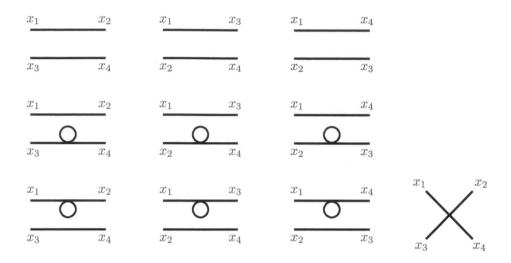

Fig. 4.1 Graphs contributing to the disconnected four-point function $\mathcal{G}(x_1, x_2, x_3, x_4)$.

Feynman rules for the $\lambda\phi^4$ theory. We can summarise our results in a few simple rules which allow us to write down Green functions directly, without the need to derive them from their generating functional. The Feynman rules for connected Green functions $G^{(n)}(x_1, \ldots, x_n)$ in coordinate space are as follows:

1. Draw all topologically different diagrams for the chosen order $\mathcal{O}(\lambda^n)$ and number of external coordinates or particles.
2. To each line connecting the points x and x' we associate a propagator $\mathrm{i}\Delta_F(x - x')$.
3. Each vertex has a factor $-\mathrm{i}\lambda$ and connects n lines for a $\lambda\phi^n$ interaction.
4. Integrate over all intermediate points.
5. Determine and add the symmetry factor.

Translation invariance of Minkowski space implies that the propagators depend like $\exp(\pm \mathrm{i}kx)$ on the position of the interaction point. Therefore the spacetime integrations result in four-momentum conservation at each vertex, $\int \mathrm{d}^4x\, \mathrm{e}^{-\mathrm{i}\sum_j k_j x} = (2\pi)^4\delta(\sum_j k_j)$. In the case of tree-level diagrams, all the four-momentum integrals of the $I = V - 1$ propagators will be eliminated by delta functions, leaving over one delta function expressing overall momentum conservation. In contrast, L integrations over loop momenta $\int \mathrm{d}^4k/(2\pi)^4$ remain in the case of loop diagrams. Accounting for the result of the spacetime integrations, we can give the Feynman rules directly in momentum space. Defining the Fourier transformed n-point function as

$$G^{(n)}(k_1, \ldots, k_n) = \int \prod_{i=1}^{n} \mathrm{d}^4x_i \, \exp\left(\mathrm{i}\sum_i k_i x_i\right) G^{(n)}(x_1, \ldots, x_n), \qquad (4.27)$$

the Feynman rules for these Green functions have the following changes:

2. To each line associate a propagator $i\Delta_F(k) = i/(k^2 - m^2 + i\varepsilon)$.

3. Fix the external momenta and impose four-momentum conservation at each vertex.

4. Integrate over all unconstrained momenta k with $\int d^4k/(2\pi)^4$. The number of independent momenta we have to integrate over equals the loop number L of the graph.

We will see in the next section using $G^{(2)}(p)$ as an example of how these rules work out.

4.3   Loop diagrams

In the Fourier-transformed Green functions $G(k_1, \ldots, k_n)$ of tree-level graphs, all integrals about the propagator momenta have been cancelled by delta functions, and $G(k_1, \ldots, k_n)$ is a mathematically well-defined distribution. In contrast, the integration over momenta in loop graphs is often divergent, requiring to regularise and to renormalise these expressions. The aim of this section is to illustrate this procedure. We concentrate first on the technicalities involved in the evaluation of these loop diagrams, before we interpret the results. We will have time to digest these examples before we will come back to the problem of renormalisation in chapter 11. The basic steps in the evaluation of simple Feynman integrals are summarised in appendix 4.A.

4.3.1   Self-energy

We consider first the only one-loop diagram contained in $Z_1[J]$: the two-point function of a scalar particle at $\mathcal{O}(\lambda)$,

$$G^{(2)}(x_1, x_2) = i\Delta_F(x_1 - x_2) - \frac{\lambda}{2}\Delta_F(0)\int d^4x\,\Delta_F(x_1 - x)\Delta_F(x - x_2). \qquad (4.28)$$

The calculation of $G^{(2)}(x_1, x_2)$ consists of three steps. First, we have to combine its two pieces into a single, modified propagator. As it stands, the expression seems to describe the propagation of two modes, a free one plus one consisting of the $\mathcal{O}(\lambda)$ correction, while $G^{(2)}(x_1, x_2)$ should correspond to the propagation of a single particle with properties modified by the self-interactions. Second, we have to calculate the loop diagram $i\Delta_F(0)$ which will turn out to be infinite. Thus the final task is the question how we should interpret this result.

   We start concentrating on the correction term, which we call the self-energy $\Sigma(x_1, x_2)$ of the scalar particle, and insert the Fourier representation of the two propagators into the integral,

$$\Sigma(x_1, x_2) = -\frac{\lambda}{2}\Delta_F(0)\int d^4x \frac{d^4p}{(2\pi)^4}\frac{d^4p'}{(2\pi)^4}\frac{e^{-ip(x_1-x)}}{p^2 - m^2 + i\varepsilon}\frac{e^{-ip'(x-x_2)}}{p'^2 - m^2 + i\varepsilon}. \qquad (4.29)$$

The d^4x integration results in $(2\pi)^4\delta(p - p')$, then one of the momentum integrations can be performed. Together this gives

$$\Sigma(x_1 - x_2) = -\frac{\lambda}{2}\Delta_F(0)\int \frac{d^4p}{(2\pi)^4}\frac{e^{-ip(x_1-x_2)}}{(p^2 - m^2 + i\varepsilon)^2}. \qquad (4.30)$$

Inserting also for the free Green function its Fourier representation, we arrive at

$$G^{(2)}(x_1 - x_2) = \int \frac{\mathrm{d}^4 p}{(2\pi)^4} \, \mathrm{e}^{-\mathrm{i}p(x_1-x_2)} \left[\frac{\mathrm{i}}{p^2 - m^2 + \mathrm{i}\varepsilon} - \frac{\frac{\lambda}{2}\Delta_F(0)}{(p^2 - m^2 + \mathrm{i}\varepsilon)^2} \right]. \qquad (4.31)$$

The Green function $G^{(2)}(p)$ in momentum space is thus given by the expression in the square bracket, which we could have written down immediately using the Feynman rules in momentum space. Next we factor out one propagator,

$$G^{(2)}(p) = \frac{\mathrm{i}}{p^2 - m^2 + \mathrm{i}\varepsilon} \left[1 + \frac{\frac{\mathrm{i}\lambda}{2}\Delta_F(0)}{p^2 - m^2 + \mathrm{i}\varepsilon} \right]. \qquad (4.32)$$

Assuming that perturbation theory is justified, the second term in the parenthesis should be small. Thus $[1 + \lambda a] = [1 - \lambda a]^{-1} + \mathcal{O}(\lambda^2)$, and we obtain

$$G^{(2)}(p) = \frac{\mathrm{i}}{p^2 - m^2 - \frac{\mathrm{i}\lambda}{2}\Delta_F(0) + \mathrm{i}\varepsilon}. \qquad (4.33)$$

The residue of the free propagator $\mathrm{i}/(p^2 - m^2 + \mathrm{i}\varepsilon)$ defines the "bare" particle mass m at zero order in λ. Switching on interactions, we continue to define the physical (or renormalised) mass m_{phys} of the scalar particle by the residue of $G^{(2)}(p)$. Thus at order λ,

$$m_{\mathrm{phys}}^2 = m^2 + \delta m^2 = m^2 + \frac{\mathrm{i}\lambda}{2}\Delta_F(0). \qquad (4.34)$$

Hence interactions shift or "renormalise" the "bare" mass m used initially in the classical Lagrangian $\mathscr{L}$. It is important to realise that such a renormalisation does not appertain to QFTs but occurs in any interacting theory. A familiar example in a classical context is the Debye screening of the electric charge in a plasma. As next step, we have to calculate (and to interpret properly)

$$\mathrm{i}\Delta_F(0) = \int \frac{\mathrm{d}^4 k}{(2\pi)^4} \frac{\mathrm{i}}{k^2 - m^2 + \mathrm{i}\varepsilon}. \qquad (4.35)$$

Since the mass correction is $\delta m^2 = \mathrm{i}\lambda\Delta_F(0)/2$, the Feynman propagator at coincident points $\Delta_F(0)$ has to be purely imaginary. Otherwise the $\lambda \phi^4$ theory would contain no stable particles.

Wick rotation.   The integrals appearing in loop graphs can be more easily integrated, if one performs a Wick rotation from Minkowski to Euclidean space. Rotating the integration contour anticlockwise to $-\mathrm{i}\infty : +\mathrm{i}\infty$ avoids both poles in the complex k_0 plane and is thus admissible. Introducing as new integration variable $\mathrm{i}k_4 = k_0$, it follows

$$\int_{-\infty}^{\infty} \mathrm{d}k_0 \frac{1}{k^2 - m^2 + \mathrm{i}\varepsilon} = \int_{-\mathrm{i}\infty}^{\mathrm{i}\infty} \mathrm{d}k_0 \frac{1}{k^2 - m^2 + \mathrm{i}\varepsilon} = \mathrm{i} \int_{-\infty}^{\infty} \mathrm{d}k_4 \frac{1}{k^2 - m^2 + \mathrm{i}\varepsilon}. \qquad (4.36)$$

We next combine $\boldsymbol{k}$ and k_4 into a new four-vector $k_E = (\boldsymbol{k}, k_4)$. Since

$$k^2 = -(|\boldsymbol{k}|^2 + k_4^2) = -k_E^2 \tag{4.37}$$

we work now (apart from the overall sign) in an Euclidean space. In particular, the denominator never vanishes and we can omit the iε. Moreover, the integrand is now spherically symmetric. Thus we have

$$i\Delta_F(0) = \int \frac{\mathrm{d}^4 k_E}{(2\pi)^4} \frac{1}{k_E^2 + m^2}. \tag{4.38}$$

As required by our interpretation $\delta m^2 = i\lambda\Delta_F(0)/2$, the propagator $\Delta_F(0)$ is imaginary. Because the relative sign of the momenta and the mass term indicates if we work in the Euclidean or Minkowski space, we will omit the index E in the following. Introducing furthermore spherical coordinates, we see that $\Delta_F(0)$ diverges quadratically for large k,

$$\lambda i\Delta_F(0) \propto \int_0^\Lambda \mathrm{d}k\, k^3 \frac{1}{k^2 + m^2} \propto \Lambda^2. \tag{4.39}$$

Dimensional regularisation. Using the integral representation

$$\frac{1}{k^2 + m^2} = \int_0^\infty \mathrm{d}s\, \mathrm{e}^{-s(k^2 + m^2)} \tag{4.40}$$

and interchanging the integrals, we can reduce the momentum integral to a Gaussian integral. Manipulations like interchanging the order of integrations or a change of integration variables in divergent expressions as Eq. (4.39) are, however, ambiguous. Before we can proceed, we have to regularise the integral, as we did introducing a cut-off function into the expression of the zero-point energy.

We will use dimensional regularisation (DR), that is, we will calculate integrals in $d = 4 - 2\varepsilon$ dimensions where they are finite. Then we find

$$i\Delta_F(0) = \int_0^\infty \mathrm{d}s \int \frac{\mathrm{d}^d k}{(2\pi)^d}\, \mathrm{e}^{-s(k^2 + m^2)} = \frac{1}{(4\pi)^{d/2}} \int_0^\infty \mathrm{d}s\, s^{-d/2} \mathrm{e}^{-sm^2}. \tag{4.41}$$

The substitution $x = sm^2$ transforms the integral into one of the standard representations of the Gamma function (see appendix 4.A for some useful formulae),

$$i\Delta_F(0) = \frac{(m^2)^{\frac{d}{2}-1}}{(4\pi)^{d/2}} \Gamma\left(1 - \frac{d}{2}\right). \tag{4.42}$$

This expression diverges for $d = 2, 4, 6, \ldots$, but is as announced finite for $d = 4 - 2\varepsilon$ and small ε. In the next step, we would like to expand the expression in a Laurent series, separating pole terms in ε and a finite remainder.

Appearance of a dimensionful scale. As the expression stands, we cannot expand the pre-factor of the Gamma function, because it is dimensionful. In order to make the factor m^{d-2} dimensionless, we should supply a new mass scale. More physically, we can understand the need for an additional dimensionful scale by the requirement that the action $S = \int \mathrm{d}^d x\, \mathscr{L}$ remains dimensionless if we deviate from $d = 4$ dimensions. From the kinetic term, we deduce that the scalar field has the mass dimension $[\phi] = d/2 - 1$. The interaction term implies then that λ acquires the dimension $[\lambda] = 4 - d$. In order to retain a dimensionless coupling constant, we introduce therefore a mass μ called the renormalisation scale as follows,

$$S_I = \int \mathrm{d}^4x \mathscr{L}_I = -\int \mathrm{d}^4x \frac{\lambda}{4!}\phi^4 \rightarrow -\mu^{4-d}\int \mathrm{d}^dx \frac{\lambda}{4!}\phi^4. \qquad (4.43)$$

Adding the factor μ^{4-d} to our previous result, we obtain

$$\lambda\mu^{4-d}\mathrm{i}\Delta_F(0) = \lambda \frac{m^2}{(4\pi)^2}\left(\frac{4\pi\mu^2}{m^2}\right)^{2-d/2}\Gamma(1 - d/2). \qquad (4.44)$$

Now we expand the dimensionless last two factors in this expression around $d = 4$ using Eq. (A.46) for the Gamma function,

$$\Gamma(1 - d/2) = \Gamma(-1 + \varepsilon) = -\frac{1}{\varepsilon} - 1 + \gamma + \mathcal{O}(\varepsilon) \qquad (4.45)$$

and

$$a^\varepsilon = \mathrm{e}^{\varepsilon\ln a} = 1 + \varepsilon\ln a + \mathcal{O}(\varepsilon^2). \qquad (4.46)$$

Note that we require the expansion of the pre-factor of the Gamma function up to $\mathcal{O}(\varepsilon^2)$ because of the pole term in (4.45). Thus the mass correction is given by

$$\lambda\mu^{4-d}\mathrm{i}\Delta_F(0) \propto m^2\left[-\frac{1}{\varepsilon} - 1 + \gamma + \mathcal{O}(\varepsilon)\right]\left[1 + \varepsilon\ln\left(\frac{4\pi\mu^2}{m^2}\right) + \mathcal{O}(\varepsilon^2)\right]. \qquad (4.47)$$

or

$$\delta m^2 = \frac{\mathrm{i}\lambda}{2}\mu^{4-d}\Delta_F(0) = \frac{\lambda}{2}\frac{m^2}{(4\pi)^2}\left[-\frac{1}{\varepsilon} - 1 + \gamma + \ln\left(\frac{m^2}{4\pi\mu^2}\right) + \mathcal{O}(\varepsilon)\right]. \qquad (4.48)$$

This expansion has allowed us to separate the correction into a divergent term $\propto 1/\varepsilon$ and a finite remainder. The latter contains an analytic part, $-1+\gamma$, and a non-analytic piece that depends on the renormalisation scale, $\ln(m^2/4\pi\mu^2)$. This result is typical for DR. First, all divergences appear in the limit $d \to 4$ as poles of the Gamma function. Second, the renormalisation scale μ enters always via Eq. (4.46) in a logarithm. Thus the only dimensionful parameter which can set the scale of the mass correction δm^2 is the mass of the particle in the loop. In the case of a theory with a single particle, the correction must have therefore the form $\delta m^2 \propto m^2$ using DR. You should contrast this behaviour with the one using as regularisation scheme an Euclidean cut-off Λ. Integrating up to momenta $\Lambda \gg m$, the particle mass m can be neglected and the correction diverges as a power-law $\delta m^2 \propto \Lambda^2$.

Let us now discuss in turn the three different kind of terms present in Eq. (4.48). First, the form of the divergent terms depends on the regularisation scheme applied. However, in any scheme we can eliminate them using Eq. (4.34), requiring that the unobservable bare mass m^2 contains the same divergent terms with the opposite sign. Second, the finite analytic terms depend also on the scheme, since we can always shift a finite part from m^2 to δm^2. In order to specify precisely δm^2 we have to fix therefore a *renormalisation scheme*, that is, a set of rules resolving these ambiguities. Finally, the finite, non-analytic terms are important predictions, which are independent of the regularisation scheme (apart from a rescaling of the arbitrary parameter μ). As we will see in chapter 9, such non-analytic terms are necessary for the unitarity of the S-matrix, or in other words that the theory preserves probability.

4.3.2 Vacuum energy density

Out of the self-energy diagram we can generate new one-loop graphs by adding or subtracting two external lines. Subtracting two lines generates a one-loop graph without external lines,[2] the "zero-point" Green function $G^{(0)}$ at order λ^0. One way to calculate this quantity is to evaluate directly $Z_0[0]$ using

$$\mathrm{Det}\, A = \exp \ln \mathrm{Det}\, A = \exp \mathrm{Tr} \ln A \tag{4.49}$$

which gives

$$Z_0[0] \propto \exp\left[-\frac{1}{2}\mathrm{Tr}\ln(\Box - m^2)\right]. \tag{4.50}$$

We postpone the question how such an expression can be evaluated and use instead another approach, recycling our result for the self-energy. Vacuum diagrams are generated by the functional $Z[J]$ setting $J = 0$,

$$\langle 0+|0-\rangle = Z_0[0] = \int \mathcal{D}\phi \exp \mathrm{i}\int \mathrm{d}^4 x \left(\frac{1}{2}\partial_\mu\phi\partial^\mu\phi - \frac{1}{2}m^2\phi^2\right). \tag{4.51}$$

We saw in Eq. (3.56) that the zero-point energy is related to the propagator at coincident points. Since we suspect a connection between vacuum diagrams and the zero-point energy, we try to relate $Z[0]$ and $\Delta_F(0)$. Taking a derivative with respect to m^2 gives

$$\frac{\partial}{\partial m^2}\langle 0+|0-\rangle = -\frac{\mathrm{i}}{2}\int \mathrm{d}^4 x\, \langle 0+|\phi(x)^2|0-\rangle = -\frac{\mathrm{i}}{2}\int \mathrm{d}^4 x\, \mathrm{i}\Delta_F(0)\langle 0+|0-\rangle. \tag{4.52}$$

The additional factor $\langle 0+|0-\rangle = \mathcal{N}^{-1}$ on the RHS takes into account that we defined the Feynman propagator with respect to a normalised vacuum. Translation invariance implies that $\langle 0+|0-\rangle$ does not depend on x. Thus we obtain

$$\frac{\partial}{\partial m^2}\ln\langle 0+|0-\rangle = -\frac{\mathrm{i}}{2}\int \mathrm{d}^4 x\, \mathrm{i}\Delta_F(0) = -\frac{\mathrm{i}}{2}\, VT\, \mathrm{i}\Delta_F(0) \tag{4.53}$$

with VT as the four-dimensional integration volume. Integrating and exponentiating the resulting formal solution, we obtain

$$\langle 0+|0-\rangle = \exp\left\{-\frac{\mathrm{i}}{2}\, VT \int \mathrm{d}m^2\, \mathrm{i}\Delta_F(0)\right\}. \tag{4.54}$$

---

[2]Although $G^{(0)}$ is often represented as a closed loop, it has also no internal line; this is in agreement with our general formula $L = n - V + 1$.

Comparing this result to $\langle 0 + |0-\rangle = \langle 0 + |\exp\left(-iHT\right)|0-\rangle = \exp\left(-i\rho VT\right)$, we see that we should associate

$$\rho = \frac{1}{2}\int dm^2\, i\Delta_F(0) \tag{4.55}$$

with the energy density of the vacuum. On the other hand, we can connect ρ to the source-free generating functionals as

$$\rho = \frac{i\ln Z[0]}{VT} = \frac{-W[0]}{VT}, \tag{4.56}$$

Thus the contribution of quantum fluctuations to the energy density of the vacuum is given by the sum of connected vacuum graphs, in accordance with (4.21).

Next we evaluate (4.55) which gives the contribution of a free scalar field to the vacuum energy density. Using our result (4.42) for the propagator, $i\Delta_F(0) = C(m^2)^{d/2-1}$, we can perform the integration over m^2,

$$\rho = C\frac{m^d}{d} - \rho_0 = \frac{m^d}{(4\pi)^{d/2}d}\,\Gamma\left(1 - \frac{d}{2}\right) - \rho_0, \tag{4.57}$$

where we introduced the integration constant ρ_0.

The energy density given by Eq. (4.57) diverges for $d = 2, 4, 6\ldots$ as $1/\varepsilon$. We can make ρ finite and equal to the observed value ρ_Λ, if we choose ρ_0 as

$$\rho_0 = \mu^{d-4}\left[\frac{1}{4}\frac{m^4}{(4\pi)^2}\frac{1}{\varepsilon} - \rho_\Lambda\right]. \tag{4.58}$$

The pre-factor μ^{d-4} ensures again that the action remains dimensionless also for $d \neq 4$.

Note that this implies that we should start off with $\mathscr{L} - \rho_0$ instead of $\mathscr{L}$. Even if we dismiss a non-zero vacuum energy in the classical Lagrangian, it will appear automatically by quantum corrections. More generally, every possible term that is not forbidden by a symmetry in $\mathscr{L}$ will show up calculating loop corrections. We have seen that we can absorb the vacuum energy density ρ into the normalisation of the path integral,

$$\int \mathcal{D}\phi\, e^{-\int d^4x\rho} = \mathcal{N}\int \mathcal{D}\phi\,.$$

Therefore, one may wonder if ρ has a real physical meaning or could be eliminated by a simple redefinition of the integration measure. The answer is no. First, we used our freedom to define the path-integral measure setting $Z_0[0] = 1$. Second, ρ depends on the parameters (masses, coupling constants) of the considered theory, but the path-integral measure should be independent of the details of the Lagrangian we integrate.

Remark 4.1: Equivalence to the zero-point energy. Performing the k^0 integral in Eq. (4.35) or using (3.28)

$$i\Delta_F(0) = \int \frac{d^3k}{(2\pi)^3 2\omega_k} = \int \frac{d^3k}{(2\pi)^3 2\sqrt{m^2 + \boldsymbol{k}^2}} \tag{4.59}$$

and integrating then with respect to m^2,

$$\rho = \frac{1}{2} \int \mathrm{d}m^2 \, \mathrm{i}\Delta_F(0) = \frac{1}{2} \int \frac{\mathrm{d}^3 k}{(2\pi)^3} \sqrt{m^2 + \boldsymbol{k}^2} \,, \tag{4.60}$$

shows that the present expression for the vacuum energy agrees with the sum over zero-point energy evaluated in Eq. (3.57). However, the results for the unrenormalised ρ differ: While Eq. (3.58) shows that $\rho \propto \Lambda^4$ using a cut-off, we have obtained $\rho \propto m^4$ in the case of dimensional regularisation. Thus in this scheme a massless particle as the photon would give a zero contribution to the cosmological constant. We will come back to this difference in chapter 26.

4.3.3 Vertex correction

Feynman amplitude. For our last example we add two external lines to the self-energy diagram. The corresponding Green function can be used to describe $2 \to 2$ scattering at $\mathcal{O}(\lambda^2)$. A scattering process corresponds, however, to a transition between an initial state at $t = -\infty$ and a final state at $t = \infty$ which contain both real, on-shell particles. In order to obtain scattering amplitudes, we should therefore replace the propagators of external lines—which describe virtual particles—by on-shell wave-functions. This rule will be derived later in chapter 9.2. For the moment, we simply anticipate that we obtain the Feynman amplitude $\mathrm{i}\mathcal{A}$ describing a scattering process using the Feynman rules for momentum space, but writing for scalar external particles simply the pre-factor of a plane wave without normalisation factor, that is simply "1". Moreover, we omit the delta function expressing the conservation of the external momenta. Thus we add to the Feynman rules in momentum space:

6. The Feynman amplitude $\mathrm{i}\mathcal{A}$ describing scattering processes is obtained omitting the factor $(2\pi)^4 \delta(\sum_i k_i - \sum_f k_f)$ expressing the conservation of the external momenta, and the propagators on external lines.

Determining the Feynman amplitude. Instead of calculating the order λ^2 term in the perturbative expansion of the generating functional $Z[J]$ we use directly the Feynman rules to obtain the Feynman amplitude for this process. According to these rules, the first steps in the calculation of the Feynman amplitude are to draw all Feynman diagrams, to find the symmetry factor and to associate then the corresponding mathematical expressions to the graphical symbols.

We first determine the symmetry factor, following the same procedure as in example 4.1. In coordinate space, we have to connect four external points (say $x_1, \ldots, x_4$) with the help of two vertices (say at x and y) which combine each four lines. An example is shown here

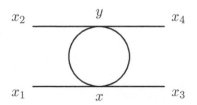

Two additional diagrams are obtained connecting x_1 with x_2 or x_4. In order to determine the symmetry factor, we consider the expression for the four-point function corresponding to the graph shown above,

$$\frac{1}{2!}\left(-i\frac{\lambda}{4!}\right)^2 \int d^4x d^4y \langle 0|T\{\phi(x_1)\phi(x_2)\phi(x_3)\phi(x_4)\phi^4(x)\phi^4(y)\}|0\rangle + (x \leftrightarrow y), \quad (4.61)$$

and count the number of possible contractions. We can connect $\phi(x_1)$ with each one of the four $\phi(x)$, and then $\phi(x_3)$ with one of the three remaining $\phi(x)$. This gives 4×3 possibilities. Another 4×3 possibilities come by the same reasoning from the upper part of the graph. The remaining pairs $\phi^2(x)$ and $\phi^2(y)$ can be combined in two possibilities. Finally, the factor $1/2!$ from the Taylor expansion is cancelled by the exchange graph. Thus the symmetry factor is

$$S = \frac{1}{2!}2!\left(\frac{4\times 3}{4!}\right)^2 2 = \frac{1}{2}. \quad (4.62)$$

Next we associate mathematical expressions with the symbols of the graphs in momentum space: We replace internal propagators by $i\Delta(k)$, external lines by 1 and vertices by $-i\lambda$. Imposing four-momentum conservation at the two vertices leaves one free loop momentum, which we call p. The momentum of the other propagator is then fixed to $p-q$, where $q^2 = s = (k_1+k_2)^2$, $q^2 = t = (k_1-k_3)^2$, and $q^2 = u = (k_1-k_4)^2$ for the three graphs[3] shown in Fig. 4.2. Thus the Feynman amplitudes in $d=4$ are at order $\mathcal{O}(\lambda^2)$

$$i\mathcal{A}_q^{(2)} = \frac{1}{2}\lambda^2 \int \frac{d^4p}{(2\pi)^4} \frac{1}{[p^2 - m^2 + i\varepsilon]} \frac{1}{[(p-q)^2 - m^2 + i\varepsilon]}. \quad (4.63)$$

The squared cms energy s and the two variables describing the momentum transfer t and u are called Mandelstam variables. For $2 \to 2$ scattering, they are connected by $s + t + u = m_1^2 + m_2^2 + m_3^2 + m_4^2$, see problem 4.4. According to the value of q^2 one calls the diagrams the s, t and u channel.

Performing again a simple counting of the powers of loop momenta, we find that the amplitude is logarithmically divergent,

$$\mathcal{A}_q^{(2)} \propto \int \frac{d^4p}{p^4} \propto \ln(\Lambda). \quad (4.64)$$

If we consider the infinite number of one-loop graphs characterised by $n = V \geq 0$, then we see that adding two external lines increases the number of propagators in the loop by one. As a result, the convergence of the loop integral improves from a quartic divergence (vacuum energy), over a quadratic divergence (self-energy energy) to a logarithmic divergence for the vertex correction. Adding two or more external lines to the vertex correction would therefore produce a finite diagram. At one-loop, the $\lambda\phi^4$ theory contains thus only three divergent Feynman graphs.

---

[3]Note that there are four additional graphs, where one of the external lines in the tree-level graph is replaced by the self-energy correction. We will later see that their effects can be neglected in this case.

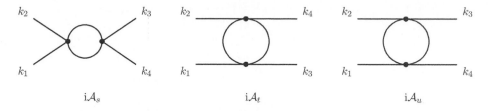

Fig. 4.2 Three graphs contributing to $\phi(k_1)\phi(k_2) \rightarrow \phi(k_3)\phi(k_4)$ at order λ^2.

Calculating the loop integral. The path to be followed in the evaluation of simple loop integrals as (4.63) can be sketched schematically as follows. Regularise the integral (and add a mass scale if you use DR). Combine then the denominators, and shift the integration variable to eliminate linear terms in the denominator by completing the square. Performing the same shift of variables in the numerator, linear terms can be dropped as they vanish after integration. Finally, Wick rotate the integrand, and reduce the integral to a known one by a suitable variable substitution. We do the last steps once in appendix 4.A where we derive a list of useful Feynman integrals which we simply look up in the future.

We start by rewriting the integral for $d = 2\omega = 4 - 2\varepsilon$ dimensions as

$$i\mathcal{A}_q^{(2)} = \frac{1}{2}\lambda^2(\mu^2)^{4-d} \int \frac{\mathrm{d}^d p}{(2\pi)^d} \frac{1}{[p^2 - m^2 + i\varepsilon]} \frac{1}{[(p-q)^2 - m^2 + i\varepsilon]}. \qquad (4.65)$$

Next we use

$$\frac{1}{ab} = \int_0^1 \frac{\mathrm{d}z}{[az + b(1-z)]^2} \qquad (4.66)$$

to combine the two denominators, setting $a = p^2 - m^2$ and $b = (p-q)^2 - m^2$,

$$\mathcal{D} \equiv az + b(1-z) = p^2 - m^2 - 2pq(1-z) + q^2(1-z). \qquad (4.67)$$

Then we eliminate the term linear in p substituting $p'^2 = [p - q(1-z)]^2$,

$$\mathcal{D} = p'^2 - m^2 + q^2 z(1-z). \qquad (4.68)$$

Since $\mathrm{d}^d p = \mathrm{d}^d p'$, we can drop the primes and find

$$i\mathcal{A}_q^{(2)} = \frac{1}{2}\lambda^2(\mu^2)^{4-d} \int_0^1 \mathrm{d}z \int \frac{\mathrm{d}^d p}{(2\pi)^d} \frac{1}{[p^2 - m^2 + q^2 z(1-z)]^2}. \qquad (4.69)$$

Performing a Wick rotation requires that $q^2 z(1-z) < m^2$ for all $z \in [0:1]$, or $q^2 < 4m^2$. The integral is of the type $I(\omega, 2)$ calculated in the appendix and equals

$$I(\omega, 2) = i \frac{1}{(4\pi)^\omega} \frac{\Gamma(2-\omega)}{\Gamma(2)} \frac{1}{[m^2 - q^2 z(1-z)]^{2-\omega}}. \qquad (4.70)$$

Inserting the result into the Feynman amplitude gives

$$A_q^{(2)} = \frac{1}{2}\lambda^2(\mu^2)^{4-d}\frac{\Gamma(2-d/2)}{(4\pi)^{d/2}} \int_0^1 dz\, [m^2 - q^2 z(1-z)]^{d/2-2} \tag{4.71a}$$

$$= \frac{\lambda^2}{32\pi^2}(\mu^2)^{2-d/2}\Gamma(2-d/2) \int_0^1 dz\, \underbrace{\left[\frac{m^2 - q^2 z(1-z)}{4\pi\mu^2}\right]}_{f}^{\frac{d}{2}-2}. \tag{4.71b}$$

In the last step, we made the function f dimensionless. Now we take the limit $\varepsilon = 2-d/2 \to 0$, expanding both the Gamma function, $\Gamma(2-d/2) = \Gamma(\varepsilon) = 1/\varepsilon - \gamma + \mathcal{O}(\varepsilon)$, and $f^{-\varepsilon}$. From

$$\frac{\lambda^2}{32\pi^2}\mu^{2\varepsilon}\left(\frac{1}{\varepsilon} - \gamma + \mathcal{O}(\varepsilon)\right)\left[1 - \varepsilon \int_0^1 dz\, \ln f + \mathcal{O}(\varepsilon^2)\right] \tag{4.72}$$

we see that all diagrams give the same divergent part, while we have to replace q^2 by the value $\{s, t, u\}$ appropriate for the three diagrams,

$$\mathcal{A} = \mathcal{A}^{(1)} + \mathcal{A}_s^{(2)} + \mathcal{A}_t^{(2)} + \mathcal{A}_u^{(2)} + \mathcal{O}(\lambda^3) \tag{4.73a}$$

$$= -\lambda\mu^\varepsilon + \frac{3\lambda^2\mu^{2\varepsilon}}{32\pi^2\varepsilon} - \frac{\lambda^2\mu^{2\varepsilon}}{32\pi^2}\left[3\gamma + F(s,m,\mu) + F(t,m,\mu) + F(u,m,\mu)\right], \tag{4.73b}$$

with

$$F(q^2, m, \mu) = \int_0^1 dz\, \ln\left[\frac{m^2 - q^2 z(1-z)}{4\pi\mu^2}\right]. \tag{4.74}$$

Note that t and u are in the physical region negative and thus the condition $q^2 < 4m^2$ is always satisfied for these two diagrams, cf. with problem 4.4. By contrast, for the s channel diagram the relation $q^2 = s > 4m^2$ holds: in this case, we have to analytically continue the result (4.74) into the physical region. We will postpone this task to chapter 9 and note for the moment only that thereby the argument of the logarithm in (4.74) changes sign. Additionally, an imaginary part of the scattering amplitude is generated.

4.3.4 Basic idea of renormalisation

The regularisation of loop integrals has introduced as a new parameter the renormalisation scale μ. As we perform perturbation theory at order λ^n, we have to connect the parameters $\{m_n, \lambda_n, \rho_n\}$ of the truncated theory with the physical ones of the full theory. This process is called renormalisation and will replace the undetermined parameter μ by a physical momentum scale relevant for the considered process.

Renormalisation of the coupling. Let us try to connect the amplitude i$\mathcal{A}$ to a physical measurement. We assume that experimentalists measured $\phi\phi \to \phi\phi$ scattering. It is sufficient that they provide us with a single value, for example with the value of the differential cross-section $d\sigma/d\Omega$ at a specific cms energy s_0 and momentum transfer t_0. The function $F(s, m, \mu)$ simplifies in the high-energy limit $s \gg m^2$ to

$$F(s, m, \mu) = \ln\left[\frac{s}{\mu^2}\right] + \text{const.} \tag{4.75}$$

The value of the bare coupling λ entering Eq. (4.73b) is not observable. This allows us to subtract the constant terms (including the $1/\varepsilon$ pole) from $\mathcal{A}$, shifting thereby only the value of λ. Introducing also the sloppy notation L for the three log terms in the square bracket, we obtain

$$\mathcal{A} = -\lambda - C\lambda^2 \big[\ln(s/\mu^2) + \ln(t/\mu^2) + \ln(u/\mu^2)\big] + \mathcal{O}(\lambda^3) \qquad (4.76a)$$

$$\equiv -\lambda - C\lambda^2 L(q^2/\mu^2) + \mathcal{O}(\lambda^3) \qquad (4.76b)$$

for $s, |t|, |u| \gg m^2$. This expression is finite but still arbitrary since it contains μ. We indicated also that our perturbative calculation is only valid up to $\mathcal{O}(\lambda^3)$ terms.

We use now the experimental measurement at the scale $q_0^2 = \{s_0.t_0, u_0\}$ to connect via

$$\mathcal{A} = -\lambda - C\lambda^2 L(q_0^2/\mu^2) \qquad (4.77)$$

the measured value λ_{phys} of the coupling to our calculation,

$$-\lambda_{\text{phys}} = -\lambda - C\lambda^2 L(q_0^2/\mu^2) + \mathcal{O}(\lambda^3). \qquad (4.78)$$

Now we solve for λ,

$$-\lambda = -\lambda_{\text{phys}} + C\lambda^2 L(q_0^2/\mu^2) + \mathcal{O}(\lambda^3) \qquad (4.79a)$$

$$= -\lambda_{\text{phys}} + C\lambda_{\text{phys}}^2 L(q_0^2/\mu^2) + \mathcal{O}(\lambda_{\text{phys}}^3). \qquad (4.79b)$$

In the second line, we could replace λ^2 by λ_{phys}^2, because their difference is of $\mathcal{O}(\lambda^3)$. Next we insert λ back into the matrix element $\mathcal{A}$ for general s and replace then again λ^2 by λ_{phys}^2,

$$\mathcal{A} = -\lambda - C\lambda^2 L(q^2/\mu^2) + \mathcal{O}(\lambda^3) \qquad (4.80a)$$

$$= -\lambda_{\text{phys}} + C\lambda_{\text{phys}}^2 L(q_0^2/\mu^2) - C\lambda_{\text{phys}}^2 L(q^2/\mu^2) + \mathcal{O}(\lambda_{\text{phys}}^3) \qquad (4.80b)$$

$$= -\lambda_{\text{phys}} - C\lambda_{\text{phys}}^2 L(q^2/q_0^2) + \mathcal{O}(\lambda_{\text{phys}}^3). \qquad (4.80c)$$

Combining the logs, the scale μ has cancelled and we find

$$\mathcal{A} = -\lambda_{\text{phys}} - \frac{\lambda_{\text{phys}}^2}{32\pi^2} \big[\ln(s/s_0^2) + \ln(t/t_0^2) + \ln(u/u_0^2)\big] + \mathcal{O}(\lambda_{\text{phys}}^3). \qquad (4.81)$$

Thus the amplitude is finite and depends only on the measured value λ_{phys} of the coupling constant and the kinematical variables s, t and u.

Running coupling. We look now from a somewhat different point of view at the problem of the apparent μ dependence of physical observables. Assume again that we have subtracted the infinite parts (and the constant term) of the amplitude $\mathcal{A}$, obtaining Eq. (4.76). We now demand that the scattering amplitude $\mathcal{A}$ as a physical observable is independent of the arbitrary scale μ, $d\mathcal{A}/d\mu = 0$. If we did a perturbative calculation up to $\mathcal{O}(\lambda^n)$, the condition $d\mathcal{A}/d\mu = 0$ can hold only up to terms $\mathcal{O}(\lambda^{n+1})$. The explicit μ dependence of the amplitude, $\partial\mathcal{A}/\partial\mu \neq 0$, can be only cancelled by a corresponding change of the parameters m and λ contained in the classical Lagrangian, converting them into "running" parameters $m(\mu)$ and $\lambda(\mu)$. Then the condition[4] $d\mathcal{A}/d\mu = 0$ becomes

---

[4]Equations of the type (4.82) that describe the change of parameters as function of the scale μ are called renormalisation group equations (RGE).

$$\left(\frac{\partial}{\partial\mu} + \frac{\partial m^2}{\partial\mu}\frac{\partial}{\partial m^2} + \frac{\partial\lambda}{\partial\mu}\frac{\partial}{\partial\lambda}\right)\mathcal{A}(s,t,m(\mu),\lambda(\mu),\mu) = 0. \tag{4.82}$$

The only explicit μ dependence of $\mathcal{A}$ is contained in the $F(q^2, m, \mu)$ functions, giving in the limit $\varepsilon \to 0$

$$\frac{\partial}{\partial\mu}\mathcal{A}(s,t,m(\mu),\lambda(\mu),\mu) = -3\frac{\lambda^2}{32\pi^2}\frac{\partial}{\partial\mu}F(q^2,m,\mu) = \frac{3\lambda^2}{16\pi^2\mu}. \tag{4.83}$$

Since the change of $m(\mu)$ and $\lambda(\mu)$ is given by loop diagrams, it includes at least an additional factor λ. Therefore the action of the derivatives ∂_{m^2} and ∂_λ on the one-loop contribution $\mathcal{A}^{(2)}$ leads to a term of $\mathcal{O}(\lambda^3)$ which can be neglected. Thus the only remaining term will be given by ∂_μ acting on the tree-level term $\mathcal{A}^{(1)}$. Note that this also holds at higher orders and ensures that $\partial_\mu\lambda$ at $\mathcal{O}(\lambda^{n+1})$ is determined by the parameters calculated at $\mathcal{O}(\lambda^n)$. Combining the two contributions we find

$$\mu\frac{\partial\lambda}{\partial\mu} = \frac{3\lambda^2}{16\pi^2} + \mathcal{O}(\lambda^3). \tag{4.84}$$

Thus the scattering amplitude $\mathcal{A}$ is independent of the scale μ, if we transform the coupling constant λ into a scale dependent "running" coupling $\lambda(\mu)$ whose evolution is given by Eq. (4.84). Since we truncate the perturbation series at a finite order, the cancellation of the scale dependence is incomplete and a residual dependence of physical quantities on μ remains. Separating variables in Eq. (4.84), we find

$$\lambda(\mu) = \frac{\lambda_0}{1 - 3\lambda_0/(16\pi^2)\ln(\mu/\mu_0)} \tag{4.85}$$

with $\lambda_0 \equiv \lambda(\mu_0)$ as initial condition. Thus the running coupling $\lambda(\mu)$ increases logarithmically for increasing μ in the $\lambda\phi^4$ theory.

Comparing (4.85) to our result for the scattering amplitude (4.81),

$$\mathcal{A} = -\lambda_0\left(1 + \frac{\lambda_0}{32\pi^2}\left[\ln(s/4m^2) + \ln(t/4m^2) + \ln(u/4m^2)\right]\right) + \mathcal{O}(\lambda_0^3), \tag{4.86}$$

we see that we can rewrite the amplitude using a symmetric point $q^2 = s = t = u$ and $\mu_0^2 = 4m^2$ as

$$\mathcal{A} = -\lambda_0\left(1 + \frac{3\lambda_0}{32\pi^2}\ln(q^2/\mu_0^2)\right) = -\lambda(q^2). \tag{4.87}$$

This shows that the q^2 dependence of the amplitude $\mathcal{A}$ in the limit $q^2 \gg m^2$ is determined completely by the scale dependence of the running coupling $\lambda(\mu)$. Therefore, we should set the renormalisation scale μ in general equal to the physical momentum scale q that characterises the considered process. We will come back to this topic in chapter 12, giving a formal definition of the running coupling.

4.A   Appendix: Evaluation of Feynman integrals

Combination of propagators.   The standard strategy in the evolution of loop integrals is the combination of the n propagator denominators into a single propagator-like denominator of higher power. One uses either Schwinger's proper-time representation

$$\frac{\mathrm{i}}{p^2 - m^2 + \mathrm{i}\varepsilon} = \int_0^\infty \mathrm{d}s \; \mathrm{e}^{\mathrm{i}s(p^2 - m^2 + \mathrm{i}\varepsilon)} \tag{4.88}$$

or the Feynman parameter integral

$$\frac{1}{x_1 \cdots x_n} = \Gamma(n) \int_0^1 \mathrm{d}\alpha_1 \cdots \int_0^1 \mathrm{d}\alpha_n \; \delta(1 - \sum_i \alpha_i) \left[\alpha_1 x_1 \cdots \alpha_n x_n \right]^{-n}$$

$$= \Gamma(n) \int_0^1 \mathrm{d}\alpha_1 \cdots \int_0^{\alpha_{n-2}} \mathrm{d}\alpha_n \; \left[\alpha_1(1 - x_1) + \alpha_2(x_1 - x_2) \cdots \alpha_n x_{n-1} \right]^{-n}. \tag{4.89}$$

In order to derive this formula for $n = 2$, consider

$$\frac{1}{b-a} \int_a^b \frac{\mathrm{d}x}{x^2} = \frac{1}{b-a} \left(\frac{1}{a} - \frac{1}{b} \right) = \frac{1}{ab} \tag{4.90}$$

for $a, b \in \mathbb{C}$. Setting $x = az + b(1 - z)$ and changing the integration variable, we obtain

$$\frac{1}{ab} = \int_0^1 \frac{\mathrm{d}z}{\left[az + b(1 - z) \right]^2}. \tag{4.91}$$

The cases $n > 2$ can be derived by induction, rewriting, for example, $1/(abc)$ as $1/(aB)$ with $B = bc$, and using the result for $n - 1$. In particular, for $n = 3$ it follows

$$\frac{1}{abc} = 2 \int_0^1 \mathrm{d}x \int_0^{1-x} \frac{\mathrm{d}y}{\left[ax + by + c(1 - x - y) \right]^3}. \tag{4.92}$$

Finally, we can generalise these formulae to expressions like $1/(a^n b^m)$ by taking derivatives with respect to a and b.

Evaluation of Feynman integrals.   We want to calculate integrals of the type

$$I_0(\omega, \alpha) = \int \frac{\mathrm{d}^{2\omega} k}{(2\pi)^{2\omega}} \frac{1}{\left[k^2 - m^2 + \mathrm{i}\varepsilon \right]^\alpha} \tag{4.93}$$

defined in Minkowski space. Performing a Wick rotation to Euclidean space and introducing spherical coordinates results in

$$I_0(\omega, \alpha) = \mathrm{i}(-1)^\alpha \int \frac{\mathrm{d}^{2\omega} k}{(2\pi)^{2\omega}} \frac{1}{\left[k^2 + m^2 \right]^\alpha} = \mathrm{i} \frac{(-1)^\alpha}{(2\pi)^{2\omega}} \Omega_{2\omega} \int_0^\infty \mathrm{d}k \frac{k^{2\omega-1}}{\left[k^2 + m^2 \right]^\alpha}, \tag{4.94}$$

where we denoted the volume $\mathrm{vol}(S^{2\omega-1})$ of a unit sphere in 2ω dimensions[5] by $\Omega_{2\omega}$. You are asked in problem 4.3 to show that $\Omega_{2\omega} = 2\pi^\omega / \Gamma(\omega)$ and thus $\Omega_4 = 2\pi^2$. Substituting

---

[5] An $n - 1$-dimensional sphere $S^{n-1}(R)$ encloses the n-dimensional volume $x_1^2 + \ldots + x_n^2 \le R^2$, while its own $n-1$-dimensional volume is given by $x_1^2 + \ldots + x_n^2 = R^2$. The volume of a unit one-sphere is a length, $\mathrm{vol}(S^1) = 2\pi$, of a unit two-sphere an area, $\mathrm{vol}(S^2) = 4\pi$, and of a unit three-sphere a volume, $\mathrm{vol}(S^3) = 2\pi^2$. If we say that the volume of a sphere is $4\pi R^3/3$, we mean in fact the volume of the three-ball $B^3(R)$, $x_1^2 + x_2^2 + x_3^2 \le R^2$ which is enclosed by the two-sphere $S^2(R)$.

$k = m\sqrt{x}$ and using the integral representation (A.29) for Euler's Beta function allows us to express the k integral as a product of Gamma functions,

$$\int_0^\infty \mathrm{d}k \, \frac{k^{2\omega-1}}{[k^2+m^2]^\alpha} = \frac{1}{2} m^{2\omega-2\alpha} \int_0^\infty \mathrm{d}x \, \frac{x^{\omega-1}}{[1+x]^\alpha} = \tag{4.95a}$$

$$\frac{1}{2} m^{2\omega-2\alpha} B(\omega, \alpha-\omega) = \frac{1}{2} m^{2\omega-2\alpha} \frac{\Gamma(\omega)\Gamma(\alpha-\omega)}{\Gamma(\alpha)} . \tag{4.95b}$$

Combining this result with $\Omega_{2\omega} = 2\pi^\omega/\Gamma(\omega)$ we obtain

$$I_0(\omega, \alpha) = \int \frac{\mathrm{d}^{2\omega}k}{(2\pi)^{2\omega}} \frac{1}{[k^2-m^2+i\varepsilon]^\alpha} = i \frac{(-1)^\alpha}{(4\pi)^\omega} \frac{\Gamma(\alpha-\omega)}{\Gamma(\alpha)} [m^2-i\varepsilon]^{\omega-\alpha} . \tag{4.96}$$

Note that m^2 can denote any function of the external momenta and masses, since we required only that it is independent of the loop momentum. We can generate additional formulae by adding first a dependence on a external momentum p^μ, shifting then the integration variable $k \to k+p$,

$$I(\omega, \alpha) = \int \frac{\mathrm{d}^{2\omega}k}{(2\pi)^{2\omega}} \frac{1}{[k^2+2pk-m^2+i\varepsilon]^\alpha} = i \frac{(-1)^\alpha}{(4\pi)^\omega} \frac{\Gamma(\alpha-\omega)}{\Gamma(\alpha)} [m^2+p^2-i\varepsilon]^{\omega-\alpha}. \tag{4.97}$$

Taking then derivatives with respect to the external momentum p^μ results in

$$I_\mu(\omega, \alpha) = \int \frac{\mathrm{d}^{2\omega}k}{(2\pi)^{2\omega}} \frac{k_\mu}{[k^2+2pk-m^2+i\varepsilon]^\alpha} = -p_\mu I(\omega, \alpha) \tag{4.98}$$

and

$$I_{\mu\nu}(\omega, \alpha) = \int \frac{\mathrm{d}^{2\omega}k}{(2\pi)^{2\omega}} \frac{k_\mu k_\nu}{[k^2+2pk-m^2+i\varepsilon]^\alpha} = \tag{4.99}$$

$$= i \frac{(-\pi)^\omega}{(2\pi)^{2\omega}} \frac{\Gamma(\alpha-\omega-1)}{\Gamma(\alpha)} \frac{p_\mu p_\nu(\alpha-\omega-1) - \frac{1}{2}\eta_{\mu\nu}(m^2+p^2)}{[m^2+p^2-i\varepsilon]^{\alpha-\omega}} . \tag{4.100}$$

Contracting both sides with $\eta^{\mu\nu}$ and using $\eta^{\mu\nu}\eta_{\mu\nu} = 2\omega$ gives

$$I_2(\omega, \alpha) = \int \frac{\mathrm{d}^{2\omega}k}{(2\pi)^{2\omega}} \frac{k^2}{[k^2+2pk-m^2+i\varepsilon]^\alpha} = i \frac{(-\pi)^\omega}{(2\pi)^{2\omega}} \frac{\Gamma(\alpha-\omega-1)}{\Gamma(\alpha)} \frac{(\alpha-2\omega-1)p^2 - \omega m^2}{[m^2+p^2-i\varepsilon]^{\alpha-\omega}} . \tag{4.101}$$

Special cases often needed are

$$I(\omega, 2) = \frac{i}{(4\pi)^\omega} \Gamma(2-\omega) (m^2+p^2-i\varepsilon)^{\omega-2}, \tag{4.102}$$

$$I_2(\omega, 2) = -\frac{i}{(4\pi)^\omega} \omega\Gamma(1-\omega) (m^2+p^2-i\varepsilon)^{\omega-1}, \tag{4.103}$$

and

$$I(2, 3) = -\frac{i}{32\pi^2} \frac{1}{m^2+p^2-i\varepsilon} . \tag{4.104}$$

Summary

(Dis-) connected n-point Green functions are generated by the functional $Z[J]$, while $iW[J] = \ln Z[J]$ generates connected Green functions. Feynman diagrams are a useful mnemonic for the perturbative expansion of $Z[J]$ and $W[J]$ which allow one to write down directly Green functions and scattering amplitudes. The three loop diagrams we calculated in the $\lambda\phi^4$ theory were infinite and had to be regularised. Renormalising the three parameters contained in the classical Lagrangian of the $\lambda\phi^4$ theory, ρ_0, m^2 and λ, eliminated the divergences and converted them into "running" quantities.

Further reading. The derivation of the Feynman rules using the graphical approach is discussed extensively in Greiner and Reinhardt (2008). Srednicki (2007) treats the $\lambda\phi^3$ theory which resembles more QED than the $\lambda\phi^4$ theory we discussed. Delamotte (2004) gives an elementary introduction into the renormalization group.

Problems

4.1 $Z[J]$ at order λ. Derive Eq. (4.10).

4.2 Vacuum diagrams. Show that vacuum diagrams in the $\lambda\phi^4$ theory are purely imaginary. Generalise the result to the ϕ^n case and give an interpretation.

4.3 Volume of n dimensional sphere. a.) Calculate the volume of the unit sphere S^{n-1} defined by $x_1^2 + \ldots + x_n^2 = 1$ in $\mathbb{R}^n$. b.) Generalise the result to arbitrary (not necessarily integer) dimensions and show that it agrees with the familiar results for $n = 1, 2$ and 3.

4.4 Mandelstam variables. Show that for $2 \to 2$ scattering $s + t + u = m_1^2 + m_2^2 + m_3^2 + m_4^2$ is valid. Express t as function of the scattering angle between $\boldsymbol{p}_1$ and $\boldsymbol{p}_3$ and derive the relation between $d\sigma/d\Omega$ and $d\sigma/dt$. Find the lower and upper limits of t and u.

4.5 Casimir effect. Repeat the calculation of the Casimir effect for a scalar field in $3 + 1$ dimensions [Hint: Convert the integration over transverse momenta into an integration over energy, or combine zeta function and dimensional regularisation.]

4.6 Renormalisation invariance of the propagator. Derive analogously to Eq. (4.82) an equation $dm^2(\mu)/d\mu = f(m^2)$ requiring that the propagator (4.33) is independent of the scale μ.

4.7 Renormalisation with an Euclidean momentum cut-off. Calculate the self-energy (mass correction) and vertex correction using an Euclidean momentum cut-off Λ. Derive the RGE equations $dm^2(\Lambda)/d\Lambda = f(m^2)$, $d\lambda(\Lambda)/d\Lambda = f(\lambda)$ and $d\rho(\Lambda)/d\Lambda = f(\rho)$ and compare them to the ones derived using DR.

4.8 Feynman amplitude (4.63). Derive (4.63) starting from the amplitude in coordinate space, associating to external incoming (outgoing) particles the factor $\exp(\mp ikx)$.

4.9 Parameter integral (4.74). Evaluate the integral (4.74).

4.10 Feynman integrals. Derive the relation (4.98).

5
Global symmetries and Noether's theorem

Emmy Noether showed in 1917 that any global continuous symmetry of a classical system described by a Lagrangian leads to a locally conserved current. We can divide such symmetries into two classes: symmetries of spacetime and internal symmetries of a group of fields. Prominent examples for the latter are the global symmetries that lead to the conservation laws of electric charge or baryon number, respectively. For a quantum system, we have to study the impact of symmetries on its generating functional. If this functional remains invariant the symmetry holds also at the quantum level. Then conserved charges exists which commute with the Hamiltonian. Analysing spacetime symmetries, we restrict ourselves in this chapter to the case when we can neglect gravity. Then spacetime is the familiar Minkowski space characterised by the Poincaré symmetry group—the product of the translation and the Lorentz group, with its corresponding conservation laws.

5.1 Internal symmetries

Up to this point, we have mainly used spacetime symmetry of Minkowski space, namely the requirement of Lorentz invariance, to deduce possible terms in the action. If we allow for more than one field, for example several scalar fields, the new possibility of *internal* symmetries arise. For instance, we can look at a theory of two massive scalar fields with quartic interactions,

$$\mathscr{L} = \frac{1}{2}\left(\partial_\mu \phi_1\right)^2 - \frac{1}{2}m_1^2\phi_1^2 - \frac{1}{4}\lambda_1\phi_1^4 + \frac{1}{2}(\partial_\mu \phi_2)^2 - \frac{1}{2}m_2^2\phi_2^2 - \frac{1}{4}\lambda_2\phi_2^4 - \frac{1}{2}\lambda_3\phi_1^2\phi_2^2. \quad (5.1)$$

In order to maintain the discrete Z_2 symmetry $\phi_i \to -\phi_i$ of the individual Lagrangians for the fields ϕ_1 and ϕ_2, we have omitted odd terms like $\phi_1\phi_2^3$. Then the theory contains five arbitrary parameters: two masses m_i and three coupling constants λ_i. For arbitrary values of these parameters, no new additional symmetry results. In nature, we find, however, often a set of particles with nearly the same mass and (partly) similar couplings. One of the first examples was suggested by Heisenberg after the discovery of the neutron, which has a mass very close to the one of the proton, $m_n \simeq m_p$. With respect to strong interactions, it is useful to view the proton and neutron as two different "isospin" states of the nucleon, similar as an electron has two spin states. An example of an exact symmetry are particles and their antiparticles, as, for example, the charged pions $\pi^\pm$ which can be combined into one complex scalar field.

Quantum Fields–From the Hubble to the Planck Scale. Michael Kachelriess. © Michael Kachelriess 2018.
Published in 2018 by Oxford University Press. DOI 10.1093/oso/9780198802877.001.0001

If we set in our case $m_1 = m_2$ and $\lambda_2 = \lambda_1$, the Lagrangian becomes invariant under the exchange $\phi_1 \leftrightarrow \phi_2$. Adding the further condition that $\lambda_3 = \lambda_1 = \lambda_2$, we arrive at

$$\mathscr{L} = \frac{1}{2}\left[(\partial_\mu \phi_1)^2 + (\partial_\mu \phi_2)^2\right] - \frac{1}{2}m^2(\phi_1^2 + \phi_2^2) - \frac{\lambda}{4}(\phi_1^2 + \phi_2^2)^2. \tag{5.2}$$

Now any orthogonal transformation $O \in O(2)$ in the two-dimensional field space $\{\phi_1, \phi_2\}$ leads to the same Lagrangian $\mathscr{L}$. In particular, the Lagrangian is invariant under a rotation $R(\alpha) \in SO(2)$ which mixes $\{\phi_1, \phi_2\}$ as

$$\begin{pmatrix} \phi_1' \\ \phi_2' \end{pmatrix} = \begin{pmatrix} \cos\alpha & \sin\alpha \\ -\sin\alpha & \cos\alpha \end{pmatrix} \begin{pmatrix} \phi_1 \\ \phi_2 \end{pmatrix}. \tag{5.3}$$

The fields transform as a vector $\boldsymbol{\phi} = \{\phi_1, \phi_2\}$, and a rotation leaves the length of this vector invariant. Generalising this to n scalar fields, $\boldsymbol{\phi} = \{\phi_1, \ldots, \phi_n\}$, we can write down immediately a theory that is invariant under transformations $\phi_a \to R_{ab}\phi_b$ where R is an element of $O(n)$,

$$\mathscr{L} = \frac{1}{2}(\partial_\mu \boldsymbol{\phi})^2 - \frac{1}{2}m^2 \boldsymbol{\phi}^2 - \frac{\lambda}{4}(\boldsymbol{\phi}^2)^2. \tag{5.4}$$

Note that the Lagrangian is only invariant under global, that is, spacetime independent rotations, since the term $\partial_\mu(R\boldsymbol{\phi})$ would breaking the invariance for $R = R(x)$.

The free Lagrangian $\mathscr{L}_0$, which is the part quadratic in the fields, is diagonal, $\mathscr{L}_0 = \mathscr{L}_0(\phi_1) + \ldots + \mathscr{L}_0(\phi_n)$. Thus the propagator $D_{ab}(x - x')$ is diagonal too, $D_{ab}(x - x') \propto \delta_{ab}$. An interaction vertex at x connects four propagators $D_{ab}(x - x_i)$. As a result of the Z_2 symmetry, an even number of ϕ_1 and ϕ_2 particles are connected at each vertex which therefore takes the form $-i\lambda(\delta_{ab}\delta_{cd} + \delta_{ac}\delta_{bd} + \delta_{ad}\delta_{bc})$.

5.2 Noether's theorem

From our experience in classical and quantum mechanics, we expect that global continuous symmetries also lead in field theory to conservation laws for the generators of the symmetry. In order to derive such a conservation law, we consider an infinitesimal change $\delta\phi_a$ of the fields that keeps by assumption $\mathscr{L}(\phi_a, \partial_\mu \phi_a)$ invariant,

$$0 = \delta\mathscr{L} = \frac{\delta\mathscr{L}}{\delta\phi_a}\delta\phi_a + \frac{\delta\mathscr{L}}{\delta\partial_\mu\phi_a}\delta\partial_\mu\phi_a. \tag{5.5}$$

Now we exchange $\delta\partial_\mu = \partial_\mu\delta$ in the second term and use then the Lagrange equations, $\delta\mathscr{L}/\delta\phi_a = \partial_\mu(\delta\mathscr{L}/\delta\partial_\mu\phi_a)$, in the first one. Then we can combine the two terms using the product rule,

$$0 = \delta\mathscr{L} = \partial_\mu\left(\frac{\delta\mathscr{L}}{\delta\partial_\mu\phi_a}\right)\delta\phi_a + \frac{\delta\mathscr{L}}{\delta\partial_\mu\phi_a}\partial_\mu\delta\phi_a = \partial_\mu\left(\frac{\delta\mathscr{L}}{\delta\partial_\mu\phi_a}\delta\phi_a\right). \tag{5.6}$$

Hence the invariance of $\mathscr{L}$ under the change $\delta\phi_a$ implies the existence of a conserved current, $\partial_\mu j^\mu = 0$, with

$$j^\mu = \frac{\delta\mathscr{L}}{\delta\partial_\mu\phi_a}\delta\phi_a. \tag{5.7}$$

If the transformation $\delta\phi_a$ leads to change in $\mathscr{L}$ that is a total four-divergence, $\delta\mathscr{L} = \partial_\mu K^\mu$, and boundary terms can be dropped, then the equations of motion remain

invariant too. The conserved current j^μ, also called Noether current, is then changed
to

$$j^\mu = \frac{\delta \mathscr{L}}{\delta \partial_\mu \phi_a} \delta \phi_a - K^\mu. \tag{5.8}$$

In Minkowski space, we can convert this differential form of a conservation law
into a global one using Gauss' theorem. Recall that this theorem allows us to convert
an n-dimensional volume integral over the divergence of a tensor field into an $n-1$-
dimensional surface integral of the tensor field,

$$\int_\Omega \mathrm{d}^4 x \, \partial_\mu X^{\mu\nu\cdots} = \int_{\partial\Omega} \mathrm{d}S_\mu X^{\mu\nu\cdots}. \tag{5.9}$$

Applied to the Noether current j^μ, we obtain assuming $|\boldsymbol{j}| \to 0$ for $|\boldsymbol{x}| \to \infty$

$$\int_\Omega \mathrm{d}^4 x \, \partial_\mu j^\mu = \int_{V(t_2)} \mathrm{d}^3 x \, j^0 - \int_{V(t_1)} \mathrm{d}^3 x \, j^0 = 0. \tag{5.10}$$

Thus the volume integral $Q = \int_V \mathrm{d}^3 x \, j^0$ over the charge density j^0 remains constant.
Often (but not always) this charge has a profound physical meaning. Finally, we note
that the conserved current j^μ is not unique, since we can add the four-divergence
$\partial_\mu K^\mu = \delta \mathscr{L}$.

Internal symmetries.   As an example, we can use the n scalar fields invariant under
the group[1] SO(n). We need the infinitesimal generators T_i of rotations,

$$\phi'_a = R_{ab}\phi_b = (1 + \alpha_i T_i + \mathcal{O}(\alpha_i^2))_{ab}\phi_b. \tag{5.11}$$

SO(n) has an antisymmetric Lie algebra with $n(n-1)/2$ generators. Thus a theory
invariant under SO(n) has $n(n-1)/2$ conserved currents. Charged pions $\pi^\pm$ are an
important application of the special case $n = 2$. We combine the two real fields ϕ_1 and
ϕ_2 into the complex field $\phi = (\phi_1 + \mathrm{i}\phi_2)/\sqrt{2}$, then the Lagrangian becomes

$$\mathscr{L} = \partial_\mu \phi^\dagger \partial^\mu \phi - m^2 \phi^\dagger \phi - \lambda(\phi^\dagger \phi)^2. \tag{5.12}$$

Now the Lagrangian $\mathscr{L}$ is invariant under the combined phase transformations $\phi \to$
$\mathrm{e}^{\mathrm{i}\vartheta}\phi$ and $\phi^\dagger \to \mathrm{e}^{-\mathrm{i}\vartheta}\phi^\dagger$. Using complex fields, the SO(2) symmetry has become an U(1)
symmetry. With $\delta\phi = \mathrm{i}\phi$ and $\delta\phi^\dagger = -\mathrm{i}\phi^\dagger$, the conserved current follows as

$$j^\mu = \mathrm{i}\left[\phi^\dagger \partial^\mu \phi - (\partial^\mu \phi^\dagger)\phi\right]. \tag{5.13}$$

The conserved charge $Q = \int \mathrm{d}^3 x \, j^0$ can be also negative and thus we cannot interpret
j^0 as the probability density to observe a ϕ particle. Instead, we should associate Q
with a conserved additive quantum number as, for example, the electric charge.

---

[1]Although the Lagrangian is invariant under the larger group O(n), we consider only the subgroup SO(n) which is continuously connected to the identity. The additional discrete transformations
contained in O(n) can be used to classify solutions of the Lagrangian, but do not lead to additional
conservation laws.

Note that Noethers theorem requires the existence of a *global* symmetry. In the cases of the conservation of electric and colour charge, the global symmetry is a consequence of an underlying *local* gauge symmetry which we will study in chapter 10 in detail. In most other cases however, such as the conservation of baryon or lepton number, the global symmetry cannot be generalised to a local one, and one speaks therefore of accidental symmetries. Such symmetries are not protected against quantum corrections and there is no reason to expect them to hold exactly. We will see later that baryon and lepton number are indeed broken.

Spacetime symmetries of Minkowski space. The Poincaré group as symmetry group of Minkowski space has ten generators. If the Lagrangian does not depend explicitly on spacetime coordinates, i.e. $\mathscr{L} = \mathscr{L}(\phi_a, \partial_\mu \phi_a)$, ten conservation laws for the fields ϕ_a follow. We consider first the behaviour of the fields ϕ_a and the Lagrangian under an infinitesimal translation $x^\mu \to x^\mu + \varepsilon \xi^\mu$. As in the case of internal symmetries, we consider only global transformations and thus ε does not depend on x. From

$$\phi_a(x^\mu) \to \phi_a(x^\mu + \varepsilon \xi^\mu) \approx \phi_a(x^\mu) + \varepsilon \xi^\nu \partial_\nu \phi_a(x^\mu) \tag{5.14}$$

we find the change

$$\delta \phi_a(x) = \xi^\mu \partial_\mu \phi_a(x) = \partial_\mu [\xi^\mu \phi_a(x)].$$

Since the Lagrange density $\mathscr{L}$ contains by assumption no explicit spacetime dependence, it will change simply as $\mathscr{L}(x^\mu) \to \mathscr{L}(x^\mu + \varepsilon \xi^\mu)$ or

$$\delta \mathscr{L}(x) = \xi^\mu \partial_\mu \mathscr{L}(x) = \partial_\mu [\xi^\mu \mathscr{L}(x)]. \tag{5.15}$$

Thus $K^\mu = \xi^\mu \mathscr{L}(x)$ and inserting both in the Noether current gives

$$j^\mu = \frac{\partial \mathscr{L}}{\partial(\partial_\mu \phi_a)} [\xi^\nu \partial_\nu \phi_a] - \xi^\mu \mathscr{L} = \xi_\nu \left[\frac{\partial \mathscr{L}}{\partial(\partial_\mu \phi_a)} \frac{\partial \phi_a}{\partial x_\nu} - \eta^{\mu\nu} \mathscr{L} \right] \equiv \xi_\nu T^{\mu\nu}, \tag{5.16}$$

where the square bracket defines the (energy–momentum) stress tensor $T^{\mu\nu}$ of the fields ϕ_a. The corresponding four conserved Noether charges are the components of the four-momentum

$$p^\nu = \int \mathrm{d}^3 x \, T^{0\nu}. \tag{5.17}$$

The conserved tensor defined by Eq. (5.16) is called the *canonical* stress tensor. The definition (5.16) does not guarantee that $T^{\mu\nu}$ is symmetric. A symmetric stress tensor, $T^{\mu\nu} = T^{\nu\mu}$, is, however, the condition for the conservation of the total angular momentum, as we will show in the next paragraph. Another reason to require this symmetry is that the stress tensor serves as source for the symmetric gravitational field. Since the Lagrange density is only determined up to a four-divergence, we can symmetrise always $T^{\mu\nu}$ adding an appropriate divergence satisfying $\partial_\lambda S^{\lambda\mu\nu} = -\partial_\lambda S^{\mu\lambda\nu}$. Thus in general, the canonical stress tensor has to be symmetrised by hand.

Example 5.1: The general expression (5.16) for the canonical stress tensor becomes for a free complex scalar field

$$T^{\mu\nu} = 2\partial^\mu \phi^\dagger \partial^\nu \phi - \eta^{\mu\nu} \mathscr{L}. \tag{5.18}$$

Thus the canonical stress tensor of a scalar field is already symmetric. Its 00 component,

$$T^{00} = \rho = \mathcal{H} = 2|\dot{\phi}|^2 - \mathcal{L} = |\dot{\phi}|^2 + |\boldsymbol{\nabla}\phi|^2 + m^2|\phi|^2 \,, \tag{5.19}$$

agrees with twice the result (3.14) for the energy density ρ of a single real scalar field. We consider now plane wave solutions to the Klein–Gordon equation, $\phi = N\exp(\mathrm{i}kx)$. If we insert $\partial_\mu \phi = \mathrm{i}k_\mu \phi$ into $\mathcal{L}$, we find $\mathcal{L} = 0$ and thus

$$T^{00} = 2N^2 k^0 k^0 \,. \tag{5.20}$$

Changing from the continuum normalisation to a box of size $V = L^3$ amounts to replace $(2\pi)^3$ by L^3. Thus the normalisation constant $N^{-2} = (2\pi)^3 2\omega$ becomes for a finite volume $N^{-2} = 2\omega V$. Thence the energy density $T^{00} = \omega/V$ agrees with the expectation for one particle with energy ω per volume V. The remaining components of $T^{\mu\nu}$ are fixed by its tensor structure,

$$T^{\mu\nu} = 2N^2 k^\mu k^\nu = \frac{k^\mu k^\nu}{\omega V}\,. \tag{5.21}$$

Since the stress tensor $T^{\mu\nu}$ is symmetric, we can find a frame in which $T^{\mu\nu}$ is diagonal with $T \propto \mathrm{diag}(\omega, v_x k_x, v_y k_y, v_z k_z)/V$. The spatial part of the stress tensor agrees with the pressure tensor of an ideal fluid, cf. problem 5.4. Thus a scalar field can be viewed as an ideal fluid with energy density ρ and pressure P, or $T^{\mu\nu} = \mathrm{diag}(\rho, P_x, P_y, P_z)$.

Angular momentum. If the tensor $T^{\mu\nu}$ is symmetric, we can construct six more conserved quantities. If we define

$$M^{\mu\nu\lambda} = x^\nu \, T^{\mu\lambda} - x^\lambda \, T^{\mu\nu}, \tag{5.22}$$

then $M^{\mu\nu\lambda}$ is conserved with respect to the index μ,

$$\partial_\mu M^{\mu\nu\lambda} = \delta^\nu_\mu \, T^{\mu\lambda} - \delta^\lambda_\mu T^{\mu\nu} = T^{\nu\lambda} - T^{\lambda\nu} = 0, \tag{5.23}$$

provided that $T^{\nu\lambda} = T^{\lambda\nu}$. In this case,

$$J^{\mu\nu} = \int \mathrm{d}^3 x \, M^{0\mu\nu} = \int \mathrm{d}^3 x \, \left[x^\mu T^{0\nu} - x^\nu T^{0\mu} \right], \tag{5.24}$$

is a globally conserved tensor. The antisymmetry of $J^{\mu\nu}$ implies that there exist six conserved charges. The three charges

$$J^{ij} = \int \mathrm{d}^3 x \, \left[x^i T^{0j} - x^j T^{0i} \right] \tag{5.25}$$

correspond to the conservation of total angular momentum, since T^{0j} is the momentum density. The remaining three charges J^{0i} express the fact that the centre-of-mass moves with constant velocity.

While $J^{\mu\nu}$ transforms as expected for a tensor under Lorentz transformations, it is not invariant under translations $x^\mu \to x^\mu + \xi^\mu$. Instead, the angular momentum changes as

$$J^{\mu\nu} \to J^{\mu\nu} + \xi^\mu p^\nu - \xi^\nu p^\mu. \tag{5.26}$$

Clearly, this is a consequence of the definition of the orbital angular momentum with respect to the centre of rotation. We therefore want to split the total angular momentum $J^{\mu\nu}$ into the orbital angular momentum $L^{\mu\nu}$ and an intrinsic part connected to a non-zero spin of the field. The latter we require to be invariant under translations. We set

$$S_\alpha = \frac{1}{2}\varepsilon_{\alpha\beta\gamma\delta}J^{\beta\gamma}u^\delta, \tag{5.27}$$

where u^α is the four-velocity of the centre-of-mass system and $\varepsilon_{\alpha\beta\gamma\delta}$ the completely antisymmetric Levi–Civita tensor. Because of the antisymmetry in $\beta\gamma$ of $\varepsilon_{\alpha\beta\gamma\delta}$, the change in (5.26) induced by a translation drops out in S_α. In the cms, $u^\alpha = (1, 0, 0, 0)$ and thus $S_0 = 0$ and $S_\alpha u^\alpha = 0$. The other components are $S_1 = J^{23}$, $S_2 = J^{31}$, and $S_3 = J^{12}$. Thus the vector S_α describes as desired the intrinsic angular momentum of a field. It is called the Pauli–Lubanski spin vector.

Example 5.2: To see that the contraction of a symmetric tensor $S_{\mu\nu}$ with an antisymmetric tensor $A_{\mu\nu}$ gives zero, consider

$$S_{\mu\nu}A^{\mu\nu} = -S_{\mu\nu}A^{\nu\mu} = -S_{\nu\mu}A^{\nu\mu} = -S_{\mu\nu}A^{\mu\nu}. \tag{5.28}$$

Here we used first the antisymmetry of $A^{\mu\nu}$, then the symmetry of $S_{\mu\nu}$, and finally exchanged the dummy summation indices. Clearly, this remains true if the tensor expression contains additional indices. Applied to the Pauli–Lubanski spin vector and recalling $p^\gamma = mu^\gamma$, its change contains terms as $\varepsilon_{\alpha\beta\gamma\delta}\xi^\beta u^\gamma u^\delta$ which are zero.

5.3 Quantum symmetries

Conserved currents. We have seen that Noether's theorem on the classical level guarantees the conservation of currents generated by global continuous symmetries. In the corresponding quantum theory, we have to study the impact of this symmetry on the generating functional Z. Since we used the equations of motion to derive Noether's theorem, current conservation holds only for classically allowed paths in field space or, in other words, for on-shell fields. Thus the action evaluated for off-shell fields is not invariant under global symmetry transformations. In the path integral, the fields are, however, only integration variables. The generating functional is therefore invariant, if we can find a field transformation $\phi_i \to \tilde{\phi}_i$ which eliminates the change of the action for off-shell fields and keeps the integration measure invariant, $\mathcal{D}\phi_i = \mathcal{D}\tilde{\phi}_i$.

Let us assume that our theory has a global symmetry under which the classical solutions transform as $\phi_a \to \phi'_a = \phi_a + \varepsilon\eta_a$. Here, ε takes the same value at all spacetime points and η_a is a function of the original fields, $\eta_a = \eta_a(\phi_a(x))$. The classically forbidden solutions will be transformed into $\tilde{\phi}_a \neq \phi'_a$. We can express $\tilde{\phi}_a$ always as

$$\phi_a \to \tilde{\phi}_a = \phi_a + \varepsilon(x)\eta_a, \tag{5.29}$$

promoting thereby $\varepsilon(x)$ to a spacetime-dependent function. To be concrete, we consider again a global U(1) symmetry for a complex scalar field. Since the transformation (5.29) is *local*, kinetic terms breaks the symmetry. Direct calculation shows that the Lagrangian (5.12) changes as

$$\delta\mathscr{L} = \mathrm{i}\left[\phi^* \partial^\mu \phi - \partial^\mu \phi^* \phi\right] \partial_\mu \varepsilon = j^\mu \, \partial_\mu \varepsilon, \tag{5.30}$$

where j^μ is the classical Noether current, cf. problem 5.8. The final result $\delta\mathscr{L} = j^\mu \, \partial_\mu \varepsilon$ holds in general.

Next we have to generalise the generating functional for a single, real scalar field given in Eq. (4.1) to a complex scalar field. We treat ϕ and ϕ^* as the two independent degrees of freedom, and add therefore also two independent sources J and J^*. Coupling them as $J\phi^* + J^*\phi$ to the fields keeps the Lagrangian real, and the generating functional becomes

$$Z[J, J^*] = \int \mathcal{D}\phi \mathcal{D}\phi^* \exp \mathrm{i} \int \mathrm{d}^4x \left(\mathscr{L} + J\phi^* + J^*\phi\right). \tag{5.31}$$

We want to calculate matrix elements of the operator representing the classical Noether current (5.8). Using the path-integral formalism, we can derive the time-ordered vacuum expectation value of a product of fields ϕ and the current operator j^μ by adding a classical external source v_μ coupled to j^μ,

$$Z[J, J^*, v_\mu] = \int \mathcal{D}\phi \mathcal{D}\phi^* \exp \mathrm{i} \int \mathrm{d}^4x \left(\mathscr{L} + J\phi^* + J^*\phi + v_\mu j^\mu\right). \tag{5.32}$$

Then we obtain the vacuum expectation value of the current as

$$\langle j^\mu(x)\rangle \equiv \langle 0|j^\mu(x)|0\rangle = \left.\frac{1}{\mathrm{i}}\frac{\delta}{\delta v_\mu(x)}\mathrm{i}W[J, J^*, v_\mu]\right|_{v_\mu = J = 0}. \tag{5.33}$$

Inverting this relation we find

$$\delta W[J, J^*, v_\mu] = \int \mathrm{d}^4x \, \langle j^\mu(x)\rangle \delta v_\mu(x). \tag{5.34}$$

We are interested in how W and Z change under a transformation of the external source v_μ. To deduce their transformation properties, it is sufficient to consider them for zero external sources J and J^*. Setting $Z[0, 0, v_\mu] \equiv Z[v_\mu]$ and choosing $\delta v_\mu(x) = -\partial_\mu \varepsilon(x)$, it follows

$$\delta W[v_\mu] = W[v_\mu - \partial_\mu \varepsilon(x)] - W[v_\mu] = -\int \mathrm{d}^4x \, \langle j^\mu(x)\rangle \partial_\mu \varepsilon(x) = \int \mathrm{d}^4x \, \partial_\mu \langle j^\mu(x)\rangle \varepsilon(x). \tag{5.35}$$

Thus $\delta W[v_\mu] = 0$ guarantees current conservation in the quantum theory, $\partial_\mu \langle j^\mu\rangle = 0$. The corresponding change of $Z[v_\mu]$ under the same transformation is

$$Z[v_\mu - \partial_\mu \varepsilon(x)] = \int \mathcal{D}\phi \mathcal{D}\phi^* \exp \mathrm{i} \int \mathrm{d}^4 x \left\{ \mathscr{L} + [v_\mu - \partial_\mu \varepsilon(x)] j^\mu \right\}. \tag{5.36}$$

We now substite $\phi_a \rightarrow \tilde{\phi}_a = \phi_a + \varepsilon(x)\eta_a$, assuming that this change of variables keeps the integration measure invariant, $\mathcal{D}\phi \mathcal{D}\phi^* = \mathcal{D}\tilde{\phi}\mathcal{D}\tilde{\phi}^*$. Recalling then that $\delta \mathscr{L} = j^\mu \, \partial_\mu \varepsilon$, we find that the generating functional is invariant,

$$Z[v_\mu - \partial_\mu \varepsilon(x)] = \int \mathcal{D}\tilde{\phi}\mathcal{D}\tilde{\phi}^* \exp \mathrm{i} \int \mathrm{d}^4 x \left(\mathscr{L}(\tilde{\phi}, \partial_\mu \tilde{\phi}) + v_\mu j^\mu \right) = Z[v_\mu]. \tag{5.37}$$

In the case of the U(1) transformation, the two phases cancel in the integration measure

$$\mathcal{D}\tilde{\phi}\mathcal{D}\tilde{\phi}^* = \prod_x \mathrm{d}\tilde{\phi}(x)\mathrm{d}\tilde{\phi}^*(x) = \prod_x \mathrm{d}\phi(x)\mathrm{d}\phi^*(x). \tag{5.38}$$

As a result, the vacuum expectation value of the electromagnetic current is conserved, $\partial_\mu \langle j^\mu \rangle = 0$.

Anomalies. The substitution $\phi_a \rightarrow \tilde{\phi}_a = \phi_a + \varepsilon(x)\eta_a$ shifts the centre of the integration at each spacetime point by the value $\varepsilon(x)\eta_a$. Such a linear shift seems harmless. Therefore it was taken for granted that the path integral remains invariant under this change and, consequently, that this approach predicts that all classical global symmetries hold also on the quantum level. It was only realised by Fujikawa in 1979 that the integration measure in the path integral may transform non-trivially under a symmetry transformation. Since the path integral is divergent, we have to regularise it and this procedure may break the classical symmetry.

If the classical symmetry is broken, one speaks of an "anomaly". The three most important examples are the trace anomaly, the chiral anomaly and the breaking of conformal invariance in string theory. We will discuss the first two cases later in some detail. The anomalous term breaking conformal invariance in string theory vanishes for a definite number of spacetime dimensions, $D = 10$ or 26, which is the reason for the prediction of extra dimensions in string theory.

Summary

Noether's theorem shows that continuous global symmetries lead classically to conservation laws. Such symmetries can be divided into spacetime and internal symmetries. Minkowski spacetime is invariant under global Poincaré transformations. The corresponding ten Noether charges are the four-momentum p^μ and the total angular momentum $J^{\mu\nu}$. Examples for conserved charges due to internal symmetries are electric and colour charge, as well as baryon or lepton number. In a quantum theory, the vacuum expectation value of a Noether current is conserved, if the symmetry transformation keeps the path-integral measure invariant.

Further reading. A more complete discussion of Noether's theorem can be found in Greiner and Reinhardt (2008) and Hill (1951).

Problems

5.1 Lagrangian for N scalar fields. The most general expression for the Lagrange density $\mathscr{L}$ of N scalar fields ϕ_i which is Lorentz invariant and at most quadratic in the fields is

$$\mathscr{L} = \frac{1}{2} A_{ij} \partial^\mu \phi_i \partial_\mu \phi_j - \frac{1}{2} B_{ij} \phi_i \phi_j - C.$$

Find the constraints on the coefficients and show that $\mathscr{L}$ can be recast into "canonical form"

$$\mathscr{L} = \frac{1}{2} \partial^\mu \phi_i \partial_\mu \phi_i - \frac{1}{2} b_i \phi_i \phi_i - C$$

by linear field redefinitions.

5.2   Lorentz invariance of charges. Show that the charge $Q = \int_V d^3 x\, j^0$ of a conserved current can be rewritten in a manifestly Lorentz invariant form as $Q(t = 0) = \int d^4 x\, j^\mu(x) \partial_\mu \vartheta(n_\mu x^\mu)$.

5.3 Stress tensor for point particles. a.) Find the stress tensor for an ensemble of N non-relativistic point particles; b) on scales L such that $\Delta N/L^3 \gg 1$, one can describe the phase space density by a smooth function $f(\boldsymbol{x}, \boldsymbol{p})$. Discuss the physical meaning of the different elements of $T^{\mu\nu}$.

5.4 Stress tensor for an ideal fluid ♣. a.) Find first the stress tensor of dust, that is, of pressureless matter. b.) Generalise this result to an ideal fluid. (Hint: The state of an ideal fluid is completely determined by its energy density ρ and its pressure P; in the rest-frame of the fluid, the pressure is isotropic $P_{ij} = P\delta_{ij}$.) c.) Compare $\partial_\mu T^{\mu\nu} = 0$ in the non-relativistic limit to the ideal fluid equations. d.) The Equation of State (EoS) is defined by $w = P/\rho$. Find the EoS for vacuum energy, that is, for the Lagrangian $\mathscr{L} = -\rho$.

5.5 Stress tensor for the electromagnetic field. Determine the energy–momentum stress tensor $T_{\mu\nu}$ of the free Maxwell field. Symmetrise $T_{\mu\nu}$, if necessary. Confirm that T^{00} corresponds to the energy density ρ. Find the trace of $T^{\mu\nu}$ and the EoS. [Possible ways: i) use Noether's theorem, ii) use Newton's law, iii) convert $\rho = (E^2 + B^2)/2$ into a relativistic law.]

5.6 Scale invariance. Consider the effect of a scale transformation $x \to e^\alpha x$ on a scalar field with a $\lambda \phi^4$ self-interaction assuming that it acts linearly on the fields, $\phi(x) \to e^{D\alpha} \phi(e^\alpha x)$. Here α and D are numbers. a) Write down first the infinitesimal version of the scale transformation, with the aim to show that $\mathscr{L}$ can be made invariant for a specific value of D, if $m = 0$. b) Find the corresponding conserved current s^μ. c) Show that the current s^μ can be written as $s^\mu = x^\nu \tilde{T}_{\mu\nu}$, where $\tilde{T}_{\mu\nu}$ is an "improved" stress tensor. [Hint: Proceed similarly as in the case of the angular momentum tensor.]

5.7 Linear Sigma model I. Show that the Lagrangian

$$\mathscr{L} = \frac{1}{2} \left[(\partial_\mu \boldsymbol{\pi})^2 + (\partial_\mu \sigma)^2 \right] - \frac{m^2}{2} (\boldsymbol{\pi}^2 + \sigma^2) \tag{5.39}$$

$$- \frac{\lambda}{4} (\boldsymbol{\pi}^2 + \sigma^2)^2$$

is invariant under the symmetry transformation $\Sigma \to \Sigma' = U \Sigma U^\dagger$, where $\Sigma \equiv \sigma + i\boldsymbol{\tau} \cdot \boldsymbol{\pi}$, $U = \exp(i\boldsymbol{\alpha} \cdot \boldsymbol{\tau}/2)$ and $\boldsymbol{\tau} = (\tau^1, \tau^2, \tau^3)$ are the Pauli matrices. Find the corresponding conserved Noether currents. Give the Feynman rules (i.e. specify propagators and vertices in momentum space) for this Lagrangian.

5.8 $\delta\mathscr{L}$ for a local U(1) transformation. Show that the change of the Lagrangian (5.12) under a local U(1) transformation can be written as (5.30).

6

Spacetime symmetries

In the previous chapter, we discussed the symmetries of Minkowski space. In this case, we could view the Poincaré group as the group generating global symmetry transformations on Minkowski space and find the resulting conservation laws. The aim of the present chapter is to extend this discussion to the case of a Riemannian manifold, that is to a curved space which looks only locally Euclidean. We will show how one can find the symmetries of such manifolds and how they determine conservation laws. Riemannian manifolds arise naturally in classical mechanics using generalised coordinates q^i, since the kinetic energy $T = a_{ik}\dot{q}^i\dot{q}^k$ defines a quadratic form a_{ik} which we can view as metric tensor on the configuration space q^i. However, the more important appearance of a (pseudo-) Riemannian manifold is in Einstein's theory of general relativity which replaces Minkowski space by a curved spacetime. Most of the mathematical structures we will introduce also have a close analogue in gauge theories which we will use later on to describe the electroweak and strong interactions.

Equivalence principle. First, we examine why one can replace the gravitational force by the curvature of spacetime discussing the equivalence principle. The idea underlying this principle emerged in the 16th century when, among others, Galileo Galilei found experimentally that the acceleration g of a test mass in a gravitational field is universal. Because of this universality, the gravitating mass m_g and the inertial mass m_i are identical in classical mechanics. While $m_i = m_g$ can be achieved for one material always by a convenient choice of units, in general there should be deviations for test bodies with differing compositions. Current limits for departures from universal gravitational attraction for different materials are, however, very tight, $|\Delta g/g| < 10^{-12}$.

As a result, gravity has, compared to the three other known fundamental interactions, the unique property that it can be switched-off locally: inside a freely falling elevator one does not feel any gravitational effects except tidal forces. The latter arise if the gravitational field is non-uniform and tries to deform the elevator. Inside a sufficiently small, freely falling system, tidal effects also play no role. Einstein promoted the equivalence of inertial and gravitating mass to the postulate of the "strong equivalence principle": in a small enough region around the centre of a freely falling coordinate system all physics is described by the laws of special relativity.

In general relativity, the gravitational force of Newton's theory that accelerates particles in an Euclidean space is replaced by a curved spacetime in which particles move force-free along geodesic lines. In particular, as in special relativity, photons still move along curves satisfying $\mathrm{d}s^2 = 0$, while all effects of gravity are now encoded in the non-Euclidean geometry of spacetime which is determined by the line element $\mathrm{d}s$

Quantum Fields–From the Hubble to the Planck Scale. Michael Kachelriess. © Michael Kachelriess 2018.
Published in 2018 by Oxford University Press. DOI 10.1093/oso/9780198802877.001.0001

or the metric tensor $g_{\mu\nu}$,

$$\mathrm{d}s^2 = g_{\mu\nu}\mathrm{d}x^\mu\mathrm{d}x^\nu . \tag{6.1}$$

Switching on a gravitational field, the metric tensor $g_{\mu\nu}$ can be transformed only locally by a coordinate change into the form $\eta_{\mu\nu} = \mathrm{diag}(1, -1, -1, -1)$. Thus we should develop the tools necessary to undertake analysis on a curved manifold $\mathcal{M}$ which geometry is described by the metric tensor $g_{\mu\nu}$.

6.1 Manifolds and tensor fields

Manifolds. A manifold $\mathcal{M}$ is any set that can be continuously parameterised. The number of independent parameters needed to specify uniquely any point of $\mathcal{M}$ is its dimension n, the parameters $x = \{x^1, \dots, x^n\}$ are called coordinates. Locally, a manifold with dimension n can be approximated by $\mathbb{R}^n$. Examples for manifolds are Lie groups, the configuration space q^i or the phase space (q^i, p_i) of classical mechanics, and spacetime in general relativity. We require the manifold to be smooth. The transitions from one set of coordinates to another one, $x^i = f(\tilde{x}^i, \dots, \tilde{x}^n)$, should be C^∞. In general, it is impossible to cover all $\mathcal{M}$ with one coordinate system that is well-defined on all $\mathcal{M}$. An example are spherical coordinate (ϑ, ϕ) on a sphere S^2, where ϕ is ill-defined at the poles. Instead one has to cover the manifold with patches of different coordinates that partially overlap.

Vector fields. A vector field $\boldsymbol{V}(x^a)$ on (a subset $\mathscr{S}$ of) $\mathcal{M}$ is a set of vectors associating to each spacetime point $x^a \in \mathscr{S}$ exactly one vector. The paradigm for such a vector field is the four-velocity $\boldsymbol{u}(\tau) = \mathrm{d}\boldsymbol{x}/\mathrm{d}\tau$ which is the tangent vector to the world-line $x(\tau)$ of a particle. Since the differential equation $\mathrm{d}\boldsymbol{x}/\mathrm{d}\sigma = \boldsymbol{X}(\sigma)$ has locally always a solution, we can find for any given $\boldsymbol{X}$ a curve $x(\sigma)$ which has $\boldsymbol{X}$ as tangent vector. Although the definition $\boldsymbol{u}(\tau) = \mathrm{d}\boldsymbol{x}/\mathrm{d}\tau$ coincides with the one familiar from Minkowski space, there an important difference. In a general manifold, we cannot imagine a vector $\boldsymbol{V}$ as an "arrow" $\overrightarrow{PP'}$ pointing from a certain point P to another point P' of the manifold. Instead, the vectors $\boldsymbol{V}$ generated by all smooth curves through P span an n-dimensional vector space at the point P called tangent space T_P. We can visualise the tangent space for the case of a two-dimensional manifold embedded in $\mathbb{R}^3$: at any point P, the tangent vectors lie in a plane $\mathbb{R}^2$ which we can associate with T_P. In general, $T_P \neq T_{P'}$ and we cannot simply move a vector $\boldsymbol{V}(x^\mu)$ to another point $\tilde{x}^\mu$. This implies in particular that we cannot add the vectors $\boldsymbol{V}(x^\mu)$ and $\boldsymbol{V}(\tilde{x}^\mu)$, if the points x^μ and $\tilde{x}^\mu$ differ. Therefore we cannot differentiate a vector field without introducing an additional mathematical structure which allows us to transport a vector from one tangent space to another.

If we want to decompose the vector $\boldsymbol{V}(x^\mu)$ into components $V^\nu(x^\mu)$, we have to introduce a basis $\boldsymbol{e}_\mu$ in the tangent space. There are two natural choices for such a basis: First, we could use Cartesian basis vectors as in a Cartesian inertial system in Minkowski space. We will follow this approach, when we discuss gravity as a gauge theory in chapter 19. Now, we will use the more conventional approach and use as basis vectors the tangential vectors along the coordinate lines x^μ in $\mathcal{M}$,

$$\boldsymbol{e}_\mu = \frac{\partial}{\partial x^\mu} \equiv \partial_\mu. \tag{6.2}$$

Here the index μ with value i of e_μ denotes the i.th basis vector $e_\mu = (0, \ldots, 1, \ldots 0)$, with a 1 at the i.th position. Using this basis, a vector can be decomposed as

$$V = V^\mu e_\mu = V^\mu \partial_\mu. \tag{6.3}$$

A coordinate change

$$x^\mu = f(\tilde{x}^1, \ldots, \tilde{x}^n), \tag{6.4}$$

or more briefly $x^\mu = x^\mu(\tilde{x}^\nu)$, changes the basis vectors as

$$e_\mu = \frac{\partial}{\partial x^\mu} = \frac{\partial \tilde{x}^\nu}{\partial x^\mu} \frac{\partial}{\partial \tilde{x}^\nu} = \frac{\partial \tilde{x}^\nu}{\partial x^\mu} \tilde{e}_\nu. \tag{6.5}$$

Therefore the vector V will be invariant under general coordinate transformations,

$$V = V^\mu \partial_\mu = \tilde{V}^\mu \tilde{\partial}_\mu = \tilde{V}, \tag{6.6}$$

if its components transform opposite to the basis vectors $e_\mu = \partial_\mu$, or

$$V^\mu = \frac{\partial x^\mu}{\partial \tilde{x}^\nu} \tilde{V}^\nu. \tag{6.7}$$

If x^μ and $\tilde{x}^\mu$ are two inertial frames in Minkowski space, we came back to Lorentz transformations $\partial x^\mu / \partial \tilde{x}^\nu = \Lambda^\mu{}_\nu$ as a special case of general coordinate transformations.

Covectors or one-forms. In quantum mechanics, we use Dirac's bracket notation to associate to each vector $|a\rangle$ a dual vector $\langle a|$ and to introduce a scalar product $\langle a | b \rangle$. If the vectors $|n\rangle$ form a basis, then the dual basis $\langle n|$ is defined by $\langle n | n' \rangle = \delta_{nn'}$. Similarly, we define a basis e^μ dual to the basis e_μ in T_P by

$$e^\mu(e_\nu) = \delta^\mu_\nu. \tag{6.8}$$

This basis can be used to form a new vector space T_P^* called the cotangent space which is dual to T_P. Its elements $\boldsymbol{\omega}$ are called covectors or one-forms,

$$\boldsymbol{\omega} = \omega_\mu e^\mu. \tag{6.9}$$

Combining a vector and a one-form, we obtain a map into the real numbers,

$$\boldsymbol{\omega}(V) = \omega_\mu V^\nu e^\mu(e_\nu) = \omega_\mu V^\mu. \tag{6.10}$$

The last equality shows that we can calculate $\boldsymbol{\omega}(V)$ in component form without reference to the basis vectors. In order to simplify notation, we will use therefore in the future simply $\omega_\mu V^\mu$; we also write $e^\mu e_\nu$ instead of $e^\mu(e_\nu)$.

Using a coordinate basis, the duality condition (6.8) is obviously satisfied, if we choose $e^\mu = \mathrm{d}x^\mu$. Then the one-form becomes

$$\boldsymbol{\omega} = \omega_\mu \mathrm{d}x^\mu. \tag{6.11}$$

Thus the familiar "infinitesimals" $\mathrm{d}x^\mu$ are actually the finite basis vectors of the cotangent space T_P^*. We require again that the transformation of the components ω_μ of a covector cancels the transformation of the basis vectors,

$$\omega_\mu = \frac{\partial \tilde{x}^\nu}{\partial x^\mu} \, \tilde{\omega}_\nu. \tag{6.12}$$

This condition guarantees that the covector itself is an invariant object, since

$$\boldsymbol{\omega} = \omega_\mu \mathrm{d}x^\mu = \frac{\partial \tilde{x}^\nu}{\partial x^\mu} \, \tilde{\omega}_\nu \frac{\partial x^\mu}{\partial \tilde{x}^\sigma} \, \mathrm{d}\tilde{x}^\sigma = \tilde{\omega}_\mu \mathrm{d}\tilde{x}^\mu = \tilde{\boldsymbol{\omega}}. \tag{6.13}$$

Covariant and contravariant tensors. Next we generalise the concept of vectors and covectors. We call a vector $\boldsymbol{X}$ also a contravariant tensor of rank one, while we call a covector also a covariant vector or covariant tensor of rank one. A general tensor of rank (n, m) is a multilinear map

$$\boldsymbol{T} = T^{\mu,\dots,\nu}_{\alpha,\dots,\beta} \underbrace{\partial_\mu \otimes \dots \otimes \partial_\nu}_{n} \otimes \underbrace{\mathrm{d}x^\alpha \otimes \dots \otimes \mathrm{d}x^\beta}_{m} \tag{6.14}$$

which components transforms as

$$\tilde{T}^{\mu,\dots,\nu}_{\alpha,\dots,\beta}(\tilde{x}) = \underbrace{\frac{\partial \tilde{x}^\mu}{\partial x^\rho} \dots \frac{\partial \tilde{x}^\nu}{\partial x^\sigma}}_{n} \underbrace{\frac{\partial x^\gamma}{\partial \tilde{x}^\alpha} \dots \frac{\partial x^\delta}{\partial \tilde{x}^\beta}}_{m} T^{\rho,\dots,\sigma}_{\gamma,\dots,\delta}(x) \tag{6.15}$$

under a coordinate change.

Metric tensor. A (pseudo-) Riemannian manifold is a differentiable manifold containing as additional structure a symmetric tensor field $g_{\mu\nu}$ which allows us to measure distances and angles. We define the scalar product of two vectors $\boldsymbol{a}(x)$ and $\boldsymbol{b}(x)$ which have the coordinates a^μ and b^μ in a certain basis $\boldsymbol{e}_\mu$ as

$$\boldsymbol{a} \cdot \boldsymbol{b} = (a^\mu \boldsymbol{e}_\mu) \cdot (b^\nu \boldsymbol{e}_\nu) = (\boldsymbol{e}_\mu \cdot \boldsymbol{e}_\nu) a^\mu b^\nu = g_{\mu\nu} a^\mu b^\nu. \tag{6.16}$$

Thus we can evaluate the scalar product between any two vectors, if we know the symmetric matrix $g_{\mu\nu}$ composed of the N^2 products of the basis vectors,

$$g_{\mu\nu}(x) = \boldsymbol{e}_\mu(x) \cdot \boldsymbol{e}_\nu(x), \tag{6.17}$$

at any point x of the manifold. This symmetric matrix $g_{\mu\nu}$ is called *metric tensor*. The manifold is called Riemannian if all eigenvalues of $g_{\mu\nu}$ are positive, and thus the scalar product defined by $g_{\mu\nu}$ is positive-definite. If the scalar product is indefinite, as in the case of general relativity, one calls the manifold pseudo-Riemannian.

In the same way, we define for the dual basis e^μ the metric $g^{\mu\nu}$ via

$$g^{\mu\nu} = e^\mu \cdot e^\nu. \tag{6.18}$$

A comparison with Eq. (6.10) shows that the metric $g^{\mu\nu}$ maps covariant vectors X_μ into contravariant vectors X^μ, while $g_{\mu\nu}$ provides a map into the opposite direction. Similarly, we can use the metric tensor to raise and lower indices of any tensor.

Next we want to determine the relation of $g^{\mu\nu}$ with $g_{\mu\nu}$. We multiply e^ρ with $e_\mu = g_{\mu\nu} e^\nu$, obtaining

$$\delta^\rho_\mu = e^\rho \cdot e_\mu = e^\rho \cdot g_{\mu\nu} e^\nu = g^{\rho\nu} g_{\mu\nu} \tag{6.19}$$

or

$$\delta^\rho_\mu = g_{\mu\nu} g^{\nu\rho}. \tag{6.20}$$

Thus the components of the covariant and the contravariant metric tensors, $g_{\mu\nu}$ and $g^{\mu\nu}$, are inverse matrices of each other. Moreover, the mixed metric tensor of rank $(1,1)$ is given by the Kronecker delta, $g^\nu_\mu = \delta^\nu_\mu$. Note that the trace of the metric tensor is therefore not -2, but

$$\mathrm{tr}(g_{\mu\nu}) = g^{\mu\mu} g_{\mu\mu} = \delta^\mu_\mu = 4, \tag{6.21}$$

because we have to contract an upper and a lower index.

6.2 Covariant derivative and the geodesic equation

Covariant derivative. In an inertial system in Minkowski space, taking the partial derivative ∂_μ maps a tensor of rank (n, m) into a tensor of rank $(n, m+1)$. Additionally, this map obeys linearity and the Leibniz product rule. We will see that in general the partial derivative in a curved space does not satisfy these rules. We therefore introduce a new derivative called covariant derivative, modified such that it fulfils these rules.

We start by considering the gradient $\partial_\mu \phi$ of a scalar ϕ. By definition, a scalar quantity does not depend on the coordinate system, $\phi(x) = \tilde\phi(\tilde x)$. Therefore its gradient transforms as

$$\partial_\mu \phi \to \tilde\partial_\mu \tilde\phi = \frac{\partial x^\nu}{\partial \tilde x^\mu} \partial_\nu \phi. \tag{6.22}$$

Thus the gradient is a covariant vector. Similarly, the derivative of a vector V transforms as a tensor,

$$\partial_\mu V \to \tilde\partial_\mu \tilde V = \frac{\partial x^\nu}{\partial \tilde x^\mu} \partial_\nu V, \tag{6.23}$$

because V is an invariant quantity. If we consider however its components $V^\mu = e^\mu \cdot V$, then the moving coordinate basis in curved spacetime, $\partial_\mu e^\nu \neq 0$, leads to an additional term in the derivative,

$$\partial_\mu V^\nu = e^\nu \cdot (\partial_\mu V) + V \cdot (\partial_\mu e^\nu). \tag{6.24}$$

The term $e^\nu \cdot (\partial_\mu V)$ transforms as a tensor, since both e^ν and $\partial_\mu V$ are tensors. This implies that the combination of the two remaining terms has to transform as tensor too, which we define as covariant derivative

$$\nabla_\mu V^\nu \equiv e^\nu \cdot (\partial_\mu \boldsymbol{V}) = \partial_\mu V^\nu - \boldsymbol{V} \cdot (\partial_\mu e^\nu). \tag{6.25}$$

The first equality tells us that we can view the covariant derivative $\nabla_\mu V^\nu$ as the projection of $\partial_\mu \boldsymbol{V}$ onto the direction e^ν.

We now expand the partial derivatives of the basis vectors as a linear combination of the basis vectors,

$$\partial_\rho e^\mu = -\Gamma^\mu{}_{\rho\sigma} e^\sigma. \tag{6.26}$$

The n^3 numbers $\Gamma^\mu{}_{\rho\sigma}$ are called (affine) connection coefficients or symbols in order to stress that they are not the components of a tensor. You are asked to derive their transformation properties in problem 6.5. Introducing this expansion into (6.25) we can rewrite the covariant derivative of a vector field as

$$\nabla_\mu V^\nu = \partial_\mu V^\nu + \Gamma^\nu{}_{\sigma\mu} V^\sigma. \tag{6.27}$$

Using $\nabla_\sigma \phi = \partial_\sigma \phi$ and requiring that the usual Leibniz rule is valid for $\phi = X_\mu X^\mu$ leads to

$$\nabla_\sigma X_\mu = \partial_\sigma X_\mu - \Gamma^\nu{}_{\mu\sigma} X_\nu \tag{6.28}$$

and to

$$\partial_\rho e_\mu = \Gamma^\sigma{}_{\mu\rho} e_\sigma. \tag{6.29}$$

For a general tensor, the covariant derivative is defined by the same reasoning as

$$\nabla_\sigma T^{\mu\cdots}{}_{\nu\cdots} = \partial_\sigma T^{\mu\cdots}{}_{\nu\cdots} + \Gamma^\mu{}_{\rho\sigma} T^{\rho\cdots}{}_{\nu\cdots} + \ldots - \Gamma^\rho{}_{\nu\sigma} T^{\mu\cdots}{}_{\rho\cdots} - \ldots \tag{6.30}$$

Note that it is the last index of the connection coefficients that is the same as the index of the covariant derivative. The *plus* sign goes together with upper (superscripts), the minus with lower indices.

Parallel transport. We say a tensor $\boldsymbol{T}$ is parallel transported along the curve $x(\sigma)$, if its components $T^{\mu\cdots}{}_{\nu\cdots}$ stay constant. In flat space, this means simply

$$\frac{\mathrm{d}}{\mathrm{d}\sigma} T^{\mu\cdots}{}_{\nu\cdots} = \frac{\mathrm{d}x^\alpha}{\mathrm{d}\sigma} \partial_\alpha T^{\mu\cdots}{}_{\nu\cdots} = 0. \tag{6.31}$$

In curved space, we have to replace the normal derivative by a covariant one. We define the directional covariant derivative along $x(\sigma)$ as

$$\frac{D}{\mathrm{d}\sigma} = \frac{\mathrm{d}x^\alpha}{\mathrm{d}\sigma} \nabla_\alpha. \tag{6.32}$$

Then a tensor is parallel transported along the curve $x(\sigma)$, if

$$\frac{D}{\mathrm{d}\sigma} T^{\mu\cdots}{}_{\nu\cdots} = \frac{\mathrm{d}x^\alpha}{\mathrm{d}\sigma} \nabla_\alpha T^{\mu\cdots}{}_{\nu\cdots} = 0. \tag{6.33}$$

Metric compatibility. Relations like $ds^2 = g_{\mu\nu}dx^\mu dx^\nu$ or $g_{\mu\nu}p^\mu p^\nu = m^2$ become invariant under parallel transport only, if the metric tensor is covariantly constant,

$$\nabla_\sigma g_{\mu\nu} = \nabla_\sigma g^{\mu\nu} = 0. \tag{6.34}$$

A connection satisfying Eq. (6.34) is called metric compatible and leaves lengths and angles invariant under parallel transport. This requirement guarantees that the local Cartesian inertial coordinate systems where the laws of special relativity are valid can be consistently connected by parallel transport using an affine connection satisfying the constraint (6.34). Note that we have already built in this constraint into our definition of the covariant derivative. If the length of a vector would not be conserved under parallel transport, then we should differentiate in (6.24) also the scalar product in $V^\mu = e^\mu \cdot \boldsymbol{V}$ leading to an additional term in Eq. (6.25).

Geodesic equation. The requirement that the affine connection is metric compatible does not fix the connection uniquely, and thus the question arises, which connection describes physics on a general spacetime? Ultimately, the combined action for gravity and matter should select the correct connection—an approach we resume in chapter 19. For the moment, we use a simple workaround which does not require the knowledge of the action of gravity. In flat space, we know that the solution to the equations of motion of a free particle is a straight line. Such a path is characterised by two properties: it is the shortest curve between the considered initial and final point, and it is the curve whose tangent vector remains constant if they are parallel transported along it. Both conditions can be generalised to curved space and the curves satisfying either one of them are called geodesics. Using the definition of a geodesic as the "straightest" line on a manifold requires as mathematical structure only the possibility to parallel transporting a tensor and thus the existence of an affine connection. In contrast, the concept of an "extremal" (shortest or longest) line between two points on a manifold relies on the existence of a metric. Requiring that these two definitions agree fixes uniquely the connection to be used in the covariant derivative.

We start by defining geodesics as the "straightest" line or an autoparallel curve on a manifold—a case which is almost trivial. The tangent vector along the path $x(\tau)$ is $u^\mu = dx^\mu/d\tau$. Then the requirement (6.33) of parallel transport for u^μ becomes

$$\frac{D}{d\tau}\frac{dx^\mu}{d\tau} = \frac{d^2x^\mu}{d\tau^2} + \Gamma^\mu{}_{\rho\sigma}\frac{dx^\rho}{d\tau}\frac{dx^\sigma}{d\tau} = 0. \tag{6.35}$$

Introducing $\dot{x}^\mu = dx^\mu/d\tau$, we obtain the geodesic equation in its standard form,

$$\ddot{x}^\mu + \Gamma^\mu{}_{\rho\sigma}\dot{x}^\rho\dot{x}^\sigma = 0. \tag{6.36}$$

Note that a possible antisymmetric part of the connection $\Gamma^\mu{}_{\rho\sigma}$ drops out of the geodesic equation, because $\dot{x}^\rho\dot{x}^\sigma$ is symmetric.

Next we derive the defining equation for a geodesics as the extremal curve between two points on a manifold. The Lagrangian of a free particle in Minkowski space, Eq. (1.53), is generalised to a curved spacetime manifold with the metric tensor $g_{\mu\nu}$ by replacing $\eta_{\mu\nu}$ with $g_{\mu\nu}$ (we set also $m = -1$),

$$L = g_{\mu\nu}\dot{x}^\mu \dot{x}^\nu. \tag{6.37}$$

The Lagrange equations are

$$\frac{\mathrm{d}}{\mathrm{d}\sigma}\frac{\partial L}{\partial(\dot{x}^\lambda)} - \frac{\partial L}{\partial x^\lambda} = 0. \tag{6.38}$$

Only the metric tensor $g_{\mu\nu}$ depends on x^μ and thus $\partial L/\partial x^\lambda = g_{\mu\nu,\lambda}\dot{x}^\mu\dot{x}^\nu$. Here we also introduced the shorthand notation $g_{\mu\nu,\lambda} = \partial_\lambda g_{\mu\nu}$ for partial derivatives. Now we use $\partial \dot{x}^\mu/\partial \dot{x}^\nu = \delta^\mu_\nu$ and apply the chain rule for $g_{\mu\nu}(x(\sigma))$, obtaining first

$$g_{\mu\nu,\lambda}\dot{x}^\mu\dot{x}^\nu = 2\frac{\mathrm{d}}{\mathrm{d}\sigma}(g_{\mu\lambda}\dot{x}^\mu) = 2(g_{\mu\lambda,\nu}\dot{x}^\mu\dot{x}^\nu + g_{\mu\lambda}\ddot{x}^\mu) \tag{6.39}$$

and then

$$g_{\mu\lambda}\ddot{x}^\mu + \frac{1}{2}(2g_{\mu\lambda,\nu} - g_{\mu\nu,\lambda})\dot{x}^\mu\dot{x}^\nu = 0. \tag{6.40}$$

Next we rewrite the second term as

$$2g_{\lambda\mu,\nu}\dot{x}^\mu\dot{x}^\nu = (g_{\lambda\mu,\nu} + g_{\lambda\nu,\mu})\dot{x}^\mu\dot{x}^\nu, \tag{6.41}$$

multiply everything by $g^{\kappa\mu}$ and arrive at our desired result,

$$\ddot{x}^\kappa + \frac{1}{2}g^{\kappa\lambda}(g_{\mu\nu,\lambda} + g_{\mu\lambda,\nu} - g_{\mu\nu,\lambda})\dot{x}^\mu\dot{x}^\nu = \ddot{x}^\kappa + \{^\kappa_{\mu\nu}\}\dot{x}^\mu\dot{x}^\nu = 0. \tag{6.42}$$

In the last step, we defined the Christoffel symbols

$$\{^\mu_{\nu\lambda}\} = \frac{1}{2}g^{\mu\kappa}(\partial_\nu g_{\kappa\lambda} + \partial_\lambda g_{\nu\kappa} - \partial_\kappa g_{\nu\lambda}). \tag{6.43}$$

They are also called Levi–Civita or Riemannian connection. A comparison with Eq. (6.36) shows that our two geodesic equations agree if we choose the Christoffel symbols as the connection. Moreover, the Christoffel symbols are symmetric in their two lower indices and, as we will show next, compatible to the metric tensor. Following standard practice, we will denote them also with $\Gamma^\lambda{}_{\mu\nu}$. In the remainder of this section, we will always use the Christoffel symbols as the affine connection.

We define[1]

$$\Gamma_{\mu\nu\lambda} = g_{\mu\kappa}\Gamma^\kappa{}_{\nu\lambda}. \tag{6.44}$$

Thus $\Gamma_{\mu\nu\lambda}$ is symmetric in the last two indices. Then it follows

$$\Gamma_{\mu\nu\lambda} = \frac{1}{2}(\partial_\nu g_{\mu\lambda} + \partial_\lambda g_{\nu\mu} - \partial_\mu g_{\nu\lambda}). \tag{6.45}$$

Adding $2\Gamma_{\mu\nu\lambda}$ and $2\Gamma_{\nu\mu\lambda}$ gives

$$\begin{aligned}
2(\Gamma_{\mu\nu\lambda} + \Gamma_{\nu\mu\lambda}) &= \partial_\nu g_{\mu\lambda} + \partial_\lambda g_{\nu\mu} - \partial_\mu g_{\nu\lambda} \\
&+ \partial_\mu g_{\nu\lambda} + \partial_\lambda g_{\mu\nu} - \partial_\nu g_{\mu\lambda} = 2\partial_\lambda g_{\mu\nu}
\end{aligned} \tag{6.46}$$

---

[1]We showed that the metric tensor can be used to raise or to lower tensor indices, but the connection Γ is not a tensor.

or

$$\partial_\lambda g_{\mu\nu} = \Gamma_{\mu\nu\lambda} + \Gamma_{\nu\mu\lambda}\,. \tag{6.47}$$

Applying the general rule for covariant derivatives, Eq. (6.30), to the metric,

$$\nabla_\lambda g_{\mu\nu} = \partial_\lambda g_{\mu\nu} - \Gamma^\kappa{}_{\mu\lambda} g_{\kappa\nu} - \Gamma^\kappa{}_{\nu\lambda} g_{\mu\kappa} = \partial_\lambda g_{\mu\nu} - \Gamma_{\nu\mu\lambda} - \Gamma_{\mu\nu\lambda}\,, \tag{6.48}$$

and inserting Eq. (6.47) shows that

$$\nabla_\lambda g_{\mu\nu} = \nabla_\lambda g^{\mu\nu} = 0\,. \tag{6.49}$$

Hence ∇_λ commutes with contracting indices,

$$\nabla_\lambda(X^\mu X_\mu) = \nabla_\lambda(g_{\mu\nu} X^\mu X^\mu) = g_{\mu\nu} \nabla_\lambda(X^\mu X^\nu) \tag{6.50}$$

and conserves the norm of vectors as announced. Thus the Christoffel symbols are symmetric and compatible with the metric. These two properties specify uniquely the connection.

Example 6.1: Calculate the Christoffel symbols of the two-dimensional unit sphere S^2.

The line element of the two-dimensional unit sphere S^2 is given by $ds^2 = d\vartheta^2 + \sin^2\vartheta\, d\phi^2$. A faster alternative to the definition (6.43) of the Christoffel coefficients is the use of the geodesic equation: From the Lagrange function $L = g_{ab}\dot{x}^a\dot{x}^b = \dot{\vartheta}^2 + \sin^2\vartheta\dot{\phi}^2$ we find

$$\frac{\partial L}{\partial \phi} = 0\,, \qquad \frac{d}{dt}\frac{\partial L}{\partial \dot{\phi}} = \frac{d}{dt}(2\sin^2\vartheta\dot{\phi}) = 2\sin^2\vartheta\ddot{\phi} + 4\cos\vartheta\sin\vartheta\dot{\vartheta}\dot{\phi}\,,$$

$$\frac{\partial L}{\partial \vartheta} = 2\cos\vartheta\sin\vartheta\dot{\phi}^2\,, \qquad \frac{d}{dt}\frac{\partial L}{\partial \dot{\vartheta}} = \frac{d}{dt}(2\dot{\vartheta}) = 2\ddot{\vartheta}\,,$$

and thus the Lagrange equations are

$$\ddot{\phi} + 2\cot\vartheta\dot{\vartheta}\dot{\phi} = 0 \qquad \text{and} \qquad \ddot{\vartheta} - \cos\vartheta\sin\vartheta\dot{\phi}^2 = 0\,.$$

Comparing with the geodesic equation $\ddot{x}^\kappa + \Gamma^\kappa{}_{\mu\nu}\dot{x}^\mu\dot{x}^\nu = 0$, we can read off the non-vanishing Christoffel symbols as $\Gamma^\phi{}_{\vartheta\phi} = \Gamma^\phi{}_{\phi\vartheta} = \cot\vartheta$ and $\Gamma^\vartheta{}_{\phi\phi} = -\cos\vartheta\sin\vartheta$. (Note that $2\cot\vartheta = \Gamma^\phi{}_{\vartheta\phi} + \Gamma^\phi{}_{\phi\vartheta}$.)

We can use also the Hamiltonian formulation for a relativistic particle. From the Lagrangian $L = \frac{1}{2}g_{\mu\nu}\dot{x}^\mu\dot{x}^\nu$ we determine first the conjugated momenta $p_\mu = \partial L/\partial\dot{x}^\mu = \dot{x}_\mu$ and then perform a Legendre transformation,

$$H(x^\mu, p_\mu, \tau) = p_\mu \dot{x}^\mu - L(x^\mu, \dot{x}^\mu, \tau) = \frac{1}{2}g^{\mu\nu}p_\mu p_\nu\,. \tag{6.51}$$

Since the Lagrangian of a free particle does not depend explicitly on the evolution parameter σ, there exists at least one conserved quantity. This conservation law, $H = $

1/2, expresses the fact that the tangent vector $\dot{x}^\mu$ has a constant norm. Hamilton equations then give

$$\dot{x}^\mu = \frac{\partial H}{\partial p_\mu} = g^{\mu\nu} p_\nu \tag{6.52}$$

and

$$\dot{p}_\mu = -\frac{\partial H}{\partial x^\mu} = -\frac{1}{2}\frac{\partial g^{\alpha\beta}}{\partial x^\mu} p_\alpha p_\beta. \tag{6.53}$$

This is a useful alternative to the standard geodesic equation. First, it makes clear that the momentum component p_μ is conserved, if the metric tensor is independent of the coordinate x^μ. Second, we can calculate $\dot{p}_\mu$ directly from the metric tensor, without knowing the Christoffel symbols. Combining the Eqs. (6.52) and (6.53) one can re-derive the standard form of the geodesic equation, cf. problem 6.7.

6.3 Integration and Gauss' theorem

Having defined the covariant derivative of an arbitrary tensor field, it is natural to ask how the inverse, the integral over a tensor field, can be defined. The short answer is that this is in general impossible. Integrating a tensor field requires summing tensors at different points in an invariant way, which is only possible for scalars. Restricting ourselves to scalar fields, we should generalise an integral like $I = \int d^4x\, \phi(x)$ valid in an Cartesian inertial frame x^μ in Minkowski space to a general spacetime with coordinates $\tilde{x}$. For a general coordinate transformation $x^\mu \rightarrow \tilde{x}^\mu$, we have to take into account that the Jacobi determinant $J = \det(\partial \tilde{x}^\mu / \partial x^\rho)$ of the transformation can deviate from one. We can express this Jacobian by the determinant $g \equiv \det(g_{\mu\nu})$ of the metric tensor as follows. Applying the transformation law of the metric tensor,

$$\tilde{g}^{\mu\nu}(\tilde{x}) = \frac{\partial \tilde{x}^\mu}{\partial x^\rho}\frac{\partial \tilde{x}^\nu}{\partial x^\sigma}\, g^{\rho\sigma}(x) \tag{6.54}$$

to the case where the x^ρ are inertial coordinates, we obtain with $g = \det(\eta_{\mu\nu}) = -1$ that

$$\det(\tilde{g}) = J^2 \det(g) = -J^2 \tag{6.55}$$

or $J = \sqrt{|\tilde{g}|}$. Thus $I = \int d^4x \sqrt{|g|}\, \phi$ is an invariant definition of an integral over a scalar field which agrees for inertial coordinates with the one known from special relativity. Now we choose as scalar ϕ the divergence of a vector field, $\phi = \nabla_\mu X^\mu$, or

$$I = \int d^4x \sqrt{|g|}\, \nabla_\mu X^\mu = \int d^4x \sqrt{|g|}\, \left(\partial_\mu X^\mu + \Gamma^\mu{}_{\lambda\mu} X^\lambda\right). \tag{6.56}$$

Our aim is to generalise Gauss' theorem (5.9). The only way that this theorem may be reconciled with (6.56) is to hope that we can express the covariant divergence as $1/\sqrt{|g|}\partial_\mu(\sqrt{|g|}X^\mu)$. In order to check this possibility, we determine first the partial derivative of the metric determinant g. As preparation, we consider the variation of a general matrix M with elements $m_{ij}(x)$ under an infinitesimal change of the coordinates, $\delta x^\mu = \varepsilon x^\mu$. It is convenient to look at the change of $\ln \det M$,

$$\delta \ln \det M \equiv \ln \det(M + \delta M) - \ln \det(M) \tag{6.57a}$$

$$= \ln \det[M^{-1}(M + \delta M)] = \ln \det[1 + M^{-1}\delta M] = \tag{6.57b}$$

$$= \ln[1 + \mathrm{tr}(M^{-1}\delta M)] + \mathcal{O}(\varepsilon^2) = \mathrm{tr}(M^{-1}\delta M) + \mathcal{O}(\varepsilon^2). \tag{6.57c}$$

In the last step, we used $\ln(1 + \varepsilon) = \varepsilon + \mathcal{O}(\varepsilon^2)$. Expressing now both the LHS and the RHS as $\delta f = \partial_\mu f \delta x^\mu$ and comparing then the coefficients of δx^μ gives

$$\partial_\mu \ln \det M = \mathrm{tr}(M^{-1}\partial_\mu M). \tag{6.58}$$

Applied to derivatives of $\sqrt{|g|}$, we obtain

$$\frac{1}{2}g^{\mu\nu}\partial_\lambda g_{\mu\nu} = \frac{1}{2}\partial_\lambda \ln g = \frac{1}{\sqrt{|g|}}\partial_\lambda(\sqrt{|g|}). \tag{6.59}$$

This expression coincides with contracted Christoffel symbols,

$$\Gamma^\mu{}_{\mu\nu} = \frac{1}{2}g^{\mu\kappa}(\partial_\mu g_{\kappa\nu} + \partial_\nu g_{\mu\kappa} - \partial_\kappa g_{\mu\nu}) = \frac{1}{2}g^{\mu\kappa}\partial_\nu g_{\mu\kappa} = \frac{1}{2}\partial_\nu \ln g = \frac{1}{\sqrt{|g|}}\partial_\nu(\sqrt{|g|}). \tag{6.60}$$

Now we can express the divergence of a vector field as

$$\nabla_\mu X^\mu = \partial_\mu X^\mu + \Gamma^\mu{}_{\lambda\mu}X^\lambda = \partial_\mu X^\mu + \frac{1}{\sqrt{|g|}}(\partial_\mu\sqrt{|g|})X^\mu = \frac{1}{\sqrt{|g|}}\partial_\mu(\sqrt{|g|}X^\mu). \tag{6.61}$$

Gauss' theorem for the divergence of a vector field follows directly,

$$\int_\Omega \mathrm{d}^4x\sqrt{|g|}\,\nabla_\mu X^\mu = \int_\Omega \mathrm{d}^4x\,\partial_\mu(\sqrt{|g|}X^\mu) = \int_{\partial\Omega} \mathrm{d}S_\mu\sqrt{|g|}\,X^\mu. \tag{6.62}$$

This implies in particular that we can drop terms like $\nabla_\mu X^\mu$ in the action, if the vector field X^μ vanishes on the boundary. Similarly, Gauss' theorem allows us to derive global conservation laws from $\nabla_\mu X^\mu = 0$ in the same way as in Minkowski space.

Next we consider the divergence of an antisymmetric tensor of rank two,

$$\nabla_\mu A^{\mu\nu} = \partial_\mu A^{\mu\nu} + \Gamma^\mu{}_{\lambda\mu}A^{\lambda\nu} + \Gamma^\nu{}_{\lambda\mu}A^{\mu\lambda} = \frac{1}{\sqrt{|g|}}\partial_\mu(\sqrt{|g|}A^{\mu\nu}). \tag{6.63}$$

Because of the antisymmetry of $A^{\mu\lambda}$ the term $\Gamma^\nu{}_{\lambda\mu}A^{\mu\lambda}$ vanishes, and we can combine the first two terms as in the vector case. This generalises to completely antisymmetric tensors of all orders. In contrast, we find for a symmetric tensor of rank two,

$$\nabla_\mu S^{\mu\nu} = \partial_\mu S^{\mu\nu} + \Gamma^\mu{}_{\lambda\mu}S^{\lambda\nu} + \Gamma^\nu{}_{\lambda\mu}S^{\mu\lambda} = \frac{1}{\sqrt{|g|}}\,\partial_\mu(\sqrt{|g|}S^{\mu\nu}) + \Gamma^\nu{}_{\lambda\mu}S^{\mu\lambda}. \tag{6.64}$$

Hence the divergence of a symmetric tensor of rank two contains an additional term $(\partial_\nu g_{\mu\lambda})S^{\mu\lambda}$ which prohibits the use of Gauss' theorem.

6.4 Symmetries of a general spacetime

In the case of a Riemannian spacetime manifold $(\mathcal{M}, g)$, we say the spacetime posseses a symmetry if it looks the same as one moves from a point P along a vector field ξ^μ to a different point $\tilde{P}$. More precisely, we mean with "looking the same" that the metric tensor transported along ξ^μ remains the same.

Such symmetries may be obvious, if one uses coordinates adapted to these symmetries. For instance, the metric may be independent from one or several coordinates. Let us assume, for example, that the metric is independent from the time coordinate x^0. Then x^0 is a cyclic coordinate, $\partial L/\partial x^0 = 0$, of the Lagrangian $L = \mathrm{d}\tau/\mathrm{d}\sigma$ of a free test particle moving in $\mathcal{M}$. With $L = \mathrm{d}\tau/\mathrm{d}\sigma$, the resulting conserved quantity $\partial L/\partial \dot{x}^0 = \mathrm{const.}$ can be written as

$$\frac{\partial L}{\partial \dot{x}^0} = g_{0\beta} \frac{\mathrm{d}x^\beta}{L\mathrm{d}\sigma} = g_{0\beta} \frac{\mathrm{d}x^\beta}{\mathrm{d}\tau} = \boldsymbol{\xi} \cdot \boldsymbol{u} \tag{6.65}$$

with $\boldsymbol{\xi} = \boldsymbol{e}_0$ and $\boldsymbol{u}$ as the four-velocity. Hence the quantity $\boldsymbol{\xi} \cdot \boldsymbol{u} = p^0/m$ is conserved along the solutions $x^\alpha(\sigma)$ of the Lagrange equation of a free particle on $\mathcal{M}$, that is along geodesics. In other words, the motion of all test particles in the corresponding spacetime conserve energy. The vector field $\boldsymbol{\xi}$ that points in the direction in which the metric does not change is called a Killing vector field.

Since we allow arbitrary coordinate systems, spacetime symmetries are, however, in general not evident by a simple inspection of the metric tensor. We say the metric is invariant moving along the Killing vector field ξ^μ, when the resulting change $\delta g^{\mu\nu}$ of the metric is zero. In order to use this condition, we have to be able to calculate how the tensor $g^{\mu\nu}$ changes as we transport it along a vector field ξ^μ. Clearly, it is sufficient to consider an infinitesimal distance. Then we can work in the approximation

$$\tilde{x}^\mu = x^\mu + \varepsilon \xi^\mu(x^\nu) + \mathcal{O}(\varepsilon^2), \qquad \varepsilon \ll 1, \tag{6.66}$$

and neglect all terms quadratic in ε.

We recall first the transformation law for a rank two tensor as the metric under an arbitrary coordinate transformation,

$$\tilde{g}^{\mu\nu}(\tilde{x}) = \frac{\partial \tilde{x}^\mu}{\partial x^\alpha} \frac{\partial \tilde{x}^\nu}{\partial x^\beta} g^{\alpha\beta}(x). \tag{6.67}$$

Applied to the transport along $\boldsymbol{\xi}$ defined in (6.66), we obtain

$$\tilde{g}^{\mu\nu}(\tilde{x}) = \frac{\partial \tilde{x}^\mu}{\partial x^\alpha} \frac{\partial \tilde{x}^\nu}{\partial x^\beta} g^{\alpha\beta}(x) = (\delta^\mu_\alpha + \varepsilon \xi^\mu_{,\alpha})(\delta^\nu_\beta + \varepsilon \xi^\nu_{,\beta}) g^{\alpha\beta}(x) \tag{6.68a}$$

$$= g^{\mu\nu}(x) + \varepsilon(\xi^{\mu,\nu} + \xi^{\nu,\mu}) + \mathcal{O}(\varepsilon^2). \tag{6.68b}$$

In order to be able to compare the new $\tilde{g}^{\mu\nu}(\tilde{x})$ with $g^{\mu\nu}(x)$, we have to express $\tilde{g}^{\mu\nu}(\tilde{x})$ as function of x. A Taylor expansion gives

$$\tilde{g}^{\mu\nu}(\tilde{x}) = \tilde{g}^{\mu\nu}(x + \varepsilon\xi) = \tilde{g}^{\mu\nu}(x) + \varepsilon \xi^\alpha \partial_\alpha \tilde{g}^{\mu\nu}(x) + \mathcal{O}(\varepsilon^2). \tag{6.69}$$

Setting equal Eq. (6.68b) and (6.69), we obtain

$$g^{\mu\nu}(x) + \varepsilon(\xi^{\mu,\nu} + \xi^{\nu,\mu}) = \tilde{g}^{\mu\nu}(x) + \varepsilon\xi^{\alpha}\partial_{\alpha}\tilde{g}^{\mu\nu}(x). \tag{6.70}$$

Thus the metric is kept invariant, if the condition

$$\xi^{\mu,\nu} + \xi^{\nu,\mu} - \xi^{\rho}\partial_{\rho}g^{\mu\nu} = 0 \tag{6.71}$$

or

$$g^{\mu\rho}\partial_{\rho}\xi^{\nu} + g^{\nu\rho}\partial_{\rho}\xi^{\mu} - \xi^{\rho}\partial_{\rho}g^{\mu\nu} = 0 \tag{6.72}$$

is satisfied. Expressing partial derivatives as covariant ones,[2] the terms containing connection coefficients cancel and we obtain the Killing equation

$$\nabla_{\mu}\xi_{\nu} + \nabla_{\nu}\xi_{\mu} = 0. \tag{6.73}$$

Its solutions $\boldsymbol{\xi}$ are the Killing vectors of the metric.

We now check that Eq. (6.73) leads to a conservation law, as required by our initial definition of a Killing vector field. We multiply the equation for geodesic motion,

$$\frac{Du^{\mu}}{d\tau} = 0, \tag{6.74}$$

by the Killing vector ξ_{μ} and use Leibniz's product rule together with the definition of the absolute derivative (6.32),

$$\xi_{\mu}\frac{Du^{\mu}}{d\tau} = \frac{d}{d\tau}(\xi_{\mu}u^{\mu}) - \nabla_{\mu}\xi_{\nu}u^{\mu}u^{\nu} = 0. \tag{6.75}$$

The second term vanishes for a Killing vector field ξ^{μ}, because the Killing equation implies the antisymmetry of $\nabla_{\mu}\xi_{\nu}$. Hence the quantity $\xi_{\mu}u^{\mu}$ is indeed conserved along any geodesics.

Example 6.2: Find all ten Killing vector fields of Minkowski space and specify the corresponding symmetries and conserved quantities.
The Killing equation $\nabla_{\mu}\xi_{\nu} + \nabla_{\nu}\xi_{\mu} = 0$ simplifies in Minkowski space to

$$\partial_{\mu}\xi_{\nu} + \partial_{\nu}\xi_{\mu} = 0. \tag{6.76}$$

Taking one more derivative and using the symmetry of partial derivatives, we arrive at

$$\partial_{\rho}\partial_{\mu}\xi_{\nu} + \partial_{\rho}\partial_{\nu}\xi_{\mu} = 2\partial_{\mu}\partial_{\rho}\xi_{\nu} = 0. \tag{6.77}$$

Integrating this equation twice, we find

$$\xi^{\mu} = \omega_{\nu}{}^{\mu}x^{\nu} + a^{\mu}. \tag{6.78}$$

The matrix $\omega^{\mu\nu}$ has to be antisymmetric in order to satisfy Eq. (6.76). Thus the Killing vector fields are determined by ten integration constants. They agree with the infinites-

---

[2]Since Eq. (6.73) is tensor equation, the previous Eq. (6.72) is also invariant under arbitrary coordinate transformations, although it contains only partial derivatives. This suggests that one can introduce the derivative of an arbitrary tensor along a vector field, called Lie derivative, without the need for a connection.

imal generators of Lorentz transformations, cf. appendix B.3.

The four parameters a^μ generate translations, $x^\mu \to x^\mu + a^\mu$, described by four Killing vector fields which can be chosen as the Cartesian basis vectors of Minkowski space,

$$\boldsymbol{T}_0 = \partial_t, \qquad \boldsymbol{T}_1 = \partial_x, \qquad \boldsymbol{T}_2 = \partial_y, \qquad \boldsymbol{T}_3 = \partial_z.$$

For a particle with momentum $p^\mu = mu^\mu$ moving along $x^\mu(\lambda)$, the existence of a Killing vector $\boldsymbol{T}^\mu$ implies

$$\frac{\mathrm{d}}{\mathrm{d}\lambda}(\boldsymbol{T}^\mu \cdot \boldsymbol{u}) = \frac{\mathrm{d}}{m\mathrm{d}\lambda}(\boldsymbol{T}^\mu \cdot \boldsymbol{p}) = 0,$$

i.e. the conservation of the four-momentum component p_μ.

Consider next the ij (=spatial) components of the Killing equation. Three additional Killing vectors are

$$\boldsymbol{J}_1 = y\partial_z - z\partial_y, \qquad \text{and cyclic permutations.} \tag{6.79}$$

The existence of Killing vectors $\boldsymbol{J}_i$ implies that $\boldsymbol{J}_i \cdot \boldsymbol{p}$ is conserved along a geodesics of particle. But

$$\boldsymbol{J}_1 \cdot \boldsymbol{p} = yp_z - zp_y = J_x$$

and thus the angular momentum around the origin of the coordinate system is conserved. The other three components satisfy the 0α component of the Killing equations ($\omega_1{}^0 = \omega_0{}^1$),

$$\boldsymbol{K}_1 = t\partial_z + z\partial_t, \qquad \text{and cyclic permutations.} \tag{6.80}$$

The conserved quantity $tp_z - zE = \text{const.}$ now depends on time and is therefore not as popular as the previous ones. Its conservation implies that the centre-of-mass of a system of particles moves with constant velocity $v_\alpha = p_\alpha/E$.

Global conservation laws. An immediate consequence of Eq. (6.61) is a covariant form of Gauss' theorem for vector fields. In particular, we can conclude from local current conservation, $\nabla_\mu j^\mu = 0$, the existence of a globally conserved charge. If the conserved current j^μ vanishes at infinity, then we obtain also in a general spacetime

$$\int_\Omega \mathrm{d}^4x \sqrt{|g|}\,\nabla_\mu j^\mu = \int_\Omega \mathrm{d}^4x \partial_\mu(\sqrt{|g|}j^\mu) = \int_{\partial\Omega} \mathrm{d}S_\mu \sqrt{|g|}\, j^\mu = 0. \tag{6.81}$$

Thus the conservation of Noether charges of internal symmetries as the electric charge, baryon number, etc., continues to hold in a curved spacetime.

Next we consider the energy–momentum stress tensor as an example of a locally conserved symmetric tensors of rank two. Now, the second term in Eq. (6.64) prevents us to convert the local conservation law into a global one. If the spacetime admits however a Killing field ξ_μ, then we can form the vector field $P^\mu = T^{\mu\nu}\xi_\nu$ with

$$\nabla_\mu P^\mu = \nabla_\mu(T^{\mu\nu}\xi_\nu) = \xi_\nu \nabla_\mu T^{\mu\nu} + T^{\mu\nu}\nabla_\mu\xi_\nu = 0. \tag{6.82}$$

Here, the first term vanishes since $T^{\mu\nu}$ is conserved and the second because $T^{\mu\nu}$ is symmetric, while $\nabla_\mu\xi_\nu$ is antisymmetric. Therefore the vector field $P^\mu = T^{\mu\nu}\xi_\nu$ is also conserved, $\nabla_\mu P^\mu = 0$, and we obtain thus the conservation of the component of the four-momentum vector in direction ξ.

Global energy conservation thus requires the existence of a time-like Killing vector field. If the metric is time-dependent, as for example in the case of the expanding universe, a time-like Killing vector field does not exist and the energy contained in a "comoving" volume changes with time.

Summary

In a curved spacetime $\mathcal{M}$ we require a connection to compare vectors at different points. The unique connection which is symmetric and compatible with the metric are the Christoffel symbols. The symmetries of a spacetime $\mathcal{M}$ are determined by its Killing vector fields ξ^μ. The momentum component parallel to ξ^μ of test particles moving in $\mathcal{M}$ is conserved. Locally conserved currents lead in general only for vector currents to globally conserved charges. In the case of locally conserved tensors (as $\nabla_\mu T^{\mu\nu} = 0$), the global conservation of the corresponding charges requires the existence of Killing vector fields.

Further reading. Landau and Lifshitz (1980) introduces classical field theory including general relativity. Carroll (2003) presents a clear introduction to differential geometry on a level accessible for physicists.

Problems

6.1 Derivative of $S^{\mu\nu}$. Express the covariant derivative $\nabla_\mu S^\mu{}_\nu$ of a symmetric tensor field without Christoffel symbols.

6.2 Lie bracket. Show that the commutator $[\boldsymbol{V}, \boldsymbol{W}]$ of two vector fields $\boldsymbol{V} = V^\mu \partial_\mu$ and $\boldsymbol{W} = W^\mu \partial_\mu$ is again a vector field.

6.3 Lie derivative. The LHS of (6.72) defines the Lie derivative of a rank two tensor. Show in the same way that the Lie derivative of a vector field $\boldsymbol{V}$ is given by $\mathscr{L}_\xi = [\boldsymbol{X}, \boldsymbol{\xi}]$, that is it agrees with the Lie bracket from problem 6.2. Generalise the Lie derivative to a general tensor field, assuming linearity and the Leibniz rule.

6.4 Conformal invariance. The condition $\tilde{g}^{\mu\nu}(\tilde{x}) = g^{\mu\nu}(\tilde{x})$ defines an isometry, that is, a distance conserving map between two spaces, $\mathrm{d}s^2 = \mathrm{d}\tilde{s}^2$. We can relax this condition to $\tilde{g}^{\mu\nu}(\tilde{x}) = \Omega^2(\tilde{x})g^{\mu\nu}(\tilde{x})$. Show that this transformation keeps the light-cone structure invariant. Derive the analogue of

(6.73) and solve it for the case of Minkowski space.

6.5 Torsion. Derive the transformation properties of a connection. Show that the difference of two connections $\Gamma^\mu{}_{\rho\sigma}$ and $\tilde{\Gamma}^\mu{}_{\rho\sigma}$ transforms as a tensor (which is called torsion).

6.6 Inertial coordinates. The equivalence principle implies that we can find in a curved spacetime locally coordinates (i.e. at a chosen point P) with $g_{\mu\nu} = \eta_{\mu\nu}$, $\partial_\rho g_{\mu\nu} = 0$ and $\Gamma^\rho{}_{\mu\nu} = 0$. Show that this requirement holds choosing as connection the Christoffel symbols.

6.7 Geodesic equation from H. Show that the Eqs. (6.52) and (6.53) imply the standard form of the geodesic equation.

6.8 Alternative Lagrangian. Use the Lagrangian (1.47) to derive the geodesic equation and the Hamiltonian. Interpret the result.

7
Spin-1 and spin-2 fields

We introduced fields transforming as tensors under coordinate general transformations. Such fields have integer spin and obey Bose–Einstein statistics. Therefore they can exist as macroscopic fields and are thus candidates to describe the electric and the gravitational force. Since both the electric and the gravitational potential $V(r)$ follow a $1/r$ law, we expect from our discussion of the Yukawa potential that the two forces are mediated by massless particles. We will find later that no interacting theory of massless particles with spin $s > 2$ exists. Therefore it is sufficient to consider the two cases $s = 1$ and $s = 2$.

7.1 Tensor fields

The momentum modes $\propto \mathrm{e}^{-ikx}$ of massive fields can be boosted to their rest-frame, where $k^\mu = (m, \mathbf{0})$. In this frame, the total angular momentum reduces to spin, and non-relativistic quantum mechanics is valid. Thus a field with spin s has $2s + 1$ spin or polarisation states. On the other hand, we can determine the spin[1] s of a field calculating the angle $s\alpha$ it is turned by a rotation $R_{ij}(\alpha)$. If we consider the transformation law of a tensor field of rank n, $\tilde{T}^{\mu_1 \cdots \mu_n} = \Lambda^{\mu_1}{}_{\nu_1} \cdots \Lambda^{\mu_n}{}_{\nu_n} T^{\nu_1 \cdots \nu_n}$, for the special case of rotations, we see that the n factors $\Lambda(\alpha)$ rotate some components of $T^{\nu_1 \cdots \nu_n}$ by the angle $n\alpha$. Therefore a tensor field of rank n has spin $s = n$. This implies that fields with spin $s \geq 1$ contain unphysical degrees of freedom. For instance, a massive spin-1 field has three and a massive spin-2 field has five polarisation states. On the other hand, a vector field A^μ has four components, and a symmetric tensor field $h^{\mu\nu}$ of rank two has ten components in $d = 4$ spacetime dimensions. The purpose of a relativistic wave equation is thus to impose the correct relativistic dispersion relation and to select the correct physical polarisation states in the chosen frame.

The first requirement is fulfilled if each component of a free field ϕ_a satisfies the Klein–Gordon equation. Additionally, we have to impose constraints f_i which eliminate the redundant components,

$$\left(\Box + m^2\right)\phi_a(x) = 0, \qquad \text{and} \quad f_i(\phi_a(x)) = 0. \tag{7.1}$$

The reason for this mismatch in the number of degrees of freedom is that in general a tensor of rank n is reducible, that is, it contains components of rank $< n$. For instance, the trace h^μ_μ of a second-rank tensor transforms clearly as a scalar. Therefore we should choose the constraints for massive fields with spin s such that all components with spin $< s$ are eliminated.

[1]See appendices B.3 and B.4 for a brief discussion.

Quantum Fields–From the Hubble to the Planck Scale. Michael Kachelriess. © Michael Kachelriess 2018.
Published in 2018 by Oxford University Press. DOI 10.1093/oso/9780198802877.001.0001

Example 7.1: An object which contains invariant subgroups with respect to a symmetry operation is called reducible. As an example, consider the reducible subgroups of a symmetric tensor $h^{\mu\nu}$ of rank two with respect to spatial rotations. Since one can boost a massive particle into its rest-frame, this is the relevant decomposition to find its spin states. We split $h^{\mu\nu}$ into a scalar h^{00}, a vector h^{0i} and a reducible tensor h^{ij},

$$h^{\mu\nu} = \begin{pmatrix} h^{00} & h^{0i} \\ h^{i0} & h^{ij} \end{pmatrix}.$$

Then we decompose h^{ij} again into its trace h^{ii} and its traceless part $h^i_j - h\delta^i_j/(d-1)$. The latter has $6 - 1 = 5$ degrees of freedom in $d = 4$, as required for a massive spin-2 field.

This problem is more severe for massless fields. We know from classical electrodynamics that the photon has only two polarisation states, and in appendix B.4 it is shown that this holds for massless fields with any spin $s > 0$. The redundant degrees of freedom of massless fields can be consistently eliminated only, if some redundancy of the field variables exists which in turn leads to a local symmetry of the field. In this chapter, we discuss the consequences of this redundancy called gauge symmetry on the level of the wave equations and their solutions for the photon and the graviton.

Tensor structure of the propagator.   We can gain some insight into the general tensor structure of the Feynman propagator for fields with spin $s > 0$ using the definition of the two-point Green function as the time-ordered vacuum expectation values of fields. In general, we can express an arbitrary solution of a free spin $s = 0, 1$ and 2 field by its Fourier components as

$$\phi(x) = \int \frac{\mathrm{d}^3k}{\sqrt{(2\pi)^3 2\omega_k}} \left[a(\boldsymbol{k}) \mathrm{e}^{-\mathrm{i}(\omega_k t - \boldsymbol{k}\boldsymbol{x})} + \mathrm{h.c.} \right], \tag{7.2}$$

$$A^\mu(x) = \sum_r \int \frac{\mathrm{d}^3k}{\sqrt{(2\pi)^3 2\omega_k}} \left[a_r(\boldsymbol{k}) \varepsilon_r^\mu(\boldsymbol{k}) \mathrm{e}^{-\mathrm{i}(\omega_k t - \boldsymbol{k}\boldsymbol{x})} + \mathrm{h.c.} \right], \tag{7.3}$$

$$h^{\mu\nu}(x) = \sum_r \int \frac{\mathrm{d}^3k}{\sqrt{(2\pi)^3 2\omega_k}} \left[a_r(\boldsymbol{k}) \varepsilon_r^{\mu\nu}(\boldsymbol{k}) \mathrm{e}^{-\mathrm{i}(\omega_k t - \boldsymbol{k}\boldsymbol{x})} + \mathrm{h.c.} \right], \tag{7.4}$$

where the momentum is on-shell, $k^\mu = (\omega_k, \boldsymbol{k})$ and r labels the spin or polarisation states. The constraints $f_i = 0$ are now conditions on the polarisation vector and tensor, respectively, which depend on $\boldsymbol{k}$. Proceeding as in the scalar case, we expect that, for example, the propagator for a vector field is given by

$$\mathrm{i}D_F^{\mu\nu}(x) = \langle 0|T\{A^\mu(x)A^{\nu*}(0)\}|0\rangle = \tag{7.5}$$

$$= \sum_r \int \frac{\mathrm{d}^3k}{(2\pi)^3 2\omega_k} \left[\varepsilon_r^\mu(\boldsymbol{k})\varepsilon_r^{\nu*}(\boldsymbol{k}) \mathrm{e}^{-\mathrm{i}kx} \vartheta(x^0) + \varepsilon_r^\mu(\boldsymbol{k})\varepsilon_r^{\nu*}(\boldsymbol{k}) \mathrm{e}^{\mathrm{i}kx} \vartheta(-x^0) \right] \tag{7.6}$$

$$= \int \frac{\mathrm{d}^4k}{(2\pi)^4} \frac{\mathcal{P}^{\mu\nu}(k)\,\mathrm{e}^{-\mathrm{i}kx}}{k^2 - m^2 + \mathrm{i}\varepsilon}. \tag{7.7}$$

The expression (7.6) is in line with the interpretation of the propagator as the probability for the creation of a particle at x with any momentum k and polarisation r, its propagation to x' followed by its annihilation. In the last step, we introduced the tensor $\mathcal{P}^{\mu\nu}(k)$ which corresponds to the sum over the polarisation states $\varepsilon_r^\mu(\mathbf{k})\varepsilon_r^{\nu*}(\mathbf{k})$. We will show that the polarisation tensors are polynomials in the momentum k, and thus Eq. (7.6) shows that $\mathcal{P}^{\mu\nu}(k)$ is even in the momentum. As a result, our discussion of causality in the scalar case applies for all tensor field, implying that these fields seen as quantum fields commute. Therefore the particles described by these fields satisfy Bose–Einstein statistics.

7.2 Vector fields

Proca and Maxwell equations. A massive vector field A^μ has four components in $d = 4$ spacetime dimensions, while it has only $2s+1 = 3$ independent spin components. Correspondingly, a four-vector A^μ transforms under a rotation as $(A^0, \boldsymbol{A})$, that is it contains a scalar and a three-vector. Therefore we have to add to the four Klein–Gordon equations for A^μ one constraint which eliminates A^0. The only linear, Lorentz-invariant possibility is

$$\left(\Box + m^2\right) A^\mu(x) = 0 \qquad \text{and} \quad \partial_\mu A^\mu = 0. \tag{7.8}$$

In momentum space, this translates into $(k^2 - m^2)A^\mu(k) = 0$ and $k_\mu A^\mu(k) = 0$. In the rest-frame of the particle, $k^\mu = (m, \mathbf{0})$, and the constraint becomes $A^0 = 0$. Hence, a field satisfying (7.8) has only three space-like components as required for a massive $s = 1$ field. We can choose the three polarisation vectors which label the three degrees of freedom in the rest-frame, for example, as the Cartesian unit vectors, $\boldsymbol{\varepsilon}_i \propto \mathbf{e}_i$.

The two equations can be combined into one equation called the Proca equation,

$$\left(\eta^{\mu\nu}\Box - \partial^\mu\partial^\nu\right) A_\nu + m^2 A^\mu = 0. \tag{7.9}$$

To show the equivalence of this equation with (7.8), we act with ∂_μ on it,

$$\left(\partial^\nu\Box - \Box\partial^\nu\right) A_\nu + m^2\partial_\mu A^\mu = m^2\partial_\mu A^\mu = 0. \tag{7.10}$$

Hence, a solution of the Proca equation automatically fulfils the constraint $\partial_\mu A^\mu = 0$ for $m^2 > 0$. On the other hand, we can neglect the second term in (7.9) for $\partial_\nu A^\nu = 0$ and obtain the Klein–Gordon equation.

We now go to the case of a massless spin-1 field which is described by Maxwell equations. In classical electrodynamics, the field-strength tensor $F_{\mu\nu}$ is an observable quantity, while the potential A_μ is merely a convenient auxiliary quantity. From the definition

$$F_{\mu\nu} = \partial_\mu A_\nu - \partial_\nu A_\mu \tag{7.11}$$

it is clear that $F_{\mu\nu}$ is invariant under the transformations

$$A_\mu(x) \to A'_\mu(x) = A_\mu(x) - \partial_\mu\Lambda(x). \tag{7.12}$$

Thus $A'_\mu(x)$ is for any $\Lambda(x)$ physically equivalent to $A_\mu(x)$, leading to the same field-strength tensor and thus, for example, to the same Lorentz force on a particle. The transformations (7.12) are called gauge transformations. Note that the mass term $m^2 A^\mu$ in the Proca equation breaks gauge invariance.

If we insert into the Maxwell equation the definition of the potential,

$$\partial_\mu F^{\mu\nu} = \partial_\mu(\partial^\mu A^\nu - \partial^\nu A^\mu) = \Box A^\nu - \partial_\mu \partial^\nu A^\mu = j^\nu, \tag{7.13}$$

we see that this expression equals the $m = 0$ limit of the Proca equation. Gauge invariance allows us to choose a potential A^μ such that $\partial_\mu A^\mu = 0$. Such a choice is called fixing the gauge, and the particular case $\partial_\mu A^\mu = 0$ is denoted as the Lorenz gauge. In this gauge, the wave equation simplifies to

$$\Box A^\mu = j^\mu. \tag{7.14}$$

Inserting then a plane wave $A^\mu \propto \varepsilon^\mu e^{ikx}$ into the free wave equation, $\Box A^\nu = 0$, we find that k is a light-like vector, while the Lorenz gauge condition $\partial_\mu A^\mu = 0$ results in $\varepsilon^\mu k_\mu = 0$. Imposing the Lorenz gauge, we can still add to the potential A^μ any function $\partial^\mu \chi$ satisfying $\Box \chi = 0$. We can use this freedom to set $A^0 = 0$, obtaining thereby $\varepsilon^\mu k_\mu = -\boldsymbol{\varepsilon} \cdot \boldsymbol{k} = 0$. Thus the photon propagates with the speed of light, is transversely polarised and has two polarisation states as expected for a massless particle.

Closely connected to the gauge invariance of electrodynamics is the fact that its source, the electromagnetic current, is conserved. The antisymmetry of $F^{\mu\nu}$, which is the basis for the symmetry (7.12), leads also to $\partial_\mu \partial_\nu F^{\mu\nu} = 0$. Thus the Maxwell equation $\partial_\mu F^{\mu\nu} = j^\nu$ implies the conservation of the electromagnetic current j^μ,

$$\partial_\mu \partial_\nu F^{\mu\nu} = \partial_\nu j^\nu = 0. \tag{7.15}$$

Propagator for massive spin-1 fields. The propagator $D_{\mu\nu}$ for a massive spin-1 field is determined by

$$\left[\eta^{\mu\nu}(\Box + m^2) - \partial^\mu \partial^\nu\right] D_{\nu\lambda}(x) = \delta^\mu_\lambda \delta(x). \tag{7.16}$$

Inserting the Fourier transformation of the propagator and of the delta function gives

$$\left[(-k^2 + m^2)\,\eta^{\mu\nu} + k^\mu k^\nu\right] D_{\nu\lambda}(k) = \delta^\mu_\lambda. \tag{7.17}$$

We will apply the tensor method to solve this equation. In this approach, we use first all tensors available in the problem to construct the required tensor of rank 2. In the case at hand, we have at our disposal only the momentum k_μ of the particle—which we can combine to $k_\mu k_\nu$—and the metric tensor $\eta_{\mu\nu}$. Thus the tensor structure of $D_{\mu\nu}(k)$ has to be of the form

$$D_{\mu\nu}(k) = A\eta_{\mu\nu} + Bk_\mu k_\nu \tag{7.18}$$

with two unknown scalar functions $A(k^2)$ and $B(k^2)$. Inserting this ansatz into (7.17) and multiplying out, we obtain

$$\left[(-k^2 + m^2)\eta^{\mu\nu} + k^\mu k^\nu\right]\left[A\eta_{\nu\lambda} + Bk_\nu k_\lambda\right] = \delta^\mu_\lambda, \tag{7.19a}$$

$$-Ak^2\delta^\mu_\lambda + Am^2\delta^\mu_\lambda + Ak^\mu k_\lambda + Bm^2 k^\mu k_\lambda = \delta^\mu_\lambda, \tag{7.19b}$$

$$-A(k^2 - m^2)\delta^\mu_\lambda + (A + Bm^2)k^\mu k_\lambda = \delta^\mu_\lambda. \tag{7.19c}$$

In the last step, we regrouped the LHS into the two tensor structures δ^μ_λ and $k^\mu k_\lambda$. A comparison of their coefficients gives then $A = -1/(k^2 - m^2)$ and

$$B = -\frac{A}{m^2} = \frac{1}{m^2(k^2 - m^2)}.$$

Thus the massive spin-1 propagator follows as

$$D^{\mu\nu}_F(k) = \frac{-\eta^{\mu\nu} + k^\mu k^\nu/m^2}{k^2 - m^2 + i\varepsilon}. \tag{7.20}$$

Note that there is a sign ambiguity, since we could have added a minus sign to the Proca equation.

Next we check this sign and our claim that the propagator $D^{ab}_F(k)$ of spin $s > 0$ fields can be obtained as sum over their polarisation states $\varepsilon^{(r)}_a$ times the scalar propagator $\Delta(k)$. As the theory is Lorentz-invariant, we can choose the frame most convenient for this comparison which is the rest-frame of the massive particle. Then $k^\mu = (m, \mathbf{0})$ and the three polarisation vectors can be chosen as the Cartesian basis vectors. Comparing then

$$-\eta^{\mu\nu} + k^\mu k^\nu/m^2 = \begin{pmatrix} 0 & 0 & 0 & 0 \\ 0 & 1 & 0 & 0 \\ 0 & 0 & 1 & 0 \\ 0 & 0 & 0 & 1 \end{pmatrix} = \sum_r \varepsilon^{\mu(r)}\varepsilon^{\nu(r)}, \tag{7.21}$$

shows that both methods agree and can be used to derive the Feynman propagator. In the latter approach, working from the RHS to the LHS of Eq. (7.21), we first derive the expression valid for the Feynman propagator in a specific frame. Then we have to rewrite the expression in an invariant way using the relevant tensors, here $\eta^{\mu\nu}$ and $k^\mu k^\nu$. Moreover, Eq. 7.21 shows that we have chosen the right sign for the propagator (7.20).

Propagator for massless spin-1 fields. As we have seen, we can set $m = 0$ in the Proca equation and obtain the Maxwell equation. The corresponding limit of the propagator (7.20) leads however to an ill-defined result. As we know that the number of degrees of freedom differs between the massive and the massless case, this is not too surprising. If we try next the limit $m \to 0$ in Eq. (7.19c), then we find

$$-Ak^2\delta^\mu_\lambda + Ak^\mu k_\lambda = \delta^\mu_\lambda. \tag{7.22}$$

This equation has for arbitrary k with $A = -1/k^2$ and $A = 0$ no solution. Moreover, the function B is undetermined. We can understand this physically, since for a massless field current conservation holds. But $\partial_\mu J^\mu(x) = 0$ implies $k_\mu J^\mu(k) = 0$ and thus the $k^\mu k^\nu$ term does not influence physical quantities: in physical measurable quantities, as, for example, $W[J]$, the propagator is always matched between conserved currents, and the longitudinal part $k^\mu k^\nu$ drops out.

We now try to construct the photon propagator from its sum over polarisation states. First we consider a linearly polarised photon with polarisation vectors $\varepsilon^{(r)}_\mu$

lying in the plane perpendicular to its momentum vector $\boldsymbol{k}$. If we perform a Lorentz boost on $\varepsilon_\mu^{(1)}$, we will find

$$\tilde{\varepsilon}_\mu^{(1)} = \Lambda^\nu{}_\mu \varepsilon_\nu^{(1)} = a_1 \varepsilon_\mu^{(1)} + a_2 \varepsilon_\mu^{(2)} + a_3 k_\mu \,, \tag{7.23}$$

where the coefficients a_i depend on the direction $\boldsymbol{\beta}$ of the boost. Thus, in general the polarisation vector will not be anymore perpendicular to $\boldsymbol{k}$. Similarly, if we perform a gauge transformation

$$A_\mu(x) \to A'_\mu(x) = A_\mu(x) - \partial_\mu \Lambda(x) \tag{7.24}$$

with

$$\Lambda(x) = -i\lambda \exp(-ikx) + \text{h.c.} \,, \tag{7.25}$$

then

$$A'_\mu(x) = (\varepsilon_\mu + \lambda k_\mu) \exp(-ikx) + \text{h.c.} = \varepsilon'_\mu \exp(-ikx) + \text{h.c.} \tag{7.26}$$

Choosing, for example, a photon propagating in z direction, $k^\mu = (\omega, 0, 0, \omega)$, we see that the gauge transformation does not affect the transverse components ε_1 and ε_2. Thus only the components of ε^μ transverse to $\boldsymbol{k}$ can have physical significance. On the other hand, the time-like and longitudinal components depend on the arbitrary parameter λ and are therefore unphysical. In particular, they can be set to zero by a gauge transformation. First, $\varepsilon'_\mu k'^\mu = 0$ implies (again for a photon propagating in z direction) $\varepsilon'_0 = -\varepsilon'_3$. From $\varepsilon'_3 = \varepsilon_3 + \lambda\omega$, we see that $\lambda = -\varepsilon_3/\omega$ sets $\varepsilon'_3 = -\varepsilon'_0 = 0$. Thus the transformation law (7.23) for the polarisation vector of a massless spin-1 particles requires the existence of the gauge symmetry (7.24). The gauge symmetry in turn implies that the massless spin-1 particle couples only to conserved currents.

We can exploit the transformation law $\varepsilon'_\mu = \varepsilon_\mu + \lambda k_\mu$ as follows: since the dependence on k_μ is nonphysical, any Feynman amplitude $\mathcal{A} = \varepsilon_\mu A^\mu$ has to vanish, if we replace the polarisation vector ε_μ of an external photon by its four-momentum, $k_\mu A^\mu = 0$. This quantum analogue of classical current conservation $k_\mu J^\mu(k) = 0$ is called "Ward identity". As an example of its application, we derive a convenient expression for the propagator of a massless vector particle. The two polarisation vectors of a photon should satisfy the normalisation $\varepsilon_\mu^{(a)*} \varepsilon^{\mu(b)} = \delta^{ab}$. For a linearly polarised photon propagating in z direction, $k^\mu = (\omega, 0, 0, \omega)$, the polarisation vectors are $\varepsilon_\mu^{(1)} = \delta_\mu^1$ and $\varepsilon_\mu^{(2)} = \delta_\mu^2$. If we perform the sum over the two polarisation states, we find

$$\sum_r \varepsilon_\mu^{(r)*} \varepsilon_\nu^{(r)} = \text{diag}\{0, 1, 1, 0\} \,. \tag{7.27}$$

If we try to rewrite this expression in an invariant way using $\eta_{\mu\nu}$ and $k_\mu k_\nu / k^2$, we fail. We cannot cancel $\eta_{00} = +1$ and $\eta_{33} = -1$ by $k_\mu k_\nu / k^2$ at the same time. We therefore introduce additionally the momentum vector $\tilde{k}^\mu = (\omega, 0, 0, -\omega)$ obtained by a spatial reflection from k^μ. This allows us to write the polarisation sum as an invariant tensor expression,

$$\sum_r \varepsilon_\mu^{(r)*} \varepsilon_\nu^{(r)} = -\eta_{\mu\nu} + \frac{k_\mu \tilde{k}_\nu + \tilde{k}_\mu k_\nu}{k\tilde{k}} \equiv -\pi_{\mu\nu} \,. \tag{7.28}$$

Current conservation, $k_\mu J^\mu(k) = 0$, implies that the second term in the polarisation sum does not contribute to physical observables. For the same reason, we can add an arbitrary term $\xi k_\mu k_\nu$. We use this freedom to eliminate the $\tilde{k}$ dependence and to set

$$J^{\mu*}\left(\sum_r \varepsilon_\mu^{(r)*}\varepsilon_\nu^{(r)}\right)J^\nu = J^{\mu*}\left(-\eta_{\mu\nu} + (1-\xi)\frac{k_\mu k_\nu}{k^2}\right)J^\nu. \tag{7.29}$$

Now we can read off the photon propagator as

$$D_F^{\mu\nu}(k) = \frac{-\eta^{\mu\nu} + (1-\xi)k^\mu k^\nu/k^2}{k^2 + i\varepsilon}. \tag{7.30}$$

A specific choice of the parameter ξ called gauge-fixing parameter corresponds to the choice of a gauge in Eq. (7.13). In particular, the Feynman gauge $\xi = 1$, which leads to a form of the propagator often most convenient in calculations, corresponds to the Lorenz gauge in Eq. (7.13). In this gauge, $\sum_{r=0}^{3}\varepsilon_\mu^{(r)*}\varepsilon_\nu^{(r)} = \text{diag}\{-1,1,1,1\}$: the propagator contains nonphysical degrees of freedom, time-like and longitudinal photons, whose contributions cancel however in physical observables. Similarly, for all other values ξ the propagator is explicitly Lorentz-invariant but contains unphysical degrees of freedom. We will see later that it is a general feature of gauge theories as electrodynamics that we have to choose between a covariant gauge which introduces unphysical degrees of freedom and a gauge which contains only the transverse degree of freedom but selects a specific frame.

Repulsive Coulomb potential by vector exchange. We consider as in the scalar case two static point charges as external sources, but now use a vector current $J^\mu = J_1^\mu(x_1) + J_2^\mu(x_2)$. Since $J^\mu = (\rho, \boldsymbol{j})$, only the zero component, $J_i^\mu = \delta_0^\mu\delta(\boldsymbol{x} - \boldsymbol{x}_i)$, contributes for a static source to $W[J]$. Moreover, we can neglect the longitudinal part $k^\mu k^\nu/m^2$ of the propagator. This is justified, since the concept of a potential energy makes only sense in the non-relativistic limit, that is, for $V \ll m$ or equivalently $r \gg 1/m$. Hence

$$W_{12}[J] = -\frac{1}{2}\int d^4x d^4x' \int \frac{d^4k}{(2\pi)^4} J_1^\mu(x)\frac{-\eta_{\mu\nu}\,e^{-ik(x-x')}}{k^2 - m^2 + i\varepsilon}J_2^\nu(x') \tag{7.31a}$$

$$= \frac{1}{2}\int dt dt' \int \frac{d^4k}{(2\pi)^4}\frac{e^{-ik^0(t-t')}e^{-i\boldsymbol{k}(\boldsymbol{x}_1-\boldsymbol{x}_2)}}{k^2 - m^2 + i\varepsilon}. \tag{7.31b}$$

Comparing with our earlier result for scalar exchange in Eq. (3.36), it becomes clear without further calculation that spin-1 exchange between equal charges is repulsive. In the limit $m \to 0$, we obtain the Coulomb potential with the correct sign for electromagnetic interactions.

7.3 Gravity

Wave equation. From Newton's law we know that gravity is fundamentally attractive and of long range. Thus the gravitational force has to be mediated by a massless particle which cannot be a spin $s = 1$ particle. Analogous to the electric field $\boldsymbol{E} = -\boldsymbol{\nabla}\phi$

we can introduce a classical gravitational field $\boldsymbol{g}$ as the gradient of the gravitational potential, $\boldsymbol{g} = -\boldsymbol{\nabla}\phi$. We obtain then $\boldsymbol{\nabla} \cdot \boldsymbol{g}(\boldsymbol{x}) = -4\pi G\rho(\boldsymbol{x})$ and as Poisson equation

$$\Delta\phi(\boldsymbol{x}) = 4\pi G\rho(\boldsymbol{x}), \tag{7.32}$$

where ρ is the mass density, $\rho = \mathrm{d}m/\mathrm{d}^3x$.

Special relativity gives us two hints of how we should transfer this equation into a relativistic framework. First, the Laplace operator Δ on the LHS is the $c \to \infty$ limit of minus the d'Alembert operator $\Box$. Second, the RHS should be the $v/c \to 0$ limit of something incorporating not only the mass density but all types of energy densities. To proceed, consider first how the mass density ρ transforms under a Lorentz transformation. An observer moving with the speed β relative to the rest-frame of the matter distribution ρ measures the energy density $\rho' = \gamma\mathrm{d}m/(\gamma^{-1}\mathrm{d}V) = \gamma^2\rho$, with $\gamma = 1/\sqrt{1 - \beta^2}$. This is the transformation law of the 00 component of a tensor of rank two and ρ as 00 component, alas the energy–momentum stress tensor $T^{\mu\nu}$.

Thus the field equation for a purely scalar theory of gravity would be

$$\Box\phi = -4\pi G T^\mu_\mu. \tag{7.33}$$

Such a theory predicts no coupling between photons and gravitation, because the trace of the stress tensor of the electromagnetic field vanishes, $T^\mu_\mu = 0$, and is therefore in contradiction to the observed gravitational lensing of light. A purely vector theory for gravity fails too, since it predicts not attraction but repulsion of two masses. Hence we are forced to consider a symmetric spin-2 field $\bar{h}_{\mu\nu}$ as mediator of the gravitational force; its source is the energy–momentum stress tensor

$$\Box\bar{h}^{\mu\nu} = -2\kappa T^{\mu\nu}. \tag{7.34}$$

The normalisation constant $\kappa \propto G_N$ has to be determined such that in the non-relativistic limit the Poisson equation (7.32) holds.

Let us consider as a warm-up the case of a massive particle. A symmetric, massive spin-2 field has ten independent components, but only $2s+1 = 5$ physical spin degrees of freedom. Thus we have to impose five constraints additional to the source-free equation

$$(\Box + m^2)\bar{h}^{\mu\nu} = 0. \tag{7.35}$$

Proceeding as in the $s = 1$ case, (7.8), we use as constraint $\partial_\mu\bar{h}^{\mu\nu} = 0$ which provides now four conditions. We can use them to set $\bar{h}^{0\mu} = 0$ in the rest-frame of the particle. We obtain the missing fifth constraint subtracting from $\bar{h}^{ij}$ the trace $\bar{h}^{ii}$ which transforms as a scalar.

We move now to the massless case considering a plane wave $\bar{h}_{\mu\nu} = \varepsilon_{\mu\nu}\exp(-\mathrm{i}kx)$. In analogy to the photon case, we expect that also the graviton has only two, transverse degrees of freedom. If we choose the plane wave propagating in the z direction, $\boldsymbol{k} = k\boldsymbol{e}_z$, then we expect that the polarisation tensor can be expressed as

$$\varepsilon^{\mu\nu} = \begin{pmatrix} 0 & 0 & 0 & 0 \\ 0 & \varepsilon_{11} & \varepsilon_{12} & 0 \\ 0 & \varepsilon_{12} & -\varepsilon_{11} & 0 \\ 0 & 0 & 0 & 0 \end{pmatrix}. \tag{7.36}$$

Here we used that the polarisation tensor has to be symmetric and traceless. The choice (7.36) is called the *transverse traceless* (TT) gauge.

Metric perturbations as a tensor field.   In the case of the photon, we could reduce the degrees of freedom from four to two, because of the redundancy implied by the gauge symmetry of electromagnetism. Moreover, the gauge symmetry lead to the conservation of the electromagnetic current. The two obvious questions to address next are which symmetry and which conservation law are connected to gravitation.

The second question is the simpler one, since we know already that in flat space $\partial_\mu T^{\mu\nu} = 0$ holds. Thus for gravity energy-momentum conservation will play the role of current conservation, implying that only a transverse gravitational wave couples to its source, $k_\mu T^{\mu\nu}(k) = 0$. In order to answer the first question, we have to consider the properties of $h^{\mu\nu}$. The equivalence principle implies that all test-particles move along the same world-line, if they are released at the same initial point and move only under the influence of the gravitational force. This universality motivated Einstein to describe the effect of gravity by the curvature of spacetime. We associate therefore the symmetric tensor field[2] $h_{\mu\nu}$ with small perturbations around the Minkowski metric $\eta_{\mu\nu}$,

$$g_{\mu\nu} = \eta_{\mu\nu} + \varepsilon h_{\mu\nu}, \qquad \varepsilon \ll 1. \tag{7.37}$$

We choose a Cartesian coordinate system x^μ and ask ourselves which transformations are compatible with the splitting (7.37) of the metric. If we consider global Lorentz transformations $\Lambda^\nu{}_\mu$, then $x'^\nu = \Lambda^\nu{}_\mu x^\mu$, and the metric tensor transforms as

$$g'_{\alpha\beta} = \Lambda^\rho{}_\alpha \Lambda^\sigma{}_\beta g_{\rho\sigma} = \Lambda^\rho{}_\alpha \Lambda^\sigma{}_\beta (\eta_{\rho\sigma} + h_{\rho\sigma}) = \eta_{\alpha\beta} + \Lambda^\rho{}_\alpha \Lambda^\sigma{}_\beta h_{\rho\sigma} = \eta'_{\alpha\beta} + \Lambda^\rho{}_\alpha \Lambda^\sigma{}_\beta h_{\rho\sigma}. \tag{7.38}$$

Since $h'_{\alpha\beta} = \Lambda^\rho{}_\alpha \Lambda^\sigma{}_\beta h_{\rho\sigma}$, we see that global Lorentz transformations respect the splitting (7.37). Thus $h_{\mu\nu}$ transforms as a rank-2 tensor under global Lorentz transformations. We can view therefore the perturbation $h_{\mu\nu}$ as a symmetric rank-2 tensor field defined on Minkowski space that satisfies the wave equation (7.34), similar as the photon field is a rank-1 tensor field fulfilling Maxwell's equations.

The splitting (7.37) is however clearly not invariant under general coordinate transformations, as they allow, for example, the finite rescaling $g_{\mu\nu} \to \Omega g_{\mu\nu}$. We restrict therefore ourselves to infinitesimal coordinate transformations,

$$\bar{x}^\mu = x^\mu + \varepsilon \xi^\mu(x^\nu) \tag{7.39}$$

with $\varepsilon \ll 1$. Then the Killing equation (6.72) simplifies to

$$h'_{\mu\nu} = h_{\mu\nu} + \partial_\mu \xi_\nu + \partial_\nu \xi_\mu, \tag{7.40}$$

because the term $\xi^\rho \partial_\rho h_{\mu\nu}$ is quadratic in the small quantities $\varepsilon h_{\mu\nu}$ and $\varepsilon \xi_\mu$ and can be neglected. Recall that the $\xi^\rho \partial_\rho h_{\mu\nu}$ term appeared, because we compared the metric tensor at different points. In its absence, it is more fruitful to view Eq. (7.40) not as a coordinate but as a gauge transformation analogous to (7.12). In this interpretation,

---

[2]We drop the bar, anticipating that $\bar{h}^{\mu\nu}$ may differ from $h^{\mu\nu}$ in Eq. (7.34). We will derive their relation, $\bar{h}_{\mu\nu} \equiv h_{\mu\nu} - \frac{1}{2}\eta_{\mu\nu}h^\alpha_\alpha$, in section 19.3.

we stay in Minkowski space and the fields $h'_{\mu\nu}$ and $h_{\mu\nu}$ describe the same physics, since the gravitational field equations do not fix uniquely $h_{\mu\nu}$ for a given source. In momentum space, Eq. (7.40) specifies how the polarisation tensor transforms under gauge transformations,

$$\varepsilon'_{\mu\nu} = \varepsilon_{\mu\nu} + \xi_\mu k_\nu + \xi_\nu k_\mu. \tag{7.41}$$

We can use this gauge freedom to eliminate four components of $\varepsilon_{\mu\nu}$. After that, we can perform another gauge transformation (7.41) using any four functions χ_μ satisfying the wave equation $\Box\chi_\mu = 0$, eliminating thereby four additional components. This justifies the use of the TT gauge.

Graviton propagator. We follow the same approach as in the derivation of the photon propagator. For a graviton propagating in z direction, $k^\mu = (\omega, 0, 0, \omega)$, we choose as the two polarisation states $\varepsilon^{(1)}_{\mu\nu}$ setting $\varepsilon_{11} = 1/\sqrt{2}$ and $\varepsilon_{12} = 0$ and $\varepsilon^{(2)}_{\mu\nu}$ setting $\varepsilon_{11} = 0$ and $\varepsilon_{12} = 1/\sqrt{2}$, respectively. They satisfy the normalisation $\varepsilon^{(a)}_{\mu\nu}\varepsilon^{\mu\nu(b)} = \delta^{ab}$. Now we should perform the sum over the two polarisation states, $\sum_r \varepsilon^{(r)}_{\mu\nu}\varepsilon^{(r)}_{\rho\sigma}$, and express the result as a linear combination of $\eta_{\mu\nu}$ and $k_\mu\tilde{k}_\nu + \tilde{k}_\mu k_\nu$. A straightforward way to do this is to combine first the ten independent quantities of the symmetric tensors in ten-dimensional vectors, $\varepsilon_{\mu\nu} \to E_a$, $\eta_{\mu\nu} \to N_a$, and $k_\mu\tilde{k}_\nu + \tilde{k}_\mu k_\nu \to K_a$, to calculate the tensor products of these vectors and to compare then the resulting $10 \otimes 10$ matrices. An alternative, shorter way is to use the requirement that the propagator is transverse in all indices, $k^\mu \sum_r \varepsilon^{(r)}_{\mu\nu}\varepsilon^{(r)}_{\rho\sigma} = \ldots = k^\sigma \sum_r \varepsilon^{(r)}_{\mu\nu}\varepsilon^{(r)}_{\rho\sigma} = 0$, because of energy–momentum conservation, $\partial_\mu T^{\mu\nu}(x) = 0$. This implies that the graviton propagator should be composed of the projection operators $\pi_{\mu\nu}$ used for the photon (cf. Eq. (7.28)) as follows

$$\sum_r \varepsilon^{(r)}_{\mu\nu}\varepsilon^{(r)}_{\rho\sigma} = A\,\pi_{\mu\nu}\pi_{\rho\sigma} + B\left[\pi_{\mu\rho}\pi_{\nu\sigma} + \pi_{\mu\sigma}\pi_{\nu\rho}\right]. \tag{7.42}$$

The last two terms have a common coefficient, since the LHS is invariant under exchanges of $\mu \leftrightarrow \nu$ or $\rho \leftrightarrow \sigma$. We fix A and B by evaluating this expression for two sets of indices. Since the only non-zero elements of $\pi_{\mu\nu}$ are $\pi_{11} = \pi_{22} = -1$, we obtain choosing, for example, $\{1212\}$ as indices

$$\sum_r \varepsilon^{(r)}_{12}\varepsilon^{(r)}_{12} = \frac{1}{2} = B\pi_{11}\pi_{22}$$

and thus $B = 1/2$. Similarly, it follows $A = -1/2$ choosing as indices $\{1111\}$, for example. Thus we found—with surprising ease—the polarisation sum required for the graviton propagator,

$$\sum_r \varepsilon^{(r)}_{\mu\nu}\varepsilon^{(r)}_{\rho\sigma} = -\frac{1}{2}\pi_{\mu\nu}\pi_{\rho\sigma} + \frac{1}{2}\pi_{\mu\rho}\pi_{\nu\sigma} + \frac{1}{2}\pi_{\mu\sigma}\pi_{\nu\rho}. \tag{7.43}$$

We continue to proceed in the same way as for the photon. Energy–momentum conservation, $k_\mu T^{\mu\nu}(k) = 0$, implies that the $k_\mu\tilde{k}_\nu + \tilde{k}_\mu k_\nu$ term in $\pi_{\mu\nu}$ does not contribute to physical observables. We drop therefore again all terms proportional to the graviton momentum k_μ,

$$T^{\mu\nu *}\left(\sum_r \varepsilon_{\mu\nu}^{(r)*}\varepsilon_{\rho\sigma}^{(r)}\right)T^{\rho\sigma} = T^{\mu\nu *}\left(-\frac{1}{2}\eta_{\mu\nu}\eta_{\rho\sigma} + \frac{1}{2}\eta_{\mu\rho}\eta_{\nu\sigma} + \frac{1}{2}\eta_{\mu\sigma}\eta_{\nu\rho}\right)T^{\rho\sigma}. \quad (7.44)$$

Thus the graviton propagator in the Feynman gauge is given by

$$D_F^{\mu\nu;\rho\sigma}(k) = \frac{\frac{1}{2}(-\eta^{\mu\nu}\eta^{\rho\sigma} + \eta^{\mu\rho}\eta^{\nu\sigma} + \eta^{\mu\sigma}\eta^{\nu\rho})}{k^2 + i\varepsilon}. \quad (7.45)$$

Other gauges are obtained by the replacement $\eta^{\mu\nu} \to \eta^{\mu\nu} - (1-\xi)k^\mu k^\nu/k^2$. In the case of gravity, the Feynman gauge $\xi = 1$ is most often called harmonic gauge, but also the names Hilbert, Loren(t)z, de Donder and confusingly many others are in use.

Attractive potential by spin-2 exchange.   We consider now the potential energy created by two point masses as external sources interacting via a tensor current $T^{\mu\nu} = T_1^{\mu\nu}(x_1) + T_2^{\mu\nu}(x_2)$. Specialising to static sources, only the zero–zero component, $T_i^{\mu\nu} = \delta_0^\mu \delta_0^\nu \delta(\boldsymbol{x} - \boldsymbol{x}_i)$, contributes to $W[J]$. Hence

$$W_{12}[J] = -\frac{1}{2}\int \mathrm{d}^4x\mathrm{d}^4x' \int \frac{\mathrm{d}^4k}{(2\pi)^4}T_1^{00}(x)\, D_{F\,00;00}(k)\, \mathrm{e}^{-ik(x-x')}\, T_2^{00}(x'). \quad (7.46)$$

Looking at the numerator of the graviton propagator, we find $-1+1+1 = 1 > 0$. Thus spin-2 exchange is attractive, as required for the force mediating gravity. Comparing Eq. (7.46) to Newton's gravitational potential, we see that the graviton couples with the strength $(8\pi G)^{1/2} \equiv \kappa^{1/2}$ to the stress tensor $T^{\mu\nu}$ of matter, cf. problem 7.5.

Helicity.   We determine now how a metric perturbation $h_{\mu\nu}$ transforms under a rotation with the angle α. We choose the wave propagating in z direction, $\boldsymbol{k} = k\boldsymbol{e}_z$, the TT gauge, and the rotation in the xy plane. Then the general Lorentz transformation $\Lambda_\mu^{\,\nu}$ becomes

$$\Lambda_\mu^{\,\nu} = \begin{pmatrix} 1 & 0 & 0 & 0 \\ 0 & \cos\alpha & \sin\alpha & 0 \\ 0 & -\sin\alpha & \cos\alpha & 0 \\ 0 & 0 & 0 & 1 \end{pmatrix}. \quad (7.47)$$

Since $\boldsymbol{k} = k\boldsymbol{e}_z$ and thus $\Lambda_\mu^{\,\nu}k_\nu = k_\mu$, the rotation affects only the polarisation tensor. We rewrite $\varepsilon'_{\mu\nu} = \Lambda_\mu^{\,\rho}\Lambda_\nu^{\,\sigma}\varepsilon_{\rho\sigma}$ in matrix notation, $\varepsilon' = \Lambda\varepsilon\Lambda^T$. It is sufficient to perform the calculation for the xy sub-matrices. The result after introducing circular polarisation states $\varepsilon_\pm = \varepsilon_{11} \pm i\varepsilon_{12}$ is

$$\varepsilon_\pm^{\prime\mu\nu} = \exp(\mp 2i\alpha)\varepsilon_\pm^{\mu\nu}. \quad (7.48)$$

The same calculation for a circularly polarised photon gives $\varepsilon_\pm^{\prime\mu} = \exp(\mp i\alpha)\varepsilon_\pm^\mu$. Any plane wave ψ which is transformed into $\psi' = \mathrm{e}^{-ih\alpha}\psi$ by a rotation of an angle α around its propagation axis is said to have helicity h. Thus if we say that a photon has spin 1 and a graviton has spin 2, we mean more precisely that electromagnetic and gravitational plane waves have helicity 1 and 2, respectively. Doing the same calculation in an arbitrary gauge, one finds that the remaining, unphysical degrees of freedom transform as helicity 1 and 0 (problem 7.8). In general, a tensor field of rank (n, m) contains states with helicity $h = 0, \ldots, n + m$. Thus we can rephrase the statement that tensor fields follow Bose–Einstein statistics as fields with integer helicity (or spin) are bosons.

7.4   Source of gravity

The dynamical energy–momentum stress tensor.   If we compare the wave equation for a photon and a graviton, then there is an important difference. The former is, in the classical limit, exact. The photon carries no charge and does not contribute to its source term. As a result, the wave equation is linear. In contrast, a gravitational wave carries energy–momentum and acts thus as its own source. The LHS of (7.34) should be therefore the limit of a more complicated equation, which we write symbolically as $G_{\mu\nu} = -\kappa T_{\mu\nu}$. The tensor $G_{\mu\nu}$ should be given as the variation of an appropriate action of gravity, called the Einstein–Hilbert action S_{EH}, with respect to the metric tensor $g_{\mu\nu}$. Even without knowing the action S_{EH}, we can derive an important conclusion. If the total action is the sum of S_{EH} and the action S_{m} including all relevant matter fields,

$$S = \frac{1}{2\kappa} S_{\text{EH}} + S_{\text{m}},$$

then the variation of the matter action S_{m} should give the stress tensor as the source of the gravitational field,

$$\frac{2}{\sqrt{|g|}} \frac{\delta S_{\text{m}}}{\delta g_{\mu\nu}} = -T^{\mu\nu}. \tag{7.49}$$

Here we included a factor $\sqrt{|g|}$ because $T^{\mu\nu}$ is a density, while the factor 2 is required to obtain agreement with the usual definition of $T^{\mu\nu}$. Since the presence of gravity implies a curved spacetime, the replacements $\{\partial_\mu, \eta_{\mu\nu}, \text{d}^4x\} \rightarrow \{\nabla_\mu, g_{\mu\nu}, \text{d}^4x\sqrt{|g|}\}$ have to performed in S_{m} before the variation is performed. The tensor $T^{\mu\nu}$ defined by this equation is called *dynamical* energy–momentum stress tensor. In order to show that this definition makes sense, we have to prove that the tensor is locally conserved, $\nabla_\mu T^{\mu\nu} = 0$, and we have to convince ourselves that this definition reproduces the standard results we know already.

Conservation of the stress tensor.   We start by proving that the dynamical energy–momentum tensor defined by Eq. (7.49) is locally conserved. We consider the change of the matter action under a variation of the metric,[3]

$$\delta S_{\text{m}} = -\frac{1}{2} \int_\Omega \text{d}^4x \sqrt{|g|}\, T^{\mu\nu} \delta g_{\mu\nu} = \frac{1}{2} \int_\Omega \text{d}^4x \sqrt{|g|}\, T_{\mu\nu} \delta g^{\mu\nu}. \tag{7.50}$$

We allow infinitesimal but otherwise arbitrary coordinate transformations,

$$\bar{x}^\mu = x^\mu + \xi^\mu(x). \tag{7.51}$$

Since the matter action S_m is a scalar, it has to be invariant under this coordinate transformation, $\delta S_m = 0$. For the resulting change $\delta g_{\mu\nu}$ in the metric we can use Eqs. (6.68b) and (6.69),

---

[3]We should view $g_{\mu\nu}$ (and not $g^{\mu\nu}$) as "the" gravitational field: in the Lagrangian of a point particle or the line element, the coordinates x^μ are contracted with $g_{\mu\nu}$. Having noted this point, we use simply the second relation in (7.50) in the future.

$$\delta g_{\mu\nu} = \nabla_\mu \xi_\nu + \nabla_\nu \xi_\mu. \tag{7.52}$$

Exploiting next that $T^{\mu\nu}$ is symmetric. we find

$$\delta S_m = -\int_\Omega d^4 x \sqrt{|g|}\, T^{\mu\nu} \nabla_\mu \xi_\nu = 0. \tag{7.53}$$

Now we apply the product rule,

$$\delta S_m = -\int_\Omega d^4 x \sqrt{|g|}\, (\nabla_\mu T^{\mu\nu}) \xi_\nu + \int_\Omega d^4 x \sqrt{|g|}\, \nabla_\mu (T^{\mu\nu} \xi_\nu) = 0. \tag{7.54}$$

The second term is a four-divergence and thus a boundary term that we can neglect. The remaining first term vanishes for arbitrary ξ_μ only, if the energy–momentum stress tensor is conserved,

$$\nabla_\mu T^{\mu\nu} = 0. \tag{7.55}$$

Hence the local conservation of energy–momentum is a consequence of the invariance of the field equations under general coordinate transformations, in the same way as current conservation follows from gauge invariance in electromagnetism. You should convince yourself that the dynamical energy–momentum stress tensor evaluated for the examples of the Klein–Gordon and the Maxwell field agrees with the symmetrised canonical stress tensor, cf. problem 7.10.

7.A Appendix: Large extra dimensions and massive gravity

Large extra dimensions. As mentioned in chapter 5, quantum corrections break the conformal invariance of string theory except we live in a world with $d = 10$ or 26 spacetime dimensions. There are two obvious answers to this result: first, one may conclude that string theory is disproven by nature or, second, one may adjust reality. Consistency of the second approach with experimental data could be achieved, if the $d - 4$ dimensions are compactified with a sufficiently small radius R, such that they are not visible in experiments sensible to wavelengths $\lambda \gg R$.

Let us check what happens to a scalar particle with mass m, if we add a fifth compact dimension y. The Klein–Gordon equation for a scalar field $\phi(x^\mu, y)$ becomes

$$(\Box_5 + m^2)\phi(x^\mu, y) = 0 \tag{7.56}$$

with the five-dimensional d'Alembert operator $\Box_5 = \Box - \partial_y^2$. The equation can be separated, $\phi(x^\mu, y) = \phi(x^\mu) f(y)$, and since the fifth dimension is compact, the spectrum of f is discrete. Assuming periodic boundary conditions, $f(x) = f(x + R)$, gives

$$\phi(x^\mu, y) = \phi(x^\mu) \cos(n\pi y/R). \tag{7.57}$$

The energy eigenvalues of these solutions are $\omega_{\boldsymbol{k},n}^2 = \boldsymbol{k}^2 + m^2 + (n\pi/R)^2$. From a four-dimensional point of view, the term $(n\pi/R)^2$ appears as a mass term, $m_n^2 = m^2 + (n\pi/R)^2$. Since we usually consider states with different masses as different particles, we see the five-dimensional particle as a tower of particles with mass m_n but otherwise identical quantum numbers. Such theories are called Kaluza–Klein theories, and the tower of particles Kaluza–Klein particles. If $R \ll \lambda$, where λ is the length-scale experimentally probed, only the $n = 0$ particle is visible and physics appears to be four-dimensional.

Since string theory includes gravity, one often assumes that the radius R of the extra-dimensions is determined by the Planck length, $R = 1/M_{\mathrm{Pl}} = (8\pi G_N)^{1/2} \sim 10^{-34}$ cm. In this case it is difficult to imagine any observational consequences of the additional dimensions. Of greater interest is the possibility that some of the extra dimensions are large,

$$R_{1,\ldots,\delta} \gg R_{\delta+1,\ldots,6} = 1/M_{\text{Pl}}.$$

Since the $1/r^2$ behaviour of the gravitational force is not tested below $d_* \sim$ mm scales, one can imagine that large extra dimensions exists that are only visible to gravity: Relating the $d = 4$ and $d > 4$ Newton's law $F \sim \frac{m_1 m_2}{r^{2+\delta}}$ at the intermediate scale $r = R$, we can derive the "true" value of the Planck scale in this model: Matching of Newton's law in 4 and $4 + \delta$ dimensions at $r = R$ gives

$$F(r = R) = G_N \frac{m_1 m_2}{R^2} = \frac{1}{M_D^{2+\delta}} \frac{m_1 m_2}{R^{2+\delta}}. \tag{7.58}$$

This equation relates the size R of the large extra dimensions to the true fundamental scale M_D of gravity in this model,

$$G_N^{-1} = 8\pi M_{\text{Pl}}^2 = R^\delta M_D^{\delta+2}, \tag{7.59}$$

while Newton's constant G_N becomes just an auxiliary quantity useful to describe physics at $r \gtrsim R$. (You may compare this to the case of weak interactions where Fermi's constant $G_F \propto g^2/m_W^2$ is determined by the weak coupling constant g and the mass m_W of the W-boson.)

Next we ask, if $M_D \sim$ TeV is possible, what would allow one to test such theories at accelerators as LHC. Inserting the measured value of G_N and $M_D = 1\,$TeV in Eq. (7.59) we find the required value for the size R of the large extra dimension as 10^{13} cm and 0.1 cm for $\delta = 1$ and 2, respectively, Thus the case $\delta = 1$ is excluded by the agreement of the dynamics of the solar system with four-dimensional Newtonian physics. The cases $\delta \geq 2$ are possible, because Newton's law is experimentally tested only for scales $r \gtrsim 1\,$mm.

Massive gravity.   Theories with extra dimensions contain often from our four-dimensional point of view a Kaluza–Klein tower of massive gravitons. Such modified theories of gravity have attracted great interest since one may hope to find an alternative explanation for the accelerated expansion of the universe. In problem 7.6, you are asked to derive the massive spin-2 propagator. As result, you should find

$$D_F^{\mu\nu;\rho\sigma}(k) = \frac{1}{2} \frac{-\frac{2}{3} G^{\mu\nu} G^{\rho\sigma} + G^{\mu\rho} G^{\nu\sigma} + G^{\mu\sigma} G^{\nu\rho}}{k^2 - m^2 + i\varepsilon}, \tag{7.60}$$

where

$$G^{\mu\nu}(k) = -\eta^{\mu\nu} + k^\mu k^\nu / m^2 \tag{7.61}$$

is the polarisation tensor for a massive spin-1 particle. Thus the nominator in the massive spin-2 propagator is as in the massless case a linear combination of the tensor products of two spin-1 polarisation tensors. However, the coefficients of the first term differ and thus the $k \to 0$ limit of the massive propagator does not agree with the massless case. In particular, the difference cannot be compensated by a rescaling of the coupling constant $\tilde{G}_N$, because it is not an overall factor. Imagine, for instance, that we determine the value of $\tilde{G}_N$ by calculating the potential energy of two non-relativistic sources like the Sun and the Earth. This requires, in massive gravity—for an arbitrarily small graviton mass—a gravitational coupling constant $\tilde{G}_N$ a factor 3/4 smaller than in the massless case. Having fixed $\tilde{G}_N$, we can predict the deflection of light by the Sun. Since the first term in the propagator couples the traces T^μ_μ of two sources, it does not contribute to the deflection of light. As a result of the reduced coupling strength, the deflection angle of light by the Sun decreases by the same factor and any non-zero graviton mass would be in conflict with observations.

When this result was first derived in 1970, its authors explained this discontinuity by the different number of degrees of freedom in the two theories: Even if the Compton wavelength of a massive graviton is larger than the observable size of the universe, and thus the Yukawa factor $\exp(-mr)$ indistinguishable from one, the additional spin states of a massive graviton may change physics. Two years later, Vainshtein realised that perturbation theory may break down in massive gravity and thus a calculation using one-graviton exchange is not reliable. More precisely, a theory of massive gravity contains an additional length scale $R_V = (GM/m^4)^{1/5}$ and for distances $r \ll R_V$ the theory has to be solved exactly.

Summary

Tensor fields satisfy second-order differential equations; their propagators are quadratic in p and thus even functions of x. As a result, tensor fields describe bosons, that is their field operators are commuting operators. Massless fields have only two, transverse degrees of freedom. A Lorentz-invariant description for such fields is only possible, if the remaining non-physical degrees of freedom are redundant. This redundancy implies that fields connected by a gauge transformation are equivalent and describe the same physics. In the case of photons, gauge symmetry implies that they couple to a conserved current, in the case of gravitons that they couple to the conserved stress tensor.

Further reading. Maggiore (2007) discusses in detail (massive) gravity as a spin-2 field in Minkowski space. The history of the gauge principle is reviewed by Jackson and Okun (2001).

Problems

7.1 Irreducible tensor components. Show that the splitting into symmetric and antisymmetric tensor components is in invariant under general coordinate transformations.

7.2 Polarisation of spin-1 particle. a.) Determine the polarisation vectors $\varepsilon_\mu^{(r)}(k)$ of a massive spin-1 particle for arbitrary k^μ. b.) Find the sum $\sum_{r=0}^3 \varepsilon_\mu^{(r)*} \varepsilon_\nu^{(r)}$ of the polarisation states of a massles spin-1 particle using the tensor method and imposing the two constraints $k_\mu \varepsilon^\mu = 0$ and $n_\mu \varepsilon^\mu = 0$, where n^μ is an arbitrary vector satisfying $n_\mu k^\mu \neq 0$.

7.3 Feynman propagator. Follow the steps from (3.25a) to (3.28) in the scalar case for a massive spin-1 propagator.

7.4 Physical states. a.) Determine the number of physical states of a photon and graviton in $d = 5$ spacetime, using that they are transverse polarised. b.) Generalise to arbitrary d.

7.5 Gravitational coupling. Determine the coupling λ in $\mathscr{L}_I = \lambda h^{\mu\nu} T_{\mu\nu}$ comparing (7.46) to Newton's gravitational potential.

7.6 Massive spin-2 propagator. Derive the propagator (7.60) of a massive spin-2 particle: Use the tensor method to write down the most general combination of tensors built from $\eta_{\mu\nu}$ and k_μ (or from $\pi_{\mu\nu}$ and k_μ), compatible with the symmetries and constraints on the polarisation tensor.

7.7 Helicity. Show that h in $\psi' = e^{-ih\alpha}\psi$ for a plane wave is the eigenvalue of the helicity operator $\boldsymbol{J}\cdot\boldsymbol{p}/|\boldsymbol{p}|$.

7.8 Scalar QED ♣. a.) Introduce in the Lagrangian (5.12) of a complex scalar field the interaction with photons via the minimal substitution, $\partial_\mu \rightarrow D_\mu = \partial_\mu + iqA_\mu$, Derive the Noether current and the current to which the photon couples (defined by $\Box A^\mu = j^\mu$). b.) Find the vertices of this theory. [Pay attention to the sign of the momentum of scalar particles.] c.) Write down the matrix element for "scalar Compton scattering" $\phi\gamma \rightarrow \phi\gamma$ and show that it is gauge invariant.

7.9 The $\phi\phi h_{\mu\nu}$ vertex. Find from the stress tensor of on-shell scalar particle the matrix elements $\langle p|' T_{\mu\nu} |p\rangle$ and determine thereby the Feynman rule for the $\phi\phi h_{\mu\nu}$ vertex.

7.10 Dynamical stress tensor. Derive the dynamical energy–momentum stress tensor $T^{\mu\nu}$ of a.) a real scalar field ϕ and b.) the photon field A^μ.

7.11 Geodesic equation from $\nabla_\mu T^{\mu\nu}$. The analogy with a pressureless fluid, $T^{\mu\nu} = \rho u^\mu u^\nu$, suggests

$$T^{\mu\nu}(\bar{x}) = \frac{m}{\sqrt{|g|}} \int d\tau \, \frac{dx^\mu}{d\tau} \, \frac{dx^\nu}{d\tau} \delta^{(4)}(\bar{x} - x(\tau))$$

as stress tensor of a point-particle moving along $x(\tau)$ with proper time τ. Show that $\nabla_\mu T^{\mu\nu}$ implies that the particle moves along a geodesic.

8
Fermions and the Dirac equation

Thus far we have discussed fields which transform as tensors under Lorentz transformations. These particles have integer spin or helicity. Since we can boost massive particles into their rest-frame, we can use our knowledge of non-relativistic quantum mechanics to anticipate that additional representations of the rotation group SO(3) and thus also of the Lorentz group SO(1, 3) exist which correspond to particles with half-integer spin. Such particles are described by anti-commuting variables that are the fundamental reason for the Pauli principle and thus the stability of matter.

8.1 Spinor representation of the Lorentz group

In order to introduce spinors we have to find the corresponding representation of the Lorentz group. As always, it is simpler to work in linear order, which in this case is the Lie algebra. The Lie algebra of the Poincaré group contains ten generators: the three generators J of rotations, the three generators K of Lorentz boosts and the four generators T of translations. The Killing vector fields V of Minkowski space generate these symmetries, and therefore the generators are given by the Killing vector fields. Thus we can use Eqs. (6.79) and (6.80) to calculate their commutation relations as (problem 8.1)

$$[J_i, J_j] = i\varepsilon_{ijk}J_k, \tag{8.1a}$$

$$[J_i, K_j] = i\varepsilon_{ijk}K_k, \tag{8.1b}$$

$$[K_i, K_j] = -i\varepsilon_{ijk}J_k. \tag{8.1c}$$

Here we followed physicist's convention and identified iV as the generators, so that they are Hermitian. Moreover, we restrict our attention to the Lorentz group which is sufficient to derive the concept of a Weyl spinor. Note that the algebra of the boost generators K is not closed. Thus in contrast to rotations, boosts do not form a subgroup of the Lorentz group.

The structure constants in all these commutation relations are $\pm\varepsilon_{ijk}$, suggesting that we can rewrite the Lorentz group as a product of two SU(2) factors. We try to decouple the two sets of generators J and K by introducing two non-Hermitian ladder operators

$$J^{\pm} = \frac{1}{2}(J \pm iK). \tag{8.2}$$

Their commutations relations are

Quantum Fields–From the Hubble to the Planck Scale. Michael Kachelriess. © Michael Kachelriess 2018.
Published in 2018 by Oxford University Press. DOI 10.1093/oso/9780198802877.001.0001

$$[J_i^+, J_j^+] = i\varepsilon_{ijk} J_k^+, \tag{8.3a}$$

$$[J_i^-, J_j^-] = i\varepsilon_{ijk} J_k^-, \tag{8.3b}$$

$$[J_i^+, J_j^-] = 0 \qquad i,j = 1,2,3. \tag{8.3c}$$

Thus $\boldsymbol{J}^-$ and $\boldsymbol{J}^+$ commute with each other and each generate an SU(2) group. The Lorentz group is[1] therefore $\sim$ SU(2) $\otimes$ SU(2), and states transforming in a well-defined way are labelled by a pair of angular momenta, (j^-, j^+), corresponding to the eigenvalues of J_z^- and J_z^+, respectively. From our knowledge of the angular momentum algebra in non-relativistic quantum mechanics, we conclude that the dimension of the representation (j^-, j^+) is $(2j^- + 1)(2j^+ + 1)$. Because of $\boldsymbol{J} = \boldsymbol{J}^- + \boldsymbol{J}^+$, the representation (j^-, j^+) contains all possible spins j in integer steps from $|j^- - j^+|$ to $j^- + j^+$.

The representation $(0,0)$ has dimension 1, transforms trivially, $\boldsymbol{J} = \boldsymbol{K} = 0$, and corresponds therefore to the scalar representation. The two smallest non-trivial representations are $\boldsymbol{J}^+ = 0$, i.e. $(j^-, 0)$ with $\boldsymbol{J}^{(1/2)} = -i\boldsymbol{K}^{(1/2)}$, and $\boldsymbol{J}^- = 0$, i.e. $(0, j^+)$ with $\boldsymbol{J}^{(1/2)} = i\boldsymbol{K}^{(1/2)}$. Both representations have spin 1/2 and dimension 2. We define therefore two types of two-component spinors,

$$\phi_L: \quad (1/2, 0), \quad \boldsymbol{J}^{(1/2)} = \boldsymbol{\sigma}/2, \quad \boldsymbol{K}^{(1/2)} = +i\boldsymbol{\sigma}/2, \tag{8.4a}$$

$$\phi_R: \quad (0, 1/2), \quad \boldsymbol{J}^{(1/2)} = \boldsymbol{\sigma}/2, \quad \boldsymbol{K}^{(1/2)} = -i\boldsymbol{\sigma}/2, \tag{8.4b}$$

which we call left-chiral and right-chiral Weyl spinors. These Weyl spinors form the fundamental representation of the Lorentz group. All higher spin states can be obtained as tensor products involving them. Their transformation properties under an (active) finite Lorentz transformation with parameters $\boldsymbol{\alpha}$ and $\boldsymbol{\eta}$ follow by exponentiating their generators as $\exp(-i\boldsymbol{J}\boldsymbol{\alpha} + i\boldsymbol{K}\boldsymbol{\eta})$ (compare to appendix B.3 for our choice of signs),

$$\phi_L \to \phi_L' = \exp\left[-\frac{i\boldsymbol{\sigma}\boldsymbol{\alpha}}{2} - \frac{\boldsymbol{\sigma}\boldsymbol{\eta}}{2}\right] \phi_L \equiv S_L \phi_L, \tag{8.5a}$$

$$\phi_R \to \phi_R' = \exp\left[-\frac{i\boldsymbol{\sigma}\boldsymbol{\alpha}}{2} + \frac{\boldsymbol{\sigma}\boldsymbol{\eta}}{2}\right] \phi_R \equiv S_R \phi_R. \tag{8.5b}$$

While the transformation matrices S_L and S_R agree for rotations, the terms describing Lorentz boosts have opposite signs. Note also that only rotations are described by a unitary transformation, while Lorentz boosts lead to a non-unitary transformation of the Weyl spinors.

We ask now if we can convert a left- into a right-chiral spinor and vice versa. Thus we should find a spinor $\tilde{\phi}_L$ constructed out of ϕ_L which transforms as $S_R\tilde{\phi}_L$. Changing S_L into S_R requires reversing the relative sign between the rotation and the boost term, which we achieve by complex conjugating ϕ_L,

$$\phi_L^{*\prime} = \left[1 + \frac{i\boldsymbol{\sigma}^*\boldsymbol{\alpha}}{2} - \frac{\boldsymbol{\sigma}^*\boldsymbol{\eta}}{2} + \dots\right] \phi_L^*. \tag{8.6}$$

Because of $\sigma_1^* = \sigma_1$, $\sigma_2^* = -\sigma_2$, $\sigma_3^* = \sigma_3$, and $\sigma_1\sigma_2 = -\sigma_2\sigma_1$, $\sigma_2\sigma_3 = -\sigma_3\sigma_2$, we obtain the desired transformation property multiplying $\phi_L^{*\prime}$ with σ_2,

---

[1] More precisely, they have the same Lie algebra and are thus locally isomorphic but differ globally.

$$\sigma_2\phi_L^{*\prime} = \sigma_2 \left[1 + \frac{\mathrm{i}(\sigma_1, -\sigma_2, \sigma_3)\boldsymbol{\alpha}}{2} - \frac{(\sigma_1, -\sigma_2, \sigma_3)\boldsymbol{\eta}}{2} + \dots \right] \phi_L^* \qquad (8.7\mathrm{a})$$

$$= \left[1 - \frac{\mathrm{i}\boldsymbol{\sigma}\boldsymbol{\alpha}}{2} + \frac{\boldsymbol{\sigma}\boldsymbol{\eta}}{2} + \dots \right] \sigma_2\phi_L^* = S_R\sigma_2\phi_L^*. \qquad (8.7\mathrm{b})$$

Thus ϕ_L and ϕ_R are connected by a non-unitary transformation, and therefore ϕ_L and ϕ_R describe different physics. Obviously, we can add to $\sigma_2\phi_L^*$ an arbitrary phase $\mathrm{e}^{\mathrm{i}\delta}$ without changing the transformation properties. We define

$$\phi_R^c \equiv -\mathrm{i}\sigma_2\phi_L^* \quad \text{and} \quad \phi_L^c \equiv \mathrm{i}\sigma_2\phi_R^*, \qquad (8.8)$$

which ensures $(\phi_L^c)^c = \phi_L$ and $(\phi_R^c)^c = \phi_R$. Later, when we discuss the coupling of a fermion to an external field, we will see that ϕ_L^c is the charge conjugated spinor of ϕ_L.

For the construction of a Lagrangian we need for the mass term scalars and for the kinetic energy vectors built out of the Weyl fields. Both the kinetic and the mass terms should be real to provide a real Lagrangian. In contrast to the real Lorentz transformation $\Lambda^\mu{}_\nu$ acting on tensor fields, the matrices $S_{L/R}$ are, however, complex and thus the Weyl fields are complex too. This suggests together with the fact that a measurement device should be the same after a rotation by 2π that observables are bilinear quantities in the fermion fields such that they transform tensorial and their eigenvalues are real.

Out of the two Weyl spinors, we can form four different products $\phi_{L/R}^\dagger\phi_{L/R}$ leading to the combinations $S_L^\dagger S_L$, $S_R^\dagger S_R$, $S_L^\dagger S_R$ and $S_R^\dagger S_L$. The rotation $\mathrm{i}\boldsymbol{\sigma}\boldsymbol{\alpha}/2$ cancels in all four products, since it enters with the same sign in S_L and S_R, and the Pauli matrices are Hermitian, $\boldsymbol{\sigma}^\dagger = \boldsymbol{\sigma}$. By contrast, the cancellation of the boost $\boldsymbol{\sigma}\boldsymbol{\eta}/2$ requires a combination of a left- and right-chiral field,

$$\phi_L^{\prime\dagger}\phi_R^\prime = \phi_L^\dagger \left[1 + \mathrm{i}\frac{\boldsymbol{\sigma}\boldsymbol{\alpha}}{2} - \frac{\boldsymbol{\sigma}\boldsymbol{\eta}}{2} + \dots \right] \left[1 - \mathrm{i}\frac{\boldsymbol{\sigma}\boldsymbol{\alpha}}{2} + \frac{\boldsymbol{\sigma}\boldsymbol{\eta}}{2} + \dots \right] \phi_R = \phi_L^\dagger\phi_R, \qquad (8.9)$$

and similarly for $\phi_R^\dagger\phi_L$. Thus $\phi_L^\dagger\phi_R$ and $\phi_R^\dagger\phi_L$ transform as Lorentz scalars, but not $\phi_L^\dagger\phi_L$ and $\phi_R^\dagger\phi_R$. So what are the transformation properties of the latter two products? Performing an infinitesimal boost along the z axis, we find

$$\phi_R^{\prime\dagger}\phi_R^\prime = \phi_R^\dagger \left[1 + \frac{\sigma_3\eta}{2} + \dots \right] \left[1 + \frac{\sigma_3\eta}{2} + \dots \right] \phi_R = \phi_R^\dagger\phi_R + \eta\phi_R^\dagger\sigma_3\phi_R. \qquad (8.10)$$

This looks like an infinitesimal Lorentz transformation of the time-like component $j^0 = \phi_R^\dagger\phi_R$ of a four-vector j^μ. If this interpretation is correct, we should be able to associate the spatial part $\boldsymbol{j}$ with $\phi_R^\dagger\boldsymbol{\sigma}\phi_R$. Checking thus how $\boldsymbol{j}$ transforms, we find using $\sigma^i\sigma^j = \delta^{ij} + \mathrm{i}\varepsilon^{ijk}\sigma^k$ that j^1 and j^2 are invariant, while j^3 transforms as

$$\phi_R^{\prime\dagger}\sigma_3\phi_R^\prime = \phi_R^\dagger \left[1 + \frac{\sigma_3\eta}{2} + \dots \right] \sigma_3 \left[1 + \frac{\sigma_3\eta}{2} + \dots \right] \phi_R = \eta\phi_R^\dagger\phi_R + \phi_R^\dagger\sigma_3\phi_R. \qquad (8.11)$$

Thus $\phi_R^\dagger\sigma^\mu\phi_R$ with $\sigma^\mu \equiv (1, \boldsymbol{\sigma})$ transforms as a four-vector, $j^0 \to j^0 + \eta j^3$ and $j^3 \to \eta j^0 + j^3$. Performing the same calculation for the left-chiral fields reproduces the same result except for an opposite sign of η. We account for this sign change setting now $\bar{\sigma}^\mu \equiv (1, -\boldsymbol{\sigma})$, so that a four-vector bilinear in ϕ_L is given by $\phi_L^\dagger \bar{\sigma}^\mu \phi_L$.

The transformations S_L and S_R that belong to the restricted Lorentz group do not mix the left- and right-chiral Weyl spinors. Consider however the effect of a parity transformation, $P\boldsymbol{x} = -\boldsymbol{x}$, on the generators $\boldsymbol{K}$ and $\boldsymbol{J}$. The velocity changes sign, $\boldsymbol{v} \to -\boldsymbol{v}$, it transforms as a polar vector, while the angular momentum $\boldsymbol{J}$ as axial vector remains invariant. Thus parity interchanges $(1/2, 0)$ and $(0, 1/2)$ and hence ϕ_L and ϕ_R, as one would expect from a left- and right-chiral object. If parity is a symmetry of the theory examined, one can therefore not consider separately the two spinors ϕ_L and ϕ_R. Instead, it proves useful to combine them into a four-spinor called Dirac (or bi-spinor)

$$\psi = \begin{pmatrix} \phi_L \\ \phi_R \end{pmatrix}. \tag{8.12}$$

Another reason to consider Dirac spinors is that the scalar terms $\phi_L^\dagger \phi_R$ and $\phi_R^\dagger \phi_L$ that qualify as mass terms combine a left- and a right-chiral field. Thus the description of a particle with such a mass term seems to require the use of both left- and right-chiral Weyl spinors. Next we will derive a field equation for this type of spinor and discuss its properties.

8.2   Dirac equation

From Weyl spinors to the Dirac equation.   We can obtain the spinor describing a particle with momentum p by boosting the one describing a particle at rest,

$$\phi_R(p) = \exp\left[\frac{\boldsymbol{\sigma}\boldsymbol{\eta}}{2}\right] \phi_R(0) = \exp\left[\frac{\eta\boldsymbol{\sigma}\boldsymbol{n}}{2}\right] \phi_R(0) = [\cosh(\eta/2) + \boldsymbol{\sigma}\boldsymbol{n} \sinh(\eta/2)]\, \phi_R(0). \tag{8.13}$$

If we replace the boost parameter η by the Lorentz factor[2] $\gamma = \cosh\eta$ and use the identities $\cosh(\eta/2) = \sqrt{(\cosh\eta + 1)/2}$ and $\sinh(\eta/2) = \sqrt{(\cosh\eta - 1)/2}$, we can express the spinor as

$$\phi_R(p) = \left[\left(\frac{\gamma + 1}{2}\right)^{1/2} + \boldsymbol{\sigma}\hat{\boldsymbol{p}}\left(\frac{\gamma - 1}{2}\right)^{1/2}\right] \phi_R(0). \tag{8.14}$$

Here $\hat{\boldsymbol{p}} = \boldsymbol{p}/|\boldsymbol{p}|$ is the unit vector in direction of $\boldsymbol{p}$. Inserting $\gamma = E/m$ and combining the two terms in the angular bracket, we arrive at

$$\phi_R(p) = \frac{E + m + \boldsymbol{\sigma}\boldsymbol{p}}{\sqrt{2m(E + m)}} \phi_R(0). \tag{8.15}$$

Similarly, we find

$$\phi_L(p) = \frac{E + m - \boldsymbol{\sigma}\boldsymbol{p}}{\sqrt{2m(E + m)}} \phi_L(0). \tag{8.16}$$

Thus ϕ_L and ϕ_R differ only by the sign of the operator $\boldsymbol{\sigma}\boldsymbol{p}$ which measures the projection of the spin $\boldsymbol{\sigma}$ on the momentum $\boldsymbol{p}$ of the particle. For a particle at rest, this

---

[2]Recall the relations $E = m \cosh\eta$ and $p = m \sinh\eta$ connecting E, p, and the rapidity η.

difference disappears and we set therefore $\phi_L(0) = \phi_R(0)$. This allows us to eliminate the zero momentum spinors, giving

$$\phi_L^R(p) = \frac{E \pm \boldsymbol{\sigma p}}{m} \phi_R^L(p). \tag{8.17}$$

In matrix form, these two equations correspond to

$$\begin{pmatrix} -m & E - \boldsymbol{\sigma p} \\ E + \boldsymbol{\sigma p} & -m \end{pmatrix} \begin{pmatrix} \phi_L(p) \\ \phi_R(p) \end{pmatrix} = \begin{pmatrix} -m & \sigma^\mu p_\mu \\ \bar{\sigma}^\mu p_\mu & -m \end{pmatrix} \begin{pmatrix} \phi_L(p) \\ \phi_R(p) \end{pmatrix} = 0. \tag{8.18}$$

We introduce the 4×4 matrices

$$\gamma^\mu = \begin{pmatrix} 0 & \sigma^\mu \\ \bar{\sigma}^\mu & 0 \end{pmatrix}. \tag{8.19}$$

Then we arrive with $\gamma^\mu p_\mu = \gamma^0 E - \boldsymbol{\gamma p}$ at the compact expression

$$(\gamma^\mu p_\mu - m)\,\psi(p) = 0. \tag{8.20}$$

Setting $p_\mu = i\partial_\mu$ we obtain the Dirac equation $(i\partial_\mu \gamma^\mu - m)\psi(x) = 0$ in coordinate space. The representation used for the Dirac spinor and the gamma matrices is called chiral or Weyl representation. Other representations can be obtained performing a unitary transformation, $U\tilde{\gamma}^\mu U^\dagger = \gamma^\mu$ and $U\tilde{\psi} = \psi$.

We can apply the tensor method to derive a definition of the gamma matrices and their properties which is independent of the considered representation. The only invariant tensor at our disposal is the metric tensor $\eta^{\mu\nu}$ and thus the gamma matrices have to satisfy $\{\gamma^\mu, \gamma^\nu\} = A\eta^{\mu\nu}$. Considering $\{\gamma^0, \gamma^0\}$ shows that $A = 2$, or

$$\{\gamma^\mu, \gamma^\nu\} = 2\eta^{\mu\nu}. \tag{8.21}$$

These anti-commutation relations define a Clifford algebra, implying

$$\left(\gamma^0\right)^2 = 1, \qquad \left(\gamma^i\right)^2 = -1 \qquad \text{and} \qquad \gamma^\mu\gamma^\nu = -\gamma^\nu\gamma^\mu \tag{8.22}$$

for $\mu \neq \nu$. The last condition shows that the Clifford algebra cannot be satisfied by normal numbers.

The definition (8.21) implies that we can apply in the usual way the metric tensor to raise or to lower the indices of the gamma matrices, $\gamma_\mu = \eta_{\mu\nu}\gamma^\nu$. Thus we can write $\gamma^\nu\partial_\nu = \gamma_\nu\partial^\nu$. Since the contraction of the gamma matrices γ^μ with a four-vector A^μ will appear frequently, we introduce the so-called Feynman slash,

$$\slashed{A} \equiv A_\mu\gamma^\mu, \tag{8.23}$$

as a useful shortcut. This notation also stresses that the gamma matrices γ^μ allow us to map a four-vector A^μ onto an element $\slashed{A}$ of the Clifford algebra which then can be applied on a spinor ψ. Although we suppress the spinor indices, you should keep in mind that the matrices $(\gamma_{ab})^\mu$ carry both tensor and spinor indices.

Dirac's way towards the Dirac equation. The Klein–Gordon equation was historically the first wave equation derived in relativistic quantum mechanics. Applied to the hydrogen atom, it failed to reproduce the correct energy spectrum. Dirac tried to derive as an alternative an equation linear in the derivatives ∂_μ. Since Lorentz invariance requires that ∂_μ has to be contracted with another object carrying the Lorentz index μ, a first-order equation has the form

$$(i\gamma^\mu \partial_\mu - m)\psi(x) = 0. \tag{8.24}$$

The main task for Dirac was to uncover the nature of the quantities γ^μ in this equation. They cannot be normal numbers, since then they would form a four-vector, specify one direction in spacetime and thus break Lorentz invariance. Multiplying the Dirac equation with $-(i\gamma^\mu \partial_\mu + m)$ and comparing the result to the Klein–Gordon equation, we find

$$-(i\gamma^\mu \partial_\mu + m)(i\gamma^\nu \partial_\nu - m)\psi = (\gamma^\mu \gamma^\nu \partial_\mu \partial_\nu + m^2)\psi = (\Box + m^2)\psi = 0. \tag{8.25}$$

Using the symmetry of partial derivatives, we can rewrite

$$\gamma^\mu \gamma^\nu \partial_\mu \partial_\nu = \frac{1}{2}\{\gamma^\mu, \gamma^\nu\}\partial_\mu \partial_\nu. \tag{8.26}$$

Remembering next the definition of the d'Alembert operator, $\Box = \eta^{\mu\nu}\partial_\mu \partial_\nu$, we re-derive that the γ^μ form a Clifford algebra.

Lagrange density. For a complex scalar field, we could rewrite after a partial integration the Lagrange density as $\mathscr{L} = -\phi^\dagger(\Box + m^2)\phi$. This expression corresponds to the Klein–Gordon operator $-(\Box + m^2)$ sandwiched between the quadratic form $\phi^\dagger\phi$, as the correspondence of the propagator and a two-point Green function requires. This suggests to try for the Dirac field as Lagrangian

$$\mathscr{L} = \psi^\dagger A(i\gamma^\mu \partial_\mu - m)\psi = \bar{\psi}(i\gamma^\mu \partial_\mu - m)\psi, \tag{8.27}$$

where we have used as quadratic from $\psi^\dagger A\psi$ with a matrix A yet to be determined. In the second step, we defined the *adjoint* spinor $\bar{\psi} = \psi^\dagger A$. Varying then the action $S[\psi, \bar{\psi}]$, we obtain

$$\delta S = \int d^4x \left\{ \delta\bar{\psi}(i\gamma^\mu \partial_\mu - m)\psi - \bar{\psi}(i\gamma^\mu \overleftarrow{\partial}_\mu + m)\delta\psi \right\}. \tag{8.28}$$

Here we made a partial integration of the $\partial_\mu \delta\psi$ term, and thus the derivative $\overleftarrow{\partial}_\mu$ acts to the left. Since we treat ψ and $\bar{\psi}$ as two independent variables, we obtain from $\delta S = 0$ two equations of motion,

$$\bar{\psi}(i\gamma^\mu \overleftarrow{\partial}_\mu + m) = 0 \quad \text{and} \quad (i\gamma^\mu \partial_\mu - m)\psi = 0. \tag{8.29}$$

Next we determine the unknown matrix A. Taking the Hermitian conjugate of the RHS of (8.29) results in

$$\psi^\dagger(-i\gamma^{\mu\dagger}\overleftarrow{\partial}_\mu - m) = 0. \tag{8.30}$$

This agrees with the LHS of (8.29), if A satisfies

$$A^{-1}\gamma^{\mu\dagger}A = \gamma^\mu. \tag{8.31}$$

One can readily check that the γ^μ matrices in the Weyl representation fulfil this relation, if we set $A = \gamma^0$. With $(\gamma^0)^2 = 1$ and $(\gamma^0)^{-1} = (\gamma^0)^\dagger = \gamma^0$, we can express this condition as

$$\gamma^{\mu\dagger} = \gamma^0\gamma^\mu\gamma^0 = \begin{cases} (\gamma^0)^2\gamma^0 = \gamma^0, \\ -\gamma^i(\gamma^0)^2 = -\gamma^i. \end{cases} \tag{8.32}$$

Thus the action principle implies that γ^0 is Hermitian, while the γ^i are anti-Hermitian matrices.

Using the gamma matrices and the Dirac spinor in the chiral representation it is straightforward to express the Dirac Lagrangian (8.27) by Weyl fields,

$$\mathscr{L} = i\phi_R^\dagger \sigma^\mu \partial_\mu \phi_R + i\phi_L^\dagger \bar{\sigma}^\mu \partial_\mu \phi_L - m(\phi_L^\dagger \phi_R + \phi_R^\dagger \phi_L). \tag{8.33}$$

This implies that the Dirac Lagrangian and the Dirac equation are invariant under Lorentz transformations because we have already checked that all ingredients of (8.33) are invariant. Note also that out of the two possible combinations of the two Lorentz scalars we found only the one invariant under parity entered the mass term. Moreover, $P(\sigma^\mu \partial_\mu) = \bar{\sigma}^\mu \partial_\mu$, and thus the combination of the kinetic energies of ϕ_L and ϕ_R is also invariant under parity.

Hamiltonian form. The Dirac equation can be transformed into Hamiltonian form by multiplying with γ^0,

$$i\partial_t \psi = H_D \psi = (-i\gamma^0\gamma^i\partial_i + \gamma^0 m)\psi. \tag{8.34}$$

Looking back at the (anti-) Hermiticity properties (8.32) of the γ^μ matrices, we see that they correspond to the one required to make the Dirac Hamiltonian Hermitian. By tradition, one rewrites H_D often with $\beta = \gamma^0$ and $\alpha^i = \gamma^0\gamma^i$ as

$$i\partial_t \psi = H_D \psi = (\boldsymbol{\alpha} \cdot \boldsymbol{p} + \beta m)\psi. \tag{8.35}$$

Considering the semi-classical limit, one sees that the matrix $\boldsymbol{\alpha}$ has the meaning of a velocity operator, see problem 8.6.

Clifford algebra and bilinear quantities. We now determine the minimal matrix representation for the Clifford algebra defined by Eq. (8.21). We find first the maximal number of independent products that we can form out of the four gamma matrices. Five obvious elements are the unit matrix $1 = (\gamma^0)^2$ and the four gamma matrices γ^μ themselves. Because of $(\gamma^\mu)^2 = \pm 1$, the remaining products should consist of γ^μ matrices with different indices. Thus the only product of four γ^μ matrices that we have to consider is $\gamma^0\gamma^1\gamma^2\gamma^3$. This combination will appear very often and thus deserves a special name. Including the imaginary unit to make it Hermitian, we define

$$\gamma^5 \equiv \gamma_5 \equiv i\gamma^0\gamma^1\gamma^2\gamma^3. \tag{8.36}$$

Because the four gamma matrices in γ^5 anti-commute, we can rewrite its definition introducing the completely antisymmetric Levi–Civita tensor $\varepsilon_{\alpha\beta\gamma\delta}$ in four dimensions as

$$\gamma^5 = \frac{\mathrm{i}}{24}\,\varepsilon_{\alpha\beta\gamma\delta}\,\gamma^\alpha\gamma^\beta\gamma^\gamma\gamma^\delta. \tag{8.37}$$

This suggests that bilinear quantities containing one γ^5 matrix transform as pseudo-tensors, that is, they change sign under a parity transformation $x \to -x$. Two important properties of the γ^5 matrix are $(\gamma^5)^2 = 1$ and $\{\gamma^\mu, \gamma^5\} = 0$.

Next we consider products of three γ^μ matrices. For instance,

$$\gamma^1\gamma^2\gamma^3 = \underbrace{\gamma^0\gamma^0}_{1}\gamma^1\gamma^2\gamma^3 = -\mathrm{i}\gamma^0\gamma^5. \tag{8.38}$$

Hence these products are equivalent to $\gamma^\mu\gamma^5$, giving us four more basis elements.

Finally, we are left with products of two different γ^μ matrices. We can associate these six products with the commutator $[\gamma^\mu, \gamma^\nu]$. Adding again for later convenience a factor i/2, we define the antisymmetric tensor $\sigma^{\mu\nu}$ as

$$\sigma^{\mu\nu} \equiv \frac{\mathrm{i}}{2}[\gamma^\mu, \gamma^\nu]. \tag{8.39}$$

The six matrices $\sigma^{\mu\nu}$ are the remaining independent elements we choose as basis for our matrix representation of the Clifford algebra. All together, the basis has dimension 16,

$$\Gamma = \{1, \gamma^5, \gamma^\mu, \gamma^5\gamma^\mu, \sigma^{\mu\nu}\}, \tag{8.40}$$

as the 4×4 matrices. Hence an arbitrary 4×4 matrix can be decomposed into a linear combination of these basis elements. Moreover, the smallest matrix representation of the Clifford algebra is given by 4×4 matrices. Some useful properties of gamma matrices are collected in appendix A.2.

Knowing the dimension of the gamma matrices, we can count the number of degrees of freedom represented by a Dirac spinor ψ. As the γ matrices and the Lorentz transformation acting on spinors are complex, the field ψ is complex too and has thus four complex degrees of freedom. We know already that the Dirac equation describes spin 1/2 particles, which come with $2s + 1 = 2$ spin degrees of freedom for a particle plus 2 for its antiparticle. Thus in this case the number of physical states matches the four components of the fields ψ. Note also the difference to the case of a complex scalar or vector field. There we introduced two complex fields $\phi^\pm = (\phi_1 \pm \mathrm{i}\phi_2)/\sqrt{2}$, which are connected by $(\phi^\pm)^* = \phi^\mp$. The real fields ϕ_1 and ϕ_2 are not mixed by Lorentz transformations and thus we count them as two real degrees of freedom.

We come now to the construction of bilinear quantities out of the Dirac spinors. Since the Lagrangian is a scalar, we know already that $\bar{\psi}\psi$ transforms as a scalar while $j^\mu = \bar{\psi}\gamma^\mu\psi$ is vector. In general, bilinear quantities are constructed as

$$\psi^\dagger\gamma^0\Gamma\psi \equiv \bar{\psi}\Gamma\psi, \tag{8.41}$$

where $\bar{\psi} = \psi^\dagger\gamma^0$ is the adjoint spinor and Γ is any of the 16 basis elements given in Eq. (8.40). In this way, the complex conjugated of a bilinear becomes

$$(\bar{\psi}\Gamma\psi')^* = (\bar{\psi}\Gamma\psi')^\dagger = \psi'^\dagger\Gamma^\dagger\gamma^0\psi = \psi'^\dagger\gamma^0\gamma^0\Gamma^\dagger\gamma^0\psi \equiv \bar{\psi}'\overline{\Gamma}\psi \tag{8.42}$$

with

$$\overline{\Gamma} \equiv \gamma^0\Gamma^\dagger\gamma^0. \tag{8.43}$$

For $\psi = \psi'$, these bilinears are real as desired. The analogue $\psi^\dagger\psi$ to the probability density $\psi^*\psi$ of the Schrödinger equation is thus the zero-component of a four-current, $\psi^\dagger\psi = \psi^\dagger\gamma^0\gamma^0\psi = \bar{\psi}\gamma^0\psi = j^0$, as one should expect in a relativistic theory.

Finally, we note that γ^0 and γ^5 are involutory matrices, that is, they satisfy the relation $A^2 = 1$. Because of $(1 \pm A)^2 = 2(1 \pm A)$, we can construct the projection operators $P_\pm = (1 \pm A)/2$, satisfying

$$P_\pm^2 = P_\pm, \qquad P_\pm P_\mp = 0, \quad \text{and} \quad P_+ + P_- = 1.$$

Thus we should be able to classify the four independent solutions of the Dirac equation with the help of $(1 \pm \gamma^0)/2$ and $(1 \pm \gamma^5)/2$, or their suitable covariant generalisations.

Lorentz transformations.   Our derivation of Weyl spinors as elements of the fundamental representation of the Lorentz group provided automatically their transformation properties under a finite Lorentz transformation. Using the Weyl representation, the transformation law for a Dirac spinor follows as

$$\psi(x) \to \psi(x') = S(\Lambda)\psi(x) = \begin{pmatrix} \phi_L(x') \\ \phi_R(x') \end{pmatrix} = \begin{pmatrix} S_L & 0 \\ 0 & S_R \end{pmatrix} \begin{pmatrix} \phi_L(x) \\ \phi_R(x) \end{pmatrix}. \tag{8.44}$$

We want to express the transformation matrix $S(\Lambda)$ by gamma matrices, such that it is representation independent and manifestly Lorentz invariant. We set

$$S(\Lambda) = \exp\left(-i\omega_{\mu\nu}J^{\mu\nu}/2\right), \tag{8.45}$$

where the antisymmetric matrix $\omega_{\mu\nu}$ parameterises the Lorentz transformation and the six generators $(J_{ab})_{\mu\nu}$ have to be determined. Since the generators are the covariant generalisation of $\boldsymbol{\sigma}$, we suspect that they are connected to $\sigma_{\mu\nu}$. Using Eq. (8.19), we obtain as explicit expression for the $\sigma^{\mu\nu}$ matrices in the Weyl representation

$$\sigma^{0i} = i\begin{pmatrix} -\sigma^i & 0 \\ 0 & \sigma^i \end{pmatrix}, \quad \text{and} \quad \sigma^{ij} = \varepsilon_{ijk}\begin{pmatrix} \sigma^k & 0 \\ 0 & \sigma^k \end{pmatrix}. \tag{8.46}$$

We split $J^{\mu\nu}$ into boosts and rotations,

$$\frac{1}{2}\omega_{\mu\nu}J^{\mu\nu} = \omega_{i0}J^{i0} + \omega_{12}J^{12} + \omega_{13}J^{13} + \omega_{23}J^{23}. \tag{8.47}$$

Identifying $\eta_i = \omega_{i0}$ and $\alpha_i = (1/2)\varepsilon_{ijk}\omega_{jk}$, we obtain $J^{\mu\nu} = \sigma^{\mu\nu}/2$. In contrast to (8.44), the expression $S(\Lambda) = \exp\left(-i\sigma_{\mu\nu}\omega^{\mu\nu}/4\right)$ is valid for any representation of the gamma matrices.

Solutions. We search for plane-wave solutions ue^{-ipx} and ve^{+ipx} of the Dirac equation with $m > 0$ and $E = p^0 = |\mathbf{p}| > 0$. The algebra is simplified, if we construct the solutions first in the rest-frame of the particle. Then $\not{p} = m\gamma^0$, and thus the use of the Dirac representation,

$$\gamma^0 = 1 \otimes \tau_3 = \begin{pmatrix} 1 & 0 \\ 0 & -1 \end{pmatrix}, \quad \text{and} \quad \gamma^i = \sigma^i \otimes i\tau_2 = \begin{pmatrix} 0 & \sigma^i \\ -\sigma^i & 0 \end{pmatrix}, \quad (8.48)$$

where γ^0 is diagonal is most convenient. Here σ_i and τ_i are the Pauli matrices, $\otimes$ denotes the tensor product, 0 and 1 are 2×2 matrices. In the Dirac representation, the γ^5 matrix is off-diagonal,

$$\gamma^5 = 1 \otimes \tau_1 = \begin{pmatrix} 0 & 1 \\ 1 & 0 \end{pmatrix}. \quad (8.49)$$

The Dirac equation becomes

$$(\not{p} - m)u = m(\gamma^0 - 1)u = 0 \quad (8.50\text{a})$$

$$(\not{p} + m)v = m(\gamma^0 + 1)v = 0. \quad (8.50\text{b})$$

The RHS shows that $(1 \pm \gamma^0)/2$ project a general spinor at rest on the subspaces of solutions with positive or negative energy. Inserting the explicit form of γ^0 into (8.50), the four solutions in the rest-frame of the particle follow as

$$u(m, +) \propto \begin{pmatrix} 1 \\ 0 \\ 0 \\ 0 \end{pmatrix}, \quad u(m, -) \propto \begin{pmatrix} 0 \\ 1 \\ 0 \\ 0 \end{pmatrix}, \quad v(m, -) \propto \begin{pmatrix} 0 \\ 0 \\ 1 \\ 0 \end{pmatrix}, \quad v(m, +) \propto \begin{pmatrix} 0 \\ 0 \\ 0 \\ 1 \end{pmatrix}.$$

$$(8.51)$$

The additional $\pm$ label should be the quantum number of a suitable operator labelling the two spin states of a Dirac particle. Note the opposite order of the spin label in the v spinor compared to u. We will see later that this choice is required by the structure of the relativistic spin operator s^μ. As an intuitive argument, we add that this labelling corresponds to our interpretation of antiparticles as particles moving backwards in time. The spinor v describes two states with negative energy, negative three-momentum $\mathbf{p}$ and negative spin s relative to u.

The solutions are orthogonal,

$$\bar{u}(p, s)u(p, s') = \mathcal{N}^2 \delta_{s,s'} \quad \text{and} \quad \bar{v}(p, s)v(p, s') = -\mathcal{N}^2 \delta_{s,s'}, \quad (8.52)$$

but not normalised to 1. Note also the minus sign introduced in $\bar{v}v$ by the corresponding minus in the (3,4) corner of γ^0. Since we know that $\psi^\dagger \psi$ is the zero component of a four-vector, the normalisation of the corresponding spinor products is

$$u^\dagger(p, s)u(p, s') = \mathcal{N}^2 \frac{E_p}{m} \delta_{s,s'} \quad \text{and} \quad v^\dagger(p, s)v(p, s') = -\mathcal{N}^2 \frac{E_p}{m} \delta_{s,s'}. \quad (8.53)$$

Summing over spins, we obtain in the rest-frame

$$\sum_s u_a(m,s)\bar{u}_b(m,s) = \begin{pmatrix} 1 & 0 \\ 0 & 0 \end{pmatrix}_{ab} \mathcal{N}^2 = \frac{1}{2}(\gamma^0 + 1)_{ab}\mathcal{N}^2, \tag{8.54}$$

$$\sum_s v_a(m,s)\bar{v}_b(m,s) = \begin{pmatrix} 0 & 0 \\ 0 & -1 \end{pmatrix}_{ab} \mathcal{N}^2 = \frac{1}{2}(\gamma^0 - 1)_{ab}\mathcal{N}^2. \tag{8.55}$$

We saw that $\gamma^0 \pm 1$ corresponds in an arbitrary frame to $(\not{p} \pm m)/m$. Thus in general these relations become

$$\Lambda_+ \equiv \sum_s u_a(p,s)\bar{u}_b(p,s) = \mathcal{N}^2\left(\frac{\not{p}+m}{2m}\right)_{ab}, \tag{8.56}$$

$$\Lambda_- \equiv -\sum_s v_a(p,s)\bar{v}_b(p,s) = \mathcal{N}^2\left(\frac{-\not{p}+m}{2m}\right)_{ab}, \tag{8.57}$$

where we defined $\Lambda_\pm$ as the projection operator on states with positive and negative energy, respectively.

The two most common normalisation conventions for the Dirac spinors are $\mathcal{N} = \sqrt{2m}$ and $\mathcal{N} = 1$. We will use the former, $\mathcal{N} = \sqrt{2m}$, which has three advantages. First, the expressions for $\Lambda_\pm$ which appear frequently become more compact. Second, spurious singularities in the limit $m \to 0$ disappear. Finally, the normalisation of fermion states and thus also the phase space volume becomes identical to the one of bosons.

The solutions of the Dirac equation for an arbitrary frame can be simplest obtained remembering $(\not{p} - m)(\not{p} + m) = p^2 - m^2 = 0$, as

$$u(p,\pm) = \frac{\not{p}+m}{\sqrt{m+E}}\, u(0,\pm) \quad \text{and} \quad v(p,\pm) = \frac{-\not{p}+m}{\sqrt{m+E}}\, v(0,\pm). \tag{8.58}$$

Here, the normalisation was fixed using (8.52).

Spin. We have seen that the Dirac equation describes a particle with helicity 1/2. Thus the $\pm$ degeneracy of the u and v spinors should correspond to the different helicity or spin states of a Dirac particle. We introduce the spin operator

$$\boldsymbol{\Sigma} = \begin{pmatrix} \boldsymbol{\sigma} & 0 \\ 0 & \boldsymbol{\sigma} \end{pmatrix}, \tag{8.59}$$

as an obvious generalisation of the non-relativistic spin matrices. This operator has the eigenvalues $\Sigma_z u(m,\pm) = \pm u(m,\pm)$ and $\Sigma_z v(m,\pm) = \mp v(m,\pm)$ and can therefore be used to classify the spin states of a Dirac particle in the rest-frame, where $[H_D,\Sigma_z] \propto [\gamma^0,\Sigma_z] = 0$. Note however, that $[H_D,\Sigma_z] \neq 0$ for $p^2 \neq m^2$, and thus the eigenvalue of Σ_z is not conserved for a moving particle. This comes not as a surprise, because the total angular momentum $\boldsymbol{L} + \boldsymbol{s}$ and not only the spin $\boldsymbol{s}$ should be conserved.

We are looking now for the relativistic generalisation of the three-dimensional spin operator $\boldsymbol{\Sigma}$. It should be a product of gamma matrices which contains in the rest-frame of the particle $\boldsymbol{\sigma}$ in the diagonal. We note first that $\gamma^5\boldsymbol{\gamma} = \begin{pmatrix} 0 & 1 \\ 1 & 0 \end{pmatrix}\begin{pmatrix} 0 & \boldsymbol{\sigma} \\ -\boldsymbol{\sigma} & 0 \end{pmatrix}$ has the required structure. Then we define the spin vector s^μ with the properties $s^2 = -1$, $s^\mu = (0, \boldsymbol{s})|_{p=m}$ and thus $s \cdot p = 0$. Since

$$\gamma^5 \not{s}|_{p=m} = -\gamma^5 \boldsymbol{s}\boldsymbol{\gamma} = \begin{pmatrix} \boldsymbol{\sigma}\boldsymbol{s} & 0 \\ 0 & -\boldsymbol{\sigma}\boldsymbol{s} \end{pmatrix}, \qquad (8.60)$$

we see that $\gamma^5 \not{s}$ measures in the rest-frame the projection of the spin along the chosen axis $\boldsymbol{s}$. Moreover, $\gamma^5 \not{s}$ commutes with the Dirac Hamiltonian, $[\gamma^5 \not{s}, \not{p}] = 0$, and has because of $(\gamma^5 \not{s})^2 = 1$ as eigenvalues ± 1. If we apply $\gamma^5 \not{s}$ on the spinor $v(p, s)$—which is easiest done in the rest-frame—

$$\gamma^5 \not{s}\, v(m, s) = \begin{pmatrix} \boldsymbol{\sigma}\boldsymbol{s} & 0 \\ 0 & -\boldsymbol{\sigma}\boldsymbol{s} \end{pmatrix} \begin{pmatrix} 0 \\ \chi \end{pmatrix} = \begin{pmatrix} 0 \\ -\boldsymbol{\sigma}\boldsymbol{s}\chi \end{pmatrix} = s_j \begin{pmatrix} 0 \\ \chi \end{pmatrix}, \qquad (8.61)$$

we see that χ_1 has the eigenvalue $s_1 = -1$, while χ_2 has the eigenvalue $s_2 = +1$. This explains the "wrong" order of the two spin states of $v(p, s)$ in (8.51). Finally, we can define a projection operator on a definite spin state by

$$\Lambda_s = \frac{1}{2} \left(1 + \gamma^5 \not{s} \right). \qquad (8.62)$$

Thus we can obtain from an arbitrary Dirac spinor ψ a state with definite sign of the energy and spin by applying the two projection operators $\Lambda_\pm$ and Λ_s.

Helicity.   An important special case of the spin operator $\gamma^5 \not{s}$ is the helicity operator $h \equiv \boldsymbol{sp}/|\boldsymbol{p}|$ which measures the projection of the spin $\boldsymbol{s} = \boldsymbol{\Sigma}/2$ on the momentum $\boldsymbol{p}$ of a particle,

$$\frac{\boldsymbol{\Sigma} \cdot \boldsymbol{p}}{2|\boldsymbol{p}|}\, \psi = h\psi. \qquad (8.63)$$

The helicity operator and the Dirac Hamiltonian commute, $[H_D, \boldsymbol{\Sigma}\boldsymbol{p}] = 0$, because there is no orbital angular momentum in the direction of $\boldsymbol{p}$. Therefore common eigenfunctions of H_D and h called helicity states can be constructed, cf. problem 8.10. Positive helicity particles are called right-handed, negative helicity particles left-handed. For a massive particle, helicity is a frame-dependent quantity. If we choose, for example, a frame with $\boldsymbol{\beta}\|\boldsymbol{p}$ and $\beta > p/m$, then the particles moves in the opposite direction and h changes sign. Since we cannot "overtake" a massless particle, helicity becomes in this case a Lorentz-invariant quantity.

Axial and vector U(1) symmetries.   Out of the 16 bilinear forms, two transform as vectors under proper Lorentz transformations, $j^\mu = \bar{\psi}\gamma^\mu\psi$ and $j_5^\mu = \bar{\psi}\gamma^5\gamma^\mu\psi$. We now want to check if these two currents are conserved. Inspection of the Lagrange density shows immediately that global U(1) transformations,

$$\psi(x) \to \psi'(x) = \mathrm{e}^{\mathrm{i}\phi}\psi(x) \qquad \text{and} \qquad \bar{\psi}(x) \to \bar{\psi}'(x) = \mathrm{e}^{-\mathrm{i}\phi}\bar{\psi}(x), \qquad (8.64)$$

keep the Lagrangian invariant, $\delta\mathscr{L} = 0$. Noether's theorem leads then to the conserved vector current $j^\mu = \bar{\psi}\gamma^\mu\psi$. In the second case, the underlying symmetry is using $\{\gamma^5, \gamma^\mu\} = 0$,

$$\psi'(x) \rightarrow e^{i\phi\gamma^5}\psi(x) \quad \text{and} \quad \bar\psi(x) \rightarrow \bar\psi'(x) = (e^{i\phi\gamma^5}\psi(x))^\dagger \gamma^0 = \bar\psi(x)e^{i\phi\gamma^5}. \quad (8.65)$$

The resulting (infinitesimal) change is

$$\delta\mathscr{L} = 2mi\bar\psi\gamma^5\psi. \quad (8.66)$$

Thus the axial or chiral symmetry $U_A(1)$ is broken by the mass term, leading to the non-conservation of the axial current j_5^μ for a massive fermion.

Chirality.   To understand this better we re-express the Dirac Lagrangian using eigenfunctions of γ^5. We can split any solution ψ of the Dirac equation into

$$\psi_L = \frac{1}{2}(1-\gamma^5)\psi \equiv P_L\psi \quad \text{and} \quad \psi_R = \frac{1}{2}(1+\gamma^5)\psi \equiv P_R\psi. \quad (8.67)$$

Since $\gamma^5\psi_L = -\psi_L$ and $\gamma^5\psi_R = \psi_R$, $\psi_{L,R}$ are eigenfunctions of γ^5 with eigenvalue ± 1. Expressing the mass term through these fields as

$$\bar\psi\psi = \bar\psi\left(P_L^2 + P_R^2\right)\psi = \psi^\dagger\left(P_R\gamma^0 P_L + P_L\gamma^0 P_R\right)\psi = \bar\psi_R\psi_L + \bar\psi_L\psi_R \quad (8.68)$$

and similarly for the kinetic term,

$$\bar\psi\slashed{\partial}\psi = \bar\psi\left(P_L^2 + P_R^2\right)\slashed{\partial}\psi = \psi^\dagger\left(P_R\gamma^0\gamma^\mu P_R + P_L\gamma^0\gamma^\mu P_L\right)\partial_\mu\psi = \bar\psi_L\slashed{\partial}\psi_L + \bar\psi_R\slashed{\partial}\psi_R, \quad (8.69)$$

the Dirac Lagrange density becomes

$$\mathscr{L} = i\bar\psi_L\slashed{\partial}\psi_L + i\bar\psi_R\slashed{\partial}\psi_R - m(\bar\psi_L\psi_R + \bar\psi_R\psi_L). \quad (8.70)$$

Comparing this expression to (8.33), we see that we can identify the Dirac fields $\psi_{L/R}$ in the chiral representation with the Weyl fields $\phi_{L/R}$ as follows:

$$\psi_L = \begin{pmatrix}\phi_L \\ 0\end{pmatrix} \quad \text{and} \quad \psi_R = \begin{pmatrix}0 \\ \phi_R\end{pmatrix}. \quad (8.71)$$

Thus the projection operators (8.67) allows us to define the left- and right-chiral components of a Dirac field in an arbitrary representation. If the mass or interaction terms treat ψ_L and ψ_R non-symmetrically, one calls them chiral fermions.

The two kinetic terms which are invariant under chiral transformations connect left- to left-chiral and right- to right-chiral fields, while the mass term mixes left- and right-chiral fields. Such a mass term is called Dirac mass. The distinction between left- and right-chiral fields is Lorentz-invariant. In terms of Weyl spinors, we saw that the Lorentz transformations S_L and S_R do not mix ϕ_L and ϕ_R,—which qualified them to form the irreducible representation of the Lorentz group. In terms of Dirac spinors, the relation $[\gamma^5, \sigma^{\mu\nu}] = 0$ guarantees that left and right chiral fields transform separately under a Lorentz transformation, $\psi'_{L/R} = S(\Lambda)\psi_{L/R}$. However, the mass term of a massive Dirac particle will mix left- and right-chiral fields as they evolve in time.

Helicity and chirality eigenstates can be seen as complementary states. The former one is a conserved, frame-dependent quantum number, while the latter is frame-independent, but not conserved. Thus helicity states are, for example, useful to describe scattering processes where the detector measures spin in a definite frame. If on the other hand the interactions of a fermion are spin-dependent, then one should choose chiral fields, since the Lagrangian should be Lorentz-invariant.

Charge conjugation. From $\mathscr{L} = \bar{\psi}(\mathrm{i}\gamma^\mu\partial_\mu - m)\psi$ and $[\mathscr{L}] = m^4$ in four dimensions, we see that the dimension of a fermion field in four dimension is $[\psi] = m^{3/2}$. Thus we can order possible couplings of a fermion to spin-1 particles according to their dimension as

$$\mathscr{L}_I = c_1 A_\mu \bar{\psi}\gamma^\mu\psi + c_2 A_\mu \bar{\psi}\gamma^5\gamma^\mu\psi + \frac{c_3}{M} F_{\mu\nu}\bar{\psi}\sigma^{\mu\nu}\psi + \dots, \tag{8.72}$$

where the coupling constants c_i are dimensionless and we introduce the mass scale M. The only coupling to the photon field with a dimensionless coupling that also respects parity is $\mathscr{L}_I = -qj^\mu A_\mu = -q\bar{\psi}\gamma^\mu\psi A_\mu$. Solving the Lagrange equations for $\mathscr{L}_0 + \mathscr{L}_I$ gives the Dirac equation including a coupling to the electromagnetic field as

$$[\mathrm{i}\gamma^\mu(\partial_\mu + \mathrm{i}qA_\mu) - m]\psi(x) = 0. \tag{8.73}$$

This corresponds to the "minimal coupling" prescription known from quantum mechanics.

Having defined the coupling to an external electromagnetic field, we can ask ourselves how the Dirac equation for a charged conjugated field ψ_c should look. In the case of a scalar particle, complex conjugation transformed a positively charged particle into a negative one and vice versa. We try the same for the Dirac equation,

$$[-\mathrm{i}\gamma^{\mu*}(\partial_\mu - \mathrm{i}qA_\mu) - m]\psi^*(x) = 0. \tag{8.74}$$

The matrix $\gamma^{\mu*}$ satisfies also the Clifford algebra. Hence we should find the unitary transformation $U^{-1}\gamma^\mu U = -\gamma^{\mu*}$ or setting $U \equiv C\gamma^0$

$$(C\gamma^0)^{-1}\gamma^\mu C\gamma^0 = -\gamma^{\mu*}. \tag{8.75}$$

If it exists, then the charge-conjugated field $\psi^c \equiv C\gamma^0\psi^*$ satisfies the Dirac equation with $-q$,

$$[\mathrm{i}\gamma^\mu(\partial_\mu - \mathrm{i}qA_\mu) - m]C\gamma^0\psi^*(x) = 0. \tag{8.76}$$

Explicit calculation shows that we may choose $C = \mathrm{i}\gamma^2\gamma^0$, see problem 8.12. In the chiral representation, $\psi^c = C\gamma^0\psi^* = \mathrm{i}\gamma^2\psi^*$ becomes

$$\psi^c = \begin{pmatrix} 0 & \mathrm{i}\sigma^2 \\ -\mathrm{i}\sigma^2 & 0 \end{pmatrix} \begin{pmatrix} \phi_L^* \\ \phi_R^* \end{pmatrix} = \begin{pmatrix} \mathrm{i}\sigma^2\phi_R^* \\ -\mathrm{i}\sigma^2\phi_L^* \end{pmatrix}, \tag{8.77}$$

which is in agreement with $\phi_L^c = \mathrm{i}\sigma^2\phi_R^*$ and $\phi_R^c = -\mathrm{i}\sigma^2\phi_L^*$ found earlier.

Example 8.1: Since the γ^2 matrix has the same form in the Dirac and the chiral representation, we find applying C on the spinors $u(p, \pm)$ and $v(p, \pm)$ immediately that

$$u^c(p, s) = C\gamma^0 u^*(p, s) = v(p, s) \quad \text{and} \quad v^c(p, s) = C\gamma^0 v^*(p, s) = u(p, s).$$

Inserted into Eq. (8.85) this implies that $S_F^T(x) = CS_F(-x)C^{-1}$.

Feynman propagator. The Green functions of the free Dirac equation are defined by

$$(i\partial\!\!\!/ - m)S(x, x') = \delta(x - x'), \tag{8.78}$$

where we omit on the RHS a unit matrix in spinor space. Translation invariance implies $S(x, x') = S(x - x')$ and, performing a Fourier transformation, the Fourier components $S(p)$ have to obey

$$(p\!\!\!/ - m)S(p) = 1. \tag{8.79}$$

After multiplication with $p\!\!\!/ + m$ and use of $a\!\!\!/^2 = \frac{1}{2}\{\gamma^\mu, \gamma^\nu\}a_\mu a_\nu = a^2$, we can solve for the propagator in momentum space,

$$iS_F(p) = i\frac{p\!\!\!/ + m}{p^2 - m^2 + i\varepsilon} = \frac{i}{p\!\!\!/ - m + i\varepsilon}, \tag{8.80}$$

where the last step is only meant as a symbolical shortcut. Here, we chose again with the $-i\varepsilon$ prescription the causal or Stückelberg–Feynman propagator for the electron and, more generally, for spin $1/2$ particles. Note also the connection to the scalar propagator Δ_F,

$$iS_F(x) = -(i\partial\!\!\!/ + m)i\Delta_F(x). \tag{8.81}$$

Example 8.2: Express the Feynman propagator as sum over the solutions $u(p, s)$ and $v(p, s)$. We follow the steps from (3.25a) to (3.28) in the scalar case, finding now

$$S_F(x) = \int \frac{d^3p}{(2\pi)^3} \int \frac{dp_0}{2\pi} \frac{(p\!\!\!/ + m)e^{-ip_0t}e^{ip\boldsymbol{x}}}{(p_0 - E_p + i\varepsilon)(p_0 + E_p - i\varepsilon)} \tag{8.82}$$

$$= \int \frac{d^3p}{(2\pi)^3} \left[-i\frac{p\!\!\!/ + m}{2E_p}e^{-iE_pt}\vartheta(x^0) + i\frac{-E_p\gamma^0 - \boldsymbol{p}\boldsymbol{\gamma} + m}{-2E_p}e^{iE_pt}\vartheta(-x^0) \right]e^{ip\boldsymbol{x}}. \tag{8.83}$$

Next we change as in the bosonic case the integration variable as $\boldsymbol{p} \to -\boldsymbol{p}$ in the second term,

$$iS_F(x) = \int \frac{d^3p}{2E_p(2\pi)^3} \left[(p\!\!\!/ + m)e^{-i(E_pt - \boldsymbol{p}\boldsymbol{x})}\vartheta(t) + (-p\!\!\!/ + m)e^{i(E_pt - \boldsymbol{p}\boldsymbol{x})}\vartheta(-t) \right]. \tag{8.84}$$

Using finally (8.56), we arrive at

$$iS_F(x) = \int \frac{d^3p}{2E_p(2\pi)^3} \sum_s \left[u(p, s)\bar{u}(p, s)e^{-i(E_pt - \boldsymbol{p}\boldsymbol{x})}\vartheta(x^0) - v(p, s)\bar{v}(p, s)e^{i(E_pt - \boldsymbol{p}\boldsymbol{x})}\vartheta(-x^0) \right].$$
$$\tag{8.85}$$

Thus the phase-space volume of fermionic states is the same as the one of bosons for the normalisation of the Dirac spinors chosen by us. The minus sign between the positive energy solution propagating forward in time and the negative energy solution propagating backward in time is a direct consequence of our $-i\varepsilon$ prescription. It implies that fermionic fields anti-commute,

$$iS_{F,ab}(x) = \langle 0|T\{\psi_a(x)\bar{\psi}_b(0)\}|0\rangle = \langle 0|\psi_a(x)\bar{\psi}_b(0)|0\rangle\vartheta(t) - \langle 0|\bar{\psi}_b(0)\psi_a(x)|0\rangle\vartheta(-t), \tag{8.86}$$

(we have added for clarity the spinor indices) and explains thereby the Pauli exclusion principle and thus the stability of matter.

Let us look back to understand why the sign appears. To simplify the discussion, we neglect the non-essential mass term. In the positive frequency term $(p^0\gamma^0 - \boldsymbol{p}\boldsymbol{\gamma})\mathrm{e}^{\mathrm{i}px}$, we pick up an additional minus relative to the bosonic case from the variable change $\boldsymbol{p} \to -\boldsymbol{p}$, resulting in $\not{p} \to -\not{p}$ and

$$S_F(x) \propto \not{p}\,\mathrm{e}^{-\mathrm{i}px}\vartheta(t) - \not{p}\,\mathrm{e}^{\mathrm{i}px}\vartheta(-t).$$

Thus the relative minus sign has its origin in the fact that the fermion propagator $S_F(p)$ is odd in the momentum, while a bosonic propagator is even. In turn, the fermion propagator is linear in the momentum, because the fermion wave equation is a first-order equation.

8.3 Quantising Dirac fermions

Spin-statistic connection. We have noted that fermionic fields should anti-commute examining the Feynman propagator in example 8.2. In a relativistic quantum field theory the spin and the statistics of a field is connected:

- The wave equations of bosons are second-order differential equations. Therefore the propagators of bosonic fields are even in the momentum p. As a result, bosonic fields commute and satisfy Bose–Einstein statistics.
- In contrast, fermions satisfy first-order differential equations and therefore the fermion propagator $S_F(p)$ is odd in p. This implies that fermions are described by anticommuting classical spinors or operators, and satisfy Fermi–Dirac statistics.

This leads to a practical and a principal question. First, the practical one: how do we implement that the classical functions which enter the path integral do anticommute? And second, does the anticommutation of fermionic variables lead to a consistent picture? In particular, is the Hamiltonian of such a theory bounded from below?

We will start to address the latter question calculating the energy density $\rho = \mathscr{H}$ of the Dirac field *assuming* that the classical spinors u and v do commute. We determine first the canonically conjugated momenta as

$$\pi = \frac{\partial \mathscr{L}}{\partial \dot{\psi}} = \mathrm{i}\bar{\psi}\gamma^0 = \mathrm{i}\psi^\dagger \tag{8.87}$$

and $\bar{\pi} = 0$. Thus the Hamilton density is

$$\mathscr{H} = \pi\dot{\psi} - \mathscr{L} = \mathrm{i}\psi^\dagger \partial_t \psi - \bar{\psi}(\mathrm{i}\gamma^\mu \partial_\mu - m)\psi = \mathrm{i}\psi^\dagger \partial_t \psi, \tag{8.88}$$

where we used the Dirac equation in the last step. To make this expression more explicit, we express ψ by plane-wave solutions,

$$\psi(x) = \sum_s \int \frac{\mathrm{d}^3 p}{\sqrt{(2\pi)^3 2E_p}} \left[b_s(p)u_s(p)\mathrm{e}^{-\mathrm{i}px} + d_s^\dagger(p)v_s(p)\mathrm{e}^{+\mathrm{i}px} \right], \tag{8.89a}$$

$$\psi^\dagger(x) = \sum_s \int \frac{\mathrm{d}^3 p}{\sqrt{(2\pi)^3 2E_p}} \left[b_s^\dagger(p)u_s^\dagger(p)\mathrm{e}^{+\mathrm{i}px} + d_s(p)v_s^\dagger(p)\mathrm{e}^{-\mathrm{i}px} \right]. \tag{8.89b}$$

Inserting these expressions into (8.88) gives schematically $(b^\dagger + d)(b - d^\dagger)$, where the relative minus sign comes from ∂_t acting on ψ. Since the spinors u and v are orthonormal, cf. (8.53), only the diagonal terms survive, $(b^\dagger + d)(b - d^\dagger) \to b^\dagger b - dd^\dagger$. Hence the energy of a Dirac field is given by

$$H = \int \mathrm{d}^3x \, \mathscr{H} = \sum_s \int \mathrm{d}^3p \, E_p \left[b_s^\dagger(p)b_s(p) - d_s(p)d_s^\dagger(p) \right]. \tag{8.90}$$

If d and $d^\dagger$ would be normal Fourier coefficients of an expansion into plane waves, the second term would be negative and the energy density of a fermion field could be made arbitrarily negative.

This conclusion is avoided if the fermion fields anticommute. In canonical quantisation, one promotes the Fourier coefficients to operators. Requiring then anti-commutation relations between the creation and annihilation operators for particles and antiparticles,

$$\{b_s(p), b_{s'}^\dagger(p')\} = \delta_{s,s'}\delta(p - p') \quad \text{and} \quad \{d_s(p), d_{s'}^\dagger(p')\} = \delta_{s,s'}\delta(p - p'), \tag{8.91}$$

compensates the sign in the second term. Repeating the discussion in section 3.3 one can show that these anti-commutation relations implement also correctly causality for fermionic fields. If we restore units then we have to add a factor $\hbar$ on the RHS of the anti-commutation relations (8.91). Since the RHS vanishes in the classical limit $\hbar \to 0$, classical spinors should anti-commute. Thus we should perform the path integral of fermionic fields over anti-commuting numbers which are called Graßmann numbers.

Graßmann variables. We now proceed to the question how we can implement the analogue of the anticommutation relations for operators into the path-integral formalism. We define a Graßmann algebra $\mathcal{G}$ requiring that for $a, b \in \mathcal{G}$ the anticommutation rules

$$\{a, a\} = \{a, b\} = \{b, b\} = 0 \tag{8.92}$$

and thus $a^2 = b^2 = 0$ are valid. Then any smooth function f of a and b can be expanded into a power-series as

$$\begin{aligned} f(a, b) &= f_0 + f_1 a + \tilde{f}_1 b + f_2 ab \\ &= f_0 + f_1 a + \tilde{f}_1 b - f_2 ba. \end{aligned} \tag{8.93}$$

Defining the derivative as acting to the right, $\partial \equiv \overrightarrow{\partial}$, we find

$$\frac{\partial f}{\partial a} = f_1 + f_2 b, \qquad \frac{\partial f}{\partial b} = \tilde{f}_1 - f_2 a, \tag{8.94}$$

and

$$\frac{\partial^2 f}{\partial a \partial b} = -\frac{\partial^2 f}{\partial b \partial a} = -f_2. \tag{8.95}$$

As integration rules for Graßmann variables, we require linearity and that the infinitesimals $\mathrm{d}a$, $\mathrm{d}b$ are also Graßmann variables,

$$\{a, \mathrm{d}a\} = \{b, \mathrm{d}b\} = \{a, \mathrm{d}b\} = \{\mathrm{d}a, b\} = \{\mathrm{d}a, \mathrm{d}b\} = 0. \tag{8.96}$$

Multiple integrals are iterated,

$$\int \mathrm{d}a\mathrm{d}bf(a,b) = \int \mathrm{d}a \left(\int \mathrm{d}bf(a,b) \right). \tag{8.97}$$

We have to determine the value of $\int \mathrm{d}a$ and $\int \mathrm{d}aa$. For the first, we write

$$\left(\int \mathrm{d}a \right)^2 = \left(\int \mathrm{d}a \right) \left(\int \mathrm{d}b \right) = \int \mathrm{d}a\mathrm{d}b = -\int \mathrm{d}b\mathrm{d}a = -\left(\int \mathrm{d}a \right)^2 \tag{8.98}$$

and find thus $\int \mathrm{d}a = 0$. We are left with $\int \mathrm{d}aa$: since there is no intrinsic scale—states are empty or occupied—we are free to set

$$\int \mathrm{d}aa = 1. \tag{8.99}$$

This implies also that there is no difference between definite and indefinite integrals for Graßmann variables. Moreover, differentiation and integration are equivalent for Graßmann variables.

Assume now that η_1 and η_2 are real Graßmann variables and $A \in \mathbb{R}$. Then

$$\int \mathrm{d}\eta_1\mathrm{d}\eta_2 \ e^{\eta_2 A \eta_1} = \int \mathrm{d}\eta_1\mathrm{d}\eta_2 \ (1 + \eta_2 A \eta_1) = \int \mathrm{d}\eta_1\mathrm{d}\eta_2 \ \eta_2 A \eta_1 = \int \mathrm{d}\eta_1 \ A\eta_1 = A. \tag{8.100}$$

Next we consider a two-dimensional integral with an antisymmetric matrix A and $\eta = (\eta_1, \eta_2)$. Then

$$A = \begin{pmatrix} 0 & a \\ -a & 0 \end{pmatrix} \tag{8.101}$$

and $\eta^T A \eta = 2a\eta_1\eta_2$. Using an arbitrary matrix would lead to the same result, since its symmetric part cancels. Expanding again the exponential gives

$$\int \mathrm{d}^2\eta \ \exp\left(\frac{1}{2}\eta^T A \eta \right) = a = (\det(A))^{1/2}. \tag{8.102}$$

An arbitrary antisymmetric matrix can be transformed into block diagonal form, where the diagonal is composed of matrices of the type (8.101). Thus the last formula holds for arbitrary n.

Finally, we introduce complex Graßmann variables $\eta = (\eta_1, \ldots, \eta_n)$ and their complex conjugates $\eta^* = (\eta_1^*, \ldots, \eta_n^*)$. For any complex matrix A,

$$\int \mathrm{d}^n\eta\mathrm{d}^n\eta^* \ \exp\left(\eta^\dagger A \eta \right) = \prod_{i=1}^{n} a_i = \det(A). \tag{8.103}$$

We can compare this to the result over commuting complex variables, $z_i = (x_i + \mathrm{i}y_i)/\sqrt{2}$ and $\bar{z}_i = (x_i - \mathrm{i}y_i)/\sqrt{2}$, with $\mathrm{d}x\mathrm{d}y = \mathrm{d}z\mathrm{d}z^*$ and

$$\int \mathrm{d}^n z \mathrm{d}^n z^* \, \exp\left(-z^\dagger A z\right) = \frac{(2\pi)^n}{\det(A)}. \tag{8.104}$$

Thus for Graßmann variables the determinant appearing in the evaluation of a Gaussian integral is in the numerator, while it is in the denominator for real or complex valued functions.

Path integral for fermions. In the bosonic case, the action $S[\phi, \pi]$ is quadratic in the canonically conjugated momenta π. The path integral over the momenta can thus be performed and we started directly with the path integral in configuration space. For a fermion, $\pi = i\psi^\dagger$, and thus the path integral in phase space is

$$Z[0] = \int \mathcal{D}\psi \mathcal{D}\bar{\psi} \, e^{iS[\psi, \bar{\psi}]} = \int \mathcal{D}\psi \mathcal{D}\bar{\psi} \, e^{i \int \mathrm{d}^4 x \, \bar{\psi}(i\slashed{\partial} - m)\psi}, \tag{8.105}$$

where we changed to $\bar{\psi}$ as integration variable. For its evaluation, we use (8.103) in the limit $n \to \infty$. Since the action is quadratic in the fields, we can perform the path integral formally,

$$Z[0] = \mathrm{Det}(i\slashed{\partial} - m) = \exp \mathrm{Tr} \ln(i\slashed{\partial} - m). \tag{8.106}$$

Using the cyclic property of the trace, we write

$$\mathrm{Tr}\ln(i\slashed{\partial} - m) = \mathrm{Tr}\ln \gamma^5 (i\slashed{\partial} - m)\gamma^5 = \mathrm{Tr}\ln(-i\slashed{\partial} - m) = \tag{8.107a}$$

$$= \frac{1}{2}[\mathrm{Tr}\ln(i\slashed{\partial} - m) + \mathrm{Tr}\ln(-i\slashed{\partial} - m)] = \frac{1}{2}[\mathrm{Tr}\ln(\square + m^2)]. \tag{8.107b}$$

Thus $Z[0] = \exp[+\mathrm{Tr}\ln(\square + m^2)/2]$. We have found the remarkable result that the zero-point energy of fermions has the opposite sign compared to the one of bosons. We arrive at the same conclusion, using anti-commutation relations $\{d_s(\boldsymbol{p}), d_{s'}^\dagger(\boldsymbol{p}')\} = \delta_{s,s'}\delta(\boldsymbol{p} - \boldsymbol{p}')$ in the Hamiltonian (8.90),

$$H = \sum_s \int \mathrm{d}^3 p \, E_p \left[b^\dagger b + d^\dagger d - \delta^{(3)}(0) \right]. \tag{8.108}$$

With $\delta^{(3)}(0) = \int \mathrm{d}^3 x/(2\pi)^3$ we see that the last term corresponds to the negative zero-point energies of a fermion.

Note that this opens the possibility that the zero-point energies of (groups of) bosons and fermions cancel exactly, provided that the degrees of freedom of fermions and bosons agree. For instance, the trace in Eq. (8.107b) includes the trace over the 4×4 matrix in spinor space, leading to a factor 4 larger results than for a single scalar. Moreover, their masses have to be the same, $m_f = m_b$. And finally their interactions have to match, so that also higher-order corrections are identical for fermions and bosons. The corresponding symmetry that guarantees automatically that the conditions i) to iii) are satisfied is called "supersymmetry". As result, the vacuum energy would be zero in an unbroken supersymmetric theory, Clearly, the second condition is most problematic, since, for example, no bosonic partner of the electron has been found (yet). Hence supersymmetry must be a broken symmetry, but as long as the mass splitting $m_f^2 - m_b^2$ between fermions and bosons is not too large, it might be still "useful".

Feynman rules. Next we add Graßmannian sources η and $\bar{\eta}$ to the action, $S[\psi, \bar{\psi}] + \bar{\eta}\psi + \bar{\psi}\eta$. Then we complete the square,

$$\bar{\psi}A\psi + \bar{\eta}\psi + \bar{\psi}\eta = (\bar{\psi} + \bar{\eta}A^{-1})A(\psi + A^{-1}\eta) - \bar{\eta}A^{-1}\eta, \tag{8.109}$$

obtaining

$$Z[\eta, \bar{\eta}] = \int \mathcal{D}\psi \mathcal{D}\bar{\psi} \; e^{i \int d^4 x d^4 x' \left[\bar{\psi}(x')A(x',x)\psi(x) + \bar{\eta}(x')\psi(x) + \bar{\psi}(x')\eta(x)\right]} =$$
$$= Z[0] \exp\left(-i \int d^4 x \, d^4 x' \; \bar{\eta}(x) S_F(x - x')\eta(x')\right). \tag{8.110}$$

Here, $A^{-1}(x, x') = S_F(x - x') = -S_F^T(x' - x)$ which corresponds to the fact that the matrix A is antisymmetric.

The propagator of a Dirac fermion is a line with an arrow representing the flow of the conserved charge which distinguishes particles and antiparticles. Thus a fermion line cannot split, and the arrow cannot change direction,

$$\xrightarrow{\quad} \atop p \qquad = iS_F(p) = \frac{i}{\not{p} - m + i\varepsilon}.$$

Next we look at possible interaction terms of a Dirac fermion with scalars and photons, restricting ourselves to dimensionless coupling constants,

$$\mathcal{L}_I = -g_s S\bar{\psi}\psi - g_a P\bar{\psi}\gamma^5\psi - q\bar{\psi}\gamma^\mu\psi A_\mu. \tag{8.111}$$

Both interaction terms of the fermion with the scalars respect parity if the field S is a true scalar and the field P a pseudo-scalar. In the non-relativistic limit, they lead to an attractive Yukawa potential between two fermions, as we discussed in section 3.2. Analogous to the $-i\lambda$ coupling in the case of a scalar self-interactions, we read off from the Lagrangian the following interaction vertices in momentum space,

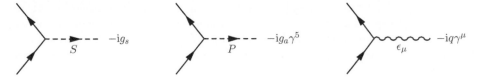

Fermion loops. A closed fermion loop with n propagators corresponds to

$$\overline{\psi(x_1)\psi(x_1)}\overline{\psi(x_2)\psi(x_2)} \cdots \overline{\psi(x_n)\psi(x_n)} \, .$$

In order to combine $\bar{\psi}(x_1)$ and $\psi(x_n)$ into $T\{\psi(x_n)\bar{\psi}(x_1)\}$, we have to anticommute $\bar{\psi}(x_1)$ with the $2n - 1$ fields $\psi(x_1) \cdots \psi(x_n)$, generating a minus sign. Thus we have to add to our set of Feynman rules that each fermion loop generates a minus sign. Another way to understand the minus sign of fermion loops is to look at the generating functional for connected graphs setting the sources to zero, $iW[0] = \ln Z[0] = \ln \det A$.

The generated graphs are closed loops with n Feynman propagators. The change from $1/\det A$ in Z for bosonic fields to $\det A$ in Z for fermionic fields implies an additional minus sign for closed fermion loops. Similarly, diagrams contributing to the same process which differ only by the exchange of two identical external fermion lines carry a relative minus sign. This applies also the exchange of a particle and antiparticle in the initial and final state (cf. also the discussion of crossing symmetry in section 9.3.1).

Furry's theorem.   What is the relation between diagrams containing fermion loops with opposite orientation in QED? A fermion loop with n external photons attached corresponds to a trace over n fermion propagators separated by gamma matrices,

$$G_1 = \mathrm{tr}[\gamma_{\mu_1} S_F(y_1, y_n)\gamma_{\mu_n} S_F(y_n, y_{n-1}) \cdots \gamma_{\mu_2} S_F(y_2, y_1)]. \tag{8.112}$$

If we insert $CC^{-1} = 1$ between all factors in the trace, use $C\gamma^\mu C^{-1} = -\gamma^{\mu T}$ and $CS_F(-x)C^{-1} = S_F^T(x)$, then we find

$$G_1 = (-1)^n \mathrm{tr}[\gamma_{\mu_1}^T S_F^T(y_n, y_1)\gamma_{\mu_n}^T S_F^T(y_{n-1}, y_n) \cdots \gamma_{\mu_2}^T S_F^T(y_1, y_2)] \tag{8.113}$$

$$= (-1)^n \mathrm{tr}[\gamma_{\mu_1} S_F(y_1, y_2) \cdots \gamma_{\mu_n} S_F(y_n, y_1)] = (-1)^n G_2. \tag{8.114}$$

Here we used $B^T A^T = (AB)^T$ in the last step. Except for the factor $(-1)^n$, the last expression corresponds to the loop G_2 with opposite orientation. Hence for an odd number of propagators the two contributions cancel, while they are equal for an even number of propagators.

Symmetry factors in QED.   The last issue we want to address in this section is the question whether the interactions in (8.111) lead to symmetry factors. We recall that drawing Feynman diagrams we should include only diagrams which are topologically distinct after the integration over internal coordinates. For instance, the two diagrams

are not topologically distinct, because a rotation around the x_1–x_2 axis interchanges them. Therefore, the corresponding two-point function $G(x_1, x_2)$ has to be invariant under a change of the orientation of the fermion loop—as it is guarantied by the Furry theorem. The two-point function $G(x_1, x_2)$ consists of two identical diagrams obtained by exchanging the integration variables y_1 and y_2. Thus the factor 2! compensates the $1/2!$ from the Taylor expansion of $\exp(\mathrm{i}\mathscr{L}_{\mathrm{int}})$, if we draw only one diagram, and its symmetry factor is 1.

Next we consider the four-point function $G(x_1, x_2, x_3, x_4)$ which describes photon–photon scattering. The four-point function $G(x_1, x_2, x_3, x_4)$ contains $4! \times 3!$ diagrams, obtained by permutating y_1, y_2, y_3, y_4 and x_2, x_3, x_4. After integration over the free y_i variables, the factor 4! compensates the $1/4!$ from the Taylor expansion of $\exp(\mathrm{i}\mathscr{L}_{\mathrm{int}})$. In configuration space, the $3! = 6$ topologically distinct diagrams shown in Fig. 8.1 remain which carry no additional symmetry factor. Thus the resulting rule for QED is very simple: we do not need symmetry factors if we draw all diagrams which are topologically distinct after the integration over internal coordinates. Fermion loops with an odd number of fermions are zero and can be omitted. Independent of the type of interaction, any fermion loop leads to an additional minus sign.

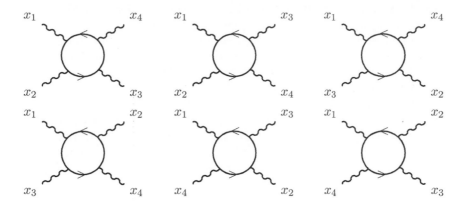

Fig. 8.1 The six topologically distinct diagrams contributing to the photon four-point function.

8.4   Weyl and Majorana fermions

Up to now we have discussed the Dirac equation, having in mind a massive particle carrying a conserved U(1) charge that allows us to distinguish particles and antiparticles. We call such particles Dirac fermions. In the SM, all particles except neutrinos carry a non-zero electric charge, are massive and are therefore Dirac fermions. In this section, we consider the case where one of these two conditions is not fulfilled.

Weyl fermions, $q \neq 0$ and $m = 0$.   The Dirac equation (8.18) in the chiral representation decouples for $m = 0$ into two equations called Weyl equations,

$$(E + \boldsymbol{\sigma}\boldsymbol{p})\phi_L(p) = 0 \quad \text{and} \quad (E - \boldsymbol{\sigma}\boldsymbol{p})\phi_R(p) = 0. \tag{8.115}$$

A fermion described by the Weyl equations is called a Weyl fermion. The correct dispersion relation, $E = |\boldsymbol{p}|$, requires that ϕ_L is an eigenstate of the helicity operator $h = \boldsymbol{\sigma}\boldsymbol{p}/(2|\boldsymbol{p}|)$ with eigenvalue $h = -1/2$, while ϕ_R has the eigenvalue $h = +1/2$. Recall also that helicity is frame-independent for a massless particle; in this case positive helicity agrees with right chirality.[3] Until the 1990s, the experimental data on neutrino masses were consistent with zero and neutrinos were incorporated into the SM as Weyl fermions. Since only left-chiral particles and right-chiral antiparticles participate in weak interactions, one set $\nu = \left(\begin{smallmatrix}\phi_L \\ 0\end{smallmatrix}\right)$, while antineutrinos were described by the CP transformed state. The lepton number L_α of the three flavours $\alpha = \{e, \mu, \tau\}$ of leptons played the role of the conserved U(1) charge that distinguishes neutrinos and antineutrinos. As result, the difference in the number of leptons and antileptons of each individual flavour was conserved. Neutrino oscillations that occur if neutrinos are massive conserve the total lepton number $L = \sum_\alpha L_\alpha$ but interchange the individual neutrino flavours L_α. Thus the observation of neutrino oscillations showed that neutrinos are not Weyl fermions.

---

[3]Most authors call $\psi_{L/R}$ and $\phi_{L/R}$ not left- and right-chiral but left- and right-handed, although this identification holds only for massless particles.

Majorana fermions, $m > 0$ and $q = 0$.   The Dirac field ψ_D has to be complex, because it transforms under the complex representation $S(\Lambda)$ of the Lorentz group. In the case of a neutral fermion, where we cannot distinguish particles and antiparticles, we should have only half the degrees of freedom of a charged Dirac field. By analogy with the scalar case, we expect that we can halve the number of degrees of the complex Dirac field by imposing a reality condition, $\psi_M = \psi_M^*$. But this condition can be Lorentz-invariant only in a special representation of the gamma matrices where $\sigma_{\mu\nu}^* = -\sigma_{\mu\nu}$ and thus $S(\Lambda)$ is real. This condition defines the Majorana representation of the gamma matrices in which all γ^μ and thus $\sigma_{\mu\nu}$ are imaginary, and the charge conjugation matrix C is the unity matrix, $C = 1$. Then the Dirac equation also becomes real and thus the time evolution preserves the reality condition. Since the spinors are real in this representation, no phase invariance $\psi(x) \to \psi'(x) = \exp(\mathrm{i}\phi)\psi(x)$ as in (8.64) can be implemented for a Majorana fermion[4] and thus they cannot carry any conserved U(1) charge.

We can halve the number of degrees of freedom of a Dirac fermion in a representation-independent way by using a self-conjugated field $\psi^c = \psi$. A fermion described by a self-conjugated field ψ_M is called a Majorana fermion and the corresponding spinor a Majorana spinor. The field operator of a Majorana field contains only one type of annihilation and creation operators,

$$\psi_M(x) = \sum_s \int \frac{\mathrm{d}^3 p}{\sqrt{(2\pi)^3 2E_p}} \left[a_s(p) u_s(p) \mathrm{e}^{-\mathrm{i}px} + a_s^\dagger(p) v_s(p) \mathrm{e}^{+\mathrm{i}px} \right]. \tag{8.116}$$

Using then

$$\psi_M^c(x) = C\gamma^0 \psi_M^*(x) = \sum_s \int \frac{\mathrm{d}^3 p}{\sqrt{(2\pi)^3 2E_p}} \left[a_s^\dagger(p) C\gamma^0 u_s^*(p) \mathrm{e}^{\mathrm{i}px} + a_s(p) C\gamma^0 v_s^*(p) \mathrm{e}^{-\mathrm{i}px} \right], \tag{8.117}$$

we confirm immediately the Majorana property $\psi_M^c(x) = \psi_M(x)$. Expressed by Weyl spinors, a Majorana spinor becomes

$$\psi^c = \psi = \begin{pmatrix} \phi_L \\ -\mathrm{i}\sigma^2 \phi_L^* \end{pmatrix} = \begin{pmatrix} \mathrm{i}\sigma^2 \phi_R^* \\ \phi_R \end{pmatrix}. \tag{8.118}$$

Thus a Majorana fermion ($m > 0$, $q = 0$) contains two degrees of freedom, which we may choose either as a left-chiral or or a right-chiral two-spinor with both helicities.

We can replace any Dirac field ψ_D by a pair of self-conjugated fields,

$$\psi_{M,1/2} = \frac{1}{\sqrt{2}} \left(\psi_D \pm \psi_D^c \right), \tag{8.119}$$

and vice versa inverting these relations. Thus describing fermions by Dirac, Weyl or Majorana spinors is merely a question of taste.

---

[4]This argument does not forbid that a Majorana fermion carries conserved charges which transform under a real representation of a symmetry group. An example are gluinos, the suggested supersymmetric partners of the gluons, which are Majorana fermions and transform under the adjoint representation of SU(3).

Dirac versus Majorana mass terms. Charge conjugated Dirac spinors were defined by

$$\psi^c = C\gamma^0\psi^* = C\bar\psi^T, \qquad \bar\psi^c = \psi^T C.$$

We define also

$$\psi_L^c \equiv (\psi_L)^c = \frac{1}{2}(1+\gamma^5)\psi^c = (\psi^c)_R, \tag{8.120}$$

which is consistent with our previous definition for Weyl spinors. As we saw, a Dirac mass term connects the left- and right-chiral components of the same field and $\psi = \psi_L + \psi_R$ is a mass eigenstate. We now use the observation that $(\psi_L)^c = (\psi^c)_R$ allows us to obtain new mass terms[5] called Majorana mass terms,

$$-\mathscr{L}_L - \mathscr{L}_R = m_L(\bar\psi_L^c\psi_L + \bar\psi_L\psi_L^c) + m_R(\bar\psi_R^c\psi_R + \bar\psi_R\psi_R^c) \tag{8.121}$$

which connect the left- and right-chiral components of charge-conjugated fields. The corresponding mass eigenstates are the self-conjugated fields

$$\chi = \psi_L + \psi_L^c = \chi^c \quad \text{and} \quad \omega = \psi_R + \psi_R^c = \omega^c \tag{8.122}$$

with $\mathscr{L}_L = -m_L\bar\chi\chi$ and $\mathscr{L}_R = -m_R\bar\omega\omega$. In the general case, both Dirac and Majorana mass terms may be present,

$$-\mathscr{L}_{DM} = m_D\bar\psi_L\psi_R + m_L\bar\psi_L^c\psi_L + m_R\bar\psi_R^c\psi_R + \text{h.c.} \tag{8.123a}$$

$$= \frac{1}{2}m_D(\bar\chi\omega + \bar\omega\chi) + m_L\bar\chi\chi + m_R\bar\omega\omega, \tag{8.123b}$$

or in matrix form

$$-\mathscr{L}_{DM} = (\bar\chi, \bar\omega) \begin{pmatrix} m_L & m_D/2 \\ m_D/2 & m_R \end{pmatrix} \begin{pmatrix} \chi \\ \omega \end{pmatrix}. \tag{8.124}$$

Physical states have a definite mass and thus we have to diagonalise the mass matrix. Its eigenvalues are

$$m_{1,2} = \frac{1}{2}\left\{(m_L + m_R) \pm \sqrt{(m_L + m_R)^2 - m_D^2}\right\} \tag{8.125}$$

and its eigenvectors

$$\eta_1 = \cos\vartheta\chi - \sin\vartheta\omega \tag{8.126a}$$

$$\eta_2 = \sin\vartheta\chi + \cos\vartheta\omega \tag{8.126b}$$

with $\tan 2\vartheta = m_D/(m_L - m_R)$. Thus the most general mass term $\mathscr{L}_{DM}$ for a four-component fermion spinor corresponds to two Majorana particles with different masses. Therefore we can view a Dirac particle as a special case of two Majorana particles with identical masses and interactions.

The *seesaw model* tries to explain why neutrinos have much smaller masses than all other particles in the standard model. Let us assume that there exist both left- and

[5]Note that terms like $\bar\psi_L^c\psi_L = \psi_L^t C\psi_L$ vanish because of $C^T = -C$, if one does not already assumes on the classical level that fields are anticommuting Graßmann variables.

right-chiral neutrinos and that they obtain Dirac masses as the other fermions, say of order $m_D \sim 100\,\text{GeV}$. The right-chiral ν_R does not participate in any SM interaction and suffers the same fate as a scalar particle. Its mass will be driven by quantum corrections to a value close to the cut-off scale used, and so we expect $m_R \gg m_D$. Moreover, in many models it is $m_L = 0$. Expanding then

$$m_{1,2} \approx \frac{1}{2}\left\{ m_R \pm m_R\sqrt{1 + m_D^2/m_R^2} \right\}, \qquad (8.127)$$

the two eigenvalues are $m_1 \approx m_D^2/(4m_R)$ and $m_2 \approx m_R$. For $m_R \sim 10^{14}\,\text{GeV}$, the light neutrino mass is in the eV or sub-eV range as required by experimental data.

Summary

The fundamental representation of the proper Lorentz group for massive particles is given by left and right-chiral Weyl spinors. These two-spinors are mixed by parity and thus one combines them into a Dirac four-spinor for parity conserving theories like electromagnetic and strong interactions.

Fermions satisfy first-order differential equations and have mass dimension $3/2$. Therefore the fermion propagator $S_F(p)$ is linear in p and thus $S_F(x)$ is an antisymmetric function in x. As a result, fermions satisfy Fermion–Dirac statistics and are described either by Graßmann variables or by anticommuting operators.

A Weyl fermion has $m = 0$ and $q \neq 0$ and satisfies the Weyl equation; its solution has two degrees of freedom, a left-chiral field with negative helicity and a right-chiral field with positive helicity. A Majorana fermion is a self-conjugated field with $m \neq 0$ and $q = 0$ which has therefore also only two degrees of freedom. It is described either by a left-chiral or a right-chiral two-spinor with both helicities. In the Majorana representation, this spinor can be chosen to be real.

Further reading.   The symmetries of the Dirac equation as well as of other relativistic wave equations are extensively discussed by Greiner (2000). More details on two-component Weyl and Majorana spinor can be found, for example, in Srednicki (2007).

Problems

8.1 Lie algebra of the Poincaré group. Derive the commutation relation (8.1) plus the missing ones involving translation using that the (Hermitian) generator T^a are connected to the Killing fields V^a by $iT^a = V^a$. Rewrite the relations in an Lorentz-invariant form.

8.2 Little group Find the Lorentz transformations $\Lambda_\mu^{\ \nu}$ which keep the momentum of a particle invariant, $\Lambda_\mu^{\ \nu} p^\mu = p^\nu$, for a a) massive particle and b) massless particle, considering an infinitesimal transformation $\omega_{\mu\nu}$. By how many parameters is $\omega_{\mu\nu}$ determined in the two cases?

8.3 Derivation of the Dirac equation. Fill in the details in the derivation (8.13) to (8.20).

8.4 g_s-factor of the electron. Square the Dirac equation with an external field A^μ and show that the spin of the electron leads to the additional interaction term $\sigma_{\mu\nu} F^{\mu\nu}$. Show that the Dirac equation predicts $H_{\text{int}} = (\boldsymbol{L} + 2\boldsymbol{s})e\boldsymbol{B}$ or $g_s = 2$ for the interaction of non-relativistic electrons with a static magnetic field.

8.5 Pauli equation. Linearise the Schrödinger equation following the logic of Eqs. (8.24–8.26) and show that the Pauli equation with $g_s = 2$ follows.

8.6 Zitterbewegung. Use Ehrenfest's theorem to show that $\boldsymbol{\alpha}$ can be interpreted as velocity operator in the semi-classical limit. Project out the negative energy solutions and show that the solution $\langle \boldsymbol{x}(t) \rangle$ contains an oscillating term.

8.7 Spin vector. i) Show that the Pauli-Lubanski spin vector (5.27), setting $J^{\mu\nu} = \sigma^{\mu\nu}$, becomes in the rest-frame $W_i = -m\Sigma_i$. ii) Show that $W_\mu s^\mu$ can be used as relativistic spin operator.

8.8 Expectation values in charge-conjugated states. a.) Show that the expectation value $\langle O \rangle \equiv \int \mathrm{d}^3 x \psi^\dagger(x) O \psi(x)$ of an operator O for a charge-conjugated state $\psi_c = \mathrm{i}\gamma^2 \psi^*$ satisfying the Dirac equation is given by $\langle O \rangle_c = \langle \psi_c | O | \psi_c \rangle = -\langle \psi | \gamma^2 O^* \gamma^2 | \psi \rangle^*$. Show that b.) $\langle \boldsymbol{p} \rangle_c = -\langle \boldsymbol{p} \rangle$ and that c.) $\langle \boldsymbol{\Sigma} \rangle_c = -\langle \boldsymbol{\Sigma} \rangle$.

8.9 Projection operators ♣. Show that the inverse of the matrix $A = \sum_i a_i P_i$, where the P_i are projection operators (i.e. $\sum_i P_i = 1$ and $P_i P_j = \delta_{ij} P_j$), is given by $A^{-1} = \sum_i a_i^{-1} P_i$.

8.10 Helicity. a.) Show that the helicity operator $h = \boldsymbol{s} \cdot \boldsymbol{p}/|\boldsymbol{p}|$ is the special case of the spin operator $\slashed{s}\gamma^5$. b.) Derive the common eigenfunctions of the Dirac Hamiltonian and h. c.) Consider these helicity states in the limit $m/E \to 0$.

8.11 Lorentz transformation and gamma's. Show that $S(\Lambda) = \exp(-\mathrm{i}\sigma_{\mu\nu}\omega^{\mu\nu}/4)$ is valid in any representation of the gamma matrices. Derive the behaviour of the gamma matrices under Lorentz transformation, $S(\Lambda)\gamma^\mu S^{-1}(\Lambda) = \Lambda^\mu_\nu \gamma'^\nu$.

8.12 Properties of C. a.) Prove the following properties of the charge conjugation matrix C, $C^T = -C$ and $C^{-1}\gamma_\mu C = -\gamma_\mu^T$. b.) Show that $\mathrm{i}\gamma^2\gamma^0$ has the required properties.

8.13 Dirac Lagrangian. Show that the Lagrangian (8.27) is not real and find a suitable generalisation.

8.14 CPT theorem. Construct all possible Lorentz scalars combining the bilinear quantities $\bar\psi\Gamma\psi$ with tensors. Find their properties under CPT transformations and show that all Lorentz scalars are CPT-even.

8.15 Majorana flip properties. Derive the properties of Majorana bilinears $\bar\xi\Gamma\eta = \pm\bar\eta\Gamma\xi$ under an exchange $\xi \leftrightarrow \eta$.

8.16 Fierz identities ♣. Derive the following identities, $\bar\sigma^\mu\sigma_\mu = 4$, $(\bar\sigma^\mu)_{ab}(\bar\sigma_\mu)_{cd} = 2(\mathrm{i}\sigma^2)_{ac}(\mathrm{i}\sigma^2)_{bd}$, and $(\mathrm{i}\sigma^2)_{ab}(\sigma^\mu)_{bc} = (\bar\sigma^{\mu T})_{ab}(\mathrm{i}\sigma^2)_{bc}$.

8.17 Fermionic vacuum energy. Calculate the contribution of a Dirac fermion to the vacuum energy density, following the steps (4.51) to (4.57) for a scalar.

8.18 Current conservation. Show that the change of the Dirac Lagrangian under a local U(1) transformation can be written as (5.30), leading to current conservation.

8.19 Linear Sigma model II. Add to the Lagrangian (5.39) nucleons $N = \binom{p}{n}$,

$$\mathscr{L} \to \mathscr{L} + \bar N \mathrm{i}\slashed\partial N + g\bar N(\sigma + \mathrm{i}\gamma^5 \boldsymbol{\tau}\boldsymbol{\pi})N.$$

It is useful to express the nucleon fields by chiral fields, N_L and N_R. a.) Show that $\mathscr{L}$ remains invariant under $\Sigma \to \Sigma' = U\Sigma U^\dagger$ and $N \to N' = UN$; find the new contribution to the Noether current. b.) Show that $\mathscr{L}$ is also invariant under $N \to N' = \exp(\mathrm{i}\boldsymbol{\beta} \cdot \boldsymbol{\tau}\gamma^5/2) N$ and $\Sigma \to \Sigma' = V^\dagger \Sigma V^\dagger$ where $V = \exp(\mathrm{i}\boldsymbol{\beta} \cdot \boldsymbol{\tau}/2)$, and $\boldsymbol{\tau}$ are again the Pauli matrices. Find the corresponding conserved Noether current.

9
Scattering processes

Most information about the properties of fundamental interactions and particles is obtained from scattering experiments. In a scattering process, the initial and final state contain widely separated particles which can be approximated as free, real particles which are on mass-shell. By contrast, n-point Green functions describe the propagation of virtual particles. In order to make contact with experiments, we have to find therefore the link between Green functions and experimental results from scattering experiments. The latter can be predicted knowing the scattering matrix S which is a unitary operator mapping an initial state at $t = -\infty$ on a final state at $t = +\infty$. First we introduce the S-matrix and show then that its unitarity restricts the analytic structure of Feynman amplitudes; in particular it implies the optical theorem. Then we derive the connection between n-point Green functions and scattering amplitudes, before we perform some explicit calculations of few tree-level processes. Finally, we consider the special case when in a scattering event additional soft particles are emitted. The relation between Feynman amplitudes and cross-sections or decay widths which is essentially the same as in non-relativistic quantum mechanics is reviewed in the appendix of this chapter.

9.1   Unitarity of the S-matrix and its consequences

A scattering process is fully described in the Schrödinger picture by the knowledge how initial states $|i, t\rangle$ at $t \to -\infty$ are transformed into final states $|f, t\rangle$ at $t \to \infty$. This knowledge is encoded in the S-matrix elements

$$|f, t = \infty\rangle = S_{fi} |i, t = -\infty\rangle. \tag{9.1}$$

An intuitive, but mathematically delicate definition of the scattering operator S is the $t \to \infty$ limit of the time-evolution operator $U(t, -t)$,

$$S = \lim_{t \to \infty} U(t, -t). \tag{9.2}$$

Thus the scattering operator S evolves an eigenstate $|n, t\rangle$ of the Hamiltonian from $t = -\infty$ to $t = +\infty$,

$$S |n, -\infty\rangle = |n, \infty\rangle. \tag{9.3}$$

The unitarity of the scattering operator, $S^\dagger S = SS^\dagger = 1$, expresses the fact that we (should) use a complete set of states for the initial and final states in a scattering process,

$$1 = \sum_n |n, +\infty\rangle \langle n, +\infty| = \sum_n S |n, -\infty\rangle \langle n, -\infty| S^\dagger = SS^\dagger. \tag{9.4}$$

Quantum Fields–From the Hubble to the Planck Scale. Michael Kachelriess. © Michael Kachelriess 2018.
Published in 2018 by Oxford University Press. DOI 10.1093/oso/9780198802877.001.0001

Optical theorem. We split the scattering operator S into a diagonal part and the transition operator T, $S = 1 + iT$, and thus

$$1 = (1 + iT)(1 - iT^\dagger) = 1 + i(T - T^\dagger) + TT^\dagger \tag{9.5}$$

or

$$iTT^\dagger = T - T^\dagger. \tag{9.6}$$

Note that in perturbation theory the LHS is $\mathcal{O}(g^{2n})$, while the RHS is $\mathcal{O}(g^n)$. Hence this equation implies a non-linear relation between the transition operator evaluated at different orders. At lowest-order perturbation theory, the LHS vanishes and T is real, $T = T^\dagger$.

We now consider matrix elements between the initial and final state,

$$\langle f| T - T^\dagger |i\rangle = T_{fi} - T_{if}^* = i \langle f| TT^\dagger |i\rangle = i \sum_n T_{fn} T_{in}^*. \tag{9.7}$$

If we set $|i\rangle = |f\rangle$, we obtain a connection between the forward scattering amplitude T_{ii} and the total cross-section σ_{tot} called the optical theorem,

$$2\Im T_{ii} = \sum_n |T_{in}|^2. \tag{9.8}$$

The optical theorem relates the attenuation of a beam of particles in the state i, $dN_i \propto -|\Im T_{ii}|^2 N_i$, to the probability that they scatter into all possible states n. Its RHS is given by the total cross-section σ_{tot} up to a factor depending on the flux of initial particles and possible symmetry factors. For the case of two particles in the initial state, comparison with Eqs. (9.149) and (9.153) from the appendix shows that

$$\Im \mathcal{A}_{ii} = 2 p_{\text{cms}} \sqrt{s}\, \sigma_{\text{tot}}. \tag{9.9}$$

Note also that the forward scattering amplitude T_{ii} means *scattering* without change in any conserved quantum number, since we already extracted the identity part, $T_{ii} = (S_{ii} - 1)/i$.

Imaginary part of the amplitude. Let us consider the Feynman amplitude $\mathcal{A}$ as a complex function of the squared centre-of-mass (c.m.) energy s. The threshold energy $\sqrt{s_0}$ in the c.m. system equals the minimal energy for which the reaction is kinematically allowed. The optical theorem implies that $\mathcal{A}$ is real for $s < s_0$ and $s \in \mathbb{R}$. Thence $s = s^*$ and $\mathcal{A}(s) = [\mathcal{A}(s)]^*$ and therefore

$$\mathcal{A}(s) = [\mathcal{A}(s^*)]^* \qquad \text{for} \quad s < s_0. \tag{9.10}$$

If $\mathcal{A}(s)$ is an analytic function, then also $[\mathcal{A}(s^*)]^*$ is analytic and we can continue this relation into the complex s plane. In particular, along the real axis we have for $s > s_0$

$$\Re \mathcal{A}(s + i\varepsilon) = \Re \mathcal{A}(s - i\varepsilon) \quad \text{and} \quad \Im \mathcal{A}(s + i\varepsilon) = -\Im \mathcal{A}(s - i\varepsilon). \tag{9.11}$$

Thus starting from s_0, the amplitude $\mathcal{A}$ has a discontinuity along the real s axis. Since the amplitude $\mathcal{A}$ should be single-valued, it has to contain a branch cut along the real s axis starting at s_0. Feynman's $m^2 - i\varepsilon$ prescription tells us then which side of the cut we should pick out as the "physical" one.

The second relation in (9.11) allows us to obtain the imaginary part of a Feynman amplitude from its discontinuity,

$$\text{disc}(\mathcal{A}) \equiv \mathcal{A}(s + i\varepsilon) - \mathcal{A}(s - i\varepsilon) = 2i\Im\mathcal{A}(s + i\varepsilon). \tag{9.12}$$

The prototype of a function having a discontinuity and a branch cut (along $\mathbb{R}^-$) is the logarithm,

$$\text{Ln}(z) = \text{Ln}(re^{i\vartheta}) = \ln(re^{i\vartheta}) + 2k\pi i = \ln(r) + (\vartheta + 2k\pi)i \tag{9.13}$$

with $\Im \ln(x + i\varepsilon) = \pi$. How does an imaginary part in a Feynman diagram arise? Comparing the relation

$$\frac{1}{x \pm i\varepsilon} = P\left(\frac{1}{x}\right) \mp i\pi\delta(x) \tag{9.14}$$

with the propagator of a virtual particle, we see that virtual particles which propagate on-shell lead to poles and to imaginary terms in the amplitude.

Example 9.1: Verify the optical theorem for $\phi\phi \to \phi\phi$ scattering in the $\lambda\phi^4$ theory at $\mathcal{O}(\lambda^2)$. The logarithmic terms in the scattering amplitude (4.73b) for $\phi\phi \to \phi\phi$ scattering at one-loop take the form

$$F(q^2, m) = \int_0^1 dz \, \ln\left[m^2 - q^2 z(1 - z)\right] \tag{9.15}$$

with $q^2 = \{s, t, u\}$. In the physical region, the relation $q^2 > 4m^2$ holds only for the s channel diagram. The argument of the logarithm becomes negative for

$$z_{1/2} = \frac{1}{2}\left[1 \pm \sqrt{1 - 4m^2/s^2}\right] = \frac{1}{2} \pm \frac{1}{2}\beta \tag{9.16}$$

with $\beta = \sqrt{1 - 4m^2/s^2}$ as the velocity of the ϕ particles in the centre-of-mass system. Using now $\Im[\ln(-q^2 - i\varepsilon)] = -\pi$, the imaginary part follows as

$$\Im(\mathcal{A}) = \pi \frac{\lambda^2}{32\pi^2} \int_{\frac{1}{2} - \frac{1}{2}\beta}^{\frac{1}{2} + \frac{1}{2}\beta} dz = \frac{\lambda^2}{32\pi}\beta. \tag{9.17}$$

The optical theorem implies thus that the total cross-section $\sigma_{\text{tot}}(\phi\phi \to \text{all})$ at $\mathcal{O}(\lambda^2)$ equals

$$\sigma_{\text{tot}} = \frac{\Im\mathcal{A}_{ii}}{2p_{\text{cms}}\sqrt{s}} = \frac{\lambda^2}{32\pi s}, \tag{9.18}$$

where we used $2p_{\text{cms}} = \sqrt{s}\beta$. On the other hand, the Feynman amplitude at tree level is simply $\mathcal{A} = -\lambda$ and thus the elastic cross-section for $\phi\phi \to \phi\phi$ scattering follows as $\sigma_{\text{el}} = \lambda^2/(32\pi s)$. At $\mathcal{O}(\lambda^2)$, the only reaction contributing to the total cross-section is elastic scattering, and thus the two cross-sections agree. Note also the treatment of the symmetry factors. In the loop diagram, the symmetry factor $S = 1/2!$ is already included, while the corresponding factor for the two identical particles in the final state is added only integrating the cross-section.

9.2   LSZ reduction formula

We defined the generating functional $Z[J] = \langle 0, \infty | 0, -\infty \rangle_J$ as the vacuum–vacuum transition amplitude in the presence of a classical source J. Thus the generating functional contains the boundary condition $\phi(x) \to 0$ for $t \to \pm\infty$. We have two options to find a bridge between S-matrix elements and the formalism we have derived up to now. One possibility is to find the connection between the Green functions derived from $Z[J]$ and S-matrix elements. Another one is to define first a new functional $Z'[J]$ with the correct boundary conditions, and then to establish the connection between $Z[J]$ and $Z'[J]$. We choose the first way, restricting ourselves for simplicity to the case of a real scalar field.

Let us start with the case of a $2 \to 2$ scattering process. We can generate a free two-particle[1] state composed of plane waves by applying two creation operators on the vacuum,

$$|\boldsymbol{k}_1, \boldsymbol{k}_2\rangle = a^\dagger(\boldsymbol{k}_1)a^\dagger(\boldsymbol{k}_2)|0\rangle. \tag{9.19}$$

We obtain localised wave packets defining new creation operators

$$a_i^\dagger = \int \mathrm{d}^3k \, f_i(\boldsymbol{k})a^\dagger(\boldsymbol{k}), \tag{9.20}$$

where $f_i(\boldsymbol{k})$ is, for example, a Gaussian centred around $\boldsymbol{k}_i$,

$$f_i(\boldsymbol{k}) \propto \exp[-(\boldsymbol{k} - \boldsymbol{k}_i)^2/(2\sigma^2)]. \tag{9.21}$$

We assume that the initial state of the scattering process at $t = -\infty$ can be described by freely propagating wave packets,

$$|i\rangle = \lim_{t \to -\infty} a_1^\dagger(t)a_2^\dagger(t)|0\rangle = |\boldsymbol{k}_1, \boldsymbol{k}_2; -\infty\rangle, \tag{9.22}$$

and similarly the final state as

$$|f\rangle = \lim_{t \to \infty} a_{1'}^\dagger(t)a_{2'}^\dagger(t)|0\rangle = |\boldsymbol{k}_{1'}, \boldsymbol{k}_{2'}; +\infty\rangle. \tag{9.23}$$

Here we changed to the Heisenberg picture, since our Green functions are time-dependent.

Our task is to connect this transition amplitude $\langle f | i\rangle$ to the corresponding four-point Green function. The latter is the time-ordered vacuum expectation value of field

---

[1]To reduce clutter, we assume $\boldsymbol{k}_1 \neq \boldsymbol{k}_2$.

operators. The first property, time-ordering, is automatically satisfied for the transition amplitude $\langle f\,|i\rangle$, since we can write

$$\langle f\,|i\rangle = \lim_{t\to\infty} \langle 0|\, a_{1'}(t)a_{2'}(t)a_1^\dagger(-t)a_2^\dagger(-t)\,|0\rangle \tag{9.24a}$$

$$= \lim_{t\to\infty} \langle 0|\, \mathrm{T}\{a_{1'}(t)a_{2'}(t)a_1^\dagger(-t)a_2^\dagger(-t)\}\,|0\rangle\,. \tag{9.24b}$$

Thus we only have to re-express the creation and annihilation operators as (projected) field operators. We define a scalar product for solutions of the Klein–Gordon equation as follows,

$$(\phi,\chi) = \mathrm{i}\int \mathrm{d}^3x\,\phi^*(x)\overleftrightarrow{\partial_0}\chi(x) \equiv \mathrm{i}\int \mathrm{d}^3x\left[\phi^*(x)\frac{\partial\chi(x)}{\partial t} - \frac{\partial\phi^*(x)}{\partial t}\chi(x)\right]. \tag{9.25}$$

Comparing this definition to Eq. (5.13), we see that the scalar product is the zero component of the conserved current j^μ. Thus the value of the scalar product (ϕ,χ) is time-independent and corresponds to the number of particles minus the number of antiparticles.

For plane-wave components with definite momentum,

$$\phi_{\boldsymbol{k}}(x) = \frac{1}{\sqrt{(2\pi)^3 2\omega_k}}\,\mathrm{e}^{-\mathrm{i}kx} = N_{\boldsymbol{k}}\,\mathrm{e}^{-\mathrm{i}kx}, \tag{9.26}$$

the scalar product is given by

$$(\phi_{\boldsymbol{k}'},\phi_{\boldsymbol{k}}) = \mathrm{i}N_{\boldsymbol{k}'}N_{\boldsymbol{k}}\int \mathrm{d}^3x\left[\mathrm{e}^{\mathrm{i}k'x}(-\mathrm{i}\omega_k)\mathrm{e}^{-\mathrm{i}kx} - \mathrm{i}\omega_{k'}\mathrm{e}^{\mathrm{i}k'x}\mathrm{e}^{-\mathrm{i}kx}\right] \tag{9.27a}$$

$$= N_{\boldsymbol{k}}^2(2\pi)^3\delta(\boldsymbol{k}-\boldsymbol{k}')2\omega_{\boldsymbol{k}}\mathrm{e}^{\mathrm{i}(\omega_k-\omega_k)t} = \delta(\boldsymbol{k}-\boldsymbol{k}'). \tag{9.27b}$$

Similarly it follows $(\phi_{\boldsymbol{k}}^*,\phi_{\boldsymbol{k}'}^*) = -\delta(\boldsymbol{k}-\boldsymbol{k}')$, while the two terms in the scalar product cancel otherwise,

$$(\phi_{\boldsymbol{k}},\phi_{\boldsymbol{k}'}^*) = (\phi_{\boldsymbol{k}}^*,\phi_{\boldsymbol{k}'}) = 0.$$

Thus we can invert the free field operator

$$\phi(x) = \int \mathrm{d}^3k\left[a(\boldsymbol{k})\phi_{\boldsymbol{k}}(x) + a^\dagger(\boldsymbol{k})\phi_{\boldsymbol{k}}^*(x)\right] = \int \frac{\mathrm{d}^3k}{\sqrt{(2\pi)^3 2\omega_k}}\left[a(\boldsymbol{k})\mathrm{e}^{-\mathrm{i}kx} + a^\dagger(\boldsymbol{k})\mathrm{e}^{+\mathrm{i}kx}\right] \tag{9.28}$$

to obtain

$$a^\dagger(\boldsymbol{k}) = -(\phi_{\boldsymbol{k}}^*,\phi) = -\mathrm{i}N_k\int \mathrm{d}^3x\,\mathrm{e}^{-\mathrm{i}kx}\overleftrightarrow{\partial_t}\phi(x). \tag{9.29}$$

Next we want to rewrite this expression in a way that shows explicitly its Lorentz invariance. Using the identity

$$a^\dagger(\boldsymbol{k},\infty) - a^\dagger(\boldsymbol{k},-\infty) = \int_{-\infty}^{\infty}\mathrm{d}t\,\frac{\partial}{\partial t}\,a^\dagger(\boldsymbol{k},t) \tag{9.30}$$

we insert first (9.29) assuming a wave-package centered at $\boldsymbol{k}_1$ and then perform the time differentiation,

$$a^\dagger(\boldsymbol{k}_1, \infty) - a^\dagger(\boldsymbol{k}_1, -\infty) = -\mathrm{i} \int \mathrm{d}^3k f_1(\boldsymbol{k}) \int \mathrm{d}^4x \, \partial_t \left(\mathrm{e}^{-\mathrm{i}kx} \overleftrightarrow{\partial_t} \phi(x) \right) \tag{9.31}$$

$$= -\mathrm{i} \int \mathrm{d}^3k f_1(\boldsymbol{k}) \int \mathrm{d}^4x \, \left(\mathrm{e}^{-\mathrm{i}kx} \partial_t^2 \phi(x) - \phi(x) \partial_t^2 \mathrm{e}^{-\mathrm{i}kx} \right),$$

where the two terms linear in ∂_t cancelled. Then we use the fact that the field is on-shell, $k^2 = m^2$, for the replacement

$$\partial_t^2 \mathrm{e}^{-\mathrm{i}kx} = (\boldsymbol{\nabla}^2 - m^2) \mathrm{e}^{-\mathrm{i}kx}.$$

Since the field is localised in space, we can perform two partial integrations moving thereby $\boldsymbol{\nabla}^2$ to the left, obtaining

$$a^\dagger(\boldsymbol{k}_1, \infty) - a^\dagger(\boldsymbol{k}_1, -\infty) = -\mathrm{i} \int \mathrm{d}^3k f_1(\boldsymbol{k}) \int \mathrm{d}^4x \, \mathrm{e}^{-\mathrm{i}kx} \left(\Box + m^2 \right) \phi(x). \tag{9.32}$$

In a free theory, $\phi(x)$ satisfies the Klein–Gordon equation and the RHS would vanish. In an interacting theory with e.g. $\mathscr{L}_I = -\lambda \phi^4 / 4!$, the RHS is, however, proportional to $\left(\Box + m^2 \right) \phi(x) = \lambda \phi^3 / 3! \neq 0$.

Having performed the partial integrations, we can forget the wave-packets, $\sigma \to 0$, and write simply

$$a^\dagger(\boldsymbol{k}, -\infty) = a^\dagger(\boldsymbol{k}, \infty) + \mathrm{i}N_{\boldsymbol{k}} \int \mathrm{d}^4x \, \mathrm{e}^{-\mathrm{i}kx} \left(\Box + m^2 \right) \phi(x). \tag{9.33}$$

Taking the Hermitian conjugate, we obtain for the annihilation operator

$$a(\boldsymbol{k}, \infty) = a(\boldsymbol{k}, -\infty) + \mathrm{i}N_{\boldsymbol{k}} \int \mathrm{d}^4x \, \mathrm{e}^{\mathrm{i}kx} \left(\Box + m^2 \right) \phi(x). \tag{9.34}$$

When we insert these expressions into $\langle f | i \rangle$, we obtain a four-point function combining the second terms from the RHS of (9.33) and (9.34). The operators $a(\boldsymbol{k}, -\infty)$ and $a^\dagger(\boldsymbol{k}, \infty)$ destroy a particle with momentum $\boldsymbol{k}$ in the initial and final state, respectively. Including these terms gives therefore only an additional contribution, if a particle propagates from $t = -\infty$ to $t = +\infty$ with its momentum unchanged, and generates thereby disconnected graphs. Hence we do not need to consider these terms, if we restrict ourselves to connected Green functions. For n particles in the initial and m particles in the final state, we obtain

$$\langle \boldsymbol{k}'_1, \ldots, \boldsymbol{k}'_m; +\infty | \boldsymbol{k}_1, \ldots, \boldsymbol{k}_n; -\infty \rangle = \mathrm{i}^{n+m} \prod_{i=1}^{n} \int \mathrm{d}^4x_i \, N_{\boldsymbol{k}_i} \, \mathrm{e}^{-\mathrm{i}k_i x_i} \left(\Box_{x_i} + m^2 \right)$$

$$\times \prod_{j=1}^{m} \int \mathrm{d}^4y_j \, N_{\boldsymbol{k}_j} \, \mathrm{e}^{\mathrm{i}k'_j y_j} \left(\Box_{y_j} + m^2 \right) \langle 0| \, \mathrm{T}\{\phi(x_1) \cdots \phi(y_m)\} \, |0\rangle. \tag{9.35}$$

This is the reduction formula of Lehmann, Symanzik and Zimmermann (LSZ). For each external particle we obtained the corresponding plane-wave component and a Klein–Gordon operator. Since the latter is the inverse of the free two-point function, we can rephrase the content of the LZS formula simply as follows: replace the two-point functions of external lines by appropriate wave-functions, that is $\phi_{\boldsymbol{k}}(x)$ for scalar particles in the initial state and $\phi_{\boldsymbol{k}}^*(x)$ for scalar particles in the final state.

Since we started from field operators in the Heisenberg picture, the matrix element is in the Heisenberg picture, too. In the Schrödinger picture, it is

$$\langle \boldsymbol{k}_1', \dots, \boldsymbol{k}_m' | \mathrm{i}T | \boldsymbol{k}_1, \dots, \boldsymbol{k}_n \rangle \equiv \mathrm{i}T_{fi},$$

where we also used $S = 1 + \mathrm{i}T$ and the fact that we neglected disconnected parts. Finally, we define the Fourier-transformed n-point function as

$$G(x_1, \dots, x_n) = \int \prod_{i=1}^{n} \frac{\mathrm{d}^4 k_i}{(2\pi)^4} \, \exp\left(-\mathrm{i}\sum_i k_i x_i\right) G(k_1, \dots, k_n). \qquad (9.36)$$

Then we obtain the LSZ reduction formula in momentum space,

$$\mathrm{i}T_{fi} = \mathrm{i}^{n+m} N_{\boldsymbol{k}_1} \cdots N_{\boldsymbol{k}_n} N_{\boldsymbol{k}_1'} \cdots N_{\boldsymbol{k}_m'} (k_1^2 - m^2) \cdots (k_n^2 - m^2)(k_1'^2 - m^2) \cdots (k_m'^2 - m^2)$$
$$\times G(k_1, \cdots, k_n, -k_1', \dots, -k_m'). \qquad (9.37)$$

The Green function $G(k_1, \cdots, k_n, -k_1', \dots, -k_m')$ is multiplied by zeros, since the external particles satisfy $k^2 = m^2$. Thus T_{fi} vanishes, except when poles $1/(k^2 - m^2)$ of $G(k_1, \cdots, k_n, -k_1', \dots, -k_m')$ cancel these zeros. In the case of external scalar particles, only their normalisation factors are left. As they are not essential for the calculation of the transition amplitudes, one includes these normalisation factors into the phase space of final-state particles and in the flux factor of initial particles. This explains our Feynman rule of replacing the scalar propagator on external lines by 1 for amplitudes in momentum space.

The derivation of the LSZ formula for particles with spin $s > 0$ proceeds in the same way. Their wave-functions additionally contain polarisation vectors $\varepsilon^\mu(k)$, tensors $\varepsilon^{\mu\nu}(k)$ or spinors $u(p)$ and $\bar{u}(p)$ and their charge conjugated states. In the case of a photon (graviton), we have to add $\varepsilon^\mu(k)$ ($\varepsilon^{\mu\nu}(k)$) in the initial state and the complex conjugated $\varepsilon^{*\mu}(k)$ ($\varepsilon^{*\mu\nu}(k)$) in the final state. In the case of Dirac fermions, we have to assign four different spinors to the four possible combinations of particle and antiparticles in the initial and final state. Having chosen $u(p)$ as particle state with an arrow along the direction of time, the simple rule that a fermion line corresponds to a complex number $\bar{\psi} \cdots \psi$ fixes the designation of the other spinors as shown in Fig. 9.1. Connecting the upper fermion lines, $\bar{u}(p') \cdots u(p)$, corresponds to the scattering $e^-(p) + X \to e^-(p') + X'$, while $\bar{v}(q) \cdots u(p)$ describes the annihilation process $e^+(q) + e^-(p) \to X + X'$. Connecting the lower fermion lines, $\bar{v}(q) \cdots v(q')$, corresponds to the scattering $e^+(q) + X \to e^+(q') + X'$, while $\bar{u}(p') \cdots v(q')$ describes the pair creation process $X + X' \to e^+(q') + e^-(p')$. Recall that the Feynman amplitude $\mathcal{A}$ is defined omitting the normalisation factor $N_p = [2\omega_p(2\pi)^3]^{-1/2}$ from all wave-functions—the splitting of S-matrix elements into Feynman amplitudes and phase space is discussed in appendix 9.A.

Wave-function renormalisation. Up to now we have pretended that we can describe the fields in the initial and final state as free particles. Although, for example, Yukawa interactions between two, by assumption, widely separated particles

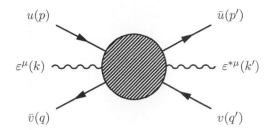

$u(p)$ $\bar{u}(p')$

$\varepsilon^\mu(k)$ $\varepsilon^{*\mu}(k')$

$\bar{v}(q)$ $v(q')$

Fig. 9.1 Feynman rules for external particles in momentum space; initial state on the left, final state on the right.

at $t = \pm\infty$ are negligibly small, self-interactions persist. These interactions lead to a renormalisation of the external wave-functions.

We can rephrase the problem as follows. If the creation operator corresponds to the one of a free theory, $a_0^\dagger(\boldsymbol{k}, -\infty)$ then it can only connect one-particle states with the vacuum. In contrast, the interacting field can also connect many-particle states to the vacuum and therefore its overlap with single-particle states is reduced,

$$a^\dagger(\boldsymbol{k}, -\infty)\,|0\rangle = \sqrt{Z}\,|\boldsymbol{k}\rangle + \sqrt{1-Z}\left\{\left|\boldsymbol{k}', \boldsymbol{k}'', \boldsymbol{k}'''\right\rangle + \ldots\right\} \tag{9.38a}$$

$$= \sqrt{Z}a_0^\dagger(\boldsymbol{k})\,|0\rangle + \sqrt{1-Z}\left\{\left|\boldsymbol{k}', \boldsymbol{k}'', \boldsymbol{k}'''\right\rangle + \ldots\right\}. \tag{9.38b}$$

Therefore the free and the interacting fields are connected by

$$\phi(x) \to \sqrt{Z}\phi_0(x) \tag{9.39}$$

for $t \to \pm\infty$, where we call the factor Z the wave-function (or the field-strength) renormalisation constant. We will show in section 11.4.2 that this factor can be extracted from the self-energy diagrams of the corresponding field. More precisely, including a factor $\sqrt{Z_k}$ for each external line of type k takes into account self-energy corrections in the external lines. Therefore it is enough to calculate the self-energy and to extract $\sqrt{Z_k}$ once; after that we can omit self-energy corrections in the external lines adding simply factors $\sqrt{Z_k}$. Finally, note that we can set $Z = 1$ in tree-level processes, since in perturbation theory $Z = 1 + \mathcal{O}(g)$ holds.

9.3 Specific processes

We now consider a few specific processes in detail. First, we derive the Klein–Nishina formula for the Compton scattering cross-section using the standard "trace method". Then we calculate polarised $e^+e^- \to \mu^+\mu^-$ and $e^+e^- \to \gamma\gamma$ scattering applying helicity methods.

9.3.1 Trace method and Compton scattering

Matrix element. The Feynman amplitude $\mathcal{A}$ of Compton scattering $e^-(p)+\gamma(k) \to e^-(p')+\gamma(k')$ at $\mathcal{O}(e^2)$ consists of the two diagrams shown in Fig. 9.2 and is given by

$$i\mathcal{A} = -ie^2 \bar{u}(p') \left[\not{\epsilon}^{*\prime} \frac{\not{p} + \not{k} + m}{(p+k)^2 - m^2} \not{\epsilon} + \not{\epsilon} \frac{\not{p} - \not{k}' + m}{(p-k')^2 - m^2} \not{\epsilon}^{*\prime} \right] u(p). \tag{9.40}$$

Since the denominator is a non-zero light-like vector, we have omitted the $i\varepsilon$. Note that the two amplitudes can be transformed into each other, replacing $\varepsilon \leftrightarrow \varepsilon^{*\prime}$ and $k \leftrightarrow -k'$. This symmetry, called crossing symmetry, relates processes where a particle is replaced by an antiparticle with negative momentum on the other side of the reaction.

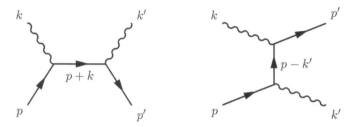

Fig. 9.2 The two Feynman diagrams contributing to Compton scattering at $\mathcal{O}(e^2)$.

We evaluate the process in the rest-frame of the initial electron. Then $p^\mu = (m, \mathbf{0})$ and choosing $\varepsilon^\mu = (0, \boldsymbol{\varepsilon})$ as well as $\varepsilon'^\mu = (0, \boldsymbol{\varepsilon}')$, it follows

$$p \cdot \varepsilon = p \cdot \varepsilon' = 0. \tag{9.41}$$

Moreover, the photons are transversely polarised,

$$k \cdot \varepsilon = k' \cdot \varepsilon' = 0, \tag{9.42}$$

and we choose real polarisation vectors. We anticommute $\not{p}$ in the numerator to the right, $\not{p}\not{\epsilon}' = 2p \cdot \varepsilon' - \not{\epsilon}'\not{p} = -\not{\epsilon}'\not{p}$ and use the Dirac equation, $\not{p}u(p) = mu(p)$. Then we simplify also the denominator using $p^2 = m^2$ and obtain

$$\mathcal{A} = -e^2 \bar{u}(p') \left[\frac{\not{\epsilon}'\not{k}\not{\epsilon}}{2p \cdot k} + \frac{\not{\epsilon}\not{k}'\not{\epsilon}'}{2p \cdot k'} \right] u(p). \tag{9.43}$$

Typically the electron target is not polarised, and the spin of the final electron is not measured. Thus we sum the squared matrix element over the final and average over the initial electron spin,

$$|\overline{\mathcal{A}}|^2 = \frac{1}{2} \sum_{s,s'} |\mathcal{A}|^2 = \frac{e^4}{2} \sum_{s,s'} \left| \bar{u}(p') \left(\frac{\not{\epsilon}'\not{k}\not{\epsilon}}{2p \cdot k} + \frac{\not{\epsilon}\not{k}'\not{\epsilon}'}{2p \cdot k'} \right) u(p) \right|^2. \tag{9.44}$$

Calculating $|\mathcal{A}|^2 = \mathcal{A}\mathcal{A}^*$ requires the knowledge of $\mathcal{A}^* = \mathcal{A}^\dagger$. Recall that we defined $\bar{\psi} = \psi^\dagger \gamma^0$ such that a general amplitude $\mathcal{A}$ composed of spinors,

$$\mathcal{A} = \bar{\psi}(p')\Gamma\psi(p) = \psi^\dagger(p')\gamma^0\Gamma\psi(p) \tag{9.45}$$

with Γ denoting a product of the basis elements given in Eq. (8.40), becomes

$$\mathcal{A}^* = \bar{\psi}(p)\gamma^0 \Gamma^\dagger \gamma^0 \psi(p') \equiv \bar{\psi}(p)\overline{\Gamma}\psi(p'). \tag{9.46}$$

Important special cases worth memorising are $\overline{\gamma^\mu} = \gamma^\mu$, $\overline{\gamma^5} = -\gamma^5$, and $\overline{\slashed{a}\slashed{b}\cdots\slashed{z}} = \slashed{z}\cdots\slashed{b}\slashed{a}$. We now write out the spinor indices,

$$|\overline{\mathcal{A}}|^2 = \frac{e^4}{2}\sum_{s,s'}\bar{u}_a(p')\left[\frac{\slashed{\varepsilon}'\slashed{k}\slashed{\varepsilon}}{2p\cdot k} + \frac{\slashed{\varepsilon}\slashed{k}'\slashed{\varepsilon}'}{2p\cdot k'}\right]_{ab} u_b(p)\bar{u}_c(p)\left[\frac{\slashed{\varepsilon}\slashed{k}\slashed{\varepsilon}'}{2p\cdot k} + \frac{\slashed{\varepsilon}'\slashed{k}'\slashed{\varepsilon}}{2p\cdot k'}\right]_{cd} u_d(p'). \tag{9.47}$$

Using the property (8.56) of the Dirac spinors, $\sum_s u_a(p,s)\bar{u}_b(p,s') = (\slashed{p}+m)_{ab}$, we obtain

$$|\overline{\mathcal{A}}|^2 = \frac{e^4}{2}\left[\slashed{p}'+m\right]_{da}\left[\frac{\slashed{\varepsilon}'\slashed{k}\slashed{\varepsilon}}{2p\cdot k} + \frac{\slashed{\varepsilon}\slashed{k}'\slashed{\varepsilon}'}{2p\cdot k'}\right]_{ab}\left[\slashed{p}+m\right]_{bc}\left[\frac{\slashed{\varepsilon}\slashed{k}\slashed{\varepsilon}'}{2p\cdot k} + \frac{\slashed{\varepsilon}'\slashed{k}'\slashed{\varepsilon}}{2p\cdot k'}\right]_{cd}. \tag{9.48}$$

Since $\slashed{p}+m$ combines one spinor in $\mathcal{A}$ and one in $\mathcal{A}^*$, the result is a trace over gamma matrices,

$$|\overline{\mathcal{A}}|^2 = \frac{e^4}{8}\text{tr}\left\{(\slashed{p}'+m)\left[\frac{\slashed{\varepsilon}'\slashed{k}\slashed{\varepsilon}}{p\cdot k} + \frac{\slashed{\varepsilon}\slashed{k}'\slashed{\varepsilon}'}{p\cdot k'}\right](\slashed{p}+m)\left[\frac{\slashed{\varepsilon}\slashed{k}\slashed{\varepsilon}'}{p\cdot k} + \frac{\slashed{\varepsilon}'\slashed{k}'\slashed{\varepsilon}}{p\cdot k'}\right]\right\}. \tag{9.49}$$

Working out some more examples of this type (e.g. in problem 9.2), you should convince yourself that each fermion line in $\mathcal{A}$ is converted into a trace in $|\overline{\mathcal{A}}|^2$. Useful identities for the evaluation of such traces are given in appendix A.2.

We simplify this trace by anticommuting identical variables such that they become neighbours. Then we can use $\slashed{a}\slashed{a} = a^2$ and thereby reduce the number of gamma matrices in each step by two. Multiplying out the terms in the trace, we obtain three contributions that we denote by

$$\text{tr}\{\ \ \} = \frac{S_1}{(p\cdot k)^2} + \frac{S_2}{(p\cdot k')^2} + \frac{2S_3}{(p\cdot k)(p\cdot k')}. \tag{9.50}$$

We consider only the first term S_1 in detail. Starting from

$$S_1 = \text{tr}\left\{(\slashed{p}'+m)\,\slashed{\varepsilon}'\slashed{k}\slashed{\varepsilon}\,(\slashed{p}+m)\,\slashed{\varepsilon}\slashed{k}\slashed{\varepsilon}'\right\} = \text{tr}\{\slashed{p}'\slashed{\varepsilon}'\slashed{\varepsilon}\,\underbrace{\slashed{k}\slashed{p}}_{2kp-\slashed{p}\slashed{k}}\,\slashed{k}\slashed{\varepsilon}\slashed{\varepsilon}'\} + m^2\text{tr}\{\slashed{\varepsilon}'\slashed{\varepsilon}\,\underbrace{\slashed{k}\slashed{k}}_{k^2=0}\,\slashed{\varepsilon}\slashed{\varepsilon}'\} =$$

$$= 2(k\cdot p)\,\text{tr}\left\{\slashed{p}'\slashed{\varepsilon}'\slashed{\varepsilon}\slashed{k}\slashed{\varepsilon}\slashed{\varepsilon}'\right\} - \text{tr}\left\{\slashed{p}'\slashed{\varepsilon}'\slashed{\varepsilon}\slashed{p}\slashed{k}\slashed{k}\slashed{\varepsilon}\slashed{\varepsilon}'\right\} \tag{9.51}$$

we arrive at an expression with only six gamma matrices. We continue the work,

$$S_1 = 2(k\cdot p)\,\text{tr}\left\{\slashed{p}'\slashed{\varepsilon}'\slashed{\varepsilon}\slashed{k}\slashed{\varepsilon}\slashed{\varepsilon}'\right\} = -2(k\cdot p)\,\text{tr}\left\{\slashed{p}'\slashed{\varepsilon}'\slashed{k}\slashed{\varepsilon}\slashed{\varepsilon}\slashed{\varepsilon}'\right\} = \tag{9.52a}$$

$$= 2(k\cdot p)\,\text{tr}\left\{\slashed{p}'\slashed{\varepsilon}'\slashed{k}\slashed{\varepsilon}'\right\} = 2(k\cdot p)\left[2(k\cdot\varepsilon')\,\text{tr}\left(\slashed{p}'\slashed{\varepsilon}'\right) - \text{tr}\left\{\slashed{p}'\slashed{\varepsilon}'\slashed{\varepsilon}'\slashed{k}\right\}\right] = \tag{9.52b}$$

$$= 8(k\cdot p)\left[2(k\cdot\varepsilon')(p'\cdot\varepsilon') + (p'\cdot k)\right], \tag{9.52c}$$

where we have used $\slashed{\varepsilon}\slashed{\varepsilon} = \slashed{\varepsilon}'\slashed{\varepsilon}' = -1$. As next step we want to eliminate the two scalar products that include p'. Four-momentum conservation implies $(p'-k)^2 = (p-k')^2$ and thus

$$p' \cdot k = p \cdot k'. \tag{9.53}$$

Multiplying the four-momentum conservation equation by ε', it follows moreover

$$p + k = p' + k' \quad \Rightarrow \quad \underbrace{\varepsilon' \cdot p}_{0} + \varepsilon' \cdot k = \varepsilon' \cdot p' + \underbrace{\varepsilon' \cdot k'}_{0}. \tag{9.54}$$

Thus our final result for S_1 is

$$S_1 = 8\,(k \cdot p)\left[2\,(k \cdot \varepsilon')^2 + k' \cdot p\right]. \tag{9.55}$$

S_2 can be obtained observing the crossing symmetry of the amplitude by the replacements $\varepsilon \leftrightarrow \varepsilon'$ and $k \leftrightarrow -k'$. The cross-term S_3 has to be calculated and we give here only the final result for the combination of the three terms, where some terms cancel

$$|\overline{\mathcal{A}}|^2 = e^4\left[\frac{\omega'}{\omega} + \frac{\omega}{\omega'} + 4(\varepsilon \cdot \varepsilon')^2 - 2\right]. \tag{9.56}$$

Cross-section. To obtain the cross-section, we have to calculate the flux factor and to perform the integration over the phase space of the final state,

$$d\sigma = \frac{1}{4I}\,(2\pi)^4\,\delta^{(4)}(P_i - P_f)|\mathcal{A}_{fi}|^2 \prod_{f=1}^{n}\frac{d^3 p_f}{2E_f(2\pi)^3} = \frac{1}{4I}\,|\mathcal{A}_{fi}|^2\,d\Phi^{(n)}, \tag{9.57}$$

with the final-state phase space $d\Phi^{(n)}$. The flux factor I in the rest system of the electron is simply

$$I \equiv v_{\text{rel}}\,p_1 \cdot p_2 = m\omega. \tag{9.58}$$

Using Eq. (2.82),

$$\frac{d^3 p'}{2E'} = d^4 p'\delta^{(4)}(p'^2 - m^2)\vartheta(p'_0), \tag{9.59}$$

the phase-space integration becomes

$$d\Phi^{(2)} = \frac{1}{(2\pi)^2}\int d\Omega_{k'}\frac{|\mathbf{k}'|^2 dk'}{2|\mathbf{k}'|}\int \frac{d^3 p'}{2E'}\,\delta^{(4)}\,(p' + k' - p - k) = \tag{9.60}$$

$$= \frac{1}{8\pi^2}\int d\Omega_{k'}|\mathbf{k}'|dk'\,\delta\left((p + k - k')^2 - m^2\right). \tag{9.61}$$

The argument of the delta function is

$$(p + k - k')^2 - m^2 = m^2 + 2p \cdot k - 2p \cdot k' - 2k \cdot k' - m^2 \tag{9.62a}$$
$$= 2m\,(\omega - \omega') - 2\omega\omega'\,(1 - \cos\vartheta) \equiv f(\omega'). \tag{9.62b}$$

In order to evaluate the delta function we have to determine the derivative $f'(\omega')$,

$$f'(\omega') = -2m - 2\omega\,(1 - \cos\vartheta), \tag{9.63}$$

and the zeros of $f(\omega')$,

$$0 = 2m \left(\omega - \omega' \right) - 2\omega\omega' \left(1 - \cos\vartheta \right) . \tag{9.64}$$

Solving for ω' gives $\omega' \left[\omega \left(1 - \cos\vartheta \right) + m \right] = m\omega$ and

$$\omega' = \frac{\omega}{1 + \frac{\omega}{m} \left(1 - \cos\vartheta \right)} . \tag{9.65}$$

This is the famous relation for the frequency shift of a photon found first experimentally in the scattering of X-rays on electrons by Compton 1921. The observed energy change of photons was crucial in accepting the quantum nature ("particle-wave duality") of photons. Combining everything, we obtain

$$d\Phi^{(2)} = \frac{1}{8\pi^2} \int d\Omega_{k'} |\omega'| d\omega' \, \delta \left(2m \left(\omega - \omega' \right) - 2\omega\omega'(1 - \cos\vartheta) \right) \tag{9.66a}$$

$$= \frac{1}{8\pi^2} \int d\Omega_{k'} \frac{\omega'}{2m \left[1 + \frac{\omega}{m} \left(1 - \cos\vartheta \right) \right]} = \frac{1}{16\pi^2 \, m\omega} \int d\Omega_{k'} \, \omega'^2 \tag{9.66b}$$

and thus as differential Klein–Nishina cross-section

$$\frac{d\sigma}{d\Omega} = \frac{1}{4m\omega} \frac{\omega'^2}{16\pi^2 \, m\omega} \left| \overline{\mathcal{A}} \right|^2 = \frac{\alpha^2}{4m^2} \frac{\omega'^2}{\omega^2} \left[\frac{\omega'}{\omega} + \frac{\omega}{\omega'} + 4(\varepsilon \cdot \varepsilon')^2 - 2 \right] \tag{9.67}$$

with $\alpha \equiv e^2/(4\pi)$. For scatterings in the forward direction, $\vartheta \to 0$ and thus $\omega' \to \omega$, the scattered photon retains (in the lab frame) its energy even in the ultra-relativistic limit $\omega \gg m$. The same holds in the classical limit, $\omega \ll m$, but now for all directions. Thus we obtain as classical limit of the Klein–Nishina formula the polarised Thomson cross-section

$$\frac{d\sigma}{d\Omega} \simeq \frac{\alpha^2}{m^2} \left(\varepsilon \cdot \varepsilon' \right)^2 = r_0^2 \left(\varepsilon \cdot \varepsilon' \right)^2 , \tag{9.68}$$

with $r_0 = \alpha/m$ as the classical electron radius.

Averaging and summing over the photon polarisation vectors is simplest, if we choose the angle between ε and ε' as ϑ. Then

$$\sum_{r,r'} \left(\varepsilon \cdot \varepsilon' \right)^2 = 1 + \cos^2\vartheta. \tag{9.69}$$

The integration over the scattering angle ϑ can be done analytically. We use $x = \cos\vartheta$ and set $\tilde{\omega} \equiv \omega/m$,

$$\sigma = \frac{\pi\alpha^2}{m^2} \int_{-1}^{1} dx \left[\frac{1}{[1 + \tilde{\omega}(1-x)]^3} + \frac{1}{1 + \tilde{\omega}(1-x)} - \frac{1 - x^2}{[1 + \tilde{\omega}(1-x)]^2} \right] \tag{9.70a}$$

$$= \frac{\pi\alpha^2}{2m^2} \left\{ \frac{1 + \tilde{\omega}}{\tilde{\omega}^3} \left[\frac{2\tilde{\omega}(1+\tilde{\omega})}{(1+2\tilde{\omega})} - \ln(1+2\tilde{\omega}) \right] + \frac{\ln(1+2\tilde{\omega})}{2\tilde{\omega}} - \frac{1 + 3\tilde{\omega}}{(1+2\tilde{\omega})^2} \right\} . \tag{9.70b}$$

Since in the electron rest-frame $s = (p+k)^2 = m^2 + 2m\omega = m^2(1+2\tilde{\omega})$, we can use $\tilde{\omega} = (s/m^2 - 1)/2$ to express σ in an explicit Lorentz-invariant form.

Approximations for the non-relativistic and the ultra-relativistic limit are

$$\sigma = \sigma_{\text{Th}} \times \begin{cases} 1 - 2\tilde{\omega} + \mathcal{O}(\tilde{\omega}^2) & \text{for } \tilde{\omega} \ll 1, \\ \frac{3}{8\tilde{\omega}} \left(\ln(2\tilde{\omega}) + \frac{1}{2} \right) + \mathcal{O}(\tilde{\omega}^{-2}) & \text{for } \tilde{\omega} \gg 1, \end{cases} \tag{9.71}$$

where the Thomson cross-section is given by $\sigma_{\text{Th}} = 8\pi\alpha^2/(3m^2)$. These approximations are shown together with the exact result in the left panel of Fig. 9.3. In the ultra-relativistic limit $s \gg m^2$, the total cross-section for Compton scattering decreases as $\sigma \propto 1/s$. On the other hand, the differential cross-section in the forward direction is constant. As a result, the relative importance of the forward region $\vartheta \sim 0$ increases for increasing s. While $d\sigma/dx$ is symmetric around $x = 0$ in the classical limit $\omega \to 0$, it becomes more and more asymmetric with a a shrinking peak around the forward region at $\vartheta \approx 0$, cf. the right panel of Fig. 9.3.

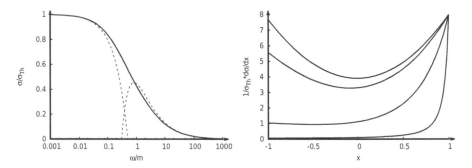

Fig. 9.3 Left: The total cross-section $\sigma/\sigma_{\text{Th}}$ as function of $\tilde{\omega}$ together with the classical and ultra-relativistic limits given in Eq. (9.71). Right: The normalised differential cross-section $\sigma_{\text{Th}}^{-1} d\sigma/dx$ as a function of $x = \cos\vartheta$ for $\tilde{\omega} = 0.01, 0.1, 1$, and 10 (from top down).

Crossing symmetry. We noticed that the two amplitudes in Compton scattering can be transformed into each other replacing $\varepsilon \leftrightarrow \varepsilon^{*\prime}$ and $k \leftrightarrow -k'$. This is an example of a general symmetry of relativistic quantum field theories called crossing symmetry. Using the Feynman rules for in- and out-going particles, it follows that matrix elements where an in-going particle is replaced by an out-going antiparticle or vice versa are related by the following substitutions,

- exchange the momentum $k \leftrightarrow -k'$;
- exchange particle and antiparticle wave-functions; thus in momentum space, $1 \leftrightarrow 1$ for spinless particles, $\varepsilon \leftrightarrow \varepsilon^{*\prime}$ for spin-1 and $u \leftrightarrow v$ for fermions.
- multiply by -1 for each exchanged fermion pair.

The additional minus for fermions is required because the spin sums of fermions and antifermions are related by

$$\sum_s u(p,s)\bar{u}(p,s) = (\slashed{p} + m) = -(\slashed{p}' - m) = -\sum_s v(p',s)\bar{v}(p',s). \tag{9.72}$$

Note that this symmetry allows us to obtain the matrix elements of different processes. For instance, we can relate the processes $e^-e^+ \to \mu^-\mu^+$ with $e^-\mu^- \to e^-\mu^-$ and $\mu^-\mu^+ \to e^-e^+$.

Following a more formal approach, one can derive the crossing symmetry not relying on perturbation theory and the Feynman rules but instead using the analytical properties of S-matrix elements. The LSZ reduction formula distinguishes in- and outgoing particles only by the sign of the momenta used in the Fourier transformation. If one can analytically continue the residue of a pole in an S-matrix element from p^0 to $-p^0$, then one converts the S-matrix for a particle with $\phi(p)$ into the one for an antiparticle with $\phi^*(-p)$. Remarkably, the basic properties of a relativistic quantum field theory, locality and causality, are sufficient to prove that this analytical continuation is possible.

Finally, note that the factor -1 for each exchanged external fermion pair implies a relative minus sign for diagrams connected by crossing which contribute to the same process. Thus there is a relative minus sign between, for example, the t and the u channel diagrams for $e^-e^- \to e^-e^-$ scattering.

9.3.2   Helicity method and polarised QED processes

Using the trace method, the number of terms that have to be calculated grows as $\sim n^2$ with the number n of diagrams. For large n, it should be therefore favourable to calculate the amplitude $\mathcal{A}(s_1, \ldots, s_f)$ for fixed polarisations s_i of the external particles. The amplitude is a complex number and can be trivially squared. An efficient way to calculate polarised amplitudes uses helicity spinors, an approach used also in most modern computer programmes for the calculation of scattering processes.

Massless fermions.   We restrict our short introduction to helicity methods to massless particles. In the case of fermions, we know that then the use of Weyl spinors in the chiral representation is most convenient,

$$u_L(p) = \begin{pmatrix} \phi_L(p) \\ 0 \end{pmatrix} \quad \text{and} \quad u_R(p) = \begin{pmatrix} 0 \\ \phi_R(p) \end{pmatrix}. \tag{9.73}$$

We do not need to consider $v_{L,R}(p)$, since they correspond to particle spinors of opposite helicity, $u_{R,L}(p)$. Moreover, two out of the four possible scalar products involving $u_{L,R}$ are zero for massless fermions,

$$\bar{u}_L(p)u_L(q) = \bar{u}_R(p)u_R(q) = 0. \tag{9.74}$$

This prompts us to introduce a bracket notation for the helicity spinors as follows

$$\bar{u}_L(p) = \langle p, \quad \bar{u}_R(p) = [p, \quad u_L(p) = p], \quad u_R(p) = p\rangle. \tag{9.75}$$

We call the quantities on the RHS *angle* and *square brackets*. The only non-zero Lorentz-invariant spinor products are given by a pair of brackets of the same type,

$$\bar{u}_L(p)u_R(q) = \langle pq\rangle, \quad \text{and} \quad \bar{u}_R(p)u_L(q) = [pq]. \tag{9.76}$$

Next we consider the tensor product of the spinors,

$$p\rangle[p = u_R(p)\bar{u}_R(p) = P_R\not{p}, \quad \text{and} \quad p]\langle p = u_L(p)\bar{u}_L(p) = P_L\not{p}. \tag{9.77}$$

These identities connect the massless spinors $p\rangle$ and $[p$ to the light-like four-vector p^μ.

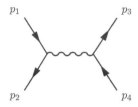

$$p_1 \qquad p_3$$
$$p_2 \qquad p_4$$

Fig. 9.4 Feynman diagrams for the process $e^- e^+ \to \mu^+ \mu^-$.

We are now in position to derive some basic properties of the brackets. First, we can connect the two types of spinor products as

$$\langle pq \rangle = \bar{u}_L(p) u_R(q) = [\bar{u}_R(q) u_L(p)]^* = [qp]^*. \tag{9.78}$$

Multiplying then $p\rangle[p$ and $q][q$ and taking the trace gives

$$\langle pq \rangle [qp] = \text{tr}\{P_R \slashed{q} P_L \slashed{p}\} = \text{tr}\{P_R \slashed{q} \slashed{p}\} = 2p \cdot q, \tag{9.79}$$

so that

$$|\langle pq \rangle|^2 = |[qp]|^2 = 2p \cdot q. \tag{9.80}$$

Next, we express the spinor products through Weyl spinors and use $u_R(p) = i\sigma^2 u_L^*(p)$,

$$\langle pq \rangle = \phi_L^\dagger(p) \phi_R(q) = \phi_{La}^*(p)(i\sigma^2)_{ab} \phi_{Lb}^*(q). \tag{9.81}$$

Then the antisymmetry of $(i\sigma^2)_{ab} = \varepsilon_{ab}$ implies

$$\langle pq \rangle = -\langle qp \rangle, \quad \text{and} \quad [pq] = -[qp]. \tag{9.82}$$

Thus the brackets are square roots of the corresponding Lorentz vector products which are antisymmetric in their two arguments. Finally, we note that the Fierz identity applied to the sigma matrices (cf. with problem 8.16),

$$(\bar{\sigma}^\mu)_{ab}(\bar{\sigma}_\mu)_{cd} = 2(i\sigma^2)_{ac}(i\sigma^2)_{bd}, \tag{9.83}$$

allows the simplification of contracted spinor expressions,

$$\langle p\gamma^\mu q \rangle \langle k\gamma_\mu \ell] = 2\langle pk \rangle [\ell q], \qquad \langle p\gamma^\mu q][k\gamma_\mu \ell \rangle = 2\langle p\ell \rangle [kq]. \tag{9.84}$$

$e^- e^+ \to \mu^- \mu^+$ **scattering.** It is now time to apply this new "bracket" formalism. We consider as first example the tree-level amplitude for $e_L^-(1) e_R^+(2) \to \mu_L^-(3) \mu_R^+(4)$ in QED, given by the single diagram shown in Fig. 9.4. As it is standard using this formalism, we consider all momenta as out-going. Then the amplitude is

$$i\mathcal{A} = (ie)^2 \frac{-i}{q^2} \, \bar{u}_L(3) \gamma^\mu u_L(4) \, \bar{u}_L(2) \gamma_\mu u_L(1) \tag{9.85a}$$

$$= \frac{ie^2}{q^2} \langle 3\gamma^\mu 4] \langle 2\gamma_\mu 1] = \frac{2ie^2}{q^2} \langle 32 \rangle [14], \tag{9.85b}$$

where we have employed the Fierz identity (9.83) in the last step. Since $\langle 32\rangle$ and $[14]$ are both square roots of

$$u = (k_2 + k_3)^2 = (k_1 + k_4)^2, \qquad (9.86)$$

we can replace them by the Mandelstam invariant u. We consider the process in the cm frame of the e^+e^- pair. With $u = -2E^2(1 + \cos\vartheta)$, and $q^2 = s = 4E^2$, the amplitude becomes

$$|\mathrm{i}\mathcal{A}|^2 = e^4(1 + \cos\vartheta)^2. \qquad (9.87)$$

You should re-derive this result using the more familiar trace formalism and compare the amount of algebra required in the two approaches (problem 9.6).

Massless gauge bosons.   In the next step, we incorporate massless gauge bosons as the photon into this framework. We claim that the polarisation vectors of a massless vector boson in the final-state can be represented as

$$\epsilon_R^{*\mu}(k) = \frac{1}{\sqrt{2}}\frac{\langle r\gamma^\mu k]}{\langle rk\rangle}, \qquad \epsilon_L^{*\mu}(k) = -\frac{1}{\sqrt{2}}\frac{[r\gamma^\mu k\rangle}{[rk]}. \qquad (9.88)$$

Here, k is the momentum of the vector boson, and r is a fixed light-like four-vector, called the reference vector, which is assumed to be not collinear with k.

Now we show that this definition makes sense. First, we note that the vectors satisfy $[\epsilon_R^*(k)]^* = \epsilon_L^*(k)$. One can also check that the polarisation vectors are correctly normalised. Moreover, the Dirac equation, $k\!\!\!/\,k] = 0$, guarantees that the polarisation vectors (9.88) are transverse,

$$k_\mu \epsilon_{R,L}^{*\mu}(k) = 0. \qquad (9.89)$$

Finally, we have to show that a change from one reference vector r to another light-like vector s corresponds to a gauge transformation and thus does not affect physics. The change of a polarisation vector under a change of reference vector $r \to s$ is

$$\varepsilon_R^{*\mu}(k;r) - \varepsilon_R^{*\mu}(k;s) = \frac{1}{\sqrt{2}}\left(\frac{\langle r\gamma^\mu k]}{\langle rk\rangle} - \frac{\langle s\gamma^\mu k]}{\langle sk\rangle}\right) \qquad (9.90a)$$

$$= \frac{1}{\sqrt{2}}\frac{1}{\langle rk\rangle\langle sk\rangle}\left\{-\langle r\gamma^\mu k]\langle ks\rangle + \langle s\gamma^\mu k]\langle kr\rangle\right\}. \qquad (9.90b)$$

Now we use first the tensor products (9.77), and then the antisymmetry of the brackets,

$$\varepsilon_R^{*\mu}(k;r) - \varepsilon_R^{*\mu}(k;s) = \frac{1}{\sqrt{2}}\frac{1}{\langle rk\rangle\langle sk\rangle}\left\{-\langle r\gamma^\mu k\!\!\!/ s\rangle + \langle s\gamma^\mu k\!\!\!/ r\rangle\right\} \qquad (9.91a)$$

$$= \frac{1}{\sqrt{2}}\frac{1}{\langle rk\rangle\langle sk\rangle}\left\{\langle s(k\!\!\!/\gamma^\mu + \gamma^\mu k\!\!\!/)r\rangle\right\} \qquad (9.91b)$$

$$= \frac{1}{\sqrt{2}}\frac{1}{\langle rk\rangle\langle sk\rangle}\langle sr\rangle\, 2k^\mu. \qquad (9.91c)$$

In the last line we have applied the Clifford algebra of Dirac matrices. Thus the difference of the polarisation vectors induced by a change of the reference vector is a function proportional to the photon momentum,

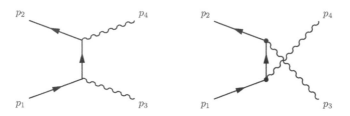

Fig. 9.5 Feynman diagrams for the process $e^-e^+ \to \gamma\gamma$.

$$\varepsilon_R^{*\mu}(k;r) - \varepsilon_R^{*\mu}(k;s) = f(r,s)k^\mu. \tag{9.92}$$

Contracted into an on-shell amplitude, $\mathcal{A} = \varepsilon^\mu \mathcal{A}_\mu$, current conservation implies that this expression vanishes. Thus we can use the most convenient reference vector s which can be chosen differently in any gauge-invariant subset of Feynman diagrams.

$e^-e^+ \to \gamma\gamma$ **scattering.** As a second example, we consider a scattering process with photons as external particles, $e^-e^+ \to \gamma\gamma$, as an illustration for the use of the polarisation vectors. We label the momenta as in Fig. 9.5, taking all momenta as out-going. Then the amplitude for this process is

$$i\mathcal{A} = (ie)^2 \langle 2 \left(\slashed{\varepsilon}(4) \frac{i(\slashed{2}+\slashed{4})}{s_{24}} \slashed{\varepsilon}(3) + \slashed{\varepsilon}(3) \frac{i(\slashed{2}+\slashed{3})}{s_{23}} \slashed{\varepsilon}(4) \right) 1], \tag{9.93}$$

where we use the shorthand $\slashed{2}+\slashed{4}$ for $\slashed{p}_2 + \slashed{p}_4$ and define $s_{ij} = (i+j)^2$.

There are four possible choices for the photon polarisations. Exchanging the momenta 3 and 4 relates the cases $\gamma_R\gamma_L$ and $\gamma_L\gamma_R$, while parity connects $\gamma_R\gamma_R$ and $\gamma_L\gamma_L$. We start showing that the latter two amplitudes are zero in the massless limit we consider. Considering $\gamma_R\gamma_R$, we choose as reference vector $r = 2$ for both polarisation vectors,

$$\varepsilon^\mu(3) = \frac{1}{\sqrt{2}} \frac{[2\gamma^\mu 3\rangle}{[23]}, \qquad \varepsilon^\mu(4) = \frac{1}{\sqrt{2}} \frac{[2\gamma^\mu 4\rangle}{[24]}. \tag{9.94}$$

Inserting the polarisation vectors into Eq. (9.93), we obtain using the Fierz identity (9.83)

$$\langle 2\gamma^\mu \varepsilon_\mu(4) \propto \langle 2\gamma^\mu[2\gamma_\mu 4\rangle = 2\langle 22\rangle[4 = 0. \tag{9.95}$$

In the last step, we used the antisymmetry of the brackets, $\langle 22\rangle = 0$. A similar cancellation occurs with $\varepsilon(3)$ and hence the entire matrix element vanishes. Parity implies then that the amplitude $\mathcal{A}(\gamma_L\gamma_L)$ vanishes too. Alternatively, we can show the same cancellation using $r = 1$ in both polarisation vectors.

Next we compute the amplitude for the case $\gamma_R\gamma_L$, choosing

$$\varepsilon^\mu(3) = \frac{1}{\sqrt{2}} \frac{[2\gamma^\mu 3\rangle}{\langle 23\rangle} \quad \text{and} \quad \varepsilon^\mu(4) = -\frac{1}{\sqrt{2}} \frac{[1\gamma^\mu 4\rangle}{[14]}. \tag{9.96}$$

Then the second diagram in Fig. 9.5 vanishes because of (9.95). Using the Fierz identity, the first diagram results in

$$i\mathcal{A} = \frac{-ie^2}{s_{24}} \frac{2 \cdot 2}{(-2)\langle 23 \rangle [14]} \langle 24 \rangle [1(\not{2} + \not{4})2][31].$$ (9.97)

Now we use the Dirac equation, $\not{2}2\rangle = 0$, and replace the vector $\not{4}_L$ by an angle bracket,

$$i\mathcal{A} = \frac{2ie^2}{s_{13}\langle 23 \rangle [14]} \langle 24 \rangle [14] \langle 42 \rangle [31] = \frac{2ie^2}{\langle 13 \rangle [31] \langle 23 \rangle} \langle 24 \rangle \langle 42 \rangle [31] = 2ie^2 \frac{\langle 24 \rangle^2}{\langle 23 \rangle^2}.$$ (9.98)

Finally, we introduce Mandelstam variables, $s_{23} = u$, $s_{13} = s_{24} = t$, reproducing the standard result

$$|i\mathcal{A}|^2 = 4e^4 \frac{t}{u} = 4e^4 \frac{1 - \cos \vartheta}{1 + \cos \vartheta}.$$ (9.99)

This short introduction into helicity methods hopefully convinced you of their efficiency. The advantage of this method over the traditional trace method increases with the number of diagrams involved, since the step $\mathcal{A} \to |\mathcal{A}|^2$ is trivial using this approach. Massive fermions can be treated using the helicity states (A.17), and efficient extensions to massive gauge bosons exist.

9.4 Soft photons and gravitons

The addition of a vertex introduces typically a factor $\alpha/\pi \sim 0.2\%$ into a QED cross-section. Thus one may hope that perturbation theory in QED converges, at least initially, reasonably fast. An exception to this rule is the emission of an additional soft or collinear photon from an external line shown in Fig. 9.6. The denominator of the additional propagator goes for $k \to 0$ to

$$\frac{1}{(p+k)^2 - m^2} \to \frac{1}{2p \cdot k} \simeq \frac{1}{2E\omega(1 - \cos \vartheta)},$$ (9.100)

where we assumed $|\boldsymbol{p}| \gg m$ in the last step. Hence the denominator can blow up in two different limits: firstly, in case of emission of soft photons, $\omega \to 0$; secondly, in case of collinear emission of photons, $\vartheta \to 0$, if the mass of the emitting particle can be neglected. We have seen in the example of Compton scattering that both soft and collinear emission correspond to the classical limit.

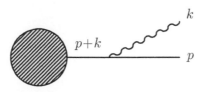

Fig. 9.6 Emission of an additional soft or collinear photon from an external line in the final state.

Universality and factorisation. The fact that a photon sees a classical current in the soft limit $k \to 0$ should lead to considerable simplifications. In particular, interference effects should disappear and the amplitude $\mathcal{A}_{n+1}$ for the emission of an additional soft photon should factorise into an universal factor $\varepsilon^\mu S_\mu$ and the amplitude $\mathcal{A}_n$ for the original process.

Let us start considering the emission of a soft photon by a spinless particle. If a scalar in the initial state with momentum p and charge q emits a photon with momentum $k \to 0$, then[2]

$$\mathcal{A}_{n+1} = q \, \frac{\varepsilon_\mu(2p^\mu + k^\mu)}{(p-k)^2 - m^2 + i\varepsilon} \, \mathcal{A}_n \to -q \, \frac{\varepsilon \cdot p}{p \cdot k - i\varepsilon} \, \mathcal{A}_n. \tag{9.101}$$

For the emission of a soft photon from a final state particle, the corresponding factor is $+q\,\varepsilon \cdot p/(p \cdot k + i\varepsilon)$. In the case of an internal line, in general no factor $(p \cdot k)^{-1}$ appears for $k \to 0$, since the virtual particle is off-shell.

For a spin-1/2 particle in the initial state, the emission of a soft photon adds the factor

$$q \, \frac{\varepsilon_\mu \bar{u}(p,s)\gamma^\mu(\slashed{p} + \slashed{k} + m)}{(p-k)^2 - m^2 + i\varepsilon} \to q \, \frac{\varepsilon_\mu \bar{u}(p,s)\gamma^\mu(\slashed{p} + m)}{-2p \cdot k + i\varepsilon} \tag{9.102}$$

to the amplitude $\mathcal{A}_n$. Now we replace $\slashed{p} + m$ by the spin sum $\sum_s u(p,s)\bar{u}(p,s)$, and use

$$\bar{u}(p,s)\gamma^\mu u(p,s') = 2E_p \, \frac{p^\mu}{p^0} \delta_{s,s'} = 2p^\mu \delta_{s,s'}. \tag{9.103}$$

This relation can be checked by direct calculation, or by noting that the current $j^\mu = \bar{u}(p,s)\gamma^\mu u(p,s)$ should become $j^\mu = (\rho, \rho\boldsymbol{v})$ in the classical limit $k \to 0$. Thus we obtain the same universal factor describing the emission of a soft photon,

$$S^\mu = -q \, \frac{p^\mu}{p \cdot k - i\varepsilon}, \tag{9.104}$$

as in the case of a scalar. Moreover, we confirmed that the amplitude indeed factorises, $\mathcal{A}_{n+1} = \varepsilon_\mu S^\mu \mathcal{A}_n \equiv \varepsilon_\mu \mathcal{A}_{n+1}^\mu$. If we allow for the emission of m soft photons from external particles with charge q_i, then

$$\mathcal{A}_{n+m}^{\mu_1 \cdots \mu_m} \to \sum_{i=1}^{m} \frac{s_i q_i \, p^{\mu_m}}{p \cdot k + i s_i \varepsilon} \, \mathcal{A}_n, \tag{9.105}$$

where the signs are $s_i = -1$ for an initial and $s_i = +1$ for a final-state particle.

We have seen that the polarisation vector $\varepsilon_\mu(k)$ of a photon does not transform as a four-vector, cf. Eq. (7.23), but acquires a term proportional to k^μ. As we exploited already at various places, amplitudes containing polarisation vectors $\varepsilon_\mu(k)$ of external photons have to vanish therefore when contracted with k^μ. Thus Eq. (9.105) implies in the limit $k \to 0$

$$k_{\mu_j} \mathcal{A}_{n+m}^{\mu_1 \cdots \mu_m} = 0 \to \sum_{i=1}^{m} s_i q_i \, \mathcal{A}_n = 0. \tag{9.106}$$

The pre-factor of $\mathcal{A}_n$ is the total charge in the final state minus the total charge in the initial state. In order to obtain a Lorentz-invariant matrix element for the soft emission

---

[2]We use the Feynman rule for a $\phi\phi A^\mu$ vertex derived in problem 7.8.

of massless spin-1 particles, we must therefore require that such particles couple to a conserved charge,

$$\sum_i q_i = \sum_f q_f.$$

Thus Lorentz invariance is sufficient to guaranty the conservation of the electromagnetic current in the low-energy limit. While this argument does not rely on gauge invariance, it tells us nothing about the behaviour of "hard" photons.

Spin $s > 1$. We can apply the same line of arguments to the emission of massless particles with spin $s > 1$. Gravitons, $s = 2$, couple to the stress tensor of matter, $\mathscr{L}_I = f h_{\mu\nu} T^{\mu\nu}$. Recalling from (5.21) the stress tensor of scalar particles, $T^{\mu\nu} = 2N^2 p^\mu p^\nu$, the universal factor becomes

$$S^{\mu\nu} = -f \frac{p^\mu p^\nu}{p \cdot k - \mathrm{i}\varepsilon}. \tag{9.107}$$

Requiring again that an amplitude containing the polarisation tensor $\varepsilon_{\mu\nu}(k)$ of external gravitons vanishes when contracted with k^μ gives the constraint

$$k_{\mu_j} \mathcal{A}_{n+m}^{\mu_1 \cdots \mu_m} = 0 \rightarrow \sum_{i=1}^m s_i f_i p_i^\nu \mathcal{A}_n = 0. \tag{9.108}$$

Now the sum $\sum_i f_i p_i^\nu$ is conserved. For $f_i \neq f_j$, a linear combination of the individual four-momenta other than the total four-momentum would be conserved in the scattering process—a condition which is not possible to satisfy in a non-trivial scattering process. Thus we have to conclude that any massless $s = 2$ particle has a *universal* coupling to all types of particle. This result can be viewed as the basis of the weak equivalence principle. Going further to $s = 3$, the universal factor becomes $S^{\mu\nu\lambda} \propto p^\mu p^\nu p^\lambda$, requiring that sums quadratic in the momenta, $\sum_i \tilde{f}_i p_i^\nu p_i^\mu$, are conserved. This is not possible for any scattering angles except $\vartheta = 0$ and $180°$, and thus no consistent theory of interacting massless particles with spin $s \geq 2$ is possible.

Bremsstrahlung. As a concrete example, we now consider the case of bremsstrahlung, that is, the emission of a real photon in the scattering of a charged particle in the Coulomb field $A^0 = -Ze/(4\pi|\boldsymbol{x}|)$ of a static nuclei with charge Ze. The S-matrix element of this process is

$$\mathrm{i}S_{fi} = 2\pi\delta(E' + \omega - E') \frac{-Ze^3}{|\boldsymbol{q}|^2} \bar{u}(p') \left[\not{\varepsilon} \frac{\not{p}' + \not{k} + m}{2p' \cdot k} \gamma^0 + \gamma^0 \frac{\not{p} - \not{k} + m}{-2p \cdot k} \not{\varepsilon} \right] u(p), \tag{9.109}$$

where $1/|\boldsymbol{q}|^2$ is the Fourier transform of A^0. Note that the external field breaks translation invariance and the momentum is not conserved. We now commute $\not{p}$ and $\not{p}'$,

$$\mathrm{i}S_{fi} \propto e^2 \bar{u}(p') \left[\frac{2\varepsilon \cdot p' - (\not{p}' - m)\not{\varepsilon} + \not{k}\not{\varepsilon}}{2p' \cdot k} \gamma^0 + \gamma^0 \frac{2\varepsilon \cdot p - \not{\varepsilon}(\not{p} - m) + \not{k}\not{\varepsilon}}{-2p \cdot k} \right] u(p), \tag{9.110}$$

such that we can use in the next step the Dirac equation. Neglecting in the soft limit the $\not{k}$ term in the numerator, we find

$$\mathrm{i}S_{fi} \propto e^2 \bar{u}(p')\gamma^0 u(p) \left[\frac{\varepsilon \cdot p'}{p' \cdot k} - \frac{\varepsilon \cdot p}{p \cdot k} \right]. \tag{9.111}$$

As we have shown in the previous paragraph in general, the amplitude factorises into the amplitude describing the "hard" process and the universal correction term. The latter consists of the two terms expected for the emission of a soft photon from an initial line with momentum p and a final line with momentum p'. The probability $\mathcal{P}$ for the emission of an additional soft photon is given integrating the square bracket over the phase space,

$$\mathrm{d}\mathcal{P}_{n+1} = \frac{\mathrm{d}\sigma_{n+1}}{\mathrm{d}\sigma_n} = \left[\frac{\varepsilon \cdot p'}{p' \cdot k} - \frac{\varepsilon \cdot p}{p \cdot k} \right]^2 \frac{\mathrm{d}^3 k}{(2\pi)^3 2\omega_k} \propto \frac{\mathrm{d}\omega_k}{\omega_k}. \tag{9.112}$$

This probability diverges for $\omega \to 0$ and therefore the process is called infrared (IR) divergent.

The resolution to this IR problem lies in the fact that soft photons with energy below the energy resolution E_{th} of the used detector are not detectable. Therefore the emission of n real soft photons with $E < E_{\mathrm{th}}$ is indistinguishable from the scattering process including virtual photons and thus these cross-sections should be added. The IR divergences in the real and virtual corrections cancel, leading to a finite result for the combined cross-section. We will discuss a detailed example of how this cancellation works in chapter 18.3.

9.A Appendix: Decay rates and cross-sections

First we establish the connection between the normalised transition matrix element $\mathcal{M}$ and the Feynman amplitude $\mathcal{A}$, where the normalisation factors of external particles are omitted. Then we derive decay rates and cross-sections describing $1 \to n$ and $2 \to n$ processes.

Normalisation. We have split the scattering operator S into a diagonal part and the transition operator T, $S = 1 + \mathrm{i}T$. Taking matrix elements, we obtain

$$S_{fi} = \delta_{fi} + (2\pi)^4 \, \delta^{(4)}(P_i - P_f)\mathrm{i}\mathcal{M}_{fi} \tag{9.113}$$

where we set also $T_{fi} = (2\pi)^4 \, \delta^{(4)}(P_i - P_f)\mathcal{M}_{fi}$.

The Feynman amplitude $\mathcal{A}$ neglects all normalisation factors of external particles, while the matrix element T_{fi} defined by (9.37) and thus $\mathcal{M}_{fi}$ contains a factor $N_{\boldsymbol{k}}$ for each external particle. Thus the transition between the matrix element $\mathcal{M}_{fi}$ and the Feynman amplitude $\mathcal{A}$ for a process with n particles in the initial and m in the final states is given by

$$\mathcal{M}_{fi} = \prod_{i=1}^{n}(2E_i V)^{-1/2} \prod_{f=1}^{m}(2E_f V)^{-1/2}\mathcal{A}_{fi}. \tag{9.114}$$

Here we changed also to a finite normalisation volume, $2E_{\boldsymbol{p}}(2\pi)^3 \to 2E_{\boldsymbol{p}}V$ what makes defining decay rates and cross-sections easier.

9.A.1 Decay rate

We consider the decay of a particle into n particles in the final state. Squaring the scattering amplitude S_{fi} for $i \neq f$ using $(2\pi)^4 \delta^{(4)}(0) = VT$ gives as the differential transition probability

$$dW_{fi} = (2\pi)^4 \, \delta^{(4)}(P_i - P_f) V T |\mathcal{M}_{fi}|^2 \prod_{f=1}^{n} \frac{V d^3 p_f}{(2\pi)^3} \, . \tag{9.115}$$

The decay rate or decay width $d\Gamma$ is the transition probability per time,

$$d\Gamma_{fi} = \lim_{T \to \infty} \frac{dW_{fi}}{T} = (2\pi)^4 \, \delta^{(4)}(P_i - P_f) V |\mathcal{M}_{fi}|^2 \prod_{f=1}^{n} \frac{V d^3 p_f}{(2\pi)^3} \, . \tag{9.116}$$

Going over to the Feynman amplitude $\mathcal{A}$ eliminates the volume factors V,

$$d\Gamma_{fi} = (2\pi)^4 \delta^{(4)}(P_i - P_f) \frac{1}{2E_i} |\mathcal{A}_{fi}|^2 \prod_{f=1}^{n} \frac{d^3 p_f}{2E_f (2\pi)^3} \, . \tag{9.117}$$

Moreover, the phase-space integrals in the final state are now Lorentz-invariant, $d^3 p_f/(2E_f)$. Introducing the n-particle phase space volume

$$d\Phi^{(n)} = (2\pi)^4 \, \delta^{(4)}(P_i - P_f) \prod_{f=1}^{n} \frac{d^3 p_f}{2E_f (2\pi)^3} \, , \tag{9.118}$$

the decay rate becomes

$$d\Gamma_{fi} = \frac{1}{2E_i} |\mathcal{A}_{fi}|^2 \, d\Phi^{(n)} \, . \tag{9.119}$$

Since both $|\mathcal{A}_{fi}|^2$ and the phase space $d\Phi^{(n)}$ are Lorentz-invariant, the decay rate $\Gamma \propto 1/E_i = 1/(\gamma_i m_i)$ shows explicitly the time dilation effect for a moving particle. Finally, we note that a symmetry factor $S = 1/n!$ has to be added to the total decay width or cross-section, if there are n identical particles in the final state.

Two-particle decays. We evaluate the two-particle phase space $d\Phi^{(2)}$ in the rest-frame of the decaying particle,

$$d\Phi^{(2)} = (2\pi)^4 \, \delta(M - E_1 - E_2) \, \delta^{(3)}(\boldsymbol{p}_1 + \boldsymbol{p}_2) \frac{d^3 p_1}{2E_1 (2\pi)^3} \frac{d^3 p_2}{2E_2 (2\pi)^3} \tag{9.120}$$

We perform the integration over $d^3 p_1$ using the momentum delta function. In the resulting expression,

$$d\Phi^{(2)} = \frac{1}{(2\pi)^2} \frac{1}{4E_1 E_2} \delta(M - E_1 - E_2) \, d^3 p_2 \, , \tag{9.121}$$

E_1 is now a function of p_2, $E_1^2 = p_2^2 + m_1^2$. Introducing spherical coordinates, $d^3 p_2 = d\Omega p_2^2 dp_2$,

$$d\Phi^{(2)} = \frac{1}{(2\pi)^2} d\Omega \int_0^{\infty} \delta(M - E_1 - E_2) \frac{p_2^2 dp_2}{4E_1 E_2} \, , \tag{9.122}$$

and evaluating the delta function with $M - E_1 - E_2 = M - x$ and $dp_2/dx = p_2 x/(E_1 E_2)$ gives

$$d\Phi^{(2)} = \frac{1}{16\pi^2} \frac{|\boldsymbol{p}'_{\text{cms}}|}{M} d\Omega \, , \tag{9.123}$$

where

$$p_{\text{cms}}^2 = \frac{\lambda(s, m_1^2, m_2^2)}{4s} = \frac{1}{4M^2} \left[M^2 - (m_1 + m_2)^2 \right] \left[M^2 - (m_1 - m_2)^2 \right] \tag{9.124}$$

is the cms momentum of the two final state particles. The Kibble function $\lambda(x, y, z)$ satisfies

$$\lambda(x, y, z) = \left[(x^2 + y^2 + z^2) - 2xy - 2yz - 2xz\right]^{1/2} \tag{9.125a}$$

$$= \left[x^2 - (\sqrt{y} + \sqrt{z})^2\right]^{1/2} \left[x^2 - (\sqrt{y} - \sqrt{z})^2\right]^{1/2} . \tag{9.125b}$$

Three-particle decays. The three-particle phase space $\mathrm{d}\Phi^{(3)}$ is in the rest-frame of the decaying particle given by

$$\mathrm{d}\Phi^{(3)} = (2\pi)^4 \, \delta(M - E_1 - E_2 - E_3) \, \delta^{(3)}(\boldsymbol{p}_1 + \boldsymbol{p}_2 + \boldsymbol{p}_3) \frac{\mathrm{d}^3 p_1}{2E_1(2\pi)^3} \frac{\mathrm{d}^3 p_2}{2E_2(2\pi)^3} \frac{\mathrm{d}^3 p_3}{2E_3(2\pi)^3} . \tag{9.126}$$

We can use again the momentum delta function to perform the integration over $\mathrm{d}^3 p_3$,

$$\mathrm{d}\Phi^{(3)} = \frac{1}{(2\pi)^5} \delta(M - E_1 - E_2 - E_3) \frac{\mathrm{d}^3 p_1 \mathrm{d}^3 p_2}{8E_1 E_2 E_3}, \tag{9.127}$$

To proceed we have to know the dependence of the matrix element on the integration variables. If there is no preferred direction (either for scalar particles or after averaging over spins), we obtain

$$\mathrm{d}\Phi^{(3)} = \frac{1}{8(2\pi)^5} \frac{4\pi p_1^2 \mathrm{d}p_1 \, 2\pi \mathrm{d}\cos\vartheta \mathrm{d}p_2}{E_1 E_2 E_3} \delta(M - E_1 - E_2 - E_3) \tag{9.128a}$$

$$= \frac{1}{32\pi^3} \frac{p_1 \mathrm{d}p_1 \, (p_1 p_2 \mathrm{d}\cos\vartheta)(p_2 \mathrm{d}p_2)}{E_1 E_2 E_3} \delta(M - E_1 - E_2 - E_3). \tag{9.128b}$$

We rewrite next the momentum integrals as energy integrals. The energy–momentum relation $E_i^2 = m_i^2 + p_i^2$ gives $E_i \mathrm{d}E_i = p_i \mathrm{d}p_i$ for $i = 1, 2$. Furthermore,

$$E_3^2 = (\boldsymbol{p}_1 + \boldsymbol{p}_2)^2 + m_3^2 = p_1^2 + p_2^2 + 2p_1 p_2 \cos\vartheta + m_3^2 \tag{9.129}$$

and thus $E_3 \mathrm{d}E_3 = p_1 p_2 \mathrm{d}\cos\vartheta$ for fixed $\boldsymbol{p}_1, \boldsymbol{p}_2$. Performing the angular integral, we obtain

$$\mathrm{d}\Phi^{(3)} = \frac{1}{32\pi^3} \mathrm{d}E_1 \mathrm{d}E_2 \mathrm{d}E_3 \delta(M - E_1 - E_2 - E_3), \tag{9.130}$$

and finally

$$\mathrm{d}\Phi^{(3)} = \frac{1}{32\pi^3} \mathrm{d}E_1 \mathrm{d}E_2. \tag{9.131}$$

The last step is only valid if the argument of the delta function is non-zero. Thus the remaining task is to determine the boundary of the integration domain. Let us choose E_1 as the outer integration variable. Then we have to determine the allowed range of E_2 for a given value of E_1. Inserting energy and momentum conservation into $E_3^2 = p_3^2 + m_3^2$, we obtain

$$(M - E_1 - E_2)^2 = m_3^2 + \boldsymbol{p}_1^2 + \boldsymbol{p}_2^2 + 2\boldsymbol{p}_1 \cdot \boldsymbol{p}_2. \tag{9.132}$$

The extrema correspond to

$$\boldsymbol{p}_1 \cdot \boldsymbol{p}_2 = \pm |\boldsymbol{p}_1||\boldsymbol{p}_2| = \pm\sqrt{(E_1^2 - m_1^2)(E_2^2 - m_2^2)}. \tag{9.133}$$

Inserting them into Eq. (9.132), we obtain the curve defining the boundary of the integration area as function of E_1 and E_2,

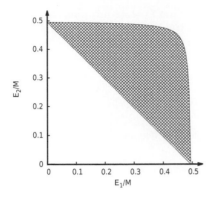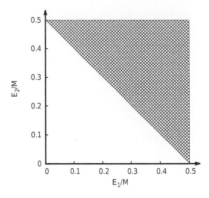

Fig. 9.7 Phase-space boundary for $m_3/M = 0.1$ (left). and $m_3 = 0$ (right); in both cases we set $m_1 = m_2 = 0$.

$$M^2 - 2M(E_1 + E_2) + 2E_1E_2 + m_1^2 + m_2^2 - m_3^2 = \pm\sqrt{(E_1^2 - m_1^2)(E_2^2 - m_2^2)}. \qquad (9.134)$$

In order to visualise the integration area more easily, we first set $m_1 = m_2 = 0$. Then the equation with the plus sign becomes $E_2 = M/2 + m_3^2/(4E_1 - 2M)$, while the equation with the minus sign simplifies to a straight line $E_2 = -E_1 + M/(2 - 2m_3^2/M^2)$. The resulting integration area is shown in Fig. 9.7 for $m_3/M = 0.1$. Setting also $m_3 = 0$, the integration area is a triangle in the E_1–E_2 plane.

If one prefers Lorentz-invariant integration variables, one can introduce the invariant mass of the pair (i,j)

$$m_{23}^2 = (p - p_1)^2 = (p_2 + p_3)^2 = M^2 - 2ME_1 + m_1^2 \qquad (9.135a)$$

$$m_{13}^2 = (p - p_2)^2 = (p_1 + p_3)^2 = M^2 - 2ME_2 + m_2^2 \qquad (9.135b)$$

$$m_{12}^2 = (p - p_3)^2 = (p_1 + p_2)^2 = M^2 - 2ME_3 + m_3^2, \qquad (9.135c)$$

Because of $m_{23}^2 + m_{13}^2 + m_{12}^2 = M^2 + m_2^2 + m_3^2$, only two out of the three variables are independent. They can be used to to replace E_1 and E_2 in Eq. (9.131).

9.A.2 Cross-sections

We now consider the interaction of two particles in the rest-frame of either particle 1 or 2. For simplicity, we consider two uniform beams. They may produce n final-state particles. The total number of such scatterings is

$$dN \propto v_{\text{Møl}} n_1 n_2 dV dt \qquad (9.136)$$

where n_i is the density of particle type $i = 1, 2$. The Møller velocity $v_{\text{Møl}}$ is a quantity which coincides in the rest-frame of particle 1 or 2 with $|\boldsymbol{v}_2|$ and $|\boldsymbol{v}_1|$, respectively. Therefore it is often denoted simply as relative velocity $\boldsymbol{v}_{\text{rel}}$. The proportionality constant in (9.136) has the dimension of an area and is called cross-section σ. We define in the rest-system of either particle 1 or 2

$$dN = \sigma v_{\text{Møl}}\, n_1 n_2\, dV dt, \qquad (9.137)$$

while we set in an arbitrary frame

$$dN = A n_1 n_2 dV dt. \qquad (9.138)$$

We determine now A. Since both dN and $dV dt = d^4 x$ are Lorentz-invariant, the expression $A n_1 n_2$ has to be Lorentz-invariant too. The densities transform as

$$n_i = n_{i,0} \gamma = n_{i,0} \frac{E_i}{m_i}, \tag{9.139}$$

and thus the expression

$$A \frac{E_1 E_2}{p_1 \cdot p_2} \tag{9.140}$$

is also Lorentz-invariant. In the rest-frame of particle 1, it becomes

$$A \frac{E_1 E_2}{E_1 E_2 - \boldsymbol{p}_1 \boldsymbol{p}_2} = A = \sigma v_{\text{Møl}}. \tag{9.141}$$

Thus we found that A in an arbitrary frame is given by

$$A = \sigma v_{\text{Møl}} \frac{p_1 \cdot p_2}{E_1 E_2}. \tag{9.142}$$

We still have to determine $v_{\text{Møl}}$. In the rest-frame 1, we have

$$p_1 \cdot p_2 = m_1 E_2 = m_1 \frac{m_2}{\sqrt{1 - v_{\text{Møl}}^2}}. \tag{9.143}$$

Thus the Møller velocity is given in general by

$$v_{\text{Møl}} = \sqrt{1 - \frac{m_1^2 m_2^2}{(p_1 \cdot p_2)^2}}. \tag{9.144}$$

Since this expression is Lorentz-invariant, we see that the notion of the Møller velocity as relative velocity is misleading.

Next we define the flux factor

$$I \equiv v_{\text{Møl}} \, p_1 \cdot p_2 = \sqrt{(p_1 \cdot p_2)^2 - m_1^2 m_2^2}. \tag{9.145}$$

Inserting (9.142) for A together with the definition of I into (9.138), we obtain

$$dN = \sigma \frac{I}{E_1 E_2 V} (n_1 V)(n_2 dV) dt. \tag{9.146}$$

Here, we regrouped the terms to make clear that after integration the total number N of scattering events is proportional to the number $N_1 = n_1 V$ and $N_2 = \int n_2 dV$ of particles of type 1 and 2, respectively. The number N of scattering events per time and per particles 1 and 2 is, however, simply the transition probability per time,

$$\frac{dW_{fi}}{T} = \frac{dN}{N_1 N_2 T} = d\sigma \frac{I}{E_1 E_2 V}. \tag{9.147}$$

Inserting the expression (9.116) for dW_{fi}, we find

$$d\sigma = \frac{E_1 E_2 V^2}{I} (2\pi)^4 \delta^{(4)}(P_i - P_f) |\mathcal{M}_{fi}|^2 \prod_{f=1}^{n} \frac{V d^3 p_f}{(2\pi)^3}. \tag{9.148}$$

Changing from the normalised matrix element $\mathcal{M}$ to the Feynman amplitude $\mathcal{A}$ introduces a factor $(2E_1 V)^{-1}(2E_2 V)^{-1}$ for the initial state and $\prod_f (2E_f V)^{-1}$ for the final state. Thus the arbitrary normalisation volume V cancels and we obtain

$$d\sigma = \frac{1}{4I} (2\pi)^4 \, \delta^{(4)}(P_i - P_f)|\mathcal{A}_{fi}|^2 \prod_{f=1}^{n} \frac{d^3 p_f}{2E_f (2\pi)^3} = \frac{1}{4I} \, |\mathcal{A}_{fi}|^2 \, d\Phi^{(n)} \tag{9.149}$$

with the final-state phase space $d\Phi^{(n)}$. The three pieces composing the differential cross-section, the flux factor I, the Feynman amplitude $\mathcal{A}$ and the final-state phase space $d\Phi^{(n)}$, are each Lorentz-invariant.

2–2 scattering. For a $1+2 \to 3+4$ scattering process, it is useful to introduce Mandelstam variables s, t and u as

$$s = (p_1 + p_2)^2 = (p_3 + p_4)^2, \tag{9.150}$$

$$t = (p_1 - p_3)^2 = (p_2 - p_4)^2, \tag{9.151}$$

$$u = (p_1 - p_4)^2 = (p_2 - p_3)^2. \tag{9.152}$$

Since $s + t + u = \sum_{i=1}^{4} m_i^2$, the scattering amplitude $\mathcal{A}$ depends only on two variables, for example $\mathcal{A}(s, t)$. In the cms, the flux factor becomes

$$I^2 = (p_1 \cdot p_2)^2 - m_1^2 m_2^2 = p_{\text{cms}}^2 (E_1 + E_2)^2 = p_{\text{cms}}^2 s. \tag{9.153}$$

Adding the expression for the two-particle phase space gives

$$\frac{d\sigma}{d\Omega} = \frac{1}{64\pi^2 s} \frac{p'_{\text{cms}}}{p_{\text{cms}}} |\mathcal{A}_{fi}|^2. \tag{9.154}$$

Here the cm momentum of the initial state is $p_{\text{cms}}^2 = \lambda(s, m_1^2, m_2^2)/(4s)$, while $p'^2_{\text{cms}} = \lambda(s, m_3^2, m_4^2)/(4s)$ is the one of the final state. Using as variable the momentum transfer t, the differential cross-section becomes

$$\frac{d\sigma}{dt} = \frac{1}{64\pi s p_{\text{cms}}^2} |\mathcal{A}_{fi}|^2, \tag{9.155}$$

where the allowed range of t has to be determined from Eq. (9.151) and $-1 \leq \cos\vartheta \leq 1$.

The optical theorem connects the imaginary part of the forward amplitude $\Im T_{ii}$ with the total cross-section as

$$\sigma_{\text{tot}} = \frac{\Im T_{ii}}{\lambda^{1/2}(s, m_1^2, m_2^2)}. \tag{9.156}$$

Summary

The LSZ reduction formula shows that S-matrix elements are obtained from connected Green functions by a replacement of the propagators on external lines with the corresponding wave-functions times the wave-function renormalisation constant $\sqrt{Z}$. Cross-sections are calculated from the squared Feynman amplitude $\mathcal{A}$, the final-state phase space $d\Phi^{(n)}$ and the flux factor I, which are all three Lorentz-invariant. Squared Feynman amplitudes can be obtained using "Casimir's trick".

If the number of diagrams increases it is more convenient to calculate directly the amplitude using helicity methods.

The amplitude for the emission of additional soft particles factorises in the amplitude of the hard process and an universal factor. Lorentz invariance requires that a massless spin-1 particle couples in the low-energy limit to a conserved charge, while a massless spin-2 particle has to couples with universal strength to the energy–momentum stress tensor.

Further reading. Sterman (1993) discusses the optical theorem and its connection to cut diagrams in more detail. The LSZ formula for particles with spin $s > 0$ is presented in Greiner and Reinhardt (2008), for example, while Bailin and Love (1993) derive S-matrix elements defining a new functional $Z'[J]$ with the correct boundary conditions. For additional information about the helicity formalisms see Haber (1994) and Peskin (2011) from which our examples are taken. Ask *et al.* (2012) provide a tutorial for several software tools useful for the calculation of scattering processes. The discussion of soft photon emission follows closely the original discussion by Weinberg (1965); for an introduction see White (2015).

Problems

9.1 Optical theorem. Consider two scalars ϕ_1 and ϕ_2 with mass m coupled to a scalar Φ with mass $M > 2m$, via $\mathscr{L}_I = g\phi_1\phi_2\Phi$. a.) Calculate the width Γ of the decay $\Phi \to \phi_1\phi_2$. b.) Draw the Feynman diagram(s) and write down the amplitude i$\mathcal{A}$ for the scattering $\phi_1(p_1)\phi_2(p_2) \to \phi_1(p_1')\phi_2(p_2')$. What is your interpretation of the behaviour of the amplitude for $s = (p_1 + p_2)^2 \to M^2$? c.) Consider the one-loop correction iΣ to the mass of Φ. Write down iΣ first for an arbitrary momentum p of Φ, then for its rest-frame, $p = (M, 0)$. Find the poles of the integrand and use the theorem of residues to perform the q^0 part of the loop integral. Use the identity (9.14) to find the imaginary part of the amplitude.

9.2 Fermion lines. Calculate in the trace formalism $|\overline{\mathcal{A}}|^2$ for the processes $e^+e^- \to e^+e^-$ and $e^+e^- \to 2\gamma$.

9.3 Gauge invariance. Show that the Compton amplitude $\mathcal{A} = \varepsilon_\mu(k)\varepsilon_\nu'(k')\mathcal{A}^{\mu\nu}$ satisfies $\varepsilon_\mu(k)\mathcal{A}^{\mu\nu} = \varepsilon_\nu'(k')\mathcal{A}^{\mu\nu} = 0$.

9.4 Identities for gamma matrices. Evaluate $\gamma^\mu \slashed{a} \gamma_\mu$, $\gamma^\mu \slashed{a}\slashed{b} \gamma_\mu$, $\mathrm{tr}[\slashed{a}\slashed{b}]$, and $\mathrm{tr}[\slashed{a}\slashed{b}\slashed{c}\slashed{d}]$.

9.5 Polarised Thomson cross-section. Derive from (9.68) the cross-section for linearly polarised photons.

9.6 Polarised cross-sections in the trace formalism. Calculate the squared matrix element of the process $e_L^-(1)e_R^+(2) \to \mu_L^-(3)\mu_R^+(4)$ using the trace formalism. (Hint: Use the spin projection operators (8.62) to obtain the required polarisations.)

9.7 Muon decay. Derive the differential decay rate of the process $\mu^-(p_1) \to e^-(p_4)\nu_\mu(p_3)\bar{\nu}_e(p_2)$, via the exchange of a W-boson described by the vertex $-\frac{ig}{\sqrt{2}}\bar{f}\gamma_\mu(1-\gamma^5)f W^\mu$. You can neglect lepton masses and use $|q^2| \ll m_W^2$ for the momentum of the propagator.

9.8 Soft photons. Consider the emission of two soft photons from the *same* external line and derive the matrix element $\mathcal{A}_{n+2}$.

9.9 Phase-space limits. Derive the phase-space limits in a three-particle decay using the invariant masses (9.135) as integration variables.

10
Gauge theories

In this chapter we discuss field theories in which the Lagrangian is invariant under a continuous group of *local* transformations in internal field space. The symmetry group of these transformations is called the gauge group and the vector fields associated to the generators of the group the gauge fields. We introduce as a first step unbroken gauge theories, that is, theories with massless gauge bosons, and defer the more complex case of broken gauge symmetries to chapters 13 and 14. The Standard Model (SM) of particle physics contains two examples for unbroken gauge theories in quantum electrodynamics (QED) and quantum chromodynamics (QCD). While QED is an abelian gauge theory based on the gauge group U(1), QCD which describes the strong interactions is a non-abelian gauge theory with group SU(3). Non-abelian gauge theories were first studied by Yang and Mills and are therefore also often called Yang–Mills theories. The structure of Yang–Mills theories has many similarities with gravity. We use this property to introduce the curvature of a spacetime as the analogue of the field strength in the Yang–Mills case.

10.1 Electrodynamics as abelian gauge theory

In classical electrodynamics, the field-strength tensor $F_{\mu\nu} = \partial_\mu A_\nu - \partial_\nu A_\mu$ is an observable quantity, while the potential A_μ is merely a convenient auxiliary quantity. From its definition as an antisymmetric tensor it is clear that $F_{\mu\nu}$ is invariant under local gauge transformations of the potentials,

$$A_\mu(x) \to A'_\mu(x) = A_\mu(x) - \partial_\mu \Lambda(x). \tag{10.1}$$

Hence $A'_\mu(x)$ is, for any smooth $\Lambda(x)$, physically equivalent to $A_\mu(x)$, leading to the same field-strength tensor and thus, for example, to the same Lorentz force on a particle.

Consider now as an example a free Dirac field $\psi(x)$ with electric charge q. We saw already that this field is invariant under *global* phase transformations $\exp[iq\Lambda] \in U(1)$, implying a conserved current $j^\mu = \bar\psi\gamma^\mu\psi$ via Noether's theorem. Can we promote this global U(1) symmetry to a local one,

$$\psi(x) \to \psi'(x) = U(x)\psi(x) = \exp[iq\Lambda(x)]\psi(x), \tag{10.2}$$

by making the phase U spacetime dependent as in (10.1)? The partial derivatives in the Dirac Lagrangian will lead to an additional term $\propto \partial_\mu U(x)$, destroying the invariance of the free Lagrangian. However, if we add a field $A_\mu(x)$ which transforms as defined

Quantum Fields–From the Hubble to the Planck Scale. Michael Kachelriess. © Michael Kachelriess 2018.
Published in 2018 by Oxford University Press. DOI 10.1093/oso/9780198802877.001.0001

in (10.1) and couples to the Noether current j^μ of the complex field as $\mathscr{L}_I = -q\,j^\mu A_\mu$, the two gauge-dependent terms will cancel. Thus local U(1) gauge invariance of the Dirac field requires the existence of a massless gauge boson and fixes its interaction with matter: the coupling of matter to photons is obtained by replacing the normal derivative by the covariant derivative,

$$\partial_\mu \to D_\mu = \partial_\mu + iqA_\mu, \tag{10.3}$$

which transforms as the matter fields,

$$D_\mu\psi(x) \to D'_\mu\psi'(x) = \{\partial_\mu + iq[A_\mu(x) - \partial_\mu\Lambda(x)]\}\exp[iq\Lambda(x)]\psi(x) = \tag{10.4}$$

$$= \exp[iq\Lambda(x)]\{\partial_\mu + iqA_\mu(x)\}\psi(x) = U(x)D_\mu\psi(x). \tag{10.5}$$

We can rewrite the gauge transformation of A_μ as

$$A_\mu(x) \to A'_\mu(x) = A_\mu(x) - \partial_\mu\Lambda(x) = A_\mu(x) - \frac{i}{q}U(x)\partial_\mu U^\dagger(x), \tag{10.6}$$

expressing the change $\delta A_\mu(x)$ through the group elements $U(x)$. Finally, we note that we can connect the field-strength tensor to the commutator of covariant derivatives,

$$[D_\mu, D_\nu]\psi = iq([\partial_\mu, A_\nu] - [\partial_\nu, A_\mu])\psi = iq\psi(\partial_\mu A_\nu - \partial_\nu A_\mu) = iq\psi F_{\mu\nu}. \tag{10.7}$$

To summarise: the invariance of complex (scalar or Dirac) fields under *global* phase transformations $\exp[iq\Lambda] \in$ U(1) implies a conserved current; promoting it to a *local* U(1) symmetry requires the existence of a massless U(1) gauge boson coupled via gauge-invariant derivatives to these fields.

10.2 Non-abelian gauge theories

10.2.1 Gauge-invariant interactions

We want to generalise now electrodynamics using as symmetry group instead of the abelian group U(1) larger groups like SO(n) or SU(n). A group like SU(n) will describe the interactions of $n^2 - 1$ gauge bosons with matter using as a single parameter the gauge coupling g. The gauge transformations will, moreover, mix fermions living in the same representation of the group, requiring these fermions to have the same interactions and the same mass if the symmetry is unbroken. In this way, non-abelian gauge theories lead to a partial *unification* of matter fields and interactions. Note the difference to an abelian symmetry. The emission of a photon does not change any quantum number (apart from the momentum) and thus does not "mix" different particles. Therefore there is also no connection between the electric charge of different particles.

The two non-abelian groups used in the SM are SU(2) for weak and SU(3) for strong interactions. Matrix representations for the fundamental representation of these two groups are the Pauli matrices, $T^a = \sigma^a/2$, and the Gell-Mann matrices, $T^a = \lambda^a/2$, respectively. Under the fundamental representation the fermions transform as doublets for SU(2), as triplets for SU(3), etc. Since the number of generators is $m =$

$n^2 - 1$ for SU(n), the group SU(2) contains three gauge bosons, while SU(3) contains eight bosons carrying strong interactions. The most important difference of these non-abelian groups compared to U(1) is that the generators $T^a \equiv T^a_{ij}$ of such groups do not commute with each other. As a result, we may expect that both the expression for the field-strength tensor, Eq. (7.11), and the transformation law for the gauge field, Eq. (7.12), becomes more complicated. In contrast, we postulate that the replacement $\partial_\mu \to D_\mu = \partial_\mu + igA_\mu$ remains valid, with the sole difference that now $A_\mu = A^a_\mu T^a$. Thus A_μ is a Lorentz vector with values in the Lie algebra of the gauge group.

We now derive the transformation laws and structure of the gauge sector, requiring that the transformation of the fermions and their interaction with the gauge field are locally invariant. A local gauge transformation

$$U(x) = \exp[ig\sum_{a=1}^{m}\vartheta^a(x)T^a] \equiv \exp[ig\vartheta(x)] \tag{10.8}$$

changes a vector of fermions ψ with components $\{\psi_1, \ldots, \psi_n\}$ as[1]

$$\psi(x) \to \psi'(x) = U(x)\psi(x). \tag{10.9}$$

Already global gauge invariance of the fermion mass term requires $m_1 = m_2 = \ldots = m_n$ and for simplicity we set $m_i = 0$. We can implement local gauge invariance, if derivatives transform in the same way as ψ. Hence we define a new covariant derivative D_μ requiring

$$D_\mu\psi(x) \to [D_\mu\psi(x)]' = U(x)[D_\mu\psi(x)]. \tag{10.10}$$

The gauge field should compensate the difference between the normal and the covariant derivative,

$$D_\mu\psi(x) = [\partial_\mu + igA_\mu(x)]\psi(x). \tag{10.11}$$

In the non-abelian case, the gauge field A_μ is a matrix that is connected to its component fields by

$$A_\mu = A^a_\mu T^a. \tag{10.12}$$

We now determine the transformation properties of D_μ and A_μ demanding that (10.9) and (10.10) hold. Combining both requirements gives

$$D_\mu\psi(x) \to [D_\mu\psi]' = UD_\mu\psi = UD_\mu U^{-1}U\psi = UD_\mu U^{-1}\psi', \tag{10.13}$$

and thus the covariant derivative transforms as $D'_\mu = UD_\mu U^{-1}$. Using its definition (10.11), we find

$$[D_\mu\psi]' = [\partial_\mu + igA'_\mu]U\psi = UD_\mu\psi = U[\partial_\mu + igA_\mu]\psi. \tag{10.14}$$

We compare now the second and the fourth term, after having performed the differentiation $\partial_\mu(U\psi)$. The result

$$[(\partial_\mu U) + igA'_\mu U]\psi = igUA_\mu\psi \tag{10.15}$$

---

[1]We suppress in the following most indices; writing them out gives, for exampe, $\psi'_i(x) = U_{ij}(x)\psi_j(x)$ with $U_{ij}(x) = \exp[ig\sum_{a=1}^{m}\vartheta^a(x)T^a_{ij}]$.

should be valid for arbitrary ψ and hence after multiplying from the right with U^{-1} we arrive at

$$A_\mu \to A'_\mu = UA_\mu U^{-1} + \frac{i}{g}(\partial_\mu U)U^{-1} = UA_\mu U^{-1} - \frac{i}{g}U\partial_\mu U^{-1}. \tag{10.16}$$

Here we also used $\partial_\mu(UU^{-1}) = 0$. In most cases, the gauge transformation U is an unitary transformation and one sets $U^{-1} = U^\dagger$. A term changing as $U(x)D_\mu(x)U^\dagger(x)$ is said to transform homogeneously, while the potential A_μ is said to transform inhomogeneously.

Example 10.1: We can determine the transformation properties of A_μ also by demanding that (10.11) defines the interaction term in a gauge invariant way. Replacing $\partial_\mu \to D_\mu$ in the free Lagrange density of fermions and inserting then $U^{-1}U = 1$ gives

$$\mathscr{L}_f + \mathscr{L}_I = i\bar{\psi}\gamma^\mu D_\mu\psi = i\bar{\psi}\gamma^\mu\partial_\mu\psi - g\bar{\psi}\gamma^\mu A_\mu\psi =$$
$$= i\bar{\psi}U^{-1}U\gamma^\mu\partial_\mu U^{-1}U\psi - g\bar{\psi}U^{-1}U\gamma^\mu A_\mu U^{-1}U\psi. \tag{10.17}$$

Using then $\psi' = U\psi$, we obtain

$$\mathscr{L}_f + \mathscr{L}_I = i\bar{\psi}'\gamma^\mu U\partial_\mu U^{-1}\psi' - g\bar{\psi}'\gamma^\mu UA_\mu U^{-1}\psi'$$
$$= i\bar{\psi}'\gamma^\mu\partial_\mu\psi' - g\bar{\psi}'\gamma^\mu\left\{UA_\mu U^{-1} - \frac{i}{g}U(\partial_\mu U^{-1})\right\}\psi'. \tag{10.18}$$

The Lagrange density $\mathscr{L}_f + \mathscr{L}_I$ is thus invariant, if the gauge field transforms as in Eq. (10.16).

Specialising to infinitesimal transformations,

$$U(x) = \exp(ig\vartheta^a(x)T^a) = 1 + ig\vartheta(x) + \mathcal{O}(\vartheta^2), \tag{10.19}$$

it follows

$$A_\mu(x) \to A'_\mu(x) = A_\mu(x) - ig[A_\mu(x), \vartheta(x)] - \partial_\mu\vartheta(x). \tag{10.20}$$

In the abelian U(1) case, the commutator term is not present and the transformation law reduces to the known $A_\mu \to A_\mu - \partial_\mu\vartheta$. For a (semi-simple) Lie group one defines

$$[T^a, T^b] = if^{abc}T^c \tag{10.21}$$

with structure constants f^{abc} that can be chosen to be completely antisymmetric. Thus

$$A^a_\mu(x) \to A^{a\prime}_\mu(x) = A^a_\mu(x) + gf^{abc}A^b_\mu(x)\vartheta^c(x) - \partial_\mu\vartheta^a(x) \tag{10.22a}$$
$$= A^a_\mu(x) - [\delta^{ac}\partial_\mu - gf^{abc}A^b_\mu(x)]\vartheta^c(x) \tag{10.22b}$$
$$\equiv A^a_\mu(x) - D^{ac}_\mu\vartheta^c(x), \tag{10.22c}$$

where the last line defines how the covariant derivative acts on the gauge fields. Comparing this expression to the general definition $D_\mu = \partial_\mu + igA^a_\mu T^a$, we see that the gauge fields live in the adjoint representation of the gauge group,[2] cf. problem 4. The infinitesimal change of the gauge fields A_μ^a is given by the covariant derivative acting on the parameters ϑ^a of the gauge transformation.

Finally, we have to derive the field-strength tensor $F_{\mu\nu} = F_{\mu\nu}^a T^a$ and the Lagrange density $\mathscr{L}_{\mathrm{YM}}$ of the gauge field. The quantity F^2 requires now additionally a summation over the group index a,

$$\mathscr{L}_{\mathrm{YM}} = -\frac{1}{4} F_{\mu\nu}^a F^{a\mu\nu} = -\frac{1}{2} \operatorname{tr} F_{\mu\nu} F^{\mu\nu}, \qquad (10.23)$$

where we assumed in the second step that the standard normalisation $\operatorname{tr} T^a T^b = \delta^{ab}/2$ for the group generators T^a holds. The last equation shows that it is sufficient for the gauge invariance of the action that the field-strength tensor transforms homogeneously,

$$F_{\mu\nu}(x) \to F'_{\mu\nu}(x) = U(x)F_{\mu\nu}(x)U^\dagger(x). \qquad (10.24)$$

There are several ways to derive the relation between $F_{\mu\nu}$ and A_μ. The field-strength tensor should be antisymmetric. Thus we should construct it out of the commutator of gauge invariant quantities that in turn should contain A_μ. An obvious try is $igF_{\mu\nu} = [D_\mu, D_\nu]$ that worked in the abelian case. Now, additionally the non-zero commutator of the gauge fields contributes,

$$F_{\mu\nu} = F_{\mu\nu}^a T^a = \frac{1}{ig}\,[D_\mu, D_\nu] = \partial_\mu A_\nu - \partial_\nu A_\mu + ig[A_\mu, A_\nu]. \qquad (10.25)$$

In components, this equation reads explicitly

$$F_{\mu\nu}^a = \partial_\mu A_\nu^a - \partial_\nu A_\mu^a - gf^{abc}A_\mu^b A_\nu^c. \qquad (10.26)$$

Remark 10.1: Antisymmetric tensors of rank n can also be seen as differential forms. We already know functions as forms of order $n = 0$ and co-vectors as forms of order $n = 1$. Since differentials $\mathrm{d}f = \partial_i f\,\mathrm{d}x^i$ of functions are forms of order $n = 1$, the $\mathrm{d}x^i$ form a basis, and one can write in general $\mathbf{A} = A_i\,\mathrm{d}x^i$. For $n > 1$, the basis has to be antisymmetrised. Hence, a two-form as the field-strength tensor is given by

$$\mathbf{F} = \frac{1}{2}\,F_{\mu\nu}\mathrm{d}x^\mu \wedge \mathrm{d}x^\nu \qquad (10.27)$$

with $\mathrm{d}x^\mu \wedge \mathrm{d}x^\nu = -\mathrm{d}x^\nu \wedge \mathrm{d}x^\mu$. Looking at $\mathrm{d}f$ suggests to define the differentiation of a form ω with coefficients w and degree n as an operation that increases its degree by one to $n + 1$,

$$\mathrm{d}\omega = \frac{1}{n!}\,(\partial_\beta w_{\alpha_1 \ldots, \alpha_n})\mathrm{d}x^\beta \wedge \mathrm{d}x^{\alpha_1} \wedge \ldots \wedge \mathrm{d}x^{\alpha_n}. \qquad (10.28)$$

Thus we have $\mathbf{F} = \mathrm{d}\mathbf{A}$. Moreover, it follows $\mathrm{d}^2\omega = 0$ for all forms. Hence we can write an abelian gauge transformation as $\mathbf{F}' = \mathrm{d}(\mathbf{A} - \mathrm{d}\Lambda) = \mathbf{F}$.

---

[2]The n complex fermion and $n^2 - 1$ real gauge fields of $\mathrm{SU}(n)$ live in different representations of the group, as already the mismatch of their number indicates, see also appendix B. Note also that the gauge transformations of the gauge fields have to be real, in contrast to the ones of the fermion fields.

10.2.2 Gauge fields as connection

There is a close analogy between the covariant derivative ∇_μ introduced for a spacetime containing a gravitational field and the gauge-invariant derivative D_μ required for a spacetime containing a gauge field. In the former case, the moving coordinate basis in curved spacetime, $\partial_\mu e^\nu \neq 0$, introduces an additional term in the derivative of vector components $V^\mu = e^\mu \cdot \mathbf{V}$. Analogously, a non-zero gauge field A^μ leads to a rotation of the basis vectors e_i in group space which in turn produces an additional term $\boldsymbol{\psi} \cdot (\partial_\mu e_i)$ performing the derivative of a $\psi_i = \boldsymbol{\psi} \cdot e_i$.

Let us rewrite our formulae such that the analogy between the covariant gauge derivative D_μ and the covariant spacetime derivative ∇_μ becomes obvious. The vector $\boldsymbol{\psi}$ of fermion fields with components $\{\psi_1, \ldots, \psi_n\}$ transforming under a representation of a gauge group can be written as

$$\boldsymbol{\psi}(x) = \psi_i(x)e_i(x). \tag{10.29}$$

We can pick out the component ψ_j by multiplying with the corresponding basis vector e_j,

$$\psi_j = \boldsymbol{\psi} \cdot e_j(x). \tag{10.30}$$

If the coordinate basis in group space depends on x^μ, then the partial derivative of ψ_i acquires a second term,

$$\partial_\mu \psi_i = (\partial_\mu \boldsymbol{\psi}) \cdot e_i + \boldsymbol{\psi} \cdot (\partial_\mu e_i). \tag{10.31}$$

We can argue, as in section 6.2, that $(\partial_\mu \boldsymbol{\psi}) \cdot e_i$ is an invariant quantity, defining therefore as gauge-invariant derivative

$$D_\mu \psi_i = (\partial_\mu \boldsymbol{\psi}) \cdot e_i = \partial_\mu \psi_i - \boldsymbol{\psi} \cdot (\partial_\mu e_i). \tag{10.32}$$

The change $\partial_\mu e_i$ of the basis vector in group space should be proportional to gA_μ. Setting

$$\partial_\mu e_i = -\mathrm{i}g(A_\mu)_{ij} e_j \tag{10.33}$$

we are back to our old notation.

Gauge loops. The correspondence between the derivatives ∇_μ and D_μ suggests that we can use the gauge field A_μ to transport fields along a curve $x^\mu(\sigma)$. In empty space, we can use the partial derivative $\partial_\mu \psi(x)$ to compare fields at different points,

$$\partial_\mu \psi(x) \propto \psi(x + \mathrm{d}x^\mu) - \psi(x). \tag{10.34}$$

If there is an external gauge field present, the field ψ is additionally rotated in group space moving it from x to $x + \mathrm{d}x$,

$$\tilde{\psi}(x + \mathrm{d}x) = \psi(x + \mathrm{d}x) + \mathrm{i}gA_\mu(x)\psi(x)\mathrm{d}x^\mu \tag{10.35a}$$
$$= \psi(x) + \partial_\mu \psi(x)\mathrm{d}x^\mu + \mathrm{i}gA_\mu(x)\psi(x)\mathrm{d}x^\mu. \tag{10.35b}$$

Then the total change is

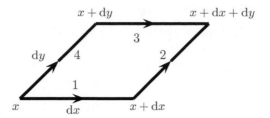

Fig. 10.1 Parallelogram used to calculate the rotation of a test field ψ_i moved along a closed loop in the presence of a non-zero gauge field A^μ.

$$\tilde{\psi}(x + \mathrm{d}x) - \psi(x) = [\partial_\mu + igA_\mu(x)]\psi(x)\mathrm{d}x^\mu = D_\mu\psi(x)\mathrm{d}x^\mu. \tag{10.36}$$

Thus we can view[3]

$$P_{\mathrm{d}x}(x) = 1 - igA_\mu(x)\mathrm{d}x^\mu \tag{10.37}$$

as an operator which allows us to transport a gauge-dependent field the infinitesimal distance from x to $x + \mathrm{d}x$.

We ask now what happens to a field $\psi_i(x)$, if we transport it along an infinitesimal parallelogram, as shown in Fig. 10.1. Calculating path 2, we find

$$\begin{aligned} P_{\mathrm{d}y}(x + \mathrm{d}x) &= 1 - igA_\nu(x + \mathrm{d}x)\mathrm{d}y^\nu \\ &= 1 - igA_\nu(x)\mathrm{d}y^\nu - ig\partial_\mu A_\nu(x)\mathrm{d}x^\mu\mathrm{d}y^\nu, \end{aligned} \tag{10.38}$$

where we Taylor expanded $A_\nu(x + \mathrm{d}x)$. Combining paths 1 and 2, we arrive at

$$\begin{aligned} P_{\mathrm{d}y}(x + \mathrm{d}x)P_{\mathrm{d}x}(x) &= [1 - igA_\nu(x)\mathrm{d}y^\nu - ig\partial_\mu A_\nu(x)\mathrm{d}x^\mu\mathrm{d}y^\nu][1 - igA_\mu(x)\mathrm{d}x^\mu] \\ &= 1 - igA_\mu(x)\mathrm{d}x^\mu - igA_\nu(x)\mathrm{d}y^\nu - ig\partial_\mu A_\nu(x)\mathrm{d}x^\mu\mathrm{d}y^\nu \\ &\quad - g^2 A_\nu(x)A_\mu(x)\mathrm{d}y^\nu\mathrm{d}x^\mu + \mathcal{O}(\mathrm{d}x^3). \end{aligned} \tag{10.39}$$

Instead of performing the calculation for a round trip $1 \to 2 \to 3 \to 4$, we evaluate next $4 \to 3$ which we then subtract from $1 \to 2$. In this way, we can reuse our result for $1 \to 2$ after exchanging labels, $A_\mu\mathrm{d}x^\mu \leftrightarrow A_\nu\mathrm{d}y^\nu$, obtaining

$$\begin{aligned} P_{\mathrm{d}x}(x + \mathrm{d}y)P_{\mathrm{d}y}(x) &= 1 - igA_\nu(x)\mathrm{d}y^\nu - igA_\mu(x)\mathrm{d}x^\mu - ig\partial_\nu A_\mu(x)\mathrm{d}x^\mu\mathrm{d}y^\nu \\ &\quad - g^2 A_\mu(x)A_\nu(x)\mathrm{d}x^\mu\mathrm{d}y^\nu + \mathcal{O}(\mathrm{d}x^3). \end{aligned} \tag{10.40}$$

The first three terms on the RHSs of (10.39) and (10.40) cancel in the result $P(\square)$ for the round trip, leaving us with

$$\begin{aligned} P(\square) &\equiv P_{\mathrm{d}y}(x + \mathrm{d}x)P_{\mathrm{d}x}(x) - P_{\mathrm{d}x}(x + \mathrm{d}y)P_{\mathrm{d}y}(x) = \\ &- ig\left\{\partial_\mu A_\nu - \partial_\nu A_\mu + ig[A_\mu, A_\nu]\right\}\mathrm{d}x^\mu\mathrm{d}y^\nu. \end{aligned} \tag{10.41}$$

Maxwell's equations inform us that the line integral of the vector potential equals the enclosed flux. The area of the parallelogram corresponds to $\mathrm{d}x^\mu\mathrm{d}y^\nu$, and the pre-factor has to be therefore the field-strength tensor. If the enclosed flux is non-zero, then $P(\square)\psi_i \neq \psi_i$ and thus the field is rotated.

---

[3]Note the sign change compared to the covariant derivative: there we pull back the field from $x + \mathrm{d}x$ to x.

10.2.3 Curvature of spacetime

Curvature and the Riemann tensor. We continue to work out the analogy between Yang–Mills theories and gravity. Both the gauge field A_μ and the connection $\Gamma^\mu{}_{\kappa\rho}$ transform inhomogeneously. Therefore we cannot use them to judge if a gauge or gravitational field is present. In the gauge case, we introduced therefore the field-strength $F_{\mu\nu}$. It transforms homogeneously and thus the statement $F_{\mu\nu}(x) = 0$ holds in any gauge. This suggests transforming (10.25) into a definition for a tensor measuring a non-zero curvature of spacetime,

$$(\nabla_\alpha \nabla_\beta - \nabla_\beta \nabla_\alpha) T^{\mu\cdots}_{\nu\cdots} = [\nabla_\alpha, \nabla_\beta] T^{\mu\cdots}_{\nu\cdots} \neq 0. \tag{10.42}$$

Thus the curvature of spacetime should be proportional to the area of a loop and the amount a tensor is rotated.

For the special case of a vector V^α we obtain with

$$\nabla_\rho V^\alpha = \partial_\rho V^\alpha + \Gamma^\alpha{}_{\beta\rho} V^\beta \tag{10.43}$$

first

$$\nabla_\sigma \nabla_\rho V^\alpha = \partial_\sigma(\partial_\rho V^\alpha + \Gamma^\alpha{}_{\beta\rho} V^\beta) + \Gamma^\alpha{}_{\kappa\sigma}(\partial_\rho V^\kappa + \Gamma^\kappa{}_{\beta\rho} V^\beta) - \Gamma^\kappa{}_{\rho\sigma}(\partial_\kappa V^\alpha + \Gamma^\alpha{}_{\beta\kappa} V^\beta). \tag{10.44}$$

The second part of the commutator follows from the simple relabelling $\sigma \leftrightarrow \rho$ as

$$\nabla_\rho \nabla_\sigma V^\alpha = \partial_\rho(\partial_\sigma V^\alpha + \Gamma^a{}_{\beta\sigma} V^\beta) + \Gamma^\alpha{}_{\kappa\rho}(\partial_\sigma V^\kappa + \Gamma^\kappa{}_{b\sigma} V^\beta) - \Gamma^\kappa{}_{\sigma\rho}(\partial_\kappa V^\alpha + \Gamma^\alpha{}_{\beta\kappa} V^\beta). \tag{10.45}$$

Now we subtract the two equations using that $\partial_\rho \partial_\sigma = \partial_\sigma \partial_\rho$ and $\Gamma^\alpha{}_{\beta\rho} = \Gamma^\alpha{}_{\rho\beta}$,

$$[\nabla_\rho, \nabla_\sigma] V^\alpha = \left[\partial_\rho \Gamma^\alpha{}_{\beta\sigma} - \partial_\sigma \Gamma^\alpha{}_{\beta\rho} + \Gamma^\alpha{}_{\kappa\rho} \Gamma^\kappa{}_{\beta\sigma} - \Gamma^\alpha{}_{\kappa\sigma} \Gamma^\kappa{}_{\beta\rho} \right] V^\beta \equiv R^\alpha{}_{\beta\rho\sigma} V^\beta. \tag{10.46}$$

The tensor $R^\alpha{}_{\beta\rho\sigma}$ is called *Riemann* or *curvature tensor*. In problem 10.7, you are asked to show that the tensor $R_{\alpha\beta\rho\sigma} = g_{\alpha\gamma} R^\gamma{}_{\beta\rho\sigma}$ is antisymmetric in the indices $\rho \leftrightarrow \sigma$, antisymmetric in $\alpha \leftrightarrow \beta$ and symmetric against an exchange of the index pairs $(\alpha\beta) \leftrightarrow (\rho\sigma)$. Therefore, we can construct out of the Riemann tensor only one non-zero tensor of rank two, contracting α either with the third or fourth index, $R^\rho{}_{\alpha\rho\beta} = -R^\rho{}_{\alpha\beta\rho}$. We define the Ricci tensor by

$$R_{\alpha\beta} = R^\rho{}_{\alpha\rho\beta} = -R^\rho{}_{\alpha\beta\rho} = \partial_\rho \Gamma^\rho{}_{\alpha\beta} - \partial_\beta \Gamma^\rho{}_{\alpha\rho} + \Gamma^\rho{}_{\alpha\beta} \Gamma^\sigma{}_{\rho\sigma} - \Gamma^\sigma{}_{\beta\rho} \Gamma^\rho{}_{\alpha\sigma}. \tag{10.47}$$

A further contraction gives the curvature scalar,

$$R = R_{\alpha\beta} g^{\alpha\beta}. \tag{10.48}$$

Example 10.2: Calculate the Ricci tensor R_{ab} and the scalar curvature R of the two-dimensional unit sphere S^2.
We have already determined the non-vanishing Christoffel symbols of the sphere S^2 as $\Gamma^\phi{}_{\vartheta\phi} = \Gamma^\phi{}_{\phi\vartheta} = \cot\vartheta$ and $\Gamma^\vartheta{}_{\phi\phi} = -\cos\vartheta \sin\vartheta$. We will show later that the Ricci tensor of a maximally symmetric space as a sphere satisfies $R_{ab} = K g_{ab}$. Since the metric is diagonal, the non-diagonal elements of the Ricci tensor are zero too, $R_{\phi\vartheta} = R_{\vartheta\phi} = 0$. We calculate with

$$R_{ab} = R^c{}_{acb} = \partial_c \Gamma^c{}_{ab} - \partial_b \Gamma^c{}_{ac} + \Gamma^c{}_{ab}\Gamma^d{}_{cd} - \Gamma^d{}_{bc}\Gamma^c{}_{ad}$$

the $\vartheta\vartheta$ component, obtaining

$$R_{\vartheta\vartheta} = 0 - \partial_\vartheta(\Gamma^\phi{}_{\vartheta\phi} + \Gamma^\vartheta{}_{\vartheta\vartheta}) + 0 - \Gamma^d{}_{\vartheta c}\Gamma^c{}_{\vartheta d} = 0 + \partial_\vartheta \cot\vartheta - \Gamma^\phi{}_{\vartheta\phi}\Gamma^\phi{}_{\vartheta\phi}$$
$$= 0 - \partial_\vartheta \cot\vartheta - \cot^2\vartheta = 1.$$

From $R_{ab} = Kg_{ab}$, we find $R_{\vartheta\vartheta} = Kg_{\vartheta\vartheta}$ and thus $K = 1$. Hence $R_{\phi\phi} = g_{\phi\phi} = \sin^2\vartheta$. The scalar curvature is (diagonal metric with $g^{\phi\phi} = 1/\sin^2\vartheta$ and $g^{\vartheta\vartheta} = 1$)

$$R = g^{ab}R_{ab} = g^{\phi\phi}R_{\phi\phi} + g^{\vartheta\vartheta}R_{\vartheta\vartheta} = \frac{1}{\sin^2\vartheta}\sin^2\vartheta + 1 \times 1 = 2.$$

We can push the analogy further by remembering that the field-strength defined in Eq. (10.25) is a matrix. Writing out the implicit matrix indices of $F_{\mu\nu}$ in Eq. (10.25) gives

$$(F_{\mu\nu})_{ij} = \partial_\mu(A_\nu)_{ij} - \partial_\nu(A_\mu)_{ij} + ig\left\{(A_\mu)_{ik}(A_\nu)_{kj} - (A_\nu)_{ik}(A_\mu)_{kj}\right\}. \tag{10.49}$$

Comparing this expression to

$$R^\alpha{}_{\beta\mu\nu} = \partial_\mu \Gamma^\alpha{}_{\beta\nu} - \partial_\nu \Gamma^\alpha{}_{\beta\mu} + \Gamma^\alpha{}_{\rho\mu}\Gamma^\rho{}_{\beta\nu} - \Gamma^\alpha{}_{\rho\nu}\Gamma^\rho{}_{\beta\mu} \tag{10.50}$$

we see that the first two indices of the Riemann tensor, α and β, correspond to the group indices ij in the field-strength tensor. This is in line with the relation of the potential $(A_\mu)_{ij}$ and the connection $\Gamma^\alpha{}_{\beta\nu}$ implied by (10.33).

10.3   Quantisation of gauge theories

10.3.1   Abelian case

We discussed in section 7.2 that we can only derive the photon propagator if we fix a gauge. Now we reconsider this problem, and ask how we should modify the Lagrange density in order to be able to obtain the photon propagator. The Lagrange density that leads to the Maxwell equation is

$$\mathscr{L} = -\frac{1}{4}F_{\mu\nu}F^{\mu\nu} = -\frac{1}{2}\left(\partial_\mu A_\nu \partial^\mu A^\nu - \partial_\mu A_\nu \partial^\nu A^\mu\right)$$
$$= \frac{1}{2}\left(A_\nu \partial_\mu \partial^\mu A^\nu - A_\nu \partial_\mu \partial^\nu A^\mu\right) = \frac{1}{2}A_\mu\left[\eta^{\mu\nu}\Box - \partial^\mu \partial^\nu\right]A_\nu = \frac{1}{2}A^\mu D^{-1}_{\mu\nu}A^\nu, \tag{10.51}$$

where we made a partial integration, dropping as usual the surface term. Deriving the photon propagator requires us to invert the term in the square bracket. Performing a Fourier transformation, we see that we should find the inverse of the operator

$$D^{-1}_{\mu\nu}(k) = k^2 P^{\mu\nu}_T(k) = k^2\left(\eta^{\mu\nu} - k^\mu k^\nu/k^2\right). \tag{10.52}$$

We have already seen that this operator projects any four-vector on the three-dimensional subspace orthogonal to k. More formally, we see that $P^{\mu\nu}_T(k)$ is a projection operator,

$$P_T^{\mu\nu} P_{T\nu}{}^\lambda = P_T^{\mu\lambda}, \tag{10.53}$$

and has thus as only eigenvalues 0 and 1. Since $P_T^{\mu\nu}(k)$ is not the unit operator, it has at least one zero eigenvalue and is thus not invertible. More precisely, its trace is

$$P_{T\mu}^\mu = \eta_{\mu\nu} P_T^{\mu\nu} = \delta_\mu^\mu - 1 = 3, \tag{10.54}$$

and thus three eigenvalues are one and one eigenvalue is zero. The latter eigenvalue corresponds to $k_\mu P_T^{\mu\nu} = 0$, as required for a projection operator on the three-dimensional subspace orthogonal to k. The orthogonal part $\delta_\nu^\mu - P_{T\nu}{}^\mu$ is given by the longitudinal projection operator $P_L^{\mu\nu} = k^\mu k^\nu / k^2$.

We can invert $D_{\mu\nu}^{-1}$, if we choose a gauge such that the subspace parallel to k is included. The simplest choice is the Lorenz gauge. Imposing this gauge on the level of the Lagrangian means adding

$$\mathscr{L} \to \mathscr{L}_{\text{eff}} = \mathscr{L} + \mathscr{L}_{\text{gf}} = \mathscr{L} - \frac{1}{2}(\partial^\mu A_\mu)^2. \tag{10.55}$$

More generally, we can add the term

$$\mathscr{L}_{\text{gf}} = -\frac{1}{2\xi}(\partial^\mu A_\mu)^2 \tag{10.56}$$

that depends on the arbitrary parameter ξ. This group of gauges is employed in the proof of the renormalisability of gauge theories and is therefore called R_ξ gauge. The combined effective Lagrange density is thus

$$\mathscr{L}_{\text{eff}} = -\frac{1}{4} F_{\mu\nu} F^{\mu\nu} - \frac{1}{2\xi}(\partial^\mu A_\mu)^2 = \frac{1}{2} A_\nu \left[\eta^{\mu\nu} \Box - \left(1 - \frac{1}{\xi} \right) \partial^\mu \partial^\nu \right] A_\mu. \tag{10.57}$$

Fourier transforming the term in the square brackets, we obtain

$$P^{\mu\nu} = -k^2 \eta^{\mu\nu} + (1 - \xi^{-1}) k^\mu k^\nu. \tag{10.58}$$

Now we split this expression into its transverse and longitudinal parts,

$$P^{\mu\nu} = - k^2 \left(P_T^{\mu\nu} + \frac{k^\mu k^\nu}{k^2} \right) + (1 - \xi^{-1}) k^\mu k^\nu \tag{10.59a}$$

$$= - k^2 P_T^{\mu\nu} - \xi^{-1} k^2 P_L^{\mu\nu}. \tag{10.59b}$$

Since $P_T^{\mu\nu}$ and $P_L^{\mu\nu}$ project on orthogonal subspaces, we obtain the inverse $P_{\mu\nu}^{-1}$ simply by inverting their prefactors, cf. problem 8.9. Thus the photon propagator in R_ξ gauge is given by

$$\mathrm{i} D_F^{\mu\nu}(k^2) = \frac{-\mathrm{i} P_T^{\mu\nu}}{k^2 + \mathrm{i}\varepsilon} + \frac{-\mathrm{i}\xi P_L^{\mu\nu}}{k^2 + \mathrm{i}\varepsilon} = \frac{-\mathrm{i}}{k^2 + \mathrm{i}\varepsilon} \left[\eta^{\mu\nu} - (1 - \xi) \frac{k^\mu k^\nu}{k^2} \right]. \tag{10.60}$$

Special cases are the Feynman gauge $\xi = 1$ and the Landau gauge $\xi = 0$, while $\xi \to \infty$ corresponds to the unitary gauge. The arbitrary, ξ-dependent part of the photon propagator vanishes in physical quantities, where it is matched between conserved currents with $\partial_\mu J^\mu(x) = 0$ or $k_\mu J^\mu(k) = 0$.

10.3.2 Non-abelian case

An important conceptional difference between abelian and non-abelian theories is that in the latter case the conserved Noether current is not gauge-invariant, cf. problem 10.1. Moreover, the non-abelian gauge transformation (10.22c) adds not only a term $\partial_\mu\vartheta$ but also mixes the gauge fields via the term $f^{abc}A_\mu^b\vartheta^c$. Therefore it is in general not guaranteed that the gauge-dependent unphysical degrees of freedom contained in the propagator (10.60) decouple and the quantisation of non-abelian theories becomes more challenging.

We consider first the two-dimensional integral

$$Z \propto \int \mathrm{d}x\mathrm{d}y \; \mathrm{e}^{\mathrm{i}S(x)}. \tag{10.61}$$

as a toy model for the generating functional of a Yang–Mills theory. Since the integration extends from $-\infty$ to ∞, the y integration does not merely change the normalisation of Z but makes the integral ill-defined. We can eliminate the dangerous y integration by introducing a delta function,

$$Z \propto \int \mathrm{d}x\mathrm{d}y \; \delta(y)\mathrm{e}^{\mathrm{i}S(x)}. \tag{10.62}$$

Since the value of y in the delta function plays no role, we can replace $\delta(y)$ by $\delta(y-f(x))$ with an arbitrary function $f(x)$. If $y = f(x)$ is the solution of $g(x,y) = 0$, we obtain with

$$\delta(g(x,y)) = \frac{\delta(y - f(x))}{|\partial g/\partial y|} \tag{10.63}$$

and assuming that $\partial g/\partial y > 0$

$$Z \propto \int \mathrm{d}x\mathrm{d}y \; \frac{\partial g}{\partial y}\delta(g)\mathrm{e}^{\mathrm{i}S(x)}. \tag{10.64}$$

Generalising this to n dimensions, we need n delta functions and have to include the Jacobian,

$$Z \propto \int \mathrm{d}^n x\mathrm{d}^n y \; \det\left(\frac{\partial g_i}{\partial y_j}\right) \prod_i \delta(g_i)\mathrm{e}^{\mathrm{i}S(x)}. \tag{10.65}$$

We now translate this toy example to the Yang–Mills case. The functions g are the gauge-fixing conditions that we choose as

$$g^a(x) = \partial^\mu A_\mu^a(x) - \omega^a(x), \tag{10.66}$$

where the $\omega^a(x)$ are arbitrary functions. The discrete index i corresponds to $\{x,a\}$, explaining why the gauge freedom results in an infinity. Although the integration measure of a compact gauge group is finite, the summation over $\mathbb{R}^4$ gives an infinite answer. Finally, we see that from the transformation law $A_\mu^a(x) \to A_\mu^{a\prime}(x) = A_\mu^a(x) - D_\mu^{ac}\vartheta^c(x)$ that the parameters ϑ^a correspond to the redundant coordinates y_i.

The generating functional for a Yang–Mills theory is thus with $\mathcal{D}A \equiv \prod_{\mu=0}^{3} \prod_{a=1}^{m} \mathcal{D}A_{\mu}^{a}$ as shortcut given by

$$Z[0] \propto \int \mathcal{D}A \, \mathrm{Det}\left(\frac{\delta g^{a}}{\delta \vartheta^{b}}\right) \prod_{x,a} \delta(g^{a}) e^{\mathrm{i}S_{\mathrm{YM}}}, \qquad (10.67)$$

where we set the sources to zero, for the moment. Our task is first to evaluate $\delta g^{a}/\delta\vartheta^{b}$ and then to transform the determinant into the Lagrangian of new, auxiliary fields such that we can use the language of Feynman diagrams to perform perturbative calculations in the usual way. Inserting into the gauge-fixing condition (10.66) the infinitesimal gauge transformation (10.22c), we obtain as change

$$\delta g^{a}(x) = -\partial^{\mu} D_{\mu}^{ab} \vartheta^{b}(x). \qquad (10.68)$$

Thus the required functional derivative is

$$\frac{\delta g^{a}(x)}{\delta \vartheta^{b}(y)} = -\partial^{\mu} D_{\mu}^{ab} \delta(x-y). \qquad (10.69)$$

We can eliminate the determinant remembering $\int \mathrm{d}\eta \mathrm{d}\bar{\eta} \, e^{\bar{\eta}A\eta} = \det A$ from Eq. (8.103), expressing the Jacobian as a path integral over Graßmann variables c^{a} and $\bar{c}^{a}$,

$$\mathrm{Det}\left[\frac{\delta g^{a}(x)}{\delta \vartheta^{b}(y)}\right] \propto \int \mathcal{D}c\mathcal{D}\bar{c} \, e^{\mathrm{i}S_{\mathrm{FP}}}. \qquad (10.70)$$

The corresponding Lagrangian is

$$\mathscr{L}_{\mathrm{FP}} = -\bar{c}^{a}\partial^{\mu}D_{\mu}^{ab}c^{b} = (\partial^{\mu}\bar{c}^{a})(D_{\mu}^{ab}c^{b}) = \partial^{\mu}\bar{c}^{a}\partial_{\mu}c^{a} + gf^{abc}\partial^{\mu}\bar{c}^{a}c^{b}A_{\mu}^{c}, \qquad (10.71)$$

where we made a partial integration and inserted the definition of the covariant derivatives acting on the gauge field, Eq. (10.22c). As a result, we have recast the determinant as the kinetic energy of complex scalar fields c^{a} that interact with the gauge fields. Since we had to use Graßmann variables c^{a} for the scalar fields, their statistics is fermionic. Clearly, such fields should be seen as a purely mathematical construct and they are therefore called Faddeev–Popov ghosts. In an abelian theory as U(1), the interaction term in Eq. (10.71) is absent and ghost fields decouple. Since then they change only the normalisation of the path integral, they can be omitted in QED.

Next we have to eliminate the $\delta(g^{a}(x))$. They contain the arbitrary functions $\omega^{a}(x)$, but the path integral does not depend on them. Thus we have the freedom to multiply with a chosen function $f(\omega^{a})$, thereby only changing the normalisation. Our aim is to generate a term $\exp(\mathrm{i}S_{\mathrm{gf}})$ after integrating over the delta functions, as in the case of QED. Choosing

$$Z \to \exp\left(-\frac{\mathrm{i}}{2\xi}\int \mathrm{d}^{4}x \, \omega^{a}(x)\omega^{a}(x)\right) Z, \qquad (10.72)$$

integrating $\prod_{x,a} \delta(g^{a}) \exp\left(-\frac{\mathrm{i}}{2\xi}\int \mathrm{d}^{4}x \, \omega^{a}(x)\omega^{a}(x)\right)$ with the help of $\delta(g^{a})$ and (10.66), we obtain as gauge-fixing term the desired

$$\mathscr{L}_{\mathrm{gf}} = -\frac{1}{2\xi}\partial^{\mu}A_{\mu}^{a}\partial^{\nu}A_{\nu}^{a}. \qquad (10.73)$$

The complete Lagrange density $\mathscr{L}_{\text{eff}}$ of a non-abelian gauge theory consists thus of four parts,

$$\mathscr{L}_{\text{eff}} = \mathscr{L}_{\text{YM}} + \mathscr{L}_{\text{gf}} + \mathscr{L}_{\text{FP}} + \mathscr{L}_{\text{s}}, \tag{10.74}$$

where the last one couples sources linearly to the fields,

$$\mathscr{L}_{\text{s}} = J^\mu A_\mu + \bar{\eta} c + \bar{c} \eta. \tag{10.75}$$

We break both $\mathscr{L}_{\text{YM}}$ and $\mathscr{L}_{\text{FP}}$ into a piece of $\mathcal{O}(g^0)$ defining the free propagator, and pieces of $\mathcal{O}(g)$ corresponding to a three gluon and a two ghost-gluon vertex, respectively, and a four gluon vertex of $\mathcal{O}(g^2)$. After a partial integration of the free part, we obtain

$$\mathscr{L}_{\text{YM}} + \mathscr{L}_{\text{FP}} = \frac{1}{2} A_\nu^a \left(\eta^{\mu\nu} \Box - \partial^\mu \partial^\nu \right) A_\mu^a - g f^{abc} A_\mu^a A_\nu^b \partial^\mu A^{c\nu} - \frac{g^2}{4} f^{abe} f^{cde} A_\mu^a A_\nu^b A^{c\mu} A^{d\nu}$$
$$- \bar{c}^a \Box c^a + g f^{abc} \partial^\mu \bar{c}^a c^b A_\mu^c. \tag{10.76}$$

The Feynman rules can now be read off after Fourier transforming into momentum space, cf. problem 10.7. Combining the resulting expression with $\mathscr{L}_{\text{gf}}$, we see that the gluon propagator is diagonal in the group indices and otherwise identical to the photon propagator in R_ξ gauge. The ghost propagator is the one of a massless scalar particle,

$$\Delta_{ab}(k) = \frac{\delta_{ab}}{k^2 + i\varepsilon}. \tag{10.77}$$

Being a fermion, a closed ghost loop introduces a minus sign, however.

Non-covariant gauges. The introduction of ghost fields can be avoided if one uses non-covariant gauges which depend on an arbitrary vector n^μ. An example used often in QED is the Coulomb or radiation gauge,

$$\partial_\mu A^\mu - n_\mu \partial^\mu (n_\nu A^\nu) = 0 \tag{10.78}$$

with $n_\mu = (1,0,0,0)$. In QCD, one employs often the set of gauges

$$n_\mu A_a^\mu = 0, \quad a = 1, \ldots, 8 \tag{10.79}$$

with the constant vector n^μ. More specifically, one calls the case $n^2 = 0$ light-cone, $n^2 < 0$ axial and $n^2 > 0$ temporal gauge. They have in common the fact that the Faddeev–Popov determinant does not depend on A_a^μ and can be absorbed in the normalisation of the path integral, cf. problem 10.9. While non-covariant gauges thus bypass the introduction of unphysical particles in loop graphs, the resulting propagators are unhandy. Moreover, they contain spurious singularities which require care. Therefore in practically all applications the use of the R_ξ gauge is advantageous.

Let us finally comment on the case of external gluons. In the case of photons, we can sum their polarisation states using $\sum_{r=0}^3 \varepsilon_\mu^{(r)*} \varepsilon_\nu^{(r)} = -\eta_{\mu\nu}$, since the nonphysical degrees of freedom cancel in physical observables. In the non-abelian case, we can use this "trick" only in the case of a single external gluon. For two or more external gluons we have to employ the polarisation sum derived in problem 7.2, since the non-abelian vertices mix physical and non-physical degrees of freedom. Since the conserved Noether current is not gauge-invariant, we cannot use the argument presented in section 7.2. Alternatively, we can use the R_ξ-gauge if we include Faddeev–Popov ghosts also as *external* particles. In order to subtract the unphysical contributions to the squared matrix elements correctly, one has to add the factor $(-1)^n$ to a term $\mathcal{A}_i \mathcal{A}_j^*$ with $2n$ Faddeev–Popov ghosts (Nachtmann, 1990).

10.A   Appendix: Feynman rules for an unbroken gauge theory

The Feynman rules for a non-broken Yang–Mills theory as QCD are given in the R_ξ gauge; for the abelian case of QED set the structure constants $f_{abc} = 0$, $T = 1$ and replace $g_s \to e q_f$, where q_f is the electric charge of the fermion in units of the elementary charge $e > 0$. The momentum flow is indicated by the thin arrow: for instance, all momenta are chosen as in-going in the triple gauge vertex (10.82).

Propagators

$$-\mathrm{i}\delta_{ab}\left[\frac{\eta_{\mu\nu}}{k^2 + \mathrm{i}\epsilon} - (1 - \xi)\frac{k_\mu k_\nu}{(k^2)^2}\right] \qquad (10.80)$$

$$\delta_{ab}\,\frac{\mathrm{i}}{k^2 + \mathrm{i}\epsilon} \qquad (10.81)$$

Triple Gauge Interactions

$$-g_s f^{abc}\big[\, \eta^{\mu\nu}(p_1 - p_2)^\rho + \eta^{\nu\rho}(p_2 - p_3)^\mu$$
$$+\eta^{\rho\mu}(p_3 - p_1)^\nu\big] \qquad (10.82)$$
$$p_1 + p_2 + p_3 = 0$$

Quartic Gauge Interactions

$$-\mathrm{i}g_s^2\Big[\, f_{eab}f_{ecd}(\eta_{\mu\rho}\eta_{\nu\sigma} - \eta_{\mu\sigma}\eta_{\nu\rho})$$
$$+f_{eac}f_{edb}(\eta_{\mu\sigma}\eta_{\rho\nu} - \eta_{\mu\nu}\eta_{\rho\sigma})$$
$$+f_{ead}f_{ebc}(\eta_{\mu\nu}\eta_{\rho\sigma} - \eta_{\mu\rho}\eta_{\nu\sigma})\Big] \qquad (10.83)$$
$$p_1 + p_2 + p_3 + p_4 = 0$$

Fermion Gauge Interactions

$$-\mathrm{i}\,g_s\gamma^\mu T^a_{ij} \qquad (10.84)$$

Ghost Interactions

$$g_s\,f^{abc}p^\mu_1$$

$$p_1 + p_2 + p_3 = 0 \qquad (10.85)$$

Summary

Requiring local symmetry under a gauge group as $SU(n)$ or $SO(n)$ specifies the self-interactions of massless gauge bosons as well as their couplings to fermions and scalars. The presence of self-interactions implies that a pure Yang–Mills theory is non-linear. The gauge-invariant derivative D_μ is the analogon to the covariant derivative ∇_μ of gravity, while the field-strength corresponds to the Riemann tensor. Both measure the rotation of a vector which is parallel-transported along a closed loop. The quantisation of Yang–Mills theories in the covariant R_ξ gauge leads to ghost particles. These fermionic scalars compensate the unphysical degrees of freedom still contained in the gauge fields A_μ using a covariant gauge-fixing condition as $\partial_\mu A^\mu = 0$.

Note also the interplay between local and global symmetries. A global symmetry transformation U maps a physical state onto a different physical state with the same properties, implying via Noether's theorem a conserved current. A local symmetry transformation $U(x)$ maps a physical state on itself, implying a redundancy in our description of the system. Since local symmetries contain global transformations as a subgroup, they imply always also the conservation of global charges via Noether's theorem.

Further reading. The Feynman rules in the appendix are taken from Romao and Silva (2012). This article contains all Feynman rules for the SM in a convention-independent notation which allows an easy comparison of references with differing conventions. Current conservation in non-abelian theories is discussed, for example in Leader and Predazzi (2013). The extension of the helicity formalism to QCD, where it leads to both phenomenological useful and theoretically interesting results, is discussed by Peskin (2011), Schwartz (2013) and Weinzierl (2016).

Problems

10.1 Non-abelian Maxwell equations. Derive the non-abelian analogue of the Maxwell equations. What are the conserved Noether currents, do gauge-invariant currents exist? Derive the constraint on the allowed gauge transformations such that the conserved charges transform covariantly.

10.2 Stress tensor. Show that the stress tensor for a single quark in the background of a classical gluon field A_μ^a is given by $T^{\mu\nu} = \frac{i}{2}\bar\psi\gamma^\mu \overleftrightarrow{D}{}^\nu \psi$.

10.3 Palatini approach. Consider the Yang–Mills action as a functional $S_{\mathrm{YM}}[A_\mu, F_{\mu\nu}]$ of the potential and the field-strength,

$$\mathscr{L} \propto F^{a\mu\nu}\left[\partial_\mu A_\nu^a - \partial_\nu A_\mu^a - g f^{abc} A_\mu^b A_\nu^c\right].$$

Derive the non-abelian Maxwell equations by varying A_μ and $F_{\mu\nu}$ independently.

10.4 Group theory. a.) Derive the Jacobi identity (B.5) and show that the structure constants $i f^{abc}$ of a Lie algebra satisfy the Lie algebra by themselves. b.) Insert $(T_A^a)_{bc} = -i f^{abc}$ for the adjoint representation into (10.11) and show that the result agrees with (10.22c). c.) Derive the relations (B.15)–(B.17).

10.5 Hyperbolic plane H^2. The line element of the Hyperbolic plane H^2 is given by $ds^2 = y^{-2}(dx^2 + dy^2)$ with $y \geq 0$. a.) Write out the geodesic equations and deduce the Christoffel symbols $\Gamma^a{}_{bc}$. b.) Calculate $R^a{}_{bcd}$ and R.

10.6 Symbolic calculations. Download the programme `differentialGeometry.py`

from the book webpage. Repeat the calculation for S^2 and H^2 using the programme.

10.7 Three and four gauge boson vertices. Derive the tensor structure $\mathbf{V}^{rst}(k_1^\rho, k_2^\sigma, k_3^\tau)$ of the three-gluon vertex by Fourier-transforming the part of $\mathscr{L}_I$ containing three gluon fields to momentum space,

$$F = \int d^4 p_1 d^4 p_2 d^4 p_3\,(2\pi)^4 \delta(p_1 + p_2 + p_2)$$
$$\times\ \mathscr{L}_I(A_\mu^a(p_1) A_\mu^b(p_2) A_\nu^c(p_3))$$

and then eliminating the fields by functional derivatives with respect to them

$$V^{rst}(k_1^\rho, k_2^\sigma, k_3^\tau) = \frac{i\delta^3 F}{\delta A_\rho^r(k_1)\,\delta A_\sigma^s(k_2)\,\delta A_\tau^t(k_3)}$$

Similarly, derive the tensor structure $\mathbf{V}^{rst}(k_1^\rho, k_2^\sigma, k_3^\tau, k_4^\lambda)$ of the four-gluon vertex. Use alternatively symmetry arguments, if possible.

10.8 Coulomb photons. Split the photon boson propagator in the Feynman gauge gauge into a transverse part and a remainder. Show that the latter contains the Fourier transform of the Coulomb potential.

10.9 Non-covariant gauge. a.) Show that the Faddeev–Popov term is independent of A_μ^a in the gauge (10.79) and, thus, can be absorbed in the normalisation of the path integral. b.) Derive the gauge boson propagator using $\mathscr{L}_{\mathrm{gf}} = \frac{1}{2\xi}(n^\mu A_\mu)^2$ as the corresponding gauge-fixing term. c.) Derive the gauge boson propagator in the Coulomb gauge.

11
Renormalisation I: Perturbation theory

We encountered three examples of divergent loop integrals discussing the $\lambda\phi^4$ theory. In these cases it was possible to subtract the infinities in such a way that we obtained finite observables which depend only on the experimentally measured values of m, λ and ρ. The aim of this and the following chapter is to obtain a better understanding of this renormalisation procedure. We will see that the $\lambda\phi^4$ theory as well as the electroweak and strong interactions of the SM are examples for renormalisable theories. For such theories, the renormalisation of the finite number of parameters contained in the classical Lagrangian is sufficient to make all observables finite at any order perturbation theory.

11.1 Overview

Why renormalisation at all? We are using perturbation theory with the free, non-interacting Lagrangian as the starting point to evaluate non-linear quantum field theories. Interactions change the parameters of the free theory, however, as we know already both from classical electrodynamics and quantum mechanics. In the former case, in 1904 Lorentz studied the connection between the measured electron mass m_{phy}, its mechanical or inertial mass m_0 and its electromagnetic self-energy m_{el} in a toy model. He described the electron as a spherically symmetric uniform charge distribution with radius r_{e}, obtaining

$$m_{\mathrm{phy}} = m_0 + m_{\mathrm{el}} = m_0 + \frac{4e^2}{5r_{\mathrm{e}}}. \tag{11.1}$$

Special relativity forces us to describe the electron as a point particle. Taking thus the limit $r_{\mathrm{e}} \to 0$, classical electrodynamics implies an infinite "renormalisation" of the "bare" electron mass m_0 by its electromagnetic self-energy m_{el}.

Another familiar example of renormalisation appears in quantum mechanics. Perturbation theory is possible, if the Hamilton operator H can be split into a solvable part $H^{(0)}$ and an interaction λV,

$$H = H^{(0)} + \lambda V, \tag{11.2}$$

and the parameter λ is small. Using then as starting point the normalised solutions $|n^{(0)}\rangle$ of $H^{(0)}$,

$$H^{(0)}|n^{(0)}\rangle = E_n^{(0)}|n^{(0)}\rangle, \tag{11.3}$$

Quantum Fields–From the Hubble to the Planck Scale. Michael Kachelriess. © Michael Kachelriess 2018.
Published in 2018 by Oxford University Press. DOI 10.1093/oso/9780198802877.001.0001

we can find the eigenstates $|n\rangle$ of the complete Hamiltonian H as a power series in λ,

$$|n\rangle = |n^{(0)}\rangle + \lambda|n^{(1)}\rangle + \lambda^2|n^{(2)}\rangle + \dots. \tag{11.4}$$

Since we started with normalised states, $\langle n^{(0)}|n^{(0)}\rangle = 1$, the new states $|n\rangle$ are no longer correctly normalised. Thus going from free (or "bare") to interacting states requires renormalising the states,

$$_R\langle n|n\rangle_R = 1 \quad \Rightarrow \quad |n\rangle_R \equiv Z^{1/2}\,|n\rangle. \tag{11.5}$$

This is a very similar problem to that we encountered introducing the LSZ formalism. In the parlance of field theory, we often continue to call this procedure wave-function renormalisation, although Z renormalises field operators.

Why regularisation at all? The familiar process of renormalisation becomes more obscure by the fact that the renormalisation constants are infinite in most quantum field theories. Mathematical manipulations such as shifting the integration variable in a divergent loop integral are only well-defined if we first convert these integrals into convergent ones. Thus we have to regularise as the first step, that is employing a method which makes our expressions finite, so that our mathematical manipulations are well-defined and we can perform the renormalisation. You should keep in mind that the two operations, regularisation and renormalisation, are logically independent. Renormalisation of the parameters in the free theory is necessary because they are changed by interactions. This change may be finite, such as the change of the photon mass in a plasma, and no regularisation is necessary.

The second question to ask is why the renormalisation constants are infinite, or in other words why do we have regularise at all? There are (at least) two possible answers to this question: either we use a bad theory as a starting point, that is, the full quantum theory defined non-perturbatively by its generating functional $Z[J]$ is ill-defined; or we employ a bad expansion scheme evaluating $Z[J]$ in perturbation theory.

Example 11.1: An example of a bad expansion is the following toy model for the $\lambda\phi^4$ interaction,

$$Z(\lambda) = \int_{-\infty}^{\infty} dx\, e^{-\frac{1}{2}x^2 - \lambda x^4} \stackrel{?}{=} \int_{-\infty}^{\infty} dx\, \left[1 - \lambda x^4 + \frac{(\lambda x^4)^2}{2!} - \dots\right] e^{-\frac{1}{2}x^2}. \tag{11.6}$$

The LHS is well-defined for $\lambda > 0$. Applying perturbation theory and summing up the first N terms of the expansion on the RHS results in an alternating series,

$$Z_N(\lambda) = \sum_{n=0}^{N} \frac{(-\lambda)^n}{n!} \int_{-\infty}^{\infty} dx\, x^{4n} e^{-x^2/2} = \sum_{n=0}^{N} a_n \lambda^n$$

with

$$a_n = \frac{(-1)^n}{n!}\, 2^{2n+1/2}\, \Gamma(2n + 1/2).$$

The coefficients a_n of this series grow like a factorial and thus the convergence radius of the expansion is zero. Plotting $Z_N(\lambda)/Z(0)$ for the first few N as function of λ, problem 11.1, you see first that adding more terms makes the expansion worse beyond a certain value $\lambda_{\max}(N)$, and second that $\lambda_{\max}(N) \to 0$ for $N \to \infty$; see Flory *et al.* (2012) for more details.

It should not be too surprising that the expansion (11.6) has a zero convergence radius. Moving from $\lambda > 0$ to $\lambda < 0$ fundamentally changes physics since the vacuum is unstable for an arbitrarily small negative λ. An interesting consequence of the failure of perturbation theory is that the complete theory may contain additional nonperturbative physics. Next we look at an example where we start from a bad theory.

Example 11.2: In problem 2.11 we discussed the scattering on a short-range potential in $d = 1$, and found that no consistent solution exists for an odd potential. We rephrase this problem now in a language close to the one used in QFT. The perturbative expansion of the S-matrix in quantum mechanics is given by

$$\langle p_f | S | p_i \rangle = 2\pi \delta(E_i - E_f) \left[\langle p_f | V | p_i \rangle + \int dp \, \langle p_f | V | p \rangle \, \frac{i}{E - p^2/2 + i\varepsilon} \, \langle p | V | p_i \rangle + \cdots \right]$$

for $p_i \neq p_f$. Recoiling on the infinitely heavy static source, the (virtual) particle in the intermediate states can have any momentum p while its energy is conserved. With $V_0(x) = c_0 \delta(x)$, it follows $\langle p_f | V_0 | p \rangle = c_0/(2\pi)$ and then the momentum integral in the second-order correction becomes

$$\left(\frac{c_0}{2\pi} \right)^2 \int dp \frac{i}{E - p^2/2 + i\varepsilon}.$$

Thus this momentum integral, and similarly those at higher orders, are well-defined. Next we set $c_0 = 0$. Using then $\delta' f = -\delta f'$, we obtain $\langle p_f | V_1 | p \rangle = i c_1 (p_f - p)/(2\pi)$ and thus the momentum integral

$$\left(\frac{c_1}{2\pi} \right)^2 \int dp \frac{i(p - p_i)(p - p_f)}{E - p^2/2 + i\varepsilon}$$

is linearly divergent. The divergence means that the scattering probability is sensitive to arbitrarily high momentum modes. We can understand this behaviour looking at the wave-function $\psi(x)$. Because the potential is odd, also $\psi(x)$ is odd and thence has to change rapidly within $|x| < a$. As result, its Fourier transform $\psi(k)$ necessarily also contains high-frequency modes.

In this simple toy model, the natural way to solve the UV divergence problem is to replace the mathematical idealisation of a delta-function-like potential by the true, smooth potential. If we either do not know the true potential or if we insist that a delta-function-like potential captures all the physics contained in a scattering process at a short-range potential, then we have to regularise the potential, replacing $V(x) = c_1 \delta'(x)$, for example by

$$V(x) = c_1 \frac{\delta(x + a) - \delta(x - a)}{2a}.$$

In this way, we eliminate high-frequency modes with $p \gg 1/a$. Repeating the computation of the transmission amplitude, we find $T \simeq iap/c_1^2$. Hence $c_R \equiv c_1^2/a$ plays

> the role of an effective coupling constant in the regularised theory. Physical observables like the transmission amplitude depend only on the single parameter c_R, if we rescale $c_1(a) \propto a^{-1/2}$. Thus this simple example from quantum mechanics exhibits the key features of a UV-divergent QFT. We regularise the theory, cutting off UV modes. Requiring the independence of physical observables from the cut-off scale, we obtain running parameters.

It is very likely that our favourite $\lambda\phi^4$ interaction suffers from both diseases. First, the expansion in λ is not convergent but results in an asymptotic series. Second, the full theory contains only the trivial $\lambda = 0$ case as a consistent solution. Even if the interacting theory may be mathematically inconsistent it can, however, be used as an effective model describing physics up to a finite energy scale.

Regularisation methods. We have already seen that the regularisation of divergent loop integrals can be done in various ways. In general, one reparameterises the integral in terms of a parameter Λ (or ε) called regulator such that the integral becomes finite for a finite value of the regulator, while the limit $\Lambda \to \infty$ (or $\varepsilon \to 0$) returns the original integral.

- We can avoid UV divergences evaluating loop integrals introducing an (Euclidean) momentum cut-off Λ. In a somewhat more sophisticated manner, we could introduce instead of a hard cut-off a smooth function which suppresses large momenta. Using Schwinger's proper-time representation (4.88) we can cut-off large momenta setting

$$\frac{1}{p^2 + m^2} \to \frac{\mathrm{e}^{-(p^2+m^2)/\Lambda^2}}{p^2 + m^2} = \int_{\Lambda^{-2}}^{\infty} \mathrm{d}s \, \mathrm{e}^{-s(p^2+m^2)} . \tag{11.7}$$

Although conceptually simple, both regularisation schemes violate all symmetries of our theory generically. This is not a principal flaw, since we should be able to recover these symmetries in the limit $\Lambda \to \infty$. However, this "recovery process" may be non-trivial to perform. Moreover, intermediate calculations become much more transparent if we can use the symmetries of the theory, and therefore these schemes are in practice not used except for the simplest cases.

- Pauli–Villars regularisation is a scheme where one adds heavy particles having the same quantum numbers and couplings as the original ones. Thus the propagator of a massless scalar particle is changed to

$$\frac{1}{k^2 + \mathrm{i}\varepsilon} \to \frac{1}{k^2 + \mathrm{i}\varepsilon} + \sum_i \frac{a_i}{k^2 - M_i^2 + \mathrm{i}\varepsilon} .$$

For $k^2 \ll M_i^2$, physics is unchanged, while for $k^2 \gg M_i^2$ and $a_i < 0$ the combined propagator scales as M_i^2/k^4 and the convergence of loop integrals improves. Since the heavy particles enter with the wrong sign, they are unphysical ghosts and serve only as a mathematical tool to regularise loop diagrams. Pauli–Villars regularisation respects the gauge invariance of QED if the heavy particles are coupled gauge invariantly to the photon.

- Lattice regularisation replaces the continuous spacetime by a discrete lattice. The finite lattice spacing a introduces a momentum cut-off, eliminating all UV divergences. Moreover the (Euclidean) path integral becomes well-defined and can be calculated numerically without the need to perform perturbation theory. Thus this approach is particularly useful in the strong-coupling regime of QCD where it has been used to calculate static quantities such as, for example, the hadron mass spectrum. Note that lattice regularisation for finite a respects gauge symmetries but spoils the translation and Lorentz symmetry of the underlying QFT. Nevertheless, one recovers a relativistic QFT in the limit $a \to 0$. A longstanding problem of lattice theory was how to implement correctly chiral fermions. This question was solved around the year 2000 and thus the SM can be defined now in a mathematically consistent, non-perturbative way as a lattice theory.

- Dimensional regularisation (DR) is the method we applied in the calculations of the one-loop diagrams of the $\lambda \phi^4$ theory. While DR has the important virtue of preserving Lorentz and gauge invariance, it is one of the least intuitive regularisation methods. We will show later that an integral without mass scale is zero in DR, for instance $\int d^d k \, k^{-2} = 0$. This example shows that the integration measure we implement using physical requirements with DR is not positive—as a mathematician would require. In problem 11.2, we examine how DR modifies the range of momentum values contributing to Feynman integrals. Using DR with fermions, we have to extend the Clifford algebra to d dimensions. A natural choice is $\mathrm{tr}(\gamma^\mu \gamma^\nu) = d\eta^{\mu\nu}$ and $\mathrm{tr}(1) = 4$. The treatment of $\gamma^5 \equiv \mathrm{i}\gamma^0 \gamma^1 \gamma^2 \gamma^3$ is problematic, however, relying heavily on $d = 4$.

- Various other regularisation methods such as zeta function regularisation or point splitting methods exist.

Even fixing a regularisation method such as DR, we can choose various renormalisation *schemes*. Four popular choices are

- on-shell renormalisation. In this scheme, we choose the subtraction such that the on-shell masses and couplings coincide with the corresponding values measured in processes with zero momentum transfer q. For instance, we define the renormalised electric charge via the Thomson limit of the Compton scattering amplitude. While this choice is very intuitive it is not practical for QCD. We will see soon that in this theory, scattering amplitudes calculated in perturbation theory become ill-defined in the limit $q^2 \to 0$.

- The momentum subtraction (MoM) scheme is a generalisation of the on-shell scheme which can be applied also to QCD. Here we subtract from the Green functions counter-terms such that the corrections are zero at a fixed space-like four-momentum $p^2 = -\mu^2$. In this way, divergences in the limit $q^2 \to 0$ are avoided.

- In the minimal subtraction (MS) scheme, we subtract only the divergent $1/\varepsilon$ poles.

- In the modified minimal subtraction $\overline{\mathrm{MS}}$ (read em-es-bar) scheme, we subtract also the $\ln(4\pi) - \gamma$ term appearing frequently. This scheme gives more compact expressions than the others and is most often used in theoretical calculations.

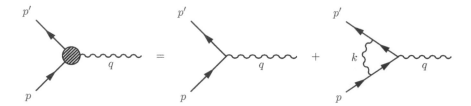

Fig. 11.1 The general vertex for the interaction of a fermion with a photon and its perturbative expansion within QED, corrections in external lines are omitted.

The main advantage of the MS and $\overline{\text{MS}}$ schemes is that they are mass-independent, that is, the subtraction terms do not depend on the particle masses. This independence simplifies the derivation of "running couplings" (cf. with the calculation of $\lambda(\mu)$ in section 4.3.4). As a drawback of the MS and $\overline{\text{MS}}$ schemes, quantities like the electron mass calculated in these schemes, m_e^{MS} or $m_e^{\overline{\text{MS}}}$, have to be translated into the physical mass m_e.

At a fixed order perturbation theory, predictions and reliability of different schemes vary for given external parameters. A simple example is the change from the MS to the $\overline{\text{MS}}$ scheme which are connected by $\tilde{\mu}^2 = 4\pi\mu^2 e^{-\gamma}$. Thus this transition is equivalent to a change of the renormalisation scale, thereby altering the size of the $\ln(\mu^2)$ term and thus the strength of the running coupling. More drastic changes result when moving from a mass-independent to a mass-dependent scheme, or comparing DR with other schemes. As a result, running couplings which are small enough to allow perturbation theory in one scheme may be prohibitive large in other schemes.

11.2   Anomalous magnetic moment of the electron

After this overview, let us move on to the calculation of the magnetic moment of the electron which is shifted by loop corrections from the tree-level value $g = 2$ you derived in problem 8.4. Apart from being the first successful loop calculation in the history of QFT, this process illustrates also several generic properties of loop graphs in renormalisable theories like QED.

Vertex function.   The tree-level interaction $\mathcal{L}_{\text{int}} = e\bar{\psi}\gamma^\mu\psi A_\mu$ between an electron and a photon corresponds in momentum space to $e\bar{u}(p')\gamma^\mu u(p)\varepsilon_\mu(q)$. Since loop integrals depend generally on the external momenta, the tree-level vertex γ^μ is modified by loop graphs such as the one shown in Fig. 11.1 and becomes a function of the momenta, $\Lambda^\mu(p, p', q)$. We want to write down the most general form of the vertex function Λ^μ for the coupling between an external electromagnetic field and an on-shell Dirac fermion, consistent with the symmetries of the problem. It is as usually convenient to apply the tensor method and to express Λ^μ as a sum of linearly independent rank-1 tensors multiplied by scalar functions.

Translation invariance implies $q = p' - p$ and thus Λ^μ is only a function of two momenta which we choose as p and p'. Since $p^2 = p'^2 = m^2$, the only non-trivial scalar variable in the problem is $p \cdot p'$. We choose to use the equivalent quantity $q^2 = (p' - p)^2$ as the variable on which the arbitrary scalar functions in our ansatz for Λ^μ depend. Next we have to form all possible vectors out of the momenta p_μ and p'_μ and the 16 basis elements (8.40) of the Clifford algebra. Restricting ourselves to QED, we have to impose additionally parity conservation what forbids the use of γ^5. Hence the most general ansatz compatible with Poincaré invariance and parity is

$$\Lambda^\mu(p, p') = A(q^2)\gamma^\mu + B(q^2)p^\mu + C(q^2)p'^\mu + D(q^2)\sigma^{\mu\nu}p_\nu + E(q^2)\sigma^{\mu\nu}p'_\nu. \quad (11.8)$$

Current conservation requires $q_\mu \Lambda^\mu(p, p') = 0$ and leads to $C = B$ and $E = -D$. Hence

$$\Lambda^\mu(p, p') = A(q^2)\gamma^\mu + B(q^2)(p^\mu + p'^\mu) + D(q^2)\sigma^{\mu\nu}q_\nu. \quad (11.9)$$

Hermiticity finally implies that A, B are real and D is purely imaginary.

Gordon decomposition. We derive now an identity that allows us to eliminate one of the three terms in Eq. (11.9), if we sandwich Λ^μ between two spinors which are on-shell. We evaluate

$$F^\mu = \bar{u}(p') \left[\slashed{p}'\gamma^\mu + \gamma^\mu \slashed{p} \right] u(p) \quad (11.10)$$

first using the Dirac equation for the two on-shell spinors, finding

$$F^\mu = 2m\bar{u}(p')\gamma^\mu u(p). \quad (11.11)$$

Second, we can use $\gamma^\mu \gamma^\nu = \eta^{\mu\nu} - i\sigma^{\mu\nu}$, obtaining

$$F^\mu = \bar{u}(p') \left[(p' + p)^\mu + i\sigma^{\mu\nu}(p' - p)_\nu \right] u(p). \quad (11.12)$$

Equating (11.11) and (11.12) gives the Gordon identity. It allows us to separate the Dirac current into a part proportional to $(p + p')^\mu$, that is with the same structure as a scalar current, and a part vanishing for $q = p' - p \to 0$ which couples to the spin of the fermion,

$$\bar{u}(p')\gamma^\mu u(p) = \bar{u}(p') \left[\frac{(p' + p)^\mu}{2m} + \frac{i\sigma^{\mu\nu}(p' - p)_\nu}{2m} \right] u(p). \quad (11.13)$$

Using the results from problem 8.4, we can identify in the non-relativistic limit the second term as contribution to the magnetic moment of the fermion.

The Gordon identity shows that the three terms in Eq. (11.9) are not independent. Depending on the context, we can therefore eliminate the most annoying term in the vertex function. We follow conventions and introduce the (real) form factors $F_1(q^2)$ and $F_2(q^2)$ by

$$\Lambda^\mu(p, p') = F_1(q^2)\gamma^\mu + F_2(q^2)\frac{i\sigma^{\mu\nu}q_\nu}{2m} = \quad (11.14a)$$

$$= F_1(q^2)\frac{(p' + p)^\mu}{2m} + [F_1(q^2) + F_2(q^2)]\frac{i\sigma^{\mu\nu}q_\nu}{2m}. \quad (11.14b)$$

The form factor F_1 is the coefficient of the electric charge, $eF_1(q^2)\gamma^\mu$, and should thus go to one for small momentum transfer, $F_1(0) = 1$. Therefore the magnetic moment of an electron is shifted by $1 + F_2(0)$ from the tree-level value $g = 2$. The deviation $a \equiv (g - 2)/2$ is called anomalous magnetic moment, the two form factors are often called electric and magnetic form factors.

Note the usefulness of the procedure to express the vertex function using only general symmetry requirements but not a specific theory for the interaction: Equation (11.14a) allows experimentalists to present their measurements using only two scalar functions which in turn can be easily compared to predictions of specific theories.

Anomalous magnetic moment.   After having discussed the general structure of the electromagnetic vertex function, we turn now to its calculation in perturbation theory for the case of QED. The Feynman diagrams contributing to the matrix element at $\mathcal{O}(e^3)$ with wave-functions as external lines are shown in Fig. 11.1, where we omit self-energy corrections in the external lines. As we will see soon, the latter do not contribute to the anomalous magnetic moment. We separate the matrix element into the tree-level part and the one-loop correction, $-ie\bar{u}(p') [\gamma^\mu + \Gamma^\mu] u(p)$. Using the Feynman gauge for the photon propagator, we obtain

$$\Gamma^\mu(p, p') = \int \frac{\mathrm{d}^4 k}{(2\pi)^4} \frac{-\mathrm{i}}{k^2 + \mathrm{i}\varepsilon} (\mathrm{i}e\gamma^\nu) \frac{\mathrm{i}}{\not{p}' + \not{k} - m + \mathrm{i}\varepsilon} \gamma^\mu \frac{\mathrm{i}}{\not{p} + \not{k} - m + \mathrm{i}\varepsilon} (\mathrm{i}e\gamma_\nu). \quad (11.15)$$

This integral is logarithmically divergent for large k,

$$\int^\Lambda \mathrm{d}k \, \frac{k^3}{k^2 k^2} \propto \ln \Lambda. \quad (11.16)$$

Before we perform the explicit calculation we want to understand if this divergence is connected to a specific kinematic configuration of the momenta. We therefore split the vertex correction into an on-shell and an off-shell part,

$$\Gamma^\mu(p, p') = \Gamma^\mu(p, p) + [\Gamma^\mu(p, p') - \Gamma^\mu(p, p)] \equiv \Gamma^\mu(p, p) + \Gamma^\mu_{\mathrm{off}}(p, p'). \quad (11.17)$$

Next we rewrite the first fermion propagator using the identity $(A + B)^{-1} = B^{-1} - B^{-1} A B^{-1} + \dots$ for small $A = p' - p$ as

$$\frac{1}{\not{p}' + \not{k} - m} = \frac{1}{\not{p} + \not{k} - m + (\not{p}' - \not{p})} = \quad (11.18\mathrm{a})$$

$$= \frac{1}{\not{p} + \not{k} - m} - \frac{1}{\not{p} + \not{k} - m} (\not{p}' - \not{p}) \frac{1}{\not{p} + \not{k} - m} + \dots. \quad (11.18\mathrm{b})$$

The first term of this expansion leads to the logarithmic divergence of the loop integral for large k. In contrast, the remainder of the expansion that vanishes for $p' - p = q \to 0$ contains additional powers of $1/k$ and is thus convergent. Hence the UV divergence is contained solely in the on-shell part of the vertex correction, while the function $\Gamma^\mu_{\mathrm{off}}(p, p') = \Gamma^\mu(p, p') - \Gamma^\mu(p, p)$ is well-behaved. Moreover, we learn from Eq. (11.14a) that the divergence is confined to $F_1(0)$, while $F_2(0)$ is finite. This is good news: the divergence is only connected to a quantity already present in the classical Lagrangian, the electric charge. Thus we can predict the function $\Gamma^\mu(p, p')$ for all values $p' \neq p$, after we have renormalised the electric charge in the limit of zero momentum transfer.

We now calculate the vertex function (11.15) explicitly. We set

$$\Gamma^\mu(q) = -\mathrm{i}e^2 \int \frac{\mathrm{d}^4 k}{(2\pi)^4} \frac{\mathcal{N}^\mu(k)}{[(p'+k)^2 - m^2]\,[(p+k)^2 - m^2]\,k^2} \tag{11.19}$$

with

$$\mathcal{N}^\mu = \gamma^\nu(\not{p}' + \not{k} + m)\gamma^\mu(\not{p} + \not{k} + m)\gamma_\nu . \tag{11.20}$$

Then we combine the propagators introducing as Feynman parameter integrals

$$\frac{1}{xyz} = 2 \int_0^1 \mathrm{d}\alpha \int_0^{1-\alpha} \mathrm{d}\beta \, \frac{1}{[z + \alpha(x-z) + \beta(y-z)]^3} = 2 \int_0^1 \mathrm{d}\alpha \int_0^{1-\alpha} \mathrm{d}\beta \, \frac{1}{D}. \tag{11.21}$$

Setting $z = k^2$, we obtain

$$D = \{k^2 + \alpha[(p'+k)^2 - m^2 - k^2] + \beta[(p+k)^2 - m^2 - k^2]\}^3. \tag{11.22}$$

The complete calculation of the vertex function (11.15) for arbitrary off-shell momenta is already quite cumbersome. In order to shorten the calculation we restrict ourselves therefore to the part contributing to the magnetic form factor $F_2(0)$. Because of

$$\Lambda^\mu(p,p') = \left[F_1(q^2) + F_2(q^2)\right]\gamma^\mu - F_2(q^2)\frac{(p'+p)^\mu}{2m} \tag{11.23}$$

we can simplify the calculation of $\mathcal{N}^\mu(k)$, throwing away all terms proportional to γ^μ which do not contribute to the magnetic moment. This also justifies why we can neglect diagrams with self-energy corrections in the external lines. Moreover, we can consider the limit that the electrons are on-shell and the momentum transfer to the photon vanishes.

Using the on-shell condition, $p^2 = p'^2 = m^2$, the two square brackets in D simplify to $2p' \cdot k$ and $2p \cdot k$, respectively,

$$D = \left\{k^2 + 2k \cdot (\alpha p' + \beta p)\right\}^3. \tag{11.24}$$

Next we eliminate the term linear in k completing the square,

$$D = \left\{\underbrace{(k + \alpha p' + \beta p)^2}_{\ell} - (\alpha p' + \beta p)^2\right\}^3 = \left\{\ell^2 - (\alpha^2 m^2 + \beta^2 m^2 + 2\alpha\beta p' \cdot p)\right\}^3. \tag{11.25}$$

Since the momentum transfer to the photon vanishes, $q^2 = 2m^2 - 2p' \cdot p \to 0$, we can replace $p' \cdot p \to m^2$ and obtain as final result for the denominator

$$D = \left\{\ell^2 - (\alpha + \beta)^2 m^2\right\}^3. \tag{11.26}$$

Now we move on to the evaluation of the numerator $\mathcal{N}^\mu(k)$. Performing the change of our integration variable from $k = \ell - (\alpha p' + \beta p)$ to ℓ, the numerator becomes

$$\mathcal{N}^\mu(\ell) = \gamma^\nu(\not{P}' + \not{\ell} + m)\gamma^\mu(\not{P} + \not{\ell} + m)\gamma_\nu \tag{11.27}$$

with $\not{P}' \equiv (1-\alpha)\not{p}' - \beta\not{p}$ and $\not{P} \equiv (1-\beta)\not{p} - \alpha\not{p}'$. Multiplying out the two brackets and ordering the result according to powers of m, we observe first that the term $\propto m^2$

leads to $\propto \gamma^\mu$ and thus does not contribute to $F_2(0)$. Next we split further the term linear in m according to powers of ℓ. The term linear in ℓ vanishes after integration, while the term $m\ell^0$ results in

$$m(\gamma^\nu \slashed{P}'\gamma^\mu\gamma_\nu + \gamma^\nu\gamma^\mu\slashed{P}\gamma_\nu) = 4m(P'^\mu + P^\mu) = 4m[(1 - 2\alpha)p'^\mu + (1 - 2\beta)p^\mu]. \quad (11.28)$$

Using the symmetry in the integration variables α and β, we can rewrite this expression as

$$\rightarrow 4m[(1 - \alpha - \beta)(p'^\mu + p^\mu)]. \quad (11.29)$$

We split the m^0 term in the same way according to the powers of ℓ. The $m^0\ell^2$ term gives a γ^μ term, the $m^0\ell$ vanishes after integration, and the $m^0\ell^0$ gives, after some work,

$$\gamma^\nu \slashed{P}'\gamma^\mu\slashed{P}\gamma_\nu \rightarrow 2m[\alpha(1 - \alpha) + \beta(1 - \beta)](p' + p)^\mu. \quad (11.30)$$

Finally, the m^0 term contributes to the anomalous magnetic moment

$$\rightarrow -2m(p' + p)^\mu[2(1 - \alpha)(1 - \beta)]. \quad (11.31)$$

Combining all terms, we find

$$\begin{aligned}
\mathcal{N}^\mu &= 4m(1 - \alpha - \beta)(p' + p)^\mu + 2m[\alpha(1 - \alpha) + \beta(1 - \beta)](p' + p)^\mu \\
&\quad - 4m(1 - \alpha)(1 - \beta)(p' + p)^\mu = \\
&= 2m[(1 - \alpha - \beta)(\alpha + \beta)](p' + p)^\mu. \quad (11.32)
\end{aligned}$$

Thus

$$\Gamma_2^\mu(0) = -2ie^2 \int \mathrm{d}\alpha\mathrm{d}\beta \int \frac{\mathrm{d}^{2\omega}\ell}{(2\pi)^{2\omega}} \frac{\mathcal{N}^\mu}{[\ell^2 - (\alpha + \beta)^2 m^2]^3}, \quad (11.33)$$

where the subscript 2 indicates that we account only for the contribution to the anomalous magnetic moment. We expressed also the loop integral in 2ω dimensions, such that we can apply the general formula derived in appendix 4.A. Using Eq. (4.104) for $I(\omega, a)$ with $\omega = 2$ and $a = 3$,

$$I(2, 3) = -\frac{i}{32\pi^2} \frac{1}{(\alpha + \beta)^2 m^2 + i\varepsilon}, \quad (11.34)$$

we obtain as expected a finite result. As last step, we perform the integrals over the Feynman parameters α and β,

$$\int_0^1 \mathrm{d}\alpha \int_0^{1-\alpha} \mathrm{d}\beta \frac{1 - \alpha - \beta}{\alpha + \beta} = \frac{1}{2}, \quad (11.35)$$

and find thus

$$\Gamma_2^\mu(0) = -\frac{e^2}{8\pi^2} \frac{1}{2m} (p' + p)^\mu. \quad (11.36)$$

Recalling Eq. (11.23), we can identify the factor $e^2/(8\pi^2)$ with the magnetic form factor $F_2(0)$. We have thus reproduced the result of the first successful calculation of a loop correction in a QFT, performed by Schwinger, and independently by Feynman

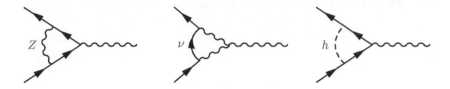

Fig. 11.2 Lowest-order electroweak corrections to the anomalous magnetic moment of fermions.

and Tomonaga, in 1948, $F_2(0) = \alpha/(2\pi)$. Together with Bethe's previous estimate of the Lamb shift in the hydrogen energy spectrum, this stimulated the view that a consistent renormalisation of QED is possible.

The currently most precise experimental value for the electron anomalous magnetic moment $a_e \equiv F_2(0)$ is

$$a_e^{\text{exp}} = 0.001\,159\,652\,180\,73 \pm 2.4 \times 10^{-10}. \tag{11.37}$$

The calculation of the universal (i.e. common to all charged leptons) QED contribution has been completed up to fourth order. An estimate of the dominant fifth-order contribution also exists,

$$\begin{aligned}
a_\ell^{\text{uni}} &= 0.5\ \left(\frac{\alpha}{\pi}\right) - 0.328\,478\,965\,579\,193\,78\ldots \left(\frac{\alpha}{\pi}\right)^2 \\
&+ 1.181\,241\,456\,587\ldots \left(\frac{\alpha}{\pi}\right)^3 - 1.9144(35)\left(\frac{\alpha}{\pi}\right)^4 + 0.0(4.6)\left(\frac{\alpha}{\pi}\right)^5 \\
&= 0.001\,159\,652\,176\,30(43)(10)(31)\cdots. \tag{11.38}
\end{aligned}$$

The three errors given in round brackets are the error from the uncertainty in α, the numerical uncertainty of the α^4 coefficient and the error estimated for the missing higher-order terms (Jegerlehner, 2007). Comparing the measured value and the prediction using QED we find an extremely good agreement. First of all, this is strong support that the methods of perturbative QFT we developed so far can be successfully applied to weakly coupled theories as QED. Second, it means that additional contributions to the anomalous magnetic moment of the electron have to be tiny.

Electroweak and other corrections. The lowest-order electroweak corrections to the anomalous magnetic moment contain in the loop virtual gauge bosons ($W^\pm$, Z) or a Higgs boson h and are shown in Fig. 11.2. We will consider the electroweak theory describing these diagrams later; for the present discussion it is sufficient to know that the weak coupling constant is $g \simeq 0.6$ and that the scalar and weak gauge bosons are much heavier than leptons, $M \gg m$.

The first diagram corresponds schematically to the expression

$$\sim g^2 \int \frac{\mathrm{d}^4 k}{(2\pi)^4} \frac{1}{k^2 - M^2} \frac{A(m^2, k)}{[(p-k)^2 - m^2]^2}. \tag{11.39}$$

As in QED, this integral has to be finite and we expect that it is dominated by momenta up to the mass M of the gauge bosons, $k \lesssim M$. Therefore its value should be proportional to $g^2 m^2/M^2$ (multiplied by a possible logarithm $\ln(M^2/m^2)$) and electroweak corrections to the anomalous magnetic moment of the electron are suppressed by a factor $(m/M)^2 \sim 10^{-10}$ compared to the QED contribution. The property that the contribution of virtual heavy particles to loop processes is suppressed in the limit $|q^2| \ll M^2$ is called "decoupling". Note the difference to the case of the mass of a scalar particle or the cosmological constant. In these examples, the loop corrections are infinite and we cannot predict these quantities. In contrast, the anomalous magnetic moment is finite but, as we include loop momenta up to infinity, depends in principal on all particles coupling to the electron, even if they are arbitrarily heavy. Only if these heavy particles "decouple", we can calculate a_e without knowing, for example, the physics at the Planck scale. Thus the decoupling property is a necessary ingredient of any reasonable theory of physics, otherwise no predictions would be possible before knowing the "theory of everything".

Clearly the contribution of heavy particles (either electroweak gauge and Higgs bosons or other not yet discovered particles) is more visible in the anomalous magnetic moment of the muon than of the electron. Moreover, a relativistic muon lives long enough that a measurement of its magnetic moment is feasible. This is one example of how radiative corrections (here evaluated at $q^2 = 0$) are sensitive to physics at higher scales M. If an observable can be measured and calculated with high enough precision, one can be sensitive to suppressed corrections of order $g^2 m^2/M^2$. Other examples are rare processes like $\mu \to e + \gamma$ or $B_s \to \mu^+ \mu^-$: these processes are suppressed by a specific property of the SM which one does not expect to hold in general. The achieved precision in measuring and calculating such processes is high enough to probe generically scales of $M \sim 100\,\text{TeV}$, that is, much higher than the mass scales that can be probed directly at current accelerators as LHC.

Finite versus divergent parts of loop corrections. We found that the vertex correction could be split into two parts

$$\Lambda^\mu(p,p') = F_1(q^2)\gamma^\mu + F_2(q^2)\frac{i\sigma^{\mu\nu}q_\nu}{2m}, \tag{11.40}$$

where the form factor $F_2(q^2)$ is finite for all q^2, while the form factor $F_1(q^2)$ diverges for $q^2 \to 0$. The important observation is that $F_2(q^2)$ corresponds to a Lorentz structure that is not present in the original Lagrangian of QED. This suggests that we can require from a "nice" theory that

- all UV divergences are connected to structures contained in the original Lagrangian, all new structures are finite. The basic divergent structures are also called "primitive divergent graphs".

- If there are no anomalies, then loop corrections respect the original (classical) symmetries. Thus, for example, the photon propagator should be at all orders transverse, respecting gauge invariance. We will see that as consequence the high-energy behaviour of the theory improves.

In such a case, we are able to hide all UV divergences in a renormalisation of the original parameters of the Lagrange density.

11.3 Power counting and renormalisability

We try to make the requirements on a "nice" theory a bit more precise. Let us consider the set of $\lambda\phi^n$ theories in $d = 4$ spacetime dimensions and check which graphs are divergent. We define the superficial degree D of divergence of a Feynman graph as the difference between the number of loop momenta in the numerator and denominator of a Feynman graph. We can restrict our analysis to those diagrams called 1P irreducible (1PI) which cannot be disconnected by cutting an internal line. All 1P-reducible diagrams can be decomposed into 1PI diagrams which do not contain common loop integrals and can be therefore analysed separately. Moreover, we are only interested in the loop integration and define therefore the *1PI Green functions*[1] as graphs where the propagators on the external lines were stripped off. In $d = 4$ spacetime dimensions, the superficial degree D of divergence of a 1PI Feynman graph is thus

$$D = 4L - 2I, \tag{11.41}$$

where L is the number of independent loop momenta and I the number of internal lines. The former contributes a factor $\mathrm{d}^4 p$, while the latter corresponds to a scalar propagator with $1/(p^2 - m^2) \sim 1/p^2$ for $p \to \infty$.

Momentum conservation at each vertex leads for an 1PI-diagram to

$$L = I - (V - 1), \tag{11.42}$$

where V is the number of vertices and the -1 takes into account the delta function leading to overall momentum conservation. The latter constrains only the external not the loop momenta. Thus

$$D = 2I - 4V + 4. \tag{11.43}$$

Each vertex connects n lines and any internal line reduces the number of external lines by two. Therefore the number E of external lines is given by

$$E = nV - 2I. \tag{11.44}$$

As result, we can express the superficial degree D by the order of perturbation theory (V), the number of external lines E and the degree n of the interaction polynomial ϕ^n,

$$D = (n - 4)V + 4 - E. \tag{11.45}$$

From this expression, we see that

- for $n < 4$, the coefficient of V is negative. Therefore only a finite number of terms in the perturbative expansion are infinite. Such a theory is called super-renormalisable, the corresponding terms in the Hamiltonian are also called relevant.
- For $n = 4$, we find $D = 4 - E$. Thus the degree of divergence is independent of the order of perturbation theory being only determined by the number of external

---

[1]Similar to their relatives, the (dis-) connected Green functions, also the 1PI Green functions can be derived from a generating functional which we will introduce in the next chapter.

Fig. 11.3 Primitive divergent diagrams in QED (without vacuum diagrams).

lines. Such theories contain an infinite number of divergent graphs, but they all correspond to a finite number of divergent structures—the so-called primitive divergent graphs. These interactions are also called marginal and are candidates for a renormalisable theory.

• Finally, for $n > 4$ the degree of divergence increases with the order of perturbation theory. As result, there exists an infinite number of divergent structures, and increasing the order of perturbation theory requires more and more input parameter to be determined experimentally. Such a theory is called non-renormalisable, the interaction irrelevant.

In particular, the $\lambda\phi^4$ theory as an example of a renormalisable theory has only three divergent structures: i) the case $E = 0$ and $D = 4$ contributes to the vacuum energy, ii) the case $E = 2$ and $D = 2$ corresponding to the self-energy, and iii) the case $E = 4$ and $D = 0$ (that is a logarithmic divergence) to the four-point function. As we saw in chapter 3, the three primitive divergent diagrams of the $\lambda\phi^4$ theory correspond to the following physical effects. Vacuum bubbles renormalise the cosmological constant. The effect of self-energy insertions is twofold: inserted in external lines it renormalises the field while self-energy corrections in internal propagators lead to a renormalisation of its mass. The vertex correction finally renormalises the coupling strength λ.

Let us move to the case of QED. Repeating the discussion, we obtain the analogue to Eq. (11.45), but accounting now for the different dimension of fermionic and bosonic fields,

$$D = 4 - B - \frac{3}{2}F, \tag{11.46}$$

where B and F count the number of external bosonic and fermionic lines, respectively. There are six different superficially divergent primitive graphs in QED shown in Fig. 11.3. The photon and the fermionic contribution to the cosmological constant ($D = 4$), the vacuum polarisation ($D = 2$), the fermion self-energy ($D = 1$), the vertex correction ($D = 0$) and light-by-light scattering ($D = 0$). Recall that Furry's theorem implies that loops with a an odd number of fermion propagator vanish in QED. Therefore we have not included in our list of primitive divergent graphs of QED the tadpole ($B = 1$ and $D = 3$) and the "photon-splitting" graph ($B = 3$ and $D = 1$).

In a theory with symmetries such as a gauge theory, the true degree of divergence can be smaller than the superficial one. For instance, light-by-light scattering corresponds to a term $\mathcal{L} \sim A^4$ that violates gauge invariance. Thus either the gauge symmetry is violated by quantum corrections or such a term is finite.

Because of the correspondence of the dimension of a field and the power of its propagator, we can connect the superficial degree of divergence of a graph to the dimension of the coupling constants at its vertices. The superficial degree $D(G)$ of divergence of a graph is connected to the one of its vertices D_v by

$$D(G) - 4 = \sum_v (D_v - 4) \tag{11.47}$$

which in turn depends as

$$D_v = \delta_v + \frac{3}{2} f_v + b_v = 4 - [g_v] \tag{11.48}$$

on the dimension of the coupling constant g at the vertex v. Here, f_v and b_v are the number of fermion and boson fields at the vertex, while δ_v counts the number of derivatives. Thus the dimension of the coupling constant plays a crucial role deciding if a certain theory is "nice" in the naive sense defined earlier. Clearly $D = 0$ or $[g] = m^0$ is the borderline case:

- If at least one coupling constant has a negative mass dimension, $[g] < 0$ and $D_v > 4$, the theory is non-renormalisable. Examples are the Fermi theory of weak interactions, $[G_F] = m^{-2}$ and gravity, $[G_N] = m^{-2}$.
- If all coupling constants have positive mass dimension, $[g] > 0$ and $D_v < 4$, the theory is super-renormalisable. An example is the $\lambda\phi^3$ theory in $D = 4$ with $[\lambda] = m^0$.
- The remaining cases, with all $[g_i] = 0$, are candidates for renormalisable theories. Examples are Yukawa interactions, $\lambda\phi^4$, Yang–Mills theories that are unbroken (QED and QCD) or broken by the Higgs mechanism (electroweak interactions).

Theories with massive bosons. We have assumed that bosonic propagators behave as $\propto 1/k^2$ for large (Euclidean) momenta k. This is true both for massive and massless scalars, while it holds only for massless particles with spin $s \geq 1$. As we have seen, the massless spin-1 and spin-2 propagators in the R_ξ gauge decrease like $\propto 1/k^2$ for large k. In contrast, the massive spin-1 propagator behaves as $D_F^{\mu\nu}(k) \propto \text{const.}$ Thus the divergences in loop diagrams are more severe for massive vector particles than for massless ones. For a massive bosonic field of spin s, the polarisation tensors contains s tensor products of $k_\mu k_\nu$ and therefore its propagator scales as $D_{\mu_1, \cdots, \mu_s; \nu_1, \cdots, \nu_s}(k) \propto k^{2s-2}$. This implies that the divergences of loop diagrams aggravate for higher spin fields. In particular, inserting additional massive propagators into a loop graph does not improve its convergence and thus a theory with massive $s > 0$ particles contains an infinite number of divergent diagrams at each loop order. Including an explicit mass term for gauge bosons leads therefore to a non-renormalisable theory. A solution to this problem is the introduction of gauge boson masses via the Higgs mechanism, which we will introduce in chapter 13.3. Combined with our finding that interacting theories of massless bosons are only possible for $s \leq 2$, we can conclude that elementary particles should have spin $s \leq 2$.

11.4 Renormalisation of the $\lambda\phi^4$ theory

We have argued that a theory with dimensionless coupling constant is renormalisable. In this case a multiplicative shift of the parameters contained in the classical

Lagrangian is sufficient to obtain finite Green functions. The simplest theory of this type in $d = 4$ is the $\lambda \phi^4$ theory for which we will discuss now the renormalisation procedure at one loop level. As a starter, we examine the general structure of the UV divergences.

11.4.1   Structure of the divergences

We learnt that the degree of divergence decreases increasing the number of external lines, since the number of propagators increases. The same effect has taking derivatives w.r.t. external momenta p,

$$\frac{\partial}{\partial \not{p}} \int \frac{\mathrm{d}^4 k}{(2\pi)^4} \frac{1}{\not{k} + \not{p} - m} = - \int \frac{\mathrm{d}^4 k}{(2\pi)^4} \frac{1}{(\not{k} + \not{p} - m)^2} \,.$$

This means that

1. we can Taylor expand loop integrals, confining the divergences in the lowest-order terms. Choosing for example $p = 0$ as expansion point in the fermion self-energy,

$$\Sigma(p) = A_0 + A_1 \not{p} + A_2 p^2 + \ldots \qquad \text{with} \qquad A_n = \frac{1}{n!} \frac{\partial^n}{\partial \not{p}^n} \Sigma(p),$$

we know that A_0 is (superficially) linear divergent. Thus A_1 can be maximally logarithmically divergent, while all other coefficients A_n are finite.
2. We could choose a different expansion point, leading to different renormalisation conditions (within the same regularisation scheme).
3. The divergences can be subtracted by local operators, that is by polynomials of the fields and their derivatives. These terms called counter-terms have for a renormalisable theory the same structure as the terms present in the classical Lagrangian. For instance, the linear divergent term A_0 can be associated to a counter-term $\delta A_0 \bar{\psi} \psi$, while $A_1 \not{p}$ corresponds to the counter-term $\delta A_1 \bar{\psi} \not{\partial} \psi$.

It is easy to show that the counter-terms are local operators at the one-loop level, where diagrams contain only one integration variable. Any loop integral $I(p)$ with superficial degree of divergence $n - 1$ becomes finite after taking n derivatives w.r.t. an external momentum p. Using a cut-off Λ as regulator, this implies that in

$$\frac{\partial^n}{\partial p^n} I(p) = f(p) + \mathcal{O}(p/\Lambda) \tag{11.49}$$

the function $f(p)$ is finite and independent of Λ, while the remainder vanishes in the limit $\Lambda \to \infty$. Integrating this expression n times we obtain

$$I(p) = F(p) + P_n(p) + \mathcal{O}(p/\Lambda), \tag{11.50}$$

where $F(p)$ is also finite and independent of Λ. The function $P_n(p)$ is an nth-order polynomial containing the integration constants. Since $F(p)$ is finite, $P_n(p)$ comprises all divergences. They are therefore the coefficients of polynomials in the external momentum p and can be subtracted by local operators, as we claimed. This argument shows also that all non-trivial analytical structures like cuts have to be contained in

$F(p)$. Moreover, choosing a different regularisation scheme or point leads to the same $f(p)$ in (11.49), and thus all the scheme dependence is contained in the polynomial $P_n(p)$. As a result, the differences caused by different schemes reside only in local terms which are absorbed in the renormalisation of the parameters.

Going to higher loop orders, non-local terms as, for example, $\ln(p^2/\mu^2)$ can be generated by sub-divergences. These are divergences connected to integration regions where one or more loop momenta are finite, while the remaining ones are send to infinity. Such terms are cancelled by counter-terms determined at lower order. A sketch of why this should be true follows:

Green functions become singular for coinciding points when the convergence factor e^{-kx} in the Euclidean Green function becomes 1. In the simplest cases as $\langle 0|\phi(x')\phi(x)|0\rangle_{x'\to x}$, the infinities are eliminated by normal ordering, that is by rewriting all creation operators on the left of the annihilation operators, cf. problem 3.6. More complicated are overlapping divergences where two or more divergent loops share a propagator. Wilson suggested to expand the product of two fields as the sum of local operators O_i times coefficient functions $C_i(x-y)$ as

$$\phi(x)\phi(y) = \sum_i C_i(x-y)O_i(x),$$

where the dependence on the relative distance is carried by the coefficients and the local operators O_i are of the type $O_i(x) = \phi(x)\partial_\mu\cdots\partial^\mu\phi(x)$. For a massless scalar field, dimensional analysis dictates that $C_i(x) \propto x^{-2+d_i}$, if the local operator O_i has dimension d_i. Note that only the unity operator has a singular coefficient function $1/x^2$ corresponding to the massless scalar propagator. Similarly we can expand products of operators,

$$O_n(x)O_m(y) = \sum_i C^i_{nm}(x-y)O_i(x),$$

where now $C^i_{nm}(x) \propto x^{-d_n-d_m+d_i}$. Thus we can use this operator product expansion (or briefly "OPE") to rewrite the overlapping divergences in terms of (singular) coefficient functions and *local* operators. Moreover, the sub-divergence occurring at order k, when $p < k$ points coincide, are eliminated by the counter-terms found at order p.

Elaborating this argument in detail, one can conclude that non-local terms due to overlapping divergences are cancelled by the counter-terms found at lower order. We will see how this works in practice in the next section, when we calculate the vacuum energy at two-loop.

11.4.2 The $\lambda\phi^4$ theory at $\mathcal{O}(\lambda)$

There are two equivalent ways to perform perturbative renormalisation. In the one called often "conventional" perturbation theory we use the "bare" (unrenormalised) parameters in the Lagrangian,

$$\mathscr{L} = \mathscr{L}_0 + \mathscr{L}_{\text{int}} = \frac{1}{2}(\partial_\mu\phi_0)^2 - \frac{1}{2}m_0^2\phi_0^2 - \frac{\lambda_0}{4!}\,\phi_0^4. \tag{11.51}$$

Then we introduce a renormalised field $\phi_R = Z_\phi^{-1/2}\phi_0$ and choose the parameters Z_ϕ, m_0 and λ_0 as functions of the regularisation parameter $(\varepsilon, \Lambda, \dots)$ such that the field ϕ_R has finite Green functions. In the following, we discuss the renormalisation procedure at the one-loop level for the Green functions of the $\lambda\phi^4$ theory in this scheme. Since any 1P-reducible diagram can be decomposed into 1PI diagrams which do not contain common loop integrals, we can restrict our analysis again to 1PI Green functions.

Mass and wave-function renormalisation.   We defined the exact or full propagator $i\Delta_F(x_1, x_2)$ in Eq. (4.2) as the path integral average of the two fields $\phi(x_1)\phi(x_2)$. Now we want to find a definition which is useful for calculations in perturbation theory. We claim that

$$[i\Delta_F(p)]^{-1} = [i\Delta_F(p)]^{-1} - \Sigma(p) = p^2 - m^2 - \Sigma(p) - i\varepsilon \qquad (11.52)$$

is the exact propagator, where the exact self-energy $\Sigma(p)$ represents the 1PI corrections to the scalar mass, $m^2_{\text{phys}} = m^2_0 + \Sigma$. Multiplying this definition from the right with $i\Delta_F(p)$ and from the left with $i\Delta_F(p)$, the so-called Dyson equation follows,

$$i\Delta_F(p) = i\Delta_F(p) + i\Delta_F(p)\Sigma(p)i\Delta_F(p) = i\Delta_F(p)\left[1 + \Sigma(p)i\Delta_F(p)\right]. \qquad (11.53)$$

Graphically, we can express this equation as

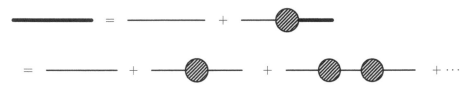

where the second line follows by iteration. Hence, $i\Delta_F(p)$ sums up the amplitudes to propagate at momentum p with $0, 1, \ldots$ self-energy Σ insertions, and corresponds therefore to the full propagator. At $\mathcal{O}(\lambda)$, we see that this relation holds comparing it to Eq. (4.33).

Next we have to show that the self-energy $\Sigma(p^2)$ is finite after renormalisation. The one-loop expression

$$-i\Sigma(p^2) = \frac{-i\lambda_0}{2} \int \frac{\mathrm{d}^4 k}{(2\pi)^4} \frac{i}{k^2 - m^2_0 + i\varepsilon} \qquad (11.54)$$

is quadratically divergent. As a particularity of the ϕ^4 theory, the p^2 dependence of the self-energy Σ shows up only at the two-loop level. We perform a Taylor expansion of $\Sigma(p^2)$ around the arbitrary point μ,

$$\Sigma(p^2) = \Sigma(\mu^2) + (p^2 - \mu^2)\Sigma'(\mu^2) + \tilde{\Sigma}(p^2), \qquad (11.55)$$

where $\Sigma(\mu^2) \propto \Lambda^2$, $\Sigma'(\mu^2) \propto \ln \Lambda$ and $\tilde{\Sigma}(p^2)$ is the finite remainder. A term linear in Λ is absent, since we cannot construct a Lorentz scalar out of p^μ. Note also $\tilde{\Sigma}(\mu^2) = 0$.

Now we insert (11.55) into (11.52),

$$\frac{i}{p^2 - m^2_0 - \Sigma(p^2) + i\varepsilon} = \frac{i}{\underbrace{p^2 - m^2_0 - \Sigma(\mu^2)}_{p^2 - \mu^2} - (p^2 - \mu^2)\Sigma'(\mu^2) - \tilde{\Sigma}(p^2) + i\varepsilon}, \qquad (11.56)$$

where we see that we can identify μ with the renormalised mass given by the pole of the propagator.

We aim at rewriting the remaining effect for $p^2 \to \mu^2 = m^2$ of the self-energy insertion, $\Sigma'(m^2)$, as a multiplicative rescaling. In this way, we could remove the divergence from the propagator by a rescaling of the field. At leading order in λ, we can write

$$\tilde{\Sigma}(p^2) = \left[1 - \Sigma'(m^2)\right] \tilde{\Sigma}(p^2) + \mathcal{O}(\lambda_0^2) \tag{11.57}$$

and thus

$$i\boldsymbol{\Delta}_F(p) = \frac{1}{1 - \Sigma'(m^2)} \frac{i}{p^2 - m^2 - \tilde{\Sigma}(p^2) + i\varepsilon} = \frac{iZ_\phi}{p^2 - m^2 - \tilde{\Sigma}(p^2) + i\varepsilon} \tag{11.58}$$

with the wave-function renormalisation constant

$$Z_\phi = \frac{1}{1 - \Sigma'(m^2)} = 1 + \Sigma'(m^2). \tag{11.59}$$

Close to the pole, the propagator is the one of a free particle with mass m,

$$i\boldsymbol{\Delta}_F(p) = \frac{iZ_\phi}{p^2 - m^2 + i\varepsilon} + \mathcal{O}(p^2 - m^2). \tag{11.60}$$

Thus the renormalisation constant Z_ϕ equals the wave-function renormalisation constant Z we had to introduce into the LSZ formalism to obtain correctly normalised states.

We define the renormalised field $\phi = Z_\phi^{-1/2}\phi_0$ such that the renormalised propagator

$$i\Delta_R(p) = \int \mathrm{d}^4 x\, e^{ipx} \langle 0| T\{\phi(x)\phi(0)\} |0\rangle = Z_\phi^{-1} i\Delta(p) = \frac{i}{p^2 - \mu^2 - \tilde{\Sigma}(p^2) + i\varepsilon} \tag{11.61}$$

is finite. Similarly, we define renormalised n-point functions by

$$G_R^{(n)}(x_1, \ldots, x_n) = \langle 0| T\{\phi(x_1) \cdots \phi_n(x_n)\} |0\rangle = Z_\phi^{-n/2} G_0^{(n)}(x_1, \ldots, x_n). \tag{11.62}$$

Since the 1PI n-point Green functions miss n field renormalisation constants compared to connected n-point Green functions, the connection between renormalised and bare 1PI n-point functions is given by

$$\Gamma_R^{(n)}(x_1, \ldots, x_n) = Z_\phi^{n/2}\, \Gamma_0^{(n)}(x_1, \ldots, x_n). \tag{11.63}$$

Coupling constant renormalisation. We can choose an arbitrary point inside the kinematical region, $s + t + u = 4\mu^2$ and $s \geq 4\mu^2$, to connect the coupling to a physical measurement at this point. For our convenience and less writing work, we choose instead the symmetric point

$$s_0 = t_0 = u_0 = \frac{4\mu^2}{3}.$$

The bare four-point 1PI Green function is (see section 4.3.3)

$$\Gamma_0^{(4)}(s,t,u) = -i\lambda_0 + \Gamma(s) + \Gamma(t) + \Gamma(u), \tag{11.64}$$

the renormalised four-point function at (s_0, t_0, u_0) is

$$\Gamma_R^{(4)}(s_0, t_0, u_0) = -i\lambda. \tag{11.65}$$

Next we expand the bare 4-point function around s_0, t_0, u_0,

$$\Gamma_0^{(4)}(s,t,u) = -i\lambda_0 + 3\Gamma(s_0) + \tilde{\Gamma}(s) + \tilde{\Gamma}(t) + \tilde{\Gamma}(u) \tag{11.66}$$

where the $\tilde{\Gamma}(x)$ are finite and zero at x_0. Now we define a vertex (or coupling constant) renormalisation constant by

$$-iZ_\lambda^{-1}\lambda_0 = -i\lambda_0 + 3\Gamma(s_0) \tag{11.67}$$

Inserting this definition in (11.66) we obtain

$$\Gamma_0^{(4)}(s,t,u) = -iZ_\lambda^{-1}\lambda_0 + \tilde{\Gamma}(s) + \tilde{\Gamma}(t) + \tilde{\Gamma}(u) \tag{11.68}$$

what simplifies at the renormalisation point to

$$\Gamma_0^{(4)}(s_0, t_0, u_0) = -iZ_\lambda^{-1}\lambda_0. \tag{11.69}$$

We use now the connection between renormalised and bare Green functions,

$$\Gamma_R^{(4)}(s,t,u) = Z_\phi^2 \Gamma_0^{(4)}(s,t,u), \tag{11.70}$$

and thus

$$-i\lambda = Z_\phi^2 Z_\lambda^{-1}(-i\lambda_0). \tag{11.71}$$

The relation between the renormalised and bare coupling in the $\lambda\phi^4$ theory is thus

$$\lambda = Z_\phi^2 Z_\lambda^{-1}\lambda_0. \tag{11.72}$$

Now we have to show that $\Gamma_R^{(4)}(s,t,u)$ is finite. Inserting (11.68) into (11.70), we find

$$\begin{aligned}
\Gamma_R^{(4)}(s,t,u) &= -iZ_\phi^2 Z_\lambda^{-1}\lambda_0 + Z_\phi^2[\tilde{\Gamma}(s) + \tilde{\Gamma}(t) + \tilde{\Gamma}(u)] \\
&= -i\lambda + Z_\phi^2[\tilde{\Gamma}(s) + \tilde{\Gamma}(t) + \tilde{\Gamma}(u)].
\end{aligned} \tag{11.73}$$

Since $Z_\phi = 1 + \mathcal{O}(\lambda^2)$ and $\tilde{\Gamma} = \mathcal{O}(\lambda^2)$, this is equivalent to

$$\Gamma_R^{(4)}(s,t,u) = -i\lambda + [\tilde{\Gamma}(s) + \tilde{\Gamma}(t) + \tilde{\Gamma}(u)] + \mathcal{O}(\lambda^3) \tag{11.74}$$

consisting only of finite expressions. This completes the proof that at one-loop order all Green functions in the $\lambda\phi^4$ theory are finite, renormalising the field ϕ, its mass and coupling constant.

Renormalised perturbation theory. In this approach, we first rescale the bare field in the classical Lagrangian by $\phi_0 = Z_\phi^{1/2}\phi$, obtaining

$$\mathscr{L} = \frac{1}{2}Z_\phi(\partial_\mu\phi)^2 - \frac{1}{2}Z_\phi m_0^2\phi^2 - \frac{\lambda_0}{4!}\,Z_\phi^2\phi^4. \tag{11.75}$$

Next we introduce the renormalised mass and coupling by $m_0^2 = Z_\phi^{-1}Z_m m^2$ and $\lambda_0 = Z_\phi^{-2}Z_\lambda\lambda$, arriving at

$$\mathscr{L} = \frac{1}{2}Z_\phi(\partial_\mu\phi)^2 - \frac{1}{2}Z_m m^2\phi^2 - \frac{\lambda}{4!}\,Z_\lambda\phi^4. \tag{11.76}$$

The renormalisation constants Z_i vanish at tree-level and allow for a perturbative expansion. Setting $Z_i = 1 + \delta_i$, we can split the Lagrangian into

$$\begin{aligned}
\mathscr{L} = {}& \frac{1}{2}(\partial_\mu\phi)^2 - \frac{1}{2}m^2\phi^2 - \frac{\lambda}{4!}\,\phi^4 \\
& + \frac{1}{2}\delta_\phi(\partial_\mu\phi)^2 - \frac{1}{2}\delta_m m^2\phi^2 - \frac{\lambda}{4!}\,\delta_\lambda\phi^4,
\end{aligned} \tag{11.77}$$

where the first line contains only renormalised quantities. The terms in the second line contain the divergent renormalisation constants, and this part is called the counter-term Lagrangian $\mathscr{L}_{\mathrm{ct}}$. An advantage of renormalised perturbation theory is that now the expansion parameter is the renormalised coupling λ. Treating $\mathscr{L}_{\mathrm{ct}}$ as a perturbation, $Z_i = 1 + \sum_{n=1}^\infty \delta_i^{(n)}$, we obtain in momentum space as additional Feynman vertices

$$\underline{\qquad\otimes\qquad} \qquad\qquad = \mathrm{i}[\delta_\phi p^2 - \delta_m m^2] \tag{11.78}$$

and

$$\times\!\!\!\!\otimes\!\!\!\!\times \qquad\qquad = -\mathrm{i}\delta_\lambda\lambda. \tag{11.79}$$

Applying renormalised perturbation theory consists of the following steps:

1. Starting from (11.76) with $n = 0$, i.e. $\delta_i^{(0)} = 0$, one derives propagator and vertices.

2. One calculates 1-loop 1PI diagrams and finds the divergent parts which determine the counter-terms $\delta_i^{(1)}$ at order $\mathcal{O}(\lambda)$. Then all other one-loop diagrams can be calculated.

3. Moving to two-loops, one generates two-loop 1PI diagrams using the Lagrangian with the one-loop counter-terms $\delta_i^{(1)}$. They are used to extract the counter-terms $\delta_i^{(2)}$ at order $\mathcal{O}(\lambda^2)$.

4. The procedure is iterated moving to higher orders.

We illustrate the use of renormalised perturbation theory with the calculation of the remaining loop diagram at $\mathcal{O}(\lambda)$, the vacuum energy density. This example of a two-loop diagram shows also how sub-divergences are cancelled by counter-terms found at lower loop order. Including the vacuum energy density, the Lagrangian (11.76) becomes $\mathscr{L} \to \mathscr{L} + \rho + \delta_\rho\rho$. This example shows also that the correct expansion parameter is the number of loops, not the power of the coupling constant.

Example 11.3: Vacuum energy density at two-loop. According to step 2, we should determine first the counter-terms in $\mathscr{L}_{\mathrm{ct}}^{(1)}$ from the already calculated one-loop 1PI diagrams. We start collecting the relevant results derived in chapter 4,

$$\rho^{(1)} = -\frac{m^4}{4(4\pi)^2}\left[\frac{1}{\varepsilon}+\ln(\mu^2/m^2)\right], \qquad \delta_\rho^{(1)} = \frac{1}{(4\pi)^2}\frac{1}{\varepsilon}, \quad \text{and} \quad \delta_m^{(1)} = \frac{\lambda}{2}\frac{1}{(4\pi)^2}\frac{1}{\varepsilon},$$

(11.80)

where we re-scaled $\mu^2 \to 4\pi\mu^2\exp(-\gamma)$. Inserting the one-loop self-energy into the two-loop expression $\rho_a^{(2)} = \lambda/8\,\Delta_F^2(0)$ results in

$$\rho_a^{(2)} = \frac{\lambda}{8}\Delta_F^2(0) = \frac{\lambda}{8}\frac{m^4}{(4\pi)^4}\left[\frac{1}{\varepsilon^2}+\frac{2}{\varepsilon}\ln\left(\frac{\mu^2}{m^2}\right)+\ln^2\left(\frac{\mu^2}{m^2}\right)\right].$$

(11.81)

Here a mixed term, combining a pole term $1/\varepsilon$ and a logarithm with argument μ^2/m^2, has appeared. In general, the logarithm will depend both on the masses of the loop particles and the external momenta p, $\ln[f(\mu^2/m^2, \mu^2/p^2)]$. Such terms cannot be subtracted by local polynomials in the momenta p as counter-terms. In a renormalisable theory, they have to therefore be cancelled by counter-terms determined at lower loop order.

In our concrete case, we have to add only the Feynman diagram generated by the counter-term $-\frac{1}{2}\delta_m^{(1)}m^2\phi^2$, since $\delta_{\bar{\lambda}}$ contributes only from the two-loop level on. This interaction generates at $\mathcal{O}(\lambda)$ the following contribution to the vacuum energy density

$$\rho_b^{(2)} = \ \text{} \ = \frac{1}{2}\delta_m^{(1)}m^2\Delta_F(0) = -\frac{\lambda}{8}\frac{m^4}{(4\pi)^4}\frac{2}{\varepsilon}\left[\frac{1}{\varepsilon}+\ln(\mu^2/m^2)\right].$$

(11.82)

Combining the two contributions, the mixed terms disappear as expected and the remaining $1/\varepsilon^2$ pole can be subtracted by the counter-term

$$\delta_\rho^{(2)} = \frac{\lambda}{8}\frac{1}{(4\pi)^4}\frac{1}{\varepsilon^2}.$$

(11.83)

Thus the two-loop contribution to the vacuum energy density is

$$\rho^{(2)} = \frac{m^4}{(4\pi)^4}\ln^2(\mu^2/m^2) = \left(\frac{m^{(1)}(\mu)}{4\pi}\right)^4.$$

(11.84)

Summary

Using a power-counting argument for the asymptotic behaviour of the free Green functions, we singled out theories with dimensionless coupling constants. Such theories with marginal interactions are renormalisable, that is they are theories with a finite number of primitive divergent diagrams. In this case, the multiplicative renormalisation of the finite number of parameters contained in the classical (effective) Lagrangian is sufficient to obtain finite Green functions at any order perturbation theory.

Further reading. The renormalisation of the $\lambda\phi^4$ theory at the two-loop level is performed, for example, by Pokorski (1987). Non-renormalisable theories are discussed by Schwartz (2013). Jegerlehner (2007) reviews the status of electroweak precision calculations.

Problems

11.1 Asymptotic expansion. Plot $Z_N(\lambda)/Z(0)$ from example 1 for the first few N as function of λ.

11.2 Dimensional regularisation. Regularisation methods modify the short-distance behaviour. Discuss how DR modifies the typical integral $I_0(\omega, \alpha)$ (cf. 4.93) performing the integral over the -2ε extra dimensions.

11.3 Dimensional regularisation and γ^5. Show that the properties $\text{tr}(\gamma^\mu\gamma^\nu) = d\eta^{\mu\nu}$, $\text{tr}(1) = 4$ and $\{\gamma^\mu, \gamma^5\} = 0$ lead to an inconsistency in $d \neq 4$ dimensions. (Hint: Consider first $d\,\text{tr}[\gamma^5] = -d\,\text{tr}[\gamma^5]$, then $d\,\text{tr}[\gamma^5\gamma_\alpha\gamma_\beta] = (4-d)\text{tr}[\gamma^5\gamma_\alpha\gamma_\beta]$ and finally $(4-d)\text{tr}[\gamma^5\gamma_\alpha\gamma_\beta\gamma_\rho\gamma_\sigma] = 0$.]

11.4 Superficial degree of divergence. Generalize the formula $D = (n-4)V + 4 - E$ for the superficial degree of divergence D of a ϕ^n theory in $d = 4$ to arbitrary d. Find the renormalisable cases.

11.5 Expanding $(A+B)^{-1}$. Show that the identity

$$\frac{1}{A+B} = \frac{1}{A} - \frac{1}{A}B\frac{1}{A} + \frac{1}{A}B\frac{1}{A}B\frac{1}{A} - \dots$$

holds for small B.

11.6 g_s-factor of gauge bosons. Derive the tree-level value of the g_s-factor for Yang–Mills gauge bosons from the non-abelian Maxwell equations.

11.7 Effective vertex for $\mu \to e\gamma$. Derive the effective vertex Γ^μ for the transition $\mu \to e\gamma$ where all three particles are on-shell and the process violates parity. Use current conservation and that the photon is on-shell to show that $\Gamma^\mu = iq^\nu\sigma_{\mu\nu}([A(q^2) + B(q^2)\gamma^5])$, where q^ν is four-momentum of the photon.

11.8 Primitive divergent diagrams of scalar QED. Find the basic primitive diagrams of scalar QED,

$$\mathscr{L} = \frac{1}{2}(D_\mu\phi)^\dagger D^\mu\phi - \frac{1}{2}m^2\phi^\dagger\phi - \frac{1}{4}F_{\mu\nu}F^{\mu\nu}$$

and their superficial degree of divergence.

11.9 Comparison of cut-off and DR. Recalculate the three basic primitive diagrams of a scalar $\lambda\phi^4$ theory using as regularisation a cut-off Λ. Find the correspondence between the coefficients of poles in DR and divergent terms in Λ.

11.10 β function of the $\lambda\phi^4$ theory. The β function determines the logarithmic change of the coupling constants. a) Consider mass independent schemes in DR and show that the β function can be written as $\beta(\lambda) \equiv \mu\partial\lambda/\partial\mu = -\varepsilon\lambda - \frac{\mu}{\tilde{Z}}\frac{d\tilde{Z}}{d\mu}\lambda$ with $\tilde{Z}^{-1} = Z_\lambda^{-1}Z_\phi^2$. b.) Show that $\tilde{Z}_\lambda^{-1} = 1 - 3\lambda/(16\pi^2\varepsilon)$ in one loop approximation and find the β-function. c.) Up to which order is the β function scheme independent? d.) Solve the differential equation for $\lambda(\mu)$.

12
Renormalisation II: Improving perturbation theory

We continue our discussion of renormalisation, introducing first the quantum action as the generating functional of 1PI Green functions. Then we apply the developed formalism to derive the Ward identities which imply, for example, that the exact photon propagator is transverse and that the renormalisation of the electric charge is universal. Next we introduce the renormalisation group equations which describe the evolution of n-point Green functions under a change of scale. These equations suggest converting the parameters contained in the classical Lagrangian into "running parameter", summing up thereby the most important corrections of an infinite set of diagrams. Finally, we introduce in the last section a non-perturbative approach based on ideas developed in solid-state physics and the renormalisation group.

12.1 Quantum action

In the classical limit, the equation of motion $\delta S[\phi]/\delta\phi = -J$ allows us to determine the source $J(x)$ which produces a given field $\phi(x)$. Our aim is to find the quantum analogue of this classical equation. Let us recall first the definition for the generating functionals of a real scalar field,

$$Z[J] = \int \mathcal{D}\phi\, \mathrm{e}^{\mathrm{i}\{S[\phi]+\int \mathrm{d}^4x\, J(x)\phi(x)\}} = \mathrm{e}^{\mathrm{i}W[J]}. \tag{12.1}$$

Then we define the classical field $\phi_c(x)$ as $\phi_c(x) = \delta W[J]/\delta J(x)$. Performing the functional derivative in its definition, we see immediately why this definition makes sense,

$$\phi_c(x) = \frac{\delta W[J]}{\delta J(x)} = \frac{1}{\mathrm{i}Z}\frac{\delta Z[J]}{\delta J(x)} = \frac{1}{Z}\int \mathcal{D}\phi\, \phi(x) \exp \mathrm{i} \int \mathrm{d}^4y(\mathscr{L} + J\phi) \tag{12.2a}$$

$$= \frac{\langle 0|\phi(x)|0\rangle_J}{\langle 0|0\rangle_J} = \langle\phi(x)\rangle_J. \tag{12.2b}$$

Thus the classical field $\phi_c(x)$ is the vacuum expectation value of the quantum field $\phi(x)$ in the presence of the source $J(x)$. Now we define the quantum[1] action $\Gamma[\phi_c]$ as the Legendre transform of $W[J]$,

---

[1] Often $\Gamma[\phi_c]$ is called the effective or quantum effective action.

Quantum Fields–From the Hubble to the Planck Scale. Michael Kachelriess. © Michael Kachelriess 2018.
Published in 2018 by Oxford University Press. DOI 10.1093/oso/9780198802877.001.0001

$$\Gamma[\phi_c] = W[J] - \int d^4x \, J(x)\phi_c(x) \equiv W[J] - \langle J\phi \rangle, \qquad (12.3)$$

where $\phi_c(x) = \delta W[J]/\delta J(x)$ should be used to replace $J(x)$ by $\phi_c(x)$ on the RHS. We compute the functional derivative w.r.t. ϕ_c of this new quantity,

$$\frac{\delta \Gamma[\phi_c]}{\delta \phi_c(y)} = \int d^4x \, \frac{\delta W}{\delta J(x)} \frac{\delta J(x)}{\delta \phi_c(y)} - \int d^4x \, \frac{\delta J(x)}{\delta \phi_c(y)} \phi_c(x) - J(y). \qquad (12.4)$$

Using the definition $\delta W/\delta J(x) = \phi_c(x)$, the first and second term cancels and we end up as desired with

$$\frac{\delta \Gamma[\phi_c]}{\delta \phi_c(y)} = -J(y). \qquad (12.5)$$

The analogy to the classical equation of motion suggests that $\Gamma[\phi_c]$ is the quantum version of the classical action. We will show next that its tree-level diagrams contain all loop corrections induced by the usual action $S[\phi]$. It is this feature that justifies the name quantum action for $\Gamma[\phi_c]$.

Expansion in $\hbar$ as a loop expansion. In order to proceed, we perform a saddle-point expansion around the classical solution ϕ_0, given by the solution to

$$\left. \frac{\delta\{S[\phi] + \langle J\phi \rangle\}}{\delta \phi} \right|_{\phi_0} = 0 \qquad (12.6)$$

or

$$\Box \phi_0 + V'(\phi_0) = J(x). \qquad (12.7)$$

We write the field as $\phi = \phi_0 + \tilde{\phi}$, that is as a classical solution with quantum fluctuations on top. Then we can approximate the path integral in $Z = \exp\{iW/\hbar\}$ as

$$Z \simeq e^{i[S[\phi_0] + \langle J\phi_0 \rangle]/\hbar} \int \mathcal{D}\tilde{\phi} \exp\left\{ \frac{i}{\hbar} \int d^4x \frac{1}{2} \left[(\partial_\mu \tilde{\phi})^2 - V''(\phi_0)\tilde{\phi}^2 \right] \right\}. \qquad (12.8)$$

We have restored Planck's constant $\hbar$ to indicate that this saddle-point expansion can be viewed as an expansion in $\hbar$. Next we want to show that the expansion in $\hbar$ corresponds to a loop expansion. We artificially introduce a parameter a into our Lagrangian so that

$$\mathcal{L}(\phi, \partial_\mu \phi, a) = a^{-1}\mathcal{L}(\phi, \partial_\mu \phi). \qquad (12.9)$$

Let us determine the power P of a in an arbitrary 1PI Feynman graph, a^P. A propagator is the *inverse* of the quadratic form in $\mathcal{L}$ and contributes thus a positive power a, while each vertex $\propto \mathcal{L}_{\text{int}}$ adds a factor a^{-1}. The number of loops in an 1PI diagram is given by $L = I - V + 1$, cf. Eq. (11.42), where I is the number of internal lines and V is the number of vertices. Putting this together we see that

$$P = I - V = L - 1 \qquad (12.10)$$

and thus[2] the power of a gives us the number of loops. We should stress that using a loop expansion does not imply a semi-classical limit, $S \gg \hbar$. Our fictitious parameter a is not small; in fact, it is 1.

---

[2]We assume here that particle masses which carry a factor $\hbar^{-1}$ can be neglected. In few applications, as, for example, calculating quantum corrections to the Newtonian potential between two masses m_1 and m_2, this assumption is not valid.

Quantum action as generating functional for 1PI Green functions. We have now all the necessary ingredients in order to show that the tree-level graphs generated by the quantum action $\Gamma[\phi]$ correspond to the complete scattering amplitudes of the corresponding action $S[\phi]$. We compare the "true" generating functional

$$Z[J] = \int \mathcal{D}\phi \, \exp\{iS + \langle J\phi \rangle\} = e^{iW[J]}, \qquad (12.11)$$

with the functional $V_a[J]$ of a fictitious field theory whose classical action S is the quantum action $\Gamma[\phi]$ of the theory (12.11) we are interested in,

$$V_a[J] = \int \mathcal{D}\phi \, \exp\left\{\frac{i}{a}\{\Gamma[\phi] + \langle J\phi \rangle\}\right\} = e^{iU_a[J]}. \qquad (12.12)$$

Additionally, we introduced the parameter a with the same purpose as in (12.9). In the limit $a \to 0$, we can perform a saddle-point expansion and the path integral is dominated by the classical path. From (12.8), we find thus

$$\lim_{a \to 0} aU_a[J] = \Gamma[\phi] + \langle J\phi \rangle = W[J], \qquad (12.13)$$

where we used the definition of the quantum action, Eq. (12.3), in the last step. The RHS is the sum of all connected Green functions of our original theory. The LHS is the classical limit of the fictitious theory $V_a[J]$: it is the sum of all connected tree graphs generated using $\Gamma[\phi]$ as action. But a connected graph which is one-particle reducible is composed of one-particle irreducible subgraphs connected by simple propagators. Hence a connected graph corresponds to a tree-level graph with 1PI subgraphs $\Gamma^{(n)}(x_1, ..., x_n)$ as (non-local) vertices. This shows that the quantum action $\Gamma[\phi]$ is the generating functional for the 1PI Green functions. Expanding $\Gamma[\phi]$ in ϕ_c gives us thus

$$\Gamma[\phi_c] = \sum_n \frac{1}{n!} \int d^4x_1 \cdots d^4x_n \Gamma^{(n)}(x_1, ..., x_n)\phi_c(x_1) \cdots \phi_c(x_n) \qquad (12.14)$$

with $\Gamma^{(n)}$ as the one-particle-irreducible Green functions. In the following, we will need only the two- and three-point functions which we construct now explicitly.

Example 12.1: Show that $\Gamma^{(2)}$ is equal to the inverse propagator or inverse two-point function, and derive the connection of $\Gamma^{(3)}$ to the connected three-point function. We write first

$$\delta(x_1 - x_2) = \frac{\delta\phi_c(x_1)}{\delta\phi_c(x_2)} = \int d^4x \, \frac{\delta\phi_c(x_1)}{\delta J(x)} \frac{\delta J(x)}{\delta\phi_c(x_2)}$$

using the chain rule. Next we insert $\phi_c(x) = \delta W/\delta J(x)$ and $J(x) = -\delta\Gamma/\delta\phi_c(x)$ to obtain

$$\delta(x_1 - x_2) = -\int d^4x \, \frac{\delta^2 W}{\delta J(x)\delta J(x_1)} \frac{\delta^2 \Gamma}{\delta\phi_c(x)\delta\phi_c(x_2)}.$$

Setting $J = \phi_c = 0$, it follows

$$\int d^4x \, iG(x, x_1)\, \Gamma^{(2)}(x, x_2) = -\delta(x_1 - x_2) \qquad (12.15)$$

or $\Gamma^{(2)}(x_1, x_2) = iG^{-1}(x_1, x_2)$. Thus the 1PI two-point function is the inverse propagator. Taking a further derivative $\delta/\delta J(x_3)$ of this relation, we have

$$\frac{\delta^3 W}{\delta J(x_1)\delta J(x_2)\delta J(x_3)} = -\frac{\delta}{\delta J(x_3)}\left(\frac{\delta^2 \Gamma}{\delta\phi_c(x_1)\delta\phi_c(x_2)}\right)^{-1}. \qquad (12.16)$$

Differentiating $MM^{-1} = 1$, we find $dM^{-1} = -M^{-1}dMM^{-1}$ for a matrix M. Applied to (12.16), we obtain

$$\frac{\delta^3 W}{\delta J(x_1)\delta J(x_2)\delta J(x_3)} = \int d^4 y_1 d^4 y_2 G^{(2)}(y_1, x_1)G^{(2)}(y_2, x_2)\frac{\delta^3 \Gamma}{\delta\phi_c(x_1)\delta\phi_c(x_2)\delta J(x_3)}. \qquad (12.17)$$

Using the chain rule, inserting $\phi_c(x_3) = \delta W/\delta J(x_3)$ and setting $J = \phi_c = 0$ gives

$$G^{(3)}(x_1, x_2, x_3) = \int d^4 y_1 d^4 y_2 d^4 y_3 G^{(2)}(y_1, x_1)G^{(2)}(y_2, x_2)G^{(2)}(y_3, x_3)\Gamma^{(3)}(x_1, x_2, x_3). \qquad (12.18)$$

Thus the connected three-point function $G^{(3)}(x_1, x_2, x_3)$ is obtained by appending propagators to the irreducible three-point vertex function $\Gamma^{(3)}(x_1, x_2, x_3)$, one for each external leg. This generalises to all $n \geq 3$ and therefore one calls the $\Gamma^{(n)}$ also *amputated* Green functions.

12.2 Ward–Takahashi identities

We now turn to QED, the simplest case of a gauge theory. Discussing the quantum version of Noether's theorem, we have already shown that the vacuum expectation value of the electromagnetic current is conserved, $\partial_\mu \langle j^\mu \rangle = 0$. Hence loop corrections respect the classical gauge symmetry and therefore we also expect that the photon remains massless. The redundancy implied by gauge invariance leads to interrelations of Green functions. In turn, such dependencies are a necessary ingredient for the quantisation of a gauge symmetry, as we can see as follows: in QED, we define wavefunction renormalisation constants for the electron and photon as

$$\psi_0 \equiv Z_2^{1/2}\psi \quad \text{and} \quad A_0^\mu \equiv Z_3^{1/2}A^\mu. \qquad (12.19)$$

Analogous to Eq. (11.72), the electric coupling is renormalised by

$$e(\mu) = Z_2 Z_3^{1/2} Z_1^{-1} e_0, \qquad (12.20)$$

where Z_1 is the charge renormalisation constant, and Z_2 and $Z_3^{1/2}$ take into account the fact that two electron fields and one photon field enter the three-point function.

As it stands, the renormalisation condition (12.20) creates two major problems. First, the factor Z_2 will vary from fermion to fermion. For instance, the wave-function renormalisation constant of a proton includes the effects of strong interactions while the one of the electron does not. As a result, it is difficult to understand why the electric charge of an electron and a proton are renormalised such that they have the same value at $q^2 = 0$. Second, we see that the renormalised covariant derivative

$$D_\mu = \partial_\mu - i\frac{Z_1}{Z_2}e(\mu)A_\mu \tag{12.21}$$

remains only gauge-invariant if $Z_1 = Z_2$. Thus it is essential that we are able to show that $Z_1 = Z_2$ holds in suitable renormalisation schemes. Clearly, this condition would then also guarantee the universality of the electric charge. In a non-abelian theory such as QCD, where we have to ensure that the gauge coupling in all terms of Eq. (10.76) remains after renormalisation the same, several constraints of the type $Z_i = Z_j$ arise.

We will proceed in two steps. First, we will show that the photon remains massless or, more technically, that the exact photon vacuum polarisation tensor is transverse. Then we will derive the Ward–Takahashi identities using the quantum action which imply in particular the relation $Z_1 = Z_2$.

Photon propagator. We start recalling the generating functional of QED,

$$Z[J^\mu, \eta, \bar{\eta}] = \int \mathcal{D}A\mathcal{D}\bar{\psi}\mathcal{D}\psi \, \exp i \int d^4x \mathcal{L}_{\text{eff}}, \tag{12.22}$$

where J^μ is a four-vector source, η and $\bar{\eta}$ are Graßmannian sources and the effective Lagrangian $\mathcal{L}_{\text{eff}}$ is composed of a classical term $\mathcal{L}_{\text{cl}}$, a gauge-fixing term $\mathcal{L}_{\text{gf}}$ and a source term $\mathcal{L}_{\text{s}}$,

$$\mathcal{L}_{\text{cl}} = -\frac{1}{4}F_{\mu\nu}F^{\mu\nu} + i\bar{\psi}\gamma^\mu D_\mu \psi - m\bar{\psi}\psi, \tag{12.23a}$$

$$\mathcal{L}_{\text{gf}} + \mathcal{L}_{\text{s}} = -\frac{1}{2\xi}(\partial_\mu A^\mu)^2 + J^\mu A_\mu + \bar{\psi}\eta + \bar{\eta}\psi. \tag{12.23b}$$

We consider the renormalised version of the Lagrangian $\mathcal{L}_{\text{eff}}$, where the renormalised covariant derivative D_μ is given by Eq. (12.21) with $Z_1 \neq Z_2$ in general. This implies that an infinitesimal gauge transformation has the form

$$A_\mu \to A_\mu' = A_\mu - \partial_\mu \Lambda \tag{12.24a}$$

$$\psi \to \psi' = \psi - i\bar{e}\Lambda\psi \quad \text{with} \quad \bar{e} \equiv \frac{Z_1}{Z_2}e. \tag{12.24b}$$

As $\mathcal{L}_{\text{cl}}$ is gauge-invariant by construction, the variation of $\mathcal{L}_{\text{eff}}$ under an infinitesimal gauge transformation consists only of

$$\delta \int d^4x \, (\mathcal{L}_{\text{gf}} + \mathcal{L}_{\text{s}}) = \int d^4x \left[\frac{1}{\xi}(\partial_\mu A^\mu)\Box\Lambda - J^\mu\partial_\mu\Lambda + i\bar{e}\Lambda(\bar{\psi}\eta - \bar{\eta}\psi)\right]. \tag{12.25}$$

Now we integrate by parts the first term twice and the second term once, to factor out the arbitrary function Λ,

$$\delta S_{\text{eff}} = \delta \int d^4x \, \mathcal{L}_{\text{eff}} = \int d^4x \left[\frac{1}{\xi}\Box(\partial_\mu A^\mu) + \partial_\mu J^\mu + i\bar{e}(\bar{\psi}\eta - \bar{\eta}\psi)\right]\Lambda. \tag{12.26}$$

Thus the variation of the generating functional $Z[J^\mu, \eta, \bar{\eta}]$ is

$$\delta Z[J^\mu, \eta, \bar{\eta}] = \int \mathcal{D}A\mathcal{D}\bar{\psi}\mathcal{D}\psi \, \exp\left(i \int d^4x \mathcal{L}_{\text{eff}}\right) i\delta \int d^4x \mathcal{L}_{\text{eff}}. \tag{12.27}$$

The fields A_μ, $\bar{\psi}$ and ψ are, however, only integration variables in the generating functional. The gauge transformation (12.24) is thus merely a change of variables which does not affect the functional $Z[J^\mu, \eta, \bar\eta]$, since the Jacobian of this transformation is one. Thus this variation has to vanish, $\delta Z[J^\mu, \eta, \bar\eta] = 0$.

If we substitute fields by functional derivatives of their sources, the change δS_{eff} can be moved outside the functional integral. Since the function $\Lambda(x)$ is arbitrary, we can drop the integration and arrive at

$$
\begin{aligned}
0 &= \left[\frac{1}{\xi} \Box \partial_\mu \frac{\delta}{\delta J_\mu} + \partial_\mu J^\mu + i\bar{e} \left(\eta \frac{\delta}{\delta \eta} - \bar\eta \frac{\delta}{\delta \bar\eta} \right) \right] \exp\{iW\} \\
&= \frac{1}{\xi} \Box \left(\partial_\mu \frac{\delta W}{\delta J_\mu} \right) + \partial_\mu J^\mu + i\bar{e} \left(\eta \frac{\delta W}{\delta \eta} - \bar\eta \frac{\delta W}{\delta \bar\eta} \right).
\end{aligned}
\tag{12.28}
$$

Differentiating this equation with respect to J_ν and then setting the sources J^μ, η and $\bar\eta$ to zero gives us our first result,

$$
-\frac{1}{\xi} \Box \left(\partial_\mu \frac{\delta^2 W}{\delta J_\mu(x) \delta J_\nu(y)} \right) = \partial_\mu \eta^{\mu\nu} \delta(x - y).
\tag{12.29}
$$

The second derivative of W w.r.t. to the vector sources J_μ is the full photon propagator $\boldsymbol{D}^{\mu\nu}(x - y)$. If we go to momentum space, we have

$$
-\frac{1}{\xi} k^2 k_\mu \boldsymbol{D}^{\mu\nu}(k) = k^\nu.
\tag{12.30}
$$

Now we split the propagator into a transverse part and a longitudinal part as in (10.60),

$$
\boldsymbol{D}^{\mu\nu} = \boldsymbol{D}_T P_T^{\mu\nu} + \boldsymbol{D}_L P_L^{\mu\nu},
\tag{12.31}
$$

with $P_T^{\mu\nu} = \eta^{\mu\nu} - k^\mu k^\nu / k^2$ and $P_L^{\mu\nu} = k^\mu k^\nu / k^2$. Then the transverse part immediately drops out and we find

$$
-\frac{1}{\xi} k^2 k_\nu \boldsymbol{D}_L(k^2) = k_\nu.
\tag{12.32}
$$

Thus the longitudinal part of the exact propagator agrees with the longitudinal part of the tree-level propagator,

$$
\boldsymbol{D}_L(k^2) = D_L(k^2) = -\frac{\xi}{k^2}.
\tag{12.33}
$$

This implies that higher-order corrections do not affect $\boldsymbol{D}_L(k^2)$. In other words, the loop correction $\Pi_{\mu\nu}$ to the photon propagator is transverse. Since we can expand all relations as power series in the coupling constant e, this holds also at any order in perturbation theory.

Ward identity $\boldsymbol{Z_1 = Z_2}$. Let us go back to the constraint for the variation of the generating functional Z under gauge transformations, Eq. (12.28). We aim to derive identities between 1PI Green functions and therefore we want to transform it into a constraint for the quantum action Γ. If we Legendre-transform $W[J, \bar{\eta}, \eta]$ into $\Gamma[A, \bar{\psi}, \psi]$,

$$\Gamma[\psi, \bar{\psi}, A_\mu] = W[\eta, \bar{\eta}, J_\mu] - \int \mathrm{d}^4 x (J_\mu A^\mu + \bar{\psi}\eta + \bar{\eta}\psi) \,, \tag{12.34}$$

we can replace the functional derivatives of W with classical fields, and the sources with functional derivatives of Γ,

$$\frac{\delta W}{\delta J_\mu} = A^\mu, \quad \frac{\delta W}{\delta \eta} = \bar{\psi}, \quad \frac{\delta W}{\delta \bar{\eta}} = \psi$$

$$\frac{\delta \Gamma}{\delta A_\mu} = -J^\mu, \quad \frac{\delta \Gamma}{\delta \bar{\psi}} = -\eta, \quad \frac{\delta \Gamma}{\delta \psi} = -\bar{\eta} \,.$$

This transforms Eq. (12.28) into

$$\frac{1}{\xi} \Box (\partial_\mu A^\mu(x)) - \partial_\mu \frac{\delta \Gamma}{\delta A_\mu(x)} + \mathrm{i}\bar{e} \left(\psi \frac{\delta \Gamma}{\delta \psi(x)} - \bar{\psi} \frac{\delta \Gamma}{\delta \bar{\psi}(x)} \right) = 0 \,, \tag{12.35}$$

a master equation from which we can derive relations between different types of Green functions. Differentiating with respect to $\psi(x_1)$ and $\bar{\psi}(x_2)$ and then setting the fields to zero gives us the most important one of the *Ward–Takahashi* identities, relating the 1PI 3-point function $\Gamma^{(3)}(x, x_1, x_2)$ to the 2-point function $\Gamma^{(2)}(x_1, x_2)$ of fermions,

$$\partial^\mu \Gamma_\mu^{(3)}(x, x_1, x_2) = \mathrm{i}\bar{e} (\Gamma^{(2)}(x, x_1) \delta(x - x_2) - \Gamma^{(2)}(x, x_2) \delta(x - x_1)) \tag{12.36}$$

or, after Fourier transforming

$$k^\mu \Gamma_\mu^{(3)}(p, k, p + k) = \bar{e} \boldsymbol{S}_F^{-1}(p + k) - \bar{e} \boldsymbol{S}_F^{-1}(p) \,. \tag{12.37}$$

Taking the limit $k^\mu \to 0$, the identity $\Gamma_\mu^{(3)}(p, 0, p) = \bar{e} \partial_\mu \boldsymbol{S}_F^{-1}(p)$ found originally by Ward follows. The Green functions in this equation are finite, renormalised quantities and thus Z_1/Z_2 has to be finite in any consistent renormalisation scheme too. In all schemes where we identify directly $e(0)$ with the measured electric charge, the finite parts of Z_1 and Z_2 also agree and thus the measured electric charge is universal.

12.3 Vacuum polarisation

The Ward identities guarantee that the gauge couplings in QED and QCD are determined solely by the loop corrections $\Pi_{\mu\nu}$ to the photon and gluon propagator, respectively. These corrections convert the classical coupling constants into running couplings. How the running coupling change as function of the scale is one of the most characteristic properties of a quantum field theory. The aim of this section is therefore to derive the one-loop corrections to the gauge boson propagator. An interpretation of these results will then be performed in the next section.

12.3.1 Vacuum polarisation in QED

Next we calculate the one-loop correction to the photon propagator, the so-called vacuum polarisation tensor $\Pi_{\mu\nu}$. In QED, only the first diagram of Fig. 12.1 contributes.

Using the Feynman rules, we obtain for the contribution of one fermion species with mass m

$$i\Pi^{\mu\nu}(q) = -(-ie)^2 \int \frac{d^4k}{(2\pi)^4} \frac{\mathrm{tr}[\gamma^\mu i(\slashed{k} + m)\gamma^\nu i(\slashed{k} + \slashed{q} + m)]}{(k^2 - m^2)[(k+q)^2 - m^2]} = -e^2 \int \frac{d^4k}{(2\pi)^4} \frac{\mathcal{N}^{\mu\nu}}{D}.$$
(12.38)

As a warm-up, we want to confirm that the vacuum polarisation tensor respects at the one-loop level gauge invariance, as we know already from Eq. (12.33). Gauge invariance implies $q_\mu \Pi^{\mu\nu}(q) = 0$ and thus the tensor structure of the vacuum polarisation tensor has to be $\Pi^{\mu\nu}(q) = (q^2 \eta^{\mu\nu} - q^\mu q^\nu)\Pi(q^2)$. Hence we have to calculate only the simpler scalar function $\Pi(q^2)$ knowing that $\Pi^{\mu\nu}(q)$ is gauge invariant.

In order to show that $q_\mu \Pi^{\mu\nu}(q) = 0$, we write first

$$q = (q + k - m) - (k - m)$$
(12.39)

and obtain

$$q_\mu \mathcal{N}^{\mu\nu} = \mathrm{tr}\{[(\slashed{q} + \slashed{k} - m) - (\slashed{k} - m)](\slashed{k} + m)\gamma^\nu(\slashed{k} + \slashed{q} + m)\}$$
(12.40a)

$$= [(q+k)^2 - m^2]\mathrm{tr}\{(\slashed{k} + m)\gamma^\nu\} - (k^2 - m^2)\mathrm{tr}\{\gamma^\nu(\slashed{k} + \slashed{q} + m)\}$$
(12.40b)

where we used the cyclic property of the trace. Employing dimensional regularisation (DR) with $d = 4 - 2\varepsilon$ in order to obtain well-defined integrals,

$$q_\mu i\Pi^{\mu\nu}(q) = -e^2 \mu^{2\varepsilon} \int \frac{d^d k}{(2\pi)^d} \left\{ \frac{\mathrm{tr}[(\slashed{k} + m)\gamma^\nu]}{(k^2 - m^2)} - \frac{\mathrm{tr}[\gamma^\nu(\slashed{k} + \slashed{q} + m)]}{(k+q)^2 - m^2} \right\},$$
(12.41)

we are allowed to shift the integration variable in one of the two terms. Thus $q_\mu \Pi^{\mu\nu}(q) = 0$ and hence the vacuum polarisation tensor at order $\mathcal{O}(e^2)$ is transverse as required by gauge invariance.

Let us pause a moment and summarise what we know already before we start with the evaluation of $\Pi(q^2)$. In our power-counting analysis we found as the superficial degree of divergence $D = 2$. This result was based on the assumption that the numerator $\mathcal{N}$ behaves as a constant. The only constant available, however, is m^2 which would lead to a mass term of the photon, $e^2 A_\mu \Pi^{\mu\nu} A_\nu \propto e^2 m^2 A_\mu \eta^{\mu\nu} A_\nu$. Thus the transversality of $\Pi^{\mu\nu}$ implies that the m^2 term in the numerator will disappear at some step of our calculation. Thereby the convergence of the polarisation tensor improves, becoming a "mild logarithmic" one.

We proceed with the explicit evaluation of $\Pi^{\mu\nu}$. Taking the trace of (12.38) and using its transversality, we find in $d = 2\omega$ dimensions

$$\Pi^\mu_\mu(q) = (q^2 \delta^\mu_\mu - q^2)\Pi(q^2) = (d-1)q^2 \Pi(q^2)$$

and

$$(d-1)q^2 i\Pi(q^2) = -e^2 \mu^{2\varepsilon} \int \frac{d^d k}{(2\pi)^d} \frac{\mathrm{tr}[\gamma^\mu(\slashed{k} + m)\gamma_\mu(\slashed{k} + \slashed{q} + m)]}{(k^2 - m^2)[(k+q)^2 - m^2]} = -e^2 \mu^{2\varepsilon} \int \frac{d^d k}{(2\pi)^d} \frac{\mathcal{N}}{D}.$$
(12.42)

We combine the two propagators introducing the Feynman parameter integral $1/(AB) = \int_0^1 dx \, [Ax + B(1-x)]^{-2}$ with $A = (k+q)^2 - m^2$ and $B = k^2 - m^2$. Thus the denominator becomes

$$\mathcal{D} = Ax + B(1-x) = k^2 + 2kqx + q^2x - m^2 = (k+qx)^2 + q^2x(1-x) - m^2. \quad (12.43)$$

Next we introduce as new integration variable $K = k + qx$,

$$\mathcal{D} = K^2 + q^2x(1-x) - m^2 = K^2 - a, \quad (12.44)$$

and $a = m^2 - q^2x(1-x) > 0$ as shortcut.

Evaluating the trace in the denominator using DR, we have to extend the Clifford algebra to $d = 2\omega$ dimensions. A natural choice is $\text{tr}(\gamma^\mu \gamma^\nu) = d\eta^{\mu\nu}$, giving with $\gamma^\mu \gamma_\mu = d$ and $\gamma^\mu \slashed{q} \gamma_\mu = (2-d)\slashed{q}$ as result for the trace

$$\mathcal{N} = \mathcal{N}^\mu_\mu = d[(2-d)k \cdot (k+q) + dm^2] = d\{(2-d)[K^2 - q^2x(1-x)] + dm^2\}. \quad (12.45)$$

In the last step, we performed the shift $k \to K = k + qx$ omitting linear terms in K that vanish after integration. Combining our results for $\mathcal{N}$ and D we arrive at

$$(d-1)q^2 i\Pi(q^2) = -e^2\mu^{2\varepsilon}d \int_0^1 dx \int \frac{d^dK}{(2\pi)^d} \frac{(2-d)K^2 + dm^2 - (2-d)q^2x(1-x)]}{(K^2 - a)^2}. \quad (12.46)$$

Finally, we use our results for the Feynman integrals $I_0(\omega, \alpha)$ and $I_2(\omega, \alpha)$ which were obtained performing a Wick rotation. The latter is possible as long as we do not pass a singularity. Since the pre-factor $x(1-x)$ of q^2 has as maximum $1/4$, this requires $q^2 < 4m^2$. We start to look for the first two terms where we expect a cancellation of the m^2 term in the numerator,

$$\int \frac{d^dK}{(2\pi)^d} \frac{(2-d)K^2 + dm^2}{(K^2 - a)^2} = (2-d)I_2(\omega, 2) + dm^2 I(\omega, 2) \quad (12.47a)$$

$$= \frac{i}{(4\pi)^\omega \Gamma(2)} \left[-2\omega(1-\omega)\Gamma(1-\omega)\,a^{\omega-1} + 2\omega m^2 \Gamma(2-\omega)\,a^{\omega-2} \right] \quad (12.47b)$$

$$= \frac{i}{(4\pi)^\omega} 2\omega\Gamma(2-\omega)(-a + m^2)a^{\omega-2} \quad (12.47c)$$

$$= \frac{4\,i}{(4\pi)^2} \Gamma(\varepsilon)\,(4\pi)^\varepsilon \frac{q^2x(1-x)}{[m^2 - q^2x(1-x)]^\varepsilon}. \quad (12.47d)$$

Hence the m^2 term dropped out of the numerator and the whole expression is proportional to q^2, as required by the LHS of (12.46). Evaluating the third term in the same way we obtain

$$-\int \frac{d^dK}{(2\pi)^d} \frac{(2-d)q^2x(1-x)}{(K^2 - a)^2} = -(2-d)q^2x(1-x)I(\omega, 2) \quad (12.48a)$$

$$= \frac{2\,i}{(4\pi)^2} \Gamma(\varepsilon)\,(4\pi)^\varepsilon \frac{q^2x(1-x)}{[m^2 - q^2x(1-x)]^\varepsilon}. \quad (12.48b)$$

Adding the two contributions, we arrive at

$$\Pi(q^2) = -\frac{8e^2}{(4\pi)^2}\,\Gamma(\varepsilon) \int_0^1 dx\,x(1-x) \left[\frac{4\pi\mu^2}{m^2 - q^2 x(1-x)}\right]^{\varepsilon}, \tag{12.49}$$

where the factor iq^2 cancelled. We included also the factor $(4\pi\mu^2)^{\varepsilon}$ into the last term, which thereby becomes dimensionless. Now we expand the Gamma function, $\Gamma(\varepsilon) = 1/\varepsilon - \gamma + \mathcal{O}(\varepsilon)$, around $d = 4 - 2\varepsilon$ and because of the resulting $1/\varepsilon$ term all other ε dependent quantities,

$$\Pi(q^2) = -\frac{e^2}{12\pi^2}\left\{\frac{1}{\varepsilon} - \gamma + \ln(4\pi) - 6\int_0^1 dx\,x(1-x)\ln\left[\frac{m^2 - q^2 x(1-x)}{\mu^2}\right]\right\}. \tag{12.50}$$

The pre-factor $x(1-x)$ has its maximum $1/4$ for $x = 1/2$. Thus the branch cut of the logarithm starts at $q^2 = 4m^2$, that is, when the virtuality of the photon is large enough that it can decay into a fermion pair of mass $2m$. This is a nice illustration of the optical theorem: the polarisation tensor is real below the pair creation threshold and acquires an imaginary part above (which equals the pair creation cross-section of a photon with mass $m^2 = q^2$, cf. problem 12.6).

The x integral can be integrated by elementary functions but we display only the result for the two limiting cases of small and large virtualities, $q^2/m^2 \to 0$ and $|q|^2/m^2 \to \infty$. In the first case, we obtain with $\ln(1-x) \simeq -x$

$$\Pi(q^2) = -\frac{e^2}{12\pi^2}\left[\frac{1}{\varepsilon} - \gamma + \ln(4\pi) + \ln(\mu^2/m^2) + \frac{q^2}{5m^2} + \dots\right], \tag{12.51}$$

while the opposite limit gives

$$\Pi(q^2) = -\frac{e^2}{12\pi^2}\left[\frac{1}{\varepsilon} - \gamma + \ln(4\pi) - \ln(|q^2|/\mu^2) + \dots\right]. \tag{12.52}$$

In the MS scheme, we obtain the renormalisation constant Z_3 for the photon field as the coefficient of the pole term,

$$Z_3 = 1 - \frac{e^2}{12\pi^2\varepsilon}. \tag{12.53}$$

More often the on-shell renormalisation scheme is used in QED. Here we require that quantum corrections to the electric charge vanish for $q^2 = 0$, that is, we choose Z_3 such that $\Pi^{\mathrm{on}}(q^2 = 0) = 0$. This is obviously achieved setting

$$\Pi^{\mathrm{on}}(q^2) = \Pi(q^2) - \Pi(0) = -\frac{e^2}{60\pi^2}\frac{q^2}{m^2} + \dots, \tag{12.54}$$

for $|q^2| \ll m^2$. This q^2 dependence leads to a modification of the Coulomb potential, which can be measured, for example, in the Lamb shift.

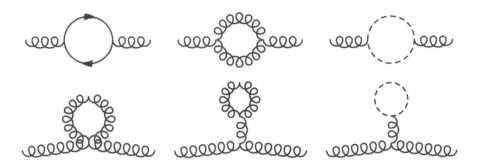

Fig. 12.1 Feynman diagrams describing vacuum polarisation in QCD at one-loop.

12.3.2   Vacuum polarisation in QCD

We only sketch the calculations in QCD, stressing the new or different points compared to QED. In Fig. 12.1, we show the various one-loop diagrams contributing to the vacuum polarisation tensor in a non-abelian theory as QCD. Most importantly, the three- and four-gluon vertices allow in addition to the quark loop now also gluon loops. Since a fermion loop has an additional minus sign, we expect that the gluon loop gives a negative contribution to the beta function. This opens up the possibility that non-abelian gauge theories are in contrast to QED asymptotically free if the number of fermion species is small enough.

Quark loop.   The vertex changes from $-\mathrm{i}e\gamma^\mu$ in QED to $-\mathrm{i}g_s T^a \gamma^\mu$ in QCD. Since the quark propagator is diagonal in the group index, a quark loop contains the factor

$$\mathrm{tr}\{T^a T^a\} = \frac{1}{2}\delta^{aa} = 4\,. \tag{12.55}$$

Thus we only have to replace $e \to 4 n_f g_s$ in the QED result, where n_f counts the number of quark flavours. For the three light quarks, u, d and s, an excellent approximation is to use $m = 0$. In contrast, the masses of the other three quarks (c, b and t) cannot be neglected when $4 m_f^2 \lesssim \mu^2$. The effect of particle masses can be approximated including in the loop only particles with mass $4 m_f^2 < \mu^2$, making n_f scale-dependent.

Loop with three-gluon vertex.   Since the three-gluon vertex connects identical particles, we have to take into account symmetry factors similar as in the case of the $\lambda\phi^4$ theory. We learnt that the imaginary part of a Feynman diagram corresponds to the propagation of real particles. Thus the imaginary part of the gluon vacuum polarisation can be connected to the total cross-section of $g \to gg$ scattering. This cross-section contains a symmetry factor $1/2!$ to account for two identical particles in the final state. Therefore the same symmetry factor should be associated to the vacuum polarisation with a gluon loop.[3] Applying the Feynman rule for the three-gluon vertex to the second diagram shown in Fig. 12.1 and using the Feynman–t'Hooft gauge, we find

---

[3]This argument does not apply to the quark loop, since cutting leads in this case to a distinguishable $\bar{q}q$ state

$$i\Pi^{\mu\nu}_{ab,\,2}(q^2) = \frac{1}{2}(-ig)^2 i^2 \int \frac{\mathrm{d}^4 k}{(2\pi)^4} \frac{\mathcal{N}^{\mu\nu}_{ab}}{(k^2 + i\varepsilon)[(k+q)^2 + i\varepsilon]} \tag{12.56}$$

with

$$\mathcal{N}^{\mu\nu}_{ab} = f_{bcd}[-\eta^{\nu\rho}(q+k)^\sigma + \eta^{\rho\sigma}(2k-q)^\nu + \eta^{\sigma\nu}(2q-k)^\rho] \\ \times f_{acd}[-\eta^\mu{}_\rho(q+k)_\sigma + \eta_{\rho\sigma}(q-2k)^\mu + \eta^\mu{}_\sigma(k-2q)_\rho]. \tag{12.57}$$

Evaluating the colour trace,

$$f^{acd} f^{cdb} = N_c \delta^{ab} \tag{12.58}$$

and extracting in the usual way the pole part using DR one obtains for $d = 4 - 2\varepsilon$

$$\Pi^{\mu\nu}_{ab,\,2}(q^2) = -\frac{N_c g^2}{32\pi^2 \varepsilon}\left(\frac{\mu^2}{q^2}\right)^{\varepsilon/2}\left(\frac{11}{3}q^\mu q^\nu - \frac{19}{6}\eta^{\mu\nu}q^2\right) + \mathcal{O}(\varepsilon^0). \tag{12.59}$$

Thus the contribution from the three-gluon vertex alone is not transverse, demonstrating again that a covariant gauge requires the introduction of ghost particles.

Ghost loop. This diagram has the same dependence on the structure constants as the previous one,

$$i\Pi^{\mu\nu}_{ab,\,3}(q^2) = -(-ig)^2 i^2 \int \frac{\mathrm{d}^4 k}{(2\pi)^4} \frac{f_{bdc}(k-q)^\nu f_{acd}k_\nu}{(k^2 + i\varepsilon)[(k+q)^2 + i\varepsilon]}, \tag{12.60}$$

and can thus be combined with the three-gluon loop. Note the extra minus sign due to the fermionic nature of the ghost particle. Evaluating the integral results in

$$\Pi^{\mu\nu}_{ab,\,3}(q^2) = \frac{N_c g^2}{32\pi^2 \varepsilon}\left(\frac{\mu^2}{q^2}\right)^{\varepsilon/2}\left(\frac{1}{3}q^\mu q^\nu + \frac{1}{6}\eta^{\mu\nu}q^2\right) + \mathcal{O}(\varepsilon^0) \tag{12.61}$$

and summing the three-gluon and ghost loops gives the expected gauge-invariant expression. Moreover, the sum has the opposite sign as the quark loop and can thus lead to the opposite behaviour of the beta function as in QED.

Four-gluon loop and tadpole diagrams. The loop with the four-gluon vertex contains a massless propagator and does not depend on external momenta,

$$i\Pi^{\mu\nu}_{ab,\,4}(q^2) \propto \int \frac{\mathrm{d}^d k}{k^2 + i\varepsilon}. \tag{12.62}$$

Our general experience with DR tells us that this graph is zero, as gluons are massless. However, this loop integral is also in DR ambiguous: for any spacetime dimension d, the integral is either UV or IR divergent. To proceed, we split therefore the integrand introducing the arbitrary mass M as

$$\frac{1}{k^2 + i\varepsilon} = \frac{1}{k^2 - M^2 + i\varepsilon} - \frac{M^2}{(k^2 + i\varepsilon)(k^2 - M^2 + i\varepsilon)}. \tag{12.63}$$

Now the IR and UV divergences are separated, and we can use $d < 2$ in the first term and $2 < d < 4$ in the second. By dimensional analysis, both terms have to be proportional to a power of the arbitrary mass M. As the LHS is independent of M, the only option is that the two terms on the RHS cancel, as an explicit calculation confirms. The remaining tadpole diagrams (5 and 6 of Fig. 12.1) vanish by the same argument. We will come back to the combined result for the coupling $g_s(Q^2)$ in QCD, after having introduced running couplings and discussing their behaviour in general.

12.4   Renormalisation group

Renormalisation group equations.   Let us consider two renormalisation schemes R and R'. In the two schemes, the renormalised field will in general differ, being $\phi_R = Z_\phi^{-1/2}(R)\phi_0$ and $\phi_{R'} = Z_\phi^{-1/2}(R')\phi_0$, respectively. Hence the connection between the two renormalised fields is

$$\phi_{R'} = \frac{Z_\phi^{-1/2}(R')}{Z_\phi^{-1/2}(R)} \, \phi_R \equiv Z_\phi^{-1/2}(R', R) \, \phi_R. \tag{12.64}$$

As both ϕ_R and $\phi_{R'}$ are finite, also $Z_\phi(R', R)$ is finite. The transformations $Z_\phi^{-1/2}(R', R)$ form a group, called the renormalisation group. If we consider

$$G_0^{(n)}(x_1, \dots, x_n) = Z_\phi^{n/2} \, G_R^{(n)}(x_1, \dots, x_n), \tag{12.65}$$

we know that the bare Green function is independent of the renormalisation scale μ. Taking a derivative with respect to μ, the LHS thus vanishes. To avoid clutter, we restrict ourselves to the simplest case of a massless theory with a single coupling g. Then the renormalised Green functions can depend (in a mass-independent scheme) only on the renormalised coupling $g(\mu)$, and we find thus

$$0 = \frac{\mathrm{d}}{\mathrm{d}\ln\mu} \left[Z_\phi^{n/2} \, G_R^{(n)}(x_1, \dots, x_n) \right] \tag{12.66a}$$

$$= \left[\frac{\partial}{\partial\ln\mu} + \frac{\partial g}{\partial\ln\mu} \frac{\partial}{\partial g} + \frac{n}{2} \frac{\partial\ln Z_\phi}{\partial\ln\mu} \right] G_R^{(n)}(x_1, \dots, x_n) \tag{12.66b}$$

$$\equiv \left[\frac{\partial}{\partial\ln\mu} + \beta \frac{\partial}{\partial g} + \frac{n}{2}\,\gamma \right] G_R^{(n)}(x_1, \dots, x_n). \tag{12.66c}$$

Here we introduced in the last step the anomalous dimension

$$\gamma(\mu) = \mu\,\frac{\partial\ln Z_\phi(\mu)}{\partial\mu} \tag{12.67}$$

of the field ϕ and the beta function

$$\beta(g) = \mu\,\frac{\partial g(\mu)}{\partial\mu}, \tag{12.68}$$

which determines the logarithmic change of the coupling constant. The beta function is often re-expressed as

$$\beta(g^2) \equiv \mu^2 \frac{\partial g^2}{\partial\mu^2} = g\beta(g) \tag{12.69}$$

or as $\beta(\alpha^2)$ with $\alpha = g^2/(4\pi)$. Equations of the type (12.66c) are called generically renormalisation group equations or briefly RGE. They come in various flavours, carrying the name of their inventors: Stückelberg–Petermann, Callan–Symanzik, Gell-Mann–Low, etc. We can use (11.63) to derive a similar RGE for the 1PI Green functions. Note that the sign difference between (11.63) and (11.62) in the power of the wave-function renormalisation constant induces a corresponding sign change in the RGE for 1PI Green functions.

Remark 12.1: In order to understand the term "anomalous dimensions", we look first at the canonical dimension of fields under a change of units. Note that in contrast to the scale transformations $x \to e^{\alpha} x$ discussed in problem 5.6, we now change all parameters including couplings and masses. The classical action is then invariant, and an n-point Green function $G(p_1, \ldots, p_n)$ can be expressed as

$$G(p_1, \ldots, p_n) = m^a g^b p_1^{c_1} \cdots p_n^{c_n}$$

though the parameters of the classical Lagrangian. Dimensional analysis constrains the exponents as $a - c_1 - \ldots - c_n = n$, and the n-point Green function scales classically as $G \to e^{n\alpha x} G$. The renormalised Green function depends, however, also on the scale μ or the subtraction point of our renormalisation scheme which we keep fixed,

$$G(p_1, \ldots, p_n) = m^a g^b p_1^{c_1} \cdots p_n^{c_n} \mu^{\gamma}.$$

Hence the renormalised n-point Green function scales as $G \to e^{(n-\gamma)\alpha x} G$ and therefore satisfies $\mu \mathrm{d}G/\mathrm{d}\mu = \gamma G$.

Knowing the two universal functions $\beta(\lambda)$ and $\gamma(\mu)$, we can calculate the change of any Green function under a change of the renormalisation scale μ. The general solution of (12.66c) can be found by the method of characteristics or by the analogy of $\mathrm{d}/\mathrm{d}\ln\mu$ with a convective derivative, cf. problem 12.5. Here we consider only the simplest case of a dimensionless observable R in a renormalisable theory which depends on a single physical momentum scale Q. Thus we assume that the coupling constant is dimensionless, and consider the limit that all masses can be neglected, $|Q^2| \gg m_i^2$. An important example is the $e^+ e^-$ annihilation cross-section into hadrons which can be made dimensionless dividing by $\sigma(e^+ e^- \to \mu^+ \mu^-)$ as reference cross-section. Then R can be only a function of the ratio Q^2/μ^2 and of $\alpha(\mu^2)$. A physical observable like R should be independent of the scale μ, or

$$0 = \mu^2 \frac{\mathrm{d}R}{\mathrm{d}\mu^2} = \left(\mu^2 \frac{\partial}{\partial \mu^2} + \beta(\alpha) \frac{\partial}{\partial \alpha} \right) R(Q^2/\mu^2, \alpha(\mu^2)) = \left(-\frac{\partial}{\partial \tau} + \beta(\alpha) \frac{\partial}{\partial \alpha} \right) R(\tau, \alpha(\mu^2)),$$

$$(12.70)$$

where we introduced $\tau = \ln(Q^2/\mu^2)$. The two differential operators compensate each other, setting

$$\tau = \int_{\alpha(\mu^2)}^{\alpha(Q^2)} \frac{\mathrm{d}x}{\beta(x)}.$$

$$(12.71)$$

The coupling $\alpha(Q^2) = \alpha(\tau, \alpha(\mu^2))$ defined by (12.71) as function of the physical momentum Q^2 is called the "running coupling". Setting the renormalisation scale equal to the physical scale, $\mu^2 = Q^2$, results in

$$R(Q^2/\mu^2, \alpha(\mu^2)) = R(1, \alpha(Q^2)) \,. \tag{12.72}$$

Hence all scale dependence in R can be absorbed into the running of $\alpha(Q^2)$. We have seen already an example of this behaviour, discussing $\phi\phi \to \phi\phi$ scattering in section 4.3.4, compare especially with Eq. (4.87).

Example 12.2: Any function of the running coupling, and thus in particular the running coupling itself, solves the homogeneous RGE. To show the latter claim, we differentiate first the definition Eq. (12.71) w.r.t. τ, obtaining

$$1 = \frac{\partial}{\partial \tau} \left[\int_{\alpha(\mu^2)}^{\alpha(Q^2)} \frac{dx}{\beta(x)} \right] = \frac{1}{\beta(\alpha(Q^2))} \frac{\partial \alpha(Q^2)}{\partial \tau} \tag{12.73}$$

or $\partial_\tau \alpha(Q^2) = \beta(\alpha(Q^2))$. Next we differentiate (12.71) w.r.t. $\alpha(\mu^2)$, obtaining

$$0 = \frac{\partial}{\partial \alpha(\mu^2)} \left[\int_{\alpha(\mu^2)}^{\alpha(Q^2)} \frac{dx}{\beta(x)} \right] = \frac{1}{\beta(\alpha(Q^2))} \frac{\partial \alpha(Q^2)}{\partial \alpha(\mu^2)} - \frac{1}{\beta(\alpha(Q^2))} \tag{12.74}$$

or

$$\frac{\partial \alpha(Q^2)}{\partial \alpha(\mu^2)} = \frac{\beta(\alpha(Q^2))}{\beta(\alpha(\mu^2))} \,. \tag{12.75}$$

Evaluating the differential operator of the homogeneous RGE (i.e. for $\gamma = 0$) for the coupling, we see that the two terms cancel as required,

$$\left(\frac{\partial}{\partial \tau} + \beta(\alpha(\mu^2)) \frac{\partial}{\partial \alpha(\mu^2)} \right) \alpha(\tau, \alpha(\mu^2)) = -\beta(\alpha(Q^2)) + \beta(\alpha_s(\mu^2)) \frac{\beta(\alpha(Q^2))}{\beta(\alpha(\mu^2))} = 0. \tag{12.76}$$

Working through problem 11.10, you found that the beta function of the $\lambda\phi^4$ theory at one loop is given by

$$\beta(\lambda) = b_1 \lambda^2 + b_2 \lambda^3 + \ldots = \frac{3\lambda^2}{16\pi^2} + \mathcal{O}(\lambda^3), \tag{12.77}$$

so that the running coupling at leading order satisfies

$$\lambda(Q) = \frac{\lambda(\mu)}{1 - 3\lambda(\mu)/16\pi^2 \ln(Q/\mu)} \,. \tag{12.78}$$

Thus our new definitions (12.71) and (12.68) reproduce the result of our more intuitive approach in section 4.3.4. As a bonus, we know now what we should choose as the argument of the coupling, at least in the simple case considered with a single physical scale Q. Expanding the running coupling,

$$\lambda(Q) = \lambda(\mu) + b_1 \ln(Q/\mu) - [b_1 \ln(Q/\mu)]^2 + \ldots . \tag{12.79}$$

we see that $\lambda(Q)$ contains arbitrary powers of $[b_1\lambda(\mu \ln(Q/\mu)]^n$, although we derived the beta function only at one-loop. The RGE ensures that the running coupling sums up the largest terms in each order perturbation theory, cf. problem 12.7. In general, the running coupling calculated at n-loop precision contains the leading $\ln^n(Q/\mu)$ terms of loop diagrams of any order. Accordingly, one speaks of LL (leading-logarithmic), NLL (next-to-leading logarithmic), NNLL, ... approximations.

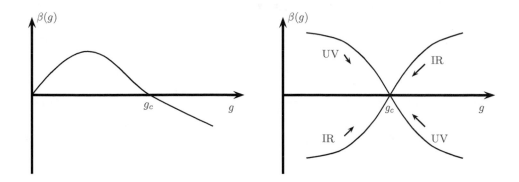

Fig. 12.2 Left: Example of a beta function with a perturbative IR and an UV fixed point. Right: General classification of UV and IR stable fixed points of the RGE flow.

Asymptotic behaviour of the beta function. The behaviour of the beta function $\beta(\mu)$ in the limit $\mu \to 0$ and $\mu \to \infty$ provides a useful classification of quantum field theories. Consider the example shown in the left panel of Fig. 12.2. This beta function has a trivial zero at zero coupling, as we expect it in any perturbative theory, and an additional zero at g_c. How does the beta function $\beta(\mu)$ evolve in the UV limit $\mu \to \infty$?

- Starting in the range $0 \le g(\mu) < g_c$ implies $\beta > 0$ and thus $\mathrm{d}g/\mathrm{d}\mu > 0$. Therefore g grows with increasing μ and the coupling is driven towards g_c.
- Starting with $g(\mu) > g_c$ implies $\beta < 0$ and thus $\mathrm{d}g/\mathrm{d}\mu < 0$. Therefore g decreases for increasing μ and we are driven again towards g_c.

Fixed points g_c approached in the limit $\mu \to \infty$ are called UV fixed points, while IR fixed points are reached for decreasing μ. The range of values $[g_1 : g_2]$ which is mapped by the RGE flow on the fixed point is called its basin of attraction.

To see what happens for $\mu \to 0$, we have only to reverse the RGE flow, $\mathrm{d}\mu \to -\mathrm{d}\mu$, and are thus driven away from g_c. If we started in $0 \le g(\mu) < g_c$, we are driven to zero, while the coupling goes to infinity for $g(\mu) > g_c$. Thus $g = 0$ is an IR fixed point. The distinction between IR and UV fixed points is sketched in the right panel of Fig. 12.2. If the beta function has several zeros, the theory consists of different phases which are not connected by the RG flow.

Looking back at our one-loop result for the beta function of the $\lambda \phi^4$ theory, we see that the theory has $\lambda = 0$ as an IR fixed point. Thus the free states we use as asymptotic initial and final states have a direct physical meaning. On the other hand, the coupling increases for $\mu \to \infty$ formally as $\lambda \to \infty$. Clearly, we cannot trust the behaviour of $\lambda(\mu)$ based on the on-loop result in the strong-coupling limit, because perturbation theory breaks down. The solution (12.78) suggests, however, that the coupling explodes already for a finite value of μ: the beta function has a pole for a finite value of μ called Landau pole where the denominator of (12.78) becomes zero.

Beta function of QED. We can derive the scale dependence of the renormalised electric charge from

$$e_0 = \mu^\varepsilon Z_3^{-1/2}\, e, \tag{12.80}$$

where we used $Z_1 = Z_2$. Then the beta function is given as

$$\beta(e) \equiv \mu \frac{\partial e}{\partial \mu} = \mu \frac{\partial}{\partial \mu} \left(\mu^{-\varepsilon} Z_3^{1/2} e_0 \right). \tag{12.81}$$

Since the bare charge e_0 is independent of μ, we have to differentiate only μ and Z_3,

$$\beta(e) = \mu \frac{\partial e}{\partial \mu} = -\varepsilon \mu^{-\varepsilon} Z_3^{1/2} e_0 + \mu \mu^{-\varepsilon} \frac{1}{2} Z_3^{-1/2} \frac{\partial Z_3}{\partial \mu} e_0 = -\varepsilon e + \frac{\mu}{2Z_3} \frac{\partial Z_3}{\partial \mu} e. \tag{12.82}$$

Inserting $Z_3 = 1 - e^2/(12\pi^2 \varepsilon)$ and thus

$$\frac{\partial Z_3}{\partial \mu} = -\frac{1}{12\pi^2 \varepsilon} \frac{2e \partial e}{\partial \mu} \tag{12.83}$$

gives

$$\beta(e) = -\varepsilon e - \frac{\mu}{12\pi^2 \varepsilon Z_3} \frac{\partial e}{\partial \mu} e^2 = -\varepsilon e - \frac{1}{12\pi^2 \varepsilon Z_3} \beta(e) \, e^2. \tag{12.84}$$

Note that Z_3 is scheme-dependent, while the beta function remains scheme-independent up to two loops for all mass-independent schemes (problem 12.10). Solving for β and neglecting higher-order terms in e^2, we find in the limit $\varepsilon \to 0$

$$\beta(e) = -\varepsilon e \left(1 - \frac{e^2}{12\pi^2 \varepsilon} \right) + \mathcal{O}(e^4) = \frac{e^3}{12\pi^2}. \tag{12.85}$$

Thus the beta function is determined by the coefficient of the pole term of Z_3. Its solution,

$$e^2(\mu) = \frac{e^2(\mu_0)}{1 - \frac{e^2(\mu_0)}{6\pi^2} \ln\left(\frac{\mu}{\mu_0}\right)} \tag{12.86}$$

shows explicitly not only the increase of e^2 with μ, but that the electric coupling has a Landau pole too. However, the scale of the Landau pole corresponds to

$$\mu = \mu_0 \exp(6\pi^2/e^2(\mu_0)) = m_e \exp(3\pi/2\alpha(m_e)) \sim 10^{56} \, \text{GeV} \gg M_{\text{Pl}} \tag{12.87}$$

and therefore has no direct physical relevance.

Alternatively, we could derive the beta function from the renormalised vacuum polarisation, using

$$\beta(e) = \frac{e}{2} \mu \frac{\partial}{\partial \mu} \Pi^{\text{MS}}(q^2), \tag{12.88}$$

where in the on-shell scheme the derivative $\partial \ln \mu^2$ should be replaced with $\partial \ln q^2$.

Beta function of QCD and asymptotic freedom. Deriving the beta function in QCD, we should evaluate

$$g(\mu) = \frac{Z_2 Z_3^{1/2}}{Z_1} g_0 = Z_3^{1/2} g_0, \tag{12.89}$$

where Z_1 is the charge renormalisation constant, and Z_2 and Z_3 are quark and gluon renormalisation constants, respectively. Without proof, we note that the generalisation of the Ward–Takahashi identities to the non-abelian case, the Slavnov–Taylor identities, ensure that $Z_1 = Z_2$. Combining all contributions to the $1/\varepsilon$ poles as one-loop contribution b_1 to the beta function of QCD gives (cf. problem 12.9)

$$\beta(\alpha_s) = \mu^2 \frac{\partial \alpha_s}{\partial \mu^2} = -\alpha_s^2 \left(b_1 + b_2 \alpha_s + b_3 \alpha_s^2 + \ldots \right) \tag{12.90}$$

with $\alpha_s \equiv g_s^2/(4\pi)$,

$$b_1 = \frac{11}{3} N_c - \frac{2}{3} n_f \tag{12.91}$$

and n_f as number of quark flavours. For $n_f < 17$, the beta function is negative and the running coupling decreases as $\mu \to \infty$. Asymptotic freedom of QCD explains the apparent paradox that protons are interacting strongly at small Q^2, while they can be described in deep-inelastic scattering as a collection of independently moving quarks and gluons.

Let us consider now the opposite limit, $\mu \to 0$. The solution of (12.90) at one loop,

$$\frac{1}{\alpha_s(\mu^2)} = \frac{1}{\alpha_s(\mu_0^2)} + b_1 \ln \left(\frac{\mu^2}{\mu_0^2} \right) \tag{12.92}$$

shows that the QCD coupling constant becomes formally infinite for a finite value of μ. We define $\Lambda_{\rm QCD}$ as the energy scale where the running coupling constant of QCD diverges, $\alpha_s^{-1}(\Lambda_{\rm QCD}^2) = 0$. Experimentally, the best measurement of the strong coupling constant has been performed at the Z resonance at LEP, giving $\alpha_s(m_Z^2) \sim 0.1184$. Thus at one-loop level,

$$\Lambda_{\rm QCD} = m_Z \exp \left(\frac{1}{2 b_1 \alpha_s(m_Z^2)} \right). \tag{12.93}$$

$\Lambda_{\rm QCD}$ depends on the renormalisation scheme and, numerically more importantly, on the number of flavours used in b_1: for instance, $\Lambda_{\rm QCD}^{\overline{\rm MS}} \simeq 220\,{\rm MeV}$ for $n_f = 3$. The fact that the running coupling provides a characteristic energy scale is called dimensional transmutation. Quantum corrections lead to the break-down of scale-invariance of classical QCD with massless quarks and to the appearance of massive bound-states, the mesons and baryons with masses of order $\Lambda_{\rm QCD}$. Note also that we are able to link exponentially separated scales by dimensional transmutation.

Coupling constant unification. While the strong coupling $\alpha_s \equiv \alpha_3$ decreases with increasing Q^2, the electromagnetic coupling $\alpha_{\rm em} \equiv \alpha_1$ increases. Since two lines in a plane meet at one point, there is a point with $\alpha_1(Q_*) = \alpha_3(Q_*)$ and one may speculate that at this point a transition to a "grand unified theory" (GUT) happens. Since the running is only logarithmic, unification happens at exponentially high scales, $Q_* \sim 10^{16}\,{\rm GeV}$, but interestingly still below the Planck scale $M_{\rm Pl}$. The problems becomes more challenging, if we add to the game the third, the weak coupling α_2. The situation in 1991 assuming the validity of the SM is shown in the left panel of Fig. 12.3. The width of the lines indicates the experimental and theoretical error, and the three couplings clearly do not meet within these errors. The right panel of the same figure assume the existence of supersymmetric partners to all SM particles with an "average mass" of around $M_{\rm SUSY} \sim 200\,{\rm GeV}$. As a result, the running changes above $Q = 1\,{\rm TeV}$, and now the three couplings meet at $2 \times 10^{16}\,{\rm GeV}$.

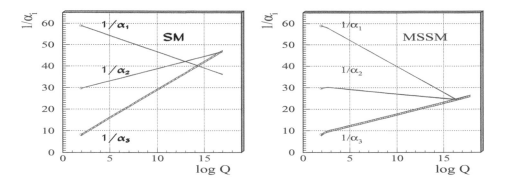

Fig. 12.3 The gauge couplings measured at low energies do not evolve towards a unified value in the SM (left), but meet assuming low-scale supersymmetry at $\simeq 2 \times 10^{16}\,{\rm GeV}$ (right), from Kazakov (2000) based on Amaldi *et al.* (1991).

12.5 Renormalisation, critical phenomena and effective theories

Overview. The behaviour thermodynamical systems exhibit close to the critical points in their phase diagram are characterised as "critical phenomenon". For a fixed number of particles, we can characterise thermodynamical systems using the free energy $F = U - TS$. Ehrenfest introduced the classification of phase transitions according to the order of the first discontinuous derivative of F with respect to any thermodynamical variable ϕ. Hence a phase transition where at least one derivative $\partial^n F/\partial \phi^n$ is discontinuous while all $\partial^{n-1} F/\partial \phi^{n-1}$ are continuous is called an nth-order phase transition.

Critical phenomena are interesting to a particle physicist for at least three reasons:

- We can learn about symmetry breaking. We should look out for ideas how we can generate mass terms without violating gauge invariance. Systems like ferromagnets show that symmetries as rotation symmetry can be broken at low energies although the Hamiltonian governing the interactions is rotation symmetric.

  Another example is a plasma. Here the screening of electric charges modifies the Coulomb potential to a Yukawa potential; the photon has three massive degrees of freedom, still satisfying gauge invariance, $k_\mu \Pi^{\mu\nu}(\omega, \boldsymbol{k}) = 0$, but with $\omega^2 - \boldsymbol{k}^2 \neq 0$.

- Experimentally one finds that close to a critical point, $T \to T_c$, the correlation length ξ diverges, while otherwise correlations are exponentially suppressed. If we consider a statistical system on a lattice, then the 2-point function of a certain order parameter ϕ scales as

$$\langle \phi_n \phi_0 \rangle \propto \exp(-n/\xi)\,,$$

where ξ is measured in multiples of the lattice spacing a. Comparing this with $|x| = na$ to the 2-point function of an Euclidean scalar field ϕ,

$$\langle \phi(x)\phi(0) \rangle \to \frac{4\pi}{|x|^2} \, \exp(-m|x|)$$

in the limit $m|x| \gg 1$, we find the correspondence $\xi = 1/(ma)$. Therefore the continuum limit $a \to 0$ is only possible for finite m, if $\xi \to \infty$.

Thus the correlation functions of a statistical system correspond for non-zero a to bare Green functions and a finite value of the regulator of the corresponding quantum field theory. The connection to renormalised Green functions ($a \to 0$ or $\Lambda \to \infty$) is only possible when the statistical system is at a critical point.

- Near a critical point, $T \to T_c$, thermodynamical systems show a universal behaviour. More precisely, they fall in different universality classes which unify systems with very different microscopic behaviour. The various universality classes can be characterised by critical exponents, that is, by the exponents γ_i with which characteristic thermodynamical quantities X_i diverge approaching T_c, i.e. $X_i = [b(T - T_c)]^{-\gamma_i}$.

  This phenomenon is similar to our realisation that, for example, two $\lambda\phi^4$ theories, one with $\lambda = 0.1$ and another one with $\lambda = 0.2$, are not fundamentally different but connected by a RGE transformation.

Landau's mean field theory. Landau suggested that the free energy F of a thermodynamical system can be expanded close to a second-order phase transition as an even series in its order parameter. Considering, for example, the magnetisation M, we can write for zero external field H the free energy as

$$F = A(T) + B(T)M^2 + C(T)M^4 + \dots. \tag{12.94}$$

We can find the possible value of the magnetisation M by minimising the free energy,

$$0 = \left.\frac{\partial F}{\partial M}\right|_T = 2B(T)M + 4C(T)M^3. \tag{12.95}$$

The variable $C(T)$ has to be positive in order that F is bounded from below. If also $B(T)$ is positive, only the trivial solution $M = 0$ exists. If however $B(T)$ is negative, two solutions with non-zero magnetisation appear. Let us use a linear approximation, $B(T) \approx b(T - T_c)$, and $C(T) \approx c$ valid close to T_c. Then

$$M = \begin{cases} 0 & \text{for } T > T_c, \\ \pm \left[\frac{b}{2c}(T_c - T)\right]^{1/2} & \text{for } T < T_c. \end{cases} \tag{12.96}$$

Note also that the ground-state breaks the $M \to -M$ symmetry of the free energy for $T < T_c$.

Representing the thermodynamical quantity M as integral of the local spin density,

$$M = \int \mathrm{d}^3 x \, s(\boldsymbol{x}), \tag{12.97}$$

we can rewrite the free energy in a way resembling the Hamiltonian of a stationary scalar field,

$$F = \int \mathrm{d}^3x \, \left[(\boldsymbol{\nabla}s)^2 + b(T - T_c)s^2 + cs^4 - \boldsymbol{H} \cdot \boldsymbol{s}\right]. \tag{12.98}$$

Here, $(\boldsymbol{\nabla}s)^2$ is the simplest ansatz leading to an alignment of spins in the continuous language. Minimising F will give us the ground-state of the system for a prescribed external field $\boldsymbol{H}(\boldsymbol{x})$ and temperature T. For small s, we can ignore the s^4 term. The spin correlation function $\langle s(\boldsymbol{x})s(\boldsymbol{0})\rangle$ is found as response to a delta function-like disturbance $H_0\delta(x)$. Using the analogy with the Yukawa potential after the substitution $m^2 \to b(T - T_c)$, the correlation function follows immediately as

$$\langle s(\boldsymbol{x})s(\boldsymbol{0})\rangle = \int \frac{\mathrm{d}^3k}{(2\pi)^3} \frac{H_0 \mathrm{e}^{\mathrm{i}\boldsymbol{k}\cdot\boldsymbol{x}}}{k^2 + b(T - T_c)} = \frac{H_0}{4\pi r} \, \mathrm{e}^{-r/\xi} \tag{12.99}$$

with

$$\xi = [b(T - T_c)]^{-1/2}. \tag{12.100}$$

Hence Landau's theory reproduces the experimentally observed behaviour $\xi \to \infty$ for $T \to T_c$. Moreover, the theory predicts as critical exponent $1/2$ for the magnetisation. Notice that the value of the exponent depends only on the polynomial assumed in the free energy, not on the underlying micro-physics. Thus another prediction of Landau's theory is an universal behaviour of thermodynamical systems close to their critical points in the dependence on $T - T_c$. Experiments show that this prediction is too strong. Thermodynamical systems fall into different universality classes, and we should try to include some micro-physics into the description of critical phenomena.

Kadanoff's block spin transformation.   Close to a critical point, collective effects play a decisive role even in case of short-range interactions. In d dimension, a particle is coupled by collective effects to $(\xi/a)^d$ particles and standard perturbative methods will certainly fail for $\xi \to \infty$. Kadanoff suggested to remove the short wavelength fluctuations by the following procedure: each step of a block spin transformation consists of i) dividing the lattice into cells of size $(2a)^d$, ii) assigning a common spin variable to the cell, and iii) of a rescaling $2a \to a$.

At each step, the number of strongly correlated spins is reduced. After n transformations, the correlation length decreases as $\xi_n = \xi/(2^n)$. When the correlation length becomes of the order of the lattice spacing, collective effects play no role: all the physics can be read off from the Hamiltonian. If the procedure is not trivial, this implies that in each step the Hamiltonian changes. In particular, the coupling constant K is changed as

$$K_2 = f(K), \qquad K_3 = f(K_2) = f(f(K)), \ldots. \tag{12.101}$$

One-dimensional Ising model.   We illustrate the idea behind Kadanoff's block spin transformation using the example of the one-dimensional Ising model. This model consists of spins with value $s_i = \pm 1$ on a line with spacing a, interacting via nearest

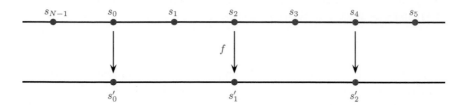

Fig. 12.4 One block spin transformation $a \to 2a$ for an one-dimensional lattice model.

neighbour interactions. We consider only the piece of six spins shown in Fig. 12.4. The corresponding partition function is

$$Z_6 = \sum_{s_{N-1}, s_0, s_1, \ldots, s_4} \exp\left[K(s_{N-1}s_0 + s_0 s_1 + \ldots s_3 s_4)\right] \tag{12.102a}$$

$$= \sum_{s_0', s_1', s_2'} \sum_{s_{N-1}, s_1, s_3} \exp\left[K(s_{N-1}s_0' + s_0' s_1 + \ldots s_3 s_4)\right], \tag{12.102b}$$

where the summation is over the two spin values ± 1 and $K = J/T$ is a dimensionless coupling constant. The step $a \to 2a$ requires to perform the sums over the unprimed spins. Expanding the exponentials using $(s_i s_j)^{2n} = 1$ gives terms like

$$\exp[K(s_0' s_1)] = 1 + K s_0' s_1 + \frac{K^2}{2!} + \frac{K^3}{3!} s_0' s_1 + \ldots \tag{12.103a}$$

$$= \cosh(K) + s_0' s_1 \sinh(K) = \cosh(K)[1 + s_0' s_1 \tanh(K)]. \tag{12.103b}$$

The terms linear in s_1 cancel in the sum and we obtain

$$\sum_{s_1 = \pm 1} \exp[K s_0' s_1] \exp[K s_1 s_1'] = 2 \cosh^2(K)[1 + \tanh^2(K) s_0' s_1']. \tag{12.104}$$

Thus the summation over the unprimed spins changes the strength of the nearest neighbourhood interaction and generates additionally a new spin-independent interaction term. We try now to rewrite the last expression in a form similar to the original one,

$$2 \cosh^2(K)[1 + \tanh^2(K) s_0' s_1'] = \exp[g(K) + K' s_0' s_1']. \tag{12.105}$$

Using (12.103b) to replace $\exp(K' s_0' s_1')$ and setting $g(K) = \ln h(K)$, we find

$$2 \cosh^2(K)[1 + \tanh^2(K) s_0' s_1'] = h(K) \cosh(K')[1 + \tanh(K') s_0' s_1']. \tag{12.106}$$

This determines the function $g = \ln h(K)$ as

$$h(K) = \frac{2 \cosh^2(K)}{\cosh(K')}, \tag{12.107}$$

while the couplings are related by

$$\tanh(K') = \tanh^2(K). \tag{12.108}$$

The summation over the other spins $s_3, s_5, \ldots$ can be performed in the same way. Thus the partition function on a lattice of size $2a$ has the same nearest-neighbour interactions with a new coupling $K' \equiv K_1$ determined by (12.108). Iterating this procedure generates a renormalisation flow with

$$\tanh(K_n) = \tanh^{2n}(K). \tag{12.109}$$

Additionally, the RGE flow generates all couplings compatible with the symmetries of the fundamental Hamiltonian. Since any operator $\mathcal{O}^n$ for $n \in \mathbb{Z}$ satisfies these symmetries if $\mathcal{O}$ does, there is an infinite number of them.

Fixed point behaviour.   In general, we will not be able to calculate the transformation function $f(K)$, but even without the knowledge of $f(K)$, we can draw some important insights from general considerations. First, the RGE equations are of the type of a heat or diffusion equation.[4] Its flow is therefore a gradient flow which has only two possible asymptotics: a runaway solution to infinity and the approach to a fixed point defined by $K_c = f(K_c)$. With $\xi_{n+1} = \xi_n/2$ and labelling the n-dependence implicitly via $\xi_n = f(K_n)$ we can write

$$\xi(f(K)) = \frac{1}{2}\,\xi(K). \tag{12.110}$$

At a fixed point $K_c = f(K_c)$ only two solutions exist,

$$\xi(K_c) = 0 \quad \text{and} \quad \xi(K_c) \to \infty. \tag{12.111}$$

The second possibility corresponds to the approach of a critical point, allowing the limit $a \to 0$ and thus the continuum limit necessary for the transition to a QFT. This point is called a critical fixed point, while the fixed point with zero correlation length is called trivial.

We can now generalise our previous discussion of the fixed point behaviour for the beta function from one to n dimensions. The general behaviour of the RGE flow can be understood from the two-dimensional example shown in Fig. 12.5. The dashed lines show surfaces of constant correlation length, including a critical surface $\xi = \infty$. Also shown are three critical fixed points (A, B, C) and a trivial one (D). In each RGE step the correlation length decreases. Thus the trivial fixed point is an attractor, that is, inside a small enough neighbourhood all points will flow towards it. Moreover, critical lines have at least one unstable direction, the one orthogonal to their surface. Even points infinitesimally close to the surface will flow away and eventually end in a trivial fixed point. Finally, also inside the critical surface stable and unstable directions exist: For instance, the fixed point B will attract all points in-between A and C ("its basin of attraction"). Universality classes of QFTs can be identified with stable critical fixed points and their basin of attraction.

Wilsonian action.   Let us now return to QFTs in the continuum limit. Wilson suggested transferring the idea of integrating out fluctuations on small scales by including only modes up to the cut-off scale $\Lambda \sim 1/a$ in the path integral. This defines an *effective field theory* which is by construction UV-finite. The corresponding effective or

---

[4] An example is given later in remark 13.1

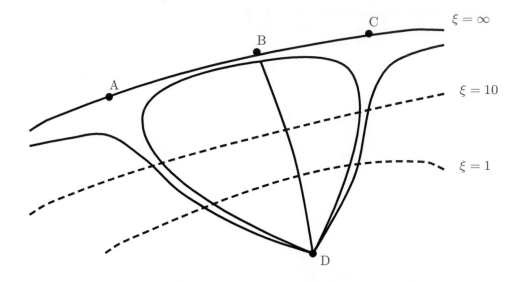

Fig. 12.5 A two-dimensional illustration of the RGE flow: A, B and C are three fixed points on the critical surface $\xi = \infty$. The fixed points A and C have a stable direction along the critical surface, while B has two unstable directions. The trivial fixed point D is a stable fixed point, attracting all points starting not on the critical surface.

Wilsonian action depends on the assumed value of the cut-off scale Λ, and we can generate a RGE flow by changing Λ.

Let us start from the Euclidean generating functional in momentum space restricted to wave numbers below a cut-off, $k \leq \Lambda$. Then we split the field modes into slow modes σ and fast modes ψ,

$$Z = \int \mathcal{D}\phi \, e^{-S[\phi]} = \int \mathcal{D}\sigma \mathcal{D}\phi \, e^{-S[\sigma+\psi]}, \tag{12.112}$$

with

$$\phi = \sigma + \psi \text{ and } \begin{cases} \sigma = 0, \text{ unless } |k| \leq \Lambda/f \\ \psi = 0, \text{ unless } \Lambda/f \leq |k| \leq \Lambda \end{cases}. \tag{12.113}$$

Next we want to integrate out the fast modes ψ,

$$e^{-S[\sigma]} = \int \mathcal{D}\psi \, e^{-S[\sigma+\psi]}, \tag{12.114}$$

thereby lowering the cut-off, $\Lambda \to \Lambda/f$. In general, we will not be able to perform this path integral. Using perturbation theory, we expand $e^{-S[\sigma+\psi]}$ and exponentiate the result. As illustrated with a toy model in problem 12.11, this procedure will renormalise the values of the parameters contained in the original Lagrangian and introduce an infinite set of irrelevant operator $\mathcal{O}_{d,i}$ with dimension $d > 5$.

Having integrated out ψ, we relabel σ as ϕ. Next we have to recover the canonical normalisation for its kinetic energy. If we rescale distances by $x \to x' = x/f$, the functional integral is again over modes $\phi(x')$ with $x > a$, complying with step iii) of the Kadanoff–Wilson prescription. Keeping the kinetic term invariant,

$$\int d^4x \, (\partial_\mu \phi)^2 = \int d^4x' \, (\partial'_\mu \phi')^2 = \frac{1}{f^2} \int d^4x \, (\partial_\mu \phi')^2, \tag{12.115}$$

requires a rescaling of the field as $\phi' = f\phi$. Let us consider now an irrelevant interaction, for example, $g_6 \phi^6$. Then

$$g_6 \int d^4x \, \phi^6 = \frac{g_6}{f^2} \int d^4x' \, \phi'^{\,6} \tag{12.116}$$

shows that the new coupling g'_6 is rescaled as $g'_6 = g_6/f^2$: As f grows and the cut-off scale Λ decreases, the value of an irrelevant coupling is driven to zero. Clearly, a relevant operator as the cosmological constant ρ or a mass term $m^2 \phi^2$ shows the opposite behaviour and grows. As result, irrelevant couplings are, in our low-energy world, suppressed and as first approximation a renormalisable theory emerges at low energies. Note that both directions of the RGE flow—towards the UV or the IR—are useful when discussing QFTs. The point of view of a RGE flow towards the IR is useful if we want to connect a theory at high energy scales to a theory valid at lower scales. An example of this approach is chiral perturbation theory where one connects QCD to an effective theory of mesons and baryons at low energies. In the opposite view, we may look, for example, at the SM as an effective theory known to be valid up to scales around TeV and ask what happens if we increase the cut-off.

Now we can generalise our earlier discussion of the two-dimensional RGE flow sketched in Fig. 12.5. The RGE flow stops at fixed points on critical surfaces of low dimension. All initial values on the critical surface inside the basin of attraction flow to the critical fixed point. Directions perpendicular to the critical surface are controlled by the irrelevant interactions; flows beginning from the surface are driven to the trivial fixed point. On the way towards $\xi \to \infty$ only the relevant and marginal interactions survive. Insisting not on $\xi \to \infty$ and keeping a finite cut-off (somewhere between TeV and M_{Pl}, depending on the limit of validity of the theory we assume) we can keep irrelevant interactions which are, however, suppressed.

Effective theories. While integrating out momentum shells in the path integral, as required in Eq. (12.114), is useful to demonstrate the concept of effective theories, it is not the method applied by practitioners. Instead, either functional RGE methods are used, for which we will give later an example in remark 13.1. Or one evaluates effective theories using the same apparatus as we have developed for renormalisable ones. In the following, we want first to illustrate how this is done and, second, to show that effective theories are predictive, despite of being non-renormalisable.

We note first that the fact that divergences are polynomial in the external momenta guarantees also for non-renormalisable theories that all divergences can be subtracted by local counter-terms. In contrast to renormalisable theories, the number of counter-terms and thus of a priori required measurements is infinite. Thus it is important to show that a truncation scheme for effective theories including loop correction exists which makes them predictive. As usually, we use as an example a real scalar field in $d = 4$ spacetime dimensions. Then an operator $\mathcal{O}_{n,m}$ consisting of n fields and m derivatives contributes to the action

$$S \approx \frac{g_{n,m}}{\Lambda^{p+q-4}} \int \mathrm{d}^4x \, \mathcal{O}_{n,m} \approx c_{n,m} \left(\frac{E}{\Lambda}\right)^{n+m-4} , \qquad (12.117)$$

where the $g_{n,m}$ are dimensionless couplings to be determined by experiment. We see again that irrelevant operators, $n + m > 4$, are suppressed at energies $E \ll \Lambda$. Clearly, an increase of the mass dimension $n + m$ results in a stronger suppression of the operator $\mathcal{O}_{n,m}$. This allows us to truncate the effective theory at some chosen dimension $D = n + m$. The truncated theory contains only a finite number of operators with unknown couplings $g_{n,m}$ which have to be determined from experiment. The predictions of the truncated theory become exact in the limit $E \to 0$. For finite $E \ll \Lambda$, we can increase the precision including higher dimensional operators, at the expense of more calculational work and additional experimental input with sufficient precision. In contrast, for $E \gtrsim \Lambda$ an infinite number of operators contribute a priori equally and the effective theory approach breaks down. Finally, note that in contrast to the Wilsonian action the value of the cut-off scale Λ is determined experimentally. For instance, calculating weak processes using a four-fermion interaction fixes the scale Λ^2 through the Fermi constant G_F.

Our estimate (12.117) holds at tree-level, and we should consider next the effect of loop corrections. To be specific, we estimate the order of the one-loop corrections induced by the operators $g_6 \phi^6/\Lambda^2$, $g_8 \phi^4 (\partial \phi)^2/\Lambda^4$, ..., to the basic $\lambda \phi^4$ vertex. Cutting the momentum integrals at Λ results in

$$\delta\lambda \approx \frac{g_6}{\Lambda^2} \int^\Lambda \frac{\mathrm{d}^4k}{(2\pi)^4} \frac{1}{k^2 - m^2} \approx \frac{c_6}{\Lambda^2} \Lambda^2 , \qquad (12.118)$$

$$\delta\lambda \approx \frac{g_8}{\Lambda^4} \int^\Lambda \frac{\mathrm{d}^4k}{(2\pi)^4} \frac{k^2}{k^2 - m^2} \approx \frac{c_8}{\Lambda^4} \Lambda^4 , \ldots . \qquad (12.119)$$

Thus all one-loop corrections are of $\mathcal{O}(1)$, since the factor $1/\Lambda^{D-4}$ contained in $\mathcal{O}_{n,m}$ is compensated by a factor Λ^{D-4} generated in the momentum integrations. Going to higher loops makes things even worse. The solution to this problem is to use a mass independent regulator as the MS or $\overline{\text{MS}}$ schemes in DR. Then the factors Λ^n originating from the momentum integration are replaces by m^n. With $E \gg m$, the loop corrections preserve thus the expansion scheme $(E/\Lambda)^{n+m-4}$ found at tree-level. As a result, the truncation of an effective theories at a chosen order D leads to a predictive theory including loop corrections at energies below the cut-off scale.

Summary

Interactions can be characterised by the asymptotic behaviour of their coupling constants. Gauge theories with a sufficiently small number of fermions are the only renormalisable interactions which are asymptotically free, that is, their running coupling constant goes to zero for $\mu \to \infty$. The scale dependence of renormalised Green functions can be interpreted as a running of coupling constants and masses.

The use of a running coupling sums up the leading logarithms of type $\ln^n(\mu^2/\mu_0^2)$, and a suitable choice of the renormalisation scale in a specific problems reduces the remaining scale dependence of perturbative results. The non-perturbative approach of Wilson provides an argument why the SM as description of our low-energy world is renormalisable: Integrating out high-energy degrees of freedom, irrelevant couplings are driven to zero and thus it is natural that a renormalisable theory emerges at low energies.

Further reading. Our discussion of the renormalisation of gauge theories left out most details. I recommend those seeking to fill the gaps to start with Ramond (1994) and Pokorski (1987). A useful textbook to learn about critical phenomena is Le Bellac (1992).

Problems

12.1 Quantum action for a free field. Show for a free scalar field ϕ that the quantum action $\Gamma[\phi_c]$ coincides with the classical action $S[\phi]$.

12.2 $\Lambda_{\rm QCD}$. Show that (12.92) can be rewritten as $\alpha_s(\mu^2) = 1/[b_1 \ln(\mu^2/\Lambda_{\rm QCD}^2)]$. Find the relation between $\Lambda_{\rm QCD}$ in the MS and $\overline{\rm MS}$ scheme.

12.3 Beta function in QED. An alternative way to derive the beta function starts from the renormalised vacuum polarisation, $\beta(e) = \frac{e}{2}\mu\frac{\partial}{\partial\mu}\Pi(q^2)$. a.) Show that using MS or $\overline{\rm MS}$ scheme reproduces our previous result. b.) Find $\Pi(q^2)$ in the MoM scheme and derive $\beta(e)$. Compare the two results, especially in the $m^2 \ll q^2$ limit.

12.4 Fermions in DR. Repeat the calculation of the vacuum polarisation $\Pi(q^2)$ using $\mathrm{tr}\{\gamma^\mu, \gamma^\nu\} = f(d)\eta^{\mu\nu}$, where $f(d = 4) = 4$ is a smooth function satisfying $f(4) = 4$.

12.5 General solution of the RGE equation. Find the general solution of the RGE equation (12.66c) using the method of characteristics or the following analogue. Coleman suggested considering the growth rate $g(x)$ of bacteria with density $\rho(x,t)$ in an one-dimensional flow $v(x)$ as analogue to the RGE,

$$\left[\frac{\partial}{\partial t} + v(x)\frac{\partial}{\partial x} - g(x)\right]\rho(x,t) = 0.$$

Find for given initial condition $\rho(x,0) = \rho_0(x)$ the time evolution of $\rho(x,t)$.

12.6 Imaginary part of the photon polarisation. a.) Derive the imaginary part of the photon polarisation $\Pi^{\rm on}(q^2)$. b.) Use the optical theorem to connect $\Im(\Pi^{\rm on})$ to the decay of a virtual photon γ^* into a fermion pair, $\gamma^* \to \bar{f}f$. [Hint: Consider $\mathrm{d}\Phi^{(2)}\sum_{s_1,s_2}\mathcal{A}_{in}^{\mu*}\mathcal{A}_{ni}^\nu$ and use the tensor method.]

12.7 Recurrence relation for the beta function. Find from the RGE for $\alpha(Q^2) = \sum_{n=0}^\infty C_n(Q^2/\mu^2)\alpha_s(\mu^2)^n$ a recurrence relation for the coefficients C_n, $C_n(x) = c_n\ln^n(x)$. Find the coefficient c_n in front of the leading log.

12.8 Effective four-fermion theory. Consider the effective theory

$$\mathscr{L} = \bar{\psi}(\mathrm{i}\slashed{\partial} - m)\psi - g/\Lambda^2(\bar{\psi}\psi)^2.$$

Calculate the mass correction δm using a) as (Euclidean) cut-off Λ, and b.) a mass-independent renormalisation scheme.

12.9 $\beta_{\rm QCD}$. Find Z_3 from the $1/\varepsilon$ poles of the gluon vacuum polarisation and determine the 1-loop contribution b_1 to the beta function of QCD.

12.10 Asymptotic freedom. In QCD, the dominance of the bosonic loop contributions to the beta function leads to asymptotic freedom. Which difference is responsible that the scalar $\lambda\phi^4$ theory is in contrast not asymptotically free?

12.11 Toy model for the effective action approach. Consider

$$Z = \int \mathrm{d}x\mathrm{d}y \exp([-(x^2+y^2+\lambda x^4+\lambda x^2 y^2)])$$

as a toy model for the generating functional of two coupled scalar fields. Integrate out the field y, assume then that λ is small:

Expand first the result and then rewrite it as an exponential. Show that this process results in a.) a renormalisation of the mass term x^2, b.) a renormalisation of the coupling term x^4, and c.) the appearance of new ("irrelevant") interactions x^n with $n \geq 6$.

12.12 Renormalisation group a.) Show that the 1D Ising model with $\tanh(K_n) = \tanh^n(K)$ is for $T \to 0$ "quasi-critical", that is, that the coupling K_n changes only logarithmically with n. b.) Assume a system which coupling changes as $K_n = \frac{1}{2}(2K)^n$. Considering a continuous change, derive its beta function. Show that the zero of the beta function coincides with the fixed point implied by K_n. What are the properties of the fixed point?

13
Symmetries and symmetry breaking

The analogy of Landau's mean field model for a ferromagnet with a scalar $\lambda\phi^4$ theory suggests that we can hide its $\phi \to -\phi$ symmetry at low temperatures, if we choose a negative mass term in the Lagrangian. Although such a choice seems at first sight unnatural, we will investigate this case in the following in detail. Our main motivation is the expectation that hiding a symmetry by choosing a non-invariant ground-state retains the "good" properties of the symmetric Lagrangian. Coupling such a scalar theory to a gauge theory, we hope to break gauge invariance in a "gentle" way which allows gauge boson masses, for example, without spoiling the renormalisability of the unbroken theory. As an additional motivation we recall that couplings and masses are not constants but depend on the scale considered. Thus it might be that the parameters determining the Lagrangian of the Standard Model at low energies originate from a more complete theory at high scales, where the squared mass parameter μ^2 is originally positive. In such a scenario, $\mu^2(Q^2)$ becomes negative only after running it down to the electroweak scale $Q = m_Z$.

13.1 Symmetry breaking and Goldstone's theorem

In the following we consider systems where the Lagrangian contains an exact symmetry which is not shared by its ground-state. Since particle masses in a field theory are determined by the ground-state, the symmetry of the Lagrangian is thus not visible in the mass spectrum of physical particles. This opens the possibility of having a gauge-invariant Lagrangian despite massive gauge bosons. If the ground-state breaks the original symmetry because one or several scalar fields acquire a non-zero vacuum expectation value, one calls this spontaneous symmetry breaking (SSB). As the symmetry is not really broken on the Lagrangian level, a perhaps more appropriate name would be "hidden symmetry".

In this and the following chapter, we discuss the case of SSB, first in general and then applied to the electroweak sector of the SM. Since the breaking of an internal symmetry should leave Poincaré symmetry intact, we can give only scalar quantities ϕ a non-zero vacuum expectation value. This excludes non-zero expectation values for fields with spin, which would single out a specific direction. On the other hand, we can construct scalar expectation values as $\langle 0|\phi|0\rangle = \langle 0|\bar{\psi}\psi|0\rangle \neq 0$ out of the product of multiple fields. In the following, we will always treat ϕ as an elementary field, but we should keep in mind the possibility that ϕ is a composite object, for example, a condensate of fermion fields, $\langle\phi\rangle = \langle\bar{\psi}\psi\rangle$, similar to the case of superconductivity.

Quantum Fields–From the Hubble to the Planck Scale. Michael Kachelriess. © Michael Kachelriess 2018.
Published in 2018 by Oxford University Press. DOI 10.1093/oso/9780198802877.001.0001

Spontaneous breaking of discrete symmetries. We start with the simplest example of a theory with a broken symmetry: a single scalar field with a discrete reflection symmetry. Consider the familiar $\lambda\phi^4$ Lagrangian, but with a negative mass term which we include into the potential $V(\phi)$,

$$\mathscr{L} = \frac{1}{2}(\partial_\mu\phi)^2 + \frac{1}{2}\mu^2\phi^2 - \frac{\lambda}{4}\phi^4 = \frac{1}{2}(\partial_\mu\phi)^2 - V(\phi). \tag{13.1}$$

The Lagrangian is for both signs of μ^2 invariant under the discrete Z_2 symmetry, $\phi \to -\phi$. The field configuration with the smallest energy is a constant field ϕ_0, chosen to minimise the potential

$$V(\phi) = -\frac{1}{2}\mu^2\phi^2 + \frac{\lambda}{4}\phi^4, \tag{13.2}$$

which has the two minima

$$\phi_0 = \pm v \equiv \pm\sqrt{\mu^2/\lambda}. \tag{13.3}$$

In quantum mechanics, we learnt that the wave-function of the ground-state for the potential $V(x) = -\frac{1}{2}\mu^2x^2 + \lambda x^4$ will be a symmetric state, $\psi(x) = \psi(-x)$, since the particle can tunnel through the potential barrier. In field theory, such a tunnelling can happen in principle too. However, the tunnelling probability is inversely proportional to the volume L^3 occupied by the system, and vanishes in the limit $L \to \infty$. In order to transform $\phi(x) = -v$ into $\phi(x) = +v$ we have to switch an infinite number of oscillators, which clearly costs an infinite amount of energy. Thus in quantum field theory, the system has to choose between the two vacua $\pm v$ and the symmetry of the Lagrangian is broken in the ground-state. Had we used the ϕ^4 Lagrangian with a positive mass term the vacuum expectation value of the field would have been zero, and the ground-state would respect the symmetry.

Quantising the theory (13.1) with the negative mass around the usual vacuum, $|0\rangle$ with $\langle 0|\phi|0\rangle = \phi_c = 0$, we find modes behaving as

$$\phi_k \propto \exp(-i\omega t) = \exp(-i\sqrt{-\mu^2 + |\mathbf{k}|^2}\, t), \tag{13.4}$$

which can grow exponentially for $|\mathbf{k}|^2 < \mu^2$. More generally, exponentially growing modes exist, if the potential is concave at the position of ϕ_c, that is, for

$$m_{\text{eff}}^2(\phi_c) = V''(\phi_c) = -\mu^2 + 3\lambda\phi_c^2 < 0 \tag{13.5}$$

or $|\phi_c| < \sqrt{\mu^2/(3\lambda)}$.

Clearly, the problem arises because we should expand the field around the ground-state v. This requires that we shift the field as

$$\phi(x) = v + \xi(x), \tag{13.6}$$

splitting it into a classical part $\langle\phi\rangle = v$ and quantum fluctuations $\xi(x)$ on top of it. Then we express the Lagrangian as function of the field ξ,

$$\mathscr{L} = \frac{\mu^4}{4\lambda} + \frac{1}{2}(\partial_\mu\xi)^2 - \frac{1}{2}(2\mu^2)\xi^2 - \mu\sqrt{\lambda}\xi^3 - \frac{\lambda}{4}\xi^4. \tag{13.7}$$

In the new variable ξ, the Lagrangian describes a scalar field with *positive* mass $m_\xi = \sqrt{2}\mu > 0$. The original symmetry is no longer apparent. Since we had to select one out of the two possible ground states, a ξ^3 term appeared and the $\phi \to -\phi$ symmetry is broken. The new cubic interaction term rises now the question, if our scalar $\lambda\phi^4$ theory is not renormalisable after SSB. As we have no corresponding counter-term at our disposal, the renormalisation of μ and λ has to cure also the divergences of the ξ^3 interaction.

Finally, we note that the contribution $\mu^4/(4\lambda)$ to the energy density of the vacuum is, in contrast to the vacuum loop diagrams generated by $Z[0]$, classical and finite. We see later that symmetries will be restored at high temperatures or at early times in the evolution of the universe. Even if we take the freedom to shift the vacuum energy density, we have either before or after SSB an unacceptable large contribution to the vacuum energy (problem 13.1).

Spontaneous breaking of continuous symmetries. Our main aim is to understand the SSB of the electroweak gauge symmetry. As next step we look therefore at a system with a global continuous symmetry. In section 5.1 we discussed the case of N real scalar fields described by the Lagrangian

$$\mathscr{L} = \frac{1}{2}\left[(\partial_\mu\boldsymbol{\phi})^2 + \mu^2\boldsymbol{\phi}^2\right] - \frac{\lambda}{4}(\boldsymbol{\phi}^2)^2. \tag{13.8}$$

Since $\boldsymbol{\phi} = \{\phi_1, \ldots, \phi_N\}$ transforms as a vector under rotations in field space, $\phi_i \to R_{ij}\phi_j$ with $R_{ij} \in \mathrm{O}(n)$, the Lagrangian is invariant under orthogonal transformations. Before we consider the general case of arbitrary N, we look at the case $N = 2$ for which the potential is shown in Fig. 13.1. Without loss of generality, we choose the vacuum pointing in the direction of ϕ_1. Thus $v = \langle\phi_1\rangle = \sqrt{\mu^2/\lambda}$ and $\langle\phi_2\rangle = 0$. Shifting the field as in the discrete case gives

$$\mathscr{L} = \frac{\mu^4}{4\lambda} + \frac{1}{2}(\partial_\mu\boldsymbol{\xi})^2 - \frac{1}{2}(2\mu^2)\xi_1^2 + \mathscr{L}_{\text{int}}, \tag{13.9}$$

and thus the two degrees of freedom of the field $\boldsymbol{\phi}$ split after SSB into one massive and one massless mode.

Since the mass matrix consists of the coefficients of the terms quadratic in the fields, the general procedure for the determination of physical masses is the following: find first the minimum of the potential $V(\boldsymbol{\phi})$. Then expand the potential up to quadratic terms,

$$V(\boldsymbol{\phi}) = V(\boldsymbol{\phi}_0) + \frac{1}{2}(\boldsymbol{\phi} - \boldsymbol{\phi}_0)_i(\boldsymbol{\phi} - \boldsymbol{\phi}_0)_j \underbrace{\frac{\partial^2 V}{\partial\phi_i\partial\phi_j}}_{M_{ij}} + \cdots. \tag{13.10}$$

The second derivatives form a symmetric matrix with elements $M_{ij} \geq 0$, because we evaluate the mass matrix by assumption at the minimum of V. Diagonalising M_{ij} gives the squared masses of the fields as eigenvalues. The eigenvectors of M_{ij} are called the mass eigenstates or physical states. Propagators and Green functions describe the evolution of fields with definite masses and should be therefore build up on these states. If the potential has $n > 0$ flat directions, the vacuum is degenerated and n massless modes appear.

$$V(\phi)$$

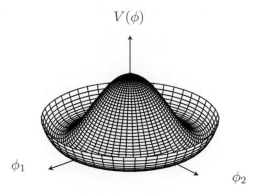

ϕ_1

ϕ_2

Fig. 13.1 Scalar potential with "Mexican hat" shape symmetric under O(2).

Looking at Fig. 13.1 suggests using polar instead of Cartesian coordinates in field space. In this way, the rotation symmetry of the potential and the periodicity of the flat direction is reflected in the variables describing the scalar fields. Introducing first the complex field $\phi = (\phi_1 + i\phi_2)/\sqrt{2}$, the Lagrangian becomes

$$\mathscr{L} = \partial_\mu \phi^\dagger \partial^\mu \phi + \mu^2 \phi^\dagger \phi - \lambda(\phi^\dagger \phi)^2. \tag{13.11}$$

Next we set

$$\phi(x) = \rho(x)e^{i\vartheta(x)} \tag{13.12}$$

and use $\partial_\mu \phi = [\partial_\mu \rho + i\rho \partial_\mu \vartheta]e^{i\vartheta}$ to express the Lagrangian in the new variables,

$$\mathscr{L} = (\partial_\mu \rho)^2 + \rho^2 (\partial_\mu \vartheta)^2 + \mu^2 \rho^2 - \lambda \rho^4. \tag{13.13}$$

Finally shifting yet again the fields as $\rho = v + \xi$ with $v = \sqrt{\mu^2/2\lambda}$, we find

$$\begin{aligned}
\mathscr{L} = &\frac{\mu^4}{4\lambda} + \frac{\mu^2}{2\lambda}(\partial_\mu \vartheta)^2 + (\partial_\mu \xi)^2 - 2\mu^2 \xi^2 - 2\mu\sqrt{2\lambda}\xi^3 - \lambda\xi^4 \\
&+ \left[\sqrt{2\mu^2/\lambda}\,\xi + \xi^2\right](\partial_\mu \vartheta)^2.
\end{aligned} \tag{13.14}$$

The phase ϑ which parameterises the flat direction of the potential $V(\vartheta, \xi)$ remained massless. This mode is called Goldstone (or Nambu–Goldstone) boson and has derivative couplings to the massive field ξ, given by the last term in Eq. (13.14). This is a general result, implying that static Goldstone bosons do not interact. Another general property of Goldstone bosons is that they carry the quantum number of the corresponding symmetry generator. They are therefore (pseudo-) scalar particles, if an internal symmetry is broken,

Let us now briefly discuss the case of general N for the Lagrangian (13.8). The lowest-energy configuration is again a constant field. The potential is minimised for any set of fields ϕ_0 that satisfies $\phi_0^2 = \mu^2/\lambda$. This equation only determines the length of the vector, but not its direction. It is convenient to choose a vacuum such that

ϕ_0 points along one of the components of the field vector. Aligning ϕ_0 with its Nth component,

$$\phi_0 = \left(0, \ldots, 0, \sqrt{\mu^2/\lambda}\right), \tag{13.15}$$

we now follow the same procedure as in the previous example. First we define a new set of fields, with the Nth field expanded around the vacuum

$$\phi(x) = (\phi^k(x), v + \xi(x)), \tag{13.16}$$

where k now runs from 1 to $N - 1$. Then we insert this, and the value $v = \sqrt{\mu^2/\lambda}$ for the vacuum expectation value into the Lagrangian, and obtain

$$\mathcal{L} = \frac{1}{2}(\partial_\mu \phi^k)^2 + \frac{1}{2}(\partial_\mu \xi)^2 - \frac{1}{2}(2\mu^2)\xi^2 + \frac{1}{4}\frac{\mu^4}{\lambda}$$
$$- \sqrt{\lambda}\mu\xi^3 - \sqrt{\lambda}\mu\xi(\phi^k)^2 - \frac{\lambda}{2}\xi^2(\phi^k)^2 - \frac{\lambda}{4}\left[(\phi^k)^2\right]^2 - \frac{\lambda}{4}\xi^4. \tag{13.17}$$

This Lagrangian describes $N - 1$ massless fields and a single massive field ξ, with cubic and quartic interactions. The $O(N)$ symmetry is no longer apparent, leaving as symmetry group the subgroup $O(N - 1)$, which rotates the ϕ^k fields among themselves. This rotation describes movements along directions where the potential has a vanishing second derivative, while the massive field corresponds to oscillations in the radial direction of V.

Goldstone's theorem. The observation that massless particles appear in theories with spontaneously broken *continuous* symmetries is a general result, known as Goldstone's theorem. The first example for such particles was suggested by Nambu in 1960. He showed that a massless quasi-particle appears in a magnetised solid, because the magnetic field breaks rotation invariance. Goldstone applied this idea to relativistic QFTs and showed that massless scalar elementary particles appear in theories with SSB. Since no massless scalar particles are known to exist, this theorem appeared to be a dead-end for the application of SSB to particle physics. Hence our task is twofold: first we should derive Goldstone's theorem and then we should find out how we can bypass the theorem applying it to gauge theories.

The theorem is obvious at the classical level. Consider a Lagrangian with a symmetry G and a vacuum state invariant under a subgroup H of G. For instance, choosing a Lagrangian invariant under $G = O(3)$ and picking out a vacuum along ϕ_3, the subgroup $H = O(2)$ of rotation around ϕ_3 keeps the vacuum invariant. Let us denote with $U(g)$ a representation of G acting on the fields ϕ and with $U(h)$ a representation of H, respectively. Since we consider constant fields, derivative terms in the fields vanish and the potential V alone has to be symmetric under G, that is,

$$V(U(g)\phi) = V(\phi). \tag{13.18}$$

Moreover, we know that the vacuum is kept invariant for all h, $\phi_0' = U(h)\phi_0$, but changes for some g, $\phi_0' \neq U(g)\phi_0$. Using the invariance of the potential and expanding $V(U(g)\phi_0)$ for an infinitesimal group transformation gives

$$V(\boldsymbol{\phi}_0) = V(U(g)\boldsymbol{\phi}_0) = V(\boldsymbol{\phi}_0) + \frac{1}{2}\left.\frac{\partial^2 V}{\partial\phi_i\partial\phi_j}\right|_0 \delta\phi_i\delta\phi_j + \ldots, \qquad (13.19)$$

where $\delta\phi_i$ denotes the resulting variation of the field. Equation (13.19) implies that

$$M_{ij}\delta\phi_i\delta\phi_j = 0. \qquad (13.20)$$

The variation $\delta\phi_i$ depends on whether the transformation belong to $U(h)$ or not. In the former case, the vacuum $\boldsymbol{\phi}_0$ is unchanged, $\delta\phi_i = 0$ and (13.20) is automatically satisfied. If on the other hand g does not belong to H, that is g is a member of the left co-set G/H, then $\delta\phi_i \neq 0$, implying that the mass matrix M_{ij} has a zero eigenvalue. It is now clear that the number of massless particles is determined by the dimensions of the two groups G and H. The number of Goldstone bosons equals $\dim(G) - \dim(H)$, or the dimension $\dim(G/H)$ of the left co-set.

Quantum case. The previous discussion was based on the classical potential. Thus we should address the question if this picture survives quantum corrections. Noether's theorem tells us that every continuous symmetry has associated to its generators g_i conserved charges Q_i. On the quantum level this means the operators Q_i commute with the Hamiltonian, $[H, Q_i] = 0$. Subtracting the vacuum energy, we have $H\,|0\rangle = 0$. If the vacuum is invariant under the symmetry Q, then $\exp(i\vartheta Q)\,|0\rangle = |0\rangle$. For the infinitesimal form of the symmetry transformation, $\exp(i\vartheta Q) \approx 1 + i\vartheta Q$, we conclude that the charge annihilates the vacuum,

$$Q\,|0\rangle = 0, \qquad (13.21)$$

or, in simpler words, the vacuum has the charge 0.

Now we came to the case we are interested in, namely that the symmetry is spontaneously broken and thus $Q\,|0\rangle \neq 0$. We first determine the energy of the state $Q\,|0\rangle$. From

$$HQ\,|0\rangle = (HQ - \underbrace{QH}_{H|0\rangle=0})\,|0\rangle = [H, Q]\,|0\rangle = 0, \qquad (13.22)$$

we see that at least another state $Q\,|0\rangle$ exists which has as the vacuum $|0\rangle$ zero energy. We represent the charge operator as the volume integral of the time-like component of the corresponding current operator,

$$Q = \int \mathrm{d}^3 x\, J^0(t, \boldsymbol{x}). \qquad (13.23)$$

The state

$$|s\rangle = \int \mathrm{d}^3 x\, \mathrm{e}^{i\boldsymbol{p}\cdot\boldsymbol{x}}\, J^0(t, \boldsymbol{x})\,|0\rangle \to Q\,|0\rangle \qquad \text{for} \quad \boldsymbol{p} \to 0 \qquad (13.24)$$

becomes in the zero-momentum limit equal to the state $Q\,|0\rangle$ we are searching for. Moreover, applying the momentum operator $\boldsymbol{P}$ on $|s\rangle$ gives (problem 13.5)

$$\boldsymbol{P}\,|s\rangle = \boldsymbol{p}\,|s\rangle. \qquad (13.25)$$

Thus the SSB of the vacuum, $Q\,|0\rangle \neq 0$, implies excitations of the system with a frequency that vanishes in the limit of long wavelengths. In the relativistic case, Goldstone's theorem predicts massless states, while in the non-relativistic case relevant for solid states the theorem predicts collective excitations with zero energy gap.

13.2   Renormalisation of theories with SSB

When we went through the SSB of the scalar field, we saw that new ϕ^3 interactions were introduced. The question then arises, are new renormalisation constants needed when a symmetry is spontaneously broken? This would make these theories non-renormalisable. We can address this questions in two ways. One possibility is to repeat our analysis of the renormalisability of the scalar theory in section 11.4.2, but now for the broken case with a negative mass term. Then we would find that the ϕ^3 term becomes finite, renormalising fields, mass and coupling as in the unbroken case. This is not unexpected, because shifting the field, which is an integration variable in the generating functional, by $\phi \to \tilde{\phi} = \phi - v$ should not affect physics. On the other hand, such a shift reshuffles the splitting $\mathscr{L} = \mathscr{L}_0 + \mathscr{L}_{\mathrm{int}}$ in our standard perturbative expansion in the coupling constant. To avoid this problem, we will use the quantum action employing a loop expansion. Additionally to not being affected by a shift of the fields, this formalism allows us to calculate the potential including all quantum corrections in the limit of constant fields.

Effective potential.   Let us start by recalling the definition of the quantum action[1]

$$\Gamma[\phi_c] = W[J] - \int \mathrm{d}^4 x'\, J(x')\phi_c(x') \equiv W[J] - \langle J\phi\rangle \,. \tag{13.26}$$

In general we will not be able to solve the quantum action. Studying SSB, we can, however, make use of a considerable simplification: the fields we are interested in are constant. Performing then a gradient expansion of the quantum action $\Gamma[\phi]$,

$$\Gamma[\phi] = \int \mathrm{d}^4 x \left[-V_{\mathrm{eff}}(\phi) + \frac{1}{2} F_2(\phi)(\partial_\mu \phi)^2 + \ldots \right], \tag{13.27}$$

only the zeroth order term $V_{\mathrm{eff}}(\phi)$ of the expansion in $(\partial_\mu \phi)^2$ survives. If we now choose the source $J(x)$ to be constant, the field $\phi(x)$ has to be uniform too, $\phi(x) = \phi$, by translation invariance. Together this implies that

$$\Gamma = -\Omega V_{\mathrm{eff}} \,, \quad \text{and} \quad -J = \frac{\delta \Gamma[\phi]}{\delta \phi} = -\Omega\, \frac{\partial V_{\mathrm{eff}}(\phi)}{\partial \phi} \,, \tag{13.28}$$

where Ω denotes the spacetime volume. In the absence of external sources, $J = 0$, Eq. (13.28) simplifies to $V'_{\mathrm{eff}}(\phi) = 0$. This is the quantum version of our old approach where we minimised the classical potential $V(\phi)$ in order to find the vacuum expectation value of ϕ. Therefore $V_{\mathrm{eff}}(\phi)$ is called the effective potential, which includes all quantum corrections to the classical potential in the limit of zero gradients.

In order to proceed, we use the fact that we know the classical potential and we assume that quantum fluctuations are small. Then we can perform a saddle-point expansion around the classical solution ϕ_0, given by the solution to $\Box \phi_0 + V'(\phi_0) = J(x)$. We split the field as $\phi = \phi_0 + \tilde{\phi}$ and approximate the path integral as

---

[1]We will suppress the subscript c on the classical field from now on and use brackets $\langle J\phi\rangle$ to indicate integration.

$$Z = e^{iW} \simeq \exp\{i\left[S[\phi_0] + \langle J\phi_0 \rangle\right]\} \int \mathcal{D}\tilde{\phi} \exp\left\{i \int d^4x \frac{1}{2}\left[(\partial_\mu\tilde{\phi})^2 - V''(\phi_0)\tilde{\phi}^2\right]\right\}.$$
$$(13.29)$$

The neglected terms are of order $\mathcal{O}(\hbar^2)$ and correspond thus to two- and higher loop contributions. The functional integral over $\tilde{\phi}$ is quadratic and is formally given by $\mathrm{Det}(\Box + V'')^{-1/2}$. Using the identity $\ln \mathrm{Det} A = \mathrm{Tr} \ln A$, we then find

$$W = S[\phi_0] + \langle J\phi_0 \rangle + \frac{i}{2}\mathrm{Tr}\ln[\Box + V''(\phi_0)] + \mathcal{O}(\hbar^2). \tag{13.30}$$

The trace implies a summation over all discrete and an integration over the continuous quantum numbers. In the case of a scalar particle, no discrete quantum numbers exist and we have to integrate the matrix element $\langle x | \ln[\Box + V'']|x\rangle$ only over spacetime. To find the eigenvalues of the operator, we insert a complete sets of plane waves,

$$\mathrm{Tr}\ln[\Box + V''] = \int d^4x \langle x|\ln[\Box + V'']|x\rangle = \int d^4x \frac{d^4k}{(2\pi)^4}\langle x|\ln[\Box + V'']|k\rangle\langle k|x\rangle =$$

$$= \int d^4x \frac{d^4k}{(2\pi)^4}\ln[-k^2 + V'']\langle x|k\rangle\langle k|x\rangle = \Omega \int \frac{d^4k}{(2\pi)^4}\ln[-k^2 + V'']. \tag{13.31}$$

Performing the Legendre transformation and using $S[\phi_0] = -\Omega V(\phi_0)$, we obtain for the effective potential $V_{\mathrm{eff}}(\phi_0)$ including the first quantum corrections

$$V_{\mathrm{eff}}(\phi_0) = V(\phi_0) - \frac{i}{2}\int \frac{d^4k}{(2\pi)^4}\ln\left[-k^2 + V''(\phi_0)\right] + \mathcal{O}(\hbar^2). \tag{13.32}$$

As an example we can use the $\lambda\phi^4$ theory. From

$$V''(\phi_0) = \mu^2 + \frac{1}{2}\lambda\phi_0^2, \tag{13.33}$$

we see that $V''(\phi_0)$ can be interpreted as an effective mass, consisting of μ^2 and the contribution $\lambda\phi_0^2/2$ due to the constant background field ϕ_0. The total effective potential at order $\mathcal{O}(\hbar^2)$ consists of the classical potential $V(\phi)$, that is the classical energy density of a scalar field with vacuum expectation value ϕ, while the first quantum correction is given by the zero-point energies[2] of a scalar particle with effective mass $V''(\phi_0)$.

Not surprisingly, the effective potential is divergent and we have to introduce counter-terms that eliminate the divergent parts. Our effective potential is then

$$V_{\mathrm{eff}}(\phi_0) = V(\phi) + \frac{1}{2}\int \frac{d^4k_E}{(2\pi)^4}\ln\left(\frac{k_E^2 + V''(\phi_0)}{k_E^2}\right) + B\phi_0^2 + C\phi_0^4 + \mathcal{O}(\hbar^2). \tag{13.34}$$

Here, we Wick-rotated the integral to Euclidean space and subtracted an infinite constant in order to make the logarithm dimensionless. (Equivalently we could have added an additional constant counter-term A renormalising the vacuum energy density.) The

---

[2]Integrating $i\Delta_F(0)$ w.r.t. m^2 reproduces the one-loop term.

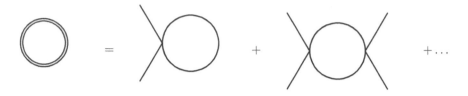

Fig. 13.2 Perturbative expansion of the one-loop effective potential $V_{\text{eff}}^{(1)}$ for the $\lambda\phi^4$ theory; all external legs have zero momentum.

integral can be evaluated in different regularisation schemes. Here we will expand the logarithm,

$$\ln\left(1+\frac{V''}{k_E^2}\right) = \sum_{n=1}^{\infty} \frac{1}{n}\left(\frac{V''}{k_E^2}\right)^n, \tag{13.35}$$

and cut-off the integral at some large momentum Λ. The first two terms of the sum will depend on the cut-off, being proportional to Λ^2 and $\ln(\Lambda^2/V'')$, respectively. Performing the integral and neglecting terms that vanish for large Λ, we obtain

$$V_{\text{eff}}(\phi_0) = V(\phi_0) + \frac{\Lambda^2}{32\pi^2}V''(\phi_0) + \frac{V''(\phi_0)^2}{64\pi^2}\ln\left(\frac{V''(\phi_0)}{\Lambda^2}\right). \tag{13.36}$$

Now we see that if we start out with a massless $\lambda\phi^4$ theory, our cut-off-dependent terms are

$$V'' = \frac{1}{2}\lambda\phi_0^2, \quad \text{and} \quad (V'')^2 = \frac{\lambda^2}{4}\phi_0^4, \tag{13.37}$$

which both can be absorbed into the counter-terms B and C by imposing appropriate renormalisation conditions.

Let us stress the important point in this result. The renormalisation of the $\lambda\phi^4$ theory using the effective potential approach is not affected by a shift of the field. We are free to use both signs of μ^2 and any value of the classical field ϕ_0 in Eq. (13.36). Independently of the sign of μ^2, we need only symmetric counter-terms, as a cubic term does not appear. We can rephrase this point as follows. If we renormalise before we shift the fields, we know that we obtain finite renormalised Green functions. Shifting the fields, however, does not change the total Lagrangian. Thus the quantum action and the effective potential are unchanged too. Consequently the theory has to stay renormalisable after SSB.

Let us now discuss what happens with a non-renormalisable theory in the effective potential approach. Including, for example, a ϕ^6 term leads to $(V'')^2 \propto \phi^8$ which requires an additional counter-term $D\phi^8$, generating in turn even higher order terms and so forth. Thus in this case an infinite number of counter-terms is needed for the calculation of $V_{\text{eff}}^{(1)}$. The reason for this behaviour becomes clear, if we look again at the series expansion of the logarithm in the one loop contribution $V_{\text{eff}}^{(1)}$,

$$V_{\text{eff}}^{(1)} = i\sum_{n=1}^{\infty}\int\frac{\mathrm{d}^4k}{(2\pi)^4}\frac{1}{2n}\left[\frac{V''(\phi_0)}{k^2}\right]^n. \tag{13.38}$$

This contribution is an infinite sum of single loops with progressively more external legs with zero-momentum attached, see Fig. 13.2 for the case of $V''(\phi_0) = \frac{1}{2}\lambda\phi_0^2$. (We added the factor i, because we returned to Minkowski space; the symmetry factor $2n$ appearing automatically in this approach accounts for the symmetry of a graph with n vertices under rotations and reflection.) As we saw, the superficial degree of divergence increases with the number of external particles for a $\lambda\phi^n$ theory and $n > 4$. Hence every single diagram in the infinite sum contained in $V_{\text{eff}}^{(1)}$ diverges for $n > 4$ and requires a counter-term of higher order. As discussed in section 12.5, we should treat non-renormalisable theories as effective theories only valid below a physical cut-off Λ. Calculating then loop corrections as in $V_{\text{eff}}^{(1)}$, we have to us a mass independent renormalisation scheme instead of a dimensionfull cut-off. An alternative is the functional RGE approach fow which we give next a brief example.

Remark 13.1: RGE flow equation for the effective potential. Let us define V_k as the effective potential cutting off the loop integrals at the scale k. We can repeat the saddle-point expansion, splitting ϕ as $\phi = \phi_0 + \psi$, where now ϕ_0 contains the modes $k < p$ and ψ the modes $p \leq k \leq \Lambda$. For simplicity, we assume that the minimum of the potential is at zero and treat the slow field ϕ_0 as uniform. Integrating out the field ψ, we obtain then

$$V_k(\phi_0) = V_\Lambda(\phi_0) + \frac{\hbar}{2} \int_k^\Lambda \frac{\mathrm{d}^4 k}{(2\pi)^4} \ln\left(\frac{k_E^2 + V''(\phi_0)}{k_E^2 + V''(0)} \right). \tag{13.39}$$

Next we differentiate w.r.t. k and obtain

$$k \frac{\partial}{\partial k} V_k(\phi_0) = -\frac{\hbar k^4}{16\pi^2} \ln\left(\frac{k_E^2 + V''(\phi_0)}{k_E^2 + V''(0)} \right), \tag{13.40}$$

that is, a differential equation describing how the effective potential changes integrating out UV modes with momentum k. Similar equations, called RGE flow equations, can be derived also for the quantum action. If we consider an asymptotically free theory, then for sufficiently high scales, $\Gamma[\phi] \simeq S[\phi]$. Thus V_k can be approximated by the classical potential, fixing our initial condition.

Finally, we note that the result for a single, real scalar generalises as

$$V_{\text{eff}}^{(1)} = (-1)^s g_i V_{\text{eff,scalar}}^{(1)} \tag{13.41}$$

to a particle with spin s and g_i internal degrees of freedom. This should come as no surprise since the one-loop expression of the effective potential sums up the logarithm of the zero-point energies.

Another proof of the Goldstone theorem. With the help of the effective potential we can give another simple proof of the Goldstone theorem. We know that the zero of the inverse propagator determines the mass of a particle. From Eq. (12.15), the exact inverse propagator in momentum space for a set of scalar fields is given by

$$\Delta_{ij}^{-1}(p^2) = \int \mathrm{d}^4 x \, e^{-ip(x-x')} \frac{\delta^2 \Gamma}{\delta\phi_i(x)\delta\phi_j(x')}. \tag{13.42}$$

Massless particles correspond to zero eigenvalues of this matrix equation for $p^2 = m^2$. If we set $\boldsymbol{p} = 0$, the fields are constant. But differentiating the quantum action w.r.t. to constant fields is equivalent to differentiating simply the effective potential,

$$\frac{\partial^2 V_{\text{eff}}}{\partial \phi_i(x) \partial \phi_j(x')} = 0. \tag{13.43}$$

Thus our previous analysis of Goldstone's theorem using the classical potential also holds in the quantum case if we replace the classical by the effective potential.

Coleman–Weinberg Problem.   We can use the effective potential to investigate whether quantum fluctuations can trigger SSB in an initially massless theory. Rewriting the effective potential (and going back to our normalisation $\lambda\phi/4!$) we have

$$V_{\text{eff}}(\phi) = \left[\frac{\Lambda^2}{64\pi^2} \lambda + B \right] \phi^2 + \left[\frac{\lambda}{4!} + \frac{\lambda^2}{(16\pi)^2} \ln \frac{\phi^2}{\Lambda^2} + C \right] \phi^4. \tag{13.44}$$

Now we impose the renormalising conditions, first

$$\left. \frac{\mathrm{d}^2 V_{\text{eff}}}{\mathrm{d}\phi^2} \right|_{\phi=0} = 0, \tag{13.45}$$

which implies that

$$B = -\frac{\lambda\Lambda^2}{64\pi^2}. \tag{13.46}$$

Renormalising the coupling constant, we have to pick a different point than $\phi = 0$, because the logarithm is ill-defined there. This means that we have to introduce a scale μ. Taking the fourth derivative and ignoring terms that are independent of ϕ, we find

$$\left. \frac{\mathrm{d}^4 V_{\text{eff}}}{\mathrm{d}\phi^4} \right|_{\phi=\mu} = \lambda = 24 \frac{\lambda^2}{(16\pi)^2} \ln \frac{\mu^2}{\Lambda^2}. \tag{13.47}$$

We can convince ourselves that this expression gives the correct beta function,

$$\beta(\mu) = \mu \frac{\partial \lambda}{\partial \mu} = \frac{3}{16\pi^2} \lambda^2 + \mathcal{O}(\lambda^3). \tag{13.48}$$

Using the complete expression for Eq. (13.47), we can determine C and obtain for the renormalised effective potential (problem 13.6)

$$V_{\text{eff}}(\phi) = \frac{\lambda(\mu)}{4!} \phi^4 + \frac{\lambda^2(\mu)}{(16\pi)^2} \phi^4 \left[\ln \frac{\phi^2}{\mu^2} - \frac{25}{6} \right] + \mathcal{O}(\lambda^3). \tag{13.49}$$

This potential has two minima outside of the origin, so it seems that SSB does indeed happen. These minima lie however outside the expected range of validity of the one-loop approximation. Rewriting the potential as $V_{\text{eff}}(\phi) = \lambda\phi^4/4!(1 + a\lambda \ln(\phi^2/\mu^2) + \dots$ suggest that we can trust the one-loop approximation only as long as $(3/32\pi^2)\lambda \ln(\phi^2/\mu^2) \ll 1$.

13.3 Abelian Higgs model

After we have shown that the renormalisability is not affected by SSB, we now try to apply this idea to a the case of a gauge symmetry. First of all because we aim to explain the masses of the W and Z bosons as consequence of SSB. And second because we saw that SSB of global symmetries leads to massless scalars which are however not observed. As SSB cannot change the number of physical degrees of freedom, we hope that each of the two diseases is the cure of the other. The Goldstone bosons which would remain massless in a global symmetry hopefully disappear becoming the required additional longitudinal degrees of freedom of massive gauge bosons in case of a spontaneously broken gauge symmetry.

The Abelian Higgs model, which is the simplest example for this mechanism, is obtained by gauging a complex scalar field theory. Introducing in the Lagrangian (13.11) the covariant derivative

$$\partial_\mu \to D_\mu = \partial_\mu + ieA_\mu \tag{13.50}$$

and adding the free Lagrangian of an $U(1)$ gauge field gives

$$\mathscr{L} = -\frac{1}{4}F_{\mu\nu}F^{\mu\nu} + (D_\mu\phi)^\dagger(D^\mu\phi) + \mu^2\phi^\dagger\phi - \lambda(\phi^\dagger\phi)^2. \tag{13.51}$$

The symmetry breaking and Higgs mechanism is best discussed changing to polar coordinates in field-space, $\phi = \rho\exp\{i\vartheta\}$. Then we insert

$$D_\mu\phi = \left[\partial_\mu\rho + i\rho(\partial_\mu\vartheta + eA_\mu)\right]e^{i\vartheta} \tag{13.52}$$

into the Lagrangian, obtaining

$$\mathscr{L} = -\frac{1}{4}F_{\mu\nu}F^{\mu\nu} + \rho^2(\partial_\mu\vartheta + eA_\mu)^2 + (\partial_\mu\rho)^2 + \mu^2\rho^2 - \lambda\rho^4. \tag{13.53}$$

The only difference to the ungauged model is the appearance of the gauge field in the prospective mass term $\rho^2(\partial_\mu\vartheta + eA_\mu)^2$. This allows us to eliminate the angular mode ϑ which shows up nowhere else by performing a gauge transformation on the field A_μ. The action of a $U(1)$ gauge transformation $A_\mu \to A'_\mu = A_\mu - \partial_\mu\Lambda$ on the original field ϕ is just a phase shift, hence ρ is unchanged and ϑ is shifted by a constant, $\vartheta \to \vartheta' = \vartheta + e\Lambda$. This means that if we consider the gauge-invariant combination

$$B_\mu = A_\mu + \frac{1}{e}\partial_\mu\vartheta \tag{13.54}$$

as new variable, we eliminate ϑ completely, as $F_{\mu\nu}(A_\mu) = F_{\mu\nu}(B_\mu)$ is gauge-invariant,

$$\mathscr{L} = -\frac{1}{4}F_{\mu\nu}F^{\mu\nu} + e^2\rho^2(B_\mu)^2 + (\partial_\mu\rho)^2 + \mu^2\rho^2 - \lambda\rho^4. \tag{13.55}$$

It is now evident that the Goldstone mode ϑ has disappeared. Eliminating the field ρ in favour of fluctuations χ around the vacuum $v = \sqrt{\mu^2/\lambda}$ by shifting the field as usual as

$$\rho = \frac{1}{\sqrt{2}}(v + \chi), \tag{13.56}$$

we find as new Lagrangian

$$
\begin{aligned}
\mathscr{L} = &-\frac{1}{4}F_{\mu\nu}F^{\mu\nu} + \frac{1}{2}M^2(B_\mu)^2 + e^2 v\chi(B_\mu)^2 + \frac{1}{2}e^2\chi^2(B_\mu)^2 \\
&+ \frac{\mu^4}{4\lambda} + \frac{1}{2}(\partial_\mu\chi)^2 - \frac{1}{2}(\sqrt{2}\mu)^2\chi^2 - \sqrt{\lambda}\mu\chi^3 - \frac{\lambda}{4}\chi^4.
\end{aligned}
\tag{13.57}
$$

As in the ungauged model we obtain a χ^3 self-interaction and a contribution to the vacuum energy density. The gauge field B_μ acquired the mass $M = ev$, therefore now having three spin degrees of freedom. The additional longitudinal one has been delivered by the Goldstone boson which in turn disappeared. The gauge field has "eaten" the Goldstone boson. We also see that the number of degrees of freedom before SSB $(2+2)$ matches the number afterwards $(3+1)$. Breaking a gauge symmetry spontaneously does not lead to massless Goldstone bosons because they become the longitudinal degree of freedom of massive gauge bosons; this phenomenon is called the Higgs effect.

The gauge transformation we used to eliminate the ϑ field corresponds to the Higgs model in the unitary gauge, where only physical particles appear in the Lagrangian. The massive gauge boson is described by the Procca Lagrangian and we know that the resulting propagator becomes constant for large momenta. Hence, this gauge is convenient for illustrating the concept of the Higgs mechanism, but not suited for loop calculations. For several reasons we expect that the renormalisability of this model is only hidden in the unitary gauge. The model before shifting the fields is renormalisable, and our discussion of SSB using the effective potential has taught us that such a shift has no impact. Moreover, we should be able to neglect v in a scattering process like $B^\mu + B^\mu \to B^\mu + B^\mu$ for $s \gg v^2$. Thus the broken theory should have the same UV behaviour as the unbroken one.

We should therefore explore alternative gauges of the same model. Avoiding the unitary gauge, we re-start using Cartesian fields $\phi = (\phi_1 + i\phi_2)/\sqrt{2}$ for the complex scalar. Then the Lagrangian is

$$
\begin{aligned}
\mathscr{L} = &-\frac{1}{4}F_{\mu\nu}F^{\mu\nu} + \frac{1}{2}\left[(\partial_\mu\phi_1 - eA_\mu\phi_2)^2 + (\partial_\mu\phi_2 + eA_\mu\phi_1)^2\right] \\
&+ \frac{\mu^2}{2}(\phi_1^2 + \phi_2^2) - \frac{\lambda}{4}(\phi_1^2 + \phi_2^2)^2.
\end{aligned}
\tag{13.58}
$$

Performing the shift due to the SSB, $\phi_1 = v + \tilde{\phi}_1$ and $\phi_2 = \tilde{\phi}_2$, the Lagrangian becomes

$$
\begin{aligned}
\mathscr{L} = &-\frac{1}{4}F_{\mu\nu}F^{\mu\nu} + \frac{1}{2}M^2A_\mu^2 + evA^\mu\partial_\mu\tilde{\phi}_2 \\
&+ \frac{1}{2}\left[(\partial_\mu\tilde{\phi}_1)^2 - 2\mu^2\tilde{\phi}_1^2\right] + \frac{1}{2}(\partial_\mu\tilde{\phi}_2)^2 + \dots,
\end{aligned}
\tag{13.59}
$$

where we have omitted interaction and vacuum terms not relevant to the discussion. As we see, the Goldstone boson $\tilde{\phi}_2$ does not disappear and it couples to the gauge field

A_μ. On the other hand, the mass spectrum of the physical particles is the same as in the unitary gauge. The degrees of freedom before and after breaking the symmetry do not match, hence there is an unphysical degree of freedom in the theory, namely that corresponding to $\tilde\phi_2$.

Gauge fixing and gauge-boson propagator. In order to make the generating functional $Z[J^\mu, J, J^*]$ of the abelian Higgs model well-defined, we have to remove the gauge freedom of the classical Lagrangian. Using the Faddeev–Popov trick to achieve this implies to add a gauge-fixing and a Faddeev–Popov ghost term to the classical Lagrangian,

$$\mathscr{L}_{\rm eff} = \mathscr{L}_{\rm cl} + \mathscr{L}_{\rm gf} + \mathscr{L}_{\rm FP} = \mathscr{L}_{\rm cl} - \frac{1}{2\xi}G^2 + \bar{c}\frac{\partial G}{\partial \vartheta}c. \tag{13.60}$$

Here $G(A^\mu, \phi) = 0$ is a suitable gauge condition, ϑ is the generator of the gauge symmetry and $c, \bar{c}$ are Graßmannian ghost fields.

In the unbroken abelian case we used as gauge condition $G = \partial_\mu A^\mu$. With the gauge transformation $A^\mu \to A^\mu - \partial^\mu \vartheta$ the ghost term becomes simply $\mathscr{L}_{\rm FP} = \bar{c}(-\Box)c$. Thus the ghost fields completely decouple from any physical particles, and the ghost term can be absorbed in the normalisation. In the present case of a theory with SSB, we want to use the Faddeev–Popov term to cancel the mixed $A^\mu \partial_\mu \phi_2$ term. Therefore we include the Goldstone boson ϕ_2 in the gauge condition,

$$G = \partial_\mu A^\mu - \xi e v \phi_2 = 0, \tag{13.61}$$

what defines the R_ξ gauge. From $\phi_2 = \partial_\mu A^\mu/(\xi ev)$ we see that the unitary gauge corresponds to $\xi \to \infty$, while we come back to $G = \partial_\mu A^\mu$ for $v \to 0$. We calculate first G^2,

$$G^2 = (\partial_\mu A^\mu)(\partial_\nu A^\nu) - 2\xi ev\phi_2 \partial_\mu A^\mu + \xi^2 e^2 v^2 \phi_2^2, \tag{13.62}$$

integrate by parts the cross-term and insert the result into $\mathscr{L}_{\rm gf}$,

$$\mathscr{L}_{\rm gf} = -\frac{1}{2\xi}G^2 = -\frac{1}{2\xi}(\partial_\mu A^\mu)^2 - ev A_\mu \partial^\mu \phi_2 - \frac{1}{2}\xi(ev)^2 \phi_2^2. \tag{13.63}$$

Now we see that the second term cancels the unwanted mixed term in $\mathscr{L}_{\rm cl}$, while a ξ dependent mass term ξM^2 for ϕ_2 appeared.

If we write out the terms in $\mathscr{L}_{\rm eff}$ quadratic in A_μ and ϕ_2,

$$\mathscr{L}_{\rm eff,2} = -\frac{1}{4}F_{\mu\nu}F^{\mu\nu} + \frac{1}{2}M^2 A_\mu^2 - \frac{1}{2\xi}(\partial_\mu A^\mu)^2 + \frac{1}{2}(\partial_\mu \phi_2)^2 - \frac{1}{2}\xi M^2 \phi_2^2, \tag{13.64}$$

we can find the boson propagator. Using the antisymmetry of $F_{\mu\nu}$ and a partial integration, we transform $F^2/4$ into standard form, $A_\mu(\eta^{\mu\nu}\Box - \partial^\mu \partial^\nu)A_\nu/2$. The part of the Lagrangian quadratic in A_μ then reads

$$\mathscr{L}_A = \frac{1}{2}A_\mu[\eta^{\mu\nu}\Box - \partial^\mu \partial^\nu]A_\nu + \frac{1}{2}A_\mu \eta^{\mu\nu}M^2 A_\nu + \frac{1}{2\xi}A_\mu \partial^\mu \partial^\nu A_\nu \tag{13.65a}$$

$$= \frac{1}{2}A_\mu[\eta^{\mu\nu}(\Box + M^2) - (1 - \xi^{-1})\partial^\mu \partial^\nu]A_\nu. \tag{13.65b}$$

To find the propagator we want to invert the term in the bracket, which we denote by $P^{\mu\nu}$. If we go to momentum space then

$$P^{\mu\nu}(k^2) = -(k^2 - M^2)\eta^{\mu\nu} + (1 - \xi^{-1})k^\mu k^\nu. \tag{13.66}$$

Next we split $P^{\mu\nu}(k^2)$ into a transverse and a longitudinal part by factoring out terms proportional to $P_T^{\mu\nu} = \eta^{\mu\nu} - k^\mu k^\nu / k^2$,

$$P^{\mu\nu} = -(k^2 - M^2)\left(P_T^{\mu\nu} + \frac{k^\mu k^\nu}{k^2}\right) + (1 - \xi^{-1})k^\mu k^\nu \tag{13.67a}$$

$$= -(k^2 - M^2)P_T^{\mu\nu} - \xi^{-1}(k^2 - \xi M^2)P_L^{\mu\nu}, \tag{13.67b}$$

with the longitudinal part given by $P_L^{\mu\nu} = k^\mu k^\nu / k^2$. We can invert the two parts separately and obtain

$$iD_F^{\mu\nu}(k^2) = \frac{-iP_T^{\mu\nu}}{k^2 - M^2 + i\varepsilon} + \frac{-i\xi P_L^{\mu\nu}}{k^2 - \xi M^2 + i\varepsilon} \tag{13.68a}$$

$$= \frac{-i}{k^2 - M^2 + i\varepsilon}\left[\eta^{\mu\nu} - (1 - \xi)\frac{k^\mu k^\nu}{k^2 - \xi M^2 + i\varepsilon}\right]. \tag{13.68b}$$

The transverse part propagates with mass M^2, while the longitudinal part propagates with mass ξM^2. The limit $\xi \to \infty$ corresponds again to the unitary gauge and $\xi = 1$ corresponds to the easier Feynman–'t Hooft gauge. For finite ξ, the propagator is proportional to $1/k^2$ as in the massless case. The Goldstone boson ϕ_2 has the usual propagator of a scalar particle, however with gauge-dependent mass ξM^2. As usual in a covariant gauge, the unphysical gauge-dependent modes have to be cancelled by ghosts, which we discuss next.

Ghosts. Using the Faddeev–Popov ansatz introduces ghosts field through the term $\mathscr{L}_{\text{FP}} = \bar{c}\,(\delta G/\delta\vartheta)\,c$ into the Lagrangian. To calculate $\delta G/\delta\vartheta$, we have to find out how the gauge-fixing condition G changes under an infinitesimal gauge transformation. Looking first at the change of the complex field,

$$\phi \to \tilde{\phi} = \phi + ie\vartheta\phi = \phi + ie\vartheta\frac{1}{\sqrt{2}}(v + \phi_1 + i\phi_2), \tag{13.69}$$

we see that the fields ϕ_1 and ϕ_2 are mixed under the gauge transformation.

$$A_\mu \to \tilde{A}_\mu = A_\mu - \partial_\mu\vartheta \tag{13.70a}$$

$$\phi_1 \to \tilde{\phi}_1 = \phi_1 - e\vartheta\phi_2 \tag{13.70b}$$

$$\phi_2 \to \tilde{\phi}_2 = \phi_2 + e\vartheta(v + \phi_1). \tag{13.70c}$$

Inserting this into the gauge-fixing condition (13.61) and differentiating with respect to the generator, we obtain

$$\frac{\delta G}{\delta\vartheta} = \frac{\delta}{\delta\vartheta}\left(\partial_\mu\tilde{A}^\mu - \xi ev\tilde{\phi}_2\right) = -\Box - \xi e^2 v(v + \phi_1). \tag{13.71}$$

Thus after spontaneous symmetry breaking, the ghost particles receive a ξ-dependent mass and interact with the Higgs field ϕ_1. To see this explicitly we insert $\delta G/\delta\vartheta$ into the ghost Lagrangian,

$$\mathscr{L}_{\mathrm{FP}} = -\bar{c}\left[\Box + \xi e^2 v(v + \phi_1)\right]c = (\partial^\mu \bar{c})(\partial_\mu c) - \xi M^2 \bar{c} c - \xi e^2 v \phi_1 \bar{c} c. \tag{13.72}$$

The second term on the RHS corresponds to the mass $\xi e^2 v^2 = \xi M^2$ for the ghost field, while the third one describes the ghost–ghost–Higgs interaction.

In summary, we have the following propagators in the R_ξ gauge, where we denote with h the physical Higgs with mass $M = ev$, with ϕ the Goldstone boson and with c the ghost:

$$\frac{i}{k^2 - M^2 + i\varepsilon}\left[-\eta_{\mu\nu} + (1 - \xi)\frac{k_\mu k_\nu}{k^2 - \xi M^2}\right] \tag{13.73}$$

$$\frac{i}{p^2 - M^2 + i\varepsilon} \tag{13.74}$$

$$\frac{i}{p^2 - \xi M^2 + i\varepsilon} \tag{13.75}$$

$$\frac{i}{p^2 - \xi M^2 + i\varepsilon} \tag{13.76}$$

Before closing this chapter, we should answer why the Goldstone theorem does not apply to the case of the Higgs model. The characteristic property of gauge theories that no manifestly covariant gauge exists which eliminates all gauge freedom is also responsible for the failure of the Goldstone theorem. In the first version of our proof, we may either choose a gauge as the Coulomb gauge. Then only physical degrees of freedom of the photon propagate, but the potential $A^0(x)$ drops only as $1/|\boldsymbol{x}|$ and the charge Q defined in (13.23) becomes ill-defined. Alternatively, we can use a covariant gauge as the Lorentz gauge. Then the charge is well-defined, but unphysical scalar and longitudinal photons exist. The Goldstone theorem does apply, but the massless Goldstone bosons do not couple to physical modes. In the second version of our proof, the effective potential for the scalar and for the gauge sector do not decouple and mix by the same reason after SSB. This invalidates our analysis including only scalar fields.

Summary

Examining spontaneous symmetry breaking of internal symmetries, we found three qualitatively different types of behaviours. For a broken global continuous symmetry, Goldstone's theorem predicts the existence of massless scalars. In the case of broken approximate symmetries, this can explain the existence of light scalar particles—an example are pions. Instances of broken global continuous symmetry which are exact seem to be not realised in nature since no massless scalar particles are observed. If we gauge the broken symmetry, the would-be massless Goldstone bosons become the longitudinal degrees of freedom required for massive spin-1 bosons. Finally, neither Noether's nor Goldstone's theorems apply to the case of discrete symmetries; therefore the breaking of discrete symmetries does not change the mass spectrum of the theory.

The effective potential is a convenient tool to study the renormalisability of spontaneously broken theories. This approach allows the calculation of all quantum corrections to the classical potential in the limit of constant fields and is invariant under a shift of fields. Thereby we could establish that renormalisability is not affected by SSB.

Further reading. Our discussion of the effective potential is based on the 1966 Erice lecture "Secret Symmetry" of Coleman (1988).

Problems

13.1 Contribution to the vacuum energy density from SSB. Calculate the difference in the vacuum energy density before and after SSB in the SM using $v = 256\,\mathrm{GeV}$ and $m_h^2 = 2\mu^2 = (125)^2\,\mathrm{GeV}^2$. Compare this to the observed value of the cosmological constant.

13.2 Scalar Lagrangian after SSB. Derive Eq. (13.9) and write down the explicit form of $\mathscr{L}_{\mathrm{int}}$.

13.3 Quantum corrections to $\langle\phi\rangle$. We implicitly assumed that quantum corrections are small enough that the field stays at the chosen classical minimum. Calculate $\langle\phi(0)^2\rangle$ for d spacetime dimensions and show that this assumption is violated for $d \le 2$.

13.4 Instability of $\langle\phi\rangle$. Calculate the imaginary part of the self-energy for a scalar field with the Lagrangian (13.1), that is, with a negative squared mass $\mu^2 < 0$. Discuss the physical interpretation.

13.5 Goldstone mode as zero mode. Show that the state $|s\rangle$ defined in Eq. (13.24) has zero energy for $\boldsymbol{k} \to 0$.

13.6 Coleman–Weinberg problem. Derive Eq. (13.49), find the minima of the potential and discuss the validity of the one-loop approximation.

13.7 Effective potential in DR. Repeat the calculation of the effective potential using DR.

14

GSW model of electroweak interactions

Fermi introduced a current–current interaction between four fermions, $\mathscr{L}_{\text{Fermi}} = G_F J^\mu(x) J_\mu^\dagger(x)/\sqrt{2}$, as explanation for the nuclear beta decay. After the discovery of parity violation in weak interactions in 1956, it was realised that these currents have a $V - A$ form, with, for example, $J_\mu(x) = \bar{\psi}_e \gamma_\mu (1 - \gamma^5) \psi_\nu + \text{h.c.}$ for the leptonic current. Using this Lagrangian, all experimental data about weak interactions known at that time could be explained. Being a dimension 6 interaction, the Fermi theory belongs, however, to the class of non-renormalisable theories according to our power counting analysis. Attempts to develop a renormalisable theory of weak interactions started therefore already in the late 1950's. A first step towards a renormalisable gauge theory for weak interactions was the introduction of "intermediate vector bosons" $W^\pm$ with mass m_W, leading to an interaction $J^\mu W_\mu^\pm$ of similar type as in QED and to a dimensionless coupling constant $g \propto m_W^2 G_F$.

Using these new vector bosons, we can write the charged current interaction in a more economical way introducing doublets of left-handed fermions,

$$\begin{pmatrix} \nu_e \\ e \end{pmatrix}_L \quad \text{and} \quad \begin{pmatrix} u \\ d \end{pmatrix}_L.$$

Associating then the bosons $W^\mp$ with the ladder operators $\tau_\pm = (\tau_1 \pm i\tau_2)/2$ (we denote the Pauli matrices in this context with τ_i), the charged current interaction of leptons becomes

$$\mathscr{L}_{CC} = \frac{-ig}{\sqrt{2}} \left(\bar{\nu}_e \ \bar{e} \right)_L (\tau^+ W_\mu^- + \tau^- W_\mu^+) \gamma^\mu \begin{pmatrix} \nu_e \\ e \end{pmatrix}_L.$$

The doublet structure suggests using SU(2) as a gauge group for weak interactions. This gauge group is often called weak isospin and denoted by $\mathrm{SU}(2)_L$ in order to stress the fact that only left-handed fermions participate in this gauge interaction, while right-handed fermions transform as singlets. If it were possible to identify τ_3 with the photon, a unification of the weak and electromagnetic forces would have been achieved. There are two major obstacles defeating this attempt:

- The generators of SU(2) are traceless, which means that the multiplets must have zero net charge.
- The currents generated by $\tau_\pm$ should have a $V - A$ structure, while the electromagnetic current has to be a pure vector current.

Quantum Fields–From the Hubble to the Planck Scale. Michael Kachelriess. © Michael Kachelriess 2018.
Published in 2018 by Oxford University Press. DOI 10.1093/oso/9780198802877.001.0001

The last argument seems to be impossible to overcome using a single gauge group. Glashow was the first to realise that nature may not always choose the most economical solution. He suggested that the gauge group of weak and electromagnetic interactions is the product of two groups, $SU(2)_L \otimes U(1)_Y$, where Y stands for hypercharge. Salam and Weinberg added the idea of SSB to this model, using the Higgs effect to break $SU(2)_L \otimes U(1)_Y$ down to $U(1)_{em}$ and to generate in a gauge-invariant way masses. In this form, the Glashow–Salam–Weinberg (GSW) model of electroweak interactions survived all experimental tests until present.

14.1 Higgs effect and the gauge sector

An $SU(2) \otimes U(1)$ gauge theory contains four gauge bosons, of which only one should remain massless. If we use as scalar field a complex $SU(2)$ doublet to break the gauge symmetry, then we add four real degrees of freedom. Three of them will become the longitudinal degrees of freedom for the three massive gauge bosons, so that just one physical Higgs field remains in this most economical version of electroweak symmetry breaking. We choose the complex scalar $SU(2)$ doublet as

$$\Phi = \begin{pmatrix} \phi^+ \\ \phi^0 \end{pmatrix} = \frac{1}{\sqrt{2}} \begin{pmatrix} \phi_1 + i\phi_2 \\ \phi_3 + i\phi_4 \end{pmatrix}. \tag{14.1}$$

The corresponding scalar Lagrangian

$$\mathscr{L} = (\partial_\mu \Phi)^\dagger (\partial^\mu \Phi) + \mu^2 \Phi^\dagger \Phi - \lambda (\Phi^\dagger \Phi)^2 \tag{14.2}$$

is invariant under both global $SU(2)$ and $U(1)$ transformations of Φ,

$$\Phi \to \exp\left\{ \frac{i\boldsymbol{\alpha} \cdot \boldsymbol{\tau}}{2} \right\} \Phi \quad \text{and} \quad \Phi \to \exp\{i\vartheta Y/2\}\Phi. \tag{14.3}$$

Here we have chosen the Pauli matrices $\boldsymbol{\tau}$ as generators $\boldsymbol{T}$ for the weak isospin transformations acting on the fundamental representation of $SU(2)$, $\boldsymbol{T} = \boldsymbol{\tau}/2$. Note that in the non-abelian case the charge is fixed by the chosen representation. In contrast, the abelian $U(1)$ charge Y of the $SU(2)$ doublet Φ can take a priori any value. We set $Y(\Phi) = 1$ with the factor $1/2$ in (14.3) added by convention.

We avoid an electrically charged vacuum by choosing the vacuum expectation value v in the ϕ^0 direction,

$$\langle 0|\Phi|0\rangle = \begin{pmatrix} 0 \\ \frac{v}{\sqrt{2}} \end{pmatrix}. \tag{14.4}$$

Electroweak symmetry breaking should leave $U_{em}(1)$ invariant. With this choice of v, we see that the combination $\mathbf{1} + \tau_3$ keeps the ground-state invariant,

$$\delta\Phi = i(\mathbf{1} + \tau_3)\Phi = i\begin{pmatrix} 2 & 0 \\ 0 & 0 \end{pmatrix}\begin{pmatrix} 0 \\ \frac{v}{\sqrt{2}} \end{pmatrix} = 0. \tag{14.5}$$

Since the hypercharge generator is the identity in the weak isospin space, we should associate thus the combination $Y + \tau_3$ with the electric charge, or $Q \propto Y + 2T_3$.

Therefore we expect that the photon is a superposition of the abelian gauge boson mediating $U(1)_Y$ and the 3-component of the non-abelian gauge boson mediating $SU(2)_L$. Applying $Q \propto Y + 2T_3$ to the upper component ϕ^+ of the Higgs doublet, we obtain the Gell-Mann–Nishijima relation $2Q = Y + 2T_3$.

Now we gauge the model, introducing covariant derivatives,

$$\partial^\mu \Phi \to D^\mu \Phi = \left(\partial^\mu + \frac{ig}{2} \boldsymbol{\tau} \cdot \boldsymbol{W}^\mu + \frac{ig'}{2} B^\mu \right) \Phi, \tag{14.6}$$

where we suppressed a unit matrix in isospin space in front of ∂^μ and B^μ. The field-strengths

$$F_a^{\mu\nu} = \partial^\mu W_a^\nu - \partial^\nu W_a^\mu - g\varepsilon_{abc} W_b^\mu W_c^\nu \tag{14.7}$$
$$G^{\mu\nu} = \partial^\mu B^\nu - \partial^\nu B^\mu, \tag{14.8}$$

correspond to the three $SU(2)_L$ gauge fields $\boldsymbol{W}^\mu$ and to the single $U_Y(1)$ field B^μ. The couplings of the two groups are g and g', respectively, and the structure constants of $SU(2)$ are the elements of the completely antisymmetric tensor ε_{abc}. The Lagrangian describing the Higgs-gauge sector is then

$$\mathscr{L} = -\frac{1}{4} F^2 - \frac{1}{4} G^2 + (D_\mu \Phi)^\dagger (D^\mu \Phi) - V(\Phi). \tag{14.9}$$

We describe SSB in this theory choosing the unitary gauge, since we are mainly interested in the mass spectrum of physical particles. Using polar coordinates for the lower component, $\phi^0 = \rho e^{i\vartheta}$, and setting $\phi^+ = \vartheta = 0$ by an $SU(2)$ gauge transformation, we are left with one physical scalar field ρ. Additionally, this gauge transformation only changes the longitudinal components of the $SU(2)$ gauge bosons, $\boldsymbol{W}^\mu \to \boldsymbol{W}'^\mu$, while the other pieces of the Lagrangian are not affected. We separate ρ into the vev $v = \sqrt{\mu^2/\lambda}$ and fluctuations $h(x)$ as

$$\Phi = \begin{pmatrix} 0 \\ \frac{1}{\sqrt{2}}(v + h(x)) \end{pmatrix} = \frac{v+h}{\sqrt{2}} \begin{pmatrix} 0 \\ 1 \end{pmatrix} \equiv \frac{v+h}{\sqrt{2}} \chi. \tag{14.10}$$

As result, we generate a mass term $m_h = \sqrt{2}\mu$ as well as trilinear and cubic self-interactions for the Higgs particle h. For the gauge bosons, inserting the part containing v into Eq. (14.9) gives as mass terms

$$\mathscr{L}_\mathrm{m} = \frac{v^2}{2} \chi^\dagger \left(\frac{g}{2} \boldsymbol{\tau} \cdot \boldsymbol{W}^\mu + \frac{g'}{2} B^\mu \right) \left(\frac{g}{2} \boldsymbol{\tau} \cdot \boldsymbol{W}_\mu + \frac{g'}{2} B_\mu \right) \chi, \tag{14.11}$$

where we have omitted the prime on $\boldsymbol{W}^\mu \to \boldsymbol{W}'^\mu$. Multiplying out the brackets and using $(\boldsymbol{\tau} \cdot \boldsymbol{W})^2 = \boldsymbol{W}^2$ and $\chi^\dagger \boldsymbol{\tau} \cdot \boldsymbol{W}^\mu \chi = -W_3^\mu$, we find

$$\mathscr{L}_\mathrm{m} = \frac{v^2}{2} \left[\frac{g^2}{4} (W_1^2 + W_2^2 + W_3^2) + \frac{g'^2}{4} B^2 - \frac{gg'}{2} W_3^\mu B_\mu \right]. \tag{14.12}$$

The neutral fields are mixed by the $W_3^\mu B_\mu$ term, and therefore we rewrite the mass term as

$$\mathscr{L}_{\mathrm{m}} = \frac{g^2 v^2}{8} (W_1^2 + W_2^2) + \frac{v^2}{8} (g W_3^\mu - g' B^\mu)^2. \tag{14.13}$$

The two fields in the first bracket are the gauge bosons appearing in the charged current interactions of the Fermi theory,

$$W^{\pm \mu} = \frac{1}{\sqrt{2}} \left(W_1^\mu \mp i W_2^\mu \right). \tag{14.14}$$

The second bracket corresponds to a new neutral massive gauge boson called Z boson, which is a mixture of the B and the W_3 fields,

$$Z^\mu = \frac{1}{\sqrt{g^2 + g'^2}} \left(g W_3^\mu - g' B^\mu \right) = \cos \vartheta_W W_3^\mu - \sin \vartheta_W B^\mu. \tag{14.15}$$

The mixing of the B and the W^3 field is parameterised by ϑ_W called the weak mixing angle. For $\vartheta_W = 0$, the mixing disappears and hypercharge and electric charge become identical.

The combination of W_3 and B orthogonal to Z does not show up in the mass Lagrangian, and corresponds therefore to the massless photon,

$$A^\mu = \frac{1}{\sqrt{g^2 + g'^2}} \left(g' W_3^\mu + g B^\mu \right) = \sin \vartheta_W W_3^\mu + \cos \vartheta_W B^\mu. \tag{14.16}$$

We can now write $\mathscr{L}_{\mathrm{m}}$ in terms of physical fields,

$$\mathscr{L}_{\mathrm{m}} = \frac{1}{2} m_W^2 W_\mu^+ W^{-\mu} + \frac{1}{2} m_Z^2 Z_\mu Z^\mu, \tag{14.17}$$

reading off the gauge boson masses, $m_W = g v / 2$, $m_Z = (g^2 + g'^2)^{1/2} v / 2 = m_W / \cos \vartheta_W$ and $m_A = 0$, as function of the still unknown values of the coupling constants g, g' and the vev v. Thus in the GSW theory, the mass ratio of the gauge bosons is fixed at tree level by the weak mixing angle, $m_W / m_Z = \cos \vartheta_W$. Because of $\Phi = (v + h) \chi / \sqrt{2}$, we can read off via $\mathscr{L}_{\mathrm{int}} = \mathscr{L}_{\mathrm{m}} (1 + h/v)$ also the tri- and quadrilinear Higgs-gauge boson interactions from the mass terms.

Couplings. The new coupling constants g and g' should be related to the electromagnetic coupling e and the weak mixing angle ϑ_W. We can find the connection between these parameters, if we replace the original fields W_3^μ and B^μ with the physical fields Z^μ and A^μ in the covariant derivative (14.6) acting on Φ by inverting Eqs. (14.15) and (14.16),

$$\begin{aligned} g W_3^\mu \tau_3 + g' B^\mu \mathbf{1} &= g(\cos \vartheta_W Z^\mu + \sin \vartheta_W A^\mu) \tau_3 + g'(- \sin \vartheta_W Z^\mu + \cos \vartheta_W A^\mu) \mathbf{1} \\ &= (g \sin \vartheta_W \tau_3 + g' \cos \vartheta_W \mathbf{1}) A^\mu + (g \cos \vartheta_W \tau_3 - g' \sin \vartheta_W \mathbf{1}) Z^\mu. \end{aligned}$$

Using then $\tan \vartheta_W = g'/g$, we obtain

$$g \frac{W_3^\mu \tau_3}{2} + g' \frac{B^\mu \mathbf{1}}{2} = \frac{1}{2} g \sin \vartheta_W (\tau_3 + 1) A^\mu + \frac{g}{2 \cos \vartheta_W} \left[\tau_3 - \sin^2 \vartheta_W (\tau_3 + 1) \right] Z^\mu. \tag{14.18}$$

Since we assigned the electric charge $q = e$ to the upper component ϕ^+, the covariant derivative $\mathcal{D}_\mu \Phi$ implies

$$e = g \sin \vartheta_W = g' \cos \vartheta_W, \tag{14.19}$$

while the lower component ϕ^0 stays, as desired, neutral. We obtain the opposite result for the lepton doublet L, if we assign the hypercharge $Y(L) = -1$ to it. Note also that the electric coupling is smaller than the weak one. Weak interactions are weak, because the mass of m_W implies a short range, not because the coupling is small.

We can use the remaining piece of the covariant derivative to derive the connection between the Fermi constant G_F and the vev v (problem 14.1), finding

$$\frac{G_F}{\sqrt{2}} = \frac{g^2}{8m_W^2} = \frac{1}{2v^2} \tag{14.20}$$

or $v \simeq 246 \, \text{GeV}$. Thus a measurement of the gauge boson masses determines the gauge couplings g and g', which in turn fix the weak mixing angle ϑ_W.

14.2 Fermions

Fermion masses. Our remaining task determining the electroweak mass spectrum at the tree level is to generate masses for the fermions. In order that the fermion mass term $m(\bar{\psi}_R \psi_L + \bar{\psi}_L \psi_R)$ can become gauge-invariant we have to promote the parameter m into a SU(2) doublet. This is easiest accomplished introducing a Yukawa coupling between the lepton doublet L, the singlet e_R and the scalar doublet Φ,

$$\mathscr{L}_Y = -y_f \left(\bar{L} \Phi e_R + \bar{e}_R \Phi^\dagger L \right). \tag{14.21}$$

This term is clearly SU(2) invariant. With $Y(\Phi) = 1$ and $Y(L) = -1$, we obtain also an $U_Y(1)$-invariant mass term, if we assign $Y(e_R) = -2$ to the lepton singlet—which is consistent with the Gell-Mann–Nishijima relation $2Q = Y + 2T_3$. The coupling generates fermion masses as well as Yukawa interactions between fermions and the Higgs. Inserting the vacuum expectation value of Φ, the mass term is

$$\mathscr{L}_m = -\frac{y_f v}{\sqrt{2}} \left[(\bar{\nu}_e \; \bar{e})_L \begin{pmatrix} 0 \\ 1 \end{pmatrix} e_R + \bar{e}_R (0 \; 1) \begin{pmatrix} \nu_e \\ e \end{pmatrix}_L \right]$$

$$= -m_f (\bar{e}_L e_R + \bar{e}_R e_L) = -m_f \bar{e} e,$$

that is, it is a normal Dirac mass term with $m_f = y_f v / \sqrt{2}$. In this way we can generate masses for the down-like fermions with $\tau_3 = -1/2$ like the electron. In order to generate masses for the other half, we must use the charge conjugated Higgs doublet, $i\tau_2 \Phi^*$. To see why this works, we recall from the discussion of spinor representations in section 8.1 that SU(2) contains two fundamental representations (ϕ_L and ϕ_R connected by $\phi_L = i\sigma^2 \phi_R^*$) that are unitary inequivalent. In the context of $SU(2)_L$, these two representations are called the fundamental 2 and the anti-fundamental 2* representations. Thus we obtain the charge conjugated Higgs doublet as $\widetilde{\Phi} \equiv \Phi^c = i\tau_2 \Phi^*$.

Field	T_3	Y	q
$\Phi = (\phi^+, \phi^0)$	$(1/2, -1/2)$	1	$(1, 0)$
$L = (\nu_e, e)_L$	$(1/2, -1/2)$	-1	$(0, -1)$
$E = e_R$	1	-2	-1
$Q = (u, d)_L$	$(1/2, -1/2)$	$1/3$	$(2/3, -1/3)$
$U = u_R$	1	$4/3$	$2/3$
$D = d_R$	1	$-2/3$	$-1/3$

Table 14.1 Quantum numbers of the scalars and fermions in the SM. The second and third fermion generations repeat these numbers.

Since in $\widetilde{\Phi}$ the neutral component sits in the upper slot, we can generate Dirac masses for up-like quarks using

$$\mathscr{L}_Y = -y_f \left(\bar{Q} \widetilde{\Phi} u_R + \bar{u}_R \widetilde{\Phi}^\dagger Q \right). \tag{14.22}$$

If desired, a corresponding term generates Dirac masses for neutrinos.

As the Yukawa couplings y_f are not predicted, the fermion masses are arbitrary parameters in the GSW theory. Theoretical prejudice suggests as the "natural" range for the values of the couplings few $\times 0.1$, which would be comparable to the gauge couplings g and g'. Even excluding neutrinos, there is instead a large hierarchy in the values of the Yukawa couplings, ranging from $y_e = \sqrt{2} m_e / v \sim 10^{-5}$ for the electron to $y_t = \sqrt{2} m_t / v \sim 1$ for the top quark.

Flavour mixing. Accounting for the second and third generation of leptons and quarks, the coupling y_f introduced in Eq. (14.21), and hence also the mass, become arbitrary 3×3 matrices, $\boldsymbol{y}$ and $\boldsymbol{m}$, in flavour space. In particular, there is no reason for $\boldsymbol{y}$ to be Hermitian or even diagonal. They can still be diagonalised by a bi-unitary transformation, that is, by

$$\boldsymbol{S}^\dagger \boldsymbol{m} \boldsymbol{T} = \boldsymbol{m}_d, \tag{14.23}$$

where $\boldsymbol{S}$ and $\boldsymbol{T}$ are two different unitary matrices and $\boldsymbol{m}_d$ is diagonal and positive. This means that the weak eigenstates $\boldsymbol{\psi} = \{\psi_e, \psi_\mu, \psi_\tau, \}$, which are diagonal in the interaction basis but have indefinite masses, can be transformed into mass eigenstates $\boldsymbol{\psi}'$ (omitting the Hermitian conjugated terms) as

$$\bar{\psi}_L \boldsymbol{m} \psi_R = \bar{\psi}_L \boldsymbol{S} \boldsymbol{S}^\dagger \boldsymbol{m} \boldsymbol{T} \boldsymbol{T}^\dagger \psi_R = \bar{\psi}'_L \boldsymbol{m}_d \psi'_R. \tag{14.24}$$

These new eigenstates will no longer be diagonal in the interaction basis, which will lead to flavour mixing. If we look at the weak current J^μ and insert the mass eigenstates,

$$J^\mu = \bar{\nu}_L \gamma^\mu e_L = \bar{\nu}'_L \gamma^\mu \boldsymbol{S}^\dagger_\nu \boldsymbol{S}_e e'_L, \tag{14.25}$$

we see that only the product $\boldsymbol{S}^\dagger_\nu \boldsymbol{S}_e \equiv \boldsymbol{U}$ will be observable. This means that we have some freedom of choice, and it is convention to move the mixing completely into the neutrino sector, setting $\boldsymbol{S}_e = \boldsymbol{1}$ and $\boldsymbol{U} = \boldsymbol{S}^\dagger_\nu$. In the same way, one shifts the mixing in the quark sector completely to the down-like quarks. One denotes the mass eigenstates as (d, s, b) and the weak eigenstates as (d', s', b'), respectively, while (u, c, t) are unmixed by definition. The two matrices U which arise due to this phenomenon are known as the PMNS matrix in the neutrino case and the CKM matrix in the quark case. Note that the mixing matrices cancel in the case of the neutral current. Because the neutral currents couple, for example, neutrinos to neutrinos, products like $\boldsymbol{S}_\nu^\dagger \boldsymbol{S}_\nu$ appear which are equal to unity due to unitarity.

Conservation laws.   We can use the global $\mathrm{U}_V(1)$ symmetry satisfied by a single Dirac fermion to define a global $\mathrm{U}(1)$ conservation law for any subgroup of fermions which is not mixed by interactions. Since electromagnetic, neutral current and strong interactions do not change the flavour, we only have to consider the weak current. In the case of quarks, the CKM matrix is non-diagonal and thus only the total number $\Delta n_q = n_q - n_{\bar{q}}$ of quarks of all flavours is conserved. This corresponds to a single global conservation law for baryon number B defined by $n_b = n_q/3$ such that a nucleon has $B = 1$.

If we set the neutrino masses to zero, then $\boldsymbol{S}_\nu$ is arbitrary. Choosing then $\boldsymbol{S}_\nu^\dagger = \boldsymbol{S}_e$ results in $\boldsymbol{U} = 1$, and thus no mixing occurs in the lepton sector. As a result, the individual lepton number L_i of each generation is conserved. These conservation laws guarantee that in perturbation theory loop corrections do not generate neutrino masses. In this sense, the "traditional" SM without neutrino masses is self-consistent: setting $m_\nu = 0$ at tree level leads to lepton number conservation L_i as additional, accidental symmetries which in turn forbid loop corrections to the neutrino masses. Accounting for the observed non-zero neutrino masses, leptons mix via the PMNS matrix and only the total lepton number is conserved, $L = L_e + L_\mu + L_\tau$. Majorana mass terms of the type $m_L(\bar{\psi}_L^c \psi_L + \bar{\psi}_L \psi_L^c)$ for neutrinos do not appear in the SM, since we cannot identify m_L in a gauge-invariant way with the vev of the Higgs field using a $d = 4$ operator. Therefore violation of lepton number as, for example, in neutrinoless double-beta decay would be a proof for physics beyond the SM.

14.3   Properties of the Higgs sector

All particles of the SM except the Higgs particle have been firmly established. The search for the Higgs has been the major goal of the Large Hadron Collider (LHC) at CERN which reported in 2012 the discovery of a scalar resonance with mass $\simeq$ 125 GeV. At the time of writing, the production and decay modes of this new scalar particle are compatible with those of the SM Higgs. Since the Higgs mass $m_h^2 = 2\lambda v^2$ depends on the unknown self-coupling λ, it is a priori a free parameter. In this section, we will sketch how one can find theoretical bounds for λ and thus the Higgs mass. Vice versa, assuming that the scalar resonance is indeed the SM Higgs, we can use its mass $m_h \simeq 125$ GeV to investigate up until which energy scale the SM can be maximally valid.

Lower bound for m_h.   Calculating the beta function of the Higgs self-coupling, we have to add to the $\lambda \left(\phi^\dagger \phi\right)^2$ self-interaction[1] the effects from all other massive SM particles. In practice, it is often sufficient to include only the massive gauge bosons, $W^\pm$ and Z, and the top quark as the only fermion with a Yukawa coupling of order one. Then one finds for the beta function

---

[1] Note the different normalisation of $V(\Phi)$ compared to the $\lambda\phi^4$ theory.

$$\frac{d\lambda}{dt} = \frac{1}{16\pi^2}\left\{12\lambda^2 + 12\lambda y_t^2 - 12y_t^4 - \frac{3}{2}\lambda\left(3g^2 + g'^2\right) + \frac{3}{16}\left[2g^4 + (g^2 + g'^2)^2\right]\right\} \quad (14.26)$$

with $t = \ln\left(\mu^2/\mu_0^2\right)$. Note the signs of the different terms: the gauge boson and the scalar loops contribute positively, while the fermion loop has a negative sign. We are looking first for a lower bound on $m_h = \sqrt{2\lambda}v$. For small values of λ, we can neglect all terms of $\mathcal{O}(\lambda)$,

$$\frac{d\lambda}{dt} \approx \frac{1}{16\pi^2}\left[-12y_t^4 + \frac{3}{16}\left(2g^4 + (g^2 + g'^2)^2\right)\right]. \quad (14.27)$$

The RHS is dominated by the Yukawa coupling of the top and thus negative. As result, the Higgs self-coupling $\lambda(\mu)$ decreases evolving to larger scales μ. As the coupling λ must be positive for the vacuum state to be stable, this results in an upper bound on $\lambda(v)$ as function of the scale Λ up to which the SM running (14.26) is valid. In general, we should solve the matrix of RGEs for λ, y_t, g and g' simultaneously, which is only numerically possible. For simplicity, we neglect therefore the scale dependence of the RHS and obtain

$$\lambda(\Lambda) - \lambda(v) = \frac{1}{16\pi^2}\left[-12y_t^4 + \frac{3}{16}\left(2g^4 + (g^2 + g'^2)^2\right)\right]\ln\left(\frac{\Lambda^2}{v^2}\right). \quad (14.28)$$

Imposing $\lambda(\Lambda) > 0$, we get as lower bound for m_h^2

$$m_h^2 > \frac{v^2}{8\pi^2}\left[12y_t^4 - \frac{3}{16}\left(2g^4 + (g^2 + g'^2)^2\right)\right]\ln\left(\frac{\Lambda^2}{v^2}\right). \quad (14.29)$$

Upper bound for m_h.   An upper bound for m_h corresponds to the case of large λ. Therefore we need to keep now only the self-coupling of the Higgs in the beta function,

$$\beta(\lambda) = \frac{d\lambda}{dt} = \frac{3\lambda^2}{4\pi^2}, \quad (14.30)$$

leading to a Landau pole at $\mu_L = \mu_0 \exp[(4\pi^2)/(3\lambda_0)]$. We should require λ to be small enough to justify perturbation theory. Including a cut-off $\Lambda < \mu_L$, and asking generously only for $\lambda < \infty$, we obtain the inequality

$$\frac{3\lambda(\mu_0)}{4\pi^2}\ln\left(\frac{\Lambda^2}{\mu_0^2}\right) < 1. \quad (14.31)$$

Using $m_h^2 = 2\lambda(v)v^2$ and $\mu_0 = v$, we find as upper bound for m_h as a function of the cut-off scale,

$$m_h^2 < \frac{8\pi^2 v^2}{3\ln\left(\Lambda^2/v^2\right)}. \quad (14.32)$$

Combining the two bounds, it is clear that the possible range of Higgs masses shrinks for increasing Λ, and disappears for a large but finite value of the cut-off. This is a sign that the electroweak theory is a trivial theory: We can keep an interacting theory

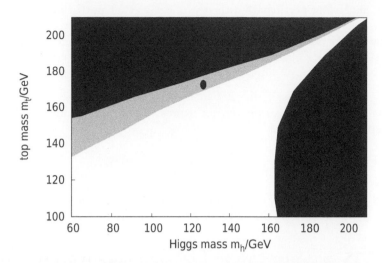

Fig. 14.1 Behaviour of $\lambda(\mu)$ at $\mu = M_{\mathrm{Pl}}$ as function of m_h and m_t using the two-loop RGE of the SM. Black (white) regions are forbidden (allowed), in the grey region the vacuum is metastable. The ellipse shows the experimentally allowed region for m_h and m_t at $3\,\sigma$; adapted from Degrassi *et al.* (2012).

only for a finite value of the cut-off, while in the limit $\Lambda \to \infty$ only the trivial solution $\lambda = 0$ is possible.

Now we turn this argument around. If the scalar particle with mass $\approx 125\,\mathrm{GeV}$ measured at the LHC is indeed the SM Higgs, at which scale Λ new physics has to appear at the latest, changing the SM RGE flow? In particular, is it possible that the SM is a valid theory up to the Planck scale M_{Pl}? The astonishing answer is that the relevant parameters, most importantly m_h and m_t, are such that the SM lies at the border between a stable and an unstable vacuum choosing as cut-off the Planck scale, $\Lambda = M_{\mathrm{Pl}}$. Figure 14.1 shows these regions obtained for $\Lambda = M_{\mathrm{Pl}}$ as function of m_h and m_t. Large m_h implies large λ and thus the SM becomes non-perturbative in the dark region on the right part of this plot. Large m_t drives λ below zero and thus the vacuum is unstable in the upper dark region. Since there is a potential barrier between the (false) vacuum at $v \simeq 246\,\mathrm{GeV}$ and the true vacuum at M_{Pl}, the universe has to tunnel through this barrier.[2] In the light-grey region the vacuum is metastable: λ is not much below zero and the tunnelling time is greater than the age of the universe. The future will show if the position of the SM in this plot happens by accident or represents an important signpost.

Perturbative unitarity. In the SM, all particle masses are given by the product of the vev v and some coupling constant. Particle masses much larger than the electroweak scale therefore require large coupling constants, leading to the breakdown of perturbation theory. More precisely, we will use that a properly normalised amplitude T has to be bounded, $|T| \leq 1$, reflecting the unitarity of the S-matrix. If this condi-

---

[2]We will estimate the decay time of a false vacuum state in section 16.2.

tion is not satisfied using perturbation theory, one says that perturbative unitarity is violated.

To see how too heavy particles within the SM violate perturbative unitarity we consider the case of $2 \to 2$ elastic scattering. The differential cross-section is

$$\frac{d\sigma_{el}}{d\Omega} = \frac{1}{64\pi^2 s} |\mathcal{A}|^2. \tag{14.33}$$

Using a partial wave decomposition we can express the Feynman amplitude for the scattering of spinless particles as

$$\mathcal{A}(s, \vartheta) = 16\pi \sum_{l=0}^{\infty} (2l + 1) P_l(\cos \vartheta) T_l, \tag{14.34}$$

where T_l is the l.th partial wave and P_l are the Legendre polynomials.[3] Inserting the partial wave expansion into (14.33) and integrating over $d\Omega = 2\pi \sin \vartheta d\vartheta$ gives for the cross-section

$$\sigma_{el} = \frac{8\pi}{s} \sum_{l=0}^{\infty} \sum_{l'=0}^{\infty} (2l + 1)(2l' + 1) T_l T_{l'}^* \int_{-1}^{1} d(\cos \vartheta) P_l(\cos \vartheta) P_{l'}(\cos \vartheta) \tag{14.35a}$$

$$= \frac{16\pi}{s} \sum_{l=0}^{\infty} (2l + 1) |T_l|^2, \tag{14.35b}$$

where we have used Eq. (A.53). From the optical theorem we obtain with $P_l(1) = 1$ on the other hand

$$\sigma_{tot} = \frac{\Im[\mathcal{A}(\vartheta = 0)]}{2p_{cms}\sqrt{s}} = \frac{16\pi}{2p_{cms}\sqrt{s}} \sum_{l=0}^{\infty} (2l + 1)\Im(T_l). \tag{14.36}$$

Below the threshold s_{th} for the production of additional particles, the elastic and the total cross-section agree, $\sigma_{el} = \sigma_{tot}$. Comparing the expressions for the two cross-sections using $p_{cms}^2 \approx s/4$ valid in the ultrarelativistic limit immediately yields the unitarity requirement

$$\Im(T_l) = |T_l|^2 = \Im(T_l)^2 + \Re(T_l)^2 \geq \Im(T_l)^2, \tag{14.37}$$

which in turn implies $|T_l| \leq 1$. Using the first and third part of Eq. (14.37) and completing the square, we arrive at

$$\left[\Im(T_l) - \frac{1}{2}\right]^2 + \Re(T_l)^2 = \frac{1}{4}. \tag{14.38}$$

Thus the partial amplitudes T_l lie on a circle of radius one-half centred at $(0, i/2)$ in the complex plane, as long as $s < s_{th}$. For the real part of the amplitude, relevant for elastic scattering, Eq. (14.38) implies the stronger bound

$$|\Re(T_l)| < \frac{1}{2}. \tag{14.39}$$

At higher energies, inelastic channels open, $\sigma_{tot} = \sigma_{el} + \sigma_{inel}$, and the partial amplitudes T_l lie inside this circle.

---

[3]The pre-factor is chosen such that $|T_l| \leq 1$, as we will see in a moment.

The simplest possible application of the bound (14.39) is the elastic scattering[4] of Higgs bosons. In lowest-order perturbation theory, the Feynman amplitude of this process is simply $i\mathcal{A}(hh \to hh) = -6i\lambda$. As the amplitude is independent of the scattering angle, it corresponds to pure s-wave scattering, $l = 0$. Thus

$$|T_0| = \frac{|\mathcal{A}|}{16\pi} = \frac{3\lambda}{8\pi} < \frac{1}{2} \tag{14.40}$$

or $m_h = \sqrt{2\lambda}v \lesssim 800\,\text{GeV}$. For larger values of m_h, the nonsensical result $|T_l| > 1/2$ indicates that perturbation theory fails. Similar findings are obtained, if one calculates the scattering of longitudinal gauge bosons as function of m_h. These considerations lead to the belief that one would either find the Higgs boson at a multi-TeV collider as the LHC or discover a new strongly interacting sector coupled to the gauge bosons.

Goldstone boson equivalence theorem. One might expect that in processes where the energy is much higher than the electroweak scale, $s \gg m_W^2$, the unbroken theory becomes a valid description. In particular, we should be able to replace a massive gauge boson by a massless transverse boson plus a massless Goldstone boson. More precisely, the equivalence theorem states that amplitudes involving longitudinal gauge bosons are equal to those of the unphysical Goldstone bosons up to corrections of $\mathcal{O}(m_W/E)$,

$$\mathcal{A}(W_L^+, W_L^-, Z_L) = \mathcal{A}(\phi^+, \phi^-, \phi_3) + \mathcal{O}(m_W/E). \tag{14.41}$$

The Higgs–Goldstone Lagrangian is

$$\begin{aligned}
\mathcal{L} = {} & \frac{1}{2}(\partial_\mu h)^2 + \frac{1}{2}\partial_\mu\phi^+ \partial^\mu\phi^- + \frac{1}{2}(\partial_\mu\phi_3)^2 - \lambda v h^2 \\
& - \lambda v h(2\phi^+\phi^- + \phi_3^2 + h^2) - \frac{1}{4}\lambda(2\phi^+\phi^- + \phi_3^2 + h^2)^2.
\end{aligned} \tag{14.42}$$

An example for this correspondence which allows to simplify calculations in the high-energy limit is discussed in problem 14.7.

14.4   Decoupling and the hierarchy problem

Decoupling. Many extensions of the SM predict particles with masses that are many orders above the weak scale. Additionally, the masses of the SM particles themselves are widely spread,

$$m_\nu \ll m_e, \ldots, m_b \ll m_W, m_Z, m_h \ll m_t.$$

Therefore it is often useful to integrate out heavy degrees of freedom. The decoupling theorem of Appelquist and Corrazone (1975) states that heavy particles can be neglected at low energies,

$$\mathcal{L}(\Phi, G, M; \phi, g, m) \xrightarrow{\text{intgr. out } \Phi} \mathcal{L}_{\text{eff}}(\phi, Z_g g, Z_m m) + \mathcal{O}\left(E/M\right),$$

if $\mathcal{L}$ and $\mathcal{L}_{\text{eff}}$ are renormalisable. In this case, all effects of the heavy particles are either suppressed or can be absorbed into a renormalisation of the parameters of the

---

[4]In this case, a symmetry factor $S = 1/2!$ has to be included both in (14.33) and (14.35b) which therefore drops out from the unitarity bound (14.39).

light sector. In order to improve our understand of the implications of this theorem, it is useful to reconsider first some of the loop corrections we met already.

Let us start with the conceptionally simplest case of a theory as QED (or QCD). At a given energy E, we can consider QED as an effective theory including only those n_f (charged) fermions with mass $m \lesssim E$. Both the effective and the full theory are renormalisable and thus the decoupling theorem applies. The behaviour of the anomalous magnetic moment illustrates this property. It represents a quantity not present in the classical Lagrangian and is therefore a finite prediction of QED; the contributions of heavy particles to the anomalous magnetic moment are suppressed as E/m. A different case is, for example, the electric charge. It is a parameter of the classical theory, and the decoupling theorem now states that all effects of heavy particles can be absorbed into its renormalisation. Since the bare parameters are not observable, there is no loss of predictivity by the fact that unknown particles of arbitrary mass contribute. Additionally to the measured value of the coupling in the classical limit, $\alpha^{\text{on}}(q^2 = 0)$, we can consider the vacuum polarisation $\Pi(q^2)$ and the resulting running of the electric charge at non-zero q^2. Here we have to distinguish which renormalisation scheme we are considering. The vacuum polarisation in the on-shell scheme—which is a physical observable via corrections to the Coulomb potential—displays decoupling as expected, $\Pi^{\text{on}}(q^2) \propto q^2/m^2 + \dots$. In contrast, in mass-independent schemes as the MS or $\overline{\text{MS}}$ scheme the pole terms and thus Z_3 do not depend by definition on the mass of the virtual particle in the loop. Thus $\Pi(q^2)$ and the beta function contain in these schemes unsuppressed contributions from arbitrarily heavy particles. This unexpected behaviour can be explained by the fact that perturbation theory in these schemes breaks down in the limit $m^2 \gg \mu^2$. Then $\Pi^{\overline{\text{MS}}}(q^2) \simeq e^2/(72\pi^2)\ln(m^2/\mu^2)$ and the large log invalidates the one-loop result. There are two ways to address this unphysical behaviour. First, one may choose instead a mass-dependent scheme as the on-shell or the MoM scheme. Alternatively, one may wish to continue to use mass-independent schemes, because the RGE are simplest in these schemes. In this case one has to match by hand the effective theory of, for example, $n_f = 1$ QED to the $n_f = 2$ effective theory at the muon threshold $q^2 = 4m_\mu^2$, requiring that the running coupling is continuous. Continuing this process, one can match finally at the top threshold the $n_f = 8$ effective theory to "full" QED.

Let us move next to cases when the decoupling theorem does not apply. An important example is the electroweak sector of the SM where decoupling does not happen in several phenomenologically important cases. For instance, sending $m_t \to \infty$ but keeping m_b finite breaks SU(2) invariance. This makes $\mathscr{L}_{\text{eff}}$ non-renormalisable, and thus the condition for the applicability of the decoupling theorem is not satisfied. As a result, electroweak loop corrections may contain terms that are proportional to the mass splitting in one doublet, as, for example, in the correction to the W mass, $\Delta m_W \propto m_t^2 - m_b^2$. Similarly, $m_h \to \infty$ results in an higgsless effective theory which is not renormalisable. Finally, the limit $m_f \to \infty$ implies also $y_f \to \infty$ which in turn leads to a violation of perturbative unitarity.

Example 14.1: Higgs production via gluon fusion. The Higgs boson couples to the massless gluons only at the one-loop level via quark loops. Apart from being the most important production process of Higgs bosons in a proton–proton collider like the LHC, this process illustrates also that heavy fermions do not decouple from the Higgs. We calculate the Feynman amplitude for the process using DR: Although we know that we will obtain a finite result, individual terms in intermediate steps are divergent (cf. Eq. (14.46)). For a fermion of mass m_f in the loop, the Feynman rules give for $i\mathcal{A} = i\mathcal{A}_1 + i\mathcal{A}_2$ with

$$= i\mathcal{A}_1 = -(-ig_s)^2 \operatorname{tr}(T^a T^b)\left(\frac{-im_f}{v}\right)$$
$$\times \int \frac{\mathrm{d}^d k}{(2\pi)^d}\, \frac{i^3 N^{\mu\nu}}{D}\, \varepsilon_\mu(p)\varepsilon_\nu(q),$$

while the exchange graph $\mathcal{A}_2$ is obtained by permutating (p,μ,a) and (q,ν,b). The denominator in $\mathcal{A}_1$ is $D = abc = (k^2 - m_f^2)[(k+p)^2 - m_f^2][(k-q)^2 - m_f^2]$. Frist we use the usual Feynman parameterisation to combine the terms,

$$\frac{1}{abc} = 2\int_0^1 \mathrm{d}x \int_0^{1-x} \frac{\mathrm{d}y}{[ax + by + c(1-x-y)]^3}. \tag{14.43}$$

We find shifting also the momentum to $k' = k + px - qy$,

$$\frac{1}{D} \to 2\int \mathrm{d}x\mathrm{d}y\, \frac{1}{[k'^2 - m_f^2 + xy m_h^2]^3}. \tag{14.44}$$

For simplicity, we evaluate the numerator assuming transverse on-shell gluons, that is dropping terms proportional to p_μ or q_ν, obtaining

$$N^{\mu\nu} = 4m_f \left[\eta^{\mu\nu}\left(m_f^2 - k^2 - \frac{m_h^2}{2}\right) + 4k^\mu k^\nu + p^\nu q^\mu\right], \tag{14.45}$$

where we also skipped the prime. Now we shift momenta in the integration, drop terms linear in k from the numerator and use $k^\mu k^\nu = k^2 \eta^{\mu\nu}/n$. The integrals can be performed using

$$I(2,3) = \int \frac{\mathrm{d}^d k}{(2\pi)^d}\, \frac{k^2}{(k^2 - C)^3} = \frac{i}{32\pi^2}(4\pi)^\varepsilon \frac{\Gamma(1+\varepsilon)}{\varepsilon}(2-\varepsilon)C^{-\varepsilon} \tag{14.46}$$

$$I_2(2,3) = \int \frac{\mathrm{d}^d k}{(2\pi)^d}\, \frac{1}{(k^2 - C)^3} = -\frac{i}{32\pi^2}(4\pi)^\varepsilon \Gamma(1+\varepsilon)C^{-1-\varepsilon}. \tag{14.47}$$

The amplitude $\mathcal{A} = 2\mathcal{A}_1$ follows as

$$\mathcal{A} = -\frac{\alpha_s m_f^2}{2\pi v}\delta_{ab}\left(\eta^{\mu\nu}\frac{m_h^2}{2} - p^\nu q^\mu\right)\int \mathrm{d}x\mathrm{d}y\left(\frac{1-4xy}{m_f^2 - xy m_h^2}\right)\varepsilon_\mu(p)\varepsilon_\nu(q). \tag{14.48}$$

Evaluating the two integrals over the Feynman parameters in the limit $m_f \gg m_h$, we find

$$\mathcal{A} = -\frac{\alpha_s}{6\pi v}\delta_{ab}\left(\eta^{\mu\nu}\frac{m_h^2}{2} - p^\nu q^\mu\right)\varepsilon_\mu(p)\varepsilon_\nu(q). \tag{14.49}$$

Thus gluon fusion $gg \to h$ is independent of the mass of the heavy fermion in the loop for $m_f \gg m_h$. Crossing symmetry implies that the amplitude for the decay $h \to gg$ has the same property. Hence we can use the decay rate $\Gamma(h \to gg)$ as a counter for all fermion species, including those with an arbitrary high mass as long as it is generated by the SM Higgs effect.

Hierarchy problem. Let us next consider loop corrections to particle masses. In the case of fermion masses, we found in the toy model of an effective four-fermion theory (cf. problem 12.8) that the mass correction δm is given by

$$\delta m = -\frac{3gm}{8\pi}\left(\frac{m}{\Lambda}\right)^2\ln(m^2/\mu^2). \tag{14.50}$$

Thus the correction vanishes both for $\Lambda \to \infty$ and for $m \to 0$. The latter behaviour is generic for fermion masses since for $m \to 0$ chiral symmetry is restored. Treating the mass term $m\bar\psi\psi$ as a perturbation, the chiral symmetry of the free Lagrangian implies that the loop corrections to m vanish for $m \to 0$. Thus the enhanced symmetry for $m = 0$ protects fermion masses against corrections of the form $\delta m \propto \Lambda$ or $\delta m \propto M$, where M is the mass of the particle in the loop. We now compare this behaviour to the case of scalar masses.

Example 14.2: Loop corrections to the scalar mass.
We consider the effect of a fermion loop on the mass of the Higgs scalar. The Yukawa interaction $\mathcal{L}_I = -yh\bar\psi\psi$ leads to

$$i\Sigma(\not p) = (-iy)^2\int\frac{d^4k}{(2\pi)^4}\frac{\text{tr}[(\not p+\not k+M)(\not k+M)]}{[(p+k)^2-M^2+i\varepsilon][k^2-M^2+i\varepsilon]}, \tag{14.51}$$

where M denotes the mass of the fermion. Evaluating the trace, $\text{tr}[(\not p+\not k+M)(\not k+M)] = 4(k^2+kp+M^2)$, combining denominators, shifting $k \to k-p(1-x)$ and dropping linear terms gives in DR

$$i\Sigma = 4y^2\mu^{4-d}\frac{d-1}{(4\pi)^{d/2}}\Gamma(1-d/2)\int_0^1 dx[M^2-p^2x(1-x)]^{d/2-1}. \tag{14.52}$$

Expanded around $d = 4 - 2\varepsilon$, we obtain

$$\Sigma = -\frac{y^2}{4\pi^2}\left[\frac{3M^2}{\varepsilon}-\frac{p^2}{2\varepsilon}+M^2-\frac{p^2}{6}+3\int_0^1 dx[p^2x(1-x)-M^2]\ln\left(\frac{M^2-p^2x(1-x)}{\tilde\mu^2}\right)\right]. \tag{14.53}$$

Thus the unrenormalised self-energy contains terms proportional to M^2. Performing on-shell renormalisation, that is requiring that the loop correction vanish at the physical Higgs mass m, $\Sigma(m^2) = \Sigma'(m^2) = 0$, we obtain in the limit $M \gg m$

$$\Sigma^{\text{on}}(p^2) = \frac{y^2}{4\pi^2}\left[\frac{(p^2-m^2)^2}{20M^2}+\mathcal{O}(m^6/M^4)\right]. \tag{14.54}$$

Thus in the on-shell scheme loop corrections of heavy particles are suppressed as $\mathcal{O}(E^2/M^2)$. In contrast, the self-energy in the mass-independent $\overline{\text{MS}}$ scheme contains unsuppressed terms $\propto M^2$. Thus the renormalised self-energy behaves similar to the beta function, showing decoupling of heavy particles if a mass-dependent scheme is used.

These results make sense and are in agreement with the decoupling theorem. The fact that we have to subtract terms of $\sim M^2 \gg m^2$ in the renormalisation process may, however, be considered as "unnatural" or "fine-tuned". But the notion of naturalness or of a fine-tuning problem only becomes well-defined in theories where the considered parameter is calculable. Imagine, for example, a more fundamental theory than the SM which predicts m_h. Thus we can calculate in this theory the Higgs mass, obtaining an expression of the type

$$m_h^2 = a\Lambda^2 + b\frac{3y_t^2}{8\pi^2}\Lambda^2 + \ldots, \tag{14.55}$$

where the coefficients $a, b, \ldots$ are finite numbers, parameterising the mass correction caused by beyond the SM graphs, the top quark, and other SM particles. If these coefficients are $\mathcal{O}(1)$, and thus m_h is of the same order as Λ, we call the theory natural. If on the other hand $\Lambda \gg m_h$, the physics of widely separated scales has to be correlated. Informed by our experience of effective theories, we consider such a behaviour as unnatural. The presence of scalars with $m \ll \Lambda$ in such theories is therefore often called "hierarchy" or "fine-tuning problem". It arises in most extensions of the SM, for instance if the SM is embedded in a GUT theory. Only two exceptions are known. Since supersymmetry connects fermions and bosons, scalars profit in these theories from the chiral symmetry of fermions. Another exception are theories where the light scalars are pseudo-Goldstone bosons. In this case, the enhanced symmetry for $m = 0$ protects the scalar masses.

Summary

The electroweak sector of the SM is described by an $\text{SU}_L(2) \otimes \text{U}_Y(1)$ gauge symmetry which is broken spontaneously to $\text{U}_{\text{em}}(1)$. Three out of the four degrees of the complex scalar SU(2) doublet are pseudo-Goldstone bosons ($\phi^\pm, \phi_3$) which become the longitudinal degrees of freedom of the three massive gauge bosons, $W^\pm$ and Z. The remaining scalar degree of freedom becomes a physical Higgs boson. The SM does not explain why the sign of μ^2 is negative. In addition, the Yukawa sector, charge quantisation and the replication of three families remain mysterious within this theory.

Further reading. An excellent source, especially for phenomenological aspects of the SM including the role of effective theories, is Donoghue *et al.* (2014). Effective theories and decoupling are also discussed by Schwartz (2013). The physics of massive neutrinos is reviewed by Valle and Romao (2015). The Feynman rules for the electroweak model in R_ξ gauge are derived, for example, in Romao and Silva (2012) or Pokorski (1987); the latter includes also counter-terms. The historical development of the electroweak theory is sketched by Quigg (2015).

Problems

14.1 Fermi constant G_F. Find the connection between the charged current in the Fermi and the GSW theory, and determine thereby the numerical value of v.

14.2 Feynman rules for Higgs-gauge sector. Derive the Feynman rules for the vertices between the scalar and the gauge bosons in unitary gauge.

14.3 Goldstone boson–fermion couplings. Derive the Feynman rules for the vertices of fermions with charged or neutral Goldstone bosons ϕ in R_ξ gauge. Use

$$\Phi(x) = \begin{pmatrix} \phi^+(x) \\ \frac{1}{\sqrt{2}}(v + h(x) + \mathrm{i}\phi_3(x)) \end{pmatrix} \tag{14.56}$$

to read off the vertices of Goldstone bosons with fermions.

14.4 Mixing matrices. Derive the number of free parameters in the quark and neutrino mixing matrices as the sum of real mixing angles ϑ_i and phases $\mathrm{e}^{\mathrm{i}\phi_i}$ for n generations of fermions. Take into account that not all phases need to be observable and distinguish between the case of Dirac and Majorana neutrinos.

14.5 Majorana neutrino masses. Find the gauge-invariant operator with the lowest dimension that generates Majorana neutrino masses.

14.6 Beta function. Draw at least one Feynman diagram contributing to each of the individual terms in Eq. (14.26).

14.7 Higgs decay $h \to W^+W^-$. Calculate the matrix element and the decay width for Higgs decay into transverse and longitudinal W-bosons, $h \to W_T^+ W_T^-$ and $h \to W_L^+ W_L^-$. Consider then the decay into Goldstone bosons, $h \to \phi^+\phi^-$, and show

that $\mathcal{M}(h \to W_L^+W_L^-) = \mathcal{M}(h \to \phi^+\phi^-) + \mathcal{O}(m_W/m_h)$.

14.8 Scattering of longitudinal gauge bosons. Consider the scattering of longitudinal gauge bosons, $W_L^+W_L^- \to W_L^+W_L^-$, which can be found to $\mathcal{O}(M_W^2/s)$ from the Goldstone boson scattering. Find the s-wave amplitude in the limit $m_h \ll s$ and $m_h \gg s$. Apply the unitarity condition (14.39) and derive the limits where perturbative unitarity breaks down and the Higgs-gauge sector of the SM becomes a strongly interacting theory.

14.9 Gluon fusion $gg \to h$. Determine the superficial degree of divergence of the amplitude for gluon fusion $gg \to h$. Why do we know that $\mathcal{A}(gg \to h)$ is finite?

14.10 Higgs decay $h \to gg$. Determine the amplitude $\mathcal{A}(h \to gg)$ for $m_f \ll m_h$ and derive an effective Lagrangian $\mathscr{L}(hgg)$ for this process.

14.11 Higgs decay $h \to \gamma\gamma$. The process $h \to \gamma\gamma$ proceeds via a fermion and W loop. a.) ♣ Show that

$$\mathcal{A}(\gamma \to \gamma h) = \frac{1}{v}\left[\sum_f m_f \frac{\partial}{\partial m_f} + m_W \frac{\partial}{\partial m_W}\right] \times \mathcal{A}(\gamma \to \gamma)$$

is valid for $m_h \ll m_W, m_f$. Look up the result for the photon vacuum polarisation in the SM and find the effective Lagrangian describing $\mathcal{A}(h \to \gamma\gamma)$. b.) ♥ Calculate the matrix element for the process $h \to \gamma\gamma$ following the example $h \to gg$.

14.12 SUSY cancellation. Consider the combined effect of the one-loop correction to the scalar mass induced by the scalar self-energy and a fermion loop. Find the condition that the two corrections cancel.

15
Thermal field theory

The Green functions we have considered so far were defined as the expectation value of field operators in a pure state, the vacuum $|0\rangle$ in the absence of real particles. Out of these Green functions, we could build up our quantities of prime interest, decay or scattering amplitudes for $1 \to n$ and $2 \to n$ particles via the LSZ formalism. In this chapter, we discuss how Green functions should be calculated for a system which is not in its ground-state but described by a density matrix ρ. Examples for such systems are the early universe or the dense, hot interior of a star. The simplest and at the same time most important cases of thermal systems are those in equilibrium. The appendix to this chapter collects a few basic formulae from statistical physics we will need later.

15.1 Overview

In equilibrium statistical physics, the partition function Z is of central importance as all thermodynamic quantities can be derived from it. In the grand canonical ensemble (where both particles and energy may be exchanged between the system and the reservoir) the partition function takes the form

$$Z(V, T, \mu_1, \mu_2, \dots) = \sum_n \langle n| \, \mathrm{e}^{-\beta H - \mu_i N_i} \, |n\rangle = \mathrm{e}^{-\beta \Omega}, \tag{15.1}$$

where $\beta \equiv 1/T$ is the inverse temperature of the system, n denotes a complete set of quantum numbers, μ_i the chemical potential and N_i the number operator for particles of type i. The Landau or grand canonical free energy

$$\Omega(V, T, \mu_i) = -T \ln Z = U - TS - \mu_i N_i = -PV \tag{15.2}$$

connects the microscopic partition function to thermodynamics. While the partition function is closely related to the generating functional of the corresponding field theory in Euclidean space, we can derive from Ω all relevant thermodynamical quantities. For instance, we obtain the pressure P from Ω as $P = \partial\Omega/\partial V|_{T,\mu_i}$. In addition, the expectation value of any observable O is given as

$$\langle O \rangle = Z^{-1} \mathrm{Tr} \left[\mathrm{e}^{-\beta H - \mu_i N_i} O \right]. \tag{15.3}$$

In the following, we will always set $\mu_i = 0$. Then the partition function $Z(N, V, T)$ determines the free (Helmholtz) energy F as $F = -T \ln Z$.

Quantum Fields–From the Hubble to the Planck Scale. Michael Kachelriess. © Michael Kachelriess 2018.
Published in 2018 by Oxford University Press. DOI 10.1093/oso/9780198802877.001.0001

Calculational approaches.   Two main approaches to calculations are used in thermal field theory:

- In the real-time formalism, one applies the formula (15.3) valid for any observable directly to Green functions, that is one evaluates

$$
\begin{aligned}
G(x_1,\ldots,x_n) &= \langle T(\phi(x_1)\cdots\phi(x_n))\rangle \\
&= Z^{-1}\mathrm{Tr}\left[\mathrm{e}^{-\beta H}\,T(\phi(x_1)\cdots\phi(x_n))\right].
\end{aligned}
\tag{15.4}
$$

The main advantage of this method is that it can be extended to non-equilibrium cases. In particular, one can investigate the time evolution of a system towards thermal equilibrium. The proper definition of the propagators becomes, however, more involved than in the vacuum.

- In the imaginary time formalism, we perform a Wick rotation from Minkowski to Euclidean space, $t \to t_E = \mathrm{i}t$, so that the transition amplitude from an initial state, $|q(t_i)\rangle$, to a final state, $|q(t_f)\rangle$, is given by

$$
\langle q(t_f)|\,\mathrm{e}^{-(t_f-t_i)H}\,|q(t_i)\rangle = \int_{q(t_i)}^{q(t_f)} \mathcal{D}q\,\mathrm{e}^{-S},
\tag{15.5}
$$

where S is now the Euclidean action. If we set the evolution time, $t_f - t_i$, equal to the inverse temperature β and integrate over all periodic paths $q(t_f) = q(t_i + \beta)$, we obtain

$$
Z = \sum_q \langle q|\,\mathrm{e}^{-\beta H}\,|q\rangle = \int_{q(t)}^{q(t+\beta)} \mathcal{D}q\,\mathrm{e}^{-S}.
\tag{15.6}
$$

We now see that we have formally connected the path integral formulation of quantum mechanics (in Euclidean space) to the partition function of statistical mechanics. In contrast to field theories at zero temperature, the partition function in the Euclidean and the resulting Euclidean Green functions are not merely mathematical tools but our main objects of interest. Since the Euclidean Green functions depend on temperature instead of time, we are not able to describe time-dependent phenomena in this approach.

Thermal Green functions.   The trace in the partition function of statistical physics implies that we have to sum over configurations connecting the same physical state at t and $t+\beta$. In the path integral corresponding to statistical mechanics, the periodicity condition $q(t) = q(t + \beta)$ for the real coordinate q is clearly the only possible choice. In contrast, fields may only be observable through bilinear quantities, as, for example, $\bar\psi\Gamma\psi$ for a fermion field. This raises the question as to whether we should require periodic or anti-periodic boundary conditions.

We start by considering thermal Green functions G_β for a free scalar field. We split the Feynman propagator into two parts, setting $G^+(x,x') = \langle\phi(x)\phi(x')\rangle$ for $t > t'$ and $G^-(x,x') = \langle\phi(x')\phi(x)\rangle$ for $t < t'$. From the Heisenberg equation for the field operator,

$$
\phi(t,\boldsymbol{x}) = \mathrm{e}^{\mathrm{i}Ht}\phi(0,\boldsymbol{x})\mathrm{e}^{-\mathrm{i}Ht},
\tag{15.7}
$$

we find inserting $1 = \mathrm{e}^{\beta H}\mathrm{e}^{-\beta H}$ into the definition (15.3) and using then (15.7),

$$G_\beta^+(t', \boldsymbol{x}'; t, \boldsymbol{x}) = \text{Tr}\,[e^{-\beta H}\phi(t', \boldsymbol{x}')\phi(t, \boldsymbol{x})]/Z \tag{15.8a}$$

$$= \text{Tr}\,[e^{-\beta H}\phi(t', \boldsymbol{x}')e^{\beta H}e^{-\beta H}\phi(t, \boldsymbol{x})]/Z \tag{15.8b}$$

$$= \text{Tr}\,[\phi(t' + \mathrm{i}\beta, \boldsymbol{x}')e^{-\beta H}\phi(t, \boldsymbol{x})]/Z \tag{15.8c}$$

$$= \text{Tr}\,[e^{-\beta H}\phi(t, \boldsymbol{x})\phi(t' + \mathrm{i}\beta, \boldsymbol{x}')]/Z = G_\beta^-(t' + \mathrm{i}\beta, \boldsymbol{x}'; t, \boldsymbol{x}). \tag{15.8d}$$

Hence Green functions satisfy $G_\beta^\pm(t', \boldsymbol{x}'; t, \boldsymbol{x}) = G_\beta^\mp(t' + \mathrm{i}\beta, \boldsymbol{x}'; t, \boldsymbol{x})$. Repeating the procedure, we obtain $G_\beta^\pm(t', \boldsymbol{x}'; t, \boldsymbol{x}) = G_\beta^\mp(t' + \mathrm{i}\beta, \boldsymbol{x}'; t, \boldsymbol{x}) = G_\beta^\pm(t' + 2\mathrm{i}\beta, \boldsymbol{x}'; t, \boldsymbol{x})$, implying periodic boundary condition for the thermal propagators of bosonic fields,

$$G_\beta(t', \boldsymbol{x}'; t, \boldsymbol{x}) = G_\beta(t' + 2\mathrm{i}\beta, \boldsymbol{x}'; t, \boldsymbol{x}). \tag{15.9}$$

We now define the Matsubara propagator $G_\beta(\tau, \boldsymbol{x})$, which is the analogue to the Feynman propagator in the imaginary-time formalism, setting the real part of the temporal argument to zero and applying time-ordering to its imaginary part $\tau = \Im(t)$,

$$G_\beta(\tau', \boldsymbol{x}'; \tau, \boldsymbol{x}) = G_\beta(\tau' + 2\beta, \boldsymbol{x}'; \tau, \boldsymbol{x}) = \text{Tr}\,\{e^{-\beta H}T_\tau[\phi(\tau', \boldsymbol{x}')\phi(\tau, \boldsymbol{x})]\}/Z. \tag{15.10}$$

The derivation (15.8d) goes through unchanged for fermionic fields,

$$S_\beta^\pm(t', \boldsymbol{x}'; t, \boldsymbol{x}) = S_\beta^\mp(t' + \mathrm{i}\beta, \boldsymbol{x}'; t, \boldsymbol{x}). \tag{15.11}$$

But now the anticommuting nature of fermionic fields, i.e. the minus sign in the definition of time-ordered product (8.85), leads to antiperiodic boundary condition for their thermal propagators,

$$S_\beta(t', \boldsymbol{x}'; t, \boldsymbol{x}) = -S_\beta(t' + 2\mathrm{i}\beta, \boldsymbol{x}'; t, \boldsymbol{x}). \tag{15.12}$$

Both periodicity conditions discretise the frequency spectrum of the thermal wavefunctions and propagator. Moreover, they constrain the set of allowed frequencies, such that bosonic fields contain only even frequencies, while fermionic fields contain only odd frequencies, cf. problem 15.2. Thus the Fourier transform of thermal fields contained in a box of size $\beta \times V$ is given by

$$\phi(t, \boldsymbol{x}) = \frac{1}{\sqrt{\beta V}} \sum_{n=-\infty}^{\infty} \sum_{\boldsymbol{p}} \phi_{n,\boldsymbol{p}}\, e^{-\mathrm{i}(\omega_n t + \boldsymbol{p}\cdot\boldsymbol{x})} \tag{15.13}$$

with $\omega_n = 2n\pi T$ for bosonic and $\omega_n = (2n+1)\pi T$ for fermionic fields, respectively, and $n \in \mathbb{Z}$. The frequencies ω_n are called Matsubara frequencies. Similarly, the Green functions for a free scalar field is given in the limit $V \to \infty$ by

$$G_\beta(t, \boldsymbol{x}) = \frac{1}{\beta} \sum_{n=-\infty}^{\infty} \int \frac{\mathrm{d}^3 p}{(2\pi)^3}\, G_n(\omega_n, \boldsymbol{p})\, e^{-\mathrm{i}(\omega_n t + \boldsymbol{p}\cdot\boldsymbol{x})} \tag{15.14}$$

with

$$G_n(\omega_n, \boldsymbol{p}) = \frac{1}{\omega_n^2 + \boldsymbol{p}^2 + m^2}. \tag{15.15}$$

Thermal Green functions and vertices come without imaginary units, because we have transformed the path integral to Euclidean time.

15.2   Scalar gas

We will illustrate the basics of thermal field theory considering the simplest example, a gas of scalar particles, evaluating its free energy density $\mathcal{F} = F/V$ as a power series in λ,

$$\mathcal{F} = \mathcal{F}_0 + \lambda \mathcal{F}_1 + \lambda^2 \mathcal{F}_2 + \dots . \tag{15.16}$$

The free energy is determined by connected vacuum diagrams. The zeroth-order contribution in perturbation theory is given by a one-loop vacuum diagram and corresponds to the Stefan–Boltzmann law valid for a free, non-interacting gas. Going on to the two-loop vacuum diagrams, we will be able to derive the first quantum correction to the Stefan–Boltzmann law.

15.2.1   Free scalar gas

In the case of a non-interacting scalar field, we can perform the path integral in the partition function Z_0, obtaining as formal solution in Euclidean space

$$\int \mathcal{D}\phi \, \mathrm{e}^{-S_0} = \mathcal{N} \, \mathrm{Det}(-\partial^2 + m^2)^{-1/2}. \tag{15.17}$$

Neglecting the temperature-independent normalisation constant $\mathcal{N}$ and using the identity $\ln \mathrm{Det} A = \mathrm{Tr} \ln A$, we arrive at

$$\beta F = -\ln Z = \frac{1}{2} \, \mathrm{Tr} \ln(-\partial^2 + m^2). \tag{15.18}$$

We evaluate the operator trace as in the case of the effective potential in section 13.2, but take into account the changes

$$\int_{-\infty}^{\infty} \mathrm{d}t \to \frac{1}{\beta} \int_0^{\beta} \mathrm{d}\tau \quad \text{and} \quad \int_{-\infty}^{\infty} \mathrm{d}k_0 \to \frac{1}{\beta} \sum_{n=-\infty}^{\infty}$$

in the completeness relation for thermal states. Inserting a complete set of plane waves, we find

$$
\begin{aligned}
\mathrm{Tr} \ln[-\partial^2 + m^2] &= \int_0^{\beta} \mathrm{d}\tau \int \mathrm{d}^3 x \langle \tau, \boldsymbol{x} | \ln[-\partial^2 + m^2] | \tau, \boldsymbol{x} \rangle \\
&= \int_0^{\beta} \mathrm{d}\tau \int \mathrm{d}^3 x \frac{1}{\beta} \sum_n \frac{\mathrm{d}^3 k}{(2\pi)^3} \langle \tau, \boldsymbol{x} | \ln[-\partial^2 + m^2] | \omega_n, \boldsymbol{k} \rangle \langle \omega_n, \boldsymbol{k} | \tau, \boldsymbol{x} \rangle \\
&= V \sum_n \frac{\mathrm{d}^3 k}{(2\pi)^3} \ln[\omega_n^2 + \boldsymbol{k}^2 + m^2].
\end{aligned} \tag{15.19}
$$

The free energy density $\mathcal{F} = F/V$ follows as

$$\beta \mathcal{F} = \frac{1}{2} \sum_{n=-\infty}^{\infty} \int \frac{\mathrm{d}^3 k}{(2\pi)^3} \ln \left[\omega_n^2 + \boldsymbol{k}^2 + m^2 \right] . \tag{15.20}$$

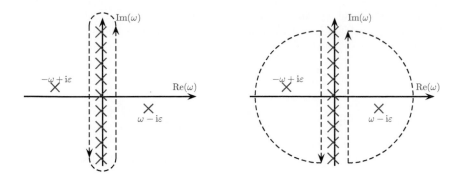

Fig. 15.1 Poles and contours in the complex ω plane used for the evaluation of the free energy $\mathcal{F}$.

In order to evaluate the sum, it is sufficient to consider

$$A = T \sum_n \ln\left[-(\mathrm{i}\omega_n)^2 + E_{\boldsymbol{k}}^2\right] \tag{15.21}$$

with $E_{\boldsymbol{k}}^2 = \boldsymbol{k}^2 + m^2$. First we eliminate the logarithm by differentiating A w.r.t. $E_{\boldsymbol{k}}$,

$$\frac{\mathrm{d}A}{\mathrm{d}E_{\boldsymbol{k}}} = -2T E_{\boldsymbol{k}} \sum_n \frac{1}{(\mathrm{i}\omega_n)^2 - E_{\boldsymbol{k}}^2}. \tag{15.22}$$

The function $\mathrm{d}A/\mathrm{d}E_{\boldsymbol{k}}$ has poles in the complex ω plane along the imaginary axis at $\omega = \mathrm{i}\omega_n = 2\pi n T\mathrm{i}$ plus two poles on the real axis at $\omega = \pm E_{\boldsymbol{k}} \mp \mathrm{i}\varepsilon$, cf. Fig. 15.1. We convert the sum into a contour integral using Cauchy's theorem in "reverse order". Because of $\coth(z) = \sinh'(z)/\sinh(z)$, we see that $\coth(z)$ has simple poles at $k\pi\mathrm{i}$ with residue 1 and $k \in \mathbb{Z}$. Thus $\coth(\beta\omega/2)$ has poles at $\mathrm{i}\omega_n$ with residues $2/\beta$, and we obtain

$$\frac{\mathrm{d}A}{\mathrm{d}E_{\boldsymbol{k}}} = -2T E_{\boldsymbol{k}} \frac{\beta}{2} \sum_n \mathrm{res}_{\omega_n} \left\{ \frac{\coth(\beta\omega/2)}{\omega_n^2 - E_{\boldsymbol{k}}^2} \right\} = \frac{E_{\boldsymbol{k}}}{2\pi\mathrm{i}} \oint_C \mathrm{d}\omega \, \frac{\coth(\beta\omega/2)}{\omega^2 - E_{\boldsymbol{k}}^2}. \tag{15.23}$$

The integrand vanishes as $1/|\omega|^2$ for $|\omega| \to \infty$, and thus we can break up the contour C into two pieces which we close at $\pm\mathrm{i}\infty$. Note that we thereby pick up a minus sign, since we change the orientation of the integration path. Now we can use Cauchy's theorem in the "normal order" to evaluate the residues of the two enclosed poles at $\pm E_{\boldsymbol{k}} \mp \mathrm{i}\varepsilon$,

$$\frac{\mathrm{d}A}{\mathrm{d}E_{\boldsymbol{k}}} = E_{\boldsymbol{k}} \sum_{\pm} \mathrm{res}_{\pm E_{\boldsymbol{k}}} \left\{ \frac{\coth(\beta\omega/2)}{\omega^2 - E_{\boldsymbol{k}}^2} \right\} = \coth(\beta E_{\boldsymbol{k}}/2) = 1 + \frac{2}{\mathrm{e}^{\beta E_{\boldsymbol{k}}} - 1}. \tag{15.24}$$

Integration gives

$$\mathcal{F} = \frac{1}{2} \int \frac{\mathrm{d}^3 k}{(2\pi)^3} \left[E_{\boldsymbol{k}} + 2T \ln(1 - \mathrm{e}^{-\beta E_{\boldsymbol{k}}}) + c \right]. \tag{15.25}$$

The integration constant c cancels against the normalisation constant of the path integral; dropping also the $T = 0$ vacuum part and taking the high-temperature limit $T \gg m$ gives

$$
\begin{aligned}
\mathcal{F} &= T \int \frac{\mathrm{d}^3 k}{(2\pi)^3} \ln(1 - \mathrm{e}^{-\beta E_k}) \\
&= \frac{T^4}{2\pi^2} \int_0^\infty \mathrm{d}x\, x^2 \ln\left(1 - \exp\{-\sqrt{x^2 + (\beta m)^2}\}\right) \simeq -\frac{\pi^2 T^4}{90} + \frac{m^2 T^2}{24}.
\end{aligned}
\tag{15.26}
$$

With $\mathcal{F} = -P$, the Stefan–Boltzmann law $P = \pi^2 T^4 / 90$ for a massless real scalar gas follows.

Example 15.1: Derive the high-temperature expansion (15.26) of the free energy $\mathcal{F}$. Performing a Taylor expansion of $\ln(1 - \exp\{-[x^2 + (\beta m)^2]^{1/2}\})$ around $\beta m = 0$, we find

$$
\int_0^\infty \mathrm{d}x\, x^2 \ln(1 - \mathrm{e}^{-[x^2 + (\beta m)^2]^{1/2}}) = \int_0^\infty \mathrm{d}x\, x^2 \ln(1 - \mathrm{e}^{-x}) + \frac{(\beta m)^2}{2} \int_0^\infty \mathrm{d}x\, \frac{x}{\mathrm{e}^x - 1} + \mathcal{O}(\beta m)^4.
$$

In the first integral we expand the logarithm,

$$
\int_0^\infty \mathrm{d}x\, x^2 \ln(1 - \mathrm{e}^{-x}) = -\sum_{n=1}^\infty \int_0^\infty \mathrm{d}x\, x^2 \mathrm{e}^{-nx} = -2\sum_{n=1}^\infty \frac{1}{n^4} = -2\zeta(4) = -2\frac{\pi^4}{90},
$$

while we use in the second integral $1/(\mathrm{e}^x - 1) = \mathrm{e}^{-x}/(1 - \mathrm{e}^x) = \sum_{n=1}^\infty \mathrm{e}^{-nx}$. With the definition (A.24) for the Gamma function, the second integral results in $\Gamma(2)\zeta(2) = \pi^2/6$.

15.2.2   Interacting scalar gas

As main application of the formalism of thermal field theory we next calculate the first quantum correction to the equation of state of a gas of scalar particles interacting with a $\lambda\phi^4$ interaction. This $\mathcal{O}(\lambda)$ correction is given by the two-loop vacuum diagram we considered already at $T = 0$; the overlapping divergences of the type $1/\varepsilon \ln(p^2/\mu^2)$ correspond now to UV divergent terms multiplied by temperature-dependent factors. Again, such terms have to be cancelled by counter-terms found at lower order.

Equation of state of a scalar gas.   The first-order contribution $\lambda\mathcal{F}_1$ to the free energy density corresponds to the two-loop vacuum diagram,

$$
\text{⚯} \quad = \frac{\lambda}{4!}\langle \phi^4 \rangle = \frac{3\lambda}{4!}\langle \phi^2 \rangle\langle \phi^2 \rangle = \frac{\lambda}{8}\Big(\Delta(0)\Big)^2,
$$

Recall from section 4.2 that the factor three accounts for the three possible ways of joining the lines to form the two loops. Alternatively, we can determine this factor using that $\langle \phi^n \rangle$ is the expectation value of ϕ^n times a Gaussian. We could now insert into this formula the Matsubara Green function derived in the imaginary time formalism. Instead, we use a more intuitive argument to obtain the propagator (for coinciding points) in the real-time formalism. At $T = 0$, we can express the propagator at coincident points as the sum over the zero-point energies,

$$\Delta(0) = \langle 0|\phi(x')\phi(x)|0\rangle_{x' \searrow x} = \int \frac{\mathrm{d}^3 k}{(2\pi)^3 2\omega_{\boldsymbol{k}}} \mathrm{e}^{-ik(x'-x)} \bigg|_{x' \searrow x} = \int \frac{\mathrm{d}^3 k}{(2\pi)^3} \frac{1}{2\omega_{\boldsymbol{k}}}. \quad (15.27)$$

The vacuum contains no real particles, and the $a_{\boldsymbol{k}}^\dagger a_{\boldsymbol{k}}$ term in $H_0 = \sum_{\boldsymbol{k}} \omega_{\boldsymbol{k}}(1/2 + a_{\boldsymbol{k}}^\dagger a_{\boldsymbol{k}})$ gives zero contribution. For $T > 0$, the expectation value of the number operator $N_{\boldsymbol{k}} = a_{\boldsymbol{k}}^\dagger a_{\boldsymbol{k}}$ is just the number distribution $n_{\boldsymbol{k}}$ of ϕ particles,

$$\Delta(t = 0, x = 0) = \int \frac{\mathrm{d}^3 k}{(2\pi)^3} \frac{1}{\omega_{\boldsymbol{k}}} \left(\frac{1}{2} + \left\langle a_{\boldsymbol{k}}^\dagger a_{\boldsymbol{k}} \right\rangle \right) = \int \frac{\mathrm{d}^3 k}{(2\pi)^3} \frac{1 + 2n_{\boldsymbol{k}}}{2\omega_{\boldsymbol{k}}}. \quad (15.28)$$

In thermal equilibrium the number density $n_{\boldsymbol{k}}$ of a scalar field is a Bose–Einstein distribution,

$$n_{\boldsymbol{k}} = \frac{1}{e^{\beta\omega_{\boldsymbol{k}}} - 1}. \quad (15.29)$$

This result can be derived directly from the periodicity condition of the Green functions (problem 15.1). Note that we can view the propagator as the sum of a vacuum part ("1/2") and a thermal part ("$n_{\boldsymbol{k}}$"). In the latter, the high-energy modes are exponentially suppressed and thus no UV divergences should appear in the temperature-dependent parts of physical observables. Thus our standard renormalisation program at $T = 0$ should apply in the same way at $T > 0$.

Continuing the derivation of the first-order correction to $\mathcal{F}$ we have

$$\frac{\lambda}{4!}\langle \phi^4 \rangle = \frac{\lambda}{8} \left[\sum_{\boldsymbol{k}} \frac{1 + 2n_{\boldsymbol{k}}}{2\omega_{\boldsymbol{k}}} \right]^2 \quad (15.30)$$

$$= \frac{\lambda}{8} \left[\underbrace{\left(\sum_{\boldsymbol{k}} \frac{1}{2\omega_{\boldsymbol{k}}} \right)^2}_{\text{vacuum}} + \underbrace{\left(\sum_{\boldsymbol{k}} \frac{n_{\boldsymbol{k}}}{\omega_{\boldsymbol{k}}} \right)^2}_{T \text{ dependent}} + \underbrace{2 \left(\sum_{\boldsymbol{k}} \frac{1}{2\omega_{\boldsymbol{k}}} \right) \left(\sum_{\boldsymbol{k}} \frac{n_{\boldsymbol{k}}}{\omega_{\boldsymbol{k}}} \right)}_{\text{vacuum} \times T \text{ dependent}} \right]. \quad (15.31)$$

The mixed term gives rise to a temperature-dependent UV divergence, which from the previous argument should not appear in our calculation. Similarly as at $T = 0$, this term should be cancelled by a one-loop counter-term. We consider therefore the one-loop self-energy Σ, which is given by

$$\Sigma = \frac{\lambda}{2} \Delta(t = 0, \boldsymbol{x} = 0) = \frac{\lambda}{2} \sum_{\boldsymbol{k}} \frac{1 + 2n_{\boldsymbol{k}}}{2\omega_{\boldsymbol{k}}}, \quad (15.32)$$

and contains the usual $T = 0$ divergence coming from the unsuppressed sum over frequencies. We use renormalised perturbation theory, adding the counter-term $\delta_m m^2 \phi^2$ to the Lagrangian, where δ_m is chosen such that m corresponds to the physical mass. Thus $\delta_m m^2$ is determined by

$$\delta_m m^2 + \frac{\lambda}{2} \sum_{\boldsymbol{k}} \frac{1}{2\omega_{\boldsymbol{k}}} = 0. \quad (15.33)$$

Because the product $\delta_m m^2 \Delta(0)$ is of $\mathcal{O}(\lambda)$, we have to add the following vacuum diagram to Ω_1,

$$\frac{1}{2}\delta_m m^2 \Delta(0,0) = \bigcirc\!\!\!\!\!\times = \frac{1}{2}\left(-\frac{\lambda}{2}\sum_k \frac{1}{2\omega_k}\right)\left(\sum_k \frac{1+2n_k}{2\omega_k}\right). \tag{15.34}$$

Comparing this expression to the troublesome mixed term, we see that they agree but have the opposite sign. Thus the one-loop subdivergence cancels the temperature dependent UV divergence at two-loop in Ω_1. As result, we obtain the consistent expression

$$\frac{\lambda}{4!}\langle \phi^4 \rangle = \frac{\lambda}{8}\left(\sum_k \frac{n_k}{\omega_k}\right)^2 + \text{vac}, \tag{15.35}$$

which we can now calculate explicitly. For simplicity, we restrict ourselves to the high-temperature limit setting $m = 0$,

$$\sum_k \frac{n_k}{\omega_k} = \int \frac{d^3k}{(2\pi)^3}\frac{1}{\omega}\frac{1}{e^{\beta\omega}-1} = \frac{T^2}{2\pi^2}\underbrace{\int_0^\infty dx \frac{x}{e^x - 1}}_{\pi^2/6} = \frac{T^2}{12}, \tag{15.36}$$

Hence our final result for $\mathcal{F}_1$ is

$$\mathcal{F}_1 = \frac{\lambda}{8}\left(\frac{T^2}{12}\right)^2 = \frac{\lambda}{1152}T^4, \tag{15.37}$$

which we may compare to the non-interacting result, $\Omega_0 = \pi^2 T^4/90$. The ratio $\Omega_1/\Omega_0 \approx 10^{-2}\lambda$ seems to indicate a fast convergence of the perturbative expansion of the pressure for any reasonable value of the coupling.

We simply quote the result to three loops or second order in λ from the literature,

$$P = \frac{\pi^2 T^4}{9}\left[\frac{1}{10} - \frac{1}{8}\frac{\lambda}{16\pi^2} + \frac{1}{8}\left(3\ln\frac{\mu}{4\pi T} + \frac{31}{35} + C\right)\left(\frac{\lambda}{16\pi^2}\right)^2\right]. \tag{15.38}$$

The parameter μ in Eq. (15.38) is not the chemical potential but as usually the renormalisation scale. As the pressure is a physical quantity, it should not depend upon such a parameter. However, the truncation of the perturbative series leads to a residual μ dependence of $\mathcal{O}(\lambda^3)$. Examining this μ dependence of P we obtain

$$\mu\frac{dP}{d\mu} = \frac{\pi^2 T^4}{9}\left[-\frac{1}{8}\frac{1}{16\pi^2}\mu\frac{d\lambda}{d\mu} + \frac{3}{8}\left(\frac{\lambda}{16\pi^2}\right)^2 + \mathcal{O}(\lambda^3)\right] = \mathcal{O}(\lambda^3), \tag{15.39}$$

where we have used $\mu d\lambda/d\mu = \beta = 3\lambda^2/16\pi^2$. Thus the dependence on the renormalisation scale μ is indeed of higher order in λ than the order of perturbation theory with which we are working. Still, we can try to minimise the remaining dependence by a suitable choice of μ. Clearly, a sensible choice in this case is $\mu = 4\pi T$, for which the logarithm vanishes. Still, any other choice is mathematically as correct as this

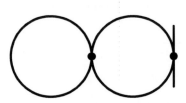 

Fig. 15.2 Left: A second-order correction to the mass. Right: Ring diagram, each external bubble corresponds to an insertion of m_{D}^2.

one. Instead of being worried about this dependence on the renormalisation scale, we may take advantage of it as follows: varying the renormalisation scale μ in a "reasonable range", say between $\mu = 2\pi T$ and $\mu = 8\pi T$, we obtain an error estimate for the missing higher-order corrections. Finally, note that $\mu \propto T$ implies that a QCD plasma becomes in the large temperature limit an asymptotically free gas of quarks and gluons. Since we know that at $T = 0$ quarks and gluons are confined in hadrons, we expect that at $T = \mathcal{O}(\Lambda_{\mathrm{QCD}})$ a phase transition from a quark–gluon gas to a gas of colourless mesons and baryons takes place. Both analytical calculations and lattice QCD simulations confirm this picture.

IR behaviour. We have found an apparently fast convergence of the perturbative expansion for small λ calculating the pressure of a scalar gas. For applications particularly interesting is the case of a massless particle and we consider now as a toy model for QCD a $\lambda\phi^4$ theory with $m = 0$.

Looking back at our previous result for the self-energy, Eq. (15.32), setting $m \to 0$ and dropping the vacuum term, we have

$$\Sigma = \frac{\lambda}{2}\Delta(t = 0, \boldsymbol{x} = 0) \to \frac{\lambda}{2}\int \frac{\mathrm{d}^3 k}{(2\pi)^3}\frac{n_{\boldsymbol{k}}}{\omega_{\boldsymbol{k}}} = \frac{\lambda}{2}\frac{T^2}{12}. \tag{15.40}$$

Thus the thermal part of the self-energy induces at first order in λ a thermal or Debye mass,

$$m_{\mathrm{D}}^2 = \frac{\lambda}{2}\frac{T^2}{12}. \tag{15.41}$$

Switching to the covariant form of the thermal propagator, Eq. (15.14), the contribution shown in the left panel of Fig. 15.2 at second order is

$$\Sigma_2 = -\frac{\lambda}{2}T\sum_n \int \frac{\mathrm{d}^3 k}{(2\pi)^3}\frac{m_{\mathrm{D}}^2}{(\omega_n^2 + \boldsymbol{k}^2)^2}. \tag{15.42}$$

While in the terms with $n \neq 0$ the ω_n act as a IR cut-off, we see that for $n = 0$ and thus $\omega_0 = 0$ the integral is proportional to $\int \mathrm{d}k/\boldsymbol{k}^2$ and thus IR divergent. If we go to higher terms in the expansion and add additional loops to the "primary loop" as shown in the right panel of Fig. 15.2, then the degree of the IR divergence increases.

The solution to this problem is to account properly for the thermal mass of the scalar particle. If we use an effective propagator which includes the Debye mass of the particle,

$$\frac{1}{\omega_n^2 + \boldsymbol{k}^2} \rightarrow \frac{1}{\omega_n^2 + \boldsymbol{k}^2 + m_{\mathrm{D}}^2}, \tag{15.43}$$

then the $n = 0$ term of Eq. (15.42) with $\omega_0 = 0$ is given by

$$\Sigma_{2,n=0} = -\frac{\lambda}{2} T \int \frac{\mathrm{d}^3 k}{(2\pi)^3} \frac{m_{\mathrm{D}}^2}{(\boldsymbol{k}^2 + m_{\mathrm{D}}^2)^2} = -\frac{\lambda^2}{4} \left(\frac{T^2}{12}\right) \left(\frac{T}{8\pi m_{\mathrm{D}}}\right) \tag{15.44}$$

and thus finite. Since the Debye mass scales as $m_{\mathrm{D}} \propto \lambda^{1/2}$, the contribution of $\Sigma_{2,n=0}$ in the perturbative expansion is not of order λ^2 but of order $\lambda^{3/2}$. Thus we obtained a term which is non-analytic in the coupling—which cannot happen, if we sum a finite number of terms. As explanation, we have to look at the expansion of the effective propagator for small m_{D} (restricting ourselves again to the $n = 0$ term),

$$\frac{1}{\boldsymbol{k}^2 + m_{\mathrm{D}}^2} = \frac{1}{\boldsymbol{k}^2} - \frac{m_{\mathrm{D}}^2}{\boldsymbol{k}^4} + \frac{m_{\mathrm{D}}^4}{\boldsymbol{k}^6} + \dots. \tag{15.45}$$

Here we can view, for example, the $m_{\mathrm{D}}^4/\boldsymbol{k}^6$ term as a ring diagram with three massless propagators $\boldsymbol{k}^{-2}$ and two factors m_{D}^2 produced by self-energy insertions. Thus including the thermal mass corresponds to summing up the infinite sum of diagrams shown in the right panel of Fig. 15.2.

We can formalise the inclusion of the Debye mass in the originally massless scalar theory as follows. We reorganise perturbation theory by adding a mass term to the free Lagrangian and subtracting it from the the interaction term,

$$\mathscr{L}_0 = \frac{1}{2} (\partial_\mu \phi)^2 - \frac{1}{2} m^2 \phi^2 \quad \text{and} \quad \mathscr{L}_{\mathrm{int}} = -\frac{\lambda}{4!} \phi^4 + \frac{1}{2} m^2 \phi^2. \tag{15.46}$$

Here we may set $m^2 = m_D^2$ or keep it as a free parameter to be determined by, for example, that the free energy is independent of this parameter, $\mathrm{d}\mathscr{F}/\mathrm{d}m^2 = 0$. This reformulation of the perturbative expansion in thermal field theories is called screened or optimised perturbation theory.

Symmetry restoration at high temperature. We have chosen the sign of the mass term such that the $\lambda\phi^4$ theory is in the unbroken phase, and the minimal energy is obtained for $\langle\phi\rangle = 0$. In this case thermal effects simply increase the effective mass of the ϕ particle,

$$V(\phi, T) = \frac{1}{2} \left(m^2 + \frac{\lambda}{24} T^2\right) \phi^2 + \frac{\lambda}{4!} \phi^4, \tag{15.47}$$

Something more interesting happens, if we consider the broken phase choosing $m^2 < 0$. Then for $T = 0$ the minimal energy is obtained for $\phi_0 = \pm\sqrt{-6m^2/\lambda}$, and the $\phi \rightarrow -\phi$ symmetry of the Lagrangian is broken. Increasing the temperature, the positive thermal mass grows. Above some critical temperature, the effective mass becomes therefore positive and the minimal energy configuration becomes $\phi_0 = 0$, cf. with Fig. 15.3. This

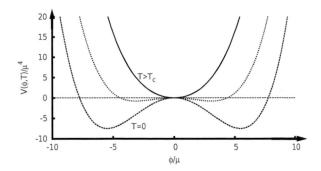

Fig. 15.3 Potential with negative mass term showing restoration of symmetry for $T > T_c$.

behaviour resembles the one we know from a ferromagnet. Below some critical temperature T_c, spontaneous magnetisation breaks rotation invariance, which above T_c is restored. In the context of cosmology, this observation suggests that symmetries that are broken today might have been unbroken in the early, hot universe.

15.A Appendix: Equilibrium statistical physics in a nut-shell

The one-particle distribution function $f(p)$ of a free gas in *kinetic equilibrium* is given by

$$f(p) = \frac{1}{\exp[\beta(E - \mu)] \pm 1} \tag{15.48}$$

where $\beta = 1/T$ is the inverse temperature, $E = \sqrt{m^2 + p^2}$, and $+1$ refers to fermions and -1 to bosons, respectively. A species X stays in kinetic equilibrium if in the reaction $X + Y \rightleftharpoons X + Y$ the energy exchange with at least one other species Y, which is in thermal equilibrium, is fast enough. The chemical potential μ is the average energy needed, if an additional X particle is added, $dU = \mu_X dN_X$. If the species X is also in *chemical equilibrium* with other species, for example via the reaction $X + \bar{X} \rightleftharpoons \gamma + \gamma$, then their chemical potentials are related by $\mu_X + \mu_{\bar{X}} = 2\mu_\gamma = 0$.

The number density n, energy density ρ and pressure P of a species X (which may be not in equilibrium) are connected to its one-particle distribution function $f(p)$ as

$$n = g \int \frac{d^3 p}{(2\pi)^3} f(p), \qquad \rho = g \int \frac{d^3 p}{(2\pi)^3} E f(p), \tag{15.49}$$

$$P = g \int \frac{d^3 p}{(2\pi)^3} \frac{p^2}{3E} f(p). \tag{15.50}$$

The factor g takes into account the internal degrees of freedom like spin or colour. Thus for a photon, a massless spin-1 particle $g = 2$, for an electron $g = 2$, a quark $g = 6$, etc.

In the non-relativistic limit $T \ll m$ or $e^{\beta(m-\mu)} \gg 1$, and thus differences between bosons and fermions disappear,

$$n = \frac{g}{2\pi^2} e^{-\beta(m-\mu)} \int_0^\infty dp\, p^2 e^{-\beta \frac{p^2}{2m}} = g \left(\frac{mT}{2\pi}\right)^{3/2} \exp[-\beta(m - \mu)], \tag{15.51}$$

Table 15.1 The number of relativistic degrees of freedom g_* present in the universe as function of its temperature T.

Temperature	new particles	$4\Delta g_*$	$4g_*$
$T < m_e$	$\gamma + \nu_i$	$4 \times (2 + 3 \times 2 \times 7/8)$	29
$m_e < T < m_\mu$	$e^\pm$	14	43
$m_\mu < T < m_\pi$	$\mu\pm$	14	57
$m_\pi < T < T_{\text{QCD}}$	$\pi^\pm, \pi^0$	12	69
$T_{\text{QCD}} < T < m_s$	$u, \bar{u}, d, \bar{d}, g$	$6 \times 14 + 4 \times 8 \times 2 - 12$	205
$m_s < T < m_c$	$s, \bar{s}$	$3 \times 14 = 42$	247
$m_c < T < m_\tau$	$c, \bar{c}$	42	289
$m_\tau < T < m_b$	$\tau^\pm$	14	303
$m_b < T < m_{W,Z}$	$b, \bar{b}$	42	345
$m_{W,Z} < T < m_h$	$W^\pm, Z$	$4 \times 3 \times 3 = 36$	381
$m_h < T < m_t$	h	4	385
$m_t < T < ?$	$t, \bar{t}$	42	427

$\rho = mn$ and $P = nT \ll \rho$. These expressions correspond to the classical Maxwell–Boltzmann statistics. The number of non-relativistic particles is exponentially suppressed, if their chemical potential is small. In the relativistic limit $T \gg m$ with $T \gg \mu$, all properties of a gas are determined by its temperature T,

$$n = \frac{g\,T^3}{2\pi^2} \int_0^\infty \mathrm{d}x \frac{x^2}{e^x \pm 1} = \varepsilon_1 \frac{\zeta(3)}{\pi^2}\, gT^3, \tag{15.52}$$

$$\rho = 3P = \frac{g\,T^4}{2\pi^2} \int_0^\infty \mathrm{d}x \frac{x^3}{e^x \pm 1} = \varepsilon_2 \frac{\pi^2}{30}\, gT^4, \tag{15.53}$$

where for bosons $\varepsilon_1 = \varepsilon_2 = 1$ and for fermions $\varepsilon_1 = 3/4$ and $\varepsilon_2 = 7/8$, respectively.

Since the energy density and the pressure of non-relativistic species are exponentially suppressed, the total energy density and the pressure of all species present in the universe can be approximated including only the relativistic ones,

$$\rho_{\text{rad}} = 3P_{\text{rad}} = \frac{\pi^2}{30}\, g_* T^4, \tag{15.54}$$

where

$$g_* = \sum_{\text{bosons}} g_i \left(\frac{T_i}{T}\right)^4 + \frac{7}{8} \sum_{\text{fermions}} g_i \left(\frac{T_i}{T}\right)^4. \tag{15.55}$$

We denote the relativistic species also collectively as radiation. Table 15.1 shows the number of relativistic degrees of freedom g_* in the SM as function of the temperature. Here T_{QCD} denotes the temperature of the QCD phase transition, above which quarks and gluons as free particles exist.

The total entropy density $s \equiv S/V$ of the universe can again approximated by the relativistic species,

$$s = \frac{2\pi^2}{45}\, g_{*S} T^3, \tag{15.56}$$

where now

$$g_{*,S} = \sum_{\text{bosons}} g_i \left(\frac{T_i}{T}\right)^3 + \frac{7}{8} \sum_{\text{fermions}} g_i \left(\frac{T_i}{T}\right)^3. \tag{15.57}$$

The entropy S is a useful quantity because it is conserved if the universe evolves adiabatically. Then the conservation of S implies that $S \propto g_{*,S} a^3 T^3 = \text{const.}$, where $a(t)$ is the scale factor describing the expansion of the universe. Thus the temperature of the universe evolves as

$$T \propto g_{*,S}^{-1/3} a^{-1}, \tag{15.58}$$

and when $g_* = \text{const.}$, the temperature scales as $T \propto 1/a$. Consider now the case that a particle species, for example, electrons, becomes non-relativistic at $T \sim m_e$. Then the particles annihilate, $e^+ e^- \to \gamma\gamma$, and their entropy is transferred to photons. Formally, $g_{*,S}$ decreases and therefore the temperature decreases for a short period less slowly than $T \propto 1/a$. Since the net number density n of particles with a conserved charge and the entropy density scale both as $\propto a^{-3}$, the ratio n/s is constant. Therefore it is often convenient to consider the time evolution of the dimensionless variable $Y = n/s$.

Summary

Thermal Green functions are (anti-) periodic functions in imaginary time, leading to discrete energies with $\omega_n = 2n\pi T$ for bosonic and $\omega_n = (2n{+}1)\pi T$ for fermionic fields, respectively. No new UV divergences appear for $T > 0$, since the thermal distribution function vanish exponentially for $E/T \to \infty$. In a plasma, even massless particles acquire a temperature dependent (Debye) mass and symmetries of the Lagrangian may be hidden at low temperatures.

Further reading. This chapter offers only a flavour of thermal field theory. The lecture notes of Blaizot (2011) are a useful starting point to learn more, before turning to textbooks dedicated to thermal field theory as Laine and Vuorinen (2016) or Kapusta and Gale (2011).

Problems

15.1 Bose–Einstein distribution. Derive (15.29) from (15.9).

15.2 Periodicity conditions. Derive the Matsubara frequencies (15.29) from the periodicity condition of thermal propagators, $G(\tau) = \pm G(\tau + \beta)$. [Hint: use (15.9).]

15.3 Asymptotic freedom. QCD processes in a quark–gluon plasma become asymptotically free for $T \gg \Lambda_{\text{QCD}}$. In contrast, the condition $s \gg \Lambda_{\text{QCD}}^2$ is not sufficient to ensure the applicability of perturbation theory at $T = 0$. Explain qualitatively the reason for this difference.

15.4 Pressure integral. a.) Show that $P = -(\partial U/\partial V)_S$. b.) Combine the quantisation condition of free particles, $p_k =$ $2\pi k/L$, with $U = V \int \mathrm{d}^3 p/(2\pi)^3 E f(p)$ and $S \propto \ln(V f(p))$ to derive the pressure integral (15.50).

15.5 Abundance for non-zero chemical potential. Show that for $\mu \ll T$, the number density of a fermion (boson) in chemical equilibrium is given by

$$\Delta n_f \equiv n_f - n_{\bar{f}} = \delta g_i \sum_i \mu_i q_i \frac{T^2}{3} \tag{15.59}$$

with $\delta = 1/2$ ($\delta = 1$).

15.6 Entropy density. Determine the entropy density s_0 of the present universe assuming that it is dominated by the entropy of CMB photons with temperature $T_0 = 2.75\,\text{K}$.

16

Phase transitions and topological defects

At the end of the last chapter we introduced the idea that spontaneously broken symmetries of field theories could be restored at high temperatures. In this case, the hot early universe would be in a symmetric phase, followed by transitions to phases with more and more broken symmetries. Phase transitions in the early universe can lead to observable consequences largely for two reasons: first, phase transitions lead often to the formation of topological defects. These defects are zero-, one- or two-dimensional extended solutions of the classical equations of motion for the Higgs-gauge sector which contain in their core the unbroken $\langle\phi\rangle = 0$ vacuum. Depending on the symmetry-breaking scale and their dimensionality, they lead to (un-) desirable cosmological consequences and can thus be used to constrain particle physics models beyond the SM. Second, the state of the universe deviates from thermodynamical equilibrium during a first-order phase transition. Thus processes like the generation of a baryon asymmetry which require out-of-equilibrium conditions may take place during first-order phase transitions.

16.1 Phase transitions

Effective potential at $T > 0$. We have seen that the ground-state of a quantum field theory including quantum fluctuations is determined by the effective potential V_{eff}. We can include additionally thermal fluctuations by studying the temperature-dependent effective potential. The latter is obtained in a rather straight-forward way in the imaginary-time formalism, where we only have to replace vacuum expectation values with thermal averages in the definition of classical fields, and to use periodic boundary conditions in the Euclidean effective action and effective potential. In particular, we can transform the $T = 0$ effective potential (13.32) defined in Euclidean space,

$$V_{\text{eff}}(\phi) = V(\phi) + \frac{1}{2} \int \frac{\mathrm{d}^4 k}{(2\pi)^4} \ln\left[k^2 + V''(\phi)\right] + \mathcal{O}(\hbar^2), \qquad (16.1)$$

into the temperature-dependent effective potential replacing the integration over the continuous energy k_0 by a summation over discrete Matsubara frequencies ω_n,

$$\beta V_{\text{eff}}(\phi) = \beta V(\phi) + \frac{1}{2} \sum_n \int \frac{\mathrm{d}^3 k}{(2\pi)^3} \ln\left[\omega_n^2 + \boldsymbol{k}^2 + V''(\phi)\right] + \mathcal{O}(\hbar^2). \qquad (16.2)$$

The sum over n is performed in the same way as in the calculation of the free energy $\mathcal{F}$ of a non-interacting scalar gas in section 15.2.1,

Quantum Fields–From the Hubble to the Planck Scale. Michael Kachelriess. © Michael Kachelriess 2018.
Published in 2018 by Oxford University Press. DOI 10.1093/oso/9780198802877.001.0001

$$\beta V_{\text{eff}}^{(1)}(\phi) = \frac{1}{2} \int \frac{\mathrm{d}^3 k}{(2\pi)^3} \left[\beta E_{\boldsymbol{k}} + 2 \ln(1 - e^{-\beta E_{\boldsymbol{k}}}) \right] \tag{16.3}$$

with $E_{\boldsymbol{k}}^2 = \boldsymbol{k}^2 + V''(\phi)$. Next we split the one-loop contribution $\beta V_{\text{eff}}^{(1)}(\phi)$ into the $T = 0$ vacuum part $V_{\text{eff}}^{(1)}(\phi, 0)$ and a temperature-dependent thermal part $V_{\text{eff}}^{(1)}(\phi, T)$,

$$V_{\text{eff}}^{(1)}(\phi) = \frac{1}{2} \int \frac{\mathrm{d}^4 k}{(2\pi)^4} \ln \left[k^2 + V''(\phi) \right] + T \int \frac{\mathrm{d}^3 k}{(2\pi)^3} \ln \left[1 - e^{-\beta E_{\boldsymbol{k}}} \right], \tag{16.4}$$

where we dropped a ϕ-independent constant.

In order to simplify the discussion we consider separately the two limiting cases that quantum or thermal fluctuations dominate. In the latter case, $V_{\text{eff}}^{(1)}(\phi, 0) \ll V_{\text{eff}}^{(1)}(\phi, T)$, we evaluate $V_{\text{eff}}^{(1)}(\phi, T)$ in the high-temperature limit[1]

$$V_{\text{eff}}^{(1)}(\phi, T) = T \int \frac{\mathrm{d}^3 k}{(2\pi)^3} \ln(1 - e^{-\beta E_{\boldsymbol{k}}}) = \frac{T^4}{2\pi^2} \int_0^\infty \mathrm{d}x \, x^2 \ln(1 - e^{-\sqrt{x^2 + a}}) \tag{16.5}$$

$$= -\frac{\pi^2 T^4}{90} + \frac{a T^4}{24} + \dots. \tag{16.6}$$

with $a = \beta^2 V''(\phi)$. For the choice $V''(\phi) = -\mu^2 + \frac{1}{2}\lambda\phi^2$, the effective potential becomes

$$V(\phi) + V_{\text{eff}}^{(1)}(\phi, T) = -\frac{1}{2}\mu^2\phi^2 + \frac{\lambda}{24}\phi^4 - \frac{\pi^2 T^4}{90} - \frac{1}{24}\mu^2 T^2 + \frac{1}{48}\lambda\phi^2 T^2 + \frac{\lambda}{12}\phi^4. \tag{16.7}$$

The terms quadratic in the field ϕ reproduce the thermal or Debye mass $m_D^2 = \lambda T^2/24$ which we found in (15.41). The minimum of $V_{\text{eff}}(\phi)$ moves smoothly away from $\phi = 0$ below the critical temperature and no barrier is formed, as shown in Fig. 15.3. Thus the phase transition is in this case of second-order. More interesting is a first-order phase transition which is shown schematically in Fig. 16.1. As the system cools down, a local minimum $\langle\phi\rangle \neq 0$ develops at the temperature T_1. The characteristic feature of a first-order transition is that in the temperature range $T_2 < T < T_1$ the two local minima at $\langle\phi\rangle = 0$ and $\langle\phi\rangle \neq 0$ coexist. Thus a potential barrier has to separate them. At the critical temperature T_c, the minimum at $\langle\phi\rangle \neq 0$ becomes the global minimum. Because of the barrier, the transition from $\langle\phi\rangle = 0$ to $\langle\phi\rangle \neq 0$ cannot proceed classically, but has to proceed via a quantum or thermal fluctuation. This tunnelling process may lead to deviations from thermal equilibrium which in turn may lead to traces observable today.

An example of a first-order phase transition is given by the SM for a small Higgs mass, $m_h \lesssim 70\,\text{GeV}$. In this (unphysical) limit, $V_{\text{eff}}(\phi, T)$ is given for large T by

$$V_{\text{eff}}(\phi, T) \simeq \frac{1}{2} a(T^2 - T_1^2)\phi^2 - \frac{1}{3} bT\phi^3 + \frac{1}{4}\lambda\phi^4 \tag{16.8}$$

with

---

[1] The derivation follows the one of Eq. (15.26); recall also that a resummation of ring diagrams may be necessary.

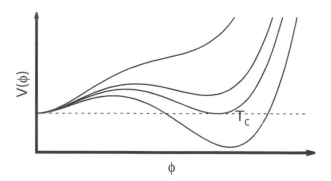

Fig. 16.1 The effective potential $V_{\text{eff}}(\phi, T)$ in case of a first-order phase transition.

$$a = \frac{3}{16}g^2 + \left(\frac{1}{2} + \frac{m_t^2}{m_h^2}\right)\lambda, \qquad b = \frac{9g^3}{32\pi}, \qquad T_1 = \frac{m_h}{2\sqrt{a}}. \qquad (16.9)$$

The critical temperature of the first-order phase transition follows as $T_c = T_1/\sqrt{1 - 2b^2/(9a\lambda)} > T_1$. Between T_c and T_1, the effective potential $V_{\text{eff}}(\phi, T)$ has two degenerate minima at $\phi = 0$ and $\phi_c = 2bT_c/(3\lambda)$ separated by a barrier. This corresponds to the effective potential sketched in Fig. 16.1.

16.2   Decay of the false vacuum

When the universe goes through a first-order phase transition as it cools down, the fields sitting in the false vacuum have to tunnel to the true minimum. The true vacuum nucleates at a localised position in spacetime, when a quantum (or thermal) fluctuation tunnels through (or above) the barrier. It forms an extending bubble which contains the energetically favoured ground-state. In the case of thermal fluctuations, this phenomenon is known to everybody from boiling water. The equivalent problem in quantum mechanics is the tunnelling through the barrier in a double-well potential $V(x)$ depicted in the left panel of Fig. 16.2. The tunnelling probability P can be calculated using the WKB method as

$$P \sim \exp\left(-\mathrm{i}\int_a^b \mathrm{d}x\sqrt{2m[V(x) - E]}\right), \qquad (16.10)$$

where a and b are the turning points of the tunnelling trajectory. In order to translate this prescription to a field theory, we rewrite the tunnelling probability P first as an Euclidean path integral. The exponent is the integral over the (imaginary) momentum of the particle, which we can express as

$$\mathrm{i}\int \mathrm{d}x\, p = \mathrm{i}\int \mathrm{d}t\, p\dot{x} = \mathrm{i}\int \mathrm{d}t\, (E + L) = \int \mathrm{d}t_E\, (-L_E). \qquad (16.11)$$

In the last step we assumed that the energy of the particle is normalised to zero and changed to Euclidean time $t_E = \mathrm{i}t$. Thus the tunnelling probability is given by the Euclidean path integral

$$P = \int \mathcal{D}x \exp(-S_E[x]) \tag{16.12}$$

with

$$S_E[x] = \int dx \left(\frac{1}{2} \frac{d^2 x}{dt_E^2} + V(x) \right). \tag{16.13}$$

From (16.10), we know that the (Minkowski) path integral is dominated by the path depicted on the left of Fig. 16.2. Since the potential V changes sign performing a Wick rotation, a classically forbidden path in Minkowski space corresponds to an allowed Euclidean path shown in the right panel of Fig. 16.2. Thus finding the tunnelling probability between two vacua amounts to finding solutions in the Euclidean theory which connect the two vacua and minimise the action.

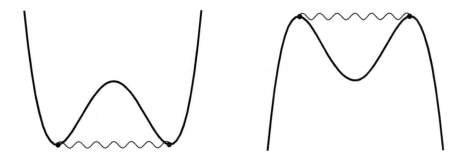

Fig. 16.2 Tunnelling through a double-hill potential in Minkowski space corresponds to the classically allowed path from one maxima to another one in Euclidean space.

Bounce solution. We now turn to the case we are interested in, namely the tunnelling of a scalar field ϕ sitting in the false, metastable vacuum a into the true vacuum b, cf. Fig. 16.3. In Euclidean space, we should find solutions of the classical action starting from b which bounce at the classical turning point a back to b. These solutions are called "bounce" or instanton, since the tunnelling happens instantaneously in Minkowski space. Moreover, these solutions should have a finite action such that they give a non-zero contribution to the semi-classical limit of the path integral. In principle, we should find all solutions with a finite action and then integrate their contribution. Since the tunnelling probability is however exponentially suppressed, the integral is dominated by the solution which minimises the action.

Solutions which are symmetric under rotations minimise the gradient energy. We use therefore an O(4) symmetric ansatz, where the real scalar field ϕ is only a function of the radial coordinate $r^2 = \boldsymbol{x}^2 + t_E^2$. Then the Euclidean action for a real scalar field becomes

$$S = 2\pi^2 \int_0^\infty dr \, r^3 \left[\frac{1}{2} \left(\frac{d\phi}{dr} \right)^2 + V(\phi) \right] \tag{16.14}$$

and the field equation simplifies to (problem 16.2)

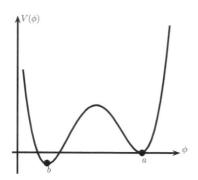 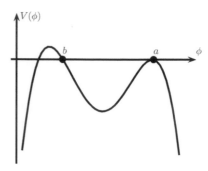

Fig. 16.3 Left: Transition from the false vacuum (a) to the true one (b) requires tunnelling through a barrier in Minkowski space. Right: In Euclidean space, a classically allowed path starts from b and bounces at a back to b; the zero point of the potential is chosen to agree with a.

$$\frac{\mathrm{d}^2\phi}{\mathrm{d}r^2} + \frac{3}{r}\frac{\mathrm{d}\phi}{\mathrm{d}r} - \frac{\mathrm{d}V}{\mathrm{d}\phi} = 0. \tag{16.15}$$

The boundary conditions $\phi \to 0$ for $r \to \infty$ and $\mathrm{d}\phi/\mathrm{d}r|_{r=0} = 0$ ensure a regular solution at $r = 0$ and a finite action. Viewing (16.15) as a mechanical problem, it describes the motion of a particle in a potential with the friction term $\frac{3}{r}\frac{\mathrm{d}\phi}{\mathrm{d}r}$. Thus the two parts $a \to b$ and $b \to a$ of the bounce are not equivalent. We should choose the starting position $\phi(t = -\infty)$ uphill from b such that at $t = +\infty$ the field comes to rest at a.

We can proceed analytically if we consider the limiting case where either the potential difference between the true and the false vacuum is much smaller or much larger than the potential barrier which separates them. We will consider here the first case, which corresponds to the so-called thin-wall approximation. This allows us to approximate the potential as

$$V(\phi) = V_0(\phi) + \mathcal{O}(\varepsilon), \tag{16.16}$$

where $V_0(\phi)$ is the symmetric potential and the difference $\varepsilon = V(\phi_+) - V(\phi_-)$ between the false and the true minima is a small parameter. For small ε, the volume energy $\propto \varepsilon r^4$ of a bubble dominates the surface energy $\propto r^3$ only, if the bubble is sufficiently large. In this case, the thickness of the bubble wall, that is, the region where $\mathrm{d}\phi/\mathrm{d}r$ deviates significantly from zero, is also much smaller than the bubble size. Thus we can neglect the $\frac{3}{r}\frac{\mathrm{d}\phi}{\mathrm{d}r}$ term in (16.15), and the field equation simplifies to the one of a one-dimensional problem,

$$\phi'' \equiv \frac{\mathrm{d}^2\phi}{\mathrm{d}r^2} = \frac{\mathrm{d}V_0}{\mathrm{d}\phi}. \tag{16.17}$$

Using the chain rule, $\mathrm{d}V_0/\mathrm{d}\phi = V_0'/\phi'$ and $(\phi'^2)' = 2\phi''\phi'$, we find after one integration

$$\frac{1}{2}\phi'^2 - V_0 = c. \tag{16.18}$$

We determine the integration constant by demanding that the contribution to the action S goes to zero for $r \to \infty$. With $\phi'(\infty) = 0$, this gives $c = -V(\phi_+)$. Separating variables in (16.18) leads then to

$$r = \int \frac{d\phi}{\sqrt{2[V_0(\phi) - V(\phi_+)]}}. \tag{16.19}$$

To gain more insight we now consider a specific potential. We choose our favourite $\lambda\phi^4$ potential,

$$V_0(\phi) = \frac{\lambda}{4} \left(\phi^2 - \eta^2\right)^2. \tag{16.20}$$

Compared to chapter 13, we shifted the potential such that the two minima at $\phi = \pm\eta = \pm\sqrt{\mu^2/\lambda}$ have the value zero, $V_0(\phi_\pm) = 0$, as required. Inserting this potential into Eq. (16.19), we find

$$r = \mp\frac{\sqrt{2}}{\eta\sqrt{\lambda}}\text{arctanh}\left(\phi/\eta\right) + r_0. \tag{16.21}$$

The integration constant r_0 determines the position of the bubble wall. Inverting (16.21) gives as field profile

$$\phi_\mp(r) = \mp\eta\tanh\left(\frac{\eta\sqrt{\lambda}}{\sqrt{2}}(r - r_0)\right). \tag{16.22}$$

The solution $\phi_\mp$ interpolates between the two vacua $\pm\eta$ of the symmetric potential $V_0(\phi)$. The thickness of the bubble wall is determined by the argument x of $\tanh(x)$ and is given by $\delta \sim \sqrt{2}/(\eta\sqrt{\lambda})$.

Knowing the solution $\phi(r)$, we can calculate the action. For $r \gg r_0$, the field is constant, $\phi \approx \eta$, and $V(\phi_+) = 0$. Therefore, the contribution $S_>$ from this region to the action is zero. For $r \ll r_0$, the field $\phi \approx -\eta$ is again constant, but the potential contributes $V(\phi_-) \simeq -\varepsilon$ and thus

$$S_< = 2\pi^2 \int_0^{r_0} dr\, r^3 \left[\frac{1}{2}(\phi')^2 + V(\phi)\right] \simeq -\frac{1}{2}\pi^2\varepsilon r_0^4. \tag{16.23}$$

Finally, the contribution from the bubble wall is

$$S_w = 2\pi^2 r_0^3 \int dr \left[\frac{1}{2}(\phi')^2 + V(\phi)\right] = 2\pi^2 r_0^3 \int dr\, 2V(\phi) =$$
$$= 2\pi^2 r_0^3 \int_{\phi_1}^{\phi_2} d\phi \sqrt{2V(\phi)} = 2\pi^2 r_0^3 I, \tag{16.24}$$

with $\phi_{1/2} \approx \phi(r_0 \mp \delta)$. The quantity

$$I = \int_{\phi_1}^{\phi_2} d\phi \sqrt{2V(\phi)} \tag{16.25}$$

has the interpretation of the surface tension of the bubble. Adding the two contributions, the total action follows as

$$S = S_< + S_w = -\frac{1}{2}\pi^2 \varepsilon r_0^4 + 2\pi^2 r_0^3 I. \tag{16.26}$$

We find the solution which gives the largest tunnelling probability minimising the action S w.r.t. r_0, resulting in the condition

$$r_0 = \frac{3I}{\varepsilon}. \tag{16.27}$$

For our choice (16.19) for the potential, the surface tension is $I = 2\mu^3/(3\lambda)$ and the action

$$S = \frac{27\pi^2 I^4}{2\varepsilon^3} = \frac{8\pi^2 \mu^{12}}{3\lambda^4 \varepsilon^3}. \tag{16.28}$$

For $\varepsilon \to 0$, the action becomes infinite. Equivalently, the tunnelling from one to another ground-state $\pm\eta$ of the symmetric potential $V_0(\phi)$ costs an infinite amount of energy, as we argued in chapter 13.

The tunnelling probability per time and volume follows then as

$$p = A \int \mathcal{D}\phi \exp(-S[\phi]), \tag{16.29}$$

where the pre-factor A is determined by $\mathrm{Det}(\Box - V''(\phi))^{-1/2}$ and its zero-modes (Callan and Coleman, 1977). In practice, the result is rather insensitive to the exact value of A, and setting $A \sim V''(\phi_0)$ is sufficient for simple estimates.

We can now justify the assumptions made. Our starting assumption that ε is small implies that r_0 is large, $r_0 \propto 1/\varepsilon$. Moreover, a small ε implies that the bubble of true vacuum is separated by a thin wall from the metastable vacuum, $\delta \sim (\varepsilon r_0)^{-1/3}$. This is the reason for the name thin-wall approximation. In order to describe the time evolution of a bubble, we have to analytically continue the O(4) symmetric solution $r_0^2 = \boldsymbol{x}^2 + t_E^2$ back to Minkowski space. From $r_0^2 = \boldsymbol{x}^2 - t^2$, we see that the bubble extends close to the speed of light for $t \gg r_0$. The resulting O(1,3) symmetry means that all inertial observers[2] will measure the same expansion law.

Scalar instantons. For the case of a massless $\lambda\phi^4$ theory we can find a class of exact solutions for the bounce (16.15), which are often called scalar instantons. For a massless particle, the classical solution should fall off as a power law at large distances. Then dimensional analysis tells us that ϕ has the dimension of an inverse length. Since the massless theory is scale invariant, the solutions should depend on an arbitrary parameter ρ characterising their size. This suggests to try the ansatz $\phi(r) \propto \rho/(r^2 + \rho^2)$ in Eq. (16.15), resulting in (problem 16.3)

$$\phi(r) = \left(\frac{8}{\lambda}\right)^{1/2} \frac{\rho}{r^2 + \rho^2}. \tag{16.30}$$

---

[2]The facts that the bubble extends with $v \sim 1$ and that the bubble is thin imply (un?)-fortunately that any potential observers will be dissolved without having the time to notice the arrival of the wall.

Thus the action of the bounce becomes

$$S = \frac{27\pi^2 I^4}{2\varepsilon^3} = \frac{8\pi^2}{3\lambda} .$$ (16.31)

Example 16.1: Decay of the metastable SM vacuum. We use $V = \lambda_{\text{eff}}\phi^4/4$ as a crude approximation for the scalar potential of the SM, where λ_{eff} is the running Higgs self-coupling at the scale $\mu = \phi$. Using the thin-wall approximation, we can check that the result (16.31) makes sense. The surface tension is $I \sim (|\lambda_{\text{eff}}|/2)^{1/2}\phi^3/3$ and $\varepsilon \sim |\lambda_{\text{eff}}|\phi^4/4$. Thus the action of the bounce becomes

$$S = \frac{27\pi^2 I^4}{2\varepsilon^3} = \frac{8\pi^2}{3|\lambda_{\text{eff}}|}$$

agreeing with (16.31). An analysis of the RGE of the SM gives for the effective Higgs boson coupling at the Planck scale $\lambda_{\text{eff}}(M_{\text{Pl}}) \approx -0.01$, using the central experimental values for m_h and m_t. For the estimate of the probability P that a bubble of the true vacuum has nucleated in the past light-cone of an observer, we can set $VT \sim t_0^4$ with t_0 as the present age of the universe. Setting the prefactor A by dimensional reasons equal to $A \sim \phi^4 \sim M_{\text{Pl}}^4$, the tunnelling probability is

$$P \sim (t_0 M_{\text{Pl}})^4 \exp\{-8\pi^2/(3|\lambda_{\text{eff}}|)\} \sim 10^{-900} .$$ (16.32)

Despite of the enormous volume factor, $(t_0 M_{\text{Pl}})^4 \sim (8\times10^{60})^4$ the tunnelling probability is exceedingly small.

Although the SM vacuum is unstable when extrapolated to the Planck scale, the probability that inside the past light-cone of the observed universe a tunnelling has happened is practically zero. This corresponds to the "metastable region" of the SM shown in Fig. 14.1.

16.3 Topological defects

A rather generic consequence of phase transitions is the formation of extended solutions to the field equations which are stable by virtue of a topological quantum number. The latter arises if the space of possible vacuum field configurations consists of two or more subspaces which cannot be connected by solutions with a finite classical action. We call two field configurations ϕ_1 and ϕ_2 topologically equivalent, if they can be transformed continuously into each other keeping the the potential energy and the Euclidean action finite during the transformation. Thus topologically equivalent field configurations form equivalence classes which can be characterised by a topological quantum number and are separated by an infinite energy barrier. Within each equivalence class, the minimal energy solution will be stable. The spatially uniform ground-state we have assumed up to now as our vacuum in perturbative calculations may thus be disconnected from other stable solutions which cannot spread out to become spatially uniform due to their non-trivial topology.

As a system cools below its critical temperature, it undergoes a transition from its symmetric state to a state with broken symmetry. While the correlation length

$\xi = \langle \phi(0)\phi(x)\rangle$ becomes formally infinite at the phase transition, the order parameter in the broken phase can take the same value only in a finite volume, restricted by the finite propagation speed of the relevant waves. In the case of the expanding universe, we expect the formation of order one topological defect per causally connected region. This process was first suggested by Kibble, and is therefore called Kibble mechanism. In a big bang model, it follows thus $\xi(t) < t$ with t as the age of the universe. The number density n of topological defects created in a phase-transition at time t is therefore bounded from below as $n > \xi^{-3} \sim t^{-3}$.

Domain walls are two-dimensional topological defects which separate three-dimensional volumes containing different vacua. Examples are ferromagnets where domains of uniform magnetisation exist. Depending on the gauge group and the pattern of symmetry breaking, topological defects with one and zero dimensions are also possible: in the first case, a one-dimensional line or string contains a vacuum different from the surrounding, while it is the second case a point-like object called monopole. We will proceed as in chapters 13 and 14, starting with a Higgs model with a single scalar field and increasing then the number of scalar fields. At the same, the dimensionality of the topological defects formed will decrease. As start, we will consider however the sine-Gordon model which is defined in $1 + 1$ spacetime dimensions.

Sine-Gordon solitons.   We have chosen in general as potential $V(\phi)$ a polynomial in the field ϕ. In the case of SSB, the periodicity of the angular variable in $\phi = \eta e^{i\vartheta}$ leads to a periodicity of the potential. Another example of a Lagrangian with a periodic potential is the sine-Gordon model defined by

$$\mathscr{L} = \frac{1}{2}(\partial_\mu \phi)^2 - V(\phi) = \frac{1}{2}(\partial_\mu \phi)^2 - \frac{a}{b^2}[1 - \cos(b\phi)] \tag{16.33}$$

with $\mu = \{0,1\}$ and two real parameters a and b. The potential $V(\phi)$ has zeros for $\phi = 2\pi n/b$. Expanding $V(\phi)$ for small ϕ,

$$\mathscr{L} = \frac{1}{2}(\partial_\mu \phi)^2 - \frac{1}{2}a\phi^2 - \frac{ab^2}{4!}\phi^4 + \ldots \tag{16.34}$$

and comparing to our usual $\lambda \phi^4$ potential, we see that they agree for small ϕ if we identify[3] $a = m^2$ and $b^2 = \lambda$. From the Lagrangian (16.33), the sine-Gordon equation

$$\frac{\partial^2 \phi}{\partial t^2} - \frac{\partial^2 \phi}{\partial x^2} + \frac{a}{b}\sin(b\phi) = 0 \tag{16.35}$$

follows. It admits static and travelling wave solutions, $\phi(x,t) = f(\pm(x - vt)) \equiv f(\pm\xi)$. You can check that

$$f(\xi) = \frac{4}{b}\arctan[\exp(\pm\gamma\sqrt{a/b}\,\xi)] \tag{16.36}$$

is a solution where $\gamma = (1 - v^2)^{-1/2}$ is the usual Lorentz factor (problem 16.4). The solution with the plus sign is called a kink, the one with the minus an anti-kink. The kink interpolates between the $n = 0$ ground-state at $x \to -\infty$ and $n = 1$ ($\phi = 2\pi/\sqrt{\lambda}$) at $x \to \infty$. The extension ℓ of the kink is determined by the argument of the arctan and is given by $\ell = \sqrt{b}/(\gamma\sqrt{a}) = \lambda/(\gamma m)$, cf. Fig. 16.4.

---

[3]Recall that a scalar field has dimension $[\phi] = (d - 2)/2$ in d spacetime dimensions. Thus the argument of $\cos(b\phi)$ is dimensionless while the ϕ^4 interaction requires a m^2 prefactor.

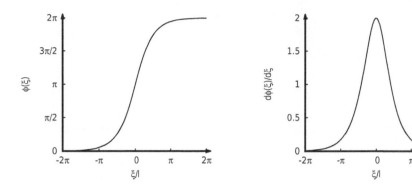

Fig. 16.4 Left: Kink solution $\phi(\xi)$ for $a = b = 1$. Right: its derivative $\mathrm{d}\phi(\xi)/\mathrm{d}\xi$.

The energy (or mass) of a static solution interpolating between $n = 0$ and $n = 1$ is

$$E = \int_{-\infty}^{\infty} \mathrm{d}x \left[\frac{1}{2} \left(\frac{\partial \phi}{\partial x} \right)^2 + V(\phi) \right] = \int_{-\infty}^{\infty} \mathrm{d}x \, 2V(\phi) = \int_{0}^{2\pi/b} \mathrm{d}\phi \, \frac{\partial x}{\partial \phi} V(\phi) =$$

$$= \int_{0}^{2\pi/b} \mathrm{d}\phi \, \sqrt{2V(\phi)} = \frac{\sqrt{2a}}{b^2} \int_{0}^{2\pi} \mathrm{d}\vartheta \, [1 - \cos(\vartheta)]^{1/2} = \frac{8m}{\lambda}. \qquad (16.37)$$

In the second step we used the field equation together with the requirement that the energy E is finite: Eq. (16.18) with $c = 0$. More generally, a static solution connecting n_1 and n_2 has the energy $E = 8|n_1 - n_2|m/\lambda$. Thus the energy spectrum of these solutions is discrete and inversely proportional to the coupling constant. The latter property implies that these solutions are not accessible in a perturbative calculation which is a power series in the coupling constant λ. In the weak coupling regime, the mass $8m/\lambda$ of the kink is much larger than the mass m of the elementary field ϕ, while in the strong coupling limit the kink is the lightest excitation.

We can imagine the sine-Gordon model as a chain of arrows, which the force $F = -\partial_x V$ tries to orient, for example, downwards. Then a kink is a solution which points from $x = -\infty \ldots - \ell$ downwards, turning at $\xi \sim 0$ by 2π, and pointing at $x \gtrsim \ell$ again downwards. To untwist the chain, we would have to turn the arrows on one of the two sides of $\xi \approx 0$ which would cost an infinite amount of energy. Therefore the winding number $\nu \equiv n_1 - n_2$ is a conserved quantum number and the solutions are stable. We can understand why the winding number is called a *topological* quantum numbers as follows: identifying the points $x = -\infty$ and $x = \infty$, we map $\mathbb{R}$ on a compact interval. Then the kink becomes a Möbius band, which cannot be smoothly transformed into a circle S^1.

A conserved quantum number implies the existence of a conserved current. The condition $E < \infty$ requires that the field approaches for $x \pm \infty$ one of the vacuum states, and thus

$$\phi(\infty) - \phi(-\infty) = \frac{2\pi}{\sqrt{\lambda}} \nu. \qquad (16.38)$$

We can rewrite this is an integral,

$$\int_{-\infty}^{\infty} \mathrm{d}x \, \partial_x \phi = \frac{2\pi}{\sqrt{\lambda}} \nu. \tag{16.39}$$

Since the solution is two-dimensional, we can set as current

$$j^\mu = \frac{\sqrt{\lambda}}{2\pi} \varepsilon^{\mu\nu} \partial_\nu \phi. \tag{16.40}$$

The antisymmetry of $\varepsilon^{\mu\nu}$ ensures then that $\partial_\mu j^\mu = 0$. The conserved charge follows with $\varepsilon^{01} = 1$ as

$$Q = \int \mathrm{d}x \, j^0 = \frac{\sqrt{\lambda}}{2\pi} \int \mathrm{d}x \, \partial_x \phi = \frac{\sqrt{\lambda}}{2\pi} [\phi(\infty) - \phi(-\infty)] = n_1 - n_2, \tag{16.41}$$

that is, it equals the winding number ν. In contrast to the conserved charges we have encountered up to now, the current is not the Noether current of a global symmetry and we did not have to use the equations of motion in its derivation. Instead, the charge has a topological origin.

The solutions of the sine-Gordon equation maintain their shape, although the equation is non-linear. Classical solutions with this property are called *solitons*. The distinctive property of the sine-Gordon solitons in $1 + 1$ spacetime dimensions is that this also holds for the asymptotic regions of a scattering event: in particular, two solitons emerge from a collision unchanged except possibly for a phase shift, although the superposition principle is not valid for non-linear equations of motion.

Domain walls. We now move to topological defects in $3 + 1$ dimensions considering two-dimensional topological defects called domain walls. Assuming that the domain wall is uniform in the $y - z$ plane, its energy (per area) is

$$E[\phi] = \int_{-\infty}^{\infty} \mathrm{d}x \left[\frac{1}{2} (\phi')^2 + V(\phi) \right]. \tag{16.42}$$

We assume that the potential energy $V(\phi)$ is shifted such that $\min(V(\phi)) = 0$. Then the energy is the sum of two (semi-) positive terms, $E[\phi] \geq 0$. Instead of integrating the equations of motion, as in Eqs. (16.17) to (16.19), we will use now an argument due to Bogomolnyi: He suggested rewriting the energy "completing the square" as

$$E[\phi] = \frac{1}{2} \int_{-\infty}^{\infty} \mathrm{d}x \left[\phi' + \sqrt{2V(\phi)} \right]^2 \pm \int_{\phi(-\infty)}^{\phi(\infty)} \mathrm{d}\tilde{\phi} \sqrt{2V(\tilde{\phi})}. \tag{16.43}$$

The second integral depends only on the boundary values of the field at infinity. For field configurations which connect the same ground-state at $x = -\infty$ and $+\infty$, this boundary term vanishes, and the minimum $E[\phi] = 0$ of the energy is attained for constant fields. If the field configurations connect different ground-states, however, then we can associate the second integral with a non-zero topological charge. The winding number in the sine-Gordon case is a specific example of such a topological charge.

A lower bound for the energy of fields with non-zero topological charge is

$$E[\phi] \geq \left| \int_{\phi(-\infty)}^{\phi(\infty)} \mathrm{d}\tilde{\phi} \sqrt{2V(\tilde{\phi})} \right|, \tag{16.44}$$

which is attained when the first integral vanishes. In this case, the field satisfies the first-order equation $\phi' = \pm\sqrt{2V(\phi)}$ which is generally much easier to solve than the original second-order equation of motion. Separating variables again, we arrive at

$$x = \pm \int \frac{\mathrm{d}\phi}{\sqrt{2V(\phi)}}. \tag{16.45}$$

To proceed, we have to choose a definite potential. For the choice (16.20), we recover the solution

$$\phi_{\mp}(x) = \mp\eta \tanh\left(\frac{\eta\sqrt{\lambda}}{\sqrt{2}}(x - x_0) \right) \tag{16.46}$$

which we used in the calculation of the bounce. The solution $\phi_{\mp}$ interpolates between the two vacua $\pm\eta$ of the symmetric potential $V_0(\phi)$ and contains at $x = x_0$ a two-dimensional plane with the unbroken vacuum $\phi = 0$. More generally, the two-dimensional surfaces which separate domains with opposite values of η can have a finite size.

Global cosmic strings. We continue our way through possible topological defects looking at the consequences of a global continuous symmetry. Thus we replace the single real field by a set of two real or one complex field $\phi = \phi_1 + i\phi_2$ with potential

$$V(\phi) = \frac{\lambda}{4}\left(\phi^{\dagger}\phi - \eta^2 \right)^2. \tag{16.47}$$

We use cylinder coordinates ρ, ξ, z, and search for static solutions using as ansatz

$$\phi = \eta e^{in\vartheta} f(\rho). \tag{16.48}$$

The phase $e^{in\vartheta}$ can be chosen arbitrarily at $\rho = 0$, unless the field is zero at the origin. In order to ensure a single-valued field, we must therefore impose the boundary condition $f(\rho) \to 0$ for $\rho \to 0$. In the opposite limit, $\rho \to \infty$, the field should approach one of its minima,

$$\phi \to \eta e^{in\vartheta} \quad \text{and} \quad f(\rho) \to 1, \tag{16.49}$$

in order to minimise the energy.

For the ansatz (16.48), the field equations are separable and we find from $\Delta\phi = \lambda(\phi^2 - \eta^2)\phi$ as differential equation for f

$$\frac{\mathrm{d}^2 f}{\mathrm{d}\xi^2} + \frac{1}{\xi}\frac{\mathrm{d}f}{\mathrm{d}\xi} - \frac{n^2}{\xi^2}f = \xi(f^2 - 1)f. \tag{16.50}$$

Here we have also introduced the new dimensionless variable $\xi = \lambda^{1/2}\eta\rho$. This differential equation together with the two boundary conditions for $\rho \to 0$ and $\rho \to \infty$ has to be solved numerically. We can, however, estimate the extension of a cosmic string by dimensional arguments. The scale of the problem is set by $\rho/\xi = \lambda^{-1/2}\eta^{-1}$. Thus the core containing the unbroken vacuum $\phi = 0$ has the radius $\mathcal{O}(\lambda^{-1/2}\eta^{-1})$, while for larger distances $\rho \gg \lambda^{-1/2}\eta^{-1}$ the field approaches the broken phase $|\phi| = \eta$. The extended solution should have a finite energy E per length L,

$$\frac{E}{L} = \int_0^\infty d\rho\rho \int_0^{2\pi} d\vartheta \left[|\boldsymbol{\nabla}\phi|^2 + V(\phi) \right] \tag{16.51a}$$

$$= \int_0^\infty d\rho\rho \int_0^{2\pi} d\vartheta \left[\partial_\rho\phi^\dagger\partial_\rho\phi + \frac{1}{\rho^2}\frac{\partial\phi^\dagger}{\partial\vartheta}\frac{\partial\phi}{\partial\vartheta} + V(\phi) \right]. \tag{16.51b}$$

The first and the last term in (16.51b) give a finite contribution to E/L, since $\partial_\rho f(\rho) \to 0$ and $V(\phi) \to 0$ for $\rho \to \infty$. In contrast, the middle term contributes $\propto \int_0^R d\rho\rho^{-1}f(\rho)$, that is, a logarithmically diverging term $\ln(R/\delta)$ to the linear energy density. The scale δ has to be determined by the typical extension of the string, and thus the energy density inside the radius R around the string is $E/L \sim \ln[R/(\lambda^{1/2}\eta)]$. If we consider the realistic case of a string network instead of the idealisation of an isolated global string then R should be given by the typical distance of strings.

Local cosmic strings.   We can avoid the (formal) problem of the infinite energy associated with a string if we gauge the model. In this case we obtain the abelian Higgs model,

$$\mathscr{L} = -\frac{1}{4}F_{\mu\nu}F^{\mu\nu} + (D_\mu\phi)^\dagger(D^\mu\phi) + \mu^2\phi^\dagger\phi - V(\phi). \tag{16.52}$$

The kinetic term $D_\mu\phi = \partial_\mu\phi + ieA_\mu\phi$ contains now two contributions which can cancel for $\rho \to \infty$. We require therefore that A_μ is a pure gauge field for $\rho \to \infty$ with

$$A_\mu \to -\frac{i}{e}\partial_\mu \ln(\phi/\eta). \tag{16.53}$$

Then $D_\mu\phi$ and $F_{\mu\nu}$ approach zero for $\rho \to \infty$, and the energy density per length of a local string is finite.

A local string carries magnetic flux Φ of the corresponding gauge field. Integrating (16.53) around a circle in a plane with $z = $ const. and large ρ gives

$$\Phi = \int \boldsymbol{B}\cdot d\boldsymbol{S} = \oint \boldsymbol{A}\cdot d\boldsymbol{l} = -\frac{i}{e}\int_0^{2\pi} d\vartheta\,\partial_\vartheta(in\vartheta) = \frac{2\pi n}{e}. \tag{16.54}$$

Thus the magnetic flux carried by local strings is quantised in units of $2\pi/e$. If we go back to cgs units, then the unit of magnetic flux becomes hc/e, determined completely by the three fundamental constants of a relativistic quantum theory. A local cosmic string with winding number n carries n flux quanta, analogous to quantised tubes of magnetic flux in a superconductor.

Global monopoles. Let us try to condense our results before we proceed. If we write complex scalar fields in terms of real ones, $\phi = \phi_1 + i\phi_2$, then we considered potentials of the type

$$V(\phi) = \frac{\lambda}{4} \left(\sum_{i=1}^{n} \phi_i^2 - \eta^2 \right)^2. \tag{16.55}$$

For $n = 1$, the possible ground-states $\phi^2 = \eta^2$ correspond to the zero-dimensional sphere S^0. The single constraint $\phi(x_1, x_2, x_3) = 0$ defines a two-dimensional surface in $\mathbb{R}^3$, which corresponds to a domain wall. The abelian Higgs model with one complex scalar doublet has $n = 2$. Now $\phi_1^2 + \phi_2^2 = \eta^2$ defines a one-dimensional sphere S^1 as manifold of the possible ground-states, while a topological defect defined by $\phi_1 = \phi_2 = 0$ is as the section of two surfaces a line. These results are summarised in the first two entries of Table 16.1.

The logical next step is to consider three scalar fields which transform under SO(3). Then the vacuum manifold is the sphere S^2 and we expect zero-dimensional topological defects which are called monopoles. We start again examining the global case. Using spherical coordinates, we search for static solutions using as ansatz

$$\phi_i = \eta h(r) \frac{x_i}{r}, \tag{16.56}$$

which satisfies the requirement

$$\phi_i \phi_i = \eta^2 \text{ for } |\boldsymbol{x}| \to \infty, \tag{16.57}$$

if $h(r) \to \pm 1$ for $r \to \infty$. Additionally, the function $h(r)$ should satisfy the boundary condition $h(r) \to 0$ for $r \to 0$ to ensure a non-singular $\phi(0)$. Estimating again the energy density at large r by simple dimensional analysis gives

$$\rho \sim \frac{1}{2}(\partial_i \phi)^2 \sim \frac{3}{2}\frac{\eta^2}{r^2} \tag{16.58}$$

and thus the energy contained inside a sphere of radius R diverges linearly, $E \approx 3\eta^2 R/2$. In contrast to the mild logarithmic divergence in case of global strings, the behaviour $E \approx 3\eta R/2$ is clearly disastrous.

The natural way-out is to consider local monopoles and to check if the gauge fields can compensate the scalar gradient energy. Before we do so we ask, however, if there is a way to ensure that the total energy of global monopoles is finite. Imagine a monopole ($h(r) = 1$ for $r \to \infty$) and an anti-monopole ($h(r) = -1$ for $r \to \infty$) pair separated by the distance d. Their fields $\phi_i(r)$ will cancel for $r \gg d$, leaving only higher multipole moments $\phi_i \sim r^{-2}$. Thus the energy density of a monopole–antimonopole

defect	n	d	homotopy group
domain wall	1	2	$\pi_0(M_0)$
cosmic string	2	1	$\pi_1(M_0)$
monopole	3	0	$\pi_2(M_0)$
texture	4	–	$\pi_3(M_0)$

Table 16.1 The number n of scalars determines the dimension d of the vacuum manifold M_0 and the dimension $3 - n$ of the hypersurface containing the unbroken vacuum.

pair scales as $\rho \sim \eta r^{-4}$ and therefore their total energy E is finite. Separating a monopole–antimonopole pair would cost an infinite amount of energy—instead a new monopole–antimonopole pair will be created as soon as the potential energy between the pair exceeds a certain threshold. We can therefore view this behaviour as a model for the confinement of coloured particles as quarks and gluons in QCD, where the potential energy $V(r)$ for distances $r \gtrsim \Lambda_{\mathrm{QCD}}^{-1}$ also scales linearly.

't Hooft–Polyakov monopoles.   The simplest model exhibiting local monopoles is the Georgi–Glashow model. It contains a SO(3) gauge field A_μ^a supplemented by a triplet of real Higgs scalars ϕ^a. Choosing a uniform vev as $\phi^a = (0, 0, v)$ leaves a residual U(1) symmetry unbroken, corresponding to rotations around the 3-axis in isospin space. Thus one can view the Georgi–Glashow model as a toy model for the electroweak sector of the SM, where the gauge field A_μ^3 plays the role of the photon and the Z boson is missing.

't Hooft and Polyakov first showed that this model contains extended classical solutions which have finite energy and correspond to local magnetic monopoles. Using the adjoint representation $(T_{\mathrm{adj}}^a)_{bc} = -\mathrm{i}f^{abc}$ for the scalar triplet of real Higgs scalars ϕ^a, the covariant derivative becomes

$$D_i \phi^a = \partial_i \phi^a - e\varepsilon^{abc} A_i^b \phi^c \tag{16.59}$$

or in vector notation $D_i \boldsymbol{\phi} = \partial_i \boldsymbol{\phi} - e\boldsymbol{A}_i \times \boldsymbol{\phi}$. Here, we assumed for simplicity that the monopole carries no electric charge, setting A_0^a. We fix again the asymptotic behaviour of the gauge fields by requiring that the kinetic term $D_i \phi^a$ vanishes for $r \to \infty$. From

$$\partial_i \phi^a \sim \eta \frac{\delta^{ai} - x^i x^a}{r^2},$$

we conclude that A_i^b should be constant, while (16.59) implies that $\boldsymbol{A}_i$ is perpendicular to $\boldsymbol{\phi}$. Evaluating $D_i \phi^a$ using $\varepsilon^{aij}\varepsilon^{akl} = \delta^{ik}\delta^{jl} - \delta^{il}\delta^{jk}$ shows that

$$A_i^a = \frac{\eta}{e}[1 - f(r)]\varepsilon^{aij}\frac{x^j}{r} \tag{16.60}$$

with $f(r) \to 0$ for $r \to \infty$ leads to the desired asymptotic behaviour of the covariant derivative. Now the regularity of the solution A_i^a at $r = 0$ requires $f(0) = 1$.

Inserting the two ansätze (16.56) and (16.60) into the energy functional of the Georgi–Glashow model and minimising the energy results in two coupled differential equations for the functions $h(r)$ and $f(r)$. In order to be able to proceed analytically, we consider only the limit $\lambda/g^2 = m_h^2/(2m_W^2) \to 0$. Then the potential energy $V(\phi)$ is negligible,

$$V(\phi) = \frac{\lambda}{4}\left(\sum_{a=1}^{3}\phi_a^2 - \eta^2\right)^2 \to 0 \qquad \text{for} \qquad \lambda \to 0, \tag{16.61}$$

if the fields satisfy the constraint (16.57). Thus the boundary conditions (16.57) remain valid in this limit. Neglecting $V(\phi)$, we can use Bogomolnyi's trick to derive an exact solution. We express the static energy as

$$E = \frac{1}{2} \int \mathrm{d}^3x \left[(B_i^a)^2 + (D_\mu \phi^a)^2 \right] = \frac{1}{2} \int \mathrm{d}^3x \left[(B_i^a \mp D_i \phi^a)^2 \pm 2 B_i^a D_i \phi^a \right] \geq 0.$$

$$(16.62)$$

Then we obtain the bound

$$E \geq \int \mathrm{d}^3x \, B_i^a D_i \phi^a \qquad (16.63)$$

which is attained, if the fields satisfy $D_i \phi^a = \pm B_i^a$. Now we only have to solve the simpler first-order equation $D_i \phi^a = \frac{1}{2} \varepsilon_{ijk} F_{jk}^a$. Its solution is called a Bogomolnyi–Parad–Sommerfield (BPS) monopole and, generalising, all solitons which minimise the classical action are denoted as BPS states or BPS solitons.

Example 16.2: Find the solution of the first-order equation $D_i \phi^a = \pm B_i^a$. We evaluate with (16.56) the LHS,

$$D_i \phi^a = \eta \frac{\delta^{ai} - x^i x^a}{r} fh + \eta \frac{x^i x^a}{r^2} h' \qquad (16.64)$$

and with (16.60) the RHS,

$$B_i^a = \frac{1}{2} \varepsilon_{ijk} F_{jk}^a = \frac{1}{2} \varepsilon_{ijk} \left[\partial_j A_k^a - \partial_k A_j^a - e \varepsilon_{abc} A_j^b A_k^c \right] \qquad (16.65a)$$

$$= \varepsilon_{ijk} \partial_j A_k^a - \frac{1}{2} e \varepsilon_{ijk} \varepsilon_{abc} A_j^b A_k^c = \frac{1}{g} \frac{r^2 \delta^{ai} - x^i x^a}{r^2} f' + \frac{1}{g} \frac{x^i x^a}{r^3} (1 - f^2). \qquad (16.65b)$$

Since the two tensors $r^2 \delta^{ai} - x^i x^a$ and $x^i x^a$ are orthogonal, the two coupled differential equations for the profile functions $h(r)$ and $f(r)$ simplify to

$$f'(r) = \pm g \eta f(r) h(r) \quad \text{and} \quad h'(r) = \frac{1}{g \eta r^2} [1 - f(r)^2]. \qquad (16.66)$$

Taking into account the boundary conditions, the solution expressed through the dimensionless variable $\xi = g \eta r$ is given by

$$h(r) = \coth(\xi) - \frac{1}{\xi} \quad \text{and} \quad f(r) = \frac{\xi}{\sinh(\xi)}. \qquad (16.67)$$

Finally, we want to determine the charge of a BPS monopole. While the massive fields fall off exponentially for $r \to \infty$, the component of F_{ij}^a connected to the massless photon decrease as a power law. Moreover, the fields (16.56) and (16.60) are practically uniform at large distances r from the centre of the monopole. Thus in a laboratory at $\boldsymbol{x}$, the field $\boldsymbol{F}_{ij}$ corresponds to the massless photon, while the two states orthogonal to $\boldsymbol{F}_{ij}$ are the massive weak gauge bosons. This implies that the part of F_{ij}^a which corresponds to the electromagnetic field is parallel to ϕ^a,

$$\mathcal{F}_{ij} = F_{ij}^a \frac{\phi^a}{v}. \qquad (16.68)$$

The magnetic part of the SU(2) field-strength tensor is

$$B_i^a = \frac{1}{2}\varepsilon_{ijk}F_{jk}^a = \varepsilon_{ijk}\partial_j A_k^a - \frac{1}{2}e\varepsilon_{ijk}\varepsilon_{abc}A_j^b A_k^c, \tag{16.69}$$

where we inserted the definition of F_{jk}^a. If we are far away from the centre of the soliton we can use the asymptotic expression for the fields, setting $f(r) = 0$ and $h(r) = 1$. Performing the differentiations (problem 16.9) we arrive at

$$B_i^a = \frac{1}{g}\frac{n_i n_a}{r^2}. \tag{16.70}$$

Thus the U(1) magnetic field is given by

$$\mathcal{B}_i = \frac{1}{g}\frac{n_i}{r^2}. \tag{16.71}$$

Comparing this to the expression $qn_i/(4\pi r^2)$ for the field of a monopole, we conclude that the magnetic charge of the BPS monopole is $q_m = 4\pi/g$.

Textures.   Finally we should comment on the case $n = 4$ which corresponds to the electroweak model. The four equations $\phi_i(x_1, x_2, x_3) = 0$ have in general no solution in $\mathbb{R}^3$. Thus regions which contain the unbroken vacuum $\phi_i = 0$ will be not formed during the electroweak phase transition. Still, correlations cannot exist beyond the horizon scale and thus non-zero gradients $\partial_\mu \phi_i$ can be produced. In the case of global textures, static solutions with positive energy density can exist. In the case of a broken gauge symmetry, gauge fields will compensate the $\partial_\mu \phi_i$ term in $D_\mu \phi_i$. Thus local textures as predicted by the SM have no observable consequences.

Homotopy groups and winding numbers.   The topological quantum numbers we have met discussing extended classical solutions can be associated with winding numbers of maps between the ground-states of a field theory and its configuration space. Let us define the ground-state or vacuum manifold M_0 of a theory as the set of all global minima $V(\phi) = 0$ of its potential,

$$M_0 = \{\phi : V(\phi) = 0\}. \tag{16.72}$$

The condition that the potential energy is finite requires that

$$\lim_{|x|\to\infty}\phi(x) = \phi \in M_0. \tag{16.73}$$

We can compactify $\mathbb{R}^n$ (where n is the number of spatial dimensions) to the sphere S^n using, for example, a stereographic projection as shown in Fig. 16.5. Then we can view the condition that the potential energy is finite as a mapping $S^n \to \mathcal{M}_0$. We ask now the question when two such mappings are topologically distinguished.

We consider only the mappings $S^1 \to S^1 \approx \mathbb{R}^2\backslash\{0\}$ which are easiest to visualise. Two closed loops with base point x_0, that is, curves $x(t)$ with $t \in [0:1]$ and $x(0) = x(1) = x_0$, are shown in Fig. 16.6. While the dashed loop is contractable, the solid

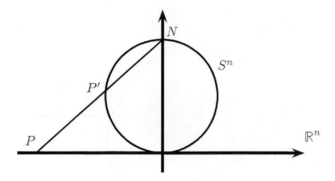

Fig. 16.5 A stereographic projection maps points $P \in \mathbb{R}^n$ onto points $P' \in S^n$. The north pole N of the sphere S^n corresponds to the sphere S_R^{n-1} at spatial infinity, $x_i^2 + \cdots + x_n^2 = R^2 \to \infty$, of $\mathbb{R}^n$.

one is wrapped once around $\{0\}$ and therefore not contractable to its base-point $\boldsymbol{x}_0$ by a continuous transformation. We define the maps

$$U^{(\nu)} = \mathrm{e}^{\mathrm{i}\vartheta\nu} = \left[U^{(1)}\right]^{\nu}, \tag{16.74}$$

which count the number of times a loop is wrapped around the origin. The integer ν is called the winding number of the map and can be rewritten as an integral,

$$\nu = \frac{\mathrm{i}}{2\pi} \int_0^{2\pi} \mathrm{d}\vartheta \, U \partial_\vartheta U^\dagger. \tag{16.75}$$

Clearly, this formula reproduces the correct values for the mappings defined earlier. Moreover, it is useful for the proof that ν is invariant under continuous transformations. It is sufficient to investigate an infinitesimal change δU. Unitarity $UU^\dagger = 1$ implies that $\delta U U^\dagger + U \delta U^\dagger = 0$ and $\delta U^\dagger = -U^\dagger \delta U U^\dagger$. As we will be interested later in the non-abelian version of the winding number, we will not use that the U are commuting complex numbers.

We calculate the variation of the integrand in the integral formula (16.75) for ν,

$$\delta(U \partial_\vartheta U^\dagger) = \delta U \partial_\vartheta U^\dagger + U \partial_\vartheta \delta U^\dagger \tag{16.76a}$$

$$= \delta U \partial_\vartheta U^\dagger - U \partial_\vartheta U^\dagger \delta U U^\dagger - U U^\dagger \partial_\vartheta \delta U U^\dagger - U U^\dagger \delta U \partial_\vartheta U^\dagger \tag{16.76b}$$

$$= -U \left[\partial_\vartheta U^\dagger \delta U + U^\dagger \partial_\vartheta \delta U\right] U^\dagger = -U \partial_\vartheta \left[U^\dagger \delta U\right] U^\dagger. \tag{16.76c}$$

Here, we inserted first $\delta U^\dagger = -U^\dagger \delta U U^\dagger$ and performed the differentiations. Then the first and fourth terms cancel, and finally we combined the remaining two terms using the product rule. In case of the abelian winding number ν of (16.75), we obtain then

$$\delta\nu = \frac{\mathrm{i}}{2\pi} \int_0^{2\pi} \mathrm{d}\vartheta \, \delta(U \partial_\vartheta U^\dagger) = -\frac{\mathrm{i}}{2\pi} \int_0^{2\pi} \mathrm{d}\vartheta \, \partial_\vartheta \left[U^\dagger \delta U\right] = 0. \tag{16.77}$$

Thus the winding number ν is an integer which is invariant under infinitesimal deformations of the loop. Since any continuous transformation can be built from infinitesimal

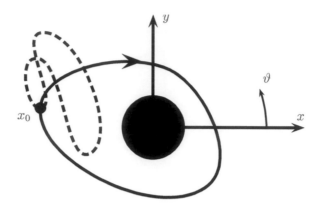

Fig. 16.6 Two loops in $\mathbb{R}^2 \backslash \{0\}$ with common base point x_0: the dashed line is contractable and has winding number $\nu = 0$, the solid loop is wrapped once around $\{0\}$ and has winding number $\nu = -1$.

ones, the maps $S^1 \to S^1$ can be divided into different equivalence classes, each charac-terised by the value $\nu \in \mathbb{Z}$. Only maps within each class can be continuously deformed into each other. Mathematicians say that such maps are homotopic. The theory of homotopy groups addresses the question into how many homotopy classes $\pi_n(X)$ the set of maps $S^n \to X$ can be divided. If $\pi_n(\mathcal{M}_0) \neq 1$, then the theory with vacuum manifold $\mathcal{M}_0$ contains stable topological defects, cf. table 16.1. Expressed in the lan-guage of homotopy groups, we have shown that $\pi_1(S^1) = \mathbb{Z}$. Since $S^1 \simeq U(1)$, this shows also that $\pi_1(U(1)) = \mathbb{Z}$.

A general results from the theory of homotopy groups says that the second ho-motopy group $\pi_2(G/H)$ of the quotient group G/H equals the first homotopy group $\pi_1(H)$ of H. If we identify the SM with H, then $H = SU(3) \otimes U(1)$ and $\pi_1(U(1)) = \mathbb{Z}$ is non-trivial. Thus the sequence

$$\pi_2(G/H) = \pi_1(H) = \pi_1(SU(3) \otimes U(1)) = \mathbb{Z} \neq 1 \tag{16.78}$$

shows that, if $U_{\mathrm{em}}(1)$ is unified at a higher scale within a larger semi-simple group G, then magnetic monopole solutions exist. Thus a grand unified theory implies that magnetic monopoles are produced at the GUT phase transition. We will see later that such monopoles would overclose the universe, leading to a very short life time of the universe. The desire to dilute the density of monopoles was a prime motivation for the invention of inflation.

Summary

Classical static solutions of theories with SSB can fall into different equivalence classes which are separated by an infinite potential energy barrier. This can lead to two-, one- or zero-dimensional topological defects which contain in their core the unbroken vacuum of the symmetric phase. Instantons or bounces are solutions of the classical field equation in Euclidean space which evolve in Euclidean time between two different vacua. If their action is finite, they describe instantaneous tunnelling from the false to the true vacuum in Minkowski space.

Further reading. A clear account of topologically non-trivial solutions of the classical field equations is given by Rubakov (2002). For a calculation of the effective potential of the SM and the resulting tunnelling probability see Espinosa (2014) and the references therein.

Problems

16.1 Order of the phase transition. Determine the critical temperature T_c of the $\lambda\phi^4$ theory from Eq. (16.7). Calculate the pressure and the heat capacity above and below T_c and confirm thereby that the phase transition is of second order.

16.2 Field equation for the bounce solution. Show that Eq. (16.15) agrees with the Klein–Gordon equation $\Delta\phi = V'(\phi)$ for a radial symmetric ϕ. (Use Eq. (6.61) to evaluate $\Delta\phi = \nabla^a\nabla_a\phi$.)

16.3 Scalar instantons. Show that the ansatz $\phi(r) = A\rho/(r^2 + \rho^2)$ solves (16.15) for the massless $\lambda\phi^4$ theory. Determine the action S.

16.4 Soliton solution of the sine-Gordon equation. Show that Eq. (16.36) solves the sine-Gordon equation and discuss the behaviour under Lorentz transformations. .

16.5 Domain wall. Estimate the extension of a domain wall using dimensional analysis.

16.6 Derrick's theorem. Consider the behaviour of the energy functional $E[\phi] = T[\phi] + V[\phi]$ for a scalar field ϕ under scale transformation $x \to \lambda x$ in D space dimensions. Show that for $D \geq 2$ no stable solutions with finite energy exist.

16.7 D_μ for the adjoint representation. Show that $D_\mu = \partial_\mu + eA_\mu$ results for SO(3) in Eq. (16.59) for the adjoint representation.

16.8 Duality for Maxwell. Show that the source-free Maxwell equations are invariant under the duality transformation $E' = E\cos\alpha + B\sin\alpha$ and $B' = -E\sin\alpha + B\cos\alpha$. Show that the invariance of the Maxwell equations with sources requires the existence of magnetic monopoles.

16.9 Magnetic field of a monopole. Derive Eq. (16.70).

17
Anomalies, instantons and axions

At the classical level, any global continuous symmetry of a system described by a Lagrangian leads to a locally conserved current. In order to decide if these symmetries survive quantisation, we have to consider if the generating functional $Z[J]$ retains the symmetries of the classical Lagrangian $\mathscr{L}$. Several examples where a quantised system does not share the symmetries of its classical counter-part have been found. As this behaviour came as a surprise, it was called "anomalous" and the non-zero terms violating the classical conservation laws on the quantum level were called "anomalies". A case we encountered already is the breaking of scale invariance (or conformal symmetry) in the process of renormalising massless theories; we will calculate this anomaly in the next chapter. The only other example of anomalous theories in $d = 4$ spacetime dimensions are models containing chiral fermions, which are theories where left- and right-chiral fermions interact differently. This anomaly is called axial or chiral anomaly and shows up in loop graphs containing chiral fermions coupled to gauge fields or gravitons. In the SM, the axial anomaly leads to three important phenomenological consequences. First, the anomaly in the electroweak currents vanishes, if the electric charges of all left-handed fermions sum up to zero. Thus the fact that electroweak Ward identities hold restricts the particle content of the theory. Second, the axial anomaly in conjunction with topologically non-trivial solutions of the Yang–Mills equations leads to the violation of CP invariance in strong interactions, in contradiction to observations. This is the so-called strong CP problem. Finally, the violation of CP is a necessary condition for the generation of a baryon asymmetry, and thus these topologically non-trivial solutions play an important role in models of baryogenesis.

The fact that a classical symmetry is broken by quantum effects is often described as "the anomaly breaks the classical symmetry". One should keep in mind that on the quantum level there is no symmetry to start with, thus there is no Goldstone boson associated with a symmetry broken by an anomaly.

17.1 Axial anomalies

Anomaly from non-invariance of the path-integral measure. The simplest model exhibiting an axial anomaly is axial electrodynamics, that is, a fermion coupled via its vector and axial current to two different gauge fields. The Lagrangian for this system reads

$$\mathscr{L} = \bar{\psi}\mathrm{i}\gamma^{\mu}(\partial_{\mu} + \mathrm{i}qV_{\mu} + \mathrm{i}gA_{\mu}\gamma^{5})\psi - \frac{1}{4}F^{2} - \frac{1}{4}G^{2} \tag{17.1}$$

with

Quantum Fields–From the Hubble to the Planck Scale. Michael Kachelriess. © Michael Kachelriess 2018.
Published in 2018 by Oxford University Press. DOI 10.1093/oso/9780198802877.001.0001

$$F_{\mu\nu} = \partial_\mu A_\nu - \partial_\nu A_\mu \quad \text{and} \quad G_{\mu\nu} = \partial_\mu V_\nu - \partial_\nu V_\nu. \tag{17.2}$$

Performing a combined $U_V(1) \otimes U_A(1)$ gauge transformation,

$$V_\mu \to V'_\mu = V_\mu - \partial_\mu \Lambda(x) \quad \text{and} \quad A_\mu \to A'_\mu = A_\mu - \partial_\mu \alpha(x), \tag{17.3}$$

induces the following change of the fermion fields,

$$\psi(x) \to \psi'(x) = \exp\{iq\Lambda(x) + ig\alpha(x)\gamma^5\}\psi(x), \tag{17.4a}$$

$$\bar\psi(x) \to \bar\psi'(x) = \bar\psi(x)\exp\{-iq\Lambda(x) + ig\alpha(x)\gamma^5\}. \tag{17.4b}$$

In order to determine the transformation properties of the fermionic path-integral measure, we introduce first eigenfunctions of the gauge-invariant Dirac operator $i\slashed{D}$,

$$i\slashed{D}\phi_n = \lambda_n \phi_n \quad \text{and} \quad \sum_n \phi_x(x)\phi_n^\dagger(y) = \delta(x-y)\,\mathbf{1}. \tag{17.5}$$

Then we can expand the fermion fields as

$$\psi(x) = \sum_n a_n \phi_n(x) \quad \text{and} \quad \bar\psi(x) = \sum_n \phi_n^\dagger(x)\bar b_n, \tag{17.6}$$

where the coefficients a_n and $\bar b_n$ are Graßmann variables. Thus we can rewrite the path-integral measure as $\mathcal{D}\psi\mathcal{D}\bar\psi = \prod_n \mathrm{d}a_n\,\mathrm{d}\bar b_n$. Next we relate the change in the fermion fields to a change in the expansion coefficients,

$$\psi'(x) = \sum_n a'_n \phi_n(x) = \sum_n a_n \exp\{iq\Lambda(x) + ig\alpha(x)\gamma^5\}\phi_n(x) \tag{17.7a}$$

$$\bar\psi'(x) = \sum_n \phi_n^\dagger(x)\bar b'_n = \sum_n \phi_n^\dagger(x)\exp\{-iq\Lambda(x) + ig\alpha(x)\gamma^5\}\bar b_n. \tag{17.7b}$$

Thus the variation introduced by a vector $U_V(1)$ gauge transformation cancels in the fermionic measure $\mathcal{D}\psi\mathcal{D}\bar\psi$, while the corresponding change under the axial $U_A(1)$ gauge transformation adds up. We can determine the latter, taking into account that the ϕ_n are an orthonormal basis, as

$$a'_m = \sum_n \int \mathrm{d}^4x\, \phi_m^\dagger(x)\exp\{ig\alpha(x)\gamma^5\}\phi_n(x)a_n \equiv C_{mn}a_n. \tag{17.8}$$

Recalling the transformation rules for Graßmann integrals, $\mathcal{D}\psi \to \mathcal{D}\psi J^{-1} = \mathcal{D}\psi[\mathrm{Det}(C)]^{-1}$, the product $\mathcal{D}\psi\mathcal{D}\bar\psi$ of the fermionic measure changes as

$$\mathcal{D}\psi\mathcal{D}\bar\psi \to \mathcal{D}\psi\mathcal{D}\bar\psi[\mathrm{Det}(C)]^{-2} = \mathcal{D}\psi\mathcal{D}\bar\psi\exp\left(-2ig\sum_n \int \mathrm{d}^4x\, \alpha(x)\mathrm{tr}\,\phi_n^\dagger(x)\gamma^5\phi_n(x)\right). \tag{17.9}$$

Here we also used the identity $\mathrm{Det}\exp(A) = \exp\mathrm{Tr}(A)$. The trace contains a product of spinors which is zero, but also a divergent sum over the eigenfunctions of $i\slashed{D}$.

Therefore we have to regularise the expression. We add as a gauge-invariant regulator the function

$$f = \exp\left\{-\lambda_n^2/M^2\right\} = \exp\left\{\slashed{D}^2/M^2\right\}, \tag{17.10}$$

which approaches zero rapidly for $\lambda_n^2 \to \infty$. The limit $f \to 1$ of the regulator corresponds to $M \to \infty$, which will allow us later an expansion of our result for

$$\mathcal{A}(x) \equiv \sum_n \mathrm{tr}\left\{\phi_n^\dagger(x)\gamma^5 \exp\left(\slashed{D}^2/M^2\right)\phi_n(x)\right\} \tag{17.11}$$

in powers of $1/M^2$. Note that we have expressed the change of the measure as an additional term $\delta\mathscr{L} = 2g\alpha(x)\mathcal{A}(x)$ in the Lagrangian.

As the vector $U_V(1)$ gauge transformation keeps the integration measure invariant, we ignore it in the following calculations. Thus we need to calculate $\slashed{D}^2$ including the axial gauge field A only,

$$\slashed{D}\slashed{D} = \gamma_\mu\gamma_\nu D^\mu D^\nu = D^2 + \frac{g}{2}\sigma_{\mu\nu}F^{\mu\nu} \tag{17.12}$$

with

$$D^2 = (\partial_\mu + igA_\mu)(\partial^\mu + igA^\mu) = \partial^2 + 2igA^\mu\partial_\mu + ig\left(\partial_\mu A^\mu\right) - g^2 A_\mu A^\mu. \tag{17.13}$$

We evaluate D^2 using for ϕ_n plane waves, $\phi_n = \mathrm{e}^{-ikx}\mathbf{1}$. Then we can replace the differentiations by momenta,

$$\exp\left(\frac{D^2}{M^2}\right)\phi_n(x) = \exp\left(-\frac{(k_\mu + gA_\mu)^2}{M^2} + \frac{ig(\partial_\mu A^\mu)}{M^2}\right)\phi_n(x). \tag{17.14}$$

Taking the continuum limit of the sum and writing out the full regulator, we obtain

$$\mathcal{A}(x) = \int \frac{\mathrm{d}^4k}{(2\pi)^4}\,\mathrm{tr}\left\{\mathrm{e}^{ikx}\gamma^5\exp\left[\slashed{D}^2/M^2\right]\mathrm{e}^{-ikx}\right\} \tag{17.15a}$$

$$= \int \frac{\mathrm{d}^4k}{(2\pi)^4}\,\mathrm{tr}\left\{\gamma^5\exp\left[-\frac{(k_\mu + gA_\mu)^2}{M^2} + \frac{ig\partial_\mu A^\mu}{M^2} + \frac{g}{2}\frac{\sigma_{\mu\nu}F^{\mu\nu}}{M^2}\right]\right\}. \tag{17.15b}$$

The second term in $\{\cdots\}$ is zero because it does not depend on k^μ and $\mathrm{tr}[\gamma^5] = 0$. Shifting variables to the dimensionless $MK_\mu = k_\mu + gA_\mu$ we obtain

$$\mathcal{A}(x) = M^4 \int \frac{\mathrm{d}^4K}{(2\pi)^4}\,\mathrm{e}^{-K^2}\mathrm{tr}\left[\gamma^5\exp\left\{\frac{g}{2}\frac{\sigma_{\mu\nu}F^{\mu\nu}}{M^2}\right\}\right]. \tag{17.16}$$

If we expand the exponential in powers of $1/M^2$, only the terms up to $\mathcal{O}(M^{-4})$ will survive in the limit $M \to \infty$. Using the antisymmetry of $F^{\mu\nu}$, we can also replace $\sigma_{\mu\nu}$ by $i\gamma_\mu\gamma_\nu$ and find then

$$\exp\left\{\ldots\right\} = 1 + \frac{ig}{2}\gamma_\mu\gamma_\nu F^{\mu\nu}\frac{1}{M^2} + \frac{1}{2}\left(\frac{ig}{2}\right)^2\gamma_\mu\gamma_\nu\gamma_\alpha\gamma_\beta F^{\mu\nu}F^{\alpha\beta}\frac{1}{M^4} + \mathcal{O}\left(\frac{1}{M^6}\right). \tag{17.17}$$

The trace properties of the gamma matrices inform us that the first two terms vanish, while the third results in a term proportional to the totally antisymmetric tensor $\varepsilon_{\mu\nu\alpha\beta}$.

Introducing the dual field-strength tensor $\tilde{F}_{\alpha\beta} = \frac{1}{2}\varepsilon_{\alpha\beta\mu\nu}F^{\mu\nu}$ and then performing the Gaussian integral over K we are left with

$$\delta S = \frac{g^2}{8\pi^2}\int \mathrm{d}^4x\,\alpha(x)\tilde{F}_{\mu\nu}F^{\mu\nu}. \tag{17.18}$$

It is sometimes stated that this result is exact, because our derivation seems not to rely on perturbation theory. However, we replaced in the evaluation of $\slashed{D}^2\phi_n$ the correct solutions ϕ_n accounting for the external gauge fields by plane waves. Therefore our derivation corresponds to an one-loop result, similar to our calculation of the effective potential in section 13.2.

In a classical theory without gauge fields, a local axial gauge transformation leads to the change

$$\mathscr{L} \to \mathscr{L} + g(\partial_\mu\alpha)\bar{\psi}\gamma^\mu\gamma^5\psi. \tag{17.19}$$

Thus the non-invariance of the measure in the path integral violates the classical conservation of the axial current,

$$\partial_\mu j_5^\mu = \partial_\mu(\bar{\psi}\gamma^\mu\gamma^5\psi) = \partial_\mu(j_R^\mu - j_L^\mu) = \frac{g^2}{8\pi^2}\tilde{F}_{\mu\nu}F^{\mu\nu}. \tag{17.20}$$

This equation is also known as the Adler–Bell–Jackiw anomaly equation.

The extra term introduced in the Lagrangian by an axial gauge transformation is proportional to $\tilde{F}_{\mu\nu}F^{\mu\nu}$ and transforms thus odd under CP. If this interaction were physically relevant, electrodynamics coupled to chiral fermions would violate CP. The $\tilde{F}_{\mu\nu}F^{\mu\nu}$ term is gauge-invariant, has dimension four and corresponds therefore to a renormalisable interaction. The reason why we have not considered it earlier is that it is a total derivative,

$$\begin{aligned}
F_{\mu\nu}\tilde{F}^{\mu\nu} &= \frac{1}{2}\varepsilon_{\mu\nu\rho\sigma}(\partial^\mu A^\nu - \partial^\nu A^\mu)(\partial^\rho A^\sigma - \partial^\sigma A^\rho) = 2\varepsilon_{\mu\nu\rho\sigma}\partial^\mu A^\nu\partial^\rho A^\sigma = \\
&= 2\varepsilon_{\mu\nu\rho\sigma}\partial^\mu(A^\nu\partial^\rho A^\sigma) = \partial^\mu(2\varepsilon_{\mu\nu\rho\sigma}A^\nu\partial^\rho A^\sigma) \equiv \partial^\mu K_\mu.
\end{aligned} \tag{17.21}$$

Since the four-divergence K_μ is not gauge-invariant, it cannot be an observable. Therefore it is not excluded that K_μ is singular, leading after integration to non-zero effects in the action. Before we discuss in more detail when it is justified to neglect a total derivative term, we will re-derive the anomaly using a diagrammatic approach.

Perturbative appearance of anomalies. Historically, the chiral anomaly was first encountered calculating the process $\pi^0 \to \gamma\gamma$ using perturbation theory. Describing the neutral pion within a non-relativistic quark picture as a $\pi^0 = (\bar{u}u + \bar{d}d)/\sqrt{2}$ state, we can view the process as

Here, the γ^5 matrix accounts for the fact that the pion is a pseudoscalar particle. The two diagrams are connected by the crossing symmetry $\kappa \leftrightarrow \lambda$, $p_1 \leftrightarrow p_2$. The total matrix element of this process is thus given by the sum

$$A_{\kappa\lambda\mu}(p_1, p_2) = S_{\kappa\lambda\mu}(p_1, p_2) + S_{\lambda\kappa\mu}(p_2, p_1), \tag{17.22}$$

where the matrix element $S_{\kappa\lambda\mu}$ describing the first diagram (neglecting coupling constants) is given by

$$S^{\kappa\lambda\mu} = -(-\mathrm{i})^3 \int \frac{\mathrm{d}^4 k}{(2\pi)^4} \mathrm{tr}\left[\gamma^\kappa \frac{\mathrm{i}}{\slashed{k} - \slashed{p}_1} \gamma^\mu \gamma^5 \frac{\mathrm{i}}{\slashed{k} + \slashed{p}_2} \gamma^\lambda \frac{1}{\slashed{k}}\right]. \tag{17.23}$$

It is sufficient to consider only massless fermions. The anomaly is connected to the UV divergences of these diagrams and in this limit masses play no role.

We check now whether the classical conservation laws for the vector and the axial current hold also at the one-loop level. Current conservation implies the following three relations in momentum space,

$$p_1^\kappa A_{\kappa\lambda\mu} = p_2^\lambda A_{\kappa\lambda\mu} = 0 \tag{17.24a}$$

$$(p_1 + p_2)^\mu A_{\kappa\lambda\mu} = 0, \tag{17.24b}$$

where (17.24a) are the equations for the conservation of the vector current, and (17.24b) for the conservation of the axial current. Crossing symmetry implies that these relations must also hold for the two individual amplitudes S.

We first check whether the axial current is conserved. In evaluating

$$I_{\kappa\lambda} \equiv (p_1 + p_2)^\mu S_{\kappa\lambda\mu} = -\int \frac{\mathrm{d}^4 k}{(2\pi)^4} \frac{\mathrm{tr}\left[\gamma_\kappa (\slashed{k} - \slashed{p}_1)(\slashed{p}_1 + \slashed{p}_2)\gamma^5 (\slashed{k} + \slashed{p}_2)\gamma_\lambda \slashed{k}\right]}{(k - p_1)^2 (k + p_2)^2 k^2} \tag{17.25}$$

we use

$$(\slashed{p}_1 + \slashed{p}_2)\gamma^5 = \slashed{p}_1 \gamma^5 - \gamma^5 \slashed{p}_2 = -(\slashed{k} - \slashed{p}_1)\gamma^5 - \gamma^5(\slashed{k} + \slashed{p}_2)$$

and $(\slashed{k} \pm \slashed{p}_i)^2 = (k \pm p_i)^2$, so that we can split the integral into two parts,

$$I_{\kappa\lambda} = \underbrace{\int \frac{\mathrm{d}^4 k}{(2\pi)^4} \frac{\mathrm{tr}\left[\gamma_\kappa \gamma^5 (\slashed{k} + \slashed{p}_2)\gamma_\lambda \slashed{k}\right]}{(k + p_2)^2 k^2}}_{A_{\kappa\lambda}(p_2)} + \underbrace{\int \frac{\mathrm{d}^4 k}{(2\pi)^4} \frac{\mathrm{tr}\left[\gamma_\kappa (\slashed{k} - \slashed{p}_1)\gamma^5 \gamma_\lambda \slashed{k}\right]}{(k - p_1)^2 k^2}}_{B_{\kappa\lambda}(p_1)}. \tag{17.26}$$

Observe that the LHS is a pseudo-tensor of rank 2, while the RHS is the sum of the two quantities $A_{\kappa\lambda}$ and $B_{\kappa\lambda}$ which each only depend on one momentum. As it is not possible to create a pseudo-tensor of rank 2 from that, the RHS must be zero. We have thus shown (ostensibly!) that the axial current is conserved.

Next we verify the conservation of the vector current, following the same line of argument as in the case of the axial current. Starting from

$$J_{\lambda\mu} \equiv p_1^\kappa S_{\kappa\lambda\mu} = -\int \frac{\mathrm{d}^4 k}{(2\pi)^4} \frac{\mathrm{tr}\left[\slashed{p}_1 (\slashed{k} - \slashed{p}_1)\gamma_\mu \gamma^5 (\slashed{k} + \slashed{p}_2)\gamma_\lambda \slashed{k}\right]}{(k - p_1)^2 (k + p_2)^2 k^2}, \tag{17.27}$$

we shift the integration variable $k' = k + p_2$ and reorder the terms in the trace, obtaining

$$
J_{\lambda\mu} = -\underbrace{\int \frac{\mathrm{d}^4 k'}{(2\pi)^4} \frac{\mathrm{tr}\left[(\slashed{k}' - \slashed{p}_1 - \slashed{p}_2)\gamma_\mu\gamma^5\slashed{k}'\gamma_\lambda\right]}{(k' - p_1 - p_2)^2 k'^2}}_{A_{\lambda\mu}(q=p_1+p_2)} + \underbrace{\int \frac{\mathrm{d}^4 k'}{(2\pi)^4} \frac{\mathrm{tr}\left[(\slashed{k}' - \slashed{p}_2\gamma_\mu\gamma^5\slashed{k}'\gamma_\lambda\right]}{k'^2(k' - p_2)^2}}_{B_{\lambda\mu}(p_2)}.
$$

$$(17.28)$$

Thus we conclude that $J_{\lambda\mu}$ also vanishes. For the $p_2^\lambda S_{\kappa\lambda\mu}$ part, we follow the same argument but we shift the integration variable by $k' = k - p_1$ instead.

We have now shown that both the axial current and the vector current are conserved. Using these results in the calculation for the decay width of a π^0 leads, however, to a width which vanishes in the limit of a massless pion. Including the small pion mass leads still to life time much longer than observed. Looking back at our calculation, we should therefore check whether all our manipulations were legitimate, although we have not regularised the divergent loop integrals. In particular, the superficial divergence of the diagrams $S_{\kappa\lambda\mu}$ is worse than the logarithmic divergences we have become accustomed to,

$$
S_{\kappa\lambda\mu} = -\int \frac{\mathrm{d}^4 k}{(2\pi)^4} \frac{\mathrm{tr}\left[\gamma_\kappa\slashed{k}\gamma_\mu\gamma^5\slashed{k}\gamma_\lambda\slashed{k}\right]}{k^6} + \text{subleading terms} \qquad (17.29)
$$

producing a linearly divergent term.

An essential ingredient for the derivation of the current conservation was the shift of our integration variable. Such a shift can only be done if the integral is properly convergent (after regularisation, if required) or only logarithmic divergent. By contrast, in the case of a linearly divergent integral, the shift $k' = k - a$

$$
\int \mathrm{d}^4 k' f(k') \stackrel{?}{=} \int \mathrm{d}^4 k f(k - a) = \int \mathrm{d}^4 k \left\{ f(k) - a_\mu \frac{\partial f}{\partial k_\mu} + \dots \right\} \qquad (17.30)
$$

changes the value of the integral. Using Gauss' theorem, we can convert the gradient term into a surface integral which will lead to a finite change of the integral because of $\mathrm{d}S_\mu \propto k^3$ and $f \propto k^{-3}$. You should show in problem 17.1 that the shift $k^\mu - a^\mu$ in (17.29) changes $S_{\kappa\lambda\mu}$ as

$$
S_{\kappa\lambda\mu} \rightarrow S'_{\kappa\lambda\mu} = S_{\kappa\lambda\mu} + \frac{1}{8\pi^2}\epsilon_{\kappa\lambda\mu\nu}a^\nu \qquad (17.31)
$$

where $\epsilon_{\kappa\lambda\mu\nu}$ is again the totally antisymmetric Levi–Civita tensor.

We look back at the corresponding proof of gauge invariance for the vacuum polarisation, Eqs. (12.39–12.41). There we used dimensional regularisation which respects gauge symmetry. Since γ^5 is not well-defined for $d \neq 4$, we cannot apply this regularisation method here. Using Pauli–Villars regularisation as an alternative would break axial symmetry, since it consists of adding massive particles. Thus, from a technical point of view, anomalies arise if no regularisation procedure exists which respects the classical symmetry.

Since the shifts of the integration variables required to obtain current conservation differ for the vector and axial current, we can absorb only one of the resulting boundary terms (17.31) by a suitably chosen counter-term. Clearly, we will choose the vector current to be conserved: otherwise electric charge conservation would be violated, while the axial current is broken by mass effects anyway. If we consider as amplitude including a counter-term

$$A_{\kappa\lambda\mu} = S_{\kappa\lambda\mu}(p_1, p_2) + S_{\lambda\kappa\mu}(p_2, p_1) + \frac{1}{4\pi^2}\epsilon_{\kappa\lambda\mu\nu}(p_1^\nu + p_2^\nu),\qquad(17.32)$$

the vector current will still be conserved, but the axial current will not

$$(p_1 + p_2)^\mu A_{\kappa\lambda\mu} = \frac{1}{2\pi^2}\epsilon_{\kappa\lambda\mu\nu}p_2^\mu p_1^\nu.\qquad(17.33)$$

The added contribution gives a non-zero contribution to the divergence of the axial current which is identical to our previous result (17.18),

$$\partial_\mu j_5^\mu = \frac{e^2}{8\pi^2}F_{\mu\nu}\tilde{F}^{\mu\nu}.\qquad(17.34)$$

Let us come back to the question whether these results are exact. Studying higher-order corrections, one can show that these corrections vanish and thus the perturbative one-loop result for the anomaly is exact. As an heuristic argument we can use the fact that adding additional propagators will reduce the superficial degree of divergence, while the anomaly is connected to linearly divergent diagrams. As a consequence, processes like $\pi^0 \to 2\gamma$ which are dominated by the anomaly can be calculated reliably in lowest-order perturbation theory, although for $Q^2 = m_\pi^2 \sim \Lambda_{\text{QCD}}^2$ the strong coupling constant $\alpha_s(Q^2)$ is certainly not small and higher-order corrections are naively expected to be large.

Cancellations of anomalies.   Anomalies may be useful when they can break accidental global symmetries like baryon and lepton number. As we will see later, this opens up the possibility of explaining why the universe consists mainly of matter. Similarly, the explicit breaking of global symmetries in the quark Lagrangian can explain why in certain cases no Goldstone bosons are observed. In contrast, symmetries and the resulting Ward identities between Green functions are crucial for the renormalisability of gauge theories. If these identities are not satisfied, physical observables in renormalisable gauges depend on the gauge-fixing parameter ξ, while Lorentz invariance is violated in unitary gauges. The excellent agreement of electroweak precision data with experiments is a strong argument that the underlying theory is renormalisable. As the V–A structure of the electroweak interactions is, however, similar to our toy model of axial electrodynamics, we can only expect that the anomalies of individual loop diagrams cancel after summing over all contributions.

Compared to axial electrodynamics, we have to consider in the GSW model both $U_Y(1)$ and SU(2) currents. Then each vertex i comes with the corresponding generator, $T_i = Y\,\mathbf{1}$ and $T_i = \sigma^a/2$, respectively. The anomaly is thus proportional to

$$A_{abc} = \text{tr}\big(T_aT_bT_c\big) + \text{tr}\big(T_aT_cT_b\big) = \text{tr}\big(T_a\{T_bT_c\}\big),\qquad(17.35)$$

where the trace is over SU(2) doublets and the second term corresponds to the exchange diagram $b \leftrightarrow c$. Concentrating on diagrams containing SU(2) currents, we have to consider $A_{abc}^{(1)} = \mathrm{tr}\big(\tau_a\{\tau_b, \tau_c\}\big)$, $A_{abc}^{(2)} = \mathrm{tr}\big(\tau_a\{Y_b, Y_c\}\big)$ and $A_{abc}^{(3)} = \mathrm{tr}\big(\tau_a\{Y_b, \tau_c\}\big)$. In the first two cases, the properties of the Pauli matrices imply that the anomaly vanishes automatically,

$$A_{abc}^{(1)} \propto \mathrm{tr}\left(\tau_a\{\tau_b, \tau_c\}\right) = 2\delta_{bc}\mathrm{tr}(\tau_a) = 0 \qquad \text{and} \qquad A_{abc}^{(2)} \propto \mathrm{tr}(\tau_a) = 0. \qquad (17.36)$$

In the third case, we use the Gell-Mann–Nishijima relation to replace the hypercharge, $Y = 2(T_3 - Q)$, obtaining

$$A_{abc}^{(3)} \propto \mathrm{tr}\left(\tau_a\{Q_b, \tau_c\}\right) = \mathrm{tr}\left(Q_b\{\tau_a, \tau_c\}\right) = 2\delta_{ab}\sum_i Q_i. \qquad (17.37)$$

Recalling that only the left-chiral fermions couple to the W boson, the condition that this anomaly vanishes is thus $\sum_i Q_i^L = 0$. If we now only look at the leptons and quarks separately, then

$$Q_e + Q_\nu = -1 \neq 0 \quad \text{and} \quad Q_u + Q_d = \frac{2}{3} - \frac{1}{3} = \frac{1}{3} \neq 0. \qquad (17.38)$$

Including both leptons and quarks where we account by the factor 3 for their colour quantum number we find as required for the anomaly cancellation

$$Q_e + Q_\nu + 3(Q_u + Q_d) = -1 + 3\frac{1}{3} = 0. \qquad (17.39)$$

The last remaining triangle contribution to the anomaly contains three $\mathrm{U}_Y(1)$ currents, $A_{abc}^{(4)} = \mathrm{tr}\big(Y_a\{Y_b, Y_c\}\big)$. Again, the anomaly vanishes, if the condition $\sum_i Q_i = 0$ is met, cf. problem 17.4. Thus chiral anomalies are cancelled within each full fermion generation. In the SM, there is no explanation for this conspiracy between the quark and lepton sector, and this has been one of the major motivations to consider GUTs. Note also that if a single member of a hypothetical fourth generation of fermions were found, anomaly cancellation would require the existence of a complete set of quarks and leptons.

We have restricted our analysis of the anomaly to an abelian model. In the non-abelian case, the field strength contains term linear and quadratic in the gauge fields. As a result, additional anomalies in square and pentagon diagrams appear. However, the absence of anomalies in the triangle diagrams guaranties also their absence in square and pentagon diagrams.

Remark 17.1: Grand unified theories. The gauge group of the SM contains four commuting generators, that is, it has rank 4. Because SU(n) has rank $n-1$, SU(5) is thus the smallest SU(n) group encompassing the SM. In the five-dimensional fundamental representation, we can set $T^a = \left(\begin{smallmatrix} \lambda^a/2 & 0 \\ 0 & 0 \end{smallmatrix}\right)$ and $T^i = \left(\begin{smallmatrix} 0 & 0 \\ 0 & \sigma^i/2 \end{smallmatrix}\right)$, with λ^a and σ^i as the Gell-Mann and Pauli matrices. Thus SU(5) contains SU(3) and SU(2) as subgroups. The remaining diagonal matrix can be chosen proportional to hypercharge, $T^0 \propto \mathrm{diag}(-2, -2, -2, 3, 3)$. Additionally, there are 12 off-diagonal matrices.

We combine three colour states of down-like quarks together with a lepton doublet in $\psi = (\bar{d}^r, \bar{d}^b, \bar{d}^g, \nu_e, e)^T$, while the remaining ten states can be fitted into an antisymmetric 5×5

matrix. From this, we can immediately draw three conclusions. First, the 12 off-diagonal matrices correspond to new gauge bosons (called X^μ and Y^μ) which interchange quarks and leptons. Thus baryon and lepton number is generically broken in GUTs, and the X^μ and Y^μ bosons have to be sufficiently heavy such that proton decay is suppressed. Second, the electric charge is diagonal and thus any multiplet has zero electric charge. Applied to $\psi = (\bar{d}^r, \bar{d}^b, \bar{d}^g, \nu_e, e)^T$, this leads to charge quantisation, $Q_{\bar{d}} = -Q_e/3$. Finally, above the energy scale M_{GUT} where the SU(5) symmetry is restored, a single gauge coupling controls all interactions. In such a picture, one expects that the SM couplings meet in one point at the scale M_{GUT}.

17.2 Instantons and the strong CP problem

The effective Lagrangian induced by the chiral anomaly is a surface term. This suggests that a term of the same structure can be obtained as the topological charge of a Yang–Mills theory, following the argument of Bogomol'nyi in Eq. (16.43). If such a topological charge exists, then an unbroken Yang–Mills theory has a non-trivial vacuum. Such a non-trivial vacuum structure has, however, only physical consequences if the corresponding classical tunnelling solutions have a finite action. We should therefore search for classical solutions of a pure, Euclidean Yang–Mills theory with $S_{\text{YM}} < \infty$. These solutions are the non-abelian analogue of the scalar bounce solution we have considered earlier.

Instantons. We define an Euclidean Yang–Mills theory by the action

$$S = \frac{1}{2} \int \mathrm{d}^4x \, \mathrm{tr}\{F_{\mu\nu}F_{\mu\nu}\}, \tag{17.40}$$

where $F_{\mu\nu} = F_{\mu\nu}^a T^a = \partial_\mu A_\nu - \partial_\nu A_\mu + ig[A_\mu, A_\nu]$ and derivatives and integrations are defined with respect to Euclidean coordinates $(x_1, x_2, x_3, x_4 = \mathrm{i}x_0)$ with $x_i \in \mathbb{R}$. Since the metric tensor is $\eta_{\mu\nu} = \delta_{\mu\nu}$, we do not need to distinguish between lower and upper indices. Note that in Euclidean space the dual of the dual field-strength tensor is again the field-strength tensor, $\tilde{\tilde{F}}_{\mu\nu} = F_{\mu\nu}$, while in Minkowski space it is its negative, $\tilde{\tilde{F}}_{\mu\nu} = -F_{\mu\nu}$. Next we define instantons as self-dual and anti-self-dual solutions,

$$\tilde{F}_{\mu\nu} = F_{\mu\nu} \quad \text{and} \quad \tilde{F}_{\mu\nu} = -F_{\mu\nu}, \tag{17.41}$$

of the classical Yang–Mills equations. Using Bogomol'nyi's trick, we can show that they correspond to the topologically non-trivial solutions with the lowest energy—if such non-trivial solutions exist. We write first

$$\mathrm{tr}\{(F_{\mu\nu} - \tilde{F}_{\mu\nu})^2\} = \mathrm{tr}\{F_{\mu\nu}^2 + \tilde{F}_{\mu\nu}^2 - 2F_{\mu\nu}\tilde{F}_{\mu\nu}\} \geq 0. \tag{17.42}$$

For the calculation of $\tilde{F}^2$ we use $\epsilon_{\mu\nu\rho\sigma}\epsilon_{\mu\nu\kappa\lambda} = 2(\delta_{\rho\kappa}\delta_{\sigma\lambda} - \delta_{\rho\lambda}\delta_{\sigma\kappa})$ and end up with $\tilde{F}^2 = F^2$. Thus

$$\mathrm{tr}\{F_{\mu\nu}F_{\mu\nu}\} \geq \mathrm{tr}\{F_{\mu\nu}\tilde{F}_{\mu\nu}\}, \tag{17.43}$$

which is minimised if F is self-dual. We obtain the same bound for an anti-self-dual solution, if we choose a plus sign in Eq. (17.42).

We first examine if QED, i.e. an abelian Yang–Mills theory, contains instantons. A finite action,

$$\left| \int_{\Omega(\tau)} \mathrm{d}^4 x \, \tilde{F} F \right| \le \left| \int_{\Omega(\tau)} \mathrm{d}^4 x \, F F \right| < \infty \tag{17.44}$$

requires that F decreases faster than τ^{-2} in the limit $\tau^2 = \boldsymbol{x}^2 + x_4^2 \to \infty$. But as a classical Yang–Mills theory contains no scale, $\tilde{F} F$ must be a polynomial in τ. Thus $F \sim \mathcal{O}(\tau^{-3})$, $A \sim \mathcal{O}(\tau^{-2})$, and then the total derivative (17.21) behaves as

$$K_\mu = 2\epsilon_{\mu\nu\rho\sigma} A_\nu \partial_\rho A_\sigma \sim \mathcal{O}(\tau^{-5}). \tag{17.45}$$

As a result, the surface term in

$$\int_\Omega \mathrm{d}^4 x \, \tilde{F} F = \int_{\partial\Omega} \mathrm{d}S_\mu K_\mu \to 0 \tag{17.46}$$

vanishes and we see that in an abelian theory $\tilde{F} F$ does not influence physical quantities. This argument justifies our usual practice of neglecting surface terms in QED.

We now turn to the non-abelian case. Then we can express $\mathrm{tr}\,\{\tilde{F}_{\mu\nu} F_{\mu\nu}\}$ again as a four-divergence,

$$\partial_\mu K_\mu = 2\,\mathrm{tr}\,\tilde{F}_{\mu\nu} F_{\mu\nu}, \tag{17.47}$$

where now

$$K_\mu = 2\varepsilon_{\mu\nu\sigma\rho}\mathrm{tr}\left(A_\nu F_{\sigma\rho} + \frac{2}{3}\mathrm{i}g A_\mu A_\sigma A_\rho \right). \tag{17.48}$$

Choosing at infinity a pure gauge field,

$$A_\mu = \frac{\mathrm{i}}{g}(\partial_\mu U)U^\dagger, \tag{17.49}$$

results in $F_{\mu\nu} = 0$ for $\tau \to \infty$ and ensures that the action is finite. On the other hand, a gauge transformation U which becomes constant for $\tau \to \infty$, that is, it depends in this limit only on the angles, gives $A \sim \mathcal{O}(\tau^{-1})$ and thus $K \sim \mathcal{O}(\tau^{-3})$. As a result, the surface integral may become non-zero. One may wonder if we can gauge away $A_\mu \propto (\partial_\mu U)U^\dagger$ on $\tau = \infty$ by performing a suitable gauge transformation $\tilde{U}$. Since $\tilde{U}$ has to be regular in all $\mathbb{R}^4$, it must be constant a $\tau = 0$ and independent of the angles. Thus $\tilde{U}$ is continuously connected to the identity and can be used only to gauge away fields A_μ in the same homotopy class. Thus the surface term $F\tilde{F}$ has only physical significance, if the gauge fields at $\tau \to \infty$ are split into non-trivial topological classes.

We next move to the specific case of SU(2). Any SU(2) matrix can be written as $U = a + \mathrm{i}\boldsymbol{b}\cdot\boldsymbol{\sigma}$ with $a^2 + |\boldsymbol{b}|^2 = 1$. Thus SU(2) is isomorphic to S^3. Compactifying the boundary $\partial\mathbb{R}^4 = \mathbb{R}^3$ of Euclidean space at large τ to S^3, the function (17.49) defines a map $S^3 \to S^3$. The question of whether non-trivial instanton solutions exist is thus equivalent to the existence of topologically non-trivial mappings $S^3 \to S^3$, what in turn requires that $\pi_3(S^3)$ is not the identity. We can generalise the winding number (16.75) from $S^1 \to S^1$ to the case $S^3 \to S^3$ as follows. If we use as coordinates the three Euler angles specifying a point on S^3, then each angle i contributes a factor

$U\partial_i U^\dagger$ as in the S^1 case. Since the winding number is a pseudo-scalar, we have to contract ijk with the Levi–Civita tensor and to take the trace over the SU(2) indices, arriving at

$$\nu = -\frac{1}{24\pi^2} \int dx_1 dx_2 dx_3 \, \varepsilon_{ijk} \text{tr} \left[(U\partial_i U^\dagger)(U\partial_j U^\dagger)(U\partial_k U^\dagger) \right]. \tag{17.50}$$

Because of $dx_i \partial_i = \tilde{d}x_i \tilde{\partial}_i$, the expression is equally valid using Cartesian coordinates. In order to show that the winding number is invariant under continuous transformations, $\delta\nu = 0$, it is sufficient to consider the variation of a single factor in the trace. Now we can profit from (16.75) where we already derived the variation of this factor in the non-abelian case as

$$\delta(U\partial_i U^\dagger) = -U\partial_i (U^\dagger \delta U) U^\dagger.$$

Inserting this relation into the integrand results in

$$E \equiv \varepsilon_{ijk} \text{tr} \left[(U\partial_i U^\dagger)(U\partial_j U^\dagger)\delta(U\partial_k U^\dagger) \right] = -\varepsilon_{ijk} \text{tr} \left[\partial_i U^\dagger U \partial_j U^\dagger U \partial_k (U^\dagger \delta U) \right]. \tag{17.51}$$

Then we perform a partial integration and use the fact that terms like $\partial_k \partial_i U^\dagger$ vanish contracted with ε_{ijk}, obtaining

$$E = \varepsilon_{ijk} \text{tr} \left[\partial_i U^\dagger \partial_k U \partial_j U^\dagger \delta U + \partial_i U^\dagger U \partial_j U^\dagger \partial_k U U^\dagger \delta U) \right]. \tag{17.52}$$

In the second term, we use $U\partial_j U^\dagger = -\partial_j U U^\dagger$ and $\partial_k U U^\dagger = -U\partial_k U^\dagger$ to symmeterise the expression in j and k,

$$E = \varepsilon_{ijk} \text{tr} \left[\partial_i U^\dagger \partial_k U \partial_j U^\dagger \delta U + \partial_i U^\dagger \partial_j U \partial_k U^\dagger \delta U \right] = 0. \tag{17.53}$$

Thus the winding number ν is invariant under infinitesimal and thus under continuous transformations.

Next we try to express the winding number as a volume integral over $\text{tr}\{\tilde{F}_{\mu\nu} F_{\mu\nu}\}$. We write (17.50) as a surface integral,

$$\nu = \frac{1}{24\pi^2} \int dS_\mu \, \varepsilon_{\mu\nu\sigma\tau} \text{tr} \left[(U\partial_\nu U^\dagger)(U\partial_\sigma U^\dagger)(U\partial_\tau U^\dagger) \right] \tag{17.54a}$$

$$= -\frac{ig^3}{24\pi^2} \int dS_\mu \, \varepsilon_{\mu\nu\sigma\tau} \text{tr} \left[A_\nu A_\sigma A_\tau \right], \tag{17.54b}$$

where we could use $U\partial_\mu U^\dagger = ig A_\mu$ because A_μ is a pure gauge field for $\tau \to \infty$. Since then also $F^a_{\sigma\rho} = 0$, only the second term in the expression (17.48) for the four-divergence K^μ survives and we obtain

$$\nu = \frac{g^2}{32\pi^2} \int dS_\mu \, K^\mu = \frac{g^2}{16\pi^2} \int d^4x \, \text{tr}\{\tilde{F}_{\mu\nu} F_{\mu\nu}\}. \tag{17.55}$$

Thus we have shown that gauge fields A^a_μ exist which are solutions of the classical Yang–Mills equations, have a finite action and fall into distinct equivalence classes characterised by the winding number[1] ν. Our remaining task is to write down an explicit form of the mappings $S^3 \to S^3$ and to show that the definition (17.50) results in a integer winding number. We choose to write $U(x)$ as

$$U^{(n)}(x) = \left(\frac{x_4 + \mathrm{i}\boldsymbol{x} \cdot \boldsymbol{\sigma}}{\tau} \right)^n \tag{17.56}$$

where x_μ is the unit vector $x_\mu = (\sin\chi\boldsymbol{e}, \cos\chi)$. Evaluation of (17.50) for $U^{(1)}(\boldsymbol{x})$ confirms then that $\nu = 1$. Now integrating $\mathrm{tr}\{F_{\mu\nu}F_{\mu\nu}\} \geq \mathrm{tr}\{F_{\mu\nu}\tilde{F}_{\mu\nu}\}$ over space, we find on the LHS twice the Euclidean action, while the RHS equals $16\pi^2\nu/g^2$. Thus the Euclidean action is bounded by

$$S \geq \frac{8\pi^2 |\nu|}{g^2}, \tag{17.57}$$

and instantons as the non-trivial solutions with the lowest energy have the action $S = 8\pi^2/g^2$.

We used for our discussion of instantons in non-abelian Yang–Mills theories the specific example of SU(2). A theorem of Brott states that for any simple Lie group G containing SU(2) the maps $S^3 \to G$ can be deformed continuously to the ones of $S^3 \to \mathrm{SU}(2)$. Thus all our results apply identically to the case of strong interactions, SU(3). In the remaining part of this chapter, we will discuss the impact of instantons on the QCD vacuum, while we postpone the electroweak case to chapter 22.

Tunnelling interpretation.   We consider the four-dimensional cylinder defined by $|x_4| \leq T$ and $|\boldsymbol{x}| \leq R$ in the limit $T, R \to \infty$. At $x_4 = -T$, we choose A_μ as a pure gauge field with winding number ν_1, and at $x_4 = T$ with ν_2. On the boundary $|\boldsymbol{x}| = R$, we choose U constant and thus A_μ is zero. Calculating the total winding number, we find

$$\frac{g^2}{16\pi^2} \int_\Omega \mathrm{d}^4x \, \mathrm{tr}\big(F\tilde{F}\big) = \frac{g^2}{32\pi^2} \int_{\partial\Omega} \mathrm{d}^3 S_\mu K^\mu =$$
$$= \frac{g^2}{32\pi^2} \int \mathrm{d}^3x \left[K^0(t = -\infty) - K^0(t = +\infty) \right] = \nu_1 - \nu_2 = \nu. \tag{17.58}$$

The minus sign appears, because of the opposite orientations of the two caps. Thus the classical solutions we have determined interpolate between a vacuum with winding number ν_1 at time $t = -\infty$ and a vacuum with winding number ν_2 at time $t = +\infty$. The two vacua are separated by a finite energy barrier, and thus the solutions describe the quantum tunnelling between different vacua. This agrees with our finding in chapter 16.2 that tunnelling solutions correspond to solutions of the Euclidean field equations. These solutions were dubbed instantons by 't Hooft, since the tunnelling they describe happens instantaneously in Minkowski space.

The ϑ vacuum.   The tunnelling interpretation indicates that the true vacuum of a pure Yang–Mills theory is the superposition of all vacua with fixed winding number

---

[1]Mathematicians call the winding number of the mapping $S^3 \to S^3$ the Pontryagin index, while mathematical physicists use often the term Chern–Simon number.

ν. Let us call these vacua $|\nu\rangle$ and the true one $|\vartheta\rangle$. Applying a gauge transformation $U \equiv U^{(1)}$ with unit winding number results in

$$U|\nu\rangle = |\nu + 1\rangle. \tag{17.59}$$

On the other hand, the Yang–Mills Hamiltonian is invariant under gauge transformations,

$$UHU^\dagger = H, \tag{17.60}$$

or $[H, U] = 0$. Thus the true vacuum $|\vartheta\rangle$ is a common eigenstate of H and U. Since the vacuum is normalised, the eigenvalue of U has to be a phase,

$$U|\vartheta\rangle = e^{i\vartheta}|\vartheta\rangle. \tag{17.61}$$

The angle ϑ is a conserved quantum number. Thus a classical Yang–Mills theory is characterised by two numbers, the coupling g and the angle ϑ.

The true vacuum $|\vartheta\rangle$ is a linear superposition of the vacua with fixed winding number n given by

$$|\vartheta\rangle = \sum_{n=-\infty}^{\infty} e^{-in\vartheta}|n\rangle, \tag{17.62}$$

since

$$U|\vartheta\rangle = \sum_{n=-\infty}^{\infty} e^{-in\vartheta}|n+1\rangle = e^{i\vartheta}\sum_{n=-\infty}^{\infty} e^{-in\vartheta}|n\rangle. \tag{17.63}$$

The matrix elements $\langle n'|H|n\rangle$ depend only on the difference $\nu = n' - n$, because (17.60) gives

$$\langle n'+1|H|n+1\rangle = \langle n'|H|n\rangle. \tag{17.64}$$

Using also the parity properties, $P|n\rangle = -|n\rangle$ and $PHP^{-1} = H$, we obtain

$$\langle n'|H|n\rangle = \langle -n'|H|-n\rangle. \tag{17.65}$$

Hence the matrix elements $\langle n'|H|n\rangle$ depend only on the absolute value $|\nu| = |n' - n|$. We conclude that instantons lead to an effective potential $V_{\text{eff}}(\vartheta)$ which is periodic and even in ϑ

$$H|\vartheta\rangle = L^3 V_{\text{eff}}(\vartheta)|\vartheta\rangle, \tag{17.66}$$

with $V_{\text{eff}}(\vartheta) = V_{\text{eff}}(\vartheta + 2\pi)$ and $V_{\text{eff}}(\vartheta) = V_{\text{eff}}(-\vartheta)$, where L^3 is the considered volume. A general argument due to Weinberg shows that points of enhanced symmetry are stationary points of the action. Thus we expect the minimum of $V_{\text{eff}}(\vartheta)$ to coincide with the CP conserving point $\vartheta = 0$.

In our definition of the path integral using the Faddeev–Popov trick, we included only gauge fields which are continuously connected with the identity. Thus our next task is to add the effect of the ϑ vacuum to the path integral. The path integral in the presence of external sources is identical to the vacuum persistence amplitude,

$$\langle \vartheta'|\vartheta\rangle_J = \sum_{n,n'} e^{i(n'\vartheta' - n\vartheta)}\langle n'|n\rangle_J. \tag{17.67}$$

Now we introduce the difference $\nu = n' - n$ so that we can rewrite the phase as $n'\vartheta' - n\vartheta = n(\vartheta' - \vartheta) + \nu\vartheta'$. But $\langle n'|n\rangle_J$ depends only on ν and thus we can perform the sum over n. This leads to a factor $\delta_{\vartheta',\vartheta}$, which expresses the fact that ϑ is conserved. Thus

$$\langle \vartheta | \vartheta \rangle_J = \sum_{\nu} e^{i\nu\vartheta} \int \mathcal{D}A^{(\nu)} e^{-S+\langle JA \rangle}, \qquad (17.68)$$

where $\mathcal{D}A^{(\nu)}$ denotes the integration over all gauge field configurations with the fixed winding number ν. Replacing ν with the help of Eq. (17.55) and introducing $\mathcal{D}A \equiv \sum_{\nu} \mathcal{D}A^{(\nu)}$ gives

$$\langle \vartheta | \vartheta \rangle_J = \int \mathcal{D}A \, e^{-S+\langle JA \rangle} \exp\left\{ i \frac{\vartheta g^2}{16\pi^2} \operatorname{tr}(F\tilde{F}) \right\}. \qquad (17.69)$$

Thus instantons induce an additional term $\mathscr{L}_\vartheta$ to the classical Yang–Mills Lagrangian,

$$\mathscr{L}_{\text{eff}} = \mathscr{L} + \frac{\vartheta g^2}{16\pi^2} \operatorname{tr}(F\tilde{F}), \qquad (17.70)$$

which depends on the arbitrary parameter $\vartheta \in [0, 2\pi[$. In order to discuss observable effects of this additional term, we have to add fermions to the pure Yang–Mills theory we discussed up to now.

Fermionic contribution to ϑ. We have derived in Eq. (17.20) the axial anomaly for an abelian gauge theory. The corresponding result for a Yang–Mills theory coupled to a single massless fermion is

$$\partial_\mu j_5^\mu = \partial_\mu (j_R^\mu - j_L^\mu) = \frac{g^2}{8\pi^2} \operatorname{tr}(F\tilde{F}). \qquad (17.71)$$

Thus the axial anomaly leads to the additional term

$$\mathscr{L}_{\text{eff}} = \mathscr{L} + \frac{\alpha n_f g^2}{8\pi^2} \operatorname{tr}(F\tilde{F}), \qquad (17.72)$$

in the effective QCD Lagrangian, if we perform a chiral $U_A(1)$ transformation $q_{L,R} \to e^{i\alpha\gamma^5} q_{L,R}$ on the n_f quark fields. This term has the same structure as the instanton contribution, and if we choose

$$\alpha = -\frac{\vartheta}{2n_f} \qquad (17.73)$$

the two terms cancel. Thus for massless quarks the ϑ parameter is unphysical and we can choose the $\vartheta = 0$ vacuum.

We can understand this, if we consider the effect of an instanton transition. Integrating Eq. (17.71) gives as change of the axial charge $Q_5 = N_R - N_L$ per massless quark flavour

$$\Delta Q_5 = Q_5(t = -\infty) - Q_5(t = +\infty) = 2\nu. \qquad (17.74)$$

Thus an instanton process $\nu = \pm 1$ changes the axial quark number by two units, creating a left-chiral and destroying a right-chiral quark and vice versa. As chirality is a conserved quantum number for massless particles, at least one of the two states connected by the instanton process can therefore not correspond to the vacuum. By contrast, for $m > 0$ the mass term mixes left- and right-chiral fields: the quark–antiquark pair can annihilate via the mass term and the states can be identified with the vacuum.

We now consider massive quarks. In this case, the conservation of the axial current is additionally to the axial anomaly explicitly broken by the quark masses, since a Dirac mass term transform as

$$m\bar{q}q \rightarrow m\bar{q}e^{2i\alpha\gamma^5}q = \cos(2\alpha)m\bar{q}q + i\sin(2\alpha)m\bar{q}\gamma^5 q \qquad (17.75)$$

under a chiral transformation $q_{L,R} \rightarrow e^{i\alpha\gamma^5}q_{L,R}$. The second term violates T (CP) invariance, as the original $\mathscr{L}_\vartheta$ we wanted to eliminate. Therefore, CP violation implied by $\mathscr{L}_\vartheta$ is a physical effect for massive quarks. If we consider n_f flavour of quarks, then the mass matrix M_{ij} will change as $M_{ij} \rightarrow e^{2i\alpha}M_{ij}$ and thus

$$\arg\det M \rightarrow \arg\det M + 2n_f\alpha. \qquad (17.76)$$

Therefore only the combination

$$\bar{\vartheta} \equiv \vartheta + \arg\det M \qquad (17.77)$$

is an observable quantity. It is a question of convenience, if we choose real mass matrices and rotate all CP violation via the axial anomaly into the ϑ term. Or if we eliminate $\mathscr{L}_\vartheta$ and transfer its effect into CP violating complex mass matrices.

Example 17.1: Two Higgs doublet model. Let assume that the Higgs doublet $\Phi_1 = (\phi^+, \phi^0)$ generates the masses for down-like fermions, while a second Higgs doublet $\Phi_2 = (\phi^0, \phi^-)$ generates the masses for up-like fermions,

$$\mathscr{L}_Y = -X_{\alpha\beta}\bar{Q}_\alpha\Phi_1 d_{R,\beta} - Y_{\alpha\beta}\bar{Q}_\alpha\Phi_2 u_{R,\beta} + \text{h.c.} \qquad (17.78)$$

with

$$\langle 0|\Phi_1|0\rangle = \frac{1}{\sqrt{2}}\begin{pmatrix} 0 \\ v_1 e^{i\delta_1} \end{pmatrix} \quad \text{and} \quad \langle 0|\Phi_2|0\rangle = \frac{1}{\sqrt{2}}\begin{pmatrix} v_2 e^{i\delta_2} \\ 0 \end{pmatrix}. \qquad (17.79)$$

The resulting quark mass matrix M_{ij}, $i,j = 1,\ldots,6$ has the determinant

$$\det(M) = \frac{(v_1 v_2)^2}{8}e^{3i(\delta_1+\delta_2)}\det(X)\det(Y) \qquad (17.80)$$

and thus

$$\arg\det M = \arg\det(XY) + 3i(\delta_1 + \delta_2). \qquad (17.81)$$

By an SU(2) gauge transformation $U = \exp(i\alpha\tau_3/2)$ we can eliminate one of the two phases. Thus a general two Higgs doublet model has one CP violating phase. In contrast, the SM uses $\Phi_2 = i\tau_2\phi^*$. Then $\delta_1 = -\delta_2$, and no CP violation arises in the Higgs sector.

We look now for observable consequences of the ϑ vacuum in the low-energy interactions of hadrons. Since the ϑ term and the axial anomaly are flavour-blind, the change c in the quark masses,

$$c \equiv \sin(2\alpha_q)m_q, \tag{17.82}$$

is the same for all quarks. In order to shift the effects of the ϑ term completely into the mass matrix of the quarks, we need also $\sum_{q=1}^{n_f} 2\alpha_q = -\vartheta$. Eliminating α_q in the limit of small ϑ gives

$$c = -\frac{\vartheta}{\sum_q m_q^{-1}}, \tag{17.83}$$

and thus the ϑ-dependent, CP-violating part of the mass term becomes

$$\mathcal{L}_m^{(\vartheta)} = -\mathrm{i}\vartheta \left(\sum_q m_q^{-1} \right)^{-1} \sum_q \bar{q}\gamma^5 q. \tag{17.84}$$

For light nucleons and mesons which consist only of u and d quarks this simplifies to

$$\mathcal{L}_m^{(\vartheta)} = -\mathrm{i}\vartheta \frac{m_u m_d}{m_u + m_d} \left(\bar{u}\gamma^5 u + \bar{d}\gamma^5 d \right). \tag{17.85}$$

This mass term generates a CP-violating effective pion–nucleon interaction which in turn leads to an electric dipole moment d_n of the neutron. The corresponding limit on d_n bounds the value of $\bar{\vartheta}$ as $|\bar{\vartheta}| \leq 2 \times 10^{-10}$. Although the value of $\bar{\vartheta}$ is a free parameter within the SM, it seems natural to ask for an explanation why $\bar{\vartheta}$ is so small.

The most straightforward explanation would be that one current quark mass is zero, that is, that one quark has no Yukawa coupling to the SM Higgs. Then (17.85) shows for $n_f = 2$ clearly that the CP-violating effect disappears. This holds also for $n_f > 2$, because $\arg \det M = 0$ if one mass eigenvalue is zero. While it has been debatable, if $m_u \sim 5\,\mathrm{MeV}$ deduced from chiral perturbation theory for the current u quark mass might be lowered to $m_u = 0$, this possibility was closed by lattice data around 2005. The question of why $\bar{\vartheta}$ is so small is called the strong CP problem of the SM.

17.3 Axions

Peceei and Quinn proposed promoting the parameter ϑ to a dynamical variable which settles automatically at its minimum zero. The basic ingredient of this proposal is a new massless pseudo-scalar field a which couples to gluons with an interaction of the same structure as the ϑ term,

$$\mathcal{L}_a = \frac{1}{2}(\partial_\mu a)^2 - \frac{g^2}{16\pi^2} \frac{a}{f_a} F\tilde{F}. \tag{17.86}$$

Since a is a pseudo-scalar, the interaction term $aF\tilde{F}$ conserves CP. Adding $\mathcal{L}_a$ to the effective Lagrangian (17.70) of QCD means that observables depend only on the combination $\vartheta - a/f_a$. If $\mathcal{L}_a$ is invariant under the shift $a \to a + \text{const.}$, then we can use this arbitrariness to absorb the ϑ parameter into a redefinition of the field a. Such a shift symmetry is typical for a Goldstone boson which has only derivative couplings. Thus the pseudo-scalar particle a should be the Goldstone boson of a spontaneously broken global symmetry, and the $aF\tilde{F}$ interaction suggests chossing this symmetry as a chiral U(1) symmetry. This symmetry is called Peceei–Quinn symmetry $U_{PQ}(1)$ and its Goldstone boson a is the axion.

Weinberg and Wilzcek realised that there is an additional twist in this proposal. The effective potential (17.66) generated via the $F\tilde{F}$ term has two effects: first, instanton effects break the shift symmetry, generating a mass term $m_a^2 = \partial^2 V_{\text{eff}}/\partial a^2$ for the axion. Thus the axion becomes a massive pseudo-Goldstone boson. Second, the vacuum expectation value of the axion field will relax to the minimum of the potential, $\partial V_{\text{eff}}/\partial a = 0$. But we have argued earlier that the minimum of $V_{\text{eff}}(\vartheta)$ is situated at $\vartheta = 0$. Thus also in the case of a massive axion the strong CP problem is solved.

Let us illustrate the key points of this idea with a simple model. We add a complex scalar field ϕ and a set of heavy fermions ψ to the SM,

$$\mathscr{L} = \mathrm{i}\bar{\psi}\slashed{\partial}\psi + \frac{1}{2}\partial_\mu\phi^\dagger\partial^\mu\phi - y_i(\bar{\psi}_L\psi_R\phi + \text{h.c.}) - V(\phi). \tag{17.87}$$

While ϕ is a SM singlet, the fermions are charged under SU(3) and U(1). The Lagrangian is invariant under the global chiral $U_{PQ}(1)$ gauge transformation

$$\phi \to \mathrm{e}^{\mathrm{i}\alpha}\phi \qquad \text{and} \qquad \psi_{L/R} \to \mathrm{e}^{\pm\mathrm{i}\alpha/2}\psi_{L/R}. \tag{17.88}$$

We now break spontaneously the Peccei–Quinn $U_{PQ}(1)$ symmetry, choosing the usual Mexican-hat potential for $V(\phi)$. Splitting ϕ into its vacuum expectation value and the fluctuating fields,

$$\phi = \frac{f_a + \rho}{\sqrt{2}}\mathrm{e}^{\mathrm{i}a/f_a}\begin{pmatrix}0\\1\end{pmatrix}, \tag{17.89}$$

generates a mass term for ρ while a remains massless. We assume that f_a is much larger than the energy scale E we are interested in, and thus we can neglect the field ρ,

$$\mathscr{L} = \frac{1}{2}(\partial_\mu a)^2 - m_i\bar{\psi}\mathrm{e}^{\mathrm{i}\gamma^5 a/f_a}\psi - V(\phi) + \mathcal{O}(E^2/f_a^2). \tag{17.90}$$

The combined expression is clearly invariant under $U_{PQ}(1)$ transformations $a \to a + \alpha f_a$. Expanding the exponential, we generate mass terms $m_i = y_i f_a/\sqrt{2}$ for the heavy fermions plus fermion–axion interactions,

$$\mathscr{L}_{\text{int}} = -m_i\bar{\psi}\psi - \mathrm{i}\frac{m_i}{f_a}a\bar{\psi}\gamma^5\psi - \dots. \tag{17.91}$$

The latter lead to a AVV triangle graph for the process $a \to 2g$ which in turn induces the desired $F\tilde{F}$ term via the axial anomaly. In the same way, an effective $a \to 2\gamma$ coupling is generated. Thus the characteristic features of an axion are its two-gluon and two-photon couplings. Moreover, the parameters of the pion and axion sector are connected by

$$m_a f_a \approx m_\pi f_\pi, \tag{17.92}$$

since the two Goldstone bosons mix via their two-gluon coupling, $a \leftrightarrow 2g \leftrightarrow \pi^0$.

Summary

The CP-odd term $\tilde{F}_{\mu\nu}F^{\mu\nu}$ is a gauge-invariant renormalisable interaction. Terms of this type are produced by instanton transitions between Yang–Mills vacua with different winding numbers and by the chiral anomaly. While we can rotate classically all CP violating phases contained in the quark mass matrices into the single CP violating phase of the CKM matrix, the chiral anomaly leads additionally to the change $\vartheta \to \bar{\vartheta} = \vartheta + \arg \det M$ in the coefficient of the $\tilde{F}_{\mu\nu}F^{\mu\nu}$ term. Since the physics origin of both contributions seem to be disconnected, it is puzzling that they sum up to $|\bar{\vartheta}| \lesssim 10^{-10}$. A possible solution is the Peceei–Quinn symmetry which promotes the parameter ϑ to the field $a = f_a \vartheta$ which settles automatically at the minimum at $\bar{\vartheta} = 0$ of the instanton potential V_{eff}.

Further reading. Instantons are discussed in more detail by Rubakov (2002). For an introduction into the physics of axions see Kawasaki and Nakayama (2013), while Cheng and Li (1988) contains a concise introduction to grand unification.

Problems

17.1 Surface term. Derive (17.31) using Gauss' theorem to convert the gradient term into a surface integral. Find the trace formula for six γ^μ and one γ^5 matrix and evaluate the trace. Transform the result into Euclidean space and evaluate it.

17.2 Anomaly for non-abelian gauge fields. Extend the model (17.1) from abelian electrodynamics to non-abelian axial gauge fields and derive the analogue of (17.20).

17.3 Pion decay width. Calculate the decay width of $\pi^0 \to 2\gamma$ taking into account the chiral anomaly. Compare the value to the measured one.

17.4 Anomaly cancellation. Show that $\text{tr}(Y_a\{Y_b, Y_c\})$ vanishes for $\sum_i Q_i = 0$.

17.5 SU(5) GUT. Argue that $Q = T_3 + Y/2 = T_3 + cT_0$ holds in SU(5) and determine c. Show that in the SU(5) symmetric limit, the relation $\sin^2 \vartheta_W = 3/8$ holds.

17.6 Euclidean YM. Derive the connection between the gauge field A_μ in Minkowski and Euclidean space. Show that the transformation leads to a real action for the gauged scalar field theory.

17.7 Four-divergence. Show that $\text{tr}(F\tilde{F})$ in the non-abelian case can be expressed by the four-divergence given in Eq. (17.48).

18
Hadrons, partons and QCD

We introduced QCD as a non-abelian gauge theory that describes strong interactions in terms of quarks and gluons. The characteristic feature of a Yang–Mills theory, asymptotic freedom, implies however that perturbation theory breaks down for small momentum transfer. The strong coupling $\alpha_s(Q^2)$ diverges in perturbation theory at a finite value, introducing $\Lambda_{\rm QCD}$ as a new mass scale. Thereby the scale invariance of QCD, classically valid in the limit of zero quark masses, is broken. Another consequence of asymptotic freedom is confinement.[1] Quarks and gluons are not observed as isolated particles, but exist only in bound-states which are colour-singlets. The aim of this chapter is to discuss how calculations in perturbation theory using quarks and gluons can be connected to experiments which observe hadrons. For instance, we aim to describe processes like $\bar{X}X \to \bar{q}q \to$ hadrons, where the initial state could, for example, be a pair of leptons or dark matter particles. We start the chapter elucidating the surprising fact that the masses of hadrons are much larger than expected from the Higgs effect.

18.1   Trace anomaly and hadron masses

The classical QCD Lagrangian is scale and conformally invariant in the limit of zero quark masses. We introduce these symmetries before we calculate the anomalous term introduced by quantum corrections which breaks these symmetries. It is this "correction" which is responsible for 95% of the mass of ordinary matter in the universe.

Scale and conformal invariance.   In chapter 7.4 we defined the dynamical energy–momentum stress tensor $T_{\mu\nu}$ as the response of the matter action $S_{\rm m}$ under an infinitesimal change of the metric tensor, $T_{\mu\nu} = (2/\sqrt{|g|})\delta S_{\rm m}/\delta g^{\mu\nu}$. This procedure implies that we temporarily leave Minkowski space, even if we are only interested in the behaviour of a Lorentz-invariant field theory. Having performed the variation $\delta g^{\mu\nu}$, we move back to Minkowski space setting $g^{\mu\nu} \to \eta^{\mu\nu}$.

We want to look at when we can extend the Poincaré group as the symmetry group of Minkowski space acting on local fields to the larger group of conformal transformations. We start considering *scale transformations* of the coordinates, $x \to x' = {\rm e}^\omega x$ with fixed ω. As a result of this coordinate transformation, the metric tensor changes as

$$g_{\mu\nu}(x) \to g_{\mu\nu}(x') = {\rm e}^{2\omega} g_{\mu\nu}(x). \tag{18.1}$$

---

[1] While confinement is an observational fact, it has not been derived from first principles and is part of one of the six open "Millennium Problems".

Quantum Fields–From the Hubble to the Planck Scale. Michael Kachelriess. © Michael Kachelriess 2018.
Published in 2018 by Oxford University Press. DOI 10.1093/oso/9780198802877.001.0001

Thus distances are rescaled by a constant factor, while angles and hence the light-cone structure given by $\mathrm{d}s^2 = 0$ are conserved. Since only the latter is important for theories without mass parameters, we expect them to be invariant under such transformations. For an infinitesimal scale change, $x \to x' = (1 + \delta\omega)x$, the metric varies as $\delta g_{\mu\nu} = 2g_{\mu\nu}\delta\omega$. The matter action S_{m} changes as a result of the variation of the metric as

$$\delta S_{\mathrm{m}} = \int \mathrm{d}^4 x \, \frac{\delta S_{\mathrm{m}}}{\delta g^{\mu\nu}} \delta g^{\mu\nu} = \frac{1}{2} \int \mathrm{d}^4 x \sqrt{|g|} \, T_{\mu\nu} \delta g^{\mu\nu} =$$
$$= \int \mathrm{d}^4 x \sqrt{|g|} \, T_{\mu\nu} g^{\mu\nu} \delta\omega = \delta\omega \int \mathrm{d}^4 x \sqrt{|g|} \, T_\mu^{\ \mu}. \tag{18.2}$$

The variation δS_{m} of the action remains unchanged, if we add a total derivative $\nabla^\mu K_\mu$ to $T_\mu^{\ \mu}$. Thus we can conclude for $\omega = \mathrm{const.}$ only that the trace of the stress tensor has to equal a total derivative,

$$T_\mu^{\ \mu} = g^{\mu\nu} \frac{2}{\sqrt{|g|}} \frac{\delta S_{\mathrm{m}}}{\delta g^{\mu\nu}} = \nabla^\mu K_\mu, \tag{18.3}$$

which may be zero. If we promote ω to a local function $\omega(x)$, the resulting spacetime-dependent transformations are called *special conformal transformations*. Now we cannot pull out $\delta\omega(x)$ from the integral and consequently the requirement $\delta S_{\mathrm{m}} = 0$ implies that the trace of the stress tensor has to vanish, $T_\mu^{\ \mu} = 0$. Clearly, conformal invariance implies scale invariance. The opposite direction holds in practically all interesting cases, but exceptions exist.

These additional symmetries should lead according to Noether's theorem to conserved currents. We recall that the stress tensor has the properties $T^{\mu\nu} = T^{\nu\mu}$ and $\partial_\mu T^{\mu\nu} = 0$. For a scale-invariant action, we can define the additional conserved current $D^\mu = T^{\mu\nu} x_\nu - K^\mu$, because then

$$\partial_\mu D^\mu = \partial_\mu (T^{\mu\nu} x_\nu - K^\mu) = T^{\mu\nu} \eta_{\mu\nu} - \partial_\mu K^\mu = T_\mu^{\ \mu} - \partial_\mu K^\mu = 0. \tag{18.4}$$

The conserved quantity D^μ is called the dilatation current, the four-divergence K^μ the virial current. For a conformally invariant action, we obtain additionally the four conserved currents

$$S^\rho = \mathrm{d}_\nu [\eta^{\mu\nu} x^2 - 2x^\mu x^\nu)] T_\mu^{\ \rho}, \tag{18.5}$$

which are parameterised by the vector d_μ. The condition that the stress tensor is traceless (and symmetric) implies the conservation of these currents,

$$\partial_\rho S^\rho = \mathrm{d}_\nu \partial_\rho [\eta^{\mu\nu} x^2 - 2x^\mu x^\nu] T_\mu^{\ \rho} = 2\mathrm{d}_\nu [x_\rho \eta^{\mu\nu} - x^\mu \delta_\rho^\nu - x^\nu \delta_\rho^\mu] T_\mu^{\ \rho} = 0. \tag{18.6}$$

In problem 6.4, we encountered the conformal Killing equation. Its solutions, the conformal Killing vector fields of Minkowski space, agree with the infinitesimal generators of the group of conformal transformations, cf. appendix B.3. This generalises our results for the Poincaré group from example 6.2.

We have already encountered theories with traceless stress tensors such as electrodynamics (or more generally pure Yang–Mills theories) and the massless Dirac field.

Thus the invariance group of these theories in Minkowski space is the 15-dimensional conformal group. Now we look at the case of a scalar field. In problem 5.6, we found that a massless scalar field is invariant under scale transformations. The trace of its stress tensor is given by

$$T_\mu{}^\mu = (\partial_\mu \phi)^2 - \delta_\mu^\mu \mathscr{L}_0 = \left(1 - \frac{d}{2}\right)(\partial_\mu \phi)^2. \tag{18.7}$$

The virial current follows using $\Box \phi = 0$ as $K_\mu = (1 - d/2)\phi \partial_\mu \phi$ and thus the action for a free massless scalar field is scale-invariant for all d. By contrast, the action is only in $d = 2$ dimensions conformally invariant, in which case the field is dimensionless and does not scale.[2].

Connection to hadron masses. The matrix element of the stress tensor $T^{\mu\nu}$ for a hadron $h(\boldsymbol{k})$ can depend only on the combination $k^\mu k^\nu$ for zero momentum transfer. Assuming hadron states normalised as $\langle h(\boldsymbol{k})|h(\boldsymbol{k})\rangle = 1$, the matrix element of the stress tensor is therefore

$$\langle h(\boldsymbol{k})|T^{\mu\nu}|h(\boldsymbol{k})\rangle = 2k^\mu k^\nu. \tag{18.8}$$

Thus the vanishing of the trace $T_\mu{}^\mu$ would imply that all hadrons are massless,

$$\langle h(\boldsymbol{k})|T_\mu{}^\mu|h(\boldsymbol{k})\rangle = 2m_h^2 = 0. \tag{18.9}$$

In reality, the masses of the light quarks are non-zero and thus the scale invariance of strong interactions is already broken on the classical level. Nevertheless, the relation (18.9) poses a problem, since the masses of the light quarks due to the Higgs effect, $m_{u,d} \approx (4 - 10)\,\mathrm{MeV}$, are much smaller than the masses of the lightest hadrons, the pions. Thus the dominant contribution to hadron masses should come from strong interactions, but Eq. (18.9) tells us that conformal invariance forbids such a contribution. The resolution to this apparent problem lies in the fact that quantum corrections spoil classical conformal invariance because we have to introduce a dimensionful parameter calculating quantum corrections.

Trace anomaly. The phenomenon that the trace of the stress tensor receives non-zero corrections due to loop effects is called trace anomaly. This effect can be analysed using the same approach as for the axial anomaly, and we therefore summarise only the main steps. If the Jacobian $\mathcal{J}$ for the conformal transformation

$$\psi(x) = \exp\{-3\omega(x)/2\}\psi'(x) \tag{18.10}$$

is not the identity, an anomalous contribution to $T_\mu{}^\mu$ in addition to the usual mass term appears,

$$\mathrm{i} \int \mathrm{d}^4 x\, \omega\, T_\mu{}^\mu = \ln \mathcal{J}/3 + \mathrm{i} \int \mathrm{d}^4 x\, \omega\, m \bar{\psi}\psi. \tag{18.11}$$

Using the same regulator in the evaluation of the Jacobian as in the case of the chiral anomaly,

---

[2]The fact that we can express the trace in (18.7) as $T_\mu{}^\mu = \partial_\mu \partial_\nu L^{\mu\nu}$ (where $L^{\mu\nu} = (2-d)/4\eta^{\mu\nu}\phi^2$ using the equations of motion) signals that we can "improve" the stress tensor adding appropriate terms such that the stress tensor becomes traceless and the action conformally invariant.

$$\mathcal{J} = \exp \omega \sum_n \int \mathrm{d}^4x \, \phi_n^*(x) \exp\left\{-\slashed{D}^2/M^2\right\} \phi_n \,, \tag{18.12}$$

the only change compared to Eq. (17.11) is the absence of the γ_5 matrix. Following the same steps but keeping track of the non-abelian part of the QCD field-strength $F_{\mu\nu}^a$, one finds

$$T_\mu^{\ \mu} = \frac{g_s^2}{48\pi^2} F_{\mu\nu}^a F^{a\mu\nu} + m\bar\psi\psi. \tag{18.13}$$

This the $\mathcal{O}(g_s^2)$ contribution to the trace anomaly. We can obtain the full non-perturbative expression for the trace anomaly by the following, more intuitive argument. We rescale first the Lagrange density by $A_\mu^a \to \bar A_\mu^a = A_\mu^a/g_s$ as

$$\mathcal{L} = -\frac{1}{4} F_{\mu\nu}^a F^{a\mu\nu} + i\bar\psi(\slashed\partial + ig_s\slashed A)\psi \to -\frac{1}{4g_s^2} \bar F_{\mu\nu}^a \bar F^{a\mu\nu} + i\bar\psi(\slashed\partial + i\slashed{\bar A})\psi. \tag{18.14}$$

If we view this Lagrange density as the renormalised Lagrange density, then g_s depends on the renormalisation scale μ and the action is not longer scale-invariant: the change δS under an infinitesimal scale transformations results solely from the dependence via the renormalised coupling,

$$\delta S = -\frac{1}{4} \int \mathrm{d}^4x \, \frac{\partial}{\partial\omega} \left(\frac{1}{4g_s^2(\mu)}\right) \bar F_{\mu\nu}^a \bar F^{a\mu\nu} \to \int \mathrm{d}^4x \, \frac{\beta(g_s)}{2g_s} F_{\mu\nu}^a F^{a\mu\nu} \,. \tag{18.15}$$

Thus the full result for the trace of the stress tensor in QCD, including the anomaly and the mass terms of the quarks that break scale invariance explicitly, is given by

$$T_\mu^{\ \mu} = \frac{\beta_{\mathrm{QCD}}^{(6)}}{2g} F_{\mu\nu}^a F^{a\mu\nu} + m_u\bar uu + m_d\bar dd + m_s\bar ss + \sum_h m_h\bar qq. \tag{18.16}$$

Since the anomaly is due to a UV divergence, it is independent of the quark masses and we should use the beta function β_{QCD} for six flavours, $b_{\mathrm{QCD}} = 11 - 2n_f/3 = 7$. We have singled out in (18.16) the contribution of the heavy quarks, $\sum_{h=c,b,t} m_h\bar qq$, since they are present in a hadron only as virtual fluctuations. Because of $m_h \gg \Lambda_{\mathrm{QCD}}$, a systematic expansion in Q^2/m_h^2 of the term $\sum_h m_h\bar qq$ is possible. At leading order, these fluctuations connect m_h via a quark loop to two gluons. Therefore one can rewrite the mass term for heavy quarks using the effective Lagrangian derived in problem 14.10 for the Higgs–gluon–gluon coupling,

$$m_h\bar qq \to \frac{2}{3} \frac{\alpha_s}{8\pi} F_{\mu\nu}^a F^{a\mu\nu}. \tag{18.17}$$

Now we see that the mass terms cancel the contribution of the heavy quarks in the beta function. As a result, we obtain for the matrix element between on-shell nucleons

$$m_N^2 = \langle N| T_\mu^{\ \mu} |N\rangle = \langle N| \frac{\beta^{(3)}}{2g_s} F_{\mu\nu}^a F^{a\mu\nu} |N\rangle + \sum_{l=u,d,s} \langle N| m_l\bar qq |N\rangle \tag{18.18}$$

with $b_{\mathrm{QCD}}^{(3)} = 9$. The matrix elements $\langle N| m_l\bar qq |N\rangle \equiv f_q^{(N)} m_N$ can be estimated using results from lattice QCD simulations. Numerical values for the so-called mass fractions

Proton		Neutron	
$f_u^{(p)}$	0.019(5)	$f_u^{(n)}$	0.013(3)
$f_d^{(p)}$	0.027(6)	$f_d^{(n)}$	0.040(9)
$f_s^{(p)}$	0.009(22)	$f_s^{(n)}$	0.009(22)

Table 18.1 The mass fractions $f_q^{(N)}$ (with errors) contributed by the u, d and s quarks to the nucleon mass deduced by Hisano *et al.* (2015) from lattice QCD simulations.

$f_q^{(N)}$ are shown in Table 18.1. Thus only $\sim 5\%$ of the mass of proton is given by the current quarks masses or, in other words, by the Higgs effect, while 95% of the proton mass is a consequence of the gluon condensate via the trace anomaly.

18.2   DGLAP equations

Experiments in the late 1960s studying electron–nucleon scattering at large momentum transfer $Q^2 \gg m_N^2$ revealed that a nucleon can be described as a collection of nearly massless, freely interacting scattering centres which share the total nucleon momentum. These point-like scattering centres were called partons and are formed by the valence quarks (which quantum numbers sum up to those of the nucleon), a sea of virtual $\bar{q}q$ pairs and gluons. Moreover, cross-sections were found in this limit to follow approximately scaling invariance. For instance, the energy spectrum of hadrons produced with energy E_h in a reaction at cm energy $\sqrt{s} \gg m_N$ can be approximated by

$$\frac{E_h}{\sigma} \frac{\mathrm{d}\sigma(s, E_h)}{\mathrm{d}E_h} \approx \frac{x}{\sigma} \frac{\mathrm{d}\sigma(x)}{\mathrm{d}x}, \tag{18.19}$$

introducing the scaling variable $x = 2E_h/\sqrt{s}$. With hindsight, we can connect these features easily to the general properties of QCD. Asymptotic freedom explains why the constituents of the nucleon behave at large momentum transfer as unbound particles. Moreover, we can treat the light quarks for $Q^2 \gg m_N^2 \gg m_q^2$ as massless, and scattering on quark and gluons therefore contains no mass scale. Hence we expect only logarithmic scaling violations, caused by the running of the coupling $\alpha_s(Q^2)$, in quantities which are not sensitive to IR singularities. As we will discuss at the end of this section, these are so-called "inclusive observables" like the e^+e^- annihilation cross-section where at fixed-order perturbation theory IR divergences from real and virtual gluon emission cancel.

In contrast, "exclusive quantities" like a differential cross-section to produce a fixed number of partons contain large IR logarithms. They are caused by the emission of soft and collinear partons, similar to the case of QED discussed in section 9.4, and spoil naive perturbation theory. What comes to our rescue, however, is the fact that these logarithms are connected to the emission of additional partons in the semi-classical regime. Therefore, we should be able to develop an approximate probabilistic picture where quantum mechanical interference terms can be neglected. Additionally, we have to add to this semi-classical picture a connection between the perturbative description using partons, valid at $Q^2 \gg \Lambda_{\mathrm{QCD}}^2$, and a description using hadrons, valid at $Q^2 \lesssim \Lambda_{\mathrm{QCD}}^2$. This connection cannot (yet) be derived from first principles and therefore one has to employ phenomenological models.

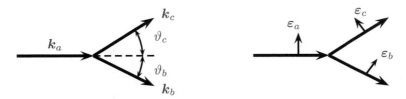

Fig. 18.1 Left: The momenta and angles used to describe the parton splitting $a \to bc$; Right: Choice of the polarisation vectors.

Our aim in this section is to describe processes like $\bar{X}X \to \bar{q}q \to$ hadrons, where the initial state could be, for example, a pair of leptons or DM particles. We can break this process into several steps. The calculation of the "hard" process $\bar{X}X \to \bar{q}q$ uses standard perturbation theory, with which we are familiar by now. Subsequently, a parton cascade develops, $q \to q + g \to q + g + g \to \dots$. In each splitting process, the virtuality t of the partons decreases, until we have to stop at $t \sim$ few $\times \Lambda^2_{\rm QCD}$ the perturbative evolution before $\alpha_s(t)$ becomes too large. Hadrons are then formed out of partons using a phenomenological model. For the calculation of this perturbative parton cascade, we need two main ingredients: An evolution equation which determines the probability that a parton evolves from t_n to t_{n+1} without splitting, and splitting functions $P_{ij}(z)$ which describe how the energy is shared between the daughter partons in a splitting process.

Splitting functions. We start by determining the splitting functions describing the time-like evolution of a parton cascade. We consider the branching of a parton a into the parton pair bc with $t \equiv k^2_a \gg k^2_b, k^2_c > 0$ in the small-angle approximation, $\vartheta = \vartheta_b + \vartheta_c \ll 1$, cf. Fig. 18.1. Defining the energy fraction taken by the parton b as $z = E_b/E_a = 1 - E_c/E_a$, we obtain

$$t = 2E_b E_c(1 - \cos\vartheta) \approx z(1-z)E^2_a\vartheta^2. \tag{18.20}$$

Conservation of transverse momentum implies $\vartheta_b E_b = \vartheta_c E_c$ or

$$\frac{\vartheta_b}{1-z} = \frac{\vartheta_c}{z}. \tag{18.21}$$

Solving (18.20) for ϑ and using (18.21) together with $\vartheta = \vartheta_b + \vartheta_c$, we find

$$\vartheta = \sqrt{\frac{t}{z(1-z)} \frac{1}{E_a}} = \frac{\vartheta_b}{1-z} = \frac{\vartheta_c}{z}. \tag{18.22}$$

To proceed, we have to fix the parton type and we consider here as an example the process of gluon splitting, $g \to gg$. The triple gluon vertex is given by

$$V_{ggg} = \mathrm{i}gf^{ABC}\varepsilon^\alpha_a\varepsilon^\beta_b\varepsilon^\gamma_c[\eta_{\alpha\beta}(k_a - k_b)_\gamma + \eta_{\beta\gamma}(k_b - k_c)_\alpha + \eta_{\gamma\alpha}(k_c - k_a)_\beta], \tag{18.23}$$

where all momenta are defined as outgoing, $k_a = -k_b - k_c$, and ε^α_a is the polarisation vector of gluon a. We assume that all three gluons are close to mass shell, $k^2 \approx 0$,

a	b	c	$F(a,b,c;z)$
in	in	in	$\frac{1-z}{z} + \frac{z}{1-z} + z(1-z)$
in	out	out	$z(1-z)$
in	in	out	$\frac{1-z}{z}$
out	out	in	$\frac{z}{1-z}$

Table 18.2 The function $F^{abc}(z)$ for different gluon polarisations.

what allows us to use transverse polarisation vectors. In order to use these conditions to simplify V_{ggg}, we replace $k_a = -k_b - k_c$ in the first and third, and $k_c = -k_a - k_b$ in the second factor, obtaining

$$V_{ggg} = -2\mathrm{i}g f^{ABC}[(\varepsilon_a \cdot \varepsilon_b)(\varepsilon_c \cdot k_b) - (\varepsilon_b \cdot \varepsilon_c)(\varepsilon_a \cdot k_b) - (\varepsilon_c \cdot \varepsilon_a)(\varepsilon_b \cdot k_c)]. \quad (18.24)$$

Then we evaluate the scalar products $\varepsilon_i \cdot k_j$ in the limit of small ϑ, choosing the two transverse polarisation states as $\varepsilon_i^{\mathrm{in}}$ in the plane spanned by $\boldsymbol{k}_b$ and $\boldsymbol{k}_c$ as shown in the right panel of Fig. 18.1 and $\varepsilon_i^{\mathrm{out}}$ perpendicular to this plane,

$$\varepsilon_i^{\mathrm{in}} \cdot \varepsilon_j^{\mathrm{in}} = \varepsilon_i^{\mathrm{out}} \cdot \varepsilon_j^{\mathrm{out}} = -1 \quad \text{and} \quad \varepsilon_i^{\mathrm{in}} \cdot \varepsilon_j^{\mathrm{out}} = \varepsilon_i^{\mathrm{out}} \cdot k_j = 0.$$

Now we express the three remaining scalar products as functions of E_a and ϑ,

$$\varepsilon_a^{\mathrm{in}} \cdot k_b = -E_b \vartheta_b = -z(1-z)E_a\vartheta, \quad (18.25)$$

$$\varepsilon_b^{\mathrm{in}} \cdot k_c = E_c\vartheta = (1-z)E_a\vartheta, \quad (18.26)$$

$$\varepsilon_c^{\mathrm{in}} \cdot k_b = -E_b\vartheta = -zE_a\vartheta. \quad (18.27)$$

Each of the three terms in V_{ggg} contains one factor ϑ. Combined with the propagator $1/t \propto 1/\vartheta^2$, this results in a collinear $1/t$ singularity of the squared amplitude. We express $|\mathcal{A}_{n+1}|^2$ as

$$|\mathcal{A}_{n+1}|^2 \approx \frac{4g^2}{t} C_A \, F(a,b,c;z)|\mathcal{A}_n|^2, \quad (18.28)$$

where the function $F(a,b,c;z)$ for the various non-zero combinations of polarisation vectors is given in table 18.2. The colour factor C_A equals for the gauge group SU(N) $C_A = \sum_{abc} f^{abc} f^{abc} = N$, see Eq. (B.14).

Example 18.1: We evaluate the vertex for the case of $\{abc\} = \{\text{in, out, out}\}$ polarisations,

$$V_{ggg} = -2\mathrm{i}g f^{abc}[(\varepsilon_a^{\mathrm{in}} \cdot \varepsilon_b^{\mathrm{out}})(\varepsilon_c^{\mathrm{out}} \cdot k_b) - (\varepsilon_b^{\mathrm{out}} \cdot \varepsilon_c^{\mathrm{out}})(\varepsilon_a^{\mathrm{in}} \cdot k_b) - (\varepsilon_c^{\mathrm{out}} \cdot \varepsilon_a^{\mathrm{in}})(\varepsilon_b^{\mathrm{out}} \cdot k_c)]$$
$$= -2\mathrm{i}g f^{abc}[0 - (-1)(-z(1-z)E_a\vartheta) - 0] = 2\mathrm{i}g f^{abc}[z(1-z)E_a\vartheta].$$

Thus the squared amplitude for this combination of polarisation states is

$$|\mathcal{A}_{n+1}|^2 = \left|\frac{V_{ggg}}{t}\mathcal{A}_n\right|^2 \approx \frac{4g^2}{t} C_A \, z(1-z)\,|\mathcal{A}_n|^2,$$

and thus $F(\text{in, out, out}; z) = z(1-z)$.

$g \to gg$	$P_{g \to g}(z) = C_A \left[\frac{1-z}{z} + \frac{z}{1-z} + z(1-z) \right]$
$g \to qq$	$P_{g \to q}(z) = n_f T_F \left[z^2 + (1-z)^2 \right]$
$q \to qg$	$P_{q \to q}(z) = C_F \left[\frac{1+z^2}{1-z} \right]$
$q \to gq$	$P_{q \to g}(z) = C_F \left[\frac{1+(1-z)^2}{z} \right]$

Table 18.3 The four unregularised splitting functions of QCD, with $C_F = 4/3$, $T_F = 1/2$, $N_C = 3$ and n_f as the number of active quark flavours.

The enhancement of the amplitude for $z \to 0$ (gluon b is soft) and $z \to 1$ (c is soft) comes from the emission of soft gluons polarised in the plane of branching. Therefore $\mathcal{A}_{n+1}$ carries some ϕ dependence. As this dependence is small and washed-out considering several successive splittings, one uses unpolarised splitting functions $P_{i \to j}(z)$ defined for $g \to gg$ by

$$P_{g \to g}(z) = \frac{1}{2} C_A \sum_{a,b,c} F(a,b,c;z) = C_A \left[\frac{1-z}{z} + \frac{z}{1-z} + z(1-z) \right]. \tag{18.29}$$

The gluon-splitting function $P_{g \to g}(z)$ is obviously symmetric under an exchange of the two gluons produced, $P_{g \to g}(z) = P_{g \to g}(1-z)$. It is called an unregularised splitting function, because it becomes infinite for $z \to 1$ (and $z \to 0$). These IR divergences can be cancelled adding the effect of virtual gluon emission, and thereby one obtains regularised splitting functions. Proceeding in the same way for the processes $g \to qq$ and $q \to qg$, you should be able to derive the other three splitting functions of QCD, $P_{g \to q}(z)$, $P_{q \to q}(z)$ and $P_{q \to g}(z)$ (problem 18.3).

We can now compute the probability for the emission of an additional parton. We compare the cross-section $\mathrm{d}\sigma_{n+1}$ and $\mathrm{d}\sigma_n$ for the process with and without emission of an additional parton. Their phase space is given by

$$\mathrm{d}\Phi_n \propto \frac{\mathrm{d}^3 k_a}{(2\pi)^3 2E_a} \tag{18.30}$$

and

$$\mathrm{d}\Phi_{n+1} \propto \frac{\mathrm{d}^3 k_b}{(2\pi)^3 2E_b} \frac{\mathrm{d}^3 k_c}{(2\pi)^3 2E_c}. \tag{18.31}$$

With $k_c = k_a - k_b$ and $\mathrm{d}^3 k_c = \mathrm{d}^3 k_a$ for fixed k_b, we obtain trading the variables $\{E_b, \vartheta_b\}$ versus $\{z, t\}$ in the small-angle approximation

$$\mathrm{d}\Phi_{n+1} = \mathrm{d}\Phi_n \frac{1}{2(2\pi)^3} \int E_b \mathrm{d}E_b \vartheta_b \mathrm{d}\vartheta_b \mathrm{d}\phi \mathrm{d}t \frac{\mathrm{d}z}{1-z} \delta(t - E_b E_c \vartheta^2) \delta(z - E_b/E_a)$$

$$= \mathrm{d}\Phi_n \frac{1}{4(2\pi)^3} \mathrm{d}t \mathrm{d}z \mathrm{d}\phi. \tag{18.32}$$

Performing the ϕ integral, we find for their ratio,

$$\frac{\mathrm{d}\sigma_{n+1}}{\mathrm{d}\sigma_n} = \frac{\mathrm{d}t}{t} \mathrm{d}z \frac{\alpha_s}{2\pi} P_{i \to j}(z). \tag{18.33}$$

This ratio is the relative probability density for the emission of an additional collinear parton with virtuality t and energy fraction z by the state n. Integrating this probability using $P(z) \sim 1/z$ for the emission of a soft gluon results in

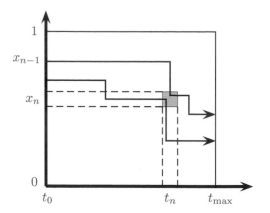

Fig. 18.2 Possible paths with $t_{n+1} > t_n$ and $x_{n+1} < x_n$ describing the space-like evolution of partons.

$d\sigma_{n+1}/d\sigma_n \propto \ln^2(t/t_{min})$. Thus the probability for the emission of an additional parton increases with t, spoiling naive perturbation theory. Moreover, the preference for the emission of collinear partons is the basis for the explanation of jets observed in hadronic final states.

Evolution equation. We want to derive an equation which determines how the number density $f_i(x,t)$ of partons of type i with a certain energy fraction x evolves by parton branching from t to $t + \delta t$. In the case of a space-like evolution, $t_{n+1} > t_n$ and $x_{n+1} < x_n$, such an equation can be used to determine the number of partons inside a hadron, if it is probed, for example, by a photon with virtuality t. The functions $f_i(x,t)$ are then called parton distribution functions. For a time-like evolution, $t_{n+1} < t_n$ and $x_{n+1} < x_n$, the same equation determines the number of partons or hadrons which are produced in a process like $e^+e^- \to$ hadrons as function of the squared cm energy s. The functions $f_i(x,t)$ are then called fragmentation or hadronisation functions.

To be concrete, we consider the space-like evolution of a single parton type. Then the change δf consists of all paths arriving from $x' > x$ and all paths leaving f, as shown in Fig. 18.2. We obtain the number of incoming paths integrating from $x' = x$ to $x' = 1$ under the constraint $x' = x/z$,

$$
\begin{aligned}
\delta f_{in}(x,t) &= \frac{\delta t}{t} \int_x^1 dx' dz \, \frac{\alpha_s}{2\pi} \, P(z) f(x',t) \delta(x - zx') \\
&= \frac{\delta t}{t} \int_0^1 \frac{dz}{z} \frac{\alpha_s}{2\pi} \, P(z) f(x/z,t).
\end{aligned}
\tag{18.34}
$$

Here, we extended the integration range from $[x,1]$ to $[0,1]$ in the second step, which is possible because of $f(x/z,t) = 0$ for $z < x$. In the same way, we find the number of leaving paths as

$$
\begin{aligned}
\delta f_{out}(x,t) &= \frac{\delta t}{t} f(x,t) \int_x^0 dx' dz \, \frac{\alpha_s}{2\pi} \, P(z) \delta(x' - zx) \\
&= \frac{\delta t}{t} f(x,t) \int_0^1 \frac{dz}{z} \frac{\alpha_s}{2\pi} \, P(z).
\end{aligned}
\tag{18.35}
$$

Combing both expressions, we obtain for the total change

$$\delta f(x,t) = \delta f_{\text{in}}(x,t) - \delta f_{\text{out}}(x,t) = \frac{\delta t}{t} \int_0^1 \mathrm{d}z \, \frac{\alpha_s}{2\pi} \, P(z) \left[\frac{f(x/z,t)}{z} - f(x,t) \right]. \quad (18.36)$$

Now we introduce the "plus prescription", defining

$$\int_0^1 \mathrm{d}x \, \frac{g(x)}{x}_+ \equiv \int_0^1 \mathrm{d}x \left[\frac{g(x) - g(1)}{x} \right], \quad (18.37)$$

for any sufficiently regular function $g(x)$, and regularised splitting functions $\hat{P}(z)$ by

$$\hat{P}(z) = P(z)_+. \quad (18.38)$$

Thus the regularised splitting function $\hat{P}(z)$ equals the unregularised splitting function $P(z)$ everywhere except at $z = 1$ where a delta function is added so that Eq. (18.37) is satisfied. For $z < 1$, both splitting functions describe the emission of real partons and allow a probabilistic interpretation. The emission of virtual partons does not change the energy and corresponds therefore to the contribution at $z = 1$. The coefficient of the delta function can be explicitly determined requiring momentum conservation, see problem 18.5, without the need to evaluate virtual processes. Rewriting the evolution equation as a differential equation, we obtain the DGLAP equation,[3],

$$t \frac{\partial f(x,t)}{\partial t} = \frac{\alpha_s(t)}{2\pi} \int_x^1 \frac{\mathrm{d}z}{z} \, \hat{P}(z) \, f(x/z,t). \quad (18.39)$$

Given the initial values on $[x_{\min} : 1]$ of the function f at a fixed scale t_0, we can calculate within perturbative QCD its evolution as function of the virtuality t. The initial values $f(x,t_0)$, however, have to be determined from measurements. These functions depend on the hadron probed, but are independent of a specific process. Thus one can determine, for example, $f_i^p(x,t)$ for a proton from a mixture of pp and e^-p reactions, and apply them then to neutrino–nucleon scattering.

Monte Carlo methods. The evolution equation (18.39) admits a probabilistic interpretation, since interference terms disappeared in the semi-classical limit. This allows a straightforward application of Monte Carlo methods to solve it. Compared to a numerical integration of (18.39), the Monte Carlo approach has the advantage of being more flexible, allowing, for example, the addition of hadronisation models and experimental cuts in straightforward manner. Moreover, this method reproduces event-by-event fluctuation which is an important ingredient for the estimation of the statistical significance of an observed signal.

We can either start from Eq. (18.36) or use as short-cut the analogy to the radioactive decay problem with a time-varying decay constant. Using the latter picture, we denote the differential probability for the decay of an atom (or the branching of

---

[3]Until the 1990s, equations of this type were called Altarelli–Parisi equations. The name evolved then from Gribov–Lipatov–AP to its present form, Dokshitzer–Gribov–Lipatov–Altarelli–Parisi or DGLAP equations.

a parton) at t by $\mathcal{P}(t)$. It equals minus the time-derivative of the survival probability which we call $\Delta(t)$. Moreover, the differential probability $\mathcal{P}(t)$ that something happens at t is proportional to the survival probability $\Delta(t)$, with a proportionality constant $F(t)$ which equals the decay rate in the radioactive decay problem,

$$\mathcal{P}(t) = -\frac{\mathrm{d}\Delta(t)}{\mathrm{d}t} = F(t)\Delta(t). \tag{18.40}$$

Integration gives

$$\Delta(t) = \Delta(0)\exp\left\{-\int_0^t \mathrm{d}t F(t)\right\}, \tag{18.41}$$

if the evolution starts at $t = 0$, and thus

$$\mathcal{P}(t) = F(t)\Delta(t) = F(t)\exp\left\{-\int_0^t \mathrm{d}t' F(t')\right\}. \tag{18.42}$$

In the case of a parton cascade, $\mathcal{P}(t)$ denotes the differential probability that a splitting happens at the virtuality t, while the probability that a parton survives, that is, it does not split between the virtuality t and $t_{\min}$, is called the Sudakov form factor $\Delta(t)$. In turn, the probability for a parton to survive between t' and t is given by the ratio $\Delta(t')/\Delta(t)$. The function $F(t)$ playing the role of the decay constant is given by the ratio $\mathrm{d}\sigma_{n+1}/\mathrm{d}\sigma_n$ from Eq. (18.33). Because of its $1/t$ singularity, we have to introduce a cut-off t_0 which should moreover be large enough that we can trust perturbative calculations. Combining all this, the differential probability distribution for the splitting of the parton i into the channels j is

$$\mathrm{d}\mathcal{P}_i(t, z) = \sum_j \frac{\mathrm{d}t}{t}\frac{\mathrm{d}z}{2\pi}\alpha_s P_{ij}(z)\frac{\Delta_i(t')}{\Delta_i(t)}. \tag{18.43}$$

Here, the sum includes all possible branching channels j, and Δ_i is given by

$$\Delta_i(t) = \exp\left[-\int_{t_{\min}}^t \frac{\mathrm{d}t'}{t'}\sum_j \int_{z_{\min}}^{z_{\max}} \frac{\mathrm{d}z}{2\pi}\alpha_s P_{ij}(z)\right]. \tag{18.44}$$

As the use of distributions like $\delta(1 - x)$ is problematic in performing numerical calculations, we have used the unregularised splitting functions $P_{ij}(z)$ given in table 18.3 together with an IR cut-off $t_{\min}$. The latter serves both as a boundary to the non-perturbative regime, $t_{\min} \gg \Lambda_{\mathrm{QCD}}^2$, and as a regulator of the IR singularities in the splitting functions,

$$z_{\min} = \sqrt{t_{\min}/t} \quad \text{and} \quad z_{\max} = 1 - \sqrt{t_{\min}/t}. \tag{18.45}$$

We can now set up a simple Monte Carlo scheme for the simulation of parton cascades. In each step $(t_n, x_n) \to (t_{n+1}, x_{n+1})$, we determine first the new t_{n+1} from the Sudakov factor as follows. We choose a random number r from an uniform distribution between $[0, 1]$. For a space-like parton cascade, we compare then this random number with the probability for the evolution from t_n to $t_{n+1} > t_n$,

$$r = \frac{\Delta(t_{n+1})}{\Delta(t_n)}, \tag{18.46}$$

and solve for t_{n+1}. If t_{n+1} is larger than the hard scale in the process considered, the cascade stops. Otherwise we determine the splitting channel j and the energy fraction $z = x_{n+1}/x_n$ according to the probability distribution $\alpha_s P_{i \to j}(z)$ and continue.

For a time-like parton cascades, where the virtuality decreases, we compare a random number r with

$$r = \frac{\Delta(t_n)}{\Delta(t_{n+1})}. \tag{18.47}$$

If $r < \Delta(t_n, t_0)$, the equation has no solution and no further splitting happens. Otherwise we solve again for the parton type and the energy fraction z and continue. As final outcome of the time-like cascade, we obtain a set of quarks and gluons whose virtualities are close to our cut-off scale $t_{\min}$.

This description of a parton cascade has been schematic in few respects. First, we have not yet specified the argument of α_s. A more careful analysis shows that choosing $p_\perp^2 = z(1-z)t$ sums up partially NLL effects. Second, in a time-like cascade coherence effects lead to an angular-ordered cascade. As a result, the evolution parameter used differs for the two cases.

Hadronisation models. The quarks and gluons produced as the final state of a parton shower have to be converted into hadrons using a phenomenological model. A method which requires no additional theoretical input is determining fragmentation functions $f_i^h(z, t_0)$ from experimental data. These functions give the probability that a parton i with energy E produces a hadron h with energy zE. Convoluting then the fragmentation functions with the parton spectra $D_i(x/z, \sqrt{s}, t_0)$ obtained from the perturbative evolution from $\sqrt{s}$ down to t_0, one obtains the hadron spectra as

$$D_h(x, \sqrt{s}) = \sum_{i=q,g} \int_x^1 \frac{dz}{z} \, D_i(x/z, \sqrt{s}, t_0) f_i^h(z, t_0). \tag{18.48}$$

An alternative to this purely phenomenological method is to develop models that try to capture basic properties of QCD like confinement. The cluster hadronisation model used first in HERWIG is based on the idea of pre-confinement; it assumes that the parton spectra at the end of the cascade resemble closely the hadron spectra. Since the P_{gg} splitting function is the most singular one, the majority of partons in the cascade are gluons. In order to create the valence quarks necessary to form mesons and baryons, one adds therefore an artificial splitting $g \to \bar{q}q$ after the parton cascade stopped at $t_{\min}$.

The Lund string model, which is another scheme often employed in Monte Carlo simulations, is based on a rather different ansatz. Lattice QCD calculations show that the potential between two static quarks can be approximated at large distance by $V(r) \approx \kappa r$ with $\kappa \approx 0.2\,\text{GeV}^2$. Thus in a confining theory like QCD, the force lines

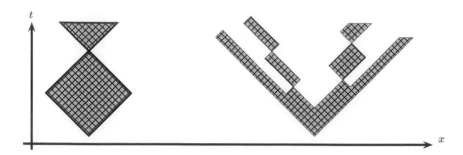

Fig. 18.3 Left: Yo-yo movement of a $\bar{q}q$ pair in the massless limit. Right: Break-up of the string into fragments.

between a $\bar{q}q$ pair are concentrated in a narrow tube connecting the pair. This tube of colour field can be viewed as a string with tension κ, trying to pull the quark pair together. The string model uses this picture to describe the hadronisation of partons into hadrons. Let us discuss the simplest case that a $\bar{q}q$ pair is created in an e^+e^- annihilation. As we will show in section 23.3, a constant acceleration $\ddot{x} = -\kappa$ leads to a hyperbola as trajectory. In the limit of massless quarks, we can approximate the hyperbola by straight line segments as shown in Fig. 18.3. Neglecting further interactions, the two quarks would oscillate back and forth in a yo-yo mode. Instead, a new $\bar{q}q$ pair will be created when the energy $\kappa \Delta x$ in the string tube is sufficiently large such that two colour singlet states can be formed, $(\bar{q}q)(\bar{q}q)$. If one adds the assumption that all breakings happen during the initial expansion phase of the yo-yo modes, a formulation of the break-up process as a probabilistic process is possible. Moreover, this process is Lorentz-invariant, since it is based on the dynamics of a relativistic string.

While the basic picture underlying the cluster hadronisation or the string model are theoretically well motivated, a rather large number of additional assumptions and parameters is required to model the momentum distributions of the various types of mesons and baryons produced and their branching ratios. These parameters are partly obtained from fits to experimental data and interfere with the properties of the basic hadronisation model. It is therefore difficult to differentiate which one of these two models describes nature better.

18.3 Corrections to e^+e^- annihilation

In this section, we examine the total annihilation cross-section of an e^+e^- pair into hadrons. While the scheme (18.43–18.44) allows us to calculate the relative fraction of final-states in a QCD process, it accounts for the effect of virtual processes only indirectly, imposing flavour conservation, $\sum_i \int dx P_{q \to q}(x) = 1$, or momentum conservation, $\sum_i \int dx\, x P_i = 1$, on the regularised splitting functions. As a consequence, we cannot calculate the impact of higher-order corrections to the *total* rate within this formalism. Instead, a perturbative QCD calculation for the total annihilation cross-section is sufficient and, as we will see, free of IR divergences. This requires the large logarithms which are present in individual processes to cancel in the total e^+e^- annihilation cross-section.

Let us first give an heuristic argument as to why the total annihilation cross-section into hadrons can be calculated using partons. The partonic process contains the time scale $1/\sqrt{s}$, while hadronisation occurs on the time scale $1/\Lambda_{\text{QCD}}$. This means that the production of partons proceeds independently of their latter hadronisation, if we consider processes with $s \gg \Lambda_{\text{QCD}}^2$. More precisely, we expect from hadronisation corrections of order Λ_{QCD}^2/s to the result of a perturbative QCD calculations using partons. We can make this argument a bit more quantitative as follows. Let us denote the amplitude to create a quark state by a hadronic QCD current $J(t)$ which evolves then into a final-state of hadrons $|h\rangle$ schematically by $\mathcal{A}_h \propto \int \mathrm{d}t \, \langle 0| \, J(t) U(t, \infty) \, |h\rangle$. The probability determining the total annihilation cross-section is then $|\mathcal{A}_h|^2$ summed over all possible hadronic states,

$$\sigma_{\text{ann}} \propto \sum_h \mathcal{A}_h \mathcal{A}_h^* \propto \sum_h \langle 0| \, J(t') U(t', \infty) \, |h\rangle \, \langle h| \, U(t, \infty) J(t) \, |0\rangle =$$
$$= \langle 0| \, J(t') U(t', \infty) U(\infty, t) J(t) \, |0\rangle = \langle 0| \, J(t') U(t', t) J(t) \, |0\rangle \,. \tag{18.49}$$

Thus unitarity, $\sum_h |h\rangle \langle h| = 1$, leads to a cancellation of the complicated long-distance physics, and only the short-distance scale $t - t' \sim 1/\sqrt{s}$ enters the total e^+e^- annihilation cross-section. As a result, such quantities can be calculated within perturbative QCD.

After these preliminaries, let us move to the specific process $e^+e^- \to$ hadrons. We split the annihilation cross-section into

$$\sigma_{\text{hadrons}} = \sigma_{q\bar{q}} + \sigma_{q\bar{q}g} + \dots \tag{18.50}$$

according to the partonic final state. Explicitly factoring out the strong coupling, the perturbative series up to α_s is

$$\sigma_{q\bar{q}} = \frac{1}{4I} \int \mathrm{d}\Phi_2 \, |\mathcal{A}_{q\bar{q}}|^2 = \frac{1}{4I} \int \mathrm{d}\Phi_2 \left[|\mathcal{A}_{q\bar{q}}^{(0)}|^2 + 2\alpha_s \Re \left(\mathcal{A}_{q\bar{q}}^{(0)} \mathcal{A}_{q\bar{q}}^{(2)*} \right) + \dots \right], \tag{18.51a}$$

$$\sigma_{q\bar{q}g} = \frac{1}{4I} \int \mathrm{d}\Phi_3 \, |\mathcal{A}_{q\bar{q}g}|^2 = \frac{1}{4I} \int \mathrm{d}\Phi_3 \left[\alpha_s |\mathcal{A}_{q\bar{q}g}^{(1)}|^2 + \dots \right]. \tag{18.51b}$$

We expect that the IR divergences of the real gluon emission process are cancelled by those of the interference term between the tree-level process and the virtual correction. At order $\mathcal{O}(\alpha_s)$, a finite total annihilation cross-section requires

$$\int \mathrm{d}\Phi_2 \, 2\Re \left(\mathcal{A}_{q\bar{q}}^{(0)} \mathcal{A}_{q\bar{q}}^{(2)*} \right) + \int \mathrm{d}\Phi_3 \, |\mathcal{A}_{q\bar{q}g}^{(1)}|^2 = \text{finite}. \tag{18.52}$$

As we have seen in the case of UV divergences, we have to apply a regularisation scheme also to IR divergences in order to make our mathematical manipulations well defined. The simplest scheme, introducing a finite gluon mass, breaks gauge invariance. We can maintain gauge invariance using DR again, now, however, in the limit $d = 2\omega = 4 + 2\varepsilon$, that is, extrapolating to more than four dimensions.

Before embarking on calculations let us discuss two features of this process that can be used to simplify the calculations. First, we can express $|\bar{\mathcal{A}}|^2 = L_{\mu\nu}H^{\mu\nu}/Q^4$ always as the product of a leptonic and hadronic tensor, $L_{\mu\nu}$ and $H_{\mu\nu}$, if we are interested only in QCD corrections. Moreover, QCD corrections affect only the hadronic tensor $H_{\mu\nu}$. Then we can write the cross-sections as

$$\sigma = \frac{1}{2Q^2}L_{\mu\nu}\frac{1}{Q^4}\int d\Phi\, H^{\mu\nu}, \qquad (18.53)$$

where we used the flux factor $4I = 4p_{\rm cms}\sqrt{s} = 2s$ and $s = Q^2$. Second, gauge invariance implies $Q_\mu L^{\mu\nu} = \cdots = Q_\nu H^{\mu\nu} = 0$, where Q_μ denotes the four-momentum of the photon or Z boson in the s channel. Hence the tensor structure of the leptonic and hadronic tensor has to be of the form $-\eta^{\mu\nu} + Q^\mu Q^\nu/Q^2$. This allows us to simplify the phase-space integration, introducing

$$\int d\Phi\, H^{\mu\nu}(Q) = \frac{1}{d-1}\left(-\eta^{\mu\nu} + \frac{Q^\mu Q^\nu}{Q^2}\right)H(Q^2) \qquad (18.54)$$

with

$$H(Q^2) = -\eta_{\mu\nu}\int d\Phi\, H^{\mu\nu}(Q) = -H_\mu^{\ \mu}(Q^2)\Phi. \qquad (18.55)$$

The last step is valid, when $H^{\mu\nu}$ depends only on Q^2, as it is the case after averaging over spins. Thus our task is reduced to the calculation of the trace of the hadronic tensor. As additional simplification, we include only the photon diagrams which give the dominant contribution to the total annihilation cross-section of an e^+e^- pair into hadrons, except close to the Z pole when $s \simeq m_Z^2$.

Tree-level process. We consider first the tree-level process $e^+e^- \to \bar{q}q$, denoting the four-momenta of the particles by $l^+, l^-, \bar{q}, q$. The leptonic and hadronic tensor, $L_{\mu\nu}$ and $H_{\mu\nu}$, are given by

$$L_{\mu\nu} = e^2[l_\mu^+ l_\nu^- + l_\mu^- l_\nu^+ - Q^2\eta_{\mu\nu}/2] \qquad (18.56)$$

and

$$H_{\mu\nu} = (e_q e)^2 4N_c[\bar{q}_\mu q_\nu + q_\mu \bar{q}_\nu - Q^2\eta_{\mu\nu}/2], \qquad (18.57)$$

where e_q denotes the quark charge in units of the positron charge e. Evaluating $L_\mu^{\ \mu} = e^2[2(l^+\cdot l^-) - Q^2 d/2] = e^2 Q^2(1-d/2)$, we find as general expression for the cross-section

$$\sigma = \frac{e^2}{4Q^4}\frac{d-2}{d-1}H(Q^2). \qquad (18.58)$$

The specific ingredient of the tree-level process is the trace of the hadronic tensor. In the limit of massless quarks, we obtain for its trace

$$H_\mu^{\ \mu} = -2(e_q e)^2 N_c(2-d)Q^2. \qquad (18.59)$$

Adding also the two-particle phase space in DR, cf. Eq. (18.82), with $p^2 = s/4 = Q^2/4$ gives for the tree-level cross-section

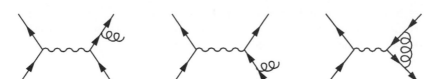

Fig. 18.4 Feynman diagrams for real gluon emission and vertex correction.

$$\sigma^{(0)} = \frac{e^2}{2Q^4} \frac{1-\varepsilon}{3-2\varepsilon} \times 4(e_q e)^2 N_c Q^2(1-\varepsilon) \times \frac{1}{4\pi} \frac{1}{2} \left(\frac{4\pi}{Q^2}\right)^\varepsilon \frac{\Gamma(1-\varepsilon)}{\Gamma(2-2\varepsilon)}$$
$$= \frac{4\pi\alpha^2}{2Q^2} e_q^2 N_c \left(\frac{4\pi\mu^2}{Q^2}\right)^\varepsilon \frac{(1-\varepsilon)^2}{3-2\varepsilon} \frac{\Gamma(1-\varepsilon)}{\Gamma(2-2\varepsilon)}. \tag{18.60}$$

Simple inspection shows that this formula reproduces the standard result for $\varepsilon \to 0$ (problem 18.7).

Real emission.   The hadronic current including the emission of an additional gluon with momentum g and polarisation vector $\varepsilon_\sigma^{(r)}$ as shown by the first two graphs in Fig. 18.4 is given by

$$J^\mu = \varepsilon_\sigma^{(r)*} J^{\mu\sigma} = -\mathrm{i}\mu^{2\varepsilon} e e_q g_s T_{ij}^a \varepsilon_\sigma^{(r)*} \bar{u}(q) \left[\gamma^\sigma \frac{\slashed{q}+\slashed{g}}{(q+g)^2}\gamma^\mu + \gamma^\mu \frac{-\slashed{\bar{q}}-\slashed{g}}{(\bar{q}+g)^2}\gamma^\sigma\right] v(\bar{q}).$$

The hadronic tensor is obtained contracting the squared currents with the polarisation vector of the gluon,

$$H^{\mu\nu} = \sum_r \varepsilon_\sigma^{(r)*} \varepsilon_\rho^{(r)} \sum J^{\mu\sigma} J^{\rho\nu*} = -J^{\mu\sigma} J_\sigma^{\ \nu*}. \tag{18.61}$$

Here we could evaluate the polarisation sum using $\sum \varepsilon_\mu^{(r)*} \varepsilon_\nu^{(r)} = -\eta_{\mu\nu}$, because only one external gluon is involved. Evaluating then the trace $H^{\mu\nu}$ gives us

$$H_\mu^{\ \mu} = C_F N_C (e e_q g_s)^2 \left[\frac{S_{qq}}{(2q\cdot g)^2} + \frac{S_{\bar{q}\bar{q}}}{(2\bar{q}\cdot g)^2} - \frac{S_{qq}+S_{\bar{q}\bar{q}}}{(2q\cdot g)(2\bar{q}\cdot g)}\right]. \tag{18.62}$$

In this expression, IR divergences have shown up as poles for $q \cdot g \to 0$ and $\bar{q} \cdot g \to 0$. With $q \cdot g = E_q E_g (1 - \cos\beta_q)$, where β_q is the quark velocity, we can identify them as a combination of soft ($E_g \to 0$) and collinear ($\beta_q \to 0$) singularities. Next we evaluate the spinor traces, finding

$$S_{qq} = S_{\bar{q}\bar{q}} = 32(1-\varepsilon)^2[(q\cdot g)(\bar{q}\cdot g)] \tag{18.63}$$

and

$$S_{q\bar{q}} + S_{\bar{q}q} = -32(1-\varepsilon)[(q\cdot q)Q^2 - 2\varepsilon(q\cdot g)(\bar{q}\cdot g)]. \tag{18.64}$$

As we have already seen discussing the splitting functions, the double pole from the propagators will be converted into a simple pole, cancelling against the factor $(q\cdot g)(\bar{q}\cdot g)$ in the traces.

Now we introduce the momentum fractions $x_i = 2p_i \cdot Q/Q^2$, which for massless quarks become, for example, $1 - x_q = 2\bar{q} \cdot g/Q^2$. Combining then H_μ^μ together with the three-particle phase space given in (18.83), we obtain

$$
\begin{aligned}
H_\mu^\mu = &\frac{C_F N_C e_q^2 \alpha \alpha_s Q^2}{\pi} \left(\frac{4\pi\mu^2}{Q^2}\right)^\varepsilon \frac{(1-\varepsilon)}{\Gamma(2-2\varepsilon)} \\
&\times \int_0^1 dx_q \int_{1-x_{\bar{q}}}^1 dx_{\bar{q}} \frac{1}{[(1-x_q)(1-x_{\bar{q}})(x_q+x_{\bar{q}}-1)]^\varepsilon} \\
&\times \left[(1-\varepsilon)\left(\frac{1-x_q}{1-x_{\bar{q}}}+\frac{1-x_{\bar{q}}}{1-x_q}\right)+\frac{2(x_q+x_{\bar{q}}-1)}{(1-x_q)(1-x_{\bar{q}})}-2\varepsilon\right].
\end{aligned}
\tag{18.65}
$$

The nested integrals become easier substituting $x_q = x$ and $x_{\bar{q}} = 1 - vx$. Solving them using the definition (A.29) of Euler's Beta function, adding the three-particle phase space (18.83) and expanding then the Gamma function around their poles, we end up with

$$
\begin{aligned}
\sigma &= \frac{2C_F N_C e_q^2 \alpha \alpha_s}{Q^2} \mu^\varepsilon \left(\frac{4\pi\mu^2}{Q^2}\right)^\varepsilon \frac{(1-\varepsilon)}{\Gamma(2-2\varepsilon)} \frac{\Gamma^3(1-\varepsilon)}{\Gamma(1-3\varepsilon)} \left(\frac{2}{\varepsilon^2}+\frac{3}{\varepsilon}+\frac{19}{2}+\mathcal{O}(\varepsilon)\right) \\
&= \sigma^{(0)} \frac{C_F \alpha_s}{2\pi} \left(\frac{4\pi\mu^2}{Q^2}\right)^\varepsilon \frac{\Gamma^2(1-\varepsilon)}{\Gamma(1-3\varepsilon)} \left(\frac{2}{\varepsilon^2}+\frac{3}{\varepsilon}+\frac{19}{2}+\mathcal{O}(\varepsilon)\right).
\end{aligned}
\tag{18.66}
$$

Here, the $1/\varepsilon^2$ term corresponds to a combined soft and collinear divergence, while the $1/\varepsilon$ term corresponds to soft singularities.

Virtual corrections. The QCD one-loop contribution to $e^+e^- \to \bar{q}q$ consists of self-energy corrections to the quark lines and the vertex correction. The interference terms between these virtual corrections and the tree-level diagram are of the same order as the real emission, cf. Eq. (18.51a), and should cancel the IR divergences found above. Since we restrict ourselves to massless quarks, the self-energy diagram contains no scale and vanishes in DR, cf. the discussion of tadpoles in section 12.3.2. Therefore we have to consider only the vertex correction. Contracting the hadronic tensor of the interference term between the vertex correction evaluated in R_ξ gauge with $\xi = 1$ and the tree-level diagram gives

$$
\begin{aligned}
-\eta_{\mu\nu} H^{\mu\nu} = &2\mathrm{i}(eqg_s\mu^{2\varepsilon})^2 \mathrm{tr}\{T^bT^b\} \int \frac{d^dk}{(2\pi)^d} \frac{1}{k^2(k+q)^2(k-\bar{q})^2} \\
&\times \mathrm{tr}\{\slashed{q}\gamma_\sigma(\slashed{k}+\slashed{q})\gamma_\mu(\slashed{k}-\slashed{\bar{q}})\gamma^\sigma\slashed{q}\gamma^\mu\}.
\end{aligned}
\tag{18.67}
$$

Combining the denominators using Feynman parameters results in

$$
\begin{aligned}
\frac{1}{k^2(k+q)^2(k-\bar{q})^2} &= \int_0^1 d\alpha \int_0^{1-\alpha} d\beta \frac{2}{[\alpha(k+q)^2+\beta(k-\bar{q})^2+(1-\alpha-\beta)k^2]^3} \\
&= \int_0^1 d\alpha \int_0^{1-\alpha} d\beta \frac{2}{[(k+\alpha q-\beta\bar{q})^2+\alpha\beta Q^2]^3}.
\end{aligned}
\tag{18.68}
$$

Now the change of variables $k \to k - \alpha q + \beta\bar{q}$ transforms the loop integral into standard form, $I(\omega, 3)$. Note that the loop integral is UV-finite, since QCD corrections cannot renormalise the $\gamma\bar{q}q$ vertex. Next we work out the trace in (18.67), and then shift variables, obtaining

$$
\begin{aligned}
\mathrm{tr}\{\cdots\} &= 8(1-\varepsilon)\{Q^4 - 4[(k\cdot q)(k\cdot\bar{q}) - 2k\cdot(q-\bar{q})Q^2 + \varepsilon k^2 Q^2\} \\
&= 8(1-\varepsilon)Q^2\{[1-\alpha-\beta+(1-\varepsilon)\alpha\beta]Q^2 - (1-\varepsilon)^2(2-\varepsilon)^{-1}k^2\}.
\end{aligned}
\tag{18.69}
$$

Here, we have omitted all terms linear in k and replaced in the second line $k^\mu k^\nu$ by $k^2\eta^{\mu\nu}/d$. Combining then the trace and the loop integral, we obtain

$$
\begin{aligned}
\sigma_V &= i\sigma_0^{(\varepsilon)}4(g_s\mu^\varepsilon)^2 C_F \int_0^1 \mathrm{d}\alpha \int_0^{1-\alpha}\mathrm{d}\beta \int \frac{\mathrm{d}^d k}{(2\pi)^d}\frac{1}{[k^2+\alpha\beta Q^2]^3} \\
&\quad \times \left[(1-\alpha-\beta+(1-\varepsilon)\alpha\beta)Q^2 - (1-\varepsilon)^2/(2-\varepsilon)k^2\right] \\
&= -\sigma_0^{(\varepsilon)}\frac{\alpha_s}{2\pi}C_F\left(\frac{4\pi\mu^2}{-Q^2}\right)^\varepsilon \Gamma(1+\varepsilon)\,I(\varepsilon),
\end{aligned}
\tag{18.70}
$$

where we decouple the parameter integrals in $I(\varepsilon)$ by the substitution $\beta = (1-\alpha)v$,

$$
\begin{aligned}
I(\varepsilon) &= \int_0^1 \mathrm{d}\alpha \int_0^{1-\alpha}\mathrm{d}\beta \frac{1}{(\alpha\beta)^\varepsilon}\left[\frac{1-\alpha-\beta+(1-\varepsilon)\alpha\beta}{\alpha\beta} - \frac{(1-\varepsilon)^2}{\varepsilon}\right] \\
&= \int_0^1 \mathrm{d}\alpha \int_0^1 \mathrm{d}v \frac{1-\alpha}{[\alpha(1-\alpha)v]^\varepsilon}\left[\frac{(1-\alpha)(1-v)+(1-\varepsilon)\alpha(1-\alpha)v}{\alpha(1-\alpha)v} - \frac{(1-\varepsilon)^2}{\varepsilon}\right] \\
&= \frac{\Gamma^2(1-\varepsilon)}{\Gamma(2-2\varepsilon)}\left(\frac{1}{\varepsilon^2}+\frac{1}{2}+\frac{1-\varepsilon}{2\varepsilon}\right) = \frac{1}{2}\frac{\Gamma^2(1-\varepsilon)}{\Gamma(1-2\varepsilon)}\left(\frac{2}{\varepsilon^2}+\frac{3}{\varepsilon}+\frac{8}{1-2\varepsilon}\right).
\end{aligned}
\tag{18.71}
$$

Inserting the result in the formula for the cross-section gives

$$
\sigma_V = -\sigma_0^{(\varepsilon)}\frac{\alpha_s}{2\pi}C_F\left(\frac{4\pi\mu^2}{-Q^2}\right)^\varepsilon\frac{\Gamma(1+\varepsilon)\Gamma^2(1-\varepsilon)}{\Gamma(1-2\varepsilon)}\left(\frac{2}{\varepsilon^2}+\frac{3}{\varepsilon}+8+\mathcal{O}(\varepsilon)\right).
\tag{18.72}
$$

Recall that only the real part of this expression enters the total cross-section.

Total cross-section. Combining the result for the real and virtual correction, we arrive at

$$
\begin{aligned}
\sigma = \sigma_0^{(\varepsilon)}\bigg\{&1 + \frac{\alpha_s}{2\pi}C_F\left(\frac{4\pi\mu^2}{Q^2}\right)^\varepsilon\frac{\Gamma^2(1-\varepsilon)}{\Gamma(1-3\varepsilon)}\left[\left(\frac{2}{\varepsilon^2}+\frac{3}{\varepsilon}+\frac{19}{2}+\mathcal{O}(\varepsilon)\right)\right. \\
&\left.+ \Re\{(-1)^\varepsilon\}\frac{\Gamma(1+\varepsilon)\Gamma(1-3\varepsilon)}{\Gamma(1-2\varepsilon)}\left(\frac{2}{\varepsilon^2}+\frac{3}{\varepsilon}-8+\mathcal{O}(\varepsilon)\right)\right]\bigg\}.
\end{aligned}
\tag{18.73}
$$

A cancellation of the poles therefore requires that the pre-factor in the second line equals one up to $\mathcal{O}(\varepsilon^3)$ terms. Using

$$
\Gamma(1+\varepsilon) = 1 - \varepsilon\gamma + \left(\frac{\pi^2}{12}+\frac{1}{2}\gamma^2\right)\varepsilon^2 + \mathcal{O}(\varepsilon^4)
\tag{18.74}
$$

we obtain

$$
\frac{\Gamma(1+\varepsilon)\Gamma(1-3\varepsilon)}{\Gamma(1-2\varepsilon)} = 1 + \frac{1}{2}(\pi\varepsilon)^2 + \mathcal{O}(\varepsilon^3).
\tag{18.75}
$$

Combined with $\Re\{(-1)^\varepsilon\} = \Re\{\exp(\mathrm{i}\pi\varepsilon)\} = 1 - (\pi\varepsilon)^2/2 + \mathcal{O}(\varepsilon^4)$, we see that the $\mathcal{O}(\varepsilon^2)$ terms do cancel as required in the combined pre-factor. The total cross-section of $e^+ e^-$ annihilation into hadrons at $\mathcal{O}(\alpha \alpha_s)$ is therefore free of IR divergences and we can perform the limit $\varepsilon \to 0$, obtaining

$$\sigma = \sigma_0 \left\{ 1 + \frac{3}{2} \frac{\alpha_s}{2\pi} C_F + \mathcal{O}(\alpha_s^2) \right\}. \tag{18.76}$$

Two obvious questions arise: first, what should we use as argument of α_s? And second, which conditions characterise in general a quantity free of IR divergences? The answer to the first question follows from our general discussion of the RGE equation in section 12.4. We can absorb all scale dependence into the running coupling $\alpha_s(\mu^2)$ choosing as scale $\mu^2 = s$. As a bonus, we sum thereby the leading logarithmic corrections of all orders perturbation theory.

IR safe observables. The second question is more complex, and we can only sketch a few general results about so-called IR safe observables. From an experimental point of view, physical observables have to be measurable. However, any experimental device has a finite energy and angular resolution, ΔE and $\Delta \vartheta$, respectively. Any soft or collinear splitting below these resolution limits cannot be measured and therefore the individual cross-sections of these unresolved processes have to be summed up. Thus any observable such as, for example, an n-jet rate should be defined such that it is invariant under the replacement $p_i \to p_j + p_k$ whenever p_j or p_k are small or collinear. While the individual rates of unresolved processes may have IR divergences, they should disappear if we use sufficiently inclusive quantities. This idea is made precise in the Kinoshita–Lee–Nauenberg theorem which states that the total rate

$$\sum_{i,f \in D} |\langle f| S |i \rangle|^2 \tag{18.77}$$

in a general quantum field theory is IR-finite. Here, $D \equiv D(p_i^\mu, \Delta E, \Delta \vartheta_i)$ denotes the set of states which are degenerated for a given energy and angular resolution with the initial and final states, respectively. In general, one has to sum thus also over all unresolved initial states, including, for example, initial bremsstrahlung. An exception is QED, where a summation over unresolved final states is sufficient to obtain IR-finite quantities, because no massless particles with non-zero electric charge exist. This also the reason why it was sufficient to sum over final-states in our example, where only the final state contained strongly interacting particles.

18.A Appendix: Phase-space integrals in DR

The two-particle phase space $d\Phi_2$ was given in Eq. (9.120) for $d = 4$ dimensions. We rewrite first this expression for general d,

$$d^{2d-2}\Phi_2 = (2\pi)^d \, \delta^{(d)}(q - p_1 - p_2) \frac{d^{d-1}p_1}{2E_1(2\pi)^{d-1}} \frac{d^{d-1}p_2}{2E_2(2\pi)^{d-1}}, \tag{18.78}$$

and integrate then over $d^{d-1}p_2$,

$$\int d^{d-1}p_2 \, \delta^{(d)}(q - p_1 - p_2) = \delta(q - E_1 - E_2). \tag{18.79}$$

The remaining integrals become

$$\frac{\mathrm{d}^{d-1}p_2}{2E_1} = \frac{1}{2} E_1^{d-3} \sin^{d-3}\vartheta_1 \sin^{d-4}\vartheta_2 \cdots \sin\vartheta_{d-3}\mathrm{d}\vartheta_1 \mathrm{d}\vartheta_2 \cdots \mathrm{d}\vartheta_{d-2}\mathrm{d}E_1, \qquad (18.80)$$

where the ϑ_i are the angles to $d-2$ coordinate axes. The spin-averaged $|\mathcal{A}|^2$ does not depend on the angles ϑ_i, and thus the integrals can be performed with the help of Eq. (A.30), giving

$$\int \frac{\mathrm{d}^{d-1}p_2}{2E_1} = 2^{d-3}\,\pi^{(d-2)/2}\,\frac{\Gamma(d/2-1)}{\Gamma(d-2)}\,\mathrm{d}E_1. \qquad (18.81)$$

Evaluating finally the energy integral as in $d=3$, we arrive at

$$\int \mathrm{d}^{2d-2}\Phi_2 = \frac{1}{4\pi}\frac{p}{\sqrt{s}}\left(\frac{\pi}{p^2}\right)^{\varepsilon}\frac{\Gamma(1-\varepsilon)}{\Gamma(2-2\varepsilon)}. \qquad (18.82)$$

Following the same procedure, one obtains for the three-particle phase space

$$\int \mathrm{d}\Phi_3 = \frac{Q^2}{2(4\pi)^3}\left(\frac{4\pi}{Q^2}\right)^{2\varepsilon}\frac{1}{\Gamma(2-2\varepsilon)}\int_0^1 \mathrm{d}x \int_{1-x}^1 \mathrm{d}y\,\frac{1}{[(1-x)(1-y)(1-z)]^{\varepsilon}}. \qquad (18.83)$$

Summary

Theories like QED or QCD with massless fermions contain no dimensionful parameters and are classically scale- and conformal-invariant. The scale invariance of these theories is broken by quantum corrections which unavoidably introduce a mass scale. This quantum effect is responsible for the bulk of hadron masses.

In the parton picture, we replace a hadron which is probed in a process with momentum transfer Q^2 by a collections of quarks and gluons. For $Q^2 \gg \Lambda_{\mathrm{QCD}}^2$, we can treat them as "free" particles which are interacting independently. Measuring the parton distribution functions $f_i(x, Q^2)$ at one scale Q^2, perturbative QCD describes via the DGLAP equation their evolution to the new scale Q'^2. In the calculation of sufficiently exclusive quantities, infrared singularities due to massless gluons and quarks cancel.

Further reading. Two very useful references covering most aspects of perturbative QCD are Dissertori *et al.* (2009) and Ellis *et al.* (2003). Our presentation follows the one given in these references, where you can find also a discussion of coherence effects and the resulting angular-ordered cascade. Effective low-energy models for QCD as well as the trace anomaly are discussed by Donoghue *et al.* (2014).

Problems

18.1 Trace anomaly. Derive the trace anomaly (18.11) following the steps in the calculation of the chiral anomaly.

18.2 Polarised gluon splitting. Verify the results in table 18.2 for the function $F^{abc}(z)$; show that all other combinations are zero.

18.3 Splitting functions. Derive the remaining splitting functions shown in table 18.2.

18.4 General DGLAP equations. Generalise the DGLAP equation from one type of parton to the realistic case of gluons and n_f quark flavours.

18.5 Regularised splitting functions. Use momentum conservation, $\int_0^1 \mathrm{d}z\, z \sum_i P_{ji}(z) = 1$, to show that the regularised splitting functions $P_{qq}(z)$ is given by $P_{qq}(z) = \frac{4}{3}\frac{1+z^2}{1-z} + A\delta(1-z)$ and determine A.

18.6 Asymptotic limit of $f_i^p(x,t)$. Find the asymptotic momentum fraction of a nucleon carried by gluons and by quarks, respectively, for $t \to \infty$.

18.7 $e^+e^- \to \bar{q}gq$ in $d=4$. a.) Show that (18.60) agrees for $\varepsilon \to 0$ with the usual result in $d=4$. b.) Derive $\mathrm{d}\sigma/\mathrm{d}x_1\mathrm{d}x_2$ for the process $e^+e^- \to \bar{q}gq$; find the total cross-section as function of a chosen energy cut-off.

18.8 $e^+e^- \to hadrons$ ♥. Fill in the missing details in the calculation of the total annihilation cross-section $e^+e^- \to hadrons$ at $\mathcal{O}(\alpha_s)$.

18.9 Expansion of the Γ function. Find in Eq. (A.46) the $\mathcal{O}(\varepsilon^2)$ terms and compare to them Eq. (18.74).

19
Gravity as a gauge theory

In this chapter, we introduce the action and the field equations of gravity, proceeding in a way which stresses the similarity of gravity and Yang–Mills theories. In particular, we determine the coupling between matter and gravity promoting the invariance of matter fields under global Lorentz transformations to a local symmetry. As a bonus, this approach allows us to describe also the gravitational interactions of fermions as well as to understand how gravity selects the connection among the many mathematically possible ones on a Riemannian manifold. We also derive the linearised Einstein equations which can be used to describe the emission and propagation of weak gravitational waves.

19.1   Vielbein formalism and the spin connection

The equivalence principle postulates that in a small enough region around the centre of a freely falling coordinate system all physics is described by the laws of special relativity. Over greater distances, gravity manifests itself as curvature of spacetime. Thus physical laws involving only quantities transforming as tensors on Minkowski space are valid on a curved spacetime performing the replacement

$$\{\partial_\mu, \eta_{\mu\nu}, \mathrm{d}^4x\} \to \{\nabla_\mu, g_{\mu\nu}, \mathrm{d}^4x\sqrt{|g|}\}\,. \tag{19.1}$$

Here, the covariant derivative ∇_μ was defined using as connection the Christoffel symbols (or Levi–Civita connection) from Eq. (6.43). We recall that the two requirements $\nabla_\rho g_{\mu\nu} = 0$ ("metric connection") and $\Gamma^\alpha{}_{\beta\gamma} = \Gamma^\alpha{}_{\gamma\beta}$ ("torsionless connection") uniquely select this connection. In the following, we seek to understand whether these conditions are a consequence of Einstein gravity or necessary additional constraints. A useful framework to address these questions is the vielbein formalism which is also necessary when including fermions into the framework of general relativity.

Vielbein formalism.   We apply the equivalence principle as a physical guide line to obtain the physical laws including gravity. More precisely, we use the fact that we can find at any point P a local inertial frame in which the physical laws become those known from Minkowski space. We demonstrate this first for the case of a scalar field ϕ. The usual Lagrange density without gravity,

$$\mathscr{L} = \frac{1}{2}\partial_\mu\phi\partial^\mu\phi - V(\phi), \tag{19.2}$$

is still valid on a general manifold $\mathcal{M}(\{x^\mu\})$ if we use at each point P locally free-falling coordinates, $\xi^a(P)$. In order to distinguish these two sets of coordinates, we

Quantum Fields–From the Hubble to the Planck Scale. Michael Kachelriess. © Michael Kachelriess 2018.
Published in 2018 by Oxford University Press. DOI 10.1093/oso/9780198802877.001.0001

label inertial coordinates by Latin letters $a, b, \ldots$ while we keep Greek indices $\alpha, \beta, \ldots$ for arbitrary coordinates. We choose the locally free-falling coordinates ξ^a to be orthonormal. Thus in these coordinates the metric is given by $\mathrm{d}s^2 = \eta_{ab}\mathrm{d}\xi^a\mathrm{d}\xi^b$ with $\eta = \mathrm{diag}(1, -1, -1, -1)$. Then the action of a scalar field including gravity is

$$S[\phi] = \int \mathrm{d}^4\xi \left[\frac{1}{2}\eta_{ab}\partial^a\phi\partial^b\phi - V(\phi)\right] \tag{19.3}$$

with $\partial_a = \partial/\partial\xi^a$. This action looks formally exactly as the one without gravity, however we have to integrate over the manifold $\mathcal{M}(\{x^\mu\})$ and all effects of gravity are hidden in the dependence $\xi^a(x^\mu)$.

We introduce now the vielbein (or for $d = 4$ tetrad) fields e^a_α by

$$\mathrm{d}\xi^m = \frac{\partial\xi^m}{\partial x^\mu}\,\mathrm{d}x^\mu \equiv \mathrm{e}^m_\mu(x)\,\mathrm{d}x^\mu. \tag{19.4}$$

Thus we can view the vielbein $\mathrm{e}^m_\mu(x)$ both as the transformation matrix between arbitrary coordinates x and inertial coordinates ξ or as a set of four vectors in $T^*_x M$. In the absence of gravity, we can find in the whole manifold coordinates such that $\mathrm{e}^m_\mu(x) = \delta^m_\mu$. The inverse vielbein e^μ_m is defined analogously by

$$\mathrm{d}x^\mu = \frac{\partial x^\mu}{\partial\xi^m}\,\mathrm{d}\xi^m \equiv \mathrm{e}^\mu_m(x)\,\mathrm{d}\xi^m. \tag{19.5}$$

The name is justified by

$$\mathrm{d}\xi^m = \mathrm{e}^m_\mu\,\mathrm{d}x^\mu = \mathrm{e}^m_\mu \mathrm{e}^\mu_n\,\mathrm{d}\xi^n \tag{19.6}$$

and thus $\mathrm{e}^m_\mu \mathrm{e}^\mu_n = \delta^m_n$. We can view the vielbein as a kind of square-root of the metric tensor, since

$$\mathrm{d}s^2 = \eta_{mn}\mathrm{d}\xi^m\mathrm{d}\xi^n = \eta_{mn}\mathrm{e}^m_\mu \mathrm{e}^n_\nu \mathrm{d}x^\mu\mathrm{d}x^\nu = g_{\mu\nu}\mathrm{d}x^\mu\mathrm{d}x^\nu \tag{19.7}$$

and hence

$$g_{\mu\nu} = \eta_{mn}\mathrm{e}^m_\mu \mathrm{e}^n_\nu. \tag{19.8}$$

Taking the determinant, we see that the volume element is

$$\mathrm{d}^4\xi = \sqrt{|g|}\mathrm{d}^4 x = \det(\mathrm{e}^m_\mu)\mathrm{d}^4 x \equiv E\mathrm{d}^4 x. \tag{19.9}$$

We can construct mixed tensors, having both Latin and Greek indices. Then Latin indices are raised and lowered by the flat metric, while Greek indices are raised and lowered by the curved metric. For instance, we can rewrite the energy–momentum stress tensor as $T_{\mu\nu} = \mathrm{e}^m_\mu T_{m\nu} = \mathrm{e}^m_\mu \mathrm{e}^n_\nu T_{mn}$.

We have now all the ingredients needed to express the action (19.3) in arbitrary coordinates x^μ of the manifold $\mathcal{M}$. We first change the derivatives,

$$\mathscr{L} = \frac{1}{2}\eta_{mn}\mathrm{e}^m_\mu \mathrm{e}^n_\nu \partial^\mu\phi\partial^\nu\phi - V(\phi) = \frac{1}{2}g_{\mu\nu}\partial^\mu\phi\partial^\nu\phi - V(\phi), \tag{19.10}$$

and then the volume element in the action,

$$S[\phi] = \int \mathrm{d}^4 x \sqrt{|g|} \left[\frac{1}{2}g_{\mu\nu}\partial^\mu\phi\partial^\nu\phi - V(\phi)\right]. \tag{19.11}$$

As it should, we reproduced the usual action of a scalar field including gravity. Note that the sole effect of gravitational interactions is contained in the metric tensor and its determinant, while the connection plays no role since $\nabla_\mu \phi = \partial_\mu \phi$. Similarly, the connection drops out of the Lagrangian of a Yang–Mills field, since its field-strength tensor is antisymmetric.

Fermions and the spin connection. We now proceed to the spin-1/2 case. In this case, the simple substitution rule $\partial_\mu \to \nabla_\mu$ cannot be used, because the connections defined by Eq. (6.26) can be applied only to objects with tensorial indices. Applying instead the vielbein formalism, we recall first the Dirac Lagrangian without gravity,

$$\mathscr{L} = \bar\psi (\mathrm{i}\gamma^\mu \partial_\mu - m)\psi \tag{19.12}$$

where $\{\gamma^\mu, \gamma^\nu\} = 2\eta^{\mu\nu}$. Performing a Lorentz transformation, $\tilde x^\nu = \Lambda^\nu{}_\mu x^\mu$, the Dirac spinor ψ transforms as

$$\tilde\psi(\tilde x) = S(\Lambda)\psi(x) = \exp\left(-\frac{\mathrm{i}}{4}\omega^{\mu\nu}\sigma_{\mu\nu}\right)\psi(x) \tag{19.13}$$

with $\omega^{\mu\nu} = -\omega^{\nu\mu}$ as the six parameters and $\sigma^{\mu\nu} = \frac{\mathrm{i}}{2}[\gamma^\mu, \gamma^\nu]$ as the six infinitesimal generators of these transformations.

Switching on gravity, we replace $x^\mu \to \xi^m$ and $\gamma^\mu \to \gamma^m$. General covariance reduces now to the requirement that in any local inertial system we have to allow independent Lorentz transformations. In other words, we promote the invariance under global, spacetime-independent Lorentz transformations Λ to the invariance under local Lorentz transformations $\Lambda(x)$. This requirement allows us to derive the correct covariant derivative, in a manner completely analogous to the Yang–Mills case. We have to compensate the term introduced by the spacetime dependence of $S(\Lambda(\xi))$ in

$$\partial_a \psi \to \tilde\partial_a \tilde\psi(\tilde\xi) = \Lambda^b{}_a \partial_b [S(\xi)\psi(\xi)] \tag{19.14}$$

by introducing a "Latin" covariant derivative,

$$\nabla_a = \mathrm{e}_a^\alpha \left(\partial_\alpha + \mathrm{i}\omega_\alpha\right), \tag{19.15}$$

and requiring the inhomogeneous transformation law

$$\omega_\alpha \to \tilde\omega_\alpha = S\omega_\alpha S^\dagger - \mathrm{i}S\partial_\alpha S^\dagger \tag{19.16}$$

for ω. As a result, the covariant derivative transforms as

$$\nabla_a \to \tilde\nabla_a = \Lambda^b{}_a S \nabla_b S^\dagger \tag{19.17}$$

and the Dirac Lagrangian is invariant under local Lorentz transformations. The connection ω_α is a matrix in spinor space. Expanding it in the basis elements $\sigma^{\mu\nu}$, we find as a more explicit expression for the covariant derivative

$$\nabla_a = \partial_a + \frac{\mathrm{i}}{2}\omega_a^{\mu\nu}\sigma_{\mu\nu} = \mathrm{e}_a^\alpha \left(\partial_\alpha + \frac{\mathrm{i}}{2}\omega_\alpha^{\mu\nu}\sigma_{\mu\nu}\right) = \mathrm{e}_a^\alpha \left(\partial_\alpha + \omega_\alpha^{\mu\nu} J_{\mu\nu}\right). \tag{19.18}$$

In the last step we replaced the infinitesimal generators $\sigma^{\mu\nu}$ specific for the spinor representation by the general generators $J^{\mu\nu}$ of Lorentz transformations chosen appropriate for the representation the ∇_a act on. In this form, the covariant derivative can be applied to a field with arbitrary spin. The Lie algebra of the Lorentz group implies that the connection $\omega_a^{\mu\nu}$ is antisymmetric in its Greek indices, if they are both up or down, $\omega_a^{\mu\nu} = -\omega_a^{\nu\mu}$.

The transformation law (19.16) of the spin connection ω_a under a Lorentz transformation S is completely analogous to the transformation properties (10.16) of a Yang–Mills field A^μ under a gauge transformation U. One should keep in mind however two important differences. First, a vector lives in a tangent space which is naturally associated with a manifold. In particular, we can associate a vector in $T_P M$ with a trajectory $x^\mu(\sigma)$ through P. Therefore we have the natural coordinate basis ∂_μ in $T_P M$ and can introduce vielbein fields in a second step. In contrast, matter fields $\psi(x)$ live in their group manifold which is attached arbitrarily at each point of the manifold, and the gauge fields act as a connection telling us how we should transport $\psi(x)$ to $\tilde\psi(x')$. Second, we associate physical particles with spin s to irreducible representations of the Poincaré group. Thus we identify fluctuations of the gauge field A_μ as the quanta of the vector field, while we associate in the case of gravity not the fluctuations of the connection but of the metric tensor $g_{\mu\nu}$ with particles.

Transition to the standard notation. We now establish the connection between the vielbein and the standard formalism, using the fact that for tensor fields the two formalisms have to agree. Inserting into the definition of the covariant derivative for the components A^a of a vector $\boldsymbol{A}$,

$$\nabla_\mu A^a = \partial_\mu A^a + \omega_{\mu\ b}^{\ a} A^b, \tag{19.19}$$

the decomposition $A^a = e_\nu^a A^\nu$ and requiring the validity of the Leibniz rule gives

$$\nabla_\mu A^a = (\nabla_\mu e_\nu^a) A^\nu + e_\nu^a (\nabla_\mu A^\nu) \tag{19.20a}$$

$$= (\nabla_\mu e_\nu^a) A^\nu + e_\nu^a \left(\partial_\mu A^\nu + \Gamma^\nu_{\ \mu\lambda} A^\lambda \right). \tag{19.20b}$$

Using

$$\partial_\mu A^a = \partial_\mu (e_\nu^a A^\nu) = e_\nu^a (\partial_\mu A^\nu) + (\partial_\mu e_\nu^a) A^\nu \tag{19.21}$$

to eliminate the second term in (19.20b), we obtain

$$\nabla_\mu A^a = (\nabla_\mu e_\nu^a) A^\nu + \partial_\mu A^a - (\partial_\mu e_\nu^a) A^\nu + e_\nu^a \Gamma^\nu_{\ \mu\lambda} A^\lambda. \tag{19.22}$$

Comparing this to Eq. (19.19) we can read off how the covariant derivative acts on an object with mixed indices,

$$\nabla_\mu e_\nu^a = \partial_\mu e_\nu^a - \Gamma^\lambda_{\ \nu\mu} e_\lambda^a + \omega_{\mu\ b}^{\ a} e_\nu^b. \tag{19.23}$$

More generally, covariant indices are contracted with the usual connection $\Gamma^\lambda_{\ \nu\mu}$ while vielbein indices are contracted with $\omega_{\mu\ b}^{\ a}$. In a moment, we will show that the covariant derivative of the vielbein field is zero, $\nabla_\mu e_\nu^a = 0$. Sometimes this property is called "tetrade postulate", but in fact it follows naturally from the definition of the vielbein field.

In order to derive an explicit formula for the spin connection $\omega_{\mu}{}^{a}{}_{b}$ we now compare the covariant derivative of a vector in the two formalisms. First, we write in a coordinate basis

$$\boldsymbol{\nabla A} = (\nabla_{\mu} A^{\nu}) \mathrm{d}x^{\mu} \otimes \partial_{\nu} = (\partial_{\mu} A^{\nu} + \Gamma^{\nu}{}_{\mu\lambda} A^{\lambda}) \mathrm{d}x^{\mu} \otimes \partial_{\nu}. \tag{19.24}$$

Next we compare this expression to the one using a mixed basis,

$$\boldsymbol{\nabla A} = (\nabla_{\mu} A^{m}) \, \mathrm{d}x^{\mu} \otimes e_{m} = (\partial_{\mu} A^{m} + \omega_{\mu}{}^{m}{}_{n} A^{n}) \, \mathrm{d}x^{\mu} \otimes e_{m} \tag{19.25a}$$

$$= [\partial_{\mu}(e_{\nu}^{m} A^{\nu}) + \omega_{\mu}{}^{m}{}_{n} e_{\lambda}^{n} A^{\lambda}] \, \mathrm{d}x^{\mu} \otimes (e_{m}^{\sigma} \partial_{\sigma}). \tag{19.25b}$$

Moving e_{m}^{σ} to the left and using the Leibniz rule as well as $e_{m}^{\sigma} e_{\nu}^{m} = \delta_{\nu}^{\sigma}$, it follows

$$\boldsymbol{\nabla A} = e_{m}^{\sigma}[e_{\nu}^{m} \partial_{\mu} A^{\nu} + A^{\nu} \partial_{\mu} e_{\nu}^{m} + \omega_{\mu}{}^{m}{}_{n} e_{\lambda}^{n} A^{\lambda}] \, \mathrm{d}x^{\mu} \otimes \partial_{\sigma} \tag{19.26a}$$

$$= [\partial_{\mu} A^{\sigma} + e_{m}^{\sigma}(\partial_{\mu} e_{\nu}^{m}) A^{\nu} + e_{m}^{\sigma} e_{\lambda}^{n} \omega_{\mu}{}^{m}{}_{n} A^{\lambda}] \, \mathrm{d}x^{\mu} \otimes \partial_{\sigma}. \tag{19.26b}$$

Relabelling indices and comparing to Eq. (19.24), we find as relation between the two connections

$$\Gamma^{\nu}{}_{\mu\lambda} = e_{m}^{\nu} \partial_{\mu} e_{\lambda}^{m} + e_{m}^{\nu} e_{\lambda}^{n} \omega_{\mu}{}^{m}{}_{n}. \tag{19.27}$$

Clearly, the connection $\Gamma^{\nu}{}_{\mu\lambda}$ is in general not symmetric, $\Gamma^{\nu}{}_{\mu\lambda} \neq \Gamma^{\nu}{}_{\lambda\mu}$. Now we are also in the position to show that the covariant derivative of the vielbein field vanishes. Multiplying Eq. (19.27) with two vielbein fields, we arrive first at

$$\omega_{\mu}{}^{m}{}_{n} = e_{\nu}^{m} e_{n}^{\lambda} \Gamma^{\nu}{}_{\mu\lambda} - e_{n}^{\lambda} \partial_{\mu} e_{\lambda}^{m}. \tag{19.28}$$

Multiplying once again with a vielbein field, we obtain

$$e_{\nu}^{n} \omega_{\mu}{}^{m}{}_{n} = e_{\nu}^{n}(e_{\sigma}^{m} e_{n}^{\lambda} \Gamma^{\sigma}{}_{\mu\lambda} - e_{n}^{\lambda} \partial_{\mu} e_{\lambda}^{m}) \tag{19.29a}$$

$$= e_{\sigma}^{m} \delta_{\nu}^{\lambda} \Gamma^{\sigma}{}_{\mu\lambda} - \delta_{\nu}^{\lambda} \partial_{\mu} e_{\lambda}^{m} = e_{\sigma}^{m} \Gamma^{\sigma}{}_{\mu\nu} - \partial_{\mu} e_{\nu}^{m}. \tag{19.29b}$$

Inserted into Eq. (19.23), the "tetrade postulate" $\nabla_{\mu} e_{\nu}^{a} = 0$ follows.

19.2   Action of gravity

Einstein–Hilbert action.   The analogy between the Yang–Mills field-strength and the Riemann tensor shown by Eq. (10.49) suggests as Lagrange density for the gravitational field

$$\mathscr{L} = -\sqrt{|g|} R_{\mu\nu\sigma\tau} R^{\mu\nu\sigma\tau}. \tag{19.30}$$

This Lagrange density has mass dimension four and would thus lead to a dimensionless gravitational coupling constant and a renormalisable theory of gravity. However, such a theory would be in contradiction to Newton's law. Hilbert chose instead the curvature scalar which has the required mass dimension $d = 6$,

$$\mathscr{L}_{\mathrm{EH}} = -\sqrt{|g|} R. \tag{19.31}$$

As we know, we can always add a constant term to the Lagrangian, $R \to R + 2\Lambda$. Such a term would imply that classically empty space has a constant vacuum energy which gravitates. The Lagrangian is a function of the metric, its first and second derivatives,[1] $\mathscr{L}_{\mathrm{EH}}(g_{\mu\nu}, \partial_\rho g_{\mu\nu}, \partial_\rho \partial_\sigma g_{\mu\nu})$. The resulting action

$$S_{\mathrm{EH}}[g_{\mu\nu}] = -\int_\Omega \mathrm{d}^4 x \sqrt{|g|}\, \{R + 2\Lambda\} \tag{19.32}$$

is a functional of the metric tensor $g_{\mu\nu}$, and a variation of the action with respect to the metric gives the field equations for the gravitational field. If we consider gravity coupled to fermions, we have to use the spin connection ω_μ in the matter Lagrangian as well as expressing $\sqrt{|g|}$ and R through Latin quantities, $\sqrt{|g|} \to E$ and $R = R_{\mu\nu} g^{\mu\nu} \to R_{mn} \eta^{mn}$.

We derive the resulting field equations for the metric tensor $g_{\mu\nu}$ directly from the action principle

$$\delta S_{\mathrm{EH}} = -\delta \int_\Omega \mathrm{d}^4 x \sqrt{|g|}(R + 2\Lambda) = -\delta \int_\Omega \mathrm{d}^4 x \sqrt{|g|}\, (g^{\mu\nu} R_{\mu\nu} + 2\Lambda) = 0. \tag{19.33}$$

We allow for variations of the metric $g_{\mu\nu}$ restricted by the condition that the variation of $g_{\mu\nu}$ and its first derivatives vanish on the boundary $\partial\Omega$,

$$\delta S_{\mathrm{EH}} = -\int_\Omega \mathrm{d}^4 x \left\{ \sqrt{|g|}\, g^{\mu\nu} \delta R_{\mu\nu} + \sqrt{|g|} R_{\mu\nu} \delta g^{\mu\nu} + (R + 2\Lambda)\, \delta\sqrt{|g|} \right\}. \tag{19.34}$$

Our task is to rewrite the first and third term as variations of $\delta g^{\mu\nu}$ or to show that they are equivalent to boundary terms. Let us start with the first term. Choosing inertial coordinates, the Ricci tensor at the considered point P becomes

$$R_{\mu\nu} = \partial_\rho \Gamma^\rho{}_{\mu\nu} - \partial_\nu \Gamma^\rho{}_{\mu\rho}. \tag{19.35}$$

Hence

$$g^{\mu\nu} \delta R_{\mu\nu} = g^{\mu\nu}(\partial_\rho \delta\Gamma^\rho{}_{\mu\nu} - \partial_\nu \delta\Gamma^\rho{}_{\mu\rho}) = g^{\mu\nu} \partial_\rho \delta\Gamma^\rho{}_{\mu\nu} - g^{\mu\rho} \partial_\rho \delta\Gamma^\nu{}_{\mu\nu}, \tag{19.36}$$

where we exchanged the indices ν and ρ in the last term. Since $\partial_\rho g_{\mu\nu} = 0$ at P, we can rewrite the expression as

$$g^{\mu\nu} \delta R_{\mu\nu} = \partial_\rho(g^{\mu\nu} \delta\Gamma^\rho{}_{\mu\nu} - g^{\mu\rho} \delta\Gamma^\nu{}_{\mu\nu}) = \partial_\rho X^\rho. \tag{19.37}$$

The quantity X^ρ is a vector, since the difference of two connection coefficients transforms as a tensor. Replacing in Eq. (19.37) the partial derivative by a covariant one promotes it therefore in a valid tensor equation,

$$g^{\mu\nu} \delta R_{\mu\nu} = \nabla_\mu V^\mu = \frac{1}{\sqrt{|g|}} \partial_\mu(\sqrt{|g|} V^\mu). \tag{19.38}$$

Thus this term corresponds to a surface term which we assume to vanish. Next we rewrite the third term using

---

[1]Recall that the Lagrange equations are modified in the case of higher derivatives which is one reason why we directly vary the action in order to obtain the field equations.

$$\delta\sqrt{|g|} = \frac{1}{2\sqrt{|g|}}\,\delta|g| = \frac{1}{2}\sqrt{|g|}\,g^{\mu\nu}\delta g_{\mu\nu} = -\frac{1}{2}\sqrt{|g|}\,g_{\mu\nu}\delta g^{\mu\nu} \qquad (19.39)$$

and obtain

$$\delta S_{\text{EH}} = -\int_\Omega \mathrm{d}^4x\sqrt{|g|}\left\{R_{\mu\nu} - \frac{1}{2}\,g_{\mu\nu}R - \Lambda\,g_{\mu\nu}\right\}\delta g^{\mu\nu} = 0. \qquad (19.40)$$

Hence the metric fulfils in vacuum the equation

$$-\frac{1}{\sqrt{|g|}}\frac{\delta S_{\text{EH}}}{\delta g^{\mu\nu}} = R_{\mu\nu} - \frac{1}{2}\,R\,g_{\mu\nu} - \Lambda g_{\mu\nu} \equiv G_{\mu\nu} - \Lambda g_{\mu\nu} = 0, \qquad (19.41)$$

where we introduced the Einstein tensor $G_{\mu\nu}$. The constant Λ is called the cosmological constant.

We consider now the combined action of gravity and matter, as the sum of the Einstein–Hilbert Lagrangian $\mathscr{L}_{\text{EH}}/2\kappa$ and the Lagrangian $\mathscr{L}_{\text{m}}$ including all relevant matter fields,

$$\mathscr{L} = \frac{1}{2\kappa}\mathscr{L}_{\text{EH}} + \mathscr{L}_{\text{m}} = -\frac{1}{2\kappa}\sqrt{|g|}(R + 2\Lambda) + \mathscr{L}_{\text{m}}. \qquad (19.42)$$

We will determine the value of the coupling constant κ in the next section, demanding that we reproduce Newtonian dynamics in the weak-field limit. We have already argued that the source of the gravitational field is the energy–momentum stress tensor which lead to the definition

$$\frac{2}{\sqrt{|g|}}\frac{\delta S_{\text{m}}}{\delta g^{\mu\nu}} = T_{\mu\nu}. \qquad (19.43)$$

Einstein's field equation follows then as

$$G_{\mu\nu} - \Lambda g_{\mu\nu} = \kappa T_{\mu\nu}. \qquad (19.44)$$

Two remarks are in order. First, the cosmological constant Λ measures the curvature of an empty classical spacetime. Moving Λ to the RHS of the Einstein equations, we see that its effect is equivalent to the stress density $T_{\mu\nu} = \kappa\Lambda g_{\mu\nu}$ or a vacuum energy density $\rho_\Lambda = \kappa\Lambda$ with EoS $w = P_\Lambda/\rho_\Lambda = -1$, see problem 19.3. Second, we assumed that the boundary term arising from $g^{\mu\nu}\delta R_{\mu\nu} = \nabla_\mu V^\mu$ vanishes, which corresponds to a zero flux of V^μ through $\partial\Omega$. In general, this flux is zero only for a compact manifold Ω (since then $\partial\Omega = 0$) or if we require that not only $\delta g_{\mu\nu}$ but also $\delta\Gamma^\kappa_{\ \mu\nu}$ vanishes on the boundary $\partial\Omega$. The second option is naturally implemented in the Palatini action principle which we consider next. Using the Einstein–Hilbert form of the action, one should add instead a boundary term which cancels the variation of $g^{\mu\nu}\delta R_{\mu\nu} = \nabla_\mu V^\mu$, for details see Poisson (2007).

Palatini action. We start from the Einstein–Hilbert Lagrangian (19.31), but consider it now as a function of the metric tensor, the connection and its first derivatives, $\mathscr{L}_{\text{EH}}(g_{\mu\nu}, \Gamma^\sigma_{\ \mu\nu}, \partial_\tau\Gamma^\sigma_{\ \mu\nu})$, while we allow an independent variation of the metric tensor and the connection in the action. We obtain the desired dependence of the Lagrangian expressing the Ricci tensor through the connection and its derivatives,

$$\mathscr{L}_{\text{EH}} = -\sqrt{|g|}g^{\mu\nu}R_{\mu\nu} = \sqrt{|g|}g^{\mu\nu}\left(\partial_\nu\Gamma^\sigma{}_{\mu\sigma} - \partial_\sigma\Gamma^\sigma{}_{\mu\nu} + \Gamma^\tau{}_{\mu\sigma}\Gamma^\sigma{}_{\tau\nu} - \Gamma^\tau{}_{\mu\nu}\Gamma^\sigma{}_{\tau\sigma}\right). \tag{19.45}$$

For simplicity, we set also $\Lambda = 0$. Then the variation with respect to the metric,

$$\delta_g S_{\text{EH}} = -\int_\Omega d^4x\, \delta\left\{\sqrt{|g|}\,g^{\mu\nu}\right\}R_{\mu\nu} = 0 \tag{19.46}$$

gives $R_{\mu\nu} = 0$. As we will show in (19.66), this condition is equivalent with the usual Einstein equations in vacuum. For the variation with respect to the connection we use first the Palatini equation,

$$\delta_\Gamma S_{\text{EH}} = -\int_\Omega d^4x\sqrt{|g|}\,g^{\mu\nu}\delta R_{\mu\nu} = -\int_\Omega d^4x\sqrt{|g|}\,g^{\mu\nu}\left[\nabla_\nu(\delta\Gamma^\sigma{}_{\mu\sigma}) - \nabla_\sigma(\delta\Gamma^\sigma{}_{\mu\nu})\right]. \tag{19.47}$$

Applying then the Leibniz rule and relabelling some indices, we find

$$\delta_\Gamma S_{\text{EH}} = -\int_\Omega d^4x\sqrt{|g|}\nabla_\nu\left[g^{\mu\nu}\delta\Gamma^\sigma{}_{\mu\sigma} - g^{\mu\sigma}\delta\Gamma^\sigma{}_{\mu\sigma}\right]$$
$$-\int_\Omega d^4x\sqrt{|g|}\left[(\nabla_\rho g^{\mu\nu})\delta\Gamma^\rho{}_{\mu\nu} - (\nabla_\nu g^{\mu\nu})\delta\Gamma^\rho{}_{\mu\rho}\right]. \tag{19.48}$$

We kept the second line, because we we do not know yet if the covariant derivative of the metric vanishes for an arbitrary connection. Next we perform a partial integration of the first two terms, converting it into a surface term which we can drop. In the remaining part we relabel indices so that we can factor out the variation of the connection,

$$\delta_\Gamma S_{\text{EH}} = \int_\Omega d^4x\sqrt{|g|}\left[\delta^\nu_\rho\nabla_\sigma g^{\mu\sigma} - \nabla_\rho g^{\mu\nu}\right]\delta\Gamma^\rho{}_{\mu\nu}. \tag{19.49}$$

We now use the fact that the connection is symmetric in the absence of fermion. Then the variation $\delta\Gamma^\rho{}_{\mu\nu}$ is also symmetric and the antisymmetric part in the square bracket drops out. Asking that $\delta_\Gamma S_{\text{EH}} = 0$ gives therefore

$$\frac{1}{2}\delta^\nu_\rho\nabla_\sigma g^{\mu\sigma} + \frac{1}{2}\delta^\mu_\rho\nabla_\sigma g^{\nu\sigma} - \nabla_\rho g^{\mu\nu} = 0 \tag{19.50}$$

or $\nabla_\sigma g^{\mu\nu} = \nabla_\sigma g_{\mu\nu} = 0$. Thus the Einstein–Hilbert action *implies* the metric compatibility of the connection.

Performing the same exercise with the Einstein–Hilbert plus the matter action considered as functional of the vielbein e^μ_m and the connection ω_μ, one finds the following: from the variation $\delta_\omega S_{\text{EH}}$ one obtains automatically a metric connection which is however in general not symmetric. The torsion is sourced by the spin density of fermions. The variation $\delta_e S_{\text{EH}}$ gives the usual Einstein equation. This result justifies the usual choice of a torsionless connection which is metric compatible. Although, for example, a star consists of a collection of individual particles carrying spin s_i, its total spin sums up to zero, $\sum_i s_i \simeq 0$, because they are uncorrelated. Thus we can describe macroscopic matter in general relativity as a classical spinless[2] point particle (or a fluid if its extension is important) leading to a symmetric connection.

19.3 Linearised gravity

We are looking for small perturbations $h_{\mu\nu}$ around the Minkowski metric $\eta_{\mu\nu}$,

$$g_{\mu\nu} = \eta_{\mu\nu} + h_{\mu\nu}, \qquad h_{\mu\nu} \ll 1. \tag{19.51}$$

These perturbations may be caused either by the propagation of gravitational waves or by the gravitational potential of a star. In the first case, experimental limits showed that one should not hope for h larger than $\mathcal{O}(h) \sim 10^{-20}$. Keeping only terms linear in h is therefore an excellent approximation. Choosing in the second case as application, for example, the spiral-in of a binary system, deviations from the Newtonian limit can become arbitrarily large. Hence one needs a systematic "post-Newtonian" expansion or has to perform a full numerical analysis to describe properly such cases.

Linearised Einstein equations in vacuum. From $\partial_\mu \eta_{\nu\rho} = 0$ and the definition

$$\Gamma^\mu{}_{\nu\lambda} = \frac{1}{2} g^{\mu\kappa}(\partial_\nu g_{\kappa\lambda} + \partial_\lambda g_{\nu\kappa} - \partial_\kappa g_{\nu\lambda}) \tag{19.52}$$

we find for the change of the connection linear in h

$$\delta\Gamma^\mu{}_{\nu\lambda} = \frac{1}{2}\eta^{\mu\kappa}(\partial_\nu h_{\kappa\lambda} + \partial_\lambda h_{\nu\kappa} - \partial_\kappa h_{\nu\lambda}) = \frac{1}{2}(\partial_\nu h^\mu_\lambda + \partial_\lambda h^\mu_\nu - \partial^\mu h_{\nu\lambda}). \tag{19.53}$$

Here we used η to raise indices which is allowed in linear approximation. Remembering the definition of the Riemann tensor,

$$R^\mu{}_{\nu\lambda\kappa} = \partial_\lambda\Gamma^\mu{}_{\nu\kappa} - \partial_\kappa\Gamma^\mu{}_{\nu\lambda} + \Gamma^\mu{}_{\rho\lambda}\Gamma^\rho{}_{\nu\kappa} - \Gamma^\mu{}_{\rho\kappa}\Gamma^\rho{}_{\nu\lambda}, \tag{19.54}$$

we see that we can neglect the terms quadratic in the connection terms. Thus we find for the change

$$\begin{aligned}
\delta R^\mu{}_{\nu\lambda\kappa} &= \partial_\lambda\delta\Gamma^\mu{}_{\nu\kappa} - \partial_\kappa\delta\Gamma^\mu{}_{\nu\lambda} \\
&= \frac{1}{2}\{\partial_\lambda\partial_\nu h^\mu_\kappa + \partial_\lambda\partial_\kappa h^\mu_\nu - \partial_\lambda\partial^\mu h_{\nu\kappa} - (\partial_\kappa\partial_\nu h^\mu_\lambda + \partial_\kappa\partial_\lambda h^\mu_\nu - \partial_\kappa\partial^\mu h_{\nu\lambda})\} \\
&= \frac{1}{2}\{\partial_\lambda\partial_\nu h^\mu_\kappa + \partial_\kappa\partial^\mu h_{\nu\lambda} - \partial_\lambda\partial^\mu h_{\nu\kappa} - \partial_\kappa\partial_\nu h^\mu_\lambda\}.
\end{aligned} \tag{19.55}$$

The change in the Ricci tensor follows by contracting μ and λ,

$$\delta R^\lambda{}_{\nu\lambda\kappa} = \frac{1}{2}\{\partial_\lambda\partial_\nu h^\lambda_\kappa + \partial_\kappa\partial^\lambda h_{\nu\lambda}) - \partial_\lambda\partial^\lambda h_{\nu\kappa} - \partial_\kappa\partial_\nu h^\lambda_\lambda\}. \tag{19.56}$$

Next we introduce $h \equiv h^\mu_\mu$, $\Box = \partial_\mu\partial^\mu$, and relabel the indices,

---

[2]The curious reader may wonder if orbital angular momentum leads to torsion. One way to see that the answer is no is to realise that one cannot define an orbital angular momentum *density* which transforms properly as a tensor, cf. Eq. (5.26f).

$$\delta R_{\mu\nu} = \frac{1}{2} \left\{ \partial_\mu \partial_\rho h_\nu^\rho + \partial_\nu \partial_\rho h_\mu^\rho - \Box h_{\mu\nu} - \partial_\mu \partial_\nu h \right\}. \tag{19.57}$$

We now rewrite all terms apart from $\Box h_{\mu\nu}$ as derivatives of the vector

$$\xi_\mu = \partial_\nu h_\mu^\nu - \frac{1}{2} \partial_\mu h, \tag{19.58}$$

obtaining

$$\delta R_{\mu\nu} = \frac{1}{2} \left\{ -\Box h_{\mu\nu} + \partial_\mu \xi_\nu + \partial_\nu \xi_\mu \right\}. \tag{19.59}$$

Looking back at the properties of $h_{\mu\nu}$ under gauge transformations, Eq. (7.40), we see that we can gauge away the second and third term. Thus the linearised Einstein equation in vacuum, $\delta R_{\mu\nu} = 0$, becomes simply

$$\Box h_{\mu\nu} = 0, \tag{19.60}$$

if the harmonic gauge

$$\xi_\mu = \partial_\nu h_\mu^\nu - \frac{1}{2} \partial_\mu h = 0 \tag{19.61}$$

is chosen. Hence the familiar wave equation holds for all ten independent components of $h_{\mu\nu}$, and the perturbations propagate with the speed of light. Inserting plane waves $h_{\mu\nu} = \varepsilon_{\mu\nu} \exp(-ikx)$ into the wave equation, one finds immediately that k is a null vector.

TT gauge. We want to re-derive our old result for the polarisation tensor describing the physical states contained in a gravitational perturbation. We consider a plane wave $h_{\mu\nu} = \varepsilon_{\mu\nu} \exp(-ikx)$. After choosing the harmonic gauge (19.61), we can still perform a gauge transformation using four functions ξ_μ satisfying $\Box \xi_\mu = 0$. We can choose them such that four components of $h_{\mu\nu}$ vanish. In the TT gauge, we set $(i = 1, 2, 3)$

$$h_{0i} = 0, \qquad h = 0. \tag{19.62}$$

The harmonic gauge condition becomes $\xi_\alpha = \partial_\beta h_\alpha^\beta$ or

$$\xi_0 = \partial_\beta h_0^\beta = \partial_0 h_0^0 = -i\omega \varepsilon_{00} e^{-ikx} = 0, \tag{19.63a}$$

$$\xi_a = \partial_\beta h_a^\beta = \partial_b h_a^b = ik^b \varepsilon_{ab} e^{ikx} = 0. \tag{19.63b}$$

Thus $\varepsilon_{00} = 0$ and the polarisation tensor is transverse, $k^b \varepsilon_{ab} = 0$. If we choose the plane wave propagating in z direction, $\boldsymbol{k} = k\boldsymbol{e}_z$, the z raw and column of the polarisation tensor vanishes too. Accounting for $h = 0$ and $\varepsilon_{\alpha\beta} = \varepsilon_{\beta\alpha}$, only two independent elements are left, and we recover our old result,

$$\varepsilon_{\alpha\beta} = \begin{pmatrix} 0 & 0 & 0 & 0 \\ 0 & \varepsilon_{11} & \varepsilon_{12} & 0 \\ 0 & \varepsilon_{12} & -\varepsilon_{11} & 0 \\ 0 & 0 & 0 & 0 \end{pmatrix}. \tag{19.64}$$

In general, one can construct the polarisation tensor in TT gauge by first setting the non-transverse part to zero and then subtracting the trace. The resulting two independent elements are (again for $\boldsymbol{k} = k\boldsymbol{e}_z$) then $\varepsilon_{11} = 1/2(\varepsilon_{xx} - \varepsilon_{yy})$ and ε_{12}.

Linearised Einstein equations with sources.   We rewrite first the Einstein equation in an alternative form where the only geometrical term on the LHS is the Ricci tensor. Because of

$$R_\mu^\mu - \frac{1}{2}\delta_\mu^\mu(R + 2\Lambda) = R - 2(R + 2\Lambda) = -R - 4\Lambda = \kappa T_\mu^\mu, \tag{19.65}$$

we can perform with $T \equiv T_\mu^\mu$ the replacement $R = -\kappa T - 4\Lambda$ in the Einstein equation and obtain

$$R_{\mu\nu} = \kappa(T_{\mu\nu} - \frac{1}{2}g_{\mu\nu}T) - g_{\mu\nu}\Lambda. \tag{19.66}$$

This form of the Einstein equations is often useful, when it is easier to calculate T that R. Note also that (19.66) informs us that an empty universe with $\Lambda = 0$ has a vanishing Ricci tensor $R_{\mu\nu} = 0$.

Now we move on to the determination of the linearised Einstein equations with sources. We found $2\delta R_{\mu\nu} = -\Box h_{\mu\nu}$. By contraction follows $2\delta R = -\Box h$. Combining both terms gives

$$\Box\left(h_{\mu\nu} - \frac{1}{2}\eta_{\mu\nu}h\right) = -2(\delta R_{\mu\nu} - \frac{1}{2}\eta_{\mu\nu}\delta R) = -2\kappa\delta T_{\mu\nu}. \tag{19.67}$$

Since we assumed an empty universe in zeroth order, $\delta T_{\mu\nu}$ is the complete contribution to the stress tensor. We omit therefore in the following the δ in $\delta T_{\mu\nu}$. Next we introduce as useful short-hand notation the "trace-reversed" amplitude as

$$\bar{h}_{\mu\nu} \equiv h_{\mu\nu} - \frac{1}{2}\eta_{\mu\nu}h\,. \tag{19.68}$$

The harmonic gauge condition becomes then

$$\partial_\mu\bar{h}_{\mu\nu} = 0 \tag{19.69}$$

and the linearised Einstein equations in the harmonic gauge follow as

$$\Box\bar{h}_{\mu\nu} = -2\kappa T_{\mu\nu}. \tag{19.70}$$

Because of $\bar{\bar{h}}_{\mu\nu} = h_{\mu\nu}$ and Eq. (19.66), we can rewrite this wave equation also as

$$\Box h_{\mu\nu} = -2\kappa\bar{T}_{\mu\nu} \tag{19.71}$$

with the trace-reversed stress tensor $\bar{T}_{\mu\nu} \equiv T_{\mu\nu} - \frac{1}{2}\eta_{\mu\nu}T$.

Newtonian limit.   The Newtonian limit corresponds to $v/c \to 0$ and thus the only non-zero element of the stress tensor becomes $T^{00} = \rho$. We will show later that the metric around a point mass can be written in the weak-field limit as

$$ds^2 = (1 + 2\Phi)dt^2 - (1 - 2\Phi)\left(dx^2 + dy^2 + dz^2\right) \tag{19.72}$$

with Φ as the Newtonian gravitational potential. Comparing this metric to Eq. (19.51), we find as metric perturbations

$$h_{00} = 2\Phi \qquad h_{ij} = 2\delta_{ij}\Phi \qquad h_{0i} = 0 \,. \tag{19.73}$$

In the static limit $\Box \to -\Delta$ and $v = 0$, and thus

$$-\Delta\left(h_{00} - \frac{1}{2}\eta_{00}h\right) = -4\Delta\Phi = -2\kappa\rho \,. \tag{19.74}$$

Hence the linearised Einstein equation has the same form as the Newtonian Poisson equation, and the constant κ equals $\kappa = 8\pi G$.

Detection principle of gravitational waves. Let us consider the effect of a gravitational wave on a free test particle that is initially at rest, $u^{\alpha} = (1,0,0,0)$. Then the geodesic equation simplifies to $\dot{u}^{\alpha} = \Gamma^{\alpha}{}_{00}$. The four relevant Christoffel symbols are in the linearised approximation, cf. Eq. (19.53),

$$\Gamma^{\alpha}{}_{00} = \frac{1}{2}(\partial_0 h_0^{\alpha} + \partial_0 h_0^{\alpha} - \partial^{\alpha}h_{00}) \,. \tag{19.75}$$

We are free to choose the TT gauge in which all component of $h_{\alpha\beta}$ appearing on the RHS are zero. Hence the acceleration of the test particle is zero and its coordinate position is unaffected by the gravitational wave: the TT gauge defines a comoving coordinate system.

The physical distance l between two test particles is given by integrating

$$\mathrm{d}l^2 = g_{ab}\mathrm{d}\xi^a\mathrm{d}\xi^b = (h_{ab} - \delta_{ab})\mathrm{d}\xi^a\mathrm{d}\xi^b \,, \tag{19.76}$$

where g_{ab} is the spatial part of the metric and $\mathrm{d}\xi$ the spatial coordinate distance between infinitesimal separated test particles. Hence the passage of a gravitational wave, $h_{\alpha\beta} \propto \varepsilon_{\alpha\beta}\cos(\omega t)$, results in a periodic change of the separation of freely moving test particles. Figure 19.1 shows that a gravitational wave exerts tidal forces, stretching and squashing test particles in the transverse plane. The relative size of the change, $\Delta L/L$, is given by the amplitude h of the gravitational wave. It is this tiny periodic change, $\Delta L/L \lesssim 10^{-21}\cos(\omega t)$, which gravitational wave experiments aim to detect. There are two basic types of gravitational wave experiments. In the first, one uses the fact that the tidal forces of a passing gravitational wave excite lattice vibrations in a solid state. If the wave frequency is resonant with a lattice mode, the vibrations might be amplified to detectable levels. In the second type of experiment, the free test particles are replaced by mirrors. Between the mirrors, a laser beam is reflected multipe times, thereby increasing the effective length L and thus ΔL, before two beams at 90° are brought to interference. As the most promising gravitational wave source the inspiral of binary systems composed of neutron stars or black holes has been suggested. In September 2015, the Advanced Laser Interferometer Gravitational-Wave Observatory (Advanced LIGO) detected such a signal for the first time (Abbott *et al.*, 2016; Coleman Miller, 2016). Additionally, a stochastic background of gravitational waves might be produced during inflation and phase transitions in the early universe.

Linearised action of gravity. We have derived the linearised wave equation describing gravitational waves directly from the Einstein equation. Now we want to obtain the linearised action of gravity. A straightforward but lengthy approach would

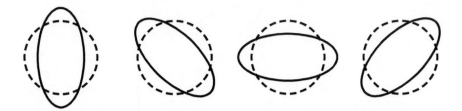

Fig. 19.1 The effect of a right-handed polarised gravitational wave on a ring of transverse test particles as function of time; the dashed line shows the state without gravitational wave.

be to expand the Einstein–Hilbert action to $\mathcal{O}(h^3)$. Instead, we profit from our knowledge of the graviton propagator, which we derived in chapter 7, cf. with Eq. (7.45),

$$D_F^{\mu\nu;\rho\sigma}(k) = \frac{\frac{1}{2}(-\eta^{\mu\nu}\eta^{\rho\sigma} + \eta^{\mu\rho}\eta^{\nu\sigma} + \eta^{\mu\sigma}\eta^{\nu\rho})}{k^2 + i\varepsilon} = \frac{P^{\mu\nu;\rho\sigma}(k)}{k^2 + i\varepsilon}. \tag{19.77}$$

The corresponding Lagrangian is as usually quadratic in the fields, $\frac{1}{2}h^{\mu\nu}P_{\mu\nu;\rho\sigma}\Box h^{\rho\sigma}$. Performing partial integrations and the contractions with the metric tensors gives us as the corresponding action

$$S = -\frac{1}{32\pi G}\int \mathrm{d}^4x \left[\frac{1}{2}(\partial_\alpha h_{\beta\gamma})^2 - \frac{1}{4}(\partial_\alpha h)^2\right]. \tag{19.78}$$

We know that the propagator of a massless particles with helicity $h > 1/2$ can be inverted only, if either the gauge freedom is completely fixed or a gauge-fixing term is added. In contrast to the TT gauge, the harmonic gauge contains still unphysical degrees of freedom. Thus the expression (19.78) equals the quadratic Einstein–Hilbert Lagrangian in the harmonic gauge plus a gauge-fixing term. The propagator derived in Eq. (7.45) corresponds to the R_ξ-gauge with $\xi = 1$, and thus the effective Lagrangian is

$$\mathscr{L}_{\text{EH}}^{(2)} + \mathscr{L}_{\text{gf}} = -\frac{1}{32\pi G}\left[\frac{1}{2}(\partial_\alpha h_{\beta\gamma})^2 - \frac{1}{4}(\partial_\alpha h)^2\right] = \mathscr{L}_{\text{EH}}^{(2)} - \left(\partial^\mu h_{\mu\nu} - \frac{1}{2}\partial_\mu h\right)^2. \tag{19.79}$$

Specialising Eq. (19.78) to the TT gauge, we obtain

$$S_{\text{EH}} = -\frac{1}{32\pi G}\int \mathrm{d}^4x \frac{1}{2}(\partial_\mu h_{ij})^2. \tag{19.80}$$

We can express an arbitrary polarisation state as the sum over the polarisation tensors for circular polarised waves,

$$h_{\mu\nu} = \sum_{a=+,-} h^{(a)}\varepsilon_{\mu\nu}^{(a)}. \tag{19.81}$$

Inserting this decomposition into (19.80) and using $\varepsilon_{\mu\nu}^{(a)}\varepsilon^{\mu\nu(b)} = \delta^{ab}$, the action becomes

$$S_{\text{EH}}^{TT} = -\frac{1}{32\pi G} \sum_a \int d^4 x \, \frac{1}{2} \left(\partial_\mu h^{(a)} \right)^2. \tag{19.82}$$

Thus the gravitational action in the TT gauge consists of two degrees of freedom, h^+ and h^-, which determine the contribution of left- and right-circular polarised waves. Apart from the pre-factor, the action is the same as the one of two scalar fields. This means that we can shortcut many calculations involving gravitational waves by using simply the corresponding results for scalar fields. We can understand this equivalence by recalling that the part of the action action quadratic in the fields just enforces the relativistic energy–momentum relation via a Klein–Gordon equation for each field component. The remaining content of (19.78) is just the rule how the unphysical components in $h^{\mu\nu}$ have to be eliminated. In the TT gauge, we have already applied this information, and thus the two scalar wave equations for $h^{(\pm)}$ summarise the Einstein equations at $\mathcal{O}(h^2)$.

Before summarising, we mention a recently found connection between gravity and gauge theories. First we note that we can express the polarisation tensor $\varepsilon_{\mu\nu}$ of a graviton as the tensor product of two polarisation vectors ε_μ of a gauge boson, $\varepsilon_{\mu\nu}^{(a)} = \varepsilon_\mu^{(a)} \varepsilon_\nu^{(a)}$. Motivated by this relation, one may wonder if one can connect amplitudes containing gravitons to those containing, for example, gluons. In general, this cannot be true since loop graphs of renormalisable and non-renormalisable theories have very different properties. However, one can connect tree-level amplitudes in gravity to the "square" of the corresponding amplitude in a non-abelian gauge theory. For instance, the vertex $V^{\mu_1\mu_2\mu_3\nu_1\nu_2\nu_3}(p_1,p_2,p_3)$ connecting three on-shell gravitons can be expressed as

$$V^{\mu_1\mu_2\mu_3\nu_1\nu_2\nu_3}(p_1,p_2,p_3) = \frac{i\kappa}{4} V^{\mu_1\mu_2\mu_3}(p_1,p_2,p_3) V^{\nu_1\nu_2\nu_3}(p_1,p_2,p_3), \tag{19.83}$$

with $V^{\nu_1\nu_2\nu_3} = \eta^{\nu_1\nu_2}(p_1-p_2)^{\nu_3} + \eta^{\nu_2\nu_3}(p_2-p_3)^{\nu_1} + \eta^{\nu_3\nu_1}(p_3-p_1)^{\nu_2}$. But $V^{\nu_1\nu_2\nu_3}$ agrees with the three-gluon vertex (10.82) after stripping off its colour factor.

Summary

The vielbein formalism allows us not only to couple fermions to gravity but it also clarifies the fact that the metric compatibility of the connection is a consequence of Einstein's general relativity. In contrast, the connection is in general not symmetric, although deviations can be neglected for macroscopic sources of gravity. The linearised Einstein equations are often sufficient to describe the propagation of gravitational waves. In the TT gauge, the Einstein–Hilbert action is proportional to the action of a massless scalar field.

Further reading. Carroll (2003) presents a clear discussion of the vielbein formalism and Cartan's structure equations. The Einstein–Hilbert action including boundary term is discussed by Poisson (2007). For an introduction to gravitational waves and their detection principle see for example Schutz (2009). The connection of tree-level amplitudes in gauge and gravity theories is discussed by Weinzierl (2016).

Problems

19.1 Dynamical stress tensor. Show that the definition of the dynamical stress tensor can be simplified to

$$T_{\mu\nu} = \frac{2}{\sqrt{|g|}} \frac{\delta S_{\mathrm{m}}}{\delta g^{\mu\nu}} = 2 \frac{\partial \mathscr{L}}{\partial g^{\mu\nu}} - g_{\mu\nu} \mathscr{L}.$$
$$(19.84)$$

19.2 Variation $\delta g_{\mu\nu}$. The variation of S_{EH} w.r.t. $g_{\mu\nu}$ will lead to different signs in Eq. (19.41). Explain why one obtains the same Einstein equation.

19.3 Cosmological constant Λ ♣. a.) Compare the stress density $T_{\mu\nu} = \kappa \Lambda g_{\mu\nu}$ of the cosmological constant to the one of an ideal fluid and determine thereby its EoS $w = P_\Lambda / \rho_\Lambda$. b.) Confirm the EoS using $U = V \rho_\Lambda$ and thermodynamics. c.) Estimate a bound on ρ_Λ using that the observable universe with size $\sim 3000\,\mathrm{Mpc}$ looks flat.

19.4 The $\phi\phi h_{\mu\nu}$ vertex. Expand the action of a scalar particle coupled to gravity,

$$S = \int \mathrm{d}^4 x \sqrt{|g|} \left(g^{\mu\nu} \partial_\mu \phi \partial_\nu \phi - \frac{1}{2} m^2 \phi^2 \right)$$

using $g_{\mu\nu} = \eta_{\mu\nu} + \lambda h_{\mu\nu}$ to first order in λ. Fix λ such that $h_{\mu\nu}$ is a canonically normalised field and find the Feynman rule for the $\phi\phi h_{\mu\nu}$ vertex. Compare with the result of problem 7.9.

19.5 Expansion of S_{EH}. Expand $g^{\mu\nu}$ and $\sqrt{|g|}$ up to $\mathcal{O}(\lambda^3)$ around Minkowski space.

Show that the $\mathcal{O}(\lambda)$ term of $\mathscr{L}_{\mathrm{EH}}$ is a total derivative which can be dropped.

19.6 Dirac-Schwarzschild Write down the Dirac equation for a fermion in the Schwarzschild metric given in Eq. (25.5).

19.7 Riemann tensor. Derive the symmetry properties of the Riemann tensor and the number of its independent components for an arbitrary number of dimensions. (Hint: Use an inertial system.)

19.8 Helicity. Show that the unphysical degrees of freedom of an electromagnetic wave transform as helicity 0, and of a a gravitational wave as helicity 0 and 1.

19.9 Gravitational wave. Consider the effect of a linearly polarised passing a transverse ring of test particles and sketch their relative movements. Combining them, obtain the effect of circular polarised gravitational waves.

19.10 Light deflection in gravity ♥. Consider the scattering of a scalar particle and a photon via graviton exchange. Calculate the scattering amplitude and the cross-section in the static limit $p^\mu = (M_\odot, \mathbf{0})$ for small-angle scattering ($k^2 = \vartheta = 0$ in the numerator) and show that it agrees with Einstein's prediction for light deflection by the Sun.

20
Cosmological models for an homogeneous, isotropic universe

The most important observational property of our universe is its homogeneity and isotropy on sufficiently large scales. Only this feature allows us to deduce from observations performed at a single point generic properties of the universe. While the homogeneity and isotropy of the universe was merely a postulate—dubbed the cosmological principle—at the beginning of the 20th century, observations of the cosmic microwave background (CMB) have shown that the early universe was isotropic at the level of 10^{-4} around us. Note that a space isotropic around at least two points is also homogeneous, while a homogeneous space is not necessarily isotropic. Baring suggestions that we live at a unique place, the early universe was therefore also homogeneous on large scales. The homogeneity of the universe implies moreover that beyond a sufficiently large scale no new types of gravitationally bound structures appear: In particular, the sequence of hierarchical structures $\rightarrow$ star clusters $\rightarrow$ galaxies $\rightarrow$ galaxy clusters $\rightarrow$ supercluster of galaxies stops and superclusters with sizes of order 100 Mpc are the largest bound systems found in the universe. For comparison, the size of the universe visible at the present epoch is of order 3000 Mpc, that is a factor 30 times larger. In this chapter, we will discuss the evolution of perfectly homogeneous and isotropic cosmological models. The question how the inhomogeneities measured in the CMB were generated and how they have evolved into the very inhomogeneous present universe will be addressed in chapter 24.

20.1 FLRW metric

Weyl's postulate. In 1923, Hermann Weyl suggested the existence of a privileged class of observers in the universe, namely those following the "average" motion of galaxies. He postulated that these observers follow time-like geodesics that never intersect. They may, however, diverge from a point in the (finite or infinite) past or converge towards such a point in the future. Weyl's postulate implies that we can find coordinates such that galaxies—and the observers residing in them—are at rest. These coordinates are called *comoving coordinates* and can be constructed as follows. One chooses first a space-like hypersurface. Through each point P in this hypersurface passes a unique world-line of a comoving observer. We choose the coordinate time such that it agrees with the proper-time of all observers, $g_{00} = 1$, and the spatial coordinate vectors such that they are constant and lie in the tangent space T_P at this point. Then $u^\alpha = \delta_0^\alpha$ and for $n^\alpha \in T_P$ it follows $n^\alpha = (0, n^i)$ and

Quantum Fields–From the Hubble to the Planck Scale. Michael Kachelriess. © Michael Kachelriess 2018.
Published in 2018 by Oxford University Press. DOI 10.1093/oso/9780198802877.001.0001

$$0 = u_\alpha n^\alpha = g_{\alpha\beta} u^\alpha n^\beta = g_{0b} n^b \,. \tag{20.1}$$

Since n^α is arbitrary, the elements g_{0b} of the metric tensor have to vanish, $g_{0b} = 0$. Hence as a consequence of Weyl's postulate we may choose the metric as

$$\mathrm{d}s^2 = \mathrm{d}t^2 - \mathrm{d}l^2 = \mathrm{d}t^2 - g_{ij}\mathrm{d}x^i\mathrm{d}x^j \,. \tag{20.2}$$

The cosmological principle constrains further the form of $\mathrm{d}l^2$. Homogeneity requires that the g_{ij} can depend on time only via a common factor $S(t)$, while isotropy requires that only the scalars $\boldsymbol{x} \cdot \boldsymbol{x} \equiv r^2$, $\boldsymbol{x} \cdot \mathrm{d}\boldsymbol{x}$ and $\mathrm{d}\boldsymbol{x} \cdot \mathrm{d}\boldsymbol{x}$ enter $\mathrm{d}l^2$. Hence the spatial part of the metric has the form

$$\mathrm{d}l^2 = C(r)(\boldsymbol{x} \cdot \mathrm{d}\boldsymbol{x})^2 + D(r)\mathrm{d}\boldsymbol{x} \cdot \mathrm{d}\boldsymbol{x} \tag{20.3a}$$

$$= C(r)r^2\mathrm{d}r^2 + D(r)[\mathrm{d}r^2 + r^2\mathrm{d}\vartheta^2 + r^2\sin^2\vartheta\mathrm{d}\phi^2]. \tag{20.3b}$$

We can eliminate the function $D(r)$ by the rescaling $r^2 \to Dr^2$. Then the line element becomes

$$\mathrm{d}l^2 = S(t)\left[B(r)\mathrm{d}r^2 + r^2\mathrm{d}\Omega\right] \tag{20.4}$$

with $\mathrm{d}\Omega = \mathrm{d}\vartheta^2 + \sin^2\vartheta\mathrm{d}\phi^2$. While the dynamical function $S(t)$ has to be determined from Einstein equations for a given matter content of the universe, we can constrain the function $B(r)$ further using the cosmological principle.

Maximally symmetric spaces. These are spaces with constant curvature, hence the Riemann tensor of such spaces can depend only on the metric tensor and a constant K specifying the curvature. The only form that respects the (anti-)symmetries of the Riemann tensor is

$$R_{\alpha\beta\gamma\delta} = K(g_{\alpha\gamma}g_{\beta\delta} - g_{\alpha\delta}g_{\beta\gamma}). \tag{20.5}$$

Then we obtain for the Ricci tensor in three spatial dimensions

$$R_{bd} = g^{ac}R_{abcd} = Kg^{ac}(g_{ac}g_{bd} - g_{ad}g_{bc}) = K(3g_{bd} - g_{bd}) = 2Kg_{bd}. \tag{20.6}$$

A final contraction gives as curvature R of a three-dimensional maximally symmetric space

$$R = g^{ab}R_{ab} = 2K\delta_a^a = 6K. \tag{20.7}$$

A comparison of Eq. (20.6) with the Ricci tensor for the metric (20.4) will fix the still unknown function $B(r)$. We proceed in the standard way. We calculate the Christoffel symbols with the help of the geodesic equation (or alternatively directly from Eq. (6.43)) and use then the definition of the Ricci tensor.

Example 20.1: Find the Christoffel symbols and the Ricci tensor for the metric (20.4). We solve the Lagrange equations for a test particle moving in (20.4),

$$L = B(r)\dot{r}^2 + r^2(\dot{\vartheta}^2 + \sin^2\vartheta\dot{\phi}^2)\,,$$

where we neglected the overall factor $S(t)$. Comparing the Lagrange equations

$$\ddot{r} - \frac{B'}{2B}\dot{r}^2 - (\dot{\vartheta}^2 + \sin^2\vartheta\dot{\phi}^2)\frac{r}{B} = 0\,, \quad \ddot{\phi} + \frac{2}{r}\dot{r}\dot{\phi} + 2\cot\vartheta\dot{\vartheta}\dot{\phi} = 0\,, \quad \ddot{\vartheta} + \frac{2\dot{r}\dot{\vartheta}}{r} - \cos\vartheta\sin\vartheta\dot{\phi}^2 = 0$$

with the geodesic equation $\ddot{x}^\kappa + \Gamma^\kappa{}_{\mu\nu}\dot{x}^\mu\dot{x}^\nu = 0$, we can read off the non-vanishing Christoffel symbols as

$$\Gamma^r{}_{rr} = -B'/(2B), \qquad \Gamma^r{}_{\phi\phi} = -r\sin^2\vartheta/B \quad \text{and} \quad \Gamma^r{}_{\vartheta\vartheta} = -r/B$$

$$\Gamma^\phi{}_{r\phi} = \Gamma^\vartheta{}_{r\vartheta} = 1/r, \qquad \Gamma^\phi{}_{\phi\vartheta} = \cot\vartheta \quad \text{and} \quad \Gamma^\vartheta{}_{\phi\phi} = -\cos\vartheta\sin\vartheta.$$

Since the metric is diagonal, the non-diagonal elements of the Ricci tensor are zero too. We calculate with

$$R_{ab} = R^c{}_{acb} = \partial_c\Gamma^c{}_{ab} - \partial_b\Gamma^c{}_{ac} + \Gamma^c{}_{ab}\Gamma^d{}_{cd} - \Gamma^d{}_{bc}\Gamma^c{}_{ad}$$

for instance the rr component as

$$R_{rr} = 0 - 0 + \Gamma^c{}_{rr}\Gamma^d{}_{cd} - \Gamma^d{}_{rc}\Gamma^c{}_{rd} = \Gamma^r{}_{rr}(\Gamma^\phi{}_{r\phi} + \Gamma^\vartheta{}_{r\vartheta}) = -\frac{B'}{rB}. \tag{20.8}$$

Similarly, we find $R_{\vartheta\vartheta} = 1 + \frac{r}{2B^2}\frac{dB}{dr} - \frac{1}{B}$ and $R_{\phi\phi} = \sin^2\vartheta R_{\vartheta\vartheta}$.

Inserting our result for the spatial Ricci tensor into Eq. (20.6), we obtain

$$R_{rr} = \frac{1}{rB}\frac{dB}{dr} = 2Kg_{rr} = 2KB \tag{20.9a}$$

$$R_{\vartheta\vartheta} = 1 + \frac{r}{2B^2}\frac{dB}{dr} - \frac{1}{B} = 2Kg_{\vartheta\vartheta} = 2Kr^2, \tag{20.9b}$$

while the $\phi\phi$ equation contains no additional information. Integration of Eq. (20.9a) gives

$$B = \frac{1}{A - Kr^2} \tag{20.10}$$

with A as integration constant. Inserting then the result into Eq. (20.9b) fixes the constant as $A = 1$. Thus we have determined the line element of a maximally symmetric 3-space with curvature K as

$$dl^2 = \frac{dr^2}{1 - Kr^2} + r^2(\sin^2\vartheta d\phi^2 + d\vartheta^2). \tag{20.11}$$

Going over to the full four-dimensional line element, we rescale for $K \neq 0$ the r coordinate by $r \to |K|^{1/2}r$. Then we absorb the factor $1/|K|$ in front of dl^2 by defining the scale factor $a(t)$ as

$$a(t) = \begin{cases} S(t)/|K|^{1/2}, & K \neq 0 \\ S(t), & K = 0. \end{cases} \tag{20.12}$$

As a result we obtain the Friedmann–Lemaître–Robertson–Walker (FLRW) metric for an homogeneous, isotropic universe,

$$ds^2 = dt^2 - a^2(t)\left[\frac{dr^2}{1 - kr^2} + r^2(\sin^2\vartheta d\phi^2 + d\vartheta^2)\right]. \tag{20.13}$$

The constant k distinguishes the cases where the three-dimensional space has negative ($k = -1$) or positive curvature ($k = 1$) or is flat ($k = 0$). Finally, we give two alternative forms of the FLRW metric that are often more useful. In the first one, we introduce a new radial variable by $r = \sin \chi$ for $k = 1$. Then $dr = \cos \chi d\chi = (1 - r^2)^{1/2} d\chi$ and the line element follows as

$$ds^2 = dt^2 - a^2(t) \left[d\chi^2 + S^2(\chi)(\sin^2 \vartheta d\phi^2 + d\vartheta^2) \right] \tag{20.14}$$

with $S(\chi) = \sin \chi = r$. Defining

$$r = S(\chi) = \begin{cases} \sin \chi & \text{for} \quad k = 1, \\ \chi & \text{for} \quad k = 0, \\ \sinh \chi & \text{for} \quad k = -1, \end{cases} \tag{20.15}$$

the metric (20.14) is valid for all three values of k.

In the second alternative, we replace the comoving time t by the conformal time $d\eta = dt/a$,

$$ds^2 = a^2(\eta) \left[d\eta^2 - \frac{dr^2}{1 - kr^2} - r^2(\sin^2 \vartheta d\phi^2 - d\vartheta^2) \right] \tag{20.16a}$$

$$= a^2(\eta) \left[d\eta^2 - d\chi^2 - S^2(\chi)(\sin^2 \vartheta d\phi^2 + d\vartheta^2) \right] . \tag{20.16b}$$

Then the metric has the same form for $k = 0$ as the one of a uniformly expanding Minkowski space. The description of light propagation becomes therefore particularly simple using the coordinates (20.16b). Since $ds^2 = 0$, a radial light-ray satisfies $d\eta = \pm d\chi$ and light-rays are straight lines at ± 45 degrees in the $\eta - \chi$ plane.

Geometry of the Friedmann–Lemaître–Robertson–Walker metric. Let us consider a sphere of fixed radius at fixed time, $dr = dt = 0$. The line element ds^2 simplifies then to $a^2(t)r^2(\sin^2 \vartheta d\phi^2 + d\vartheta^2)$, which is the usual line element of a sphere S^2 with radius $ra(t)$. The area of a sphere is $A = 4\pi(ra(t))^2 = 4\pi[S(\chi)a(t)]^2$ and the circumference of a circle is $L = 2\pi ra(t)$. Hence these quantities agree for $a = 1$ with the ones in an Euclidean space. By contrast, the radial distance between two points (r, ϑ, ϕ) and $(r + dr, \vartheta, \phi)$ is $dl = a(t)dr/\sqrt{1 - kr^2}$. Thus the radius of a sphere centered at $r = 0$ is

$$l = a(t) \int_0^r \frac{dr'}{\sqrt{1 - kr'^2}} = a(t) \times \begin{cases} \arcsin(r) & \text{for} \quad k = 1, \\ r & \text{for} \quad k = 0, \\ \text{arcsinh}\,(r) & \text{for} \quad k = -1. \end{cases} \tag{20.17}$$

Using χ as coordinate, the same result follows immediately,

$$l = a(t) \int_0^{\chi(r)} d\chi = a(t)\chi. \tag{20.18}$$

Hence for $k = 0$, that is a flat space, one obtains the usual result $L/l = 2\pi$, while for $k = 1$ (spherical geometry) it is $L/l = 2\pi r/\arcsin(r) < 2\pi$ and for $k = -1$ (hyperbolic geometry) $L/l = 2\pi r/\text{arcsinh}\,(r) > 2\pi$. For $k = 0$ and $k = -1$, l is unbounded, while for $k = +1$ there exists a maximal distance $l_{\max}(t)$. Hence the first two cases correspond to open spaces with an infinite volume, while the latter is a closed space with finite volume.

Hubble's law.   Lemaître, Hubble and other astronomers found empirically that the spectral lines of "distant" galaxies are redshifted, $z = \Delta\lambda/\lambda_0 > 1$, proportional to their distance d,

$$z = H_0 d. \tag{20.19}$$

If this redshift is interpreted as Doppler effect, $z = \Delta\lambda/\lambda_0 = v_r$, then the recession velocity of galaxies follows as

$$v = H_0 d. \tag{20.20}$$

The restriction "distant galaxies" means more precisely that the peculiar motion of galaxies caused by the gravitational attraction of nearby galaxy clusters should be small compared to the Hubble flow, $H_0 d \gg v_{\text{pec}} \approx \text{few} \times 100\,\text{km/s}$. Note that the interpretation of v as recession velocity is problematic. The validity of such an interpretation is certainly limited to $v \ll 1$. The parameter H_0 is called Hubble constant and has the value $H_0 \simeq (67.7 \pm 0.8)\,\text{km/s/Mpc}$. We will see soon that the Hubble law Eq. (20.20) is an approximation valid for $z \ll 1$. In general, the Hubble constant is not constant but depends on time, $H = H(t)$, and we will call it therefore the Hubble parameter for $t \neq t_0$.

We can derive Hubble's law as a consequence of the homogeneity of space. Consider this law as a vector equation with us at the centre of the coordinate system,

$$\boldsymbol{v} = H\boldsymbol{d}. \tag{20.21}$$

What does a different observer at the position $\boldsymbol{d}'$ see? She has the velocity $\boldsymbol{v}' = H\boldsymbol{d}'$ relative to us. We are assuming that velocities are small and thus

$$\boldsymbol{v}'' \equiv \boldsymbol{v} - \boldsymbol{v}' = H(\boldsymbol{d} - \boldsymbol{d}') = H\boldsymbol{d}'', \tag{20.22}$$

where $\boldsymbol{v}''$ and $\boldsymbol{d}''$ denote the position relative to the new observer. Hence a linear relation between v and d as Hubble law is the only relation compatible with homogeneity and thus the "cosmological principle" for $v \ll 1$.

Lemaître's redshift formula.   A non-static universe possesses no time-like Killing vector field and thus we expect that the energy of a particle is affected by the expansion of space. In order to derive this change, we consider a massive particle moving along a geodesic $x(\tau)$ with four-velocity u^α. An observer with four-velocity U^α will measure as energy E of the massive particle $E = mu^\beta U_\beta$. Along the trajectory $x(\tau)$, this energy E changes as

$$\frac{\mathrm{d}}{\mathrm{d}\tau} E = u^\alpha \nabla_\alpha E = m u^\alpha \nabla_\alpha \left(u^\beta U_\beta \right) = m u^\alpha u^\beta \nabla_\alpha U_\beta, \tag{20.23}$$

where we used the geodesic equation $u^\alpha \nabla_\alpha u^\beta = 0$ in the last step. Evaluating the derivative using the Christoffel symbols for the FLRW metric and for comoving observers with $U^\alpha = (1, \boldsymbol{0})$ gives (cf. problem 20.4)

$$\nabla_\alpha U_\beta = \frac{\dot{a}}{a}(g_{\alpha\beta} - U_\alpha U_\beta). \tag{20.24}$$

Inserting the derivative, we find with $u_\alpha u^\alpha = 1$

$$\frac{\mathrm{d}}{\mathrm{d}\tau} E = m\frac{\dot{a}}{a}\left(1 - E^2/m^2\right) = -\frac{\dot{a}}{a}\left(\frac{E^2 - m^2}{m}\right), \tag{20.25}$$

or

$$-\frac{\mathrm{d}a}{a} = m\frac{\mathrm{d}t}{\mathrm{d}\tau}\frac{\mathrm{d}E}{E^2 - m^2} = \frac{E\mathrm{d}E}{E^2 - m^2} = \frac{1}{2}\frac{\mathrm{d}E(E^2 - m^2)}{E^2 - m^2} = \frac{\mathrm{d}p}{p}. \tag{20.26}$$

Thus the momentum of a particle is inverse proportionally to the scale factor of the universe, $p \propto 1/a$. An intuitive explanation of this result is that the expansion of the universe stretches all length scales of unbound systems, including the wavelength of free particles. As a consequence, the kinetic energy of non-relativistic particles goes quadratically to zero, and hence peculiar velocities relative to the Hubble flow are strongly damped by the expansion of the universe.

We can derive now Hubble's law by a Taylor expansion of the scale factor $a(t)$,

$$a(t) = a(t_0) + (t - t_0)\dot{a}(t_0) + \frac{1}{2}(t - t_0)^2\ddot{a}(t_0) + \ldots \tag{20.27a}$$

$$= a(t_0)\left[1 + (t - t_0)H_0 - \frac{1}{2}(t - t_0)^2 q_0 H_0^2 + \ldots\right], \tag{20.27b}$$

where we introduced the Hubble constant

$$H_0 \equiv \frac{\dot{a}(t_0)}{a(t_0)} \quad \text{and} \quad q_0 \equiv -\frac{\ddot{a}(t_0)a(t_0)}{\dot{a}^2(t_0)}, \tag{20.28}$$

the so-called deceleration parameter. If the expansion is slowing down, $\ddot{a} < 0$ and $q_0 > 0$. Hubble's law follows as an approximation for small redshift. For not too large time-differences, we can use the expansion Eq. (20.27a) and write

$$1 - z \simeq \frac{1}{1 + z} = \frac{a(t)}{a_0} \simeq 1 + (t - t_0)H_0. \tag{20.29}$$

Hence Hubble's law, $z = (t_0 - t)H_0 = d/cH_0$, is valid as long as $z \simeq H_0(t_0 - t) \ll 1$. Deviations from its linear form arises for $z \gtrsim 1$ and can be used to determine q_0.

Proper distance. In an expanding universe, the distance to an object depends on the expansion history, that is, the behaviour of the scale factor $a(t)$ between the time of emission t of the observed light signal and its reception at t_0. From the metric (20.14) we can define the (radial) coordinate distance

$$\chi = \int_t^{t_0} \frac{\mathrm{d}t}{a(t)} \tag{20.30}$$

as well as the proper distance $d = g_{\chi\chi}\chi = a(t)\chi$. The proper distance d corresponds to the physical distance between two points on a hypersurface $t = \text{const.}$ However, it is only for a static metric a directly measurable quantity and cosmologists therefore use other operationally defined measures for the distance. The two most important examples are the luminosity and the angular diameter distances.

Luminosity distance. The luminosity distance d_L is defined such that the inverse-square law between the luminosity L of a source at the distance d and the received energy flux $\mathcal{F}$ is valid,

$$d_L = \left(\frac{L}{4\pi\mathcal{F}}\right)^{1/2}. \tag{20.31}$$

Assume now that a (isotropically emitting) source with luminosity $L(t)$ and comoving coordinate χ is observed at t_0 by an observer at O. The cut at O through the forward light-cone emitted at t_e by the source defines a sphere S^2 with proper area

$$A = 4\pi a^2(t_0)S^2(\chi). \tag{20.32}$$

Additionally, we have to take into account that the frequency of a single photon is redshifted, $\nu_0 = \nu_e/(1+z)$, and that the arrival rate of photons is reduced by the same factor due to time-dilation. Hence the received flux is

$$\mathcal{F}(t_0) = \frac{1}{(1+z)^2}\frac{L(t_e)}{4\pi a_0^2 S^2(\chi)} \tag{20.33}$$

and the luminosity distance in a FLRW universe follows as

$$d_L = (1+z)\,a_0 S(\chi). \tag{20.34}$$

Note that d_L depends via χ on the expansion history of the universe between t_e and t_0.

Angular diameter distance. Instead of basing a distance measurement on standard candles, one may use standard rods with known proper length l whose angular diameter $\Delta\vartheta$ can be observed. Then we define the angular diameter distance as

$$d_A = \frac{l}{\Delta\vartheta}. \tag{20.35}$$

From the angular part of the FLRW metric it follows that $l = a(t_e)S(\chi)\Delta\vartheta$ and thus

$$d_A = a(t_e)S(\chi) = a(t_0)\frac{a(t_e)}{a(t_0)}S(\chi) = \frac{a_0 S(\chi)}{1+z}. \tag{20.36}$$

At small distances, $z \ll 1$, the angular diameter and the luminosity distance agree by construction, while for large redshift the difference increases as $(1+z)^2$.

Observable are not the coordinates χ or r, but the redshift z of a galaxy. Differentiating $1+z = a_0/a(t)$, we obtain

$$dz = -\frac{a_0}{a^2}\,da = -\frac{a_0}{a^2}\frac{da}{dt}\,dt = -(1+z)H dt. \tag{20.37}$$

Denoting by t_0 the present age of the universe, we can find its age t at redshift z for a given $H(z)$ as

$$t_0 - t = \int_t^{t_0} dt = \int_z^0 \frac{dz}{H(z)(1+z)}. \tag{20.38}$$

Inserting the relation (20.37) into Eq. (20.30), we find the coordinate χ of a galaxy at redshift z as

$$\chi = \int_t^{t_0} \frac{\mathrm{d}t}{a(t)} = \frac{1}{a_0} \int_0^z \frac{\mathrm{d}z}{H(z)}. \tag{20.39}$$

For small redshift $z \ll 1$, we can use the expansion (20.27b) to approximate

$$\chi = \int_t^{t_0} \frac{\mathrm{d}t}{a_0} [1 - (t - t_0)H_0 + \ldots]^{-1} \tag{20.40a}$$

$$\approx \frac{1}{a_0} [(t - t_0) + \frac{1}{2}(t - t_0)^2 H_0 + \ldots] = \frac{1}{a_0 H_0} [z - \frac{1}{2}(1 + q_0)z^2 + \ldots]. \tag{20.40b}$$

In practice, one observes only the luminosity within a certain frequency range instead of the total (or bolometric) luminosity. A correction for this effect requires the knowledge of the intrinsic source spectrum.

20.2 Friedmann equations

We could determine the metric of a homogeneous and isotropic universe by purely geometrical considerations except for the value of $k \in \{-1, 0, +1\}$ and the unknown function $a(t)$. Since the scale factor is a dynamical quantity, it has to be determined by the Einstein equations. In order to write down these equations, we have to fix a model for the stress tensor $T_{\mu\nu}$ of the universe. A surprisingly generic ansatz for the matter content of the universe which is consistent with Weyl's postulate is an ideal fluid. Although an ideal fluid is characterised by only one free parameter—the equation of state (EoS) $w = P/\rho$ fixing the ratio of its pressure P and its energy density ρ—it can describe[1] both macrophysical systems (a fluid of galaxies) and elementary particles (e.g. a gas of photon or a fluid of cold dark matter particles). The various components of an ideal fluid evolve independently and thus the total stress tensor is simply the sum of the individual contributions.

In problem 5.4 you derived as the stress tensor of an ideal fluid in Minkowski space

$$T_{\mu\nu} = (\rho + P)u_\mu u_\nu - P\eta_{\mu\nu}. \tag{20.41}$$

Clearly, the homogeneity and isotropy of the FLRW metric imply that the stress tensor has the same symmetries and thus ρ and P can be only functions of time. Replacing $\eta_{\mu\nu}$ by $g_{\mu\nu}$ and choosing the comoving coordinates of Eq. (20.13) gives

$$g_{00} = 1, \quad g_{11} = -\frac{a^2}{1 - kr^2}, \quad g_{22} = -a^2 r^2 \sin^2 \vartheta \quad \text{and} \quad g_{33} = -a^2 r^2. \tag{20.42}$$

Since the FLRW metric is diagonal, the elements of the inverse metric are simply given by $g^{\mu\mu} = 1/g_{\mu\mu}$. Thus the stress tensor follows for a comoving observer as

$$T_{00} = \rho g_{00} \quad \text{and} \quad T_{ij} = P\delta_{ij}. \tag{20.43}$$

[1]Although deviations from this idealisation are small and happen only in specific phases during the evolution of the universe, they are crucial in explaining the observed amount of relic particles as baryons and dark matter. This topic will be introduced in the next chapter.

Next we have to determine the still missing Christoffel symbols $\Gamma^\sigma{}_{\mu\nu}$, see problem 20.1. The result is

$$\Gamma^i{}_{0i} = \frac{\dot{a}}{a} \quad \text{and} \quad \Gamma^0{}_{ij} = -\frac{\dot{a}}{a}\delta_{ij} \tag{20.44}$$

from which the non-zero components of the Ricci tensor follow as

$$R_{00} = -3\frac{\ddot{a}}{a}g_{00} \quad \text{and} \quad R_{ij} = \frac{a\ddot{a} + 2\dot{a}^2 + 2k}{a^2}\delta_{ij}. \tag{20.45}$$

The last ingredient needed for the Einstein equations is the curvature scalar,

$$R = g^{\mu\nu}R_{\mu\nu} = -6\frac{a\ddot{a} + \dot{a}^2 + k}{a^2}. \tag{20.46}$$

All the quantities appearing in the gravitational field equation are proportional to the metric tensor and thus the symmetries of the FLRW metric lead to only two independent equations of motion. The time–time component of the Einstein equation gives

$$3\frac{\dot{a}^2 + k}{a^2} = \kappa\rho + \Lambda, \tag{20.47}$$

while the space–space part results in

$$\frac{2a\ddot{a} + \dot{a}^2 + 2k}{a^2} = -\kappa P + \Lambda. \tag{20.48}$$

Using $\kappa = 8\pi G$, we obtain from Eq. (20.47) the Friedmann equation,

$$H^2 \equiv \left(\frac{\dot{a}}{a}\right)^2 = \frac{8\pi}{3}G\rho - \frac{k}{a^2} + \frac{\Lambda}{3}, \tag{20.49}$$

while the "acceleration equation" follows from (20.48) after eliminating $\ddot{a}$ with (20.47) as

$$\frac{\ddot{a}}{a} = \frac{\Lambda}{3} - \frac{4\pi G}{3}(\rho + 3P). \tag{20.50}$$

This equation determines the (de-) acceleration of the universe as a function of its matter and energy content. "Normal" matter is characterised by $\rho > 0$ and $P \geq 0$. Thus a static solution is impossible for a universe with $\Lambda = 0$. Such a universe is decelerating and since today $\dot{a} > 0$, $\ddot{a}$ was always negative and there was a "big bang".

We define the *critical density* ρ_{cr} as the density for which the spatial geometry of the universe is flat. From $k = 0$, it follows

$$\rho_{\text{cr}} = \frac{3H_0^2}{8\pi G} \tag{20.51}$$

and thus ρ_{cr} is uniquely fixed by the value of H_0. One "hides" this dependence by introducing h,

$$H_0 = 100\,h\,\text{km}/(\text{s Mpc}).$$

Then one can express the critical density as function of h,

$$\rho_{\rm cr} = 2.77 \times 10^{11} h^2 M_\odot/\text{Mpc}^3 = 1.88 \times 10^{-29} h^2 \text{g/cm}^3 = 1.05 \times 10^{-5} h^2 \text{GeV/cm}^3.$$

Thus a flat universe with $h = 1$ requires an energy density of ~ 10 protons per cubic metre. We define the energy fraction Ω_i of the different players in cosmology as their energy density relative to the critical density, $\Omega_i = \rho/\rho_{\rm cr}$. We will often include the cosmological constant Λ as another contribution to the energy density ρ via

$$\frac{8\pi}{3} G \rho_\Lambda = \frac{\Lambda}{3}. \tag{20.52}$$

Thereby one recognises also that the cosmological constant acts as a constant energy density with magnitude

$$\rho_\Lambda = \frac{\Lambda}{8\pi G} \qquad \text{or} \qquad \Omega_\Lambda = \frac{\Lambda}{3H_0^2}. \tag{20.53}$$

We can understand the consequences of the cosmological constant better by replacing Λ by $(8\pi G)\rho_\Lambda$ in the acceleration equation. Comparing then the effect of normal matter and of the Λ term on the acceleration,

$$\frac{\ddot{a}}{a} = \frac{8\pi G}{3} \rho_\Lambda - \frac{4\pi G}{3}(\rho + 3P). \tag{20.54}$$

we recognise that Λ is equivalent to matter with an EoS $w_\Lambda = P/\rho = -1$, as we showed already in problem 19.3. Using ρ_Λ instead of Λ corresponds on the level of the Einstein equations to a reshuffling of the cosmological constant from the geometry to the matter side, $\kappa T_{\mu\nu} + g_{\mu\nu}\Lambda \to \kappa \tilde{T}_{\mu\nu}$. Thus the observed value of the cosmological constant $\rho_{\rm obs}$ includes in addition to ρ_Λ both the classical contribution $V(\phi_0) \neq 0$ of all scalar potentials whose minima are not at zero and the (renormalised) quantum vacuum fluctuations of all matter fields. The borderline between an accelerating and decelerating universe is given by $\rho = -3P$ or $w = -1/3$. The condition $\rho < -3P$ violates the so-called strong energy condition for "normal" matter in equilibrium. An accelerating universe requires therefore a positive cosmological constant or a dominating form of matter that is not in equilibrium. As a phenomenological generalisation of the cosmological constant, one calls any form of matter *dark energy* which leads in the present epoch to an EoS parameter $w \approx -1$.

Relativistic species today are photons and possibly one of the three SM neutrinos. Their energy contribution is much smaller than the one from non-relativistic matter (stars, gas and cold dark matter). Thus the pressure term in the acceleration equation can be neglected at the present epoch. Measuring ρ, $\dot{a}/a$ and $\ddot{a}/a$ therefore fixes the geometry of the universe. For a long time, cosmology was thus described as the quest for two numbers, H_0 and q_0.

Thermodynamics. With $dS = 0$ the first law of thermodynamics becomes for a ideal fluid simply $dU = TdS - PdV = -PdV$. Considering a fixed comoving volume,

$$d(\rho a^3) = -Pd(a^3), \tag{20.55}$$

and dividing by dt,

$$a\dot{\rho} + 3(\rho + P)\dot{a} = 0, \tag{20.56}$$

we obtain

$$\dot{\rho} = -3(\rho + P)H. \tag{20.57}$$

This result could be also derived from $\nabla_\alpha T^{\alpha\beta} = 0$, cf. problem 8. Since $\nabla_\alpha T^{\alpha\beta} = 0$ is built-in in the Einstein equations, the three equations (20.49), (20.50) and (20.57) are not independent.

Scale-dependence of different energy forms.   The dependence of different energy forms as function of the scale factor a can derived from energy conservation, $dU = -PdV$, if an EoS $P = P(\rho) = w\rho$ is specified. For $w = \text{const.}$, it follows

$$d(\rho a^3) = -3Pa^2 da \tag{20.58}$$

or eliminating P

$$\frac{d\rho}{da}a^3 + 3\rho a^2 = -3w\rho a^2. \tag{20.59}$$

Separating the variables,

$$-3(1 + w)\frac{da}{a} = \frac{d\rho}{\rho}, \tag{20.60}$$

we can integrate and obtain

$$\rho \propto a^{-3(1+w)} = \begin{cases} a^{-3} & \text{for matter} \quad (w = 0), \\ a^{-4} & \text{for radiation} \quad (w = 1/3), \\ \text{const.} & \text{for } \Lambda \quad (w = -1). \end{cases} \tag{20.61}$$

The obtained scaling can be understood from heuristic arguments. The kinetic energy of non-relativistic matter is negligible, $kT \ll m$. Thus $\rho = nm \gg nT = P$ and non-relativistic matter is pressure-less, $w = 0$. The mass m is constant and $n \propto 1/a^3$, hence ρ is just diluted by the expansion of the universe, $\rho \propto 1/a^3$. Radiation is not only diluted but the energy of a single particle is additionally redshifted, $E \propto 1/a$. Thus the energy density of radiation scales as $\propto 1/a^4$. Alternatively, one can use that $\rho \propto T^4$ and $T \propto \langle E \rangle \propto 1/a$. Finally, the cosmological constant Λ acts by definition as an energy density $\rho_\lambda = \Lambda/(8\pi G)$ that is constant in time, independent from a possible expansion or contraction of the universe. Note also that cosmologists call relativistic particles radiation while they include in matter only non-relativistic particles.

Let us rewrite the Friedmann equation for the present epoch as

$$\frac{k}{a_0^2} = H_0^2\left(\frac{8\pi G}{3H_0^2}\rho_0 + \frac{\Lambda}{3H_0^2} - 1\right) = H_0^2\left(\Omega_{\text{tot},0} - 1\right). \tag{20.62}$$

We express the curvature term for arbitrary times through $\Omega_{\text{tot},0}$ and the redshift z as

$$\frac{k}{a^2} = \frac{k}{a_0^2}(1 + z)^2 = H_0^2(\Omega_{\text{tot},0} - 1)(1 + z)^2. \tag{20.63}$$

Note that the curvature term $(\Omega_{\text{tot},0} - 1)(1 + z)^2$ decreases at a slower rate than radiation or matter. Observations indicate that the universe is close to flat, and thus the curvature term can be safely neglected in the early universe. This behaviour poses the question of why the universe is (nearly) flat, since for generic initial conditions one would expect $|\Omega_{\text{tot},0} - 1|/\Omega_{\text{tot}} \gg 1$. Dividing the Friedmann equation (20.49) by $H_0^2 = 8\pi G \rho_{\text{cr}}/3$, we obtain

$$\frac{H^2(z)}{H_0^2} = \sum_i \Omega_i(z) - (\Omega_{\text{tot},0} - 1)(1+z)^2 \tag{20.64a}$$

$$= \Omega_{\text{rad},0}(1+z)^4 + \Omega_{\text{m},0}(1+z)^3 + \Omega_\Lambda - (\Omega_{\text{tot},0} - 1)(1+z)^2. \tag{20.64b}$$

This expression allows us to calculate the age of the universe (20.38), distances (20.34), etc., for a given cosmological model, that is, specifying the energy fractions $\Omega_{i,0}$ and the Hubble parameter H_0 at the present epoch.

20.3 Evolution of simple cosmological models

Cosmological models with one energy component. We consider a flat universe, $k = 0$, with one dominating energy component with EoS $w = P/\rho = \text{const}$. With $\rho = \rho_{\text{cr}} (a/a_0)^{-3(1+w)}$, the Friedmann equation becomes

$$\dot{a}^2 = \frac{8\pi}{3} G \rho a^2 = H_0^2 a_0^{3+3w} \, a^{-(1+3w)}, \tag{20.65}$$

where we inserted the definition of $\rho_{\text{cr}} = 3H_0^2/(8\pi G)$. Separating variables we obtain

$$a_0^{-(3+3w)/2} \int_0^{a_0} da \, a^{(1+3w)/2} = H_0 \int_0^{t_0} dt = t_0 H_0 \tag{20.66}$$

and hence the age of the universe follows as

$$t_0 H_0 = \frac{2}{3+3w} = \begin{cases} 2/3 & \text{for matter} \quad (w = 0), \\ 1/2 & \text{for radiation} \quad (w = 1/3), \\ \to \infty & \text{for } \Lambda \quad (w = -1). \end{cases} \tag{20.67}$$

While a universe with only a cosmological constant Λ has no "beginning", models with $w > -1$ needed a finite time to expand from the initial singularity $a(t = 0) = 0$ to the current size a_0. Since for $t \to 0$ the temperature and density formally diverge, one calls these cases often (hot) big bang models. We should expect that classical gravity breaks down when $\rho \sim M_{\text{Pl}}^4$. As long as $a \propto t^\alpha$ with $\alpha < 1$, most time elapsed during the last fractions of $t_0 H_0$. Hence our result for the age of the universe does not depend on unknown physics close to the big bang as long as $w > -1/3$.

If we integrate Eq. (20.66) to the arbitrary time t, we obtain as time-dependence of the scale factor

$$a(t) \propto t^{2/(3+3w)} = \begin{cases} t^{2/3} & \text{for matter} \quad (w = 0), \\ t^{1/2} & \text{for radiation} \quad (w = 1/3), \\ \exp(t) & \text{for } \Lambda \quad (w = -1). \end{cases} \tag{20.68}$$

The special case with an exponential growing scale factor for $w = -1$ is called de Sitter spacetime.

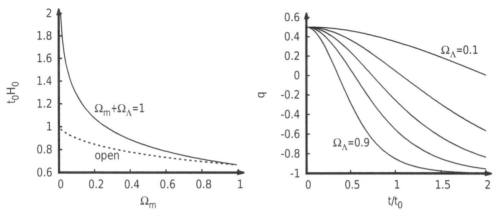

Fig. 20.1 Left: The product $t_0 H_0$ for an open universe containing only matter (dotted line) and for a flat cosmological model with $\Omega_\Lambda + \Omega_m = 1$ (solid line). Right: The deceleration parameter q as function of t/t_0 for the ΛCDM model and various values for Ω_Λ (0.1, 0.3, 0.5, 0.7 and 0.9 from the top to the bottom).

Age problem of the universe. The age of a matter-dominated universe expanded around $\Omega_0 = 1$ is given by (problem 20.12)

$$t_0 = \frac{2}{3H_0} \left[1 - \frac{1}{5}(\Omega_0 - 1) + \ldots \right]. \tag{20.69}$$

Globular cluster ages require $t_0 \geq 13$ Gyr. Using $\Omega_0 = 1$ leads to $h \leq 0.50$. Thus a flat universe with $t_0 = 13$ Gyr without cosmological constant requires a value of H_0 which is too small compared with observations. Choosing $\Omega_m = 0.3$ increases the age by just 14%, cf. also with the left panel of Fig. 20.1.

We derive the age t_0 of a flat universe with $\Omega_m + \Omega_\Lambda = 1$ in the next section as

$$\frac{3t_0 H_0}{2} = \frac{1}{\sqrt{\Omega_\Lambda}} \ln \frac{1 + \sqrt{\Omega_\Lambda}}{\sqrt{1 - \Omega_\Lambda}}. \tag{20.70}$$

Requiring $H_0 \geq 65$ km/s/Mpc and $t_0 \geq 13$ Gyr means that the function on the RHS should be larger than $\simeq 1.3$ or $\Omega_\Lambda \geq 0.55$. Thus the lower limit $t_0 \geq 13$ Gyr on the age of the universe together with a lower limit on the Hubble constant is sufficient to deduce a non-zero value of the cosmological constant.

The ΛCDM model. We consider as approximation to the late universe a flat model containing as its only two components pressure-less matter and a cosmological constant, $\Omega_m + \Omega_\Lambda = 1$. Thus the curvature term in the Friedmann equation and the pressure term in the deceleration equation play no role and we can hope to solve these equations for $a(t)$. Multiplying the deceleration equation (20.50) by 2 and adding it to the Friedmann equation (20.49), we eliminate ρ_m,

$$2\frac{\ddot{a}}{a} + \left(\frac{\dot{a}}{a}\right)^2 = \Lambda.$$ (20.71)

Next we rewrite first the LHS and then the RHS as total time derivatives. With

$$\frac{\mathrm{d}}{\mathrm{d}t}(a\dot{a}^2) = \dot{a}^3 + 2a\dot{a}\ddot{a} = \dot{a}a^2\left[\left(\frac{\dot{a}}{a}\right)^2 + 2\frac{\ddot{a}}{a}\right]$$ (20.72)

we obtain

$$\frac{\mathrm{d}}{\mathrm{d}t}(a\dot{a}^2) = \dot{a}a^2\Lambda = \frac{1}{3}\frac{\mathrm{d}}{\mathrm{d}t}(a^3)\Lambda.$$ (20.73)

Integrating is now trivial,

$$a\dot{a}^2 = \frac{\Lambda}{3}a^3 + C.$$ (20.74)

The integration constant can be determined most easily setting $a(t_0) = 1$ and comparing the Friedmann equation (20.49) with (20.74) for $t = t_0$ as $C = 8\pi G\rho_{\mathrm{m},0}/3$. Now we introduce the new variable $x = a^{3/2}$. Then

$$\frac{\mathrm{d}a}{\mathrm{d}t} = \frac{\mathrm{d}x}{\mathrm{d}t}\frac{\mathrm{d}a}{\mathrm{d}x} = \frac{\mathrm{d}x}{\mathrm{d}t}\frac{2x^{-1/3}}{3},$$ (20.75)

and we obtain as a new differential equation

$$\dot{x}^2 - \Lambda x^2/4 + C/3 = 0.$$ (20.76)

Inserting the solution $x(t) = A\sinh(\sqrt{\Lambda}t/2)$ of the homogeneous equation fixes the constant A as $A = \sqrt{3C/\Lambda}$. We can express A also by the current values of Ω_i as $A = \Omega_m/\Omega_\Lambda = (1 - \Omega_\Lambda)/\Omega_\Lambda$. Hence the time-dependence of the scale factor is

$$a(t) = A^{1/3}\sinh^{2/3}(\sqrt{3\Lambda}t/2).$$ (20.77)

The time scale of the expansion is set by $t_\Lambda = 2/\sqrt{3\Lambda}$. The present age t_0 of the universe follows by setting $a(t_0) = 1$ as

$$t_0 = t_\Lambda \mathrm{arctanh}\left(\sqrt{\Omega_\Lambda}\right).$$ (20.78)

The deceleration parameter $q = -\ddot{a}/aH^2$ is an important quantity for observational tests of the ΛCDM model. We first calculate the Hubble parameter

$$H(t) = \frac{\dot{a}}{a} = \frac{2}{3t_\Lambda}\coth(t/t_\Lambda)$$ (20.79)

and then find

$$q(t) = \frac{1}{2}[1 - 3\tanh^2(t/t_\Lambda)].$$ (20.80)

The limiting behaviour of q corresponds with $q = 1/2$ for $t \to 0$ and $q = -1$ for $t \to \infty$ as expected to the one of a flat $\Omega_m = 1$ and an $\Omega_\Lambda = 1$ universe. More interesting is the transition region and, as shown in the right panel of Fig. 20.1, the transition from a decelerating to an accelerating universe happens for $\Omega_\Lambda = 0.7$ at $t \approx 0.55t_0$. This can be easily converted to redshift, $z_* = a(t_0)/a(t_*) - 1 \approx 0.7$, that is directly measured in supernova observations. The fact that the acceleration has set in recently, i.e. that z_* is of order one, has been coined the coincidence problem.

Remark 20.1: Evolution of topological defects. We have argued that topological defects are produced during cosmological phase transition via the Kibble mechanism. In order to understand the impact of topological defects on the evolution of the universe, one has to determine their EoS. The simplest case are magnetic monopoles which are created with negligible velocities. Hence their energy density behaves as $\rho \sim 1/a^3$, and their abundance Ω increases during the radiation-dominated phase. As a result, they tend to dominate the energy density of the universe, leading to its early collapse, cf. exercise 20.15. Next we consider how the energy density of a network of domain walls evolves in an expanding universe. If we model them for simplicity as a set of static infinitely extended domain walls parallel to the yz plane, then the expansion of the universe along the y and z directions does not change the number density of domain walls. Therefore, the energy density of a network of domain walls evolves as $\rho \propto 1/a$. Simulations of dynamical networks of domain walls confirm approximately this behaviour. Interestingly, the equation of state of domain walls is thus negative, $w = -2/3$, leading to an accelerated expansion of the universe. However, even a single wall inside the Hubble radius containing, for example, the electroweak vacuum would overclose the universe. Thus broken discrete symmetries are in general in conflict with cosmological observations. The remaining option, a network of cosmic strings, is viable if the string scale is below $\mu \lesssim (10^{-4} M_{\mathrm{Pl}})^2$.

20.4   Horizons

An important consequence of the finite speed of light and the expansion of the universe is the possibility that regions of spacetime may be inaccessible for a given observer, either at the present time, or perhaps for all time. The borderline between the accessible and the inaccessible parts of the universe is termed a horizon.

Particle horizon.   We define the particle horizon d_p of a comoving observer O as the surface of the region in the past causally connected to O. Because of the homogeneity of the FLRW metric, it is sufficient to consider an observer at $\chi = 0$. The causally connected region is limited by light-rays, and thus bounded by

$$\chi_p(t_*) = \eta(t) - \eta(t_*) = \int_{t_*}^{t} \frac{\mathrm{d}t}{a(t)} = \int_0^{a(t)} \frac{\mathrm{d}a}{a\dot{a}}. \tag{20.81}$$

Here, the lower integration limit t_* is zero for models starting with a big bang. and $t_* \to -\infty$ otherwise. If the integral diverges, then the observer O can receive light-signals from an arbitrary spatial point in the past for sufficiently early t_*. If, however, the integral converges then $\chi_p(t_*)$ defines a particle horizon. The physical size of the horizon is given by $d_p(t) = a(t)\chi_p(t_*)$. If the model has a big bang origin, $a(0) = 0$, then also $z \to \infty$ for $t_* \to 0$ and thus the particle horizon corresponds to a surface of infinite redshift.

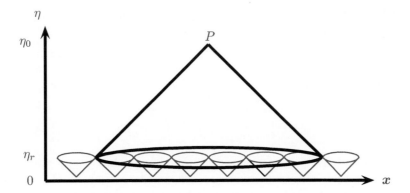

Fig. 20.2 Causal structure of a spacetime with big bang: particle horizons at the time η_r are shown in grey together with the event horizon of point P (black cone).

The second form of the integral shows that a finite particle horizon exists if $\ddot{a} < 0$, that is, if the model was decelerating for $t \to t_*$. We can show this explicitly, if a single form of matter dominates. Then $a(t) = a_0(t/t_0)^\alpha$, and thus

$$\chi_p(t_0) = \frac{1}{a_0} \int_0^t \mathrm{d}t \left(\frac{t}{t_0}\right)^{-\alpha} = \frac{t_0}{a_0(1-\alpha)} \tag{20.82}$$

for $\alpha < 1$. The physical size of the horizon follows as $d_p(t_0) = t_0/(1-\alpha)$. A particle horizon exists if $\alpha = 2/[3(1+w)] > 1$ or $w > -1/3$ and $q > 0$, that is for an decelerated expansion of the universe.

Differentiating the definition (20.81) gives $\mathrm{d}\chi_p/\mathrm{d}t = 1/a(t)$, and thus the particle horizon grows in an expanding universe. As time goes by, new objects become visible[2] after they cross the horizon and their redshift decreases from $z = \infty$. Note that in a universe with $w > -1/3$, these objects were never before in causal contact with the observer O. In such models, the observed homogeneity and isotropy of the universe on large scales is actually a very puzzling fact called the horizon problem. The best observational probe of anisotropies are the photons of the cosmic microwave background (CMB) which scattered at redshift $z \simeq 1300$ the last time. Their intensity is the one of a black-body, with the same temperature in all directions except for fluctuations $\delta T/T \sim 10^{-5}$.

Hubble scale. The Hubble scale (which is often called the Hubble horizon) is defined by $d_H(t) = H^{-1}(t)$. Since H^{-1} is of the same order as the curvature scalar R, the Hubble scale is a useful measure on which scales spacetime curvature can be neglected and a simpler Newtonian analysis using an inertial coordinate system is possible. Moreover, the Hubble scale approximately equals the particle horizon for decelerating cosmological models, with $H^{-1} = (3+3w)/2\,t$. It is therefore often used as a substitute

---

[2]Since photons can interact, the boundary of the visible universe is given more precisely by the "surface of last scattering", which is however comparable, see problem 20.13.

for the particle horizon. Note, however, that the difference between the two scales become very important for $w \to -1$.

Event horizon. Although the particle horizon grows in an expanding universe, there may be galaxies which may be never be seen by an observer. This is the case if the integral in Eq. (20.81) converges for $t \to t_{max}$. We define the event horizon by

$$\chi_p(t_*) = \int_{t_*}^{t_{max}} \frac{dt}{a(t)} = \int_0^{a(t)} \frac{da}{a\dot{a}} , \qquad (20.83)$$

where t_{max} is either infinite or corresponds to the big crunch, $a(t_{max}) = 0$, that is, the end of the universe in a future singularity. The existence of an event horizon in the limit $t_{max} \to \infty$ suggests that space expands faster than the speed of light, which is often termed as "superluminal expansion". Note that this notion is misleading, since the definition of the relative velocity of two observers O and O' relies on the existence of an (approximate) inertial system connecting the two. For distances $\gtrsim 1/H$, when the "relative velocity" becomes superluminal, such a connection is not possible and thus their relative velocity is not defined.

In summary, the particle horizon is the maximal distance from which we can receive signals, while the event horizon defines the maximal distance to which we can send signals. The horizon problem of a radiation or matter-dominated universe is illustrated schematically in Fig. 20.2. Particle horizons, i.e. causally connected regions, at the time of recombination η_r are shown in red. The event horizon (black cone) of the observer at the point P is much larger at the time of recombination and contains therefore many apparently causally disconnected patches.

Summary

A homogeneous and isotropic universe is described by an FLRW metric. This family of spacetimes is completely determined by the scale factor $a(t)$ and the parameter k distinguishing between hyperbolic, flat, or spherical 3-spaces. The time evolution of these models is specified, if at a given time t the Hubble parameter $H(t)$, the abundances Ω_i of all relevant matter forms and their equation of state $w_i = P_i/\rho_i$ is known. Observational data are well described by the ΛCDM model with $\Omega_\Lambda \simeq 0.7$, $\Omega_m \simeq 0.3$ and $H_0 \simeq 70$km/s/Mpc.

Further reading. For a discussion of important observations in cosmology and their interpretation see e.g. Coles and Lucchin (2002). The properties of simple cosmological models are discussed in more detail by Hobson *et al.* (2006).

Problems

20.1 Connection coefficients. Calculate the missing connection coefficient $\Gamma^\sigma{}_{\mu\nu}$ using a.) the geodesic equation (6.36) and b.) the definition (6.43) for the FLRW metric.

20.2 Comoving observers. Show that a) a particle with $\mathrm{d}x^i/\mathrm{d}\lambda = 0$ in the FLRW metric stays at rest; b.) a clock at rest shows the coordinate time.

20.3 Redshift evolution of P^μ. Solve the geodesic equation for P^μ for $k = 0$ and show that $P^i \propto a^{-2}$. Explain the difference to the redshift formula.

20.4 Derivative of a vector field. Calculate the covariant derivative of a vector field $\boldsymbol{V}$ in the FLRW metric. Show that for $V^\alpha = (1, \boldsymbol{0})$ Eq. (20.24) holds.

20.5 Redshift. Derive Lemaître's redshift formula for massless particles using that they propagate along null geodesics, $\mathrm{d}s^2 = 0$.

20.6 Planck distribution. Show that a photon gas in thermal equilibrium stays in equilibrium, with its temperature redshifting as $T' = T/(1 + z)$. How does this generalises to massive bosons?

20.7 Intensity. Assume that sources with constant comoving number density emit photons with the energy spectrum dN/dE. What is the resulting intensity $I_\gamma(E)$ of photons observed at earth?

20.8 Energy equation Derive $\dot\rho = -3(\rho + P)H$ from $\nabla_a T^{ab} = 0$.

20.9 Closed universe Rewrite the FLRW metric using conformal time $\mathrm{d}\eta = \mathrm{d}t/a$. Assume a closed matter-dominated universe for which $a(\eta) = \Omega/[2H_0(\Omega - 1)^{3/2}](1 - \cos\eta)$. i.) Draw a $\eta - \chi$ spacetime diagram for a closed matter-dominated universe including big bang, big crunch and the past light-cone of an observer at the origin at the moment of maximal expansion. ii.) Are there parts of the universe the observer can never see? iii.) Can an observer traverse the entire universe in the time between big bang and big crunch?

20.10 Temperature–time relation. Derive the relation between temperature and time in the early (radiation-dominated) universe using $\rho = g\pi^2 T^4/30$ as expression for the energy density of a gas with g relativistic degrees of freedom. What is the temperature at $t = 1\,\mathrm{s}$?

20.11 Matter-radiation equilibrium. Derive the redshift z_{eq} when matter and radiation contributed equally to the energy density of the universe (for their present day values see appendix A.1).

20.12 Matter with curvature. Find the evolution of a universe with $\Omega_m + \Omega_k = 1$. Derive Eq. (20.69).

20.13 Last scattering surface. Compare the physical size of the particle horizon in a radiation-dominated universe to the radius of the surface of last scattering of photons at redshift $z \sim 1300$. Find the angular size of causally connected regions on the last scattering surface.

20.14 Steady-state theory. The steady-state theory of the universe is based on the "perfect cosmological principle": the universe presents the same aspect from any place at any time. a.) Calculate the rate $\dot n$ of spontaneous matter creation required for such a model in an expanding universe (with $H = 70\mathrm{km/s/Mpc}$ and $\rho = \rho_{\mathrm{cr}} = 3H_0^2/(8\pi G)$). b.) How behaves the scale factor $a(t)$ as function of time? c.) What is the mean-age of galaxies in this model?

20.15 Magnetic monopole problem. Assume that at $T = M_{\mathrm{GUT}} = 10^{16}$ GeV one magnetic monopole with mass $M_{\mathrm{GUT}}/\alpha_{\mathrm{GUT}}$ per horizon is created. Determine the subsequent evolution of the universe.

21
Thermal relics

At large redshifts, we can treat the different contributions to the total energy density of the universe as nearly homogeneous fluids whose momenta are increasing as $p \propto 1/a$ for $a \to 0$. This energy increase has three main effects. First, bound states like atoms, nuclei and hadrons are dissolved when the temperature reaches their binding energy, $T \gtrsim E_b$. Second, particles with mass m_X can be produced in reactions like $\gamma\gamma \to \bar{X}X$, when $T \gtrsim 2m_X$. Finally, most scattering rates $\Gamma = n\sigma v$ increase faster than the expansion rate of the universe for $t \to 0$, since the number density n scales as $n \propto T^3$ for relativistic particles, while the expansion rate goes as $H \propto \rho_{\rm rad}^{1/2} \propto T^2$. Therefore, reactions that have become "frozen in" today may have been important in the early universe. As a result, the early universe consisted of a plasma containing particles with $m_X \lesssim T$ in a state of thermal equilibrium—except those particles that are very weakly interacting as, for example, axions.

In the previous chapter, we have assumed that the evolution of the universe proceeds adiabatically. While this is an excellent approximation during most of its evolution, deviations are crucial for the explanation of the present universe. Without such deviations, the abundance of all particles with masses smaller than the present temperature of the CMB, $m \gg T_{\rm CMB} \simeq 2 \times 10^{-4}\,{\rm eV}$ would be exponentially suppressed, or a non-zero chemical potential should ensure their survival. In particular, the abundance of baryons, of light elements from helium up to lithium, and probably of dark matter are connected to deviations from thermal equilibrium. Their present abundance can be calculated knowing their reaction rates and the expansion history of the universe, and we will develop the required formalism in this chapter.

21.1 Boltzmann equation

The evolution of the phase space distribution function of a thermal relic in the expanding universe can be modelled by a Boltzmann equation. As first step in its derivation, we recall Liouville's theorem and the resulting collisionless Boltzmann equation in Minkowski space, before we generalise this equation to a spacetime described by an FLRW metric.

Liouville's theorem. We consider first the evolution of a classical many-particle system in $6n$-dimensional phase space $\omega_j \equiv (q^j, p_j)$ with $j = 1, \ldots, 3n$. The phase-space density $f(q^j, p_j, t)$ determines the probability to find the system in the state $\omega = (q^j, p_j)$ at time t. Conservation of the particle number leads to a conservation law for f,

$$\frac{\partial f}{\partial t} + \frac{\partial}{\partial \omega_i}(f\dot{\omega}_i) = 0. \tag{21.1}$$

Quantum Fields–From the Hubble to the Planck Scale. Michael Kachelriess. © Michael Kachelriess 2018.
Published in 2018 by Oxford University Press. DOI 10.1093/oso/9780198802877.001.0001

We first use Hamilton's equations (1.22) to replace $\dot{\omega}$ with ω,

$$\frac{\partial}{\partial \omega_i}(f\dot{\omega}_i) = \frac{\partial}{\partial q^i}(f\dot{q}^i) + \frac{\partial}{\partial p_i}(f\dot{p}_i) = \frac{\partial}{\partial q^i}\left(f\frac{\partial H}{\partial p_i}\right) - \frac{\partial}{\partial p_i}\left(f\frac{\partial H}{\partial q^i}\right), \qquad (21.2)$$

and then that the mixed second derivatives of H commute, $\partial q^i \partial p_i = \partial p_i \partial q^i$,

$$\frac{\partial}{\partial \omega_i}(f\dot{\omega}_i) = \frac{\partial f}{\partial q^i}\frac{\partial H}{\partial p_i} - \frac{\partial f}{\partial p_i}\frac{\partial H}{\partial q^i} = \{f, H\} = \dot{q}^i\frac{\partial f}{\partial q^i} + \dot{p}_i\frac{\partial f}{\partial p_i}. \qquad (21.3)$$

Inserting these results into the conservation law (21.1), Liouville's theorem follows

$$\frac{\mathrm{d}f}{\mathrm{d}t} = \frac{\partial f}{\partial t} + \{f, H\} = \frac{\partial f}{\partial t} + \frac{\partial f}{\partial q^i}\dot{q}^i + \frac{\partial f}{\partial p_i}\dot{p}_i = 0. \qquad (21.4)$$

Thus the phase-space density $f(q^j, p_j, t)$ of a Hamiltonian system stays constant along a trajectory in phase space.

Boltzmann equation in Minkowski space. Particles in a thermal plasma can exchange energy via elastic collisions, and can be created and destroyed in inelastic collisions. We can split the time-evolution of such a system into the free movement of particles along spacetime geodesics interrupted by collisions, if these scatterings are caused by short-range interactions. Without scatterings, the $6n$-dimensional phase-space function is separable; it is the n-dimensional product of single-particle distribution functions. Therefore it is sufficient to describe the time-evolution of a collisionless many-particle system in the simpler six-dimensional phase space $\omega \equiv (\boldsymbol{q}, \boldsymbol{p})$. In a second step, we will account for interactions by adding a collision term.

Now the six-dimensional phase-space density $f(\boldsymbol{q}, \boldsymbol{p}, t)$ determines the density $\mathrm{d}N = f(\boldsymbol{x}, \boldsymbol{p})\mathrm{d}^3 x \mathrm{d}^3 p$ of particles in the state $\omega = (\boldsymbol{q}, \boldsymbol{p})$ at time t. We can also view this system as a fluid with $\dot{\omega}$ as the fluid velocity and f as its density. If the number of particles is not changed by interactions, then again a conservation law for f is valid,

$$\frac{\partial f}{\partial t} + \frac{\partial}{\partial \omega_i}(f\dot{\omega}_i) = \frac{\partial f}{\partial t} + \frac{\partial f}{\partial q^i}\dot{q}^i + \frac{\partial f}{\partial p_i}\dot{p}_i = 0. \qquad (21.5)$$

This is the collisionless Boltzmann equation in Minkowski space.

Boltzmann equation for an FLRW metric. Next we apply the Boltzmann equation to the case of a universe described by an FLRW metric. Particles move along geodesics which we parameterise by the parameter λ. Then the collisionless Boltzmann equation states that the phase-space density $f(\boldsymbol{x}, \boldsymbol{p}, t)$ stays constant along all trajectories, $\mathrm{d}f/\mathrm{d}\lambda = 0$. In order to evaluate the total derivative $\mathrm{d}f/\mathrm{d}t = 0$, we have to fix the dependence $\lambda(t)$. Using the choice of $\mathrm{d}\lambda = \mathrm{d}t/E$ as affine parameter (such that $\mathrm{d}x^\mu/\mathrm{d}\lambda = p^\mu$) and assuming an isotropic and uniform distribution of matter in space, $f(\boldsymbol{x}, \boldsymbol{p}, t) = f(p, t)$, we find the collisionless Boltzmann equation for an FLRW metric as

$$\frac{\mathrm{d}f}{\mathrm{d}\lambda} = \left(\frac{\mathrm{d}t}{\mathrm{d}\lambda}\frac{\partial}{\partial t} + \frac{\mathrm{d}\boldsymbol{p}}{\mathrm{d}\lambda}\frac{\partial}{\partial \boldsymbol{p}}\right)f(p, t) = E\left(\frac{\partial}{\partial t} - H\boldsymbol{p}\frac{\partial}{\partial \boldsymbol{p}}\right)f(p, t) = 0. \qquad (21.6)$$

In the second step we used also the redshift formula for a free particle,

$$\frac{\mathrm{d}\boldsymbol{p}}{\mathrm{d}\lambda} = \frac{\mathrm{d}\boldsymbol{p}}{\mathrm{d}t}\frac{\mathrm{d}t}{\mathrm{d}\lambda} = -H\boldsymbol{p}\,E. \tag{21.7}$$

Adding the effect of collisions, the total time derivative of the phase-space density f becomes non-zero,

$$\frac{\mathrm{d}f}{\mathrm{d}\lambda} = E\left(\frac{\partial}{\partial t} - H\boldsymbol{p}\frac{\partial}{\partial\boldsymbol{p}}\right)f(p,t) = C(p,t), \tag{21.8}$$

and is determined by the collision term $C(p,t)$.

Most often, we are not interested in the momentum dependence of $f(p,t)$ but only in the total density $n(t)$ of the relic particles. Dividing by the energy E and taking the first moment of the Boltzmann equation, we arrive at

$$g\int\frac{\mathrm{d}^3p}{(2\pi)^3}\left(\frac{\partial}{\partial t} - H\boldsymbol{p}\frac{\partial}{\partial\boldsymbol{p}}\right)f(p,t) = g\int\frac{\mathrm{d}^3p}{(2\pi)^3}\frac{1}{E_p}C(p,t). \tag{21.9}$$

The first term on the LHS is the time derivative of the number density n of relic particles, as a comparison to the general definition in Eq. (15.49) shows. In the second term, we use the isotropy of the particle distribution and perform a partial integration. The boundary term $p^3 f(p,t)|_0^\infty$ vanishes,[1] and thus we obtain

$$\frac{\partial}{\partial t}n + 3Hn = g\int\frac{\mathrm{d}^3p}{(2\pi)^3}\frac{1}{E_p}C(p). \tag{21.10}$$

We can express the LHS as the change of the comoving number density a^3n,

$$\frac{\mathrm{d}(a^3n)}{\mathrm{d}t} = a^3\,g\int\frac{\mathrm{d}^3p}{(2\pi)^3}\frac{1}{E_p}C(p) \equiv a^3\frac{\mathrm{d}n_{\mathrm{col}}}{\mathrm{d}t}, \tag{21.11}$$

and thus the RHS is the net change $a^3\dot{n}_{\mathrm{col}}$ of the comoving particle density due to interactions.

Let us determine $\dot{n}_{\mathrm{col}}$ for the case that the relic particle interacts via decays and inverse decays, $1 \leftrightarrow 1'2'\cdots n'$, with other particles in the thermal plasma. Recall first Eq. (9.117), that gives the number of decays $1 \to 1'2'\cdots n'$ per time of a given particle 1 with momentum $\boldsymbol{p}_1$ at temperature $T = 0$. This rate should be multiplied with $f_1(\boldsymbol{p})$ to get the number of decays per time of all relic particles of type 1 with momentum $\boldsymbol{p}_1$. Moreover, in a thermal bath, we have to account for the Pauli blocking of fermions and the stimulated emission of bosons by adding the factors $1 \pm f_{i'}(\boldsymbol{p}_{i'})$ to the final-state phase space. The inverse reaction $1'2'\cdots n' \to 1$ is proportional to the number of plasma particles and contains a factor $1 \pm f_1(\boldsymbol{p})$ in the final phase space. Combining everything, we have

$$\frac{\mathrm{d}n_1^{\mathrm{col}}(\boldsymbol{p}_1)}{\mathrm{d}t} = -\frac{1}{2E_1}\int\prod_{f=1}^{n'}\frac{\mathrm{d}^3p_f}{2E_f(2\pi)^3}(2\pi)^4\delta^{(4)}(p_1 - P_f)$$
$$\times\left[f_1(1\pm f_{1'})\cdots(1\pm f_{n'})|\widetilde{\mathcal{A}_{fi}}|^2 - f_{1'}\cdots f_{n'}(1\pm f_1)|\widetilde{\mathcal{A}_{if}}|^2\right], \tag{21.12}$$

---

[1] A finite energy density requires that $f(p)$ decreases faster than $1/p^4$ for large p.

with $P_f = \sum_{i'} p_{i'}^{\mu}$, $f_i \equiv f(\boldsymbol{p}_i)$, while $|\widetilde{\mathcal{A}_{fi}}|^2 = \sum_{s_1,\dots,s_{n'}} |\mathcal{A}_{fi}|^2$ denotes the squared Feynman amplitude summed over both final and initial internal degrees of freedom. If necessary, symmetry factors $S = 1/n!$ should be added for n identical particles in the initial or final state. As final step, we have to integrate the differential rate $\dot{n}_1^{\text{col}}(\boldsymbol{p}_1)$ over $\mathrm{d}^3 p_1/(2\pi)^3$ to obtain the total collision rate $\dot{n}_1^{\text{col}}$ per physical volume, finding as Boltzmann equation

$$\frac{1}{a^3} \frac{\mathrm{d}(a^3 n_1^{\text{col}})}{\mathrm{d}t} = -\int \frac{\mathrm{d}^3 p_1}{2E_1(2\pi)^3} \prod_{f=1}^{n} \frac{\mathrm{d}^3 p_f}{2E_f(2\pi)^3} (2\pi)^4 \delta^{(4)}(p_1 - P_f)$$

$$\times \left[f_1(1 \pm f_{1'}) \cdots (1 \pm f_{n'}) |\widetilde{\mathcal{A}_{fi}}|^2 - f_{1'} \cdots f_{n'}(1 \pm f_1) |\widetilde{\mathcal{A}_{if}}|^2 \right].$$

(21.13)

It is instructive to tailor this equation to the simplest case of a $1 \leftrightarrow 2 + 3$ process. Then the decay widths scale as $\Gamma_{1 \to 2+3} \propto g_1^{-1} \sum_{s_1, s_2, s_3} |\mathcal{A}_{1 \to 2+3}|^2$ and $\Gamma_{2+3 \to 3} \propto g_2^{-1} g_3^{-1} \sum_{s_1, s_2, s_3} |\mathcal{A}_{2+3 \to 1}|^2$, respectively. Thus we can rewrite the Boltzmann equation as

$$\frac{1}{a^3} \frac{\mathrm{d}(a^3 n_1^{\text{col}})}{\mathrm{d}t} = -n_1 \langle \Gamma_{1 \to 2+3} \rangle + n_2 n_3 \langle \Gamma_{2+3 \to 1} \rangle,$$

(21.14)

where the factors g_i entered the densities n_i and $\langle \Gamma \rangle$ are the rates thermally averaged over the gamma factors m_i/E_i. Going over from a decay $1 \leftrightarrow 1'2' \cdots n'$ to a scattering process $12 \cdots n \leftrightarrow 1'2' \cdots m'$, we should add to Eq. (21.13) the corresponding phase space factors

$$\int \frac{\mathrm{d}^3 p_2}{2E_i(2\pi)^3} f_2(\boldsymbol{p}_2) \quad \text{and} \quad \int \frac{\mathrm{d}^3 p_2}{2E_i(2\pi)^3} (1 - f_2(\boldsymbol{p}_2))$$

to the first and second term in the square bracket of Eq. (21.13), respectively.

In most cases of interest, the collision rate can be simplified drastically. First, we can set $|\mathcal{A}_{fi}|^2 \simeq |\mathcal{A}_{if}|^2$, because CP violation is generically small.[2] At tree level, this relation holds exactly. Second, kinetic decoupling happens typically much later than chemical decoupling. Consider, for example, a heavy relic particle X scattering with neutrinos as part of the thermal bath. Then, elastic collisions $X\nu \to X\nu$ may keep the distribution function of X still close to the one in thermal equilibrium, although inelastic collisions like $\bar{\nu}\nu \to \bar{X}X$ became ineffective below $T \sim m_X$. As a result, deviations of the relic density from a thermal equilibrium distribution in the time between chemical and kinetic decoupling can be simply parameterised by a time-dependent chemical potential $\mu(t)$. Third, we can assume that the densities are small enough such that the distribution functions can be approximated by Maxwell–Boltzmann distributions,

$$n(t) = e^{\beta(t)\mu(t)} n_{\text{eq}}(t) = e^{\beta(t)\mu(t)} g \int \frac{\mathrm{d}^3 p}{2E(2\pi)^3} \exp\left(-\beta(t)E\right).$$

(21.15)

Since this corresponds to the classical limit, we can also neglect the factors for Pauli blocking and stimulated emission, setting $1 \pm f_i \simeq 1$. Finally, we often (but not always)

---

[2]We postpone the effect of CP violation to chapter 22.

assume that in the early universe any asymmetry in the number of particles and antiparticles is zero. If $\mu(0) = \bar{\mu}(0) = 0$, then $\mu(t) = \bar{\mu}(t)$ also at later times.

We now employ the first three approximations but keep $\mu_i \neq 0$. Specialising also to the case of $12 \to 34$ scatterings, the second line in (21.13) becomes

$$\left[f_1 f_2 (1 \pm f_3)(1 \pm f_4)|\widetilde{\mathcal{A}_{fi}}|^2 - f_3 f_4 (1 \pm f_1)(1 \pm f_2)|\widetilde{\mathcal{A}_{if}}|^2 \right] \tag{21.16}$$

$$\simeq e^{-\beta(E_1 + E_2)} \left(e^{\beta(\mu_1 + \mu_2)} - e^{\beta(\mu_3 + \mu_4)} \right) |\widetilde{\mathcal{A}_{fi}}|^2 \tag{21.17}$$

$$= e^{-\beta(E_1 + E_2)} \left[\frac{n_1 n_2}{n_1^{eq} n_2^{eq}} - \frac{n_3 n_4}{n_3^{eq} n_4^{eq}} \right] |\widetilde{\mathcal{A}_{fi}}|^2 . \tag{21.18}$$

Here we used also energy conservation, $E_1 + E_2 = E_3 + E_4$, before we eliminated in the final step the chemical potentials in favour of the number densities, using $n_i = e^{\beta \mu_i} n_i^{eq}$. Now we define the thermally averaged cross-section $\langle \sigma v \rangle$ as

$$\langle \sigma v \rangle \equiv \frac{1}{n_1^{eq} n_2^{eq}} \prod_{i=1}^{4} \int \frac{d^3 p_i}{2 E_i (2\pi)^3} \, e^{-\beta(E_1 + E_2)} \, (2\pi)^4 \delta^{(4)}(P_i - P_f)|\widetilde{\mathcal{A}_{fi}}|^2 . \tag{21.19}$$

Here the velocity v in $\langle \sigma v \rangle$ is the relative or Møller velocity, $v \equiv v_{\text{Møl}}$. Its sole purpose is to cancel the corresponding factor $v_{\text{Møl}}$ in the flux factor I contained in the $T = 0$ cross-section σ, compare with Eq. (9.149). As a result, the integration on the RHS is over Lorentz-invariant factors that do contain no explicit factor $v_{\text{Møl}}$. Employing the definition (21.19) for $\langle \sigma v \rangle$, we arrive at a rather compact expression for the Boltzmann equation,

$$\frac{1}{a^3} \frac{d(a^3 n_1)}{dt} = -n_1^{eq} n_2^{eq} \langle \sigma v \rangle \left[\frac{n_1 n_2}{n_1^{eq} n_2^{eq}} - \frac{n_3 n_4}{n_3^{eq} n_4^{eq}} \right] . \tag{21.20}$$

Note that we managed to reduce the integro-differential equation (21.8) to an ordinary differential equation. This equation, together with the initial condition $n_1 \simeq n_{eq}$ for $T \to \infty$, determines $n_1(t)$ for a given thermal cross-section $\langle \sigma v \rangle$.

Important applications of this equation are the freeze-out of the baryon asymmetry and the dark matter abundance, the formation of light elements in big bang nucleosynthesis (BBN), and the recombination of electrons and protons into hydrogen. While the structure of the Boltzmann equation is the same for all theses processes, the energy scales involved vary from few eV to scales possibly as high as 10^{16} GeV. Correspondingly, the microphysics required as input to describe recombination consists of atomic physics, while BBN depends on weak interactions between nucleons and neutrinos as well as on nuclear physics. By contrast, the freeze-out of DM is a problem involving mainly particle physics. Therefore we will pick out this problem for a discussion of the freeze-out mechanism which is at work in all the four processes. After that, we will only sketch the main points of BBN. We postpone discussing baryogenesis until the next chapter.

21.2 Thermal relics as dark matter

A wealth of observational data suggests that a viable dark matter (DM) candidate has to be non-baryonic and should be non-relativistic, at least from the time of matter–radiation equilibrium on (Lisanti, 2016). As none of the particles in the SM has the required properties, DM has to belong to a new sector of particles beyond the SM. The various particles X proposed as DM candidates can be divided in two main sub-categories: thermal relics were in chemical equilibrium with the thermal plasma at least once during the history of the universe, while non-thermal relics have either sufficiently small interactions or a high enough mass m_X never to be produced efficiently by processes like $e^-e^+ \to XX$. Next we consider thermal relics, while we come back to the topic of non-thermal production of relics later.

21.2.1 Abundance of thermal relics

We assume that the thermal relic X and their annihilation products are symmetric, that is, that no asymmetry in the number density of particles and antiparticles exist. Moreover, we assume that the annihilation products are in thermal equilibrium. Then the Boltzmann equation simplifies to,

$$\frac{1}{a^3}\frac{\mathrm{d}(a^3 n_X)}{\mathrm{d}t} = -\langle\sigma_{\mathrm{ann}}v\rangle\left[n_X^2 - (n_X^{\mathrm{eq}})^2\right]. \qquad (21.21)$$

In the calculation of $\langle\sigma_{\mathrm{ann}}v\rangle$ a sum over all relevant final states should be included. Knowing $\langle\sigma_{\mathrm{ann}}v\rangle$ as function of temperature, this equation can be numerically integrated using as initial condition $n_X \simeq n_X^{\mathrm{eq}}$ for a large enough initial temperature T. In order to obtain simple numerical estimates and gaining some insight which physical parameters determine the final abundance, we will develop instead the more intuitive *Gamov criterion*: it states that a reaction becomes ineffective, when its rate Γ drops below the expansion rate H of the universe. Note that H defines also the typical time scale for changes in the temperature T. Thus $\Gamma \ll H \propto \dot{T}/T$ implies that the particle cannot longer maintain an equilibrium distribution. For the important case of a power law expansion, the condition $\Gamma \ll H$ means that the time $1/\Gamma$ between a reaction becomes larger than the age $t \propto 1/H(t)$ of the universe. In the case of annihilations, we thus have to determine the "freeze-out" time t_{f} defined by $\Gamma_{\mathrm{A}}(t_{\mathrm{f}}) = H(t_{\mathrm{f}})$ with $\Gamma_{\mathrm{A}} = n_{\mathrm{eq}}\langle\sigma_{\mathrm{ann}}v\rangle$ as the annihilation rate. At later times, the comoving density of the relic particle is then constant.

Freeze-out of thermal relic particles. Without interactions, the number density n_X of a particle species X is only diluted by the expansion of space, $n_X \propto a^{-3}$. It is convenient to account for this trivial expansion effect by dividing n_X through the entropy density[3] s, that is, to use the dimensionless quantity $Y = n_X/s$. The equilibrium abundance Y_{eq} expressed as function of $x = T/m_X$ is given for vanishing chemical potential $\mu_X = 0$ by

---

[3]Some formulae from equilibrium statistics are collected in appendix 15.A.

$$Y_{\rm eq} = \frac{n_X}{s} = \begin{cases} \frac{45}{2\pi^4}\left(\frac{\pi}{8}\right)^{1/2}\frac{\varepsilon_1 g_X}{g_{*S}}\,x^{3/2}\exp(-x) = 0.145\,\frac{\varepsilon_1 g_X}{g_{*S}}\,x^{3/2}\exp(-x) & \text{for } x \gg 3, \\ \frac{45\zeta(3)}{2\pi^4}\frac{\varepsilon g_X}{g_{*S}} = 0.278\,\frac{\varepsilon g_X}{g_{*S}} & \text{for } x \ll 3 \end{cases}$$

$$(21.22)$$

with $\varepsilon = 3/4$ ($\varepsilon = 1$) for fermions (bosons). If the particle X is in chemical equilibrium, its abundance is determined for $T \gg m$ by its contribution to the total number of degrees of freedom of the plasma, while $Y_{\rm eq}$ is exponentially suppressed for $T \ll m$.

In an expanding universe, one may expect that the reaction rate Γ for processes like $e^+e^- \leftrightarrow \bar{X}X$ drops below the expansion rate H mainly for two reasons. First, cross-sections may decrease with energy as, for example, in weak processes with $\sigma \propto s \propto T^2$ for $s \lesssim m_W^2$. Second, the density n_X decreases at least as $n_X \propto T^3$. When a reaction rate drops below the expansion rate it becomes ineffective or "freezes out". Around this freeze-out time x_f, the true abundance Y starts to deviate from the equilibrium abundance $Y_{\rm eq}$ and becomes constant, $Y(x) \simeq Y_{\rm eq}(x_f)$ for $x \gtrsim x_f$. This behaviour is illustrated in Fig. 21.1.

Next we rewrite the evolution equation for $n_X(t)$ using the dimensionless variables Y and x. We assume that the freeze-out occurs during the radiation-dominated epoch. Thus $\rho_{\rm rad} \propto 1/a^4$, $H = 1/(2t)$ and the curvature term k/R^2 can be neglected. Then the Friedmann equation simplifies to $H^2 = (8\pi/3)G\rho$ with $\rho = g_*\pi^2/30T^4$, or

$$\frac{1}{2t} = H = 1.66\sqrt{g_*}\,\frac{T^2}{M_{\rm Pl}} \propto x^{-2}.$$

$$(21.23)$$

Here we introduced also the Planck mass $M_{\rm Pl} = 1/\sqrt{G_N} \simeq 1.2 \times 10^{19}$ GeV. Changing then from $n_X = sY$ to Y, we eliminate the $3Hn_X$ term obtaining

$$\frac{dY}{dx} = -\frac{sx}{H}\langle\sigma_{\rm ann}v\rangle\left(Y^2 - Y_{\rm eq}^2\right).$$

$$(21.24)$$

Finally we recast the Boltzmann equation in a form that makes our intuitive Gamov criterion explicit,

$$\frac{x}{Y_{\rm eq}}\frac{dY}{dx} = -\frac{\Gamma_A}{H}\left[\left(\frac{Y}{Y_{\rm eq}}\right)^2 - 1\right]$$

$$(21.25)$$

with $\Gamma_A = n_{\rm eq}\langle\sigma_{\rm ann}v\rangle$. The relative change of Y is controlled by the factor Γ_A/H times the deviation from equilibrium. The evolution of $Y = n_X/s$ is shown schematically in Fig. 21.1. As the universe expands and cools down, n_X decreases at least as a^{-3}. Therefore, the annihilation rate quenches and is not longer sufficiently large to keep the particle in chemical equilibrium. As a result, the abundance freezes-out, that is, the ratio n_X/s stays constant. For the discussion of approximate solutions to this equation, it is convenient to distinguish, according to the freeze-out temperature, hot dark matter (HDM) with $x_f \ll 3$, cold dark matter (CDM) with $x_f \gg 3$ and the intermediate case of warm dark matter with $x_f \sim 3$.

Abundance of hot dark matter. For $x_f \ll 3$, freeze-out occurs when the particle is still relativistic and $Y_{\rm eq}$ is not changing with time. Thus the asymptotic value of the abundance, $Y(x \to \infty) \equiv Y_\infty$, is given by the equilibrium value at freeze-out,

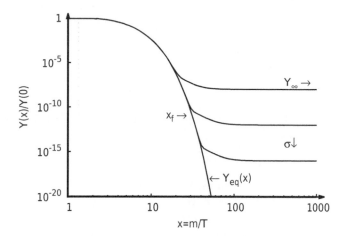

Fig. 21.1 Illustration of the freeze-out process: For $x \gtrsim x_{\mathrm{f}}$, the abundance Y becomes constant, $T \to Y_\infty$; for increasing $\langle \sigma_{\mathrm{ann}} v \rangle$, the final abundance Y_∞ decreases.

$$Y_\infty = Y_{\mathrm{eq}}(x_{\mathrm{f}}) = 0.278 \, \frac{g_{\mathrm{eff}}}{g_{*S}}, \tag{21.26}$$

and the only temperature-dependence is contained in g_{*S}. The number density today is then

$$n_0 = s_0 Y_\infty = 2970 \, Y_\infty \mathrm{cm}^{-3} = 825 \, \frac{g_{\mathrm{eff}}}{g_{*S}} \, \mathrm{cm}^{-3}, \tag{21.27}$$

where we used for the present entropy density $s_0 \simeq 2891/\mathrm{cm}^3$ derived in problem 15.6. Although an HDM particle was relativistic at freeze-out it is now non-relativistic if its mass is $m \gg 3\mathrm{K} \simeq 0.2\,\mathrm{meV}$. In this case its energy density is simply $\rho_0 = m s_0 Y_\infty$ and its abundance $\Omega h^2 = \rho_0 / \rho_{\mathrm{cr}}$ is given by

$$\Omega h^2 = 7.8 \times 10^{-2} \, \frac{m}{\mathrm{eV}} \, \frac{g_{\mathrm{eff}}}{g_{*S}} \,. \tag{21.28}$$

Hence the abundance of HDM particles with mass $m \gtrsim 10\,\mathrm{eV}$ exceeds the observed total abundance of matter, $\Omega_m h^2 \simeq 0.1$.

Abundance of cold dark matter. For CDM with $x_{\mathrm{f}} \ll 3$, freeze-out occurs when the particles are already non-relativistic and Y_{eq} is exponentially changing with time. Thus the main problem is to find x_{f}, for later times we use $Y_\infty \approx Y(x_{\mathrm{f}})$, i.e. the equilibrium value at freeze-out. We parameterise the temperature-dependence of the annihilation cross-section as $\langle \sigma_{\mathrm{ann}} v \rangle = \sigma_0 (T/m)^n = \sigma_0 / x^n$ which corresponds to an expansion in $v_{\mathrm{Møl}}^{2n}$. For simplicity, we consider only the case of s-wave annihilation for CDM, $n = 0$. Then the Gamov criterion becomes with $H = 1.66 \sqrt{g_*} \, T^2 / M_{\mathrm{Pl}}$ and $\Gamma_A = n_{\mathrm{eq}} \langle \sigma_{\mathrm{ann}} v \rangle$

$$g \left(\frac{m T_{\mathrm{f}}}{2\pi} \right)^{3/2} \exp(-m/T_{\mathrm{f}}) \, \sigma_0 = 1.66 \sqrt{g_*} \, \frac{T_{\mathrm{f}}^2}{M_{\mathrm{Pl}}} \tag{21.29}$$

or

$$x_{\mathrm{f}}^{-1/2} \exp(x_{\mathrm{f}}) = 0.038 \frac{g}{\sqrt{g_*}} M_{\mathrm{Pl}} m \sigma_0 \equiv C. \tag{21.30}$$

To obtain an approximate solution, we neglect first in

$$\ln C = -\frac{1}{2} \ln x_{\mathrm{f}} + x_{\mathrm{f}} \tag{21.31}$$

the slowly varying term $\ln x_{\mathrm{f}}$. Inserting next $x_{\mathrm{f}} \approx \ln C$ into Eq. (21.31) to improve the approximation then gives

$$x_{\mathrm{f}} = \ln C + \frac{1}{2} \ln(\ln C). \tag{21.32}$$

For a DM particle with thermal annihilation cross-section $\sigma_0 = 3 \times 10^{-26} \mathrm{cm}^3/\mathrm{s}$ the freeze-out temperature changes slowly from $x_{\mathrm{f}} \simeq 23$ for $m = 10\,\mathrm{GeV}$ to $x_{\mathrm{f}} \simeq 28$ for $m = 1\,\mathrm{TeV}$. The relic abundance for CDM follows from $n(x_{\mathrm{f}}) = 1.66\sqrt{g_*}\,T_{\mathrm{f}}^2/(\sigma_0 M_{\mathrm{Pl}})$ and $n_0 = n(x_{\mathrm{f}})[R(x_{\mathrm{f}})/R_0]^3 = n(x_{\mathrm{f}})[g_{*,f}/g_{*,0}][T_0/T(x_{\mathrm{f}})]^3$ as

$$\rho_0 = mn_0 \approx 10 \frac{x_{\mathrm{f}} T_0^3}{\sqrt{g_{*,f}} \sigma_0 M_{\mathrm{Pl}}} \tag{21.33}$$

or

$$\Omega_{\mathrm{CDM}} h^2 = \frac{mn_0}{\rho_{\mathrm{cr}}} \approx \frac{1 \times 10^{-28} \mathrm{cm}^3/\mathrm{s}}{\sigma_0} x_{\mathrm{f}}. \tag{21.34}$$

Thus the abundance of a CDM particle is inverse proportionally to its annihilation cross-section, since a more strongly interacting particle stays longer in equilibrium. Note that the explicit dependence on the freeze-out temperature T_{f} in the factors $n(x_{\mathrm{f}})$ and $[a(x_{\mathrm{f}})/a_0]^3$ cancelled, leading only to an implicit dependence of the abundance via $g_{*,\mathrm{f}}$ on T_{f}. Moreover, the abundance depends only logarithmically on the mass m via Eq. (21.32).

The observed value $\Omega_{\mathrm{CDM}} h^2 \simeq 0.1$ implies $\sigma_0 \simeq 3 \times 10^{-26} \mathrm{cm}^3/\mathrm{s} = 1 \times 10^{-36} \mathrm{cm}^2$. Such cross-sections are typical for a weakly interacting massive particles, and thus such CDM particles are called WIMPs. The numerical coincidence of the annihilation cross-section of a thermal relic with a typical weak interaction cross-section in the SM has been called by *aficionados* of this scenario the WIMP miracle.

Unitarity limit for m_X. We saw that the abundance of a CDM particle scales approximately inversely with its annihilation cross-section. Therefore the observed abundance of DM implies a lower bound on $\langle \sigma v \rangle$. We show in this section that the unitarity of the S-matrix implies $\langle \sigma v \rangle \propto 1/m_X^2$ and derive thereby an upper bound on its mass m_X.

We repeat the steps from Eq. (14.33) to (14.39) which lead to the unitary bound on partial waves amplitudes, but take into account that the final states differs from the initial one, $p_{\mathrm{cms}} \neq p'_{\mathrm{cms}}$. Comparing

$$\sigma_{\mathrm{tot}} = \frac{16\pi}{s} \frac{p'_{\mathrm{cms}}}{p_{\mathrm{cms}}} \sum_{l=0}^{\infty} (2l+1)|T_l|^2 \equiv \sum_{l=0}^{\infty} \sigma_l, \tag{21.35}$$

to the result using the optical theorem (14.36) gives then the unitarity constraint

$$\frac{2p'_{\text{cms}}}{\sqrt{s}}\, |T_l| \leq 1\,.\tag{21.36}$$

Inserting this constraint for T_l bounds the contribution of the l.th partial wave to the total cross-section σ_l as

$$\sigma_l = \frac{4\pi}{p^2_{\text{cms}}}(2l+1)\,.\tag{21.37}$$

For the non-relativistic scattering of two particles with mass m_X, we can use $p_{\text{cms}} \simeq v_{\text{Møl}} m_X/2$, giving as unitarity limit

$$\sigma_l v_{\text{Møl}} \leq (2l+1)\frac{16\pi}{v_{\text{Møl}} m_X^2}\,.\tag{21.38}$$

A bound stronger by a factor four applies to the annihilation cross-section, $\sigma_{\text{ann}} = \sigma_{\text{tot}} - \sigma_{\text{el}}$. In the case of particles with spin, we have to divide this result by the number of spin degrees of freedom of the initial particles (Griest and Kamionkowski, 1990; Hui, 2001). For a CDM particle, only the $l = 0$ and $l = 1$ partial waves give a sizeable contribution to the total annihilation cross-section. Using then $\Omega_X h^2 \leq \Omega_{\text{CDM}} h^2 = 0.11$ and $v_{\text{Møl}}^2 = x_{\text{f}}/6$ and $x_{\text{f}} \approx 30$ we obtain an upper limit of $m_X \lesssim 20\,\text{TeV}$ for any stable particle that was once in thermal equilibrium.

21.2.2 Annihilation cross-section in the non-relativistic limit

Since WIMPs freeze-out typically at $x_{\text{f}} \sim 20$, it is useful to consider the WIMP annihilation cross-section in the non-relativistic limit $x \gg 1$. This is even more true for annihilations today in the Milky Way where WIMPs have velocities of order $v \sim 220\,\text{km/s} \sim 10^{-3}$. An expansion in $x \propto v^2$ corresponds to a partial wave expansion. Therefore we can either project out from the general Feynman amplitude the contribution to the first partial waves $l = 0, 1, \ldots$, or we can perform an expansion in v^2 of the thermally averaged cross-section. Such an expansion contains even powers of the Møller velocity,

$$\langle \sigma_{\text{ann}} v \rangle = \sigma_{\text{ann}}^{(s)} + \sigma_{\text{ann}}^{(l)} \langle v \rangle^2 + \sigma_{\text{ann}}^{(p)} \langle v \rangle^4 + \ldots,\tag{21.39}$$

assuming that the Feynman amplitude is not singular in the limit $v \to 0$. The occurrence of singularities signals that the initial state can form a bound state. For instance, at low energies, the process $e^+ e^- \to n\gamma$ should not be described as the annihilation of a free electron and a positron but as the annihilation of its bound-state positronium. More generally, the small relative velocities of CDM particles, $\beta = v \ll 1$, means that factors g^2/β or $\ln(g^2/\beta)$ can lead to a breakdown of perturbation theory. This effect was first studied by Sommerfeld for Coulomb interactions, and the resulting boost of reaction rates is therefore called Sommerfeld enhancement.

The enhancement of the annihilation cross-section can be calculated non-relativistically and is, neglecting bound-state effects, characterised by two parameters: the mass ratio $\varepsilon = M/m_X$ of the exchange and the DM particle determines, if the annihilation proceeds in the Coulomb ($\varepsilon \ll 1$) or in the Yukawa ($\varepsilon \gg 1$) regime, while the ratio $x = g_{\text{eff}}^2/\beta$ of the squared effective coupling constant and the velocity determines, if factors g_{eff}^2/β or $\ln(g_{\text{eff}}^2/\beta)$ lead to a breakdown of perturbation theory. Here, the effective coupling constant g_{eff} includes all prefactors in front of the Yukawa potential, as, for example, mixing matrix elements. In the Coulomb case, the Sommerfeld factor $\mathcal{R}$ as ratio of the perturbative and non-perturbative annihilation cross-section is given by

$$\mathcal{R} = \frac{\sigma_{\text{np}}}{\sigma_{\text{pert}}} \sim \frac{\eta}{1 - \exp(-\eta)} \tag{21.40}$$

with $\eta = \pm g_{\text{eff}}^2/(2\beta)$ (Landau and Lifshitz, 1981).

Projection operators and the non-relativistic limit.   We now give an example how one can obtain the non-relativistic expansion of a Feynman amplitude by projecting out the the partial waves $l = 0, 1, \ldots$. In many models, the DM particle is a Majorana fermion which leads to some particular features which the example will also illustrate.

We use spinors in the Dirac representation (8.51) to describe a Majorana fermion pair at rest. The adjoint antiparticle spinors are $\bar{v}(m, -) = \sqrt{2m}(0, 0, 0, -1)$ and $\bar{v}(m, +) = \sqrt{2m}(0, 0, 1, 0)$. Since we have a pair of indistinguishable fermions in the initial state, we have to antisymmetrise the initial state. We compute the antisymmetrised two-particle state $u(m, s_1)\bar{v}(m, s_2) - u(m, s_2)\bar{v}(m, s_1)$ for different spin configurations: If the two spins are parallel, then the result is zero,

$$u(m, -)\bar{v}(m, -) - u(m, -)\bar{v}(m, -) = u(m, +)\bar{v}(m, +) - u(m, +)\bar{v}(m, +) = 0. \tag{21.41}$$

For antiparallel spins, we obtain

$$u(m, +)\bar{v}(m, -) - u(m, -)\bar{v}(m, +) = 2m \begin{pmatrix} 0 & 1 \\ 0 & 0 \end{pmatrix} \equiv \psi_1 \tag{21.42}$$

and

$$u(m, -)\bar{v}(m, +) - u(m, +)\bar{v}(m, -) = -2m \begin{pmatrix} 0 & 1 \\ 0 & 0 \end{pmatrix} \equiv \psi_2. \tag{21.43}$$

The two states ψ_1 and ψ_2 are linearly dependent and we can combine them into

$$\Phi = \frac{1}{\sqrt{2}}(\psi_1 - \psi_2) = 2\sqrt{2}m \begin{pmatrix} 0 & 1 \\ 0 & 0 \end{pmatrix}. \tag{21.44}$$

Next we want to rewrite the expression for Φ which is valid in the rest-frame of the two Majorana fermions into a Lorentz invariant way. We express first $\begin{pmatrix} 0 & 1 \\ 0 & 0 \end{pmatrix}$ by gamma matrices,

$$\Phi = 2\sqrt{2}m \frac{1}{2}(1 + \gamma^0)\gamma^5 = \sqrt{2}(m + P\!\!\!/ /2)\gamma^5. \tag{21.45}$$

In the second step we introduced the total momentum $P = (p_1 + p_2) = (2m, \mathbf{0})$ of the fermion pair and replaced $m\gamma^0$ with $P\!\!\!/ /2$.

Let us now illustrate the usefulness of this method with a concrete example. We consider the annihilation of two Majorana fermions χ with mass m_χ into a fermion pair $f\bar{f}$ with mass m_f via the exchange of a scalar $\tilde{f}_L$ with mass M. Their interaction is given by

$$\mathscr{L} = g_L\bar{\chi}P_L f\tilde{f}_L + \text{h.c.} = g_L\bar{\chi}P_L f\tilde{f}_L + g_L\bar{f}P_R\chi\tilde{f}_L. \tag{21.46}$$

For simplicity, we consider the limit that the mass M of the exchanged scalar is much larger than the DM mass, $M \gg m_\chi$. Then the interaction becomes an effective four-fermion interaction, and the Feynman amplitude simplifies to

$$\mathcal{A} = \frac{g_L^2}{M^2}\bar{v}_{\bar{f}}(p_4)P_R[u_\chi(m_\chi,s_1)\bar{v}_\chi(m_\chi,s_2) - u_\chi(m_\chi,s_2)\bar{v}_\chi(m_\chi,s_1)]P_L u_f(p_3) \tag{21.47a}$$

$$= \frac{\sqrt{2}g_L^2}{M^2}\bar{v}_{\bar{f}}(p_4)P_R(m_\chi + \not{P}/2)\gamma^5 P_L u_f(p_3). \tag{21.47b}$$

The m_χ term vanishes because of $P_L P_R = 0$. Using $P = p_3+p_4$ and the Dirac equation, we obtain $p_3 + p_4 = -2m_f$ and thus

$$\mathcal{A} = -\frac{\sqrt{2}m_f g_L^2}{M^2}\bar{v}_{\bar{f}}(p_4)\gamma^5 P_L u_f(p_3). \tag{21.48}$$

Thus the amplitude is proportional to m_f and the annihilation into light fermions is strongly suppressed. In addition to obtaining this insight with little effort, the remaining amplitude is easier to evaluate and no expansion in $v_{\text{Møl}}$ needs be performed.

Let us now try to understand why the amplitude is proportional to m_f. The initial wave-function $|\Phi\rangle = |L,S\rangle$ of the identical fermion pair has to be antisymmetric. For zero relative velocity, the orbital angular momentum L is zero and thus $|L\rangle$ is symmetric. Therefore the spin wave-function $|S\rangle$ has to be antisymmetric, $|S\rangle = |\uparrow,\downarrow\rangle$ or $|S\rangle = |\downarrow,\uparrow\rangle$. Thus the total spin is zero, and the pair of Majorana fermions is in a ${}^1S_0$ state. Consequently, $\Phi = \sqrt{2}(m + \not{P}/2)\gamma^5$ acts as a projection operator which inserted between two arbitrary spinors extracts the ${}^1S_0$ state or the s-wave contribution to the annihilation amplitude. If the produced fermion pair were massless, the final state would be either $f_L\bar{f}_R$ or $f_R\bar{f}_L$. For instance, the Majorana fermion pair could annihilate into a left-handed electron and a right-handed positron. The electron–positron pair is produced back-to-back and therefore its total spin is $S = 1$. Since the Dirac mass term connects left- and right-chiral fields, the chirality flip required to create an $S = 0$ state leads to an amplitude proportional to m_f. Thus for massless fermions, the annihilation cross-section has to vanish for $v_{\text{Møl}} \to 0$. Alternatively, we can allow for a non-zero relative velocity and consider that the orbital angular momentum is one. Thus $|L\rangle$ is antisymmetric, and from the two states ${}^3P_0$ and ${}^3P_1$ only the latter is allowed. Now it is possible to produce the fermion pair without a helicity flip, but the amplitude will be proportional to $v_{\text{Møl}}$.

21.2.3 Detection of WIMPs

Searches for WIMPs rely on the assumption that they interact with SM particles. Such searches can be divided into three main categories. Indirect detection uses the annihilation process $X + X \to \text{SM} + \text{SM}$ into SM particles, while direct detection relies on the elastic scattering $X + \text{SM} \to X + \text{SM}$ of WIMPs on SM particles. Finally, one can search at accelerators for the production $SM + SM \rightarrow X + X$ of WIMPs by colliding normal matter.

Cosmology connects the measured CDM abundance with the required annihilation cross-section of a thermal relic as $\langle \sigma_{\text{ann}} v \rangle \simeq 3 \times 10^{-26} \, \text{cm}^3/\text{s}$. Moreover, we know that the three processes used in the different search categories are related by crossing symmetry. This leads to the question of how well we can constrain the possible signal strength to be expected in the WIMP scenario in these three channels. Let us consider first indirect detection where one can use the annihilation of WIMPs in the centre of the Sun, the halo of the Milky Way or other galaxies. In all cases, typical WIMP velocities are much smaller than the ones during freeze-out. For instance, as typical WIMP velocity relevant for the annihilation of DM in the Milky Way we can use the rotation velocity of the Sun around the centre of the Galaxy, $v \sim 220 \, \text{km/s} \sim 10^{-3}$. As a result, the thermal annihilation cross-section today is—in the absence of non-perturbative effects—only bounded from above by the one in the early universe. If annihilations are dominated by the p-wave contribution, the thermal cross-section $\langle \sigma_{\text{ann}} v \rangle$ relevant for indirect searches could be six orders of magnitude smaller than in the early universe. On the other hand, the annihilation cross-section at small velocities may be enhanced via the Sommerfeld effect compared to the cosmological one.

The connection is even less tight in the case of direct and accelerator searches. In these searches we test mainly the couplings of the WIMP to the first generation of quarks and leptons, while the annihilation cross-section sums up all relevant channels. In many models, the main final states of WIMP annihilations are heavy fermions, gauge and higgs bosons. As a result, the elastic cross-section on nucleons or the production via proton–proton scattering could be suppressed. Moreover, WIMPs are produced at accelerators as ultrarelativistic particles, probing again a different kinematical regime than annihilations in the early universe.

Direct detection. A direct detection experiment aims to measure the nuclear recoil, when a WIMP scatters on a nucleus in a detector. Let us assume that the WIMP interacts via the exchange of a gauge or Higgs boson with the nucleus. The momentum transfer in such a reaction is small, $q^2 \lesssim 100 \, \text{keV}^2$, cf. problem 21.4. Therefore, the exchanged virtual particles do not resolve the quark and gluon content of a nucleon but interact with the whole nucleon.[4] Instead of the couplings to quarks, we have therefore to know the effective coupling of a gauge or Higgs boson to a nucleon. The small momentum transfer implies also that one can integrate out the intermediate virtual particles, constructing an effective Lagrangian for the interactions between the DM particle and quarks and gluons. Within a given DM model, this step is lengthy but straight-forward and gives an effective Lagrangian containing higher-dimensional operators.

Let us illustrate this procedure with an example. We assume that the WIMP is a fermion and write down its interactions with light quarks and gluons. For our purposes, it is sufficient to consider as an example the contribution of scalar operators to the effective Lagrangian,

---

[4]Since the momentum transfer is comparable to the nuclear size, they interact (partly) coherently with the whole nucleus. Thus in an additional step, nuclear physics effects have to be incorporated.

$$\mathscr{L}_{\text{eff}} = C_S^q \bar{X} X m_q \bar{q} q + C_S^g \frac{\alpha_s}{\pi} \bar{X} X F_{\mu\nu}^a F^{a\mu\nu}. \tag{21.49}$$

All the information about the physics integrated out is contained in the coefficients C_S^N. We obtain the matrix elements of these effective operators between nucleon states following the same strategy as in section 18.1 discussing the trace anomaly. We use Eq. (18.18) at leading order in α_s,

$$T_\mu^\mu = -\frac{9}{8} \frac{\alpha_s}{\pi} F_{\mu\nu}^a F^{a\mu\nu} + \sum_{q=u,d,s} m_q \bar{q} q, \tag{21.50}$$

together with the mass fractions $f_q^{(N)}$ calculated in lattice QCD. Evaluating then T_μ^μ between nucleon states $|N\rangle$ and using $\langle N|T_\mu^\mu|N\rangle = m_N$, we obtain

$$\langle N|\frac{\alpha_s}{\pi} F_{\mu\nu}^a F^{a\mu\nu}|N\rangle = -\frac{8}{9} m_N f_G^{(N)} \tag{21.51}$$

with $f_G^{(N)} \equiv 1 - \sum_{q=u,d,s} f_q^{(N)}$. This determines the effective operators at the scale $q^2 \simeq m_N$ and allows one to calculate scattering rates for given coefficients C_S^N. The effective Lagrangian (21.49) is, however, defined at the scale corresponding to the mass scale of the virtual particles integrated out. Thus as a final step, one has to derive the RGE of these effective operators and to evolve the coefficients C_S^N down to the scale m_N.

Indirect detection. The average density of DM in the Milky Way is increased by a factor $\sim 10^5$ compared to the extragalactic space. Therefore the annihilation rate of DM can become again appreciable inside the Milky Way, and in particular in regions where DM is strongly accumulated. The secondaries of DM annihilations will be the stable particle of the standard model, that is, photons, neutrinos, electrons and protons. The challenge for indirect detection consists of disentangling these annihilation products from the background of high-energy particles produced by astrophysical sources. Two channels provide a rather unique signature: DM annihilations into two photons, or a photon and Z, lead to line features which cannot be mimicked by a astrophysical background. Since a WIMP is by assumption neutral, this process can proceed however only via loop graphs and it thus suppressed by a factor $(\alpha/\pi)^2$.

Another unique signature are high-energy neutrinos produced by WIMPs that accumulate, for example, in the Sun. Here, the directional signal together with the fact that neutrinos produced by astrophysical processes have energies $\lesssim$ GeV provides the distinctive signature. In all other cases, a detailed knowledge of the spectral shape of antimatter fluxes, both for the background produced, for example, in pp collisions, and in DM annihilations, is required. This requires the knowledge of strong interactions at small virtualities and relies on the use of Monte Carlo methods, as described in section 18.2.

21.3 Big bang nucleosynthesis

Big bang nucleosynthesis (BBN) is controlled mainly by two parameters: the mass difference between protons and neutrons, $\Delta \equiv m_n - m_p \simeq 1.29\,\text{MeV}$, and the freeze-out temperature T_f of reactions converting protons into neutrons and vice versa. Since the binding energy per nucleon has a large peak for ^{4}He, essentially all free neutrons are bound into helium. Heavier elements (except ^{7}Li) are produced later by stellar fusion, where the densities are sufficiently high that e.g. the triple α process $3\,^4\text{He} \to\,^{12}\text{C}$ can bridge the gap of missing tightly bound nuclei between ^{4}He and ^{12}C.

Equilibrium distributions.   In the non-relativistic limit $T \ll m$, the number density of a nuclear species with mass number A and charge Z is given by

$$n_A = g_A \left(\frac{m_A T}{2\pi}\right)^{3/2} \exp[\beta(\mu_A - m_A)].   \tag{21.52}$$

In chemical equilibrium, $\mu_A = Z\mu_p + (A - Z)\mu_n$ and we can eliminate μ_A by inserting the equivalent expression of (21.52) for protons and neutrons,

$$e^{\beta\mu_A} = \exp[\beta(Z\mu_p + (A - Z)\mu_n)] = \frac{n_p^Z n_n^{A-Z}}{2^A} \left(\frac{2\pi}{m_N T}\right)^{3A/2} \exp[\beta(Zm_p + (A - Z)m_n)].   \tag{21.53}$$

Here and in the following we can set in the pre-factors $m_p \simeq m_n \simeq m_N$ and $m_A \simeq A m_N$, keeping the exact masses only in the exponentials. Inserting this expression for $\exp(\beta\mu_A)$ together with the definition of the binding energy of a nucleus, $B_A = Zm_p + (A - Z)m_n - m_A$, we obtain

$$n_A = g_A \left(\frac{2\pi}{m_N T}\right)^{3(A-1)/2} \frac{A^{3/2}}{2^A}\, n_p^Z n_n^{A-Z} \exp(\beta B_A).   \tag{21.54}$$

The mass fraction X_A contributed by a nuclear species is

$$X_A = \frac{A n_A}{n_B} \quad \text{with} \quad n_B = n_p + n_n + \sum_i A_i n_{A_i} \quad \text{and} \quad \sum_i X_i = 1.   \tag{21.55}$$

Next we introduce the baryon–photon ratio $\eta = n_B/n_\gamma$ as variable. With $n_p^Z n_n^{A-Z}/n_N = X_p^Z X_n^{A-Z} n_N^{A-1}$ and $\eta \propto T^3$ and thus $n_B^{A-1} \propto \eta^{A-1} T^{3(A-1)}$, we have

$$X_A \propto \left(\frac{T}{m_N}\right)^{3(A-1)/2} \eta^{A-1} X_p^Z X_n^{A-Z} \exp(\beta B_A).   \tag{21.56}$$

The fact that $\eta \simeq 6 \times 10^{-10} \ll 1$, that is, that the number of photons per baryon is extremely large, means that nuclei with $A > 1$ are much less abundant and that nucleosynthesis takes place later than naively expected.

Let us consider the particular case of deuterium in Eq. (21.56),

$$\frac{X_D}{X_p X_n} = \frac{24\zeta(3)}{\sqrt{\pi}} \left(\frac{T}{m_N}\right)^{3/2} \eta \exp(\beta B_D)   \tag{21.57}$$

with $B_D = 2.23$ MeV. The start of nucleosynthesis could be defined approximately by the condition $X_D/(X_p X_n) = 1$ in Eq. (21.57), what results in $T_{\text{NS}} \simeq 0.07$ MeV. Figure 21.2 shows the results, if the Eqs. (21.56) together with $\sum_i X_i = 1$ are solved for the lightest and stablest nuclei: In thermal equilibrium, essentially all free neutrons will bind to ^{4}He at temperatures $T \lesssim 0.2$ MeV. The formation of heavier elements is strongly suppressed because of their much smaller binding energy per nucleon. Moreover, the Coulomb barrier will prevent the production of nuclei with $Z \gg 1$.

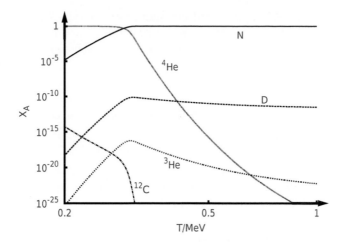

Fig. 21.2 Relative equilibrium abundance $X_D/(X_p X_n)$ of deuterium as function of temperature T (left) and equilibrium mass fractions of nucleons, D, ^{3}He, ^{2}He and ^{12}C (right).

Neutron abundance. If we do not aim at the calculation of the (small) abundance of elements other than ^{4}He, we have to solve only the Boltzmann equation for the neutron abundance. Nucleons are inter-converted by weak processes as $n \leftrightarrow p + e^- + \nu_e$. For an estimate of the freeze-out temperature of weak interactions, we can use the Gamov criteria. The cross-section of processes like $n \leftrightarrow p + e^- + \nu_e$ or $e^+ e^- \leftrightarrow \bar{\nu}\nu$ is $\sigma \approx G_F^2 E^2$. If we approximate the energy of all particle species by their temperature T, their velocity by c and their density by $n \approx T^3$, then the interaction rate of weak processes is

$$\Gamma \approx \langle v \sigma n_\nu \rangle \approx G_F^2 T^5. \tag{21.58}$$

In the radiation-dominated epoch, $\Gamma(T_{\text{fr}}) = H(T_{\text{fr}})$ gives as freeze-out temperature T_{fr} of weak processes

$$T_{\text{fr}} \approx \left(\frac{1.66 \sqrt{g_*}}{G_F^2 M_{\text{Pl}}} \right)^{1/3} \approx 1 \text{MeV} \tag{21.59}$$

with $g_* = 10.75$. Thus weak interaction freeze out before nucleosynthesis starts. Because of $T_{\text{fr}} \approx 1$ MeV, we can treat nucleons in the non-relativistic limit. Then their relative equilibrium abundance is given by the Boltzmann factor $\exp(-\Delta/T)$ for $T \gtrsim T_{\text{fr}}$. Hence at freeze-out, the ratio of their equilibrium distributions is given by

$$\frac{n_n^{\text{eq}}}{n_p^{\text{eq}}} = \exp\left(-\frac{\Delta}{T_{\text{fr}}}\right). \tag{21.60}$$

We consider in Eq. (21.13) the reaction $n + e^- \leftrightarrow p + \nu_e$, using for the leptons equilibrium distributions,

$$\frac{1}{a^3} \frac{\mathrm{d}(a^3 n_n)}{\mathrm{d}t} = n_l^{\text{eq}} \langle \sigma v \rangle \left[\frac{n_p n_n^{\text{eq}}}{n_p^{\text{eq}}} - n_n \right]. \tag{21.61}$$

Next we replace the ratio $n_n^{\mathrm{eq}}/n_p^{\mathrm{eq}}$ by the Boltzmann factor $\exp(-\beta\Delta)$. Moreover, the loss term $n_l^{\mathrm{eq}} n_n \langle \sigma v \rangle$ equals the neutron–proton scattering rate Γ_{np}. Changing also from the variable n_n to the neutron mass fraction X_n, we arrive at

$$\frac{\mathrm{d}X_n}{\mathrm{d}x} = \Gamma_{np} \left[(1 - X_n)\mathrm{e}^{-\beta\Delta} - X_n \right]. \qquad (21.62)$$

Introducing again the dimensionless $x = \Delta/T$ as evolution variable, we obtain

$$\frac{\mathrm{d}X_n}{\mathrm{d}x} = \frac{x\Gamma_{np}}{H(1)} \left[\mathrm{e}^{-x} - X_n(1 + \mathrm{e}^{-x}) \right] \qquad (21.63)$$

with $H(1) \simeq 1.13/\mathrm{s}$. The neutron-proton conversion rate can be connected to the neutron life time $\tau_n \simeq 886.7\mathrm{s}$ as (problem 21.6),

$$\Gamma_{np} = \frac{255}{\tau_n x^5} \left(12 + 6x + x^2 \right). \qquad (21.64)$$

Now Eq. (21.63) can be integrated and the result shows that X_n freezes-out below $0.5\,\mathrm{MeV}$, approaching the asymptotic value $X_n \simeq 0.15$.

Until now we have treated the neutron as stable. We can include into our simple picture the effect of neutron decays by adding the factor $\exp(-t/\tau_n)$ to the neutron abundance X_n. If we use the fact that nucleosynthesis starts at $T_{\mathrm{NS}} \simeq 0.07\,\mathrm{MeV}$, $t_{\mathrm{NS}} \simeq 270\,\mathrm{s}$, then $\exp(-t/\tau_n) \simeq 0.74$. As result, the mass fraction of neutrons which can be fused into helium is $X_n = 0.11$ and thus the helium fraction $X_4 = 2X_n = 0.22$. Numerical calculations that include the full nuclear reaction network lead to $X_4 = 0.24$, and thus our simple estimate is only 10% away from the true value. A fit to these numerical results shows that the helium abundance depends logarithmically on η_b,

$$X_4 = 0.226 + 0.013 \ln(\eta_b/10^{-10}). \qquad (21.65)$$

Thus the helium abundance alone can be used to determine the baryon–photon ratio as $\eta_b \simeq \text{few} \times 10^{-10}$. A comparison of the predicted with the observed abundance of deuterium and lithium allows then for a consistency check of the BBN picture. An independent determination of Ω_b using CMB observations leads to $\eta_b \simeq 6.2 \times 10^{-10}$. The success of BBN can be used to limit, for example, the injection of energy through particle decays which could destroy light elements.

Summary

Boltzmann equations are an important tool to describe processes as diverse as the evolution of the DM density, BBN nucleosynthesis or recombination. The Gamov criterion states that processes freeze out when their rate becomes smaller than the Hubble rate. The mass of any thermal relic is bounded by $\lesssim 20\,\mathrm{TeV}$. The abundance of a CDM particle with $\langle \sigma v \rangle \simeq 3 \times 10^{-26}\mathrm{cm}^3/\mathrm{s}$ corresponds to the observed one, $\Omega_{\mathrm{CDM}} = 0.2$. BBN explains successfully the abundance of light elements like D and ^{4}He, and fixes thereby also η_b.

Further reading. A classic presentation of the freeze-out mechanism is given in Kolb and Turner (1994). A more detailed analytical treatment of BBN is presented by Mukhanov (2005) and Gorbunov and Rubakov (2011*b*), who also discuss recombination. Dropping the assumption of uniformity, Boltzmann-type equations can be also used to describe the evolution of perturbations, a topic which is treated in detail, for example, by Dodelson (2003).

Problems

21.1 Boltzmann equation. Consider the phase space density $f(x^\mu, p^\mu)$ as function of the four-vectors x^μ and p^μ and calculate $df/d\lambda$. Use the geodesic equation for $dp^\mu/d\lambda$ and show that the resulting equation agrees with (21.6).

21.2 Kinetic decoupling. Consider a fourth generation Dirac neutrino as a hypothetical DM particle. a.) Calculate its annihilation cross-section into SM fermions and estimate its abundance as function of its mass. b.) Calculate the time of kinetic decoupling assuming that the reaction $\bar\nu_4\nu_i \to \bar\nu_4\nu_i$ dominates the energy exchange.

21.3 Relative velocity at freeze out. Find the average relative velocity of two CDM particles annihilating at a given x_f.

21.4 Direct detection of WIMPs. Find the maximal energy a WIMP with mass M and velocity $v = 220$km/s transfers to a nucleon in a scattering events. What is the rate of such events in a detector with mass 1 t using a "typical" weak cross-section?

21.5 Baryon abundance from freeze-out. Calculate the expected baryon abundance for zero chemical potential.

21.6 Rate of weak reactions. Derive the amplitude for $\nu_e + n \leftrightarrow p + e^-$ in the limit $s \ll m_W^2$ and express the coupling constant via the neutron life time. Find the cross-section in vacuum and the rate (21.64) of the process in a thermal plasma.

21.7 Effective Lagrangian. Find the values C_i^N for a Majorana fermion interacting via the Lagrangian (21.46) with nucleons.

21.8 Additional relativistic species. Estimate the effect of additional relativistic species on BBN.

21.9 Recombination. Find the equilibrium abundance of free electrons. Describe recombination as the freeze-out of the process $p + e^- \leftrightarrow H + \gamma$ using for the recombination rate $\langle \sigma v \rangle \simeq 9.78(\alpha/m_e)^2 (\beta E_0)^{1/2} \ln(\beta E_0)$ with $E_0 = 13.6$ eV, and the ionisation rate $\beta = \langle v\sigma \rangle (m_e T/(2\pi))^{3/2} \exp(-E_0/T)$.

22
Baryogenesis

We have seen that BBN determines the baryon–photon ratio as $\eta = n_B/n_\gamma \simeq 6 \times 10^{-10}$. In problem 21.5, we calculated the baryon abundance in the usual freeze-out formalism for zero chemical potential. Since the $\bar{p}p$ annihilation cross-section is large, nucleons freeze-out very late ($x_f \sim 44$) when their density is already strongly suppressed. As a result, the baryon–photon ratio in a baryon symmetric world would be $\eta \simeq 7 \times 10^{-20}$ which is much smaller than the observed value. This implies that at temperatures above the freeze-out a tiny surplus of one quark per 10^{10} quarks and antiquarks existed. Therefore the proper definition of the baryon–photon ratio is $\eta = (n_b - n_{\bar{b}})/n_\gamma$ and the challenge is to explain the origin of the asymmetry between baryons and antibaryons. Astronomers do not observe the photons from the processes $e^+e^- \to 2\gamma$ and $\bar{p}p \to X\gamma$ that would occur at the boundaries of matter–antimatter domains. Using these observational limits, one can conclude that the whole observable universe consists of matter. Moreover, an inflationary period in the early universe eliminates any pre-existing baryon asymmetry, forcing us to explain the observed baryon asymmetry dynamically. We will see that such explanations require necessarily physics beyond the SM.

22.1 Sakharov conditions and the SM

Sakharov conditions for baryogenesis. In 1967 Sakharov developed the first model which contained the three ingredients necessary for the dynamically generation of a non-zero baryon number. These so-called Sakharov conditions for baryogenesis are

1. violation of baryon number B,

2. violation of the discrete symmetries C and CP, and

3. departure from thermal equilibrium.

The first condition, the non-conservation of baryon number B, is obviously necessary, if the universe should evolve from a state with $n_B = 0$ to $n_B > 0$. We can understand the second condition from the transformation properties of the baryon number operator B under C and CP. Because of $\text{C}B\text{C}^{-1} = -B$ and $\text{CP}B(\text{CP})^{-1} = -B$, the (thermal) expectation value of B has to vanish, if C and CP are symmetries of the model considered. More explicitly, we can express the expectation value of B in this case with $A = \{\text{C}, \text{CP}\}$ as

Quantum Fields–From the Hubble to the Planck Scale. Michael Kachelriess. © Michael Kachelriess 2018.
Published in 2018 by Oxford University Press. DOI 10.1093/oso/9780198802877.001.0001

$$\langle B \rangle = Z^{-1}\mathrm{Tr}\left[e^{-\beta H}\, B\right] = Z^{-1}\mathrm{Tr}\left[AA^{-1}e^{-\beta H}\, B\right] \tag{22.1a}$$

$$= Z^{-1}\mathrm{Tr}\left[e^{-\beta H}\, A^{-1}BA\right] = -Z^{-1}\mathrm{Tr}\left[e^{-\beta H}\, B\right] = -\langle B \rangle, \tag{22.1b}$$

where we used $[H, A] = 0$ going from the first to the second line. Thus $\langle B \rangle = 0$, if $[H, C] = 0$ or $[H, \mathrm{CP}] = 0$. By the same token, we can show that a departure from thermal equilibrium is required. Since any unitary Lorentz-invariant quantum field theory is invariant under CPT, we have $[H, \mathrm{CPT}] = 0$ but also $\mathrm{CPT}B(\mathrm{CPT})^{-1} = -B$. We can avoid the conclusion $\langle B \rangle = 0$ only, if Eq. (22.1a), or in other words thermal equilibrium, does not hold.

CP violation. Another consequence of CPT invariance is that CP violation is equivalent to T violation. The latter is an anti-unitary operator, $\mathrm{T}^{-1}\mathrm{i}\mathrm{T} = -\mathrm{i}$, and therefore complex parameters in the Lagrangian lead to T and thus CP violation. The only complex parameters contained in the SM are the phases of the fermion mixing matrices. In order to see that such phases result indeed in CP violation, we express the charged current interaction for the example of leptons by mass eigenstates,

$$\mathscr{L}_{\mathrm{CC}} = \frac{g}{\sqrt{2}}\bar{\nu}_{L,i}\gamma^{\mu}U_{ij}e_{L,j}W_{\mu}^{+} + \mathrm{h.c.} \tag{22.2}$$

$$= \frac{g}{\sqrt{2}}\left[\bar{\nu}_{L,i}\gamma^{\mu}U_{ij}e_{L,j}W_{\mu}^{+} + \bar{e}_{L,j}\gamma^{\mu}U_{ji}^{*}\nu_{L,i}W_{\mu}^{-}\right]. \tag{22.3}$$

We choose the rest-frame of the W-boson so that $W_0^{\pm} = 0$. A CP transformation transforms the current $\bar{\nu}_{L,i}\gamma e_{L,j}$ into $\bar{e}_{L,j}\gamma\nu_{L,i}$ and the W_i^{+} into a W_i^{-}. It also exchanges the left- and right-circular polarisations of the W and transforms all arguments $x^{\mu} = (t, \boldsymbol{x})$ of the fields into $x^{\mu} = (t, -\boldsymbol{x})$. The latter two effects are harmless, since we integrate over x^{μ} and sum over the two polarisations of the W in the action. Thus the combined effect of CP on the first term in (22.3) is given by

$$\frac{g}{\sqrt{2}}\bar{\nu}_{L,i}\gamma^{\mu}U_{ij}e_{L,j}W_{\mu}^{+} \rightarrow \frac{g}{\sqrt{2}}\bar{e}_{L,j}\gamma^{\mu}U_{ji}\nu_{L,i}W_{\mu}^{-}. \tag{22.4}$$

The CP-transformed term is identical to the Hermitian conjugated of the original term, if the mixing matrix is real, $\boldsymbol{U} = \boldsymbol{U}^{*}$. In this case, a CP transformation simply exchanges the first and the second term in (22.3). If, however, the mixing matrix is complex, the CP-transformed Lagrangian differs from the original one, and CP is violated.

In general, the complex parameters required for CP violation can arise in two ways. First, interactions may contain physical phases, as in the case of the CKM and the MNSP matrices in the SM. Second, vacuum expectation values may contain physical phases. This option, called spontaneous CP violation, requires at least two Higgs doublets, as we discussed in remark 17.1. Enlarging the Higgs sector is therefore an efficient way to add greater CP violation to the SM.

Sphaleron transitions and $B - L$ violation. Recall from section 17.2 the effect of an instanton transition on the divergence of the axial current,

$$\partial_{\mu}j_A^{\mu} = \partial_{\mu}(j_R^{\mu} - j_L^{\mu}) = \frac{g^2}{16\pi^2}\,\mathrm{tr}(F_{\mu\nu}\tilde{F}^{\mu\nu}). \tag{22.5}$$

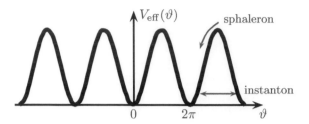

Fig. 22.1 Instanton versus sphaleron transition between different ϑ vacua.

Strong interactions couple equally to left- and right-chiral fermions, and therefore the sole effect of an instanton transition is a chirality flip of the fermion. In contrast, an electroweak instanton couples only to left-chiral fermions,

$$\sum_i \partial_\mu j^\mu_{L,i} = -n_g \frac{g^2}{16\pi^2} \operatorname{tr}(W_{\mu\nu}\tilde{W}^{\mu\nu}) = \Delta\nu, \qquad (22.6)$$

where n_g counts the number of generations, i stands for all components of the left-chiral SU(2) doublets, $L = (\nu_l, l)$ with $l = \{e, \mu\,\tau\}$ and the corresponding nine quark doublets. Now an instanton transition changes the fermion number of the left-chiral fields. More precisely, a transition which changes the winding number by one unit, $\Delta\nu = 1$, changes the fermion number in each doublet by one unit and thus the total fermion number by 12. Since each quark carries baryon number 1/3, the changes in the lepton flavour and baryon number are connected by

$$\Delta L_e = \Delta L_\mu = \Delta L_\tau = \frac{1}{3}\Delta L = \frac{1}{3}\Delta B \qquad (22.7)$$

or $\Delta L = \Delta B = n_g = 3$ for $\Delta\nu = 1$. Thus we see that the combination $B - L$ is conserved in the SM even in the presence of instanton transitions, while both the baryon and the lepton number are broken. Adding conservation of electric charge and colour, the states connected by an instanton are fixed. In particular, a transition with $\Delta\nu = 1$ creates the state

$$|0\rangle \rightarrow \left| u_L u_L d_L e_L^- + c_L c_L s_L \mu_L^- + t_L t_L b_L \tau_L^- \right\rangle,$$

where the three quarks of each generations form a colour singlet. Since mass terms and Higgs interactions mix left- and right-chiral fields, the resulting change in the number of left-chiral fermions is transferred to the right-chiral fermions.

The differential probability p per time and volume of a tunnelling process connecting two vacua separated by one winding follows from (17.57) as

$$p = \xi m_W^{-4} \exp(-S) = \xi m_W^{-4} \exp\left(-\frac{8\pi^2}{g^2}\right) \approx \frac{10^{-160}}{m_W^4}, \qquad (22.8)$$

where the pre-factor ξ is a dimensionless function of order 1. Thus at zero temperature, the effects of electroweak instantons are completely negligible. In the early universe however, we should take into account not only quantum but also thermal fluctuations.

We include the latter replacing the tunnelling factor $\exp(-S)$ by the Boltzmann factor $\exp(-E/T)$. The classical field configuration[1] connecting the top of the barrier with the vacuum which has the smallest free energy is called sphaleron, cf. Fig. 22.1. Its energy in the broken phase is $E_{\mathrm{sp}} \sim m_W(T)/\alpha_W$, where $m_W(T)$ is the temperature-dependent mass of the W. Thus the probability that a thermal fluctuation crosses the barrier is $\propto \exp(-2E_{\mathrm{sp}}(T)/T)$. At temperatures $T \gtrsim m_W$, thermal fluctuations are larger than the barrier height between the vacua and the baryon number violating processes should proceed unsuppressed. At these temperatures, the SM is in the unbroken phase and the only relevant length scale is the analogue of the Debye mass m_D for a gauge boson, $m_D \sim \alpha_W T$. Thus we expect that the sphaleron rate[2] is given in the high-temperature limit by

$$p \sim \xi(\alpha_W T)^4 \,. \tag{22.9}$$

Comparing the rate per particle, $\Gamma \sim \xi \alpha_W^4 T$, with the Hubble rate, wee see that the $B + L$ violating sphaleron transition are (with $\xi \sim 0.1$) up to $T_{\mathrm{sp}} \sim 10^{12}\,\mathrm{GeV}$ in equilibrium. As the universe cools down, sphaleron processes go out of equilibrium around $T \sim 100\,\mathrm{GeV}$, when the rate becomes Boltzmann suppressed.

Connecting B and $B - L$. In thermal equilibrium, a linear combination of $B+aL$ has to vanish. The exact value of a can be determined using that the universe is neutral, that is, that the conserved charges should be zero. Before the electroweak phase transition, all three gauge forces are long-range. The corresponding charges have therefore to be zero (or better negligibly small), in order not to outweigh the gravitational forces. The only other conserved charge is $B - L$. It is sufficient to consider the hypercharge Y and $B - L$. Setting $\mu \equiv \mu_{B-L}$, a particle of type i has the chemical potential

$$\mu_i = \mu(B_i - L_i) + \mu_Y Y_i/2 = -\mu_{\bar{i}}. \tag{22.10}$$

Using the quantum number assignments of table 14.1, it follows, for example,

$$\mu_{h^+} = \mu_{h^0} = \frac{1}{2}\mu_Y \quad \text{and} \quad \mu_{u_L} = \mu_{d_L} = \frac{\mu}{3} + \frac{\mu_Y}{6}, \tag{22.11}$$

with analogous relations for the other members of the first fermion generation. Since the asymmetry is small, $\mu_i \ll T$, we can use Eq. (15.59) obtaining

$$\Delta n_i \equiv n_i - n_{\bar{i}} = \delta \sum_i g_i \mu_i q_i \frac{T^2}{3} \tag{22.12}$$

with $\delta = 2$ for bosons and $\delta = 1$ for fermions. The condition that the hypercharge of the plasma vanishes becomes

$$\sum_i \Delta n_i Y_i = n_g \left(\frac{5}{3}\mu_Y + \frac{4}{3}\mu \right) + \frac{1}{2}n_h \mu_Y = 0, \tag{22.13}$$

---

[1]The potential energy is a functional of the field configurations, $V_{\mathrm{eff}}[W^\mu, \phi]$, and thus the figure represents a cut through an infinite dimensional field space; the maximum is in fact a saddle-point.

[2]Our dimensional argument does not account for an additional factor α_W arising in perturbative calculations due to IR dynamics, but is nevertheless with $\xi \approx 0.1$ a good approximation for the sphaleron rate determined in lattice calculation.

where we summed over n_g fermion generations and n_h Higgs doublets. Now we can solve for μ, eliminate then this variable in the relations for Δn_i, arriving at

$$n_B = \frac{1}{3}(\Delta n_{u_L} + \Delta n_{u_R} + \Delta n_{d_L} + \Delta n_{d_R}) = -\mu_Y \left(\frac{1}{2}n_g + \frac{1}{4}n_h\right)\frac{T^3}{3}, \qquad (22.14)$$

$$n_L = \Delta n_{\nu_L} + \Delta n_{e_L} + \Delta n_{e_R} = \mu_Y \left(\frac{7}{8}n_g + \frac{9}{16}n_h\right)\frac{T^3}{3}. \qquad (22.15)$$

Subtraction gives an expression for n_{B-L}. Eliminating T^3 with the help of (22.14), we find finally

$$n_B = \frac{24 + 4n_h}{66 + 13n_h} n_{B-L} = a n_{B-L} \qquad (22.16)$$

with $a = 28/79$ in the SM. Because the $B + L$ violating sphaleron transitions are in thermal equilibrium above $T \sim 100\,\text{GeV}$, we therefore have to modify the first Sakharov condition: any asymmetry proportional to $B + L$ will relax to zero, while the part proportional to $B - L$ will lead a final baryon asymmetry. In particular, a non-zero baryon number arises too, if we generate at $T \gg 100\,\text{GeV}$ an asymmetry only in the lepton number.

Toy model. Let us consider a simple model, where we couple a heavy scalar X via Yukawa interactions to four SM fermions f_i,

$$\mathscr{L}_X = g_X^{12} X \bar{f}_2 f_1 + g_X^{34} X \bar{f}_4 f_3 + \text{h.c.} \qquad (22.17)$$

To be concrete, we choose the scalar as a heavy leptoquark X with electric charge $q(X) = -4/3$ and the two decay modes as

$$X \to \bar{u}\bar{u}, \qquad r \qquad (22.18)$$

$$X \to e^- d, \qquad 1 - r, \qquad (22.19)$$

where r is the corresponding branching ratio. The baryon number of the first decay mode is $B = -2/3$, while the one of the second mode is $B = 1/3$. Therefore we cannot assign a baryon number to X and thus baryon number is violated. Next we require that in these decays C is violated. Then the charge-conjugated decay modes have different branching ratios $\bar{r} \neq r$,

$$\bar{X} \to uu, \qquad \bar{r} \qquad (22.20)$$

$$\bar{X} \to e^+ \bar{d}, \qquad 1 - \bar{r}. \qquad (22.21)$$

The resulting change ΔB of the baryon number B per decay of a $X, \bar{X}$ pair is thus

$$\Delta B = -\frac{2}{3}r + \frac{1}{3}(1 - r) + \frac{2}{3}\bar{r} - \frac{1}{3}(1 - \bar{r}) = \bar{r} - r, \qquad (22.22)$$

that is, proportional to the amount of C violation. If we consider these processes at tree level, then the decay widths are $\Gamma(X \to \bar{f}_2 f_1) = |g_X^{12}|^2 I_X$ and $\Gamma(\bar{X} \to \bar{f}_2 f_1) = |g_X^{12*}|^2 I_{\bar{X}}$ where $I_X = I_{\bar{X}}$ is a real kinematical factor determined by the masses, $I =$

$I(m_X, m_i, m_j)$. Thus at tree level, CP violating effects are absent, $r = \bar{r}$, the second Sakharov condition is not satisfied, and thus no net baryon number is generated.

This example shows that for successful baryogenesis two additional requirements have to be satisfied. First, the kinematical terms I should be complex. The required imaginary part can be generated in a loop graph if one of the virtual particles can become on-shell. This can happen, if the mass of the decaying particle is to be larger than the sum of the fermion masses in the loop. Second, at least two heavy particles have to interfere, such that the decay widths contain the complex combination $g_X g_Y^*$ instead of the real $|g_X|^2$. Adding therefore

$$\mathscr{L}_Y = g_Y^{12} Y \bar{f}_2 f_1 + g_Y^{34} Y \bar{f}_4 f_3 + \text{h.c.}$$

to the Lagrangian produces at the one-loop a potentially baryon number-violating term,

$$\Delta B \propto \text{Im}(g_Y^{12*} g_X^{12} g_Y^{34*} g_X^{34})\text{Im}(I_{XY}). \tag{22.23}$$

22.2 Baryogenesis in out-of-equilibrium decays

Sakharov's third condition for baryogenesis, a departure from thermal equilibrium, can be satisfied during phase transitions or when a particle species is out of chemical equilibrium. In this section, we examine the conceptionally simpler second possibility, considering out-of-equilibrium decays of unstable heavy particle X which violate baryon number.

Boltzmann equation for decays. As the first step, we simplify the Boltzmann equation for decays $X \to bb$ and its inverse reaction. Here, the unstable heavy X and Y particles may be out of chemical equilibrium, while the light decay products b are in equilibrium. In a GUT scenario for baryogenesis, we might identify the X particles with the lepto quark-like gauge bosons with electric charge $q = 4/3$. Finally, we assume $M_Y \gg M_X$ such that the heavier Y particle contributes only as virtual state in the loop correction. Then the decays $X \to bb$ corresponds to the annihilation term β and the inverse decays $bb \to X$ to the production term ψ in the Boltzmann equation. Using detailed balance to relate decays and inverse decays, we find

$$\frac{dn_X}{dt} + 3H n_X = -\Gamma_D(n_X - n_{X,\text{eq}}). \tag{22.24}$$

Let us assume that the relevant decays of the X particle are to two bb particles and to two $\bar{b}\bar{b}$ antiparticles, with $B = +1/2$ and $B = -1/2$, respectively. We denote the squared Feynman amplitudes as

$$|\mathcal{A}(X \to bb)|^2 = |\mathcal{A}(\bar{b}\bar{b} \to X)|^2 = \frac{1}{2}(1 + \varepsilon)|A_0|^2, \tag{22.25}$$

$$|\mathcal{A}(X \to \bar{b}\bar{b})|^2 = |\mathcal{A}(bb \to X)|^2 = \frac{1}{2}(1 - \varepsilon)|A_0|^2. \tag{22.26}$$

Then the asymmetry per decay of one X particle is given by

$$\varepsilon_X = \sum_f B_f \frac{\Gamma(X \to f) - \Gamma(\bar{X} \to \bar{f})}{\Gamma_{\text{tot}}} = \frac{\frac{1}{2}(1+\varepsilon)|A_0|^2 - \frac{1}{2}(1-\varepsilon)|A_0|^2}{\frac{1}{2}(1+\varepsilon)|A_0|^2 + \frac{1}{2}(1-\varepsilon)|A_0|^2} \equiv \varepsilon. \tag{22.27}$$

We derive in the same way the Boltzmann equations for the number density of the b and $\bar{b}$ particles. However, we have to include additionally to the decays $2 \to 2$ scattering processes that may wash-out the created baryon asymmetry.[3] Subtracting the two Boltzmann equations and dividing by 2, we obtain as equation for the baryon number density $n_B = n_b - n_{\bar{b}}$,

$$\frac{\mathrm{d}n_B}{\mathrm{d}t} + 3Hn_B = \varepsilon \Gamma_{\mathrm{D}}(n_X - n_{X,\mathrm{eq}}) - \Gamma_{\mathrm{D}} n_B \frac{n_{X,\mathrm{eq}}}{n_\gamma} - 2n_B n_n \langle \sigma v \rangle. \tag{22.28}$$

The only term which can be positive and can thus drive the baryon asymmetry, $\varepsilon \Gamma_{\mathrm{D}}(n_X - n_{X,\mathrm{eq}})$, shows clearly the three Sakharov conditions. It is zero in case of thermal equilibrium, $n_X = n_{X,\mathrm{eq}}$, or if C, CP, or B are not violated, or in other wordes, if $\varepsilon = 0$. The second term accounts for the inverse decays of the X boson, where we used detailed balance to relate decays and inverse decays. Finally, the third part includes $2 \to 2$ baryon number-violating scattering processes. Here, a subtle point arises: the imaginary part of this processes corresponds to the product of an inverse decay and a decay, which we have already taken into account. Thus we should include only the real, off-shell part of the scattering process.

Next we change to dimensionless variables, introducing $x = m_X/T$ and $Y_i = n_i/s$. We define as measure for the departure from equilibrium

$$K \equiv \left. \frac{\Gamma_D(x)}{2H(x)} \right|_{x=1} = \frac{\alpha M_{\mathrm{Pl}}}{3.3 g_*^{1/2} m_X}. \tag{22.29}$$

Inserting all this, the new Boltzmann equations can be written as

$$\frac{\mathrm{d}Y_X}{\mathrm{d}x} = -Kx\gamma_D \left(Y_X - Y_X^{\mathrm{eq}} \right) \tag{22.30a}$$

$$\frac{\mathrm{d}Y_B}{\mathrm{d}x} = \varepsilon Kx\gamma_D \left(Y_X - Y_X^{\mathrm{eq}} \right) - Kx\gamma_B Y_B \tag{22.30b}$$

with $\gamma_D = \Gamma_D(x)/\Gamma_D(1)$ and $\gamma_B = g_* Y_X^{\mathrm{eq}} \gamma_D + 2n_\gamma \langle \sigma v \rangle / \Gamma_D(1)$. The baryon asymmetry Y_B is driven by the departure from equilibrium, $\Delta = Y_X - Y_X^{\mathrm{eq}}$ and damped by inverse decays and $2 \to 2$ scatterings. Changing to Δ as variable, we can rewrite Eq. (22.30) as

$$\frac{\mathrm{d}\Delta}{\mathrm{d}x} = -\frac{\mathrm{d}Y_X^{\mathrm{eq}}}{\mathrm{d}x} - Kx\gamma_D\Delta, \tag{22.31a}$$

$$\frac{\mathrm{d}Y_B}{\mathrm{d}x} = \varepsilon Kx\gamma_D\Delta - Kx\gamma_B Y_B. \tag{22.31b}$$

Integrating these first-order equations results in

$$\Delta(x) = \Delta(x_0) \exp \left[-\int_{x_0}^x \mathrm{d}z \, zK\gamma_D(z) \right] - \int_{x_0}^x \mathrm{d}z X'_{\mathrm{eq}}(z) \exp \left[\int_z^x \mathrm{d}z' \, z'K\gamma_D(z') \right] \tag{22.32}$$

---

[3] The existence of such processes is implied by unitarity, cf. for an example Fig. 22.2

and

$$
\begin{aligned}
Y_B(x) = Y_B(x_0) \exp\left[-\int_{x_0}^x \mathrm{d}z\, z K \gamma_B(z)\right] \\
+ \varepsilon K \int_{x_0}^x \mathrm{d}z \gamma_D(z) \Delta(z) \exp\left[-\int_z^x \mathrm{d}z'\, z' K \gamma_B(z')\right].
\end{aligned}
\tag{22.33}
$$

The most interesting limiting case is $K \ll 1$, when the out-of-equilibrium condition is satisfied. Then the exponentials are of order 1, and the abundances become with $\gamma_D(x) \simeq 1$ for $x \gg 1$

$$
Y_X(x) \simeq Y_X(0) \exp\left[-Kx^2/2\right], \tag{22.34a}
$$
$$
Y_B(x) = \varepsilon[X_X(0) - X(z)] \simeq \varepsilon X(0). \tag{22.34b}
$$

Thus the X particles decay around $x \approx K^{-1/2}$, resulting in a baryon asymmetry ε/g_*.

GUT baryogenesis. Since GUT theories unify quarks and lepton, they also contain gauge bosons X_μ and Y_μ similar to the lepto-quark discussed in our toy model. In the simplest GUT theory, SU(5), $B - L$ is conserved and thus any non-zero B will be washed out by sphaleron processes. In GUT theories based on larger groups such as SO(10), $B - L$ is broken and baryogenesis based on out-of-equilibrium decays is in principle possible. However, the necessary temperatures, $T \gtrsim M_{\mathrm{GUT}} \sim 10^{16}$ GeV, are larger than the maximal temperature the universe is reheated after inflation. A possible solution to this problem is sketched in section 24.3.

Leptogenesis. This is an attractive model for baryogenesis which connects the smallness of neutrino masses with the creation of the baryon asymmetry. In a first step, L violating decays of heavy right-handed neutrinos generate $L \neq 0$. Latter, sphaleron processes that conserve only $B - L$ convert the lepton asymmetry into a baryon asymmetry.

In the seesaw model, we extend the SM by three right-chiral neutrinos $N_{R\alpha}$ with $\alpha = \{e, \mu, \tau\}$. Their interactions are described by

$$
\mathscr{L} = M_\alpha \bar{N}_{R\alpha}^c N_{R\alpha} + (y_{\alpha\beta} \bar{l}_{L\alpha} \Phi N_{R\beta} + \text{h.c.}). \tag{22.35}
$$

Since we include Majorana masses, lepton number is violated in the decays of the right-chiral neutrinos,

$$
N_{R\alpha} \to l_{L\beta} + \bar{\Phi} \quad \text{and} \quad N_{R\alpha} \to l_{L\beta}^c + \Phi. \tag{22.36}
$$

If there is enough CP violation and the out-of-equilibrium condition is satisfied, these decays can result in a L asymmetry. We first check the latter condition, comparing the decay width $\Gamma \sim |y_{\alpha 1}|^2 M_1/(16\pi)$ of the lightest right-chiral neutrino N_{R1} with mass M_1 to the Hubble rate $H(T)$ at $T = M_1$. This results in the bound $M_1 \gtrsim 10^{14}$ GeV for a Yukawa coupling of $\mathcal{O}(y) = 0.1$. Numerical calculations show that the decoupling can happen later, which relaxes the bound to $M_1 \gtrsim 10^{12}$ GeV.

As sources for CP violation, we have to consider the interference terms of the tree-level decays (22.36) with one-loop corrections. We have to require $N_j \neq N_1$ as virtual

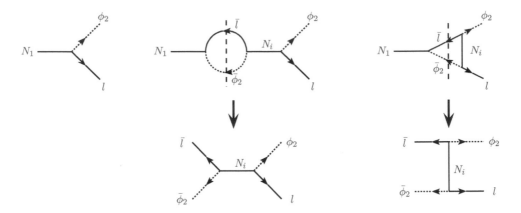

Fig. 22.2 Top: Feynman diagrams for the tree-level decay and the one loop corrections which source CP violation. Bottom: Scattering processes with $\Delta L = 2$.

particle, while the presence of the light particles l and Φ ensures that an imaginary part is created. Additionally to the vertex correction, these requirements are also satisfied by the self-energy insertion, cf. the top of Fig. 22.2. Cutting the loops as indicated leads automatically to scattering processes which violate the lepton number by two units, $\Delta L = 2$. These scattering processes will wash out the generated L asymmetry, if they are in equilibrium. They are described by the effective interaction

$$\mathscr{L} = \frac{1}{M} ll\Phi\Phi, \tag{22.37}$$

which generates after electroweak symmetry breaking Majorana masses $m_\nu = v^2/M$ for the three light neutrinos via the seesaw mechanism. For dimensional reasons, the rate $\Gamma(\Delta L = 2)$ of these processes is proportional to

$$\Gamma(\Delta L = 2) \sim \frac{T^3}{M^2} \sim \frac{T^3 \sum_i m_i^2}{v^4}. \tag{22.38}$$

These processes are not effective, if their rate is smaller than the Hubble rate in the range $T_{\text{ew}} < T < M_1$. This implies a limit on the masses of the light neutrinos,

$$\sum_{i=e,\mu,\tau} m_i^2 \lesssim (0.2\,\text{eV})^2, \tag{22.39}$$

which is comparable to the upper limit from neutrinoless double-beta decay and from cosmology.

22.3 Baryogenesis in phase transitions

The most economical model for baryogenesis is the attempt to use the SM only: It contains C, CP as well as B violation and hence the first two of the three Sakharov conditions are satisfied. Although electroweak interactions rates are fast compared to the Hubble rate at $T \sim m_W$ and we can thus not use out-of-equilibrium decays, a deviation from thermal equilibrium may occur during the electroweak phase transition. In practice, however, the amount of CP violation in the CKM matrix is too small and the phase transition is only a smooth cross-over. It is still tempting to ask if electroweak baryogenesis is possible for a slightly modified SM, adding a second Higgs doublet, for example, since such models are testable at the LHC.

Recall that in a first-order transition, the two minima at the critical temperature T_c are separated by a potential barrier. While the universe cools below T_c, it will be trapped for some time in the false minimum. As a result of tunnelling through the potential barrier, nucleation of bubbles containing the true vacuum starts. Initially, their surface tension is too large and the bubbles will collapse. When the temperature lowers further, the volume energy gain overweights the surface tension and the bubbles start to grow. As the bubbles expand, the expectation value $\langle \phi \rangle$ of the Higgs field changes from $\langle \phi \rangle = 0$ to $\langle \phi \rangle = v(T)$ when the bubble passes the considered point. It is this change of the order parameter $v(T)$ which provides the necessary departure from equilibrium.

Baryogenesis in phase transitions is a complicated dynamical process and we can offer only a sketch of the basic physics involved. Let us estimate first the relative size of the relevant scales:

- The sphaleron rate changes from $\Gamma_{\rm sp} \sim \xi \alpha_W^4 T \sim 10^{-6}T$ outside the bubble to $\Gamma_{\rm sp} \sim \exp(-E_{\rm sp}/T) \sim \exp(-m_W(T)/\alpha_W T)$ in the broken phase contained in the expanding bubble. Requiring that the generated baryon number is not washed out in the broken phase, requires $m_W(T)/T = gv(T)/2T \gg 1$ or, in other words, a large jump in the order parameter $v(T)$ of the phase transition.

- The time scale for kinetic equilibration is given by strong or electroweak Coulomb scatterings with rate close to the plasma temperature, $\Gamma_{\rm th} \sim 0.1T$.

- The time scale for the change of the Higgs vev when the bubble passes depends on the bubble speed v_b and the width of the bubble wall δ. Typically, one finds $\Gamma_{\rm v} \sim \dot{\phi}/\phi \sim v_b/\delta \sim (0.01 - 0.1)T$.

Hence the rate of baryon number-violating processes is always out of equilibrium near the wall, $\Gamma_{\rm sp} \ll \Gamma_{\rm v}$. Depending on the relative size of $\Gamma_{\rm th}$ and $\Gamma_{\rm v}$ one distinguishes two regimes:

- The adiabatic thick-wall regime, $\Gamma_{\rm th} \gg \Gamma_{\rm v}$, where the plasma is in quasi-static equilibrium with time-dependent chemical potentials.

- The non-adiabatic thin-wall regime, $\Gamma_{\rm th} \ll \Gamma_{\rm v}$. Here, the individual CP violating transmission and reflection of particles at the bubble wall has to be calculated. These processes differ for L and R chiral fermions, because CP is violated: therefore there is an excess of $f_l + \bar{f}_R$ compared to $f_R + \bar{f}_L$ in front of the bubble wall. Sphaleron interaction are effective only in the unbroken phase outside the bubble. Other processes exchange L to R. Since sphalerons react only with f_L, the rate for

$$e_L^- \mu_L^- \tau_L^- \rightarrow u_L u_L d_L + c_L c_L s_L + t_L t_L b_L,$$

is slower than for

$$e_R^+ \mu_R^+ \tau_R^+ \rightarrow \bar{u}_R \bar{u}_R \bar{d}_R + \bar{c}_R \bar{c}_R \bar{s}_R + \bar{t}_R \bar{t}_R \bar{b}_R,$$

because there are more $\bar{f}_R$ than f_L. The change of the baryon number in the two reactions is opposite, and thus a difference in their rates results in the creation of a baryon asymmetry. Since the wall is moving, part of the created baryons will end up inside the broken phase where wash-out processes are not active. It is this part which survives and creates the final baryon asymmetry.

Adiabatic thick-wall regime. Let us consider now the opposite regime in a bit more detail. We use as a toy model the Yukawa interactions between one fermion doublet Q_L, a singlet q_R and two Higgs doublets. Their vev's are spacetime-dependent and contain one physical phase which we can choose as $\phi_1 = v_1 e^{i\delta}$ and $\phi_2 = v_2 e^{-i\delta}$. The Yukawa interactions contain terms of the type

$$\mathcal{L}_Y = -y_f v_1 e^{i\delta} \bar{d}_L d_R + \text{h.c.} \tag{22.40}$$

Both v_1 and δ vary across the bubble wall. Since the bubble is expanding, v_1 and δ and thus also the Lagrangian are time-dependent. We can eliminate the time-dependence of $\mathcal{L}_Y$ caused by $\delta(t)$ performing the time-dependent rotation $d_R \to e^{-i\delta(t)} d_R$. This induces an additional term in the kinetic energy of right-chiral quarks,

$$i\bar{d}_R \slashed{\partial} d_R \to i\bar{d}_R \slashed{\partial} d_R + \dot{\delta} \bar{d}_R \gamma^0 d_R. \tag{22.41}$$

As a result, the Hamiltonian changes as

$$H \to H - \dot{\delta} \int \mathrm{d}^3 x \, \bar{d}_R \gamma^0 d_R = H - \dot{\delta} N_R. \tag{22.42}$$

Electroweak processes which interchange left- and right-chiral quarks are fast compared to $\dot{\delta}$. In contrast, the sphaleron rate is slower than $\dot{\delta}$ and we neglect this rate in the first step. In this approximation, the baryon number $B = N_R + N_L$ is zero and we have to add to (22.42) a chemical potential,

$$H \to H - \dot{\vartheta} N_R - \mu_B (N_R + N_L). \tag{22.43}$$

The resulting effective chemical potentials for right- and left-chiral quarks differ, and using again (15.59), we obtain

$$\Delta n_B = n_{B,R} + n_{B,L} = \left[(\mu_B + \dot{\delta}) + 2\mu_B \right] \frac{T^2}{6}, \tag{22.44}$$

where the factor 2 accounts for the two components of the fermion doublet. Setting $\Delta n_B = 0$, we find

$$\mu_B = -\frac{1}{3} \dot{\delta}. \tag{22.45}$$

Next we include the effect of the non-zero sphaleron rate. Using $\Delta B = 1$ and $\Delta F = \mu_B \Delta B = \mu_B$ as change of the free energy F and the baryon number B per sphaleron process, we find applying detailed balance

$$\frac{\mathrm{d} n_B}{\mathrm{d} t} = -\frac{\Delta F \, \Delta B}{T} \Gamma_{\text{sp}} = -\frac{\mu_B}{T} \Gamma_{\text{sp}} = \frac{1}{3} \frac{\dot{\delta}}{T} \Gamma_{\text{sp}}. \tag{22.46}$$

In the last step, we used also that the change of μ_B is mainly driven by $\dot{\delta}$, while $\Gamma_{\rm sp}$ is slow. Integrating this relation, we obtain the baryon number density generated by a passing bubble wall,

$$n_B = \frac{1}{3T} \int \mathrm{d}t \, \dot{\delta}\Gamma_{\rm sp}. \tag{22.47}$$

For an order of magnitude estimate we can use that the sphaleron rate drops from $\Gamma_{\rm sp} \approx \xi\alpha_W^4 T \sim 10^{-6}T$ to zero. Then we find for the ratio of baryon and entropy density

$$\frac{n_B}{s} \approx \frac{45\xi\alpha_W^4 T\Delta\vartheta}{6\pi^2 g_*} \sim 10^{-8}\Delta\vartheta. \tag{22.48}$$

This simple estimate shows that successful electroweak baryogenesis is possible, if the SM is extended such that there is sufficient CP violation and the phase transition is strong enough. While the first condition is relatively easy too satisfy the second one requires additional particles which generically should be discovered at the LHC.

Summary

The dynamical generation of a baryon asymmetry is only possible, if the Sakharov conditions (violation of B, of C and CP, and departure from thermal equilibrium) are satisfied. The SM contains neither a sufficiently strong source of CP violation, nor leads to a departure from thermal equilibrium in the early universe. Thus baryogenesis requires necessarily physics beyond the SM. Electroweak baryogenesis has the virtue of being testable at accelerators, while leptogenesis is supported by the observation that the light neutrino masses required to avoid wash-out of the baryon asymmetry fall in the experimentally allowed range.

Further reading. A more detailed discussion of baryogenesis which includes also the Affleck–Dine mechanism can be found in Gorbunov and Rubakov (2011b). The connection between leptogensis and neutrino masses is reviewed in Buchmüller *et al.* (2005). Baryogenesis is an application where loop corrections involving thermal fields are essential. This requires to go beyond the Boltzmann equation, since its derivation is based on the distribution function in classical phase space; for some approaches in this direction see Anisimov *et al.* (2011).

Problems

22.1 CP transformation. Derive (22.4) from the properties of the fields (operator) under C and P.

22.2 Relic abundance of antimatter. Derive the Boltzmann equation for the abundance of antiprotons in the presence of a non-zero chemical potential. Estimate the relic abundance of antiprotons and positrons using $\eta = 6 \times 10^{-10}$.

22.3 Leptogenesis. Check the out-of-equilibrium condition in the decays of right-chiral neutrinos. Derive the mass-limit on light neutrinos from $\Delta L = 2$ scatterings.

23

Quantum fields in curved spacetime

Our treatment of quantum field theory has been restricted up to now to *inertial* observers and detectors in *Minkowski* space. This combination is very peculiar for two reasons: first, all Minkowski space is covered by a time-like Killing vector field. Second, no event horizons exist for inertial observers in this spacetime. The existence of a unique time-like Killing vector field ∂_t which has as eigenfunctions the modes $e^{-i\omega t}$ implies that all inertial observers can agree how to split positive and negative frequency modes. This splitting selects in turn the standard Minkowski vacuum $|0\rangle_M$. As a result, the vacuum states defined by different inertial observers agree, and no inertial detector will register particles in the vacuum state $|0\rangle_M$. In this chapter we will consider more general situations. A case of special interest is the expanding universe described by the FLRW metric, where no time-like Killing vector field exists. Intuitively, we expect that a time-dependent gravitational field, in analogy to an electric field, can create particles. Our analysis confirms this expectation, but also teaches us that the concept of a "particle" becomes dubious in a non-stationary metric. Even more astonishingly, we will find that particle creation can also occur in the case of a *static* spacetime, if an event horizon exists. Such a horizon obstructs the construction of an unique time-like Killing vector field and corresponds to a surface of infinite redshift. As a result, a *thermal* spectrum of particles is created close to the horizon.

23.1 Conformal invariance and scalar fields

Conformally flat spacetimes. The problem of quantising field theories in curved spacetimes simplifies considerably, if we apply the following two restrictions. First, we consider only conformally flat spacetimes, that is, spacetimes which are connected by a conformal transformation to Minkowski space,

$$g_{\mu\nu}(x) = \Omega^2(x)\eta_{\mu\nu}(x) = e^{2\omega(x)}\eta_{\mu\nu}(x). \tag{23.1}$$

Note that conformal transformations $g_{\mu\nu}(x) \to \tilde{g}_{\mu\nu}(x) = \Omega^2(x)g_{\mu\nu}(x)$ of the *metric* are not equivalent to conformal transformations of the *coordinates*, $x \to \tilde{x} = e^{\omega(x)}x$, which we considered in chapter 18.1. In the latter case, the argument of the metric tensor in the LHS and the RHS in (23.1) would differ, see also problem 23.1. Recall also that a coordinate transformation $x \to \tilde{x}(x)$ only relabels the spacetime points, but does not affect physics, since we require that any action S is invariant under general coordinate transformations. By contrast, a conformal transformation of the metric shrinks and stretches the Riemannian manifold $\{\mathcal{M}, g_{\mu\nu}\}$ into another manifold $\{\tilde{\mathcal{M}}, \tilde{g}_{\mu\nu}\}$.

Quantum Fields–From the Hubble to the Planck Scale. Michael Kachelriess. © Michael Kachelriess 2018.
Published in 2018 by Oxford University Press. DOI 10.1093/oso/9780198802877.001.0001

Conformal transformations change distances, but keep angles invariant. Thus the causal structure of two conformally related spacetimes is identical. In particular, light-rays also propagate in conformally flat spacetimes along straight lines at ± 45 degrees to the time axis. Important examples for conformally flat spacetimes are the flat FLRW metric and all two-dimensional spacetimes.

Remark 23.1: Geometric quantities derived for the metric $g_{\mu\nu}(x)$ are connected to those of a conformally transformed metric $\tilde{g}_{\mu\nu}(x) = \Omega^2(x)g_{\mu\nu}(x) = e^{2\omega(x)}g_{\mu\nu}(x)$ as

$$\tilde{\Gamma}^\mu{}_{\alpha\beta} = \Gamma^\mu{}_{\alpha\beta} + \Omega^{-1}\left[\delta^\mu_\alpha\partial_\beta\Omega + \delta^\mu_\beta\partial_\alpha\Omega - g_{\alpha\beta}g^{\mu\nu}\partial_\nu\Omega\right], \tag{23.2a}$$

$$\tilde{R}_{\mu\nu} = R_{\mu\nu} - g_{\mu\nu}\Box\omega - (d-2)\nabla_\mu\nabla_\nu\omega + (d-2)\nabla_\mu\omega\nabla_\nu\omega - (d-2)g_{\mu\nu}\nabla^\lambda\omega\nabla_\lambda\omega, \tag{23.2b}$$

$$\tilde{R} = \Omega^{-2}\left[R - 2(d-1)\Box\omega - (d-1)(d-2)\nabla^\mu\omega\nabla_\mu\omega\right], \tag{23.2c}$$

as one can check by direct (but tedious) computation.

Conformal invariance.   Another simplification occurs if we consider field theories which are conformally invariant in Minkowski space. Recall that such theories contain no dimensions-full parameter and satisfy $T_\mu{}^\mu = 0$. In a curved spacetime, we call such a theory conformally or Weyl invariant.[1] The action

$$S[\phi, g^{\mu\nu}] = \int \mathrm{d}^d x\sqrt{|g|}\,\mathscr{L}(\phi, \partial_\mu\phi, g^{\mu\nu}) \tag{23.3}$$

of such a theory is invariant under a conformal transformation of the metric,

$$S[\phi, g^{\mu\nu}] = S[\tilde{\phi}, \tilde{g}^{\mu\nu}] = \int \mathrm{d}^d x\sqrt{\tilde{g}}\,\mathscr{L}(\tilde{\phi}, \partial_\mu\tilde{\phi}, \tilde{g}^{\mu\nu}), \tag{23.4}$$

if we rescale the field according to its canonical dimension, that is, $\phi(x) \to \tilde{\phi}(x) = \Omega^{(2-d)/2}(x)\phi(x)$ for a boson. Therefore the equations of motion for the field $\tilde{\phi}$ using the metric $\tilde{g}_{\mu\nu}$ are the same as those for the field ϕ using the metric $g_{\mu\nu}$. This allows us to relate the quantisation of a Weyl invariant theory in a conformally flat space to the known problem of quantising the field in Minkowski space.

Conformal invariance of a scalar field.   We have experienced that most calculations in Minkowski space for a scalar field are considerably less involved than for fields with non-zero spin. This holds true also in curved spacetimes, except for one aspect. While massless Dirac and Yang–Mills fields are classically conformal-invariant, this is not the case for a scalar field minimally coupled to gravity. As discussed in section 18.1 the trace $T_\mu{}^\mu$ of the stress tensor for a massless scalar field is given by

$$T_\mu{}^\mu = -\left(1 - \frac{d}{2}\right)\Box\phi^2 \tag{23.5}$$

and vanishes only in $d = 2$, when the field ϕ is dimensionless.

---

[1]H. Weyl suggested in 1918 the combined scale transformation $\tilde{g}_{\mu\nu}(x) = e^{2\omega(x)}g_{\mu\nu}(x)$ and $\tilde{A}_\mu(x) = e^{\omega(x)}A_\mu(x)$ in an attempt to unify the gravitational and electromagnetic field.

However, we have the freedom to improve the action or the stress tensor by appropriate terms which do not affect the equations of motion or the generators of the Poincaré algebra, respectively. We pursue the second approach in problem 21.4. Here we ask if we can make the scalar action Weyl-invariant, while retaining the usual equations of motion in Minkowski space.

The latter constraint is taken into account if we only modify the coupling of the scalar field to gravity, that is, if we abandon the substitution rule $\{\partial_\mu, \eta_{\mu\nu}, d^4x\} \to \{\nabla_\mu, g_{\mu\nu}, d^4x\sqrt{|g|}\}$. While this rule is implied by the strong equivalence principle, there are two reasons to expect deviations. First, we already know that even within Einstein gravity non-zero torsion can exist. In this case, it is not possible to eliminate locally all effects of gravity by introducing a local inertial system. Second, we expect that quantum corrections will add all renormalisable coupling terms between the scalar and the gravitational field even if we set them to zero at tree level. Thus we should ask ourselves which additional renormalisable coupling terms between the scalar and the gravitational field exist. Since $[R] = m^2$ and thus also $[R_{\mu\nu}] = [R_{\mu\nu\rho\sigma}] = m^2$, the only dimensionless additional interaction term is a linear coupling of the curvature scalar R to ϕ^2. Such a coupling $R\phi^2/2$ acts as a curvature dependent, additional mass term for the scalar field. Now we ask how the action[2]

$$S = \int d^dx\sqrt{|g|} \left[\frac{1}{2}g^{\alpha\beta}\partial_\alpha\phi\partial_\beta\phi - \frac{1}{2}(m^2 - \xi R)\phi^2 \right], \tag{23.6}$$

where ξ parameterises the coupling to the curvature scalar, transforms under a conformal transformation of the metric,

$$g_{\mu\nu}(x) \to \tilde{g}_{\mu\nu}(x) = \Omega^2(x)g_{\mu\nu}(x). \tag{23.7}$$

From our discussion of scale transformations in Minkowski space, we know that a bosonic field in d spacetime dimensions scales as

$$\phi(x) \to \tilde{\phi}(x) = \Omega^{(2-d)/2}(x)\phi(x) = \Omega^D(x)\phi(x). \tag{23.8}$$

This leads for $d > 2$ clearly to a non-trivial transformation of the kinetic term which has to be compensated by the non-trivial transformation of the scalar curvature term. The transformation (23.7) implies

$$g^{\mu\nu} \to \tilde{g}^{\mu\nu} = \Omega^{-2}g^{\mu\nu} \quad \text{and} \quad \sqrt{|g|} \to \sqrt{\tilde{g}} = \Omega^d\sqrt{|g|}. \tag{23.9}$$

As a result, the action S_0 obtained setting $\xi = 0$ changes as

$$S_0 \to \tilde{S}_0 = \frac{1}{2}\int d^dx\sqrt{|g|}\,\Omega^d \left[\Omega^{-2}g^{\alpha\beta}\nabla_\alpha(\Omega^D\phi)\nabla_\beta(\Omega^D\phi) - m^2\Omega^{2D}\phi^2 \right] \tag{23.10}$$

or

---

[2]Since $g_{\alpha\beta}$ and $g^{\alpha\beta}$ transform inversely, we have to distinguish between $\partial_\mu\phi$ and $\partial^\mu\phi$. Convention is to use only $g^{\alpha\beta}$ and to write all fields and derivatives with lower indices.

$$\tilde{S}_0 = \frac{1}{2} \int d^d x \sqrt{|g|} \left[g^{\alpha\beta} \nabla_\alpha \phi \nabla_\beta \phi + 2\Omega^{d-2+D} g^{\alpha\beta} \nabla_\alpha \phi (\nabla_\beta \Omega^D) \phi \right. \tag{23.11a}$$

$$\left. + \Omega^{d-2} \phi^2 g^{\alpha\beta} (\nabla_\alpha \Omega^D)(\nabla_\beta \Omega^D) - m^2 \Omega^{2D} \phi^2 \right]$$

$$= \frac{1}{2} \int d^d x \sqrt{|g|} \left[g^{\alpha\beta} \partial_\alpha \phi \partial_\beta \phi - D(\Box \omega) \phi^2 + D^2 (\nabla \omega)^2 \phi^2 - m^2 \Omega^{2D} \phi^2 \right]. \tag{23.11b}$$

Taking into account the transformation rule (23.2c) for the scalar curvature R,

$$\sqrt{|g|} \, R\phi^2 \to \sqrt{|g|} \, [R - 2(d-1)\Box\omega - (d-1)(d-2)\nabla^\mu \omega \nabla_\mu \omega]\phi^2, \tag{23.12}$$

we find that the scalar action is invariant choosing $m = 0$ and

$$\xi = \xi_d \equiv \frac{d-2}{4(d-1)}. \tag{23.13}$$

A scalar field with $\xi = \xi_d$ is called conformally coupled to gravity, while the choice $\xi = 0$ is called minimally coupled to gravity. Note that for $d = 2$, that is when ϕ is dimensionless and does not scale under conformal transformations, minimal and conformal coupling agree.

Scalar field equation in an FLRW background. Next we derive the equation of motion for a scalar field with arbitrary ξ in an expanding universe described by the flat FLRW metric. Choosing as coordinates $\{t, \boldsymbol{x}\}$, we have $g_{\mu\nu} = \text{diag}(1, -a^2, -a^2, -a^2)$, $g^{\mu\nu} = \text{diag}(1, -a^{-2}, -a^{-2}, -a^{-2})$, and $\sqrt{|g|} = a^3$. Varying the action

$$S = \int d^4 x \, a^3 \left\{ \frac{1}{2} \dot{\phi}^2 - \frac{1}{2a^2} (\boldsymbol{\nabla}\phi)^2 - V(\phi) \right\} \tag{23.14}$$

gives

$$\delta S = \int d^4 x \, a^3 \left\{ \dot{\phi} \delta\dot{\phi} - \frac{1}{a^2} (\boldsymbol{\nabla}\phi) \cdot \delta(\boldsymbol{\nabla}\phi) - V_{,\phi} \delta\phi \right\} \tag{23.15a}$$

$$= \int d^4 x \left\{ -\frac{d}{dt} (a^3 \dot{\phi}) + a\boldsymbol{\nabla}^2 \phi - a^3 V_{,\phi} \right\} \delta\phi \tag{23.15b}$$

$$= \int d^4 x \, a^3 \left\{ -\ddot{\phi} - 3H\dot{\phi} + \frac{1}{a^2} \boldsymbol{\nabla}^2 \phi - V_{,\phi} \right\} \delta\phi, \tag{23.15c}$$

setting $V_{,\phi} \equiv dV/d\phi$. Thus the Klein–Gordon equation for a scalar field with the potential $V(\phi) = (m^2 + \xi R)\phi^2/2$ in a flat FLRW background is

$$\ddot{\phi} + 3H\dot{\phi} - \frac{1}{a^2} \boldsymbol{\nabla}^2 \phi + (m^2 + \xi R)\phi = 0. \tag{23.16}$$

The term $3H\dot{\phi}$ acts in an expanding universe as a friction term for the oscillating ϕ field. Moreover, the gradient of ϕ is also suppressed for increasing a; this term can be therefore often neglected.

Next we want to rewrite this equation as the one for an harmonic oscillator with a time-dependent oscillation frequency. We introduce first the conformal time $d\eta = dt/a$.

Remark 23.2: For a power law-like behaviour of the scale factor, $a(t) \propto t^p$, the conformal time η evolves as $\eta = \int dt\, t^{-p} \propto t^{1-p}$ and thus $a(\eta) \propto \eta^{\frac{p}{1-p}}$. In particular, η scales as $a(\eta) \propto \tau$ in the radiation dominated ($p = 1/2$) and as $a(\eta) \propto \tau^2$ in the matter dominated epoch ($p = 2/3$). For a de Sitter phase, $a(t) \propto e^{Ht}$, we obtain

$$\eta = \int dt\, e^{-Ht} = -H^{-1}e^{-Ht} + \eta_0 = -(aH)^{-1} + \eta_0.$$

Setting $\eta_0 = 0$ results in $a(\eta) = -1/(H\eta)$. Inflation is defined as the phase in the early universe with an accelerated expansion, $\ddot{a} > 0$. Using the convention $\eta_0 = 0$, inflation ends at $\eta = 0$, followed by the standard big bang evolution for $\eta > 0$.

Then we change the derivatives of the field,

$$\dot{\phi} = \frac{d\phi}{dt} = \frac{d\phi}{d\eta}\frac{d\eta}{dt} = \frac{1}{a}\phi', \quad \text{and} \quad \ddot{\phi} = \frac{1}{a}\frac{d}{d\eta}\left(\frac{1}{a}\phi'\right) = \frac{1}{a^2}\phi'' - \frac{a'}{a^3}\phi', \tag{23.17}$$

and also express the Hubble parameter as function of η,

$$H = \frac{\dot{a}}{a} = \frac{a'}{a^2} \equiv \frac{\mathscr{H}}{a}. \tag{23.18}$$

Inserting these expressions into Eq. (23.16) and multiplying with a^2 gives

$$\phi'' + 2\mathscr{H}\phi' - \nabla^2\phi + a^2 V_{,\phi} = 0. \tag{23.19}$$

Performing then a Fourier transformation, $\phi(\eta, \boldsymbol{x}) = \sum_k \phi_k(\eta) e^{i\boldsymbol{k}\boldsymbol{x}}$, we obtain

$$\phi_k'' + 2H\phi_k' + [k^2 + (m^2 + \xi R)a^2]\phi_k = 0. \tag{23.20}$$

Note that $\boldsymbol{k}$ is the comoving wave number. Since the proper distance varies as $\boldsymbol{x} \propto a$, the physical momentum is $\boldsymbol{p} = \boldsymbol{k}/a$.

Finally, we can eliminate the friction term $2H\phi_k'$ by introducing the auxiliary field $\chi_k(\eta) = a(\eta)\phi_k(\eta)$. Then we obtain a harmonic oscillator equation for χ_k,

$$\chi_k'' + \omega_k^2 \chi_k = 0, \tag{23.21}$$

with the time-dependent frequency

$$\omega_k^2(\eta) = k^2 + (m^2 + \xi R)a^2 - \frac{a''}{a}. \tag{23.22}$$

For a massless conformally coupled ($\xi = 1/6$) scalar field, the frequency is independent of the expansion of the universe, $\omega_k^2(\eta) = k^2$, cf. problem 21.3.

Now we choose the special case of a de Sitter universe as an approximation for the inflationary phase of the early universe. Moreover, we consider a minimally coupled scalar field with negligible mass. Combining then $a = -1/(H\eta)$ and $a'' = -2/(H\eta^3)$, or

$$\frac{a''}{a} = \frac{2}{\eta^2},$$
(23.23)

the wave equation simplifies to

$$\chi_k'' + \left(k^2 - \frac{2}{\eta^2} \right) \chi_k = 0.$$
(23.24)

We examine first the short and the long-wavelength limit. In the first case, $k \gg |1/\eta|$, the field equation is conformally equivalent to the one in Minkowski space, with solution

$$\chi_k(\eta, \boldsymbol{x}) = \frac{1}{\sqrt{2k}} (A_k e^{-\mathrm{i}kx} + B_k e^{\mathrm{i}kx}).$$
(23.25)

Here we factored out a normalisation factor $1/\sqrt{2k} \equiv 1/\sqrt{2\omega_k}$. With $\eta = -1/(aH)$, we can rewrite the short-wavelength condition as $|k|/a \gg H^{-1}$. Thus the comoving wavelength of the particle is much shorter than the comoving Hubble radius aH, or equivalently, its physical wavelength is much shorter than the Hubble radius. Therefore these solutions are called subhorizon modes.

In the opposite limit, we find $a''\chi_k = a\chi_k''$ which has as a growing solution $\chi_k \propto a$ and thus $\phi_k = \mathrm{const}$. Thus modes with wavelengths larger than the horizon are "frozen in" and do not oscillate. They are called superhorizon modes. The complete solution is given by Hankel functions $H_{3/2}(\eta)$,

$$\chi_k(\eta, \boldsymbol{x}) = A_k \frac{e^{-\mathrm{i}kx}}{\sqrt{2k}} \left(1 - \frac{\mathrm{i}}{k\eta} \right) + B_k \frac{e^{\mathrm{i}kx}}{\sqrt{2k}} \left(1 + \frac{\mathrm{i}}{k\eta} \right).$$
(23.26)

We could now set out for the quantisation of the scalar field χ. If we ignore the time-dependence and quantise the scalar field with the time-dependent mass term (23.22) in the standard way, we will obtain different vacua and different Fock spaces at different times t. As a result, a state which was empty at time t will contain in general particles at time t'. Thus the time-dependent gravitational field can excite modes χ_k, supplying energy and leading to particle production. We will postpone the quantisation of a scalar field in a FLRW metric to the next chapter where this equation will play a prominent role. Before that we will introduce some formalism needed and discuss two conceptionally simpler examples.

23.2 Quantisation in curved spacetimes

As in Minkowski space, we can use both canonical quantisation or the path-integral approach to quantise classical field theories in curved spacetimes. The latter method is particularly useful if we are interested in quantum corrections to the stress tensor: its expectation value for the quantum field ϕ in the background of a classical gravitational field $g^{\mu\nu}$ is

$$\langle T_{\mu\nu} \rangle = \frac{\int \mathcal{D}\phi \, T_{\mu\nu} \, e^{\mathrm{i}S[\phi, g^{\mu\nu}]}}{\int \mathcal{D}\phi \, e^{\mathrm{i}S[\phi, g^{\mu\nu}]}}.$$
(23.27)

Note that now the gravitational field $g^{\mu\nu}$ plays the usual role of a classical source. Inserting the definition (7.49) of the dynamical stress tensor and recalling that the denominator in (23.27) is the generating functional $Z = \exp(\mathrm{i}W)$ leads to

$$\langle T_{\mu\nu} \rangle = \frac{1}{Z[g^{\mu\nu}]} \frac{2}{\mathrm{i}\sqrt{|g|}} \frac{\delta}{\delta g^{\mu\nu}} Z[g^{\mu\nu}] = \frac{2}{\sqrt{|g|}} \frac{\delta W[g^{\mu\nu}]}{\delta g^{\mu\nu}}. \tag{23.28}$$

Having calculated $\langle T_{\mu\nu} \rangle$, one could aim at solving the Einstein equations in the semi-classical limit, replacing $T_{\mu\nu}$ by $\langle T_{\mu\nu} \rangle$. In this way, one takes into account two effects: first, the gravitational background can produce particles. Second, it changes the zero-point energies of the ϕ vacuum, analogous to the Casimir effect or vacuum polarisation.

One of the main advantages of this approach is that it is based on a local quantity, $\langle T_{\mu\nu} \rangle = \langle T_{\mu\nu}(x) \rangle$. Thus if we can show in a specific frame that, for example, $\langle T_{\mu\nu}(x) \rangle = 0$, then any observer will agree on that. By contrast, we will see that the expectation value $\langle \tilde{0}|N|\tilde{0} \rangle$ for the number of particles measured in a specific vacuum $|\tilde{0}\rangle$ depends on the trajectory of the considered detector. Therefore the concept of particle number and production is not a local one. This implies in particular that we can address the question of where or when a particle is created only in an approximate way. An essential ingredient in both approaches is the vacuum state to be used in the calculation of expectation values. We will therefore concentrate first on this question, and then apply the simpler canonical quantisation formalism to specific examples.

23.2.1 Bogolyubov transformations

Ambiguity of the vacuum. A basic ingredient of the canonical quantisation procedure is to split the fields into positive frequencies propagating forward in time, and negative frequencies propagating backwards. Since we associate annihilation operators with negative frequency modes and creation operators with positive frequency modes, this splitting defines the vacuum. As Minkowski space contains a time-like Killing vector field ∂_t which has as eigenfunctions the modes $\mathrm{e}^{-\mathrm{i}\omega t}$ with positive eigenvalues ω, the vacuum is invariant under Lorentz transformation. All observers in inertial frames agree on the choice of the vacuum and thus also on one- and many-particle states.

In curved spacetimes no inertial system can be globally extended to cover the whole manifold. No unique definition of the vacuum is possible and thus the notion of particle number becomes observer-dependent, which in turn implies the creation of particles. Using the correspondence $g^{\mu\nu}(x) \to J(t)$, we can illustrate this behaviour with the simple example of a harmonic oscillator which is driven during a finite time interval $0 < t < T$ by an external force $J(t)$.

Example 23.1: Excitation of a driven harmonic oscillator. In the notation of chapter 2.4, the Hamiltonian of a driven harmonic oscillator is given by

$$H(\phi, \pi) = \frac{1}{2}\pi^2 + \frac{1}{2}\omega^2\phi^2 - J\phi. \tag{23.29}$$

We assume that the classical external source $J(t)$ acts only in the finite time interval $0 < t < T$. We keep the definition (2.60) of the annihilation and creation operators, from which we find now as equation of motion

$$\dot{a} = -\mathrm{i}\omega a + \frac{\mathrm{i}}{2\omega}\, J(t). \tag{23.30}$$

Specifying the initial value before we apply the external force as $a(t = 0) \equiv a_\mathrm{in}$ results in

$$a(t) = a_\mathrm{in}\mathrm{e}^{-\mathrm{i}\omega t} + \frac{\mathrm{i}}{2\omega}\int_0^t \mathrm{d}t'\, J(t')\mathrm{e}^{-\mathrm{i}\omega t'}. \tag{23.31}$$

For $t > T$, we set $a(t) \equiv \mathrm{e}^{-\mathrm{i}\omega t} a_\mathrm{out} \equiv \mathrm{e}^{-\mathrm{i}\omega t}(a_\mathrm{in} + J_0)$. Our aim is to express the in-states in term of the out-states. We set $|0_\mathrm{in}\rangle = \sum_{n=0}^\infty c_n\, |n_\mathrm{out}\rangle$, where the coefficients c_n have to be determined. Acting with a_out on the in-vacuum gives

$$a_\mathrm{out}\,|0_\mathrm{in}\rangle = \sum_{n=0}^\infty c_n a_\mathrm{out}\,|n_\mathrm{out}\rangle = \sum_{n=0}^\infty \sqrt{n}c_n\,|(n-1)_\mathrm{out}\rangle = \sum_{n=0}^\infty \sqrt{n+1}c_{n+1}\,|n_\mathrm{out}\rangle, \tag{23.32}$$

where we relabelled $n \to n+1$ in the last step. On the other hand, applying $a_\mathrm{in} + J_0 = a_\mathrm{out}$ on the in-vacuum results in

$$(a_\mathrm{in} + \tilde{J})\,|0_\mathrm{in}\rangle = J_0\,|0_\mathrm{in}\rangle = \sum_{n=0}^\infty J_0 c_n\,|n_\mathrm{out}\rangle. \tag{23.33}$$

Comparing the two expressions, we obtain $c_{n+1} = J_0 c_n/\sqrt{n+1}$ and thus $c_n = J_0^n c_0/\sqrt{n!}$. Requiring that the vacuum is normalised, we can determine c_0 as $|c_0| = \exp(-\frac{1}{2}|J_0|^2)$. Thus

$$|0_\mathrm{in}\rangle = \exp(-\frac{1}{2}|J_0|^2 + J_0 a_\mathrm{out}^\dagger)\,|0_\mathrm{out}\rangle, \tag{23.34}$$

up to an undetermined phase. The driving force $J(t)$ converted the in-vacuum into a coherent state which contains all n-particle states with the amplitude c_n.

Our corresponding task in field theory is to find a mapping between field operators defined with respect to different vacua. The relation between the two sets of field operators is a special case of a Bogolyubov transformation. We will first discuss this transformation in general.

Bogolyubov transformation. We defined a scalar product for solutions of the Klein–Gordon equation in Eq. (9.25). To simplify the notation, we normalise the solution in a box of finite volume, obtaining a discrete spectrum. Then the scalar product for plane waves becomes

$$(\varphi_i, \varphi_j) = \delta_{ij}, \quad \text{and} \quad (\varphi_i^*, \varphi_j^*) = -\delta_{ij} \tag{23.35}$$

and zero otherwise,

$$(\varphi_i, \varphi_j^*) = (\varphi_i^*, \varphi_j) = 0. \tag{23.36}$$

If we quantise the scalar field (23.21) with its time-dependent frequency at different times t and $\tilde{t}$, we will obtain different vacua and different Fock spaces,

$$a_i\,|0\rangle = 0 \quad \forall i, \qquad \tilde{a}_i\,|\tilde{0}\rangle = 0 \quad \forall i.$$

We now search for the connection between the two sets of annihilation and creation operators. We can express the field at any time through the two sets of creation and annihilation operators,

$$\phi(x) = \sum_i \left[a_i \varphi_i(x) + a_i^\dagger \varphi_i^*(x) \right] = \sum_j \left[\tilde{a}_j \tilde{\varphi}_j(x) + \tilde{a}_j^\dagger \tilde{\varphi}_j^*(x) \right]. \tag{23.37}$$

Both sets of solutions, $\{\varphi_j(x), \varphi_j^*(x)\}$ and $\{\tilde{\varphi}_j(x), \tilde{\varphi}_j^*(x)\}$, form a complete basis. Thus we can decompose any basis vector $\tilde{\varphi}_j(x)$ as

$$\tilde{\varphi}_j(x) = \sum_i \left[\alpha_{ji} \varphi_i(x) + \beta_{ji} \varphi_i^*(x) \right]. \tag{23.38}$$

The unknown matrices α_{ij} and β_{ij} are called Bogolyubov coefficients. Using the orthogonality relations (23.35) and (23.36), we can determine the Bogolyubov coefficients as

$$(\varphi_k, \tilde{\varphi}_j) = \sum_i \left[\alpha_{ij} (\varphi_k, \varphi_i) + \beta_{ij}^\dagger (\varphi_k, \varphi_i^*) \right] = \alpha_{jk} \tag{23.39}$$

and $\beta_{jk} = -(\varphi_k^*, \tilde{\varphi}_j)$. In the reverse direction, we find in the same way

$$\varphi_i(x) = \sum_j \left[\alpha_{ji}^* \tilde{\varphi}_j(x) - \beta_{ji}^\dagger \tilde{\varphi}_j^*(x) \right]. \tag{23.40}$$

Rewriting Eq. (23.38) and its complex conjugated expression in matrix form gives

$$\begin{pmatrix} \tilde{\varphi} \\ \tilde{\varphi}^* \end{pmatrix} = \begin{pmatrix} \alpha & \beta \\ \beta^* & \alpha^* \end{pmatrix} \begin{pmatrix} \varphi \\ \varphi^* \end{pmatrix} = U \begin{pmatrix} \varphi \\ \varphi^* \end{pmatrix}. \tag{23.41}$$

Since both bases are orthonormal, the matrix U is unitary,

$$\alpha_{ij} \alpha_{jk}^* - \beta_{ij} \beta_{jk}^* = \delta_{ik}. \tag{23.42}$$

The transformation properties of the annihilation and creation operator follow as

$$\left(a, a^\dagger \right) = \left(\tilde{a}, \tilde{a}^\dagger \right) U. \tag{23.43}$$

If $\beta_{ij} \neq 0$, time-evolution mixes positive and negative frequency modes. As a result, the vacuum $|\tilde{0}\rangle$ evaluated with the number operator $N_i = a_i^\dagger a_i$ will contain particles,

$$\langle \tilde{0} | N_i | \tilde{0} \rangle = \sum_j |\beta_{ji}|^2. \tag{23.44}$$

The corresponding energy density ρ in the continuum limit is

$$\rho_{\boldsymbol{k}} = \int \frac{d^3 k'}{(2\pi)^3} \, \omega_{k'} |\beta_{\boldsymbol{k}, \boldsymbol{k}'}|^2. \tag{23.45}$$

The Bogolyubov coefficients β should therefore decrease faster than k^{-2} to ensure a finite energy density.

We will see that the presence of horizons lead to a thermal flux of particles. For a thermal spectrum with temperature T, the Bogolyubov coefficients have to satisfy the condition

$$|\alpha_{ji}|^2 = e^{\omega_i/T}|\beta_{ji}|^2. \tag{23.46}$$

Then the unitarity condition (23.42) of the Bogolyubov coefficients gives

$$\langle \tilde{0}|N_i|\tilde{0}\rangle = \sum_j |\beta_{ji}|^2 = \frac{1}{e^{\omega_i/T} - 1}. \tag{23.47}$$

23.2.2 Choosing the vacuum state

Having set up the formalism of Bogolyubov transformations, we have to fill the formalism with physics: since there is no unique vacuum, we have to decide case by case which is the physically relevant one. Additionally, we need a scheme for the calculation of the Bogolyubov coefficients. The first problem, the choice of a physically sensible vacuum, simplifies if the spacetime has "useful" symmetries. This includes in particular the case that the spacetime is conformally flat. spacetimes which approach asymptotically Minkowski space for $\eta \to \pm\infty$ are also useful toy models. For such models, we can apply our standard formalism to construct the in and out Fock space. We therefore study this case first, using a model which has the further virtue of being analytically solvable.

Solvable model. We now illustrate the production of particles in an expanding universe for a specific, exactly solvable model. In this two-dimensional model, the line element is given by

$$ds^2 = C(\eta)\left[d\eta^2 - dx^2\right] \tag{23.48}$$

and the scale factor changes as

$$C(\eta) \equiv a^2(\eta) = A + B\tanh(\rho\eta) \tag{23.49}$$

with $A > B$ and $\rho > 0$. For $\eta \to \pm\infty$, the scale factor approaches the constant value $C(\eta) \to A \pm B$ and thus the metric approaches asymptotically Minkowski space. In this limit we should be able to apply our standard approach to define vacua and one-particle states.

Using the separation ansatz $\phi_k(\eta, x) = \chi_k(\eta)e^{ikx}/(2\pi)^{1/2}$ in the scalar field equation with $\xi = 0$ gives

$$\chi_k'' + \left[k^2 + C(\eta)m^2\right]\chi_k = 0. \tag{23.50}$$

Thus asymptotic in and out states have the frequency

$$\omega_{\text{in}} = \sqrt{k^2 + m^2(A - B)} \tag{23.51}$$

$$\omega_{\text{out}} = \sqrt{k^2 + m^2(A + B)} \tag{23.52}$$

and approach

$$\chi_k(\eta) \to \frac{e^{-i(\omega_{\text{in}}\eta - kx)}}{\sqrt{4\pi\omega_{\text{in}}}} \quad \text{and} \quad \chi_k(\eta) \to \frac{e^{-i(\omega_{\text{out}}\eta - kx)}}{\sqrt{4\pi\omega_{\text{out}}}} \tag{23.53}$$

for $\eta \to \mp\infty$, respectively. The complete solutions are given by hypergeometric functions; we could find them either using a computer algebra program or performing the substitution $\xi = 1 + \tanh(\rho\eta)$. The important point is that one can relate the in and out solutions

$$\chi_k^{\text{in}}(\eta, x) = \alpha_k \chi_k^{\text{out}}(\eta, x) + \beta_k \chi_{-k}^{\text{out}*}(\eta, x) \qquad (23.54)$$

using the linear transformation properties of these functions. The coefficients β_k determine the Bogolyubov coefficients, $\beta_{kk'} = \delta_{-k,k'}\beta_k$, and are given by

$$|\beta_k|^2 = \frac{\sinh^2(\pi\omega_-/\rho)}{\sinh(\pi\omega_{\text{in}}/\rho)\sinh(\pi\omega_{\text{out}}/\rho)} \qquad (23.55)$$

where $\omega_- = (\omega_{\text{in}} - \omega_{\text{out}})/2$.

Using a variation of the Gamov criterion, particle creation should be controlled by the ratio of the expansion rate $H = \dot{C}/(2C) \sim \rho$ and ω. High-frequency modes (or subhorizon modes using the language of the previous section) with $\rho/\omega \ll 1$ should be not affected by the expansion and behave as in Minkowski space. Expanding $|\beta_k|^2$ for small ρ, we see that particle production is *exponentially* suppressed in this limit, $|\beta_k|^2 \to \exp(-2\pi\omega_-/\rho)$.

Adiabatic vacuum. In cases of practical interest such as the FLRW metric, the spacetime does not approach asymptotically Minkowski space. In this case, we need an approximation scheme for the calculation of the Bogolyubov coefficients together with a suitable definition for the vacuum at an arbitrary intermediate time. Two particular choices for the vacuum state are the instantaneous and the adiabatic vacuum. The first one defines the vacuum as the state of lowest energy at each moment of time. This implies that one uses the time-dependent $\omega_k(\eta)$ in the usual Minkowski modes. This scheme overpredicts the effect of particle production, as the analogue of a (quantum) mechanical pendulum with variable length makes clear: the number of quanta E/ω is an invariant, if the length of the pendulum (i.e. ω) changes adiabatically (Landau and Lifshitz, 1981). This invariance is however not taken into account choosing the instantaneous vacuum.

The definition of the adiabatic vacuum is motivated by the requirement that high-frequency modes should not be affected by the expansion of the universe. In particular, the model (23.49) suggests that in this limit particle production is exponentially suppressed. Motivated by the WKB approximation, we express the positive mode functions as

$$\chi_k^+(\eta) = \frac{1}{\sqrt{W_k(\eta)}} \exp\left(i \int^\eta d\eta\, W_k(\eta)\right), \qquad (23.56)$$

while the negative modes functions are given by $\chi_k^-(\eta) = \chi_k^{+*}(\eta)$. We can implement the idea that high-frequency modes are unaffected by the expansion of the universe requiring the asymptotic expansion of W_k as

$$W_k(\eta) = \omega_k(\eta)\left[1 + \delta_2(\eta)\omega_k^{-2} + \delta_4(\eta)\omega_k^{-4} + \ldots\right]. \qquad (23.57)$$

An adiabatic vacuum (of order n) is defined by choosing the initial conditions for the exact solution of $\chi_k'' + \omega_k^2 \chi_k = 0$ such that only the positive modes (23.56) (with W_k evaluated at order δ_n) are present at the initial time η_0. This guarantees that particle production is exponentially suppressed in the adiabatic, high-frequency limit.

At lowest order in the asymptotic expansion, $W_k(\eta) = \omega_k(\eta)$, we can obtain a numerical approximation as follows: In general, the field modes can be expressed as

$$\chi_k(\eta) = \frac{\alpha_k(\eta)}{\sqrt{2\omega_k(\eta)}} \exp\left[-i \int^\eta d\eta\, \omega_k(\eta)\right] + \frac{\beta_k(\eta)}{\sqrt{2\omega_k(\eta)}} \exp\left[i \int^\eta d\eta\, \omega_k(\eta)\right], \quad (23.58)$$

where $\alpha_k(\eta)$ and $\beta_k(\eta)$ are the "instantaneous" Bogolyubov coefficient. The mode equation $\chi_k'' + \omega_k^2 \chi_k = 0$ is satisfied, if the coefficients satisfy

$$\alpha_k' = \frac{\omega_k'}{2\omega_k} \exp\left(2i \int d\eta\, \omega_k\right) \beta_k \qquad (23.59a)$$

$$\beta_k' = \frac{\omega_i'}{2\omega_k} \exp\left(-2i \int d\eta\, \omega_k\right) \alpha_k. \qquad (23.59b)$$

Choosing the adiabatic vacuum at the time η_0 implies the initial conditions $\alpha_k(\eta_0) = 1$ and $\beta_k(\eta_0) = 0$. Neglecting a quadratic term, we obtain as closed expression for the Bogolyubov coefficients

$$\beta_k(\eta) \simeq \int^\eta d\eta' \frac{\omega_k'}{2\omega_k} \exp\left(-2i \int^{\eta'} d\eta''\, \omega_k\right). \qquad (23.60)$$

Conformal vacuum. For a conformally flat spacetime, we can connect the solutions of a conformally invariant theory to those of Minkowski space,

$$g_{\mu\nu}(x) = \Omega^{-2}(x)\eta_{\mu\nu}. \qquad (23.61)$$

In the case of a conformally coupled scalar, the wave equation $(\Box - \xi_d R)\phi = 0$ becomes

$$\Box\tilde{\phi} = \eta_{\mu\nu}\partial_\mu\partial_\nu(\Omega^D \phi), \qquad (23.62)$$

where we used the fact that in Minkowski space $\tilde{R} = 0$ and the scaling law (23.8) for bosonic fields. The field $\tilde{\phi}$ has the usual Minkowski Fourier modes,

$$\tilde{\varphi}_k(x) = [2\omega(2\pi)^{d-1}]^{-1/2} e^{-ikx} \qquad (23.63)$$

which are eigenfunctions of the conformal Killing vector $\boldsymbol{\eta} = (1, \mathbf{0})$,

$$\frac{\partial}{\partial\eta}\tilde{\varphi}_k(x) = -i\omega_{\boldsymbol{k}}\tilde{\varphi}_k(x) \qquad (23.64)$$

with $\omega_{\boldsymbol{k}} > 0$. Reversing the scaling law (23.8), $\tilde{\phi} = \Omega^{-D}\phi$, we obtain the solution and the Green function in the spacetime $(\mathcal{M}, g_{\mu\nu})$ as

$$\phi(x) = \Omega^{-D}(x) \sum_i \left[a_i\tilde{\varphi}_i(x) + a_i^\dagger\tilde{\varphi}_i^*(x)\right] \qquad (23.65)$$

$$\Delta_F(x, x') = \Omega^{-D}(x)D_F(x, x')\Omega^{-D}(x'), \qquad (23.66)$$

where the $\tilde{\varphi}_i$ are given by (23.63). The vacuum of the field ϕ defined by $a_i\,|0\rangle = 0$ is called conformal vacuum. The second relation for the propagator follows immediately from the definition of the Green function as time-ordered product of fields. Note also the correspondence to the variable substitution $u_k(\eta) = a(\eta)\phi_k(\eta) = \Omega(\eta)\phi_k(\eta)$ which transformed the scalar field equation in $d = 4$ into a Minkowskian form.

Since modes which are positive eigenmodes of Eq. (23.64) at one time remain positive for all times, no mixing between positive and negative frequency modes occurs. As result, particle creation of conformally invariant fields in a conformally flat spacetime is absent. The exactly solvable model confirms this behaviour, since $m \to 0$ results in $\omega_- \to 0$ and no particle production for conformally coupled massless scalar occurs. Phenomenologically important cases of massless particles are the photon and the graviton. In the first case, the conformal invariance of the Maxwell equation implies that electromagnetic fields cannot be generated during inflation, unless a mechanism which breaks gauge invariance and generates a mass term for the photon is invoked. In the case of the gravitational field $g_{\mu\nu}$, the scaling law $\tilde{g}_{\mu\nu} = \Omega^{(2-d)/2} g_{\mu\nu}$ of a bosonic field is in conflict with the conformal transformation law $\tilde{g}_{\mu\nu} = \Omega^2 g_{\mu\nu}$ for all d. Therefore gravitons are generated in an expanding universe.

23.3   Accelerated observers and the Unruh effect

Particle production can be divided into two different cases: in the first, the spacetime is time-dependent (e.g. via the scale factor $a(t)$) and can perform "work" and thus create particles. The second, perhaps more intriguing case, is the emission of a thermal spectrum of particles close to a horizon. We will consider in this section the second case, investigating the simplest case of an accelerated observer in Minkowski space.

Uniformly accelerated observer.   In the rest-frame of an uniformly accelerated observer, the four-acceleration is given by $a^\alpha = \ddot{x}^\alpha = (0, \boldsymbol{a})$ with $|\boldsymbol{a}| = a = \text{const}$. We can convert this condition into a covariant form, writing

$$\eta_{\alpha\beta}\ddot{x}^\alpha \ddot{x}^\beta = -a^2. \tag{23.67}$$

In order to determine the trajectory $x^\alpha(\tau)$ of the accelerated observer, it is convenient to change to light-cone coordinates,

$$u = t - x \quad \text{and} \quad v = t + x. \tag{23.68}$$

Here, we assume that the trajectory is contained in the t–x plane; in the following we will suppress the transverse coordinates y and z. Forming the differentials du and dt, we see that the line element in the new coordinates is $ds^2 = du\,dv$. The normalisation condition $\eta_{\alpha\beta}\dot{x}^\alpha \dot{x}^\beta = 1$ of the four-velocity becomes therefore $\dot{u}\dot{v} = 1$, while the acceleration equation (23.67) results in $\ddot{u}\ddot{v} = -a^2$. Differentiating then $\dot{u} = 1/\dot{v}$, we obtain $\ddot{u} = -\ddot{v}/\dot{v}^2$ or

$$\frac{\ddot{v}}{\dot{v}} = \pm a. \tag{23.69}$$

Integrating results in

$$v(\tau) = \frac{A}{a}\exp(a\tau) + C \tag{23.70}$$

and, using $\dot{u} = 1/\dot{v}$, in

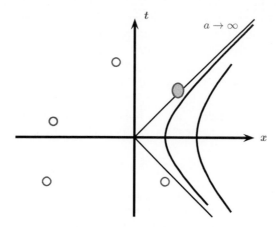

Fig. 23.1 Trajectories of uniformly accelerated observers together with the horizon and some vacuum fluctuations.

$$u(\tau) = -\frac{1}{Aa} \exp(-a\tau) + D. \tag{23.71}$$

Going back to the original Cartesian coordinates, we obtain

$$t(\tau) = \frac{1}{a} \sinh(a\tau) \quad \text{and} \quad x(\tau) = \frac{1}{a} \cosh(a\tau), \tag{23.72}$$

where we set the integration constants $A = 1$ and $C = D = 0$, which selects the trajectory with $t(0) = 0$ and $x(0) = 1/a$. Two trajectories for finite acceleration a are shown together with the limiting curve $a \to \infty$ in Fig. 23.1.

Exponential redshift. Later we will discuss gravitational particle production as the effect of a non-trivial Bogolyubov transformation between different vacua. Before we apply this formalism, we will examine the basis of this physical phenomenon in a classical picture. As a starter, we want to derive the formula for the relativistic Doppler effect. Consider an observer who is moving with constant velocity v relative to the Cartesian inertial system $x^\mu = (t, x)$ where we neglect the two transverse dimensions. We can parameterise the trajectory of the observer as

$$x^\mu(\tau) = (t(\tau), x(\tau)) = (\tau\gamma, \tau\gamma v), \tag{23.73}$$

where γ denotes its Lorentz factor. A monochromatic wave of a scalar, massless field $\phi(k) \propto \exp[-i\omega(t - x)]$ will be seen by the moving observer as

$$\phi(\tau) \equiv \phi(x^\mu(\tau)) \propto \exp\left[-i\omega\tau\,(\gamma - \gamma v)\right] = \exp\left[-i\omega\tau\sqrt{\frac{1 - v}{1 + v}}\right]. \tag{23.74}$$

Thus this simple calculation reproduces the usual Doppler formula, where the frequency ω of the scalar wave is shifted as

$$\omega' = \sqrt{\frac{1 - v}{1 + v}}\,\omega. \tag{23.75}$$

Next we apply the same method to the case of an accelerated observer. Then $t(\tau) = a^{-1}\sinh(a\tau)$ and $x(\tau) = a^{-1}\cosh(a\tau)$. Inserting this trajectory again into a monochromatic wave with $\phi(k) \propto \exp(-i\omega(t-x))$ now gives

$$\phi(\tau) \propto \exp\left[-\frac{i\omega}{a}\left[\sinh(a\tau) - \cosh(a\tau)\right]\right] = \exp\left[\frac{i\omega}{a}\exp(-a\tau)\right] \equiv e^{-i\vartheta}. \qquad (23.76)$$

Thus an accelerated observer does not see a monochromatic wave, but a superposition of plane waves with varying frequencies. Defining the instantaneous frequency by

$$w(\tau) = \frac{d\vartheta}{d\tau} = \omega\,\exp(-a\tau), \qquad (23.77)$$

we see that the phase measured by the accelerated observer is exponentially redshifted. As next step, we want to determine the power spectrum $P(\nu) = |\phi(\nu)|^2$ measured by the observer, for which we have to calculate the Fourier transform $\phi(\nu)$.

Example 23.2: Determine the Fourier transform of the wave $\phi(\tau)$.
Substituting $y = \exp(-a\tau)$ in

$$\phi(\nu) = \int_{-\infty}^{\infty} d\tau\,\phi(\tau)e^{i\nu\tau} = \int_{-\infty}^{\infty} d\tau \exp\left(\frac{i\omega}{a}\exp(-a\tau)\right) e^{i\nu\tau} \qquad (23.78)$$

gives

$$\phi(\nu) = \frac{1}{a}\int_0^{\infty} dy\, y^{-i\nu/a-1}\, e^{i(\omega/a)y}. \qquad (23.79)$$

On the other hand, we can rewrite the integral representation (A.24) of the Gamma function as

$$\int_0^{\infty} dt\, t^{z-1}e^{-bt} = b^{-z}\,\Gamma(z) = \exp(-z\ln b)\,\Gamma(z) \qquad (23.80)$$

for $\Re(z) > 0$ and $\Re(b) > 0$. Comparing these two expressions, we see that they agree setting $z = -i\nu/a + \varepsilon$ and $b = -i\omega/a + \varepsilon$. Here we added an infinitesimal positive real quantity $\varepsilon > 0$ to ensure the convergence of the integral. In order to determine the correct phase of b^{-z}, we have rewritten this factor as $\exp(-z\ln b)$ and have used

$$\ln b = \lim_{\varepsilon \to 0} \ln\left(-\frac{i\omega}{a} + \varepsilon\right) = \ln\left|\frac{\omega}{a}\right| - \frac{i\pi}{2}\mathrm{sign}(\omega/a). \qquad (23.81)$$

Thus the Fourier transform $\phi(\nu)$ is given by

$$\phi(\nu) = \frac{1}{a}\left(\frac{\omega}{a}\right)^{i\nu/a}\Gamma(-i\nu/a)e^{\pi\nu/(2a)}. \qquad (23.82)$$

The Fourier transform $\phi(\nu)$ contains negative frequencies,

$$\phi(-\nu) = \phi(\nu)e^{-\pi\nu/a} = \frac{1}{a}\left(\frac{\omega}{a}\right)^{i\nu/a}\Gamma(-i\nu/a)e^{-\pi\nu/(2a)}. \qquad (23.83)$$

Using the reflection formula of the Gamma function for imaginary arguments,

$$\Gamma(ix)\Gamma(-ix) = \frac{\pi}{x\sinh(\pi x)}, \qquad (23.84)$$

we find the power spectrum at negative frequencies as

$$P(-\nu) = \frac{\pi}{a^2} \frac{e^{-\pi\nu/a}}{(\nu/a)\sinh(\pi\nu/a)} = \frac{\beta}{\nu} \frac{1}{e^{\beta\nu} - 1} \tag{23.85}$$

with $\beta = 2\pi/a$. Remarkably, the dependence on the frequency ω of the scalar wave—still present in the Fourier transform $\phi(\nu)$—has dropped from the negative frequency part of the power spectrum $P(-\nu)$ which corresponds to a thermal Planck law with temperature $T = 1/\beta = a/(2\pi)$.

The occurrence of negative frequencies is the classical analogue for the mixing of positive and negative frequencies in the Bogolyubov method. Therefore we expect that on the quantum level a uniformly accelerated detector will measure a thermal Planck spectrum with temperature $T = 1/\beta = a/(2\pi)$. This phenomenon is called Unruh effect and $T = a/(2\pi)$ the Unruh temperature.

Rindler spacetime. Recall that the trajectory of an accelerated observer is given by

$$t(\tau) = \frac{1}{a}\sinh(a\tau) \quad \text{and} \quad x(\tau) = \frac{1}{a}\cosh(a\tau). \tag{23.86}$$

It describes one branch of the hyperbola $x^2 - t^2 = a^{-2}$, cf. Fig. 23.1.

Our aim is to determine the vacuum experienced by the uniformly accelerated observer. As a first step, we have to find a frame $\{\xi, \chi\}$ comoving with the observer. In this frame, the observer is at rest, $\chi(\tau) = 0$, and the coordinate time ξ agrees with the proper time, $\xi = \tau$. Introducing comoving light-cone coordinates,

$$\tilde{u} = \xi - \chi \quad \text{and} \quad \tilde{v} = \xi + \chi, \tag{23.87}$$

these conditions become

$$\tilde{u}(\tau) = \tilde{v}(\tau) = \tau. \tag{23.88}$$

Moreover, we choose the comoving coordinates such that the metric is conformally flat,

$$ds^2 = \Omega^2(\xi, \chi)(d\xi^2 - d\chi^2) = \Omega^2(\tilde{u}, \tilde{v})d\tilde{u}d\tilde{v}. \tag{23.89}$$

Next we have to relate the comoving coordinates $\{\tilde{u}, \tilde{v}\}$ to Minkowski coordinates $\{t, x\}$. Since $d\tilde{u}^2$ and $d\tilde{v}^2$ are missing in the line element, the functions $u(\tilde{u}, \tilde{v})$ and $v(\tilde{u}, \tilde{v})$ can depend only on one of their two arguments. We can set therefore $u(\tilde{u})$ and $v(\tilde{v})$. Expressing the four-acceleration $\dot{u}$ as

$$\frac{du}{d\tau} = \frac{du}{d\tilde{u}}\frac{d\tilde{u}}{d\tau}, \tag{23.90}$$

inserting $\dot{u} = -au$ and $\dot{\tilde{u}} = 1$ we arrive at

$$-au = \frac{du}{d\tilde{u}}. \tag{23.91}$$

Separating variables and integrating we end up with $u = C_1 e^{-a\tilde{u}}$. In the same way, we find $v = C_2 e^{a\tilde{v}}$. Since the line element has to agree along the trajectory with the proper-time, $ds^2 = d\tau^2$, the two integration constants C_1 and C_2 have to satisfy the constraint $-a^2 C_1 C_2 = 1$. Choosing $C_1 = -C_2$, the desired relation between the two sets of coordinates becomes

$$u = -\frac{1}{a} e^{-a\tilde{u}} \quad \text{and} \quad v = \frac{1}{a} e^{a\tilde{v}}, \tag{23.92}$$

or using Cartesian coordinates,

$$t = \frac{1}{a} e^{a\xi} \sinh(a\xi) \quad \text{and} \quad x = \frac{1}{a} e^{a\chi} \cosh(a\chi). \tag{23.93}$$

The spacetime described by the coordinates defining the comoving frame of the accelerated observer,

$$ds^2 = e^{2a\chi}(d\xi^2 - d\chi^2), \tag{23.94}$$

is called Rindler spacetime. It is locally equivalent to Minkowski space but differs globally. If we vary the Rindler coordinates over their full range, $\xi \in \mathbb{R}$ and $\chi \in \mathbb{R}$, then we cover only the one quarter of Minkowski space with $x > |t|$. Thus for an accelerated observer an event horizon exist: Evaluating on a hypersurface of constant comoving time, $\xi = \text{const.}$, the physical distance from $\chi = -\infty$ to the observer placed at $\chi = 0$ gives

$$d = \int_{-\infty}^{0} d\chi \sqrt{|g_{\chi\chi}|} = \frac{1}{a}. \tag{23.95}$$

This corresponds to the coordinate distance between the observer and the horizon in Minkowski coordinates.

Unruh effect.   We have found that the Fourier spectrum of a classical wave seen by an accelerated observer contains negative frequencies which exhibit a thermal spectrum. Now we want to discuss this phenomenon which is called the Unruh effect at the quantum level. In order to simplify the calculation, we consider the simplest case of a massless scalar field in $1+1$ dimensions. Then the action is conformally invariant,

$$S = \frac{1}{2} \int dt dx \sqrt{|g|}\, \eta^{\alpha\beta} \nabla_\alpha \phi \nabla_\beta \phi = \frac{1}{2} \int d\xi d\chi \sqrt{|g|}\, g^{\alpha\beta} \nabla_\alpha \phi \nabla_\beta \phi \tag{23.96}$$

and the resulting wave equation has the same form for an an inertial observer using u, v and an acclcerated observer using $\tilde{u}, \tilde{v}$ (light-cone) coordinates,

$$\frac{\partial^2 \phi}{\partial u \partial v} = \frac{\partial^2 \phi}{\partial \tilde{u} \partial \tilde{v}} = 0. \tag{23.97}$$

The corresponding solutions are

$$\phi(t, x) = f(u) + g(v) \quad \text{and} \quad \phi(\xi, \chi) = f(\tilde{u}) + g(\tilde{v}), \tag{23.98}$$

where f and g are arbitrary smooth functions specifying the wave form.

In the overlap region $x > |t|$, we can quantise the field using either set of coordinates,

$$\phi(x) = \int_0^\infty \frac{d\omega}{\sqrt{(2\pi)2\omega}} \left[a_\omega e^{-i\omega u} + a_\omega^\dagger e^{i\omega u} \right] + \text{left-movers} \tag{23.99a}$$

$$= \int_0^\infty \frac{d\Omega}{\sqrt{(2\pi)2\Omega}} \left[b_\Omega e^{-i\Omega \tilde{u}} + b_\Omega^\dagger e^{i\Omega \tilde{u}} \right] + \text{left-movers}. \tag{23.99b}$$

Here we wrote down explicitly only the right-moving modes. Because of $u(\tilde{u})$ and $v(\tilde{v})$, the two sets of modes propagate independently and we can consider them separately.

The vacuum defined by $b_i |0_R\rangle = 0$ is called the Rindler vacuum, while the usual Minkowski vacuum is defined by $a_i |0_M\rangle = 0$. One may wonder which one of the two vacua is the "better" one. First, the Rindler coordinates cover only part of the Minkowski spacetime; they, and as a result also the Rindler vacuum $|0_R\rangle$, are singular on the horizon. Using the two different representation for the field ϕ, its energy density follows as

$$\rho = \langle 0_M | (\partial_u \phi)^2 |0_M\rangle = \langle 0_R | (\partial_{\tilde{u}} \phi)^2 |0_R\rangle. \tag{23.100}$$

Now we compare the expectation value of $(\partial_u \phi)^2$ for the two vacua,

$$\rho = \langle 0_R | (\partial_u \phi)^2 |0_R\rangle = \left(\frac{\partial \tilde{u}}{\partial u} \right)^2 \langle 0_R | (\partial_{\tilde{u}} \phi)^2 |0_R\rangle = \frac{1}{(au)^2} \langle 0_M | (\partial_u \phi)^2 |0_M\rangle. \tag{23.101}$$

Since the expectation value in the Minkowski vacuum is well-behaved, the Rindler vacuum diverges for $u \to 0$. More precisely, we see that the contribution of the left-movers to the energy density of the Rindler vacuum diverges at the future horizon $u = 0$. Similarly, the right-movers add an infinite energy density at the past horizon $v = 0$. While the Rindler vacuum is thus not able to describe physics close to the horizon, the corresponding set of field operators should be used to calculate the response of an uniformly accelerated detector to the Minkowski vacuum.

We now express in Eq. (23.99b) the operator b_Ω using the Bogolyubov relation, and compare then the coefficients of the positive frequency part,

$$\frac{1}{\sqrt{2\omega}} e^{-i\omega u} = \int_0^\infty \frac{d\Omega'}{\sqrt{2\Omega'}} \left[e^{-i\Omega' \tilde{u}} \alpha_{\Omega'\omega} - e^{-i\Omega' \tilde{u}} \beta_{\Omega'\omega}^* \right]. \tag{23.102}$$

Next we multiply with $e^{-i\Omega \tilde{u}}$ and integrate over $\tilde{u}$. Performing then the trivial Ω' integral on the RHS, we arrive at

$$\int_0^\infty d\tilde{u} \, e^{\mp i\omega u + i\Omega \tilde{u}} = \begin{cases} \alpha_{\Omega\omega} \\ \beta_{\Omega\omega} \end{cases}, \tag{23.103}$$

which has the same form as (23.78). Hence the Bogolyubov coefficients satisfy the condition for a thermal spectrum,

$$|\alpha_{\Omega\omega}|^2 = \exp(2\pi\Omega/a) |\beta_{\Omega\omega}|^2. \tag{23.104}$$

The expectation value of the number operator valid for the accelerated observer in the Minkowski vacuum becomes

$$\left\langle \tilde{N}_\Omega \right\rangle = \left\langle 0_M \right| b_\Omega^\dagger b_\Omega \left| 0_M \right\rangle = \int d\omega |\beta_{\omega\Omega}|^2. \tag{23.105}$$

For $\omega = \Omega$, the normalisation condition (23.42) becomes in the continuum limit $\int_0^\infty d\omega \left(|\alpha_{\Omega\omega}|^2 - |\beta_{\Omega\omega'}|^2 \right) = \delta(0)$. Using also (23.104) we arrive at

$$\left\langle \tilde{N}_\Omega \right\rangle = \delta(0) \frac{1}{\exp(2\pi\Omega/a) - 1}. \tag{23.106}$$

Identifying the factor $\delta(0)$ as usually with the volume, we obtain for the number density of scalar particles detected by an accelerated observer in the Minkowski vacuum

$$\langle \tilde{n}_\Omega \rangle = \frac{1}{\exp(2\pi\Omega/a) - 1}. \tag{23.107}$$

Since energy is conserved in Minkowski space, you should be worried about this thermal flux measured in an accelerated detector. A tempting answer is that this energy is delivered by the external agent which accelerates the detector. Figure 23.1 suggests, however, a different interpretation: Minkowski space vacuum fluctuations that are crossing the horizon of an accelerated observer are experienced by the observer as real, thermal fluctuations. For instance, the vacuum fluctuation shown in Figure 23.1 as a filled ellipse is seen for an accelerated observer as a real particle existing from $\xi = -\infty$ to $\xi = \infty$. Similarly, all other vacuum fluctuation crossing the horizon are interpreted as the creation and annihilation of real particles. Thus the horizon seems to be equipped with a thermal atmosphere, which temperature increases the closer an accelerated observer approaches it. In contrast, for any Minkowski observer these fluctuations are the usual "harmless" vacuum fluctuations.

We close this section with a remark on the topology of the Rindler and Minkowski spacetimes. Both spacetimes are flat and agree locally in their overlapping region. They differ only from a global point of view, and thus their topology should disagree. Performing a Wick rotation, the hyperbola $x^2 - t^2 = a^{-2}$ defining the horizon becomes a circle. Hence the Rindler spacetime has the topology $S^1 \times \mathbb{R}^3$. We can view this periodicity as another indication for the presence of a thermal spectrum. Its temperature T equals the inverse of the circumference $2\pi/a$ of the event horizon in the compactified dimension.

Summary

The definition of the vacuum and of the number of particles in a spacetime without time-like Killing vector field is ambiguous and dependent on the observer. Field operators defined with respect to different vacua are related by a Bogolyubov transformation; the coefficients relating positive and negative frequencies in two different vacua determine the amount of particle production. In spacetimes without time-like Killing vector fields, all SM particles except photons and gluons are produced by the changing metric. In a spacetime with an event horizon, a thermal spectrum of particles is created close to the horizon.

Further reading. Mukhanov and Winitzki (2007) give a pedagogical introduction of quantum effects in gravity, including path integral methods and the calculation of the conformal anomaly. Two exhaustive discussions of quantum field theory in curved spacetime are the books by Birrell and Davies (1982) and by Parker and Toms (2009).

Problems

23.1 Conformal transformation. Show that a conformal transformation $\tilde{g}^{\mu\nu}(x) = \Omega^2(x)g^{\alpha\beta}(x)$ of the metric is not a coordinate transformation for the special case of constant Ω.

23.2 Conformal transformation properties ♥. Derive the connection (23.2) between the geometrical quantities for two metrics which are connected by a conformal transformation.

23.3 Auxiliary field χ. a.) Show that the substitution $\chi_k(\eta) = a(\eta)\phi_k(\eta)$ gives (23.21). b.) Show that for a massless, conformally coupled scalar field the oscillation frequency (23.22) becomes $\omega_k^2(\eta) = k^2$. c.) Derive the action for the field $\chi(\eta, \boldsymbol{x})$.

23.4 Stress tensor of scalar field. Calculate the dynamical stress tensor for the scalar action (23.6) and show that $T_\mu{}^\mu = 0$ for

$\xi = \xi_d$. (Hint: You can partially recycle the calculation for the Einstein–Hilbert action; evaluate the only new term in an inertial system.)

23.5 Maxwell. a.) Show that the Maxwell action is conformally invariant. b.) Use that the evolution of the scale factor $a(t)$ in the FLRW metric can be seen as a conformal transformation to derive the redshift formula.

23.6 Reflection formula. Derive the reflection formula (23.84) from the definition of the Gamma function.

23.7 Unruh temperature. Estimate the Unruh temperature for a.) a proton accelerated in the LHC, b.) a quark moving during hadronisation in a potential with tension $\kappa = 1\,\text{GeV/fm}$.

24
Inflation

Observational data show that we live in a universe which is nearly flat, $|\Omega_k| \equiv |\Omega_{\text{tot}} - 1| \lesssim 0.1\%$. A flat universe, however, is an unstable fixed point of the time evolution of a radiation or matter-dominated universe. This raises the question about the "naturalness" of the initial conditions of our universe. Can we explain why the universe is flat? A similar problem is posed by the formation of structures. Gravitational collapse can enhance initial fluctuations but is not able to create them. Thus in an initially uniform FLRW universe no structures like galaxies would evolve. A successful cosmological theory should answer, therefore, how the primordial fluctuations were created which have served as seeds for the observed structures today.

Proposing a inflationary phase in the early universe is a very successful attempt to explain these and other observations which are puzzling in the traditional big bang picture. Inflation is, however, not a specific theory but from the point of view of particle physics resembles rather a paradigm. Very different models can lead to an observationally indistinguishable inflationary phase in the early universe. In order to be specific, we will use in most of our discussion as prototype for inflation a single scalar field, the *inflaton*, equipped with one of our standard potentials, $V(\phi) = m^2\phi^2/2$ or $V(\phi) = \lambda\phi^4/4$.

24.1 Motivation for inflation

According to the standard big bang scenario, the universe evolved looking backwards to $t \to 0$ from a matter dominated into an radiation-dominated epoch until it reached the initial singularity at $t = 0$. This standard big bang model leads to several shortcomings:

- *Causality* or *horizon problem*. Why are causally disconnected regions of the universe homogeneous, as the isotropy of the CMB shows? This problem arises because the (particle) horizon grows like the cosmic time t, but the scale factor a increases in the radiation or matter-dominated epoch only as $t^{2/3}$ or $t^{1/2}$, respectively. Thus for any scale L contained today completely inside the horizon, there exists a time $t < t_0$ where L crossed the horizon. A solution to the horizon problem requires that a grows faster than t. Since $a \propto t^{2/[3(1+w)]}$, this demands $w < -1/3$ or $\ddot{a} > 0$, that is, an accelerated expansion of the universe.

- *Flatness problem*. The curvature term in the Friedmann equation scales as k/a^2 and thus decreases slower than the matter ($\propto 1/a^3$) and radiation ($1/a^4$) terms. Let us rewrite the Friedmann equation as

Quantum Fields–From the Hubble to the Planck Scale. Michael Kachelriess. © Michael Kachelriess 2018.
Published in 2018 by Oxford University Press. DOI 10.1093/oso/9780198802877.001.0001

$$\frac{k}{a^2} = H^2 \left(\frac{8\pi G}{3H^2}\rho - 1\right) = H^2 \left(\Omega_{\text{tot}} - 1\right). \tag{24.1}$$

The LHS scales as $(1+z)^2$, the squared Hubble parameter in the matter-dominated epoch as $(1 + z)^3$, and in the radiation-dominated epoch as $(1 + z)^4$. Classical gravity is supposed to be valid until the Planck scale M_{Pl}. Throughout most of time the universe was radiation-dominated, so we can estimate the redshift at the Planck time as $1 + z_{\text{Pl}} = (t_0/t_{\text{Pl}})^{1/2} \approx 10^{30}$. Thus if today the deviation from flatness is $|\Omega_{\text{tot}} - 1| \lesssim 1\%$, then it had too be extremely small at the Planck time, $|\Omega_{\text{tot}} - 1| \lesssim 10^{-2}/(1 + z_{\text{Pl}})^2 \approx 10^{-62}$. Taking the time-derivative of

$$|\Omega_{\text{tot}} - 1| = \frac{|k|}{H^2 a^2} = \frac{|k|}{\dot{a}^2} \tag{24.2}$$

gives

$$\frac{\mathrm{d}}{\mathrm{d}t} |\Omega_{\text{tot}} - 1| = \frac{\mathrm{d}}{\mathrm{d}t} \frac{|k|}{\dot{a}^2} = -\frac{2|k|\ddot{a}}{\dot{a}^3} < 0 \tag{24.3}$$

for $\ddot{a} > 0$. Hence $|\Omega_{\text{tot}} - 1|$ increases if the universe decelerates ($\dot{a}$ decreases), and decreases if the universe accelerates ($\dot{a}$ increases). Thus again a phase with $\ddot{a} > 0$ (or $w < -1/3$) may avoid this problem.

- *Magnetic monopole problem.* Grand unified theories (GUT) predict the existence of magnetic monopoles with masses $M_{\text{GUT}}/\alpha_{\text{GUT}}$. If they are produced via the Kibble mechanism during the GUT symmetry breaking, they would overclose the universe, cf. problem 20.15.
- The standard big bang model contains no source for the *initial fluctuations* required for structure formation.

Classical gravity breaks down around $t \approx t_{\text{Pl}}$ or $T \approx M_{\text{Pl}}$, and one may wonder if these problems can be avoided by an appropriate modification of gravity above the Planck scale. This possibility seems to be rather contrived, because for most predictions the time interval between the singularity and t_{Pl} is negligible. Therefore setting $t_{\text{Pl}} = 0$ seems to be a good approximation, see also problem 24.1.

Solution by inflation. Inflation is a modification of the standard big bang model where a phase of accelerated expansion in the very early universe is introduced. While the initial singularity in a big bang model happens at $t = 0$ or $\eta = 0$, an inflationary phase adds "additional" time at $\eta < 0$ (cf. the remark 23.2). During this phase, the universe behaves close to a de Sitter universe with $w = -1$, $H = \text{const.}$, and $a(t) = a_0 \exp(Ht)$. If this phase prevails long enough it solves the horizon, flatness, and the monopole problem and generates as bonus density fluctuations:

- The exponential growth of the scale factor, $a(t_{\text{f}})/a(t_{\text{i}}) = \mathrm{e}^{(t_{\text{f}} - t_{\text{i}})H} \gg t_{\text{f}}/t_{\text{i}}$, blows up a small, at time t_{i} causally connected region to superhorizon scales.
- Similarly, the growth of the scale factor solves the flatness problem, driving $|\Omega_{\text{tot}} - 1| \propto \mathrm{e}^{-2Ht}$ exponentially towards zero.
- Magnetic monopoles or other superheavy relics are diluted as $n \propto \exp(-3Ht)$. At the end of a sufficiently long inflationary phase, their density is therefore practically zero.

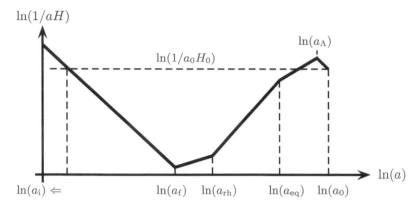

Fig. 24.1 Evolution of the present Hubble scale as function of time; the condition $(a_0 H_0)^{-1} < (a_i H_i)^{-1}$ sets a lower limit on a_f/a_i.

- Inflation blows up quantum fluctuation to astronomical scales, generating initial fluctuation without scale, $P_0(k) \propto k^{-1+\varepsilon}$ with $\varepsilon \ll 1$, as required by observations.

Figure 24.1 illustrates how the comoving Hubble scale $1/(aH)$ evolves in a universe with an inflationary phase. During inflation, $1/(aH)$ decreases, because $\ddot{a} > 0$ is equivalent to

$$\frac{\mathrm{d}}{\mathrm{d}t}\left(\frac{1}{aH}\right) = \frac{\mathrm{d}}{\mathrm{d}t}(\dot{a})^{-1} = -\frac{\ddot{a}}{(\dot{a})^2}. \tag{24.4}$$

More precisely, $w \simeq -1$ and $H \simeq$ const. imply that a increases exponentially. Thus the comoving Hubble radius decreases as $\ln(1/aH) = -\ln(a) + $const. during inflation. As a result, physical scales k which had been previously on superhorizon distance come in causal contact. Inflation ends at a_f, and the comoving Hubble scale starts to grow. During the intermediate phase $a_f < a < a_{\rm rh}$ the universe goes through a phase of reheating before the standard big bang evolution starts at $a > a_{\rm rh}$. The value of $a_{\rm rh}$ is determined by the temperature to which the universe is reheated and is very model-dependent.

Conditions for successful inflation. We can rewrite the condition for accelerated expansion $\ddot{a} > 0$ as

$$\frac{\mathrm{d}}{\mathrm{d}t}\left(\frac{1}{aH}\right) = -\frac{\dot{a}H + a\dot{H}}{(aH)^2} = -\frac{1}{a}(1 - \varepsilon_H) \qquad \text{with} \qquad \varepsilon_H \equiv -\frac{\dot{H}}{H^2}. \tag{24.5}$$

The condition $\varepsilon_H \ll 1$ ensures that w is close to -1 (problem 24.3). Next, as measure for the expansion we define the number N of e-foldings during inflation, $N = \ln(a_2/a_1)$. Using then $\mathrm{d}N = \mathrm{d}\ln(a) = H\mathrm{d}t$, we can express ε_H as

$$\varepsilon_H = -\frac{\dot{H}}{H^2} = -\frac{\mathrm{d}\ln H}{\mathrm{d}N}. \tag{24.6}$$

The relative change of ε_H should be small per Hubble time H^{-1} so that inflation can persist during a sufficiently large number of e-foldings. This motivates us to introduce as a second parameter

$$\eta_H \equiv \frac{\mathrm{d}\ln\varepsilon_H}{\mathrm{d}N} = \frac{1}{H}\frac{\dot{\varepsilon}_H}{\varepsilon_H}, \tag{24.7}$$

which should also be small, $|\eta_H| \ll 1$. Thus we can quantify the conditions for successful inflation by $\varepsilon_H \ll 1$ and $|\eta_H| \ll 1$. These two conditions guarantee that the time evolution of the scale factor is for a sufficiently long time close to the one in a de Sitter universe.

Example 24.1: How much inflation is needed?
We can find the minimal number of e-folding to solve the horizon problem by comparing the comoving Hubble radius today to the one at the beginning of inflation (cf. Fig. 24.1),

$$(a_0 H_0)^{-1} < (a_i H_i)^{-1}. \tag{24.8}$$

We neglect the reheating phase and approximate the complete evolution of the universe for $a > a_{\mathrm{rh}}$ as radiation dominated. With $H \propto 1/a^2$, we find

$$\frac{a_0 H_0}{a_f H_f} = \frac{a_0}{a_f}\left(\frac{a_f}{a_0}\right)^2 = \frac{a_f}{a_0} = \frac{T_0}{T_f} \approx 10^{-28}, \tag{24.9}$$

where we used $T_f = 10^{15}$ GeV as the temperature of the universe at beginning of the standard big bang evolution for the numerical estimate. Thus

$$(a_i H_i)^{-1} > (a_0 H_0)^{-1} \sim 10^{28}(a_f H_f)^{-1}. \tag{24.10}$$

With $H_i \approx H_f$, $N \approx 65$ e-foldings are required to solve the horizon problem for our choice of T_f.

24.2 Inflation in the homogeneous limit

We assume in the following that a single scalar field, the *inflaton*, is responsible for the inflationary phase in the early universe. We know that displacing the minimum of a scalar potential from zero generates a cosmological constant, which in turn leads to accelerated expansion. In a successful inflationary model, we have to convert this static picture into a dynamical process, since inflation has to start and to stop. In particular, the end of inflation has to be successfully connected to the smooth big bang picture, which is called often the "graceful exit problem".

Recall also our discussion of scalar fields with non-zero vev in chapter 13. There we split the scalar field $\phi(x)$ into a classical part $\langle\phi\rangle \equiv \phi_0$ and quantum fluctuations $\delta\phi(x)$ on top of it. Now we consider the more general case where also the classical field depends on time,

$$\phi(x) = \phi_0(t) + \delta\phi(x), \tag{24.11}$$

and its dynamics is governed by the Einstein equations. Clearly, this problem cannot be solved in general. Observations of the CMB show, however, that the early universe was very homogeneous, $\delta T/T \sim 10^{-5}$, and thus we can expect that lowest-order perturbation theory in $\delta\phi(x)$ should work reliable. We already considered the evolution of $\phi_0(t)$ in an FLRW background. Now we analyse under which conditions this evolution leads successfully to inflation before we calculate the fluctuations $\delta\phi(x)$.

Equation of state of a scalar field condensate. A scalar field ϕ sitting at the minimum of its potential $V(\phi)$ has the desired EoS $w = -1$ to drive accelerated expansion. The simplest dynamical model for an inflationary phase is a single, scalar field which is initially displaced from its minimum. A necessary condition for accelerated expansion is $w < -1/3$, and thus we should find the EoS of a scalar field ϕ evolving in an FLRW background.

The stress tensor for a scalar field with $\mathscr{L} = \frac{1}{2}g^{\mu\nu}\nabla_\mu\phi\nabla_\nu\phi - V(\phi)$ follows from

$$T_{\mu\nu} = 2\frac{\partial\mathscr{L}}{\partial g^{\mu\nu}} - g_{\mu\nu}\mathscr{L} = \nabla_\mu\phi\nabla_\nu\phi - g_{\mu\nu}\left[\frac{1}{2}g^{\rho\sigma}\nabla_\rho\phi\nabla_\sigma\phi - V(\phi)\right], \qquad (24.12)$$

where we used the relation (19.84) derived in problem 19.1. We can describe the scalar field also as an ideal fluid. Equating the two expressions for the stress tensor gives

$$T_{\mu\nu} = \nabla_\mu\phi\nabla_\nu\phi - g_{\mu\nu}\mathscr{L} \overset{!}{=} (\rho + P)u_\mu u_\nu - Pg_{\mu\nu}. \qquad (24.13)$$

Comparing the two independent tensor structures we can identify $P = \mathscr{L}$ and

$$\nabla_\mu\phi\nabla_\nu\phi = (\rho + P)u_\mu u_\nu. \qquad (24.14)$$

Contracting the indices with $g^{\mu\nu}$, remembering $u_\mu u^\mu = 1$ and using $\nabla_\mu\phi\nabla^\mu\phi = 2L + 2V$ results in

$$\rho = P + 2V. \qquad (24.15)$$

Now we have to calculate only the energy density $\rho = T_{00}$ in order to determine the (isotropic) pressure P and the equation of state $w = P/\rho$. In an FLRW background, the energy density of the field ϕ is given by

$$\rho = T_{00} = \dot{\phi}^2 - \left[\frac{1}{2}\dot{\phi}^2 - \frac{1}{2a^2}(\boldsymbol{\nabla}\phi)^2 - V(\phi)\right] = \frac{1}{2}\dot{\phi}^2 + \frac{1}{2a^2}(\boldsymbol{\nabla}\phi)^2 + V(\phi). \qquad (24.16)$$

Thus the pressure[1] follows as

$$P = \frac{1}{2}\dot{\phi}^2 + \frac{1}{2a^2}(\boldsymbol{\nabla}\phi)^2 - V(\phi). \qquad (24.17)$$

If we require the field ϕ to respect the symmetries of the FLRW background then ϕ has to be homogeneous and the $(\boldsymbol{\nabla}\phi)^2$ term vanishes. As result, the equation of state simplifies to

$$w = \frac{P}{\rho} = \frac{\dot{\phi}^2 - 2V(\phi)}{\dot{\phi}^2 + 2V(\phi)} \in [-1:1]. \qquad (24.18)$$

Thus a classical scalar field may act as dark energy, $w < -1/3$, leading to an accelerated expansion of the universe. A necessary condition is that the field is "slowly rolling", that is, that its kinetic energy is sufficiently smaller than its potential energy, $\dot{\phi}^2 < 2V/3$.

---

[1]For an alternative definition of the pressure see problem 24.5.

Slow-roll conditions for the potential. We can integrate $\dot{a} = aH$ for an arbitrary time evolution of H,

$$a(t) = a(t_0) \exp\left(\int \mathrm{d}t H(t)\right). \tag{24.19}$$

The number N of e-foldings is connected to the evolution of H and ϕ as

$$N = \ln \frac{a_2}{a_1} = \int \mathrm{d}t\, H(t) = \int \frac{\mathrm{d}\phi}{\dot{\phi}} H(t). \tag{24.20}$$

The potential energy can dominate only long enough if $\ddot{\phi}$ is small. Therefore we can approximate the field equation $\ddot{\phi} + 3H\dot{\phi} + V' = 0$ as $\dot{\phi} \approx -V'/(3H)$. Then using the Friedmann equation $H^2 = 8\pi G V/3$, it follows

$$N = -\int \mathrm{d}\phi\, \frac{3H^2}{V'} = -\int \mathrm{d}\phi\, \frac{8\pi G V}{V'} = -\int \frac{\mathrm{d}\phi}{\widetilde{M}_{\mathrm{Pl}}} \frac{V}{\widetilde{M}_{\mathrm{Pl}} V'}, \tag{24.21}$$

where we introduced the reduced Planck mass $\widetilde{M}_{\mathrm{Pl}} \equiv (8\pi G)^{-1/2}$. Successful inflation requires $N \gtrsim 60 \gg 1$ and thus we require as slow-roll parameter for the potential

$$\varepsilon_V \equiv \frac{1}{2}\left(\frac{\widetilde{M}_{\mathrm{Pl}} V'}{V}\right)^2 \ll 1. \tag{24.22}$$

Hence the inflaton potential should be flat and its value $V(\phi)$ should be large. An additional constraint on the curvature V'' follows by differentiating V'/V and then using $\varepsilon_V \ll 1$ as

$$\eta_V \equiv \widetilde{M}_{\mathrm{Pl}}^2 \frac{V''}{V} \ll 1. \tag{24.23}$$

This defines the second slow-roll condition which requires that the curvature V'' of the potential measured in Planck units is small compared to the value of the potential $V(\phi)$.

Example 24.2: The arguable least exotic model for inflation uses a single scalar field ϕ with potential $V = \frac{1}{2}m^2\phi^2$. Then the two slow-roll parameters coincide and are given by

$$\varepsilon_V = \eta_V = \frac{2\widetilde{M}_{\mathrm{Pl}}^2}{\phi^2}. \tag{24.24}$$

Thus we see that inflation in this model requires trans-Planckian field values, $\phi_{\mathrm{i}} \gg \sqrt{2}\widetilde{M}_{\mathrm{Pl}}$. (With a certain understatement, one calls in this context field values "large" for $\phi \gtrsim \widetilde{M}_{\mathrm{Pl}}$ and small otherwise.) Combining (24.21) and (24.22), we can express the number of e-foldings as

$$N = \int_{\phi_{\mathrm{i}}}^{\phi_{\mathrm{f}}} \frac{\mathrm{d}\phi}{\widetilde{M}_{\mathrm{Pl}}} (2\varepsilon_V)^{-1/2} = \frac{\phi_{\mathrm{i}}^2}{4\widetilde{M}_{\mathrm{Pl}}^2} - \frac{1}{2}, \tag{24.25}$$

where we used $\max\{\varepsilon_V, \eta_V\} = 1$ as condition for the end of inflation. Solving the horizon and flatness problem with inflation requires $N \sim 60$ e-foldings. Thus the largest scales observed in the CMB, $N \sim 60$, correspond to field values $\phi = 2\sqrt{N_{\mathrm{CMB}}} \sim 15 M_{\mathrm{Pl}}$.

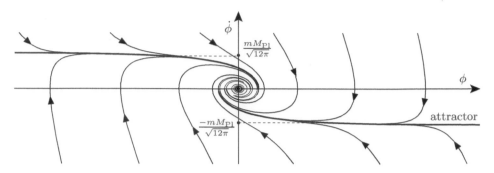

Fig. 24.2 Phase portrait for the $V = m^2\phi^2/2$ potential, from Mukhanov (2005).

Classical general relativity should be valid as long as the energy density ρ satisfies $\rho \sim V \ll M_{\rm pl}^4$. This allows field values as large as $\phi_{\rm i} \sim 10^3$ for $m = 10^{15}$ GeV, and as maximal number of e-folding $N \sim 10^6$. Therefore it is natural to expect that the true number of e-foldings is much larger than 60. In this case, deviations from flatness would be extremely small.

Hamilton–Jacobi equations and phase portraits. The evolution of the scale factor a and the field ϕ in single-field inflation is determined by

$$H^2 = \frac{8\pi G}{3}\rho - \frac{k}{a^2} = \frac{8\pi G}{3}\left(\frac{1}{2}\dot\phi^2 + V\right) - \frac{k}{a^2}, \tag{24.26a}$$

$$\frac{\ddot a}{a} = \frac{8\pi G}{3}\left(V - \dot\phi^2\right), \tag{24.26b}$$

$$\ddot\phi + 3H\dot\phi + V_{,\phi} = 0. \tag{24.26c}$$

If the field $\phi(t)$ is a monotonic function of the time t, then we can replace t by $\phi(t)$ as evolution variable. We differentiate first $H = \dot a/a$, obtaining $\dot H = \ddot a/a - H^2$. Inserting then (24.26a) and (24.26b) as well as setting $k = 0$, we arrive at

$$\dot H = -4\pi G \dot\phi^2 \tag{24.27}$$

or

$$H_{,\phi} = -4\pi G \dot\phi. \tag{24.28}$$

Now we can eliminate $\dot\phi$ in (24.26a), obtaining

$$H_{,\phi}^2 - 12\pi G H^2 = -32\pi^2 G^2 V. \tag{24.29}$$

The last two equations are equivalent to the usual two Friedmann equations. They have the virtue of showing that choosing the potential $V(\phi)$ according to (24.29), single-field inflation is flexible enough to describe an arbitrary evolution of the Hubble parameter H. Moreover, we can use them to connect the two sets of slow-roll parameters, see problem 24.3.

Next, we want to check how generic the slow-roll conditions in the specific case of a $V = \frac{1}{2}m^2\phi^2$ potential occur. Inserting the Friedmann equation (24.26a) and the explicit form of the potential into the Klein–Gordon equation (24.26c) gives

$$\ddot{\phi} + [12\pi G(\dot{\phi} + m^2\phi^2)]^{1/2}\dot{\phi} + m^2\phi = 0. \tag{24.30}$$

Since this second-order differential equation contains no explicit time-dependence, it can be reduced to a first-order equation eliminating $\ddot{\phi}$ with the help of

$$\ddot{\phi} = \dot{\phi}\,\frac{\mathrm{d}\dot{\phi}}{\mathrm{d}\phi}. \tag{24.31}$$

The result,

$$\frac{\mathrm{d}\dot{\phi}}{\mathrm{d}\phi} = -\frac{[12\pi G(\dot{\phi} + m^2\phi^2)]^{1/2}\dot{\phi} + m^2\phi}{\dot{\phi}}, \tag{24.32}$$

allows to plot the phase portrait shown in Fig. 24.2. We observe two lines at $\dot{\phi}_a = \pm m M_{\mathrm{Pl}}/\sqrt{12\pi}$ that attract all trajectories starting at sufficiently large field values ϕ. A field that starts its evolution with $|\dot{\phi}| \gg \dot{\phi}_a$ loses fast its kinetic energy $\dot{\phi}$, moving to the attractor line. Close to the attractor line, it evolves with $|\ddot{\phi}| \sim 0$, that is it satisfies the slow-roll condition and can drive inflation. Around $\phi_f \approx \sqrt{2}\widetilde{M}_{\mathrm{Pl}}$, this condition is violated: the trajectory leaves the attractor line and spirals towards the origin. This final stage corresponds to coherent oscillations of the inflaton around its minimum and leads to the reheating of the universe. Thus the phase portrait in Fig. 24.2 illustrates that the potential $V = m^2\phi^2/2$ implies an inflationary phase for sufficiently large initial field values ϕ.

Models for inflation. Up to now we have discussed the, for a particle physicist, perhaps most natural option that a scalar field drives inflation. Moreover, we have restricted our attention to the case of a *single* scalar field. Single-field models can be characterised by two parameters, for example, the width and the height of the potential. Generically, we can divide these models into large and small field models as shown in Fig. 24.3. In the first class, inflation requires trans-Planckian field values as, for example, for the $m^2\phi^2$ or the $\lambda\phi^4$ models. The potential of these models has positive curvature, $V'' > 0$, and thus $\varepsilon_V > 0$.

The trans-Planckian field values required to start inflation clearly lead to the question of how the inflaton was displaced from its equilibrium position. A suggestion by Linde is "chaotic inflation". The inflaton field ϕ acquires random values due to quantum fluctuations. In a region of initial size $1/M_{\mathrm{Pl}}^4$ with $\phi \gg M_{\mathrm{Pl}}$, inflation starts and produces a homogeneous patch inside the universe. Other regions do not undergo inflation at all, or with only a few e-foldings. Thus on scales much larger than our successfully inflated patch, the universe is very inhomogeneous. In a variant, called "stochastic inflation", quantum fluctuations disturb the classical slow-roll trajectory so strongly that the volume filled with large quantum fluctuations $\phi \gg M_{\mathrm{Pl}}$ grows exponentially. As a result, new patches of inflating "miniverses" are generated continuously, leading to an eternal self-reproduction of the inflationary universe. A controversial question in these types of models is how generic is the observed universe, and how such a statement can be made precise.

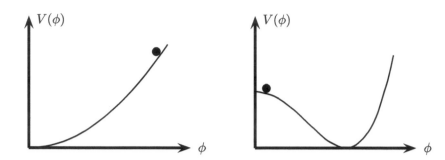

Fig. 24.3 Typical potential of a large field (left) and of a small field model (right) for inflation.

Typical examples for a small field model are potentials like $V(\phi) = \lambda(\eta^2 - \phi^2)^2/4$ or of the Coleman–Weinberg type. Such potentials are generically much flatter than those of large field models. They often are connected to SSB, and the inflaton sits initially at the unstable equilibrium position $\phi = 0$. The potential has negative curvature, $V'' < 0$, and thus $\varepsilon_V < 0$. The idea of small field models implies that we choose parameters such that inflation starts at sub-Planckian field values, and thus, for example, $\lambda\phi_{\rm f}^4 \ll V_0 \lesssim M_{\rm Pl}^4$ for $V(\phi) = V_0 - \lambda\phi^4/4$. Inflation may be realised in this class of models as follows. Before the start of inflation, the universe is at high temperature in a potentially inhomogeneous state. The symmetry of the temperature-dependent effective potential $V_{\rm eff}(\phi, T)$ is restored. Thus the field ϕ sits initially at $\phi = 0$. As the universe expands, it cools down and the size of the temperature-dependent corrections becomes smaller, $V(\phi, T) \simeq V(\phi, 0)$. Finally, the symmetry is broken, the field starts to roll down the potential and inflation starts.

Using a scalar field as the driving force of inflation, a natural question to address is if we can identify the inflaton with a Higgs field. During the 1980s, one tried to connect the GUT phase transition and GUT Higgs fields to inflation. However, combining the slow-roll conditions and the size of density fluctuations (which will be discussed in the after next section) restricts the potential severely. Generically, loop corrections destroy the flatness of the potential if the inflation is not extremely weakly coupled to the SM fields. Therefore, it seems natural to consider the inflaton as a gauge singlet. The discovery of the SM Higgs has created nevertheless interest in the question if the SM Higgs can act as inflaton. First, we know that the Higgs potential flattens for large values of the renormalisation scale. Second, its coupling $\xi\phi^2 R$ to the curvature scalar is unconstrained. A large enough number of e-foldings can be achieved, if the coupling ξ is large, $\xi \approx 50\,000$. Such a term flattens the Higgs potential below $M_{\rm Pl}/\sqrt{\xi}$ sufficiently to lead to slow-roll inflation. This scenario faces however two problems: first, perturbative unitarity is violated below $M_{\rm Pl}$, requiring the existence of new degrees of freedom. Thus predictions in "Higgs inflation" depend on the unknown UV completion. More severely, we have seen that the SM Higgs potential (for the values of m_h and m_t currently favoured) develops an instability below $M_{\rm Pl}$. Thus the SM Higgs cannot be the main agent of inflation but may play some role during inflation.

The range of options widens drastically as soon as one uses several inflaton fields, reducing at the same time, however, the predictive power of the models. We comment here only briefly on another option, namely abandoning the idea that the inflaton is a fundamental field. In particular, during the very early universe higher derivative terms in the gravitational action may have played an important role. As a specific possibility, one can modify gravity by generalising the Einstein–Hilbert action as $\mathscr{L}_{\mathrm{EH}} = R \rightarrow f(R)$. Here, the function $f(R)$ should be chosen such that observational constraints are obeyed in the $R \rightarrow 0$ limit, while for large R modified gravity may lead to inflation. An example of this approach is the Starobinsky model proposed in 1979 that uses $f(R) = R - R^2/(6M^2)$. It represents the first working theory of inflation and is still in excellent agreement with data. Neglecting other matter fields, this theory is equivalent to standard gravity with a scalar field. Changing first the metric as $g_{\mu\nu} \rightarrow g_{\mu\nu}/\chi$ with $\chi \equiv \partial f(R)/\partial R$, and using then $\chi = \exp[4\sqrt{\pi}\phi/(\sqrt{3}M_{\mathrm{Pl}})]$ in order to obtain a canonically normalised kinetic term gives the scalar potential

$$V(\phi) = \frac{3M^2 M_{\mathrm{Pl}}^2}{32\pi^2} \left(1 - 1/\chi\right)^2 = \frac{3M^2 M_{\mathrm{Pl}}^2}{32\pi^2} \left[1 - \exp\left(\frac{4\sqrt{\pi}\phi}{\sqrt{3}M_{\mathrm{Pl}}}\right)\right]^2. \tag{24.33}$$

Thus one can analyse the Starobinsky model using $V(\phi)$ and standard gravity. However, the transformation $g_{\mu\nu} \rightarrow g_{\mu\nu}/\chi$ induces couplings of gravitational strength between ϕ and all other SM fields. These additional gravitational couplings indicate that it is more natural to see this class of models as a modification of gravity.

In order to distinguish between these various possibilities, we have to work out the fluctuations predicted by these models. Prior to that, we consider first the transition from the inflationary phase to the standard radiation dominated universe.

24.3 Reheating and preheating

The inflationary period ends when the EoS becomes larger than $w = -1/3$ and the expansion slows down. In practice, we can define as the end of inflation the time when one of the slow-roll parameters becomes order 1. At this point, the universe became empty and cold; all its energy was contained in the inflaton field. Reheating is a collective term for all the mechanisms by which this energy is transferred to a thermal state of ordinary matter, initiating the usual radiation dominated epoch of the universe. One distinguishes between "perturbative reheating" where the inflaton transfers its energy via perturbative two- or three-particle decays and "preheating" where matter fields coupled to the inflaton develop exponentially growing instabilities.

Perturbative reheating. We determine first the EoS valid when the energy density of the universe is dominated by the oscillating inflaton. Assuming a polynomial as inflaton potential, $V \propto \phi^n$, we can use the virial theorem $\langle T \rangle = -n/2\langle V \rangle$ to obtain the EoS averaged over several oscillations,

$$w = \frac{P}{\rho} = \frac{\langle T \rangle - \langle V \rangle}{\langle T \rangle + \langle V \rangle} = \frac{n - 2}{n + 2}. \tag{24.34}$$

Here we neglected the friction factor $3H$ which is by assumption small. Thus during reheating, the universe expands as matter-dominated for $n = 2$ and as radiation-dominated for $n = 4$.

Initially, the reheating process of the universe was modelled as perturbative two-particle decays of the inflaton ϕ. For instance, we can consider decays into bosons χ and fermions ψ via the interaction terms $\mathscr{L}_{\text{int}} = -g_1 v \phi \chi^2 - g_2 \phi \bar{\psi} \psi$. The self-energy of the inflaton obtains an imaginary part for $q^2 \geq \min\{4m_\chi^2, 4m_\psi^2\}$ which is connected to the total decay width Γ_{tot} via the optical theorem (cf. problem 9.1) by

$$\Im[\Sigma(m)] = m \Gamma_{\text{tot}}(\phi). \tag{24.35}$$

We can include this imaginary part of the self-energy into the Klein–Gordon equation, setting $m^2 \to m^2 + i m \Gamma_{\text{tot}}(\phi)$,

$$\ddot{\phi} + 3H\dot{\phi} + (m^2 + i m \Gamma_{\text{tot}})\phi = 0. \tag{24.36}$$

The inflaton has to be very weakly coupled to other fields, $\Gamma_{\text{tot}} \ll \phi$, and the friction term has to be small too, $H \ll m$. Therefore we can approximate the inflaton evolution as

$$\phi(t) = \Phi(t)\sin(mt) \tag{24.37}$$

with the time-dependent amplitude

$$\Phi(t) = \phi_{\text{i}} \exp\left[-\frac{1}{2}\int dt\,(3H + \Gamma_{\text{tot}})\right]. \tag{24.38}$$

We can view the oscillating classical field $\phi(t)$ as a coherent wave of ϕ particles with zero momentum and number density $n(t) = \rho/m \sim m\phi^2(t)/2$.

To be concrete, we fix the potential as $V = m^2\phi^2/2$. Then we find in the limit $\Gamma_{\text{tot}} \ll H$ with $H = 2/(3t)$

$$\Phi(t) = \phi_{\text{f}}\frac{t_{\text{f}}}{t} = \frac{1}{\sqrt{3\pi}t}\frac{M_{\text{Pl}}}{m}, \tag{24.39}$$

where we used $\phi_{\text{f}} = \sqrt{2}\widetilde{M}_{\text{Pl}} = M_{\text{Pl}}/\sqrt{4\pi}$ as the inflaton value at the end of inflation, $t_{\text{f}} = 2H_{\text{f}}/3$ and $H_{\text{f}} = V(\phi_f)$. This implies $\rho \sim 1/a^3$ as expected for $n = 2$ (problem 24.6). In the opposite limit, $\Gamma_{\text{tot}} \gg H$, the amplitude (24.38) follows the usual decay law,

$$\Phi(t) = \phi_{\text{i}} \exp(-\Gamma_{\text{tot}}t/2). \tag{24.40}$$

Defining the reheating time t_{rh} by $3\Gamma_{\text{tot}} = H$, the inflaton energy density at that time is

$$\rho_{\text{rh}} = \frac{\Gamma_{\text{tot}}^2 M_{\text{Pl}}^2}{24\pi} \overset{!}{=} \frac{\pi^2}{30}g_* T_{\text{rh}}^4. \tag{24.41}$$

In the second step, we assumed that reheating occurs instantaneously. In this approximation, the reheating temperature T_{rh} is therefore

$$T_{\text{rh}} = \left(\frac{5}{4\pi^3 g_*}\right)^{1/4}(\Gamma_{\text{tot}}M_{\text{Pl}})^{1/2} \simeq 0.14\left(\frac{100}{g_*}\right)^{1/4}(\Gamma_{\text{tot}}M_{\text{Pl}})^{1/2} \tag{24.42}$$

$$\ll (mM_{\text{Pl}})^{1/2} \simeq 10^{15}\ \text{GeV},$$

where we used $\Gamma_{\text{tot}} \ll m$ and as inflaton mass $m \approx 10^{-6}M_{\text{Pl}}$.

Using instead the estimate from problem 24.7 for the maximal possible decay width of the inflaton, $\Gamma_{\text{tot}} \lesssim 10^8\,\text{GeV}$, the instantaneous reheating temperature is of order $10^{13}\,\text{GeV}$. It is thus considerably smaller than the GUT energy and thus GUT baryogenesis seems to be impossible in this picture. Moreover, in any realistic model, reheating will be delayed and thus the maximal temperature the universe reaches after inflation will be below $10^{13}\,\text{GeV}$. In this case, even leptogenesis seems to be only marginally possible. A way out is the non-perturbative scenario of preheating we discuss next, where in a first stage particles are resonantly produced before they thermalise by perturbative processes.

Preheating. We consider again a light scalar field χ coupled to the inflaton ϕ via the interaction $\mathscr{L}_{\text{int}} = -g\phi^2\chi^2$, but now treat ϕ as a classical field and study the evolution of the quantum field χ in this background. Then the field modes χ_k satisfy

$$\ddot{\chi}_k + 3H\dot{\chi}_k + \left(\frac{k^2}{a^2} + g^2\phi^2\right)\chi_k = 0. \tag{24.43}$$

When the inflaton ϕ starts to oscillate, it can transfer energy to the χ field. In particular, modes χ_k which are resonant with the ϕ oscillations can became unstable, leading to an exponential growth of their occupation number. This phenomenon is called parametric resonance.

In order to illustrate the basis of the mechanism, we transform (24.43) into an equation resembling the Schrödinger equation for an electron in a periodic potential. Rescaling $u_k = a^{3/2}\chi_k$ gives the oscillator equation $\ddot{u}_k + \omega_k^2 u_k = 0$ with energy

$$\omega_k^2 = \frac{k^2}{a^2} + g^2\phi^2 - \frac{3}{4}(2\dot{H} + 3H^2). \tag{24.44}$$

Now we introduce as new time variable $z = mt$ as well as $q^2 = g^2\Phi^2/(4m^2)$ and $A_k = 2q + k^2/(ma)^2$, where $\Phi(t)$ satisfies (24.38). Then we arrive at

$$\frac{\mathrm{d}^2 u_k}{\mathrm{d}z^2} + (A_k - 2q\cos(2z) + \Delta)u_k = 0, \tag{24.45}$$

where we lumped the unimportant factors into

$$\Delta = \frac{m_\chi^2}{m^2} - \frac{3}{4m^2}(2\dot{H} + 3H^2). \tag{24.46}$$

Using $H = 2/(3t)$, and thus $2\dot{H} + 3H^2 = 0$, we see that the second term in Δ vanishes. Moreover, we can assume that the χ field is light relative to the inflaton, $m_\chi/m \ll 1$, and thus Δ can be neglected altogether. If the fields evolve fast compared to the expansion of the universe, one may neglect also the a dependence in A_k and q. The differential equation (24.45) then has constant coefficients and becomes a Mathieu equation. Its solutions have the form $u_k \propto \exp(iv_k z)$, and are therefore unstable for $\Im(v_k) < 0$. The wave bands with $\Im(v_k) < 0$ correspond to forbidden energy bands of the equivalent Schrödinger equation for an electron in a periodic potential. For $q \gtrsim 2A_k$, the bands of unstable wave numbers k become large, and occupy most of k

space. The production of particles is efficient, if ω_k changes non-adiabatically, that is, if $\dot{\omega}_k \gg \omega_k^2$. This happens when the effective mass of χ, $m_\chi(t) = g\phi(t)$, becomes zero for $\phi(t) = 0$.

Example 24.3: Broad resonance as scattering. We can break the evolution of $u_k(t)$ into adiabatic pieces where the modes evolve as $u_k \propto \exp(\pm i \int dt \omega_k)$ and the particle number is conserved, and "scattering events" when the inflaton crosses zero, $\phi(t_i) = 0$. Thus the field modes satisfy for $t \neq t_i$ Eq. (23.58) with constant Bogolyubov coefficients and energies,

$$u_k(t) = \frac{\alpha_k}{\sqrt{2\omega_k}} \exp\left[-i \int^t dt' \, \omega_k\right] + \frac{\beta_k}{\sqrt{2\omega_k}} \exp\left[i \int^t dt' \, \omega_k\right]. \tag{24.47}$$

At t_i, the incoming field is scattered into an outgoing field with new Bogolyubov coefficients. Let us assume that the incoming state $t < t_i$ contained no particles. Then we know that the rate of particle production is given by $|\beta_k|^2$. Close to $\phi = 0$, we can approximate $g^2\phi^2(t) \approx g^2\Phi^2 m^2(t - t_j)^2 \equiv k_*^4(t - t_j)^2$ and thus the oscillator equation $\ddot{u}_k + \omega_k^2 u_k = 0$ becomes

$$\ddot{u}_k + \left[k^2/a^2 + k_*^4(t - t_j)^2\right] u_k = 0. \tag{24.48}$$

Now we rescale momenta, $\kappa = k/a$, and time, $\tau = k_*(t - t_j)$, mapping thereby our problem on the calculation of the tunnelling probability through the one-dimensional potential $V = -\tau^2$,

$$\frac{d^2 u_k}{d\tau^2} + \left(\kappa^2 + \tau^2\right) u_k = 0. \tag{24.49}$$

The analytical solution to this problem is known. Alternatively, one can employ the WKB approximation to find $n_k = |\beta_k|^2 = \exp(-\pi\kappa^2)$, cf. with problem 24.8.

Thus particle production is efficient, if $\kappa^2 \lesssim 1$ or $(k/a)^2 \lesssim k_*^2 = gm\Phi$. The analysis for a non-empty initial state shows that generally particles can be created or destroyed, depending on the value of the relative phase between initial and final state. If one includes the expansion of the universe, the relative phases behave nearly as random variables. In this limit, one finds that in 75% of scatterings particle are created. The number of created particles therefore grows exponentially with time.

At some point this growth has to slow down and finally stop. Generically, this already happens when still most energy is contained in the inflaton field. One possible reason is the back-reaction of the produced particles on the inflaton field. The term $\langle\chi^2\rangle\phi^2$ acts as additional mass term for the inflaton, changes for $\langle\chi^2\rangle \sim m^2$ the dynamics and shuts off particle production via preheating. Thus the initial phase of preheating has to be followed by perturbative reheating and thermalisation. Note also that the back-reaction of the particle production on the evolution of the scale factor couples the evolution of χ, ϕ and the Hubble parameter H.

Remark 24.1: Superheavy dark matter. Particles with masses up to $\sim 100T_{\rm rh}$ could be produced during preheating. Another option is their gravitational production during the end of the inflation, as discussed in section 23.2. While the first possibility is strongly model-dependent, the second one relies only on gravitational interactions and is therefore universal. If such particles are stable (or have life times large compared to the age of the universe), this opens the possibility of superheavy dark matter (SHDM). They should have weak enough interactions with SM particles not to thermalise, in order to avoid the unitarity limit (21.38). This ensures also that SHDM behaves as cold dark matter.

The energy density of the particle type X with mass M_X at the end of inflation can be calculated using Eq. (23.60). The numerical results are well described by

$$\rho_X \simeq 10^{-3} M_X^4 \left(\frac{M_X}{H(t_{\rm f})}\right)^{-3/2} \exp\left(-2M_X/H(t_{\rm f})\right), \qquad (24.50)$$

where $H(t_{\rm f})$ denotes the Hubble parameter at the end of inflation. Since the modes generated are non-relativistic, the relative abundance of SHDM has grown as $T_0/T_{\rm rh}$ since reheating. The present abundance follows therefore setting also $H(t_{\rm f})$ equal to the inflaton mass m as

$$\Omega_{X,0} \simeq 3 \times 10^{-7} \left(\frac{T_{\rm rh}}{T_0}\right) \left(\frac{m}{M_{\rm Pl}}\right)^2 \left(\frac{M_X}{m}\right)^{5/2} \exp\left(-2M_X/m\right). \qquad (24.51)$$

Since $T_{\rm rh}/T_0$ is large, the initial abundance of SHDM has to be tiny and a stable particle with mass $M_X \sim 10^{13}$ GeV would have today an abundance of order 1.

24.4 Generation of perturbations

We move now on to the study of the fluctuations $\delta\phi(x)$. First we neglect that they back-react via the Einstein equations on the spacetime and calculate their evolution in a fixed FLRW metric. Allowing in the next step for deviations from the FLRW metric means that no preferred foliation of the spacetime in time and space exists. This arbitrariness in the coordinate choice requires a careful definition of observables, so that an unphysical gauge-dependence of observables is avoided. Having defined suitable gauge invariant variables, we then solve the coupled equations for perturbation in the linear theory.

24.4.1 Fluctuations in a fixed FLRW background

We consider fluctuations of the inflaton field ϕ around its classical, uniform but time-dependent average value,

$$\phi(\boldsymbol{x}, t) = \phi_0(t) + \delta\phi(\boldsymbol{x}, t). \qquad (24.52)$$

Inserting this splitting into the Klein–Gordon equation (23.16) with $\xi = 0$ and a general potential $V(\phi)$ gives six terms,[2]

$$\ddot{\phi} + \partial_t^2 \delta\phi + 3H(\dot{\phi} + \delta\dot{\phi}) - \frac{1}{a^2}\boldsymbol{\nabla}^2 \delta\phi + V'(\phi_0 + \delta\phi) = 0. \qquad (24.53)$$

First we evaluate the potential term V', expanding it around the classical field ϕ_0,

$$V'(\phi_0 + \delta\phi) = V'(\phi_0) + V''(\phi_0)\delta\phi = V'(\phi_0) + m_{\rm eff}^2 \delta\phi. \qquad (24.54)$$

---

[2]Here, a prime on the potential, V', denotes its derivative with respect to its argument, not to conformal time.

Here we also introduced $m_{\text{eff}}^2 = V''(\phi_0)$ as an effective mass term for the fluctuations $\delta\phi$. Taking into account that the classical field ϕ_0 separately satisfies the field equation (23.16) gives as equation for the fluctuations

$$\left[\frac{\partial^2}{\partial t^2} - \frac{1}{a^2}\nabla^2 + 3H\frac{\partial}{\partial t} + m_{\text{eff}}^2 \right] \delta\phi = 0. \tag{24.55}$$

Next we perform a Fourier expansion of the fluctuations, $\delta\phi(\boldsymbol{x}, t) = V^{-1}\sum_k \phi_k(t)e^{i\boldsymbol{kx}}$, with k as comoving wave number. Since the proper distance varies as $a\boldsymbol{x}$, the physical momentum is then given by $\boldsymbol{p} = \boldsymbol{k}/a$. Inserting the expansion into (24.55), we obtain

$$\ddot{\phi}_k + 3H\dot{\phi}_k + \left(\frac{k^2}{a^2} + m_{\text{eff}}^2 \right) \phi_k = 0. \tag{24.56}$$

Thus the fluctuations basically obey the same equation as the average field, with the effective mass term as the only difference. Introducing the auxiliary field $\chi_k(\eta) = a(\eta)\phi_k(\eta)$ again, we arrive at

$$\chi_k'' + \left(k^2 - \frac{a''}{a} + m_{\text{eff}}^2 \right) \chi_k = 0. \tag{24.57}$$

The effective mass term $m_{\text{eff}}^2 = V_{,\phi\phi}$ is much smaller than the squared Hubble parameter H during slow-roll, since

$$\frac{V_{,\phi\phi}}{H^2} \approx \frac{3\widetilde{M}_{\text{Pl}}V_{,\phi\phi}}{V} = 3\eta_V \ll 1. \tag{24.58}$$

This implies that we can neglect $V_{,\phi\phi}$ relative to a''/a, because

$$\frac{a''}{a} = \frac{2}{\eta^2} = 2a^2H^2 \ll a^2 V_{,\phi\phi} \tag{24.59}$$

is valid in the slow-roll regime. Hence the modes $\chi_k(\eta)$ are also described by

$$\chi_k(\eta) = A_k \frac{e^{-i\omega_k\eta}}{\sqrt{2\omega_k}}\left(1 - \frac{i}{k\eta} \right) + B_k \frac{e^{i\omega_k\eta}}{\sqrt{2\omega_k}}\left(1 + \frac{i}{k\eta} \right). \tag{24.60}$$

In order to find the spectrum of fluctuations, we should quantise χ and determine its two-point function. In the limit $\eta \to -\infty$, the auxiliary field χ resembles a Minkowski field, and thus we can write down immediately the field operator in this limit,

$$\chi(\eta, \boldsymbol{x}) = \int \frac{d^3k}{\sqrt{2\omega_k(2\pi)^3}} \left[A_k a_k e^{-i(\omega_k\eta - \boldsymbol{kx})} + B_k a_k^\dagger e^{i(\omega_k\eta - \boldsymbol{kx})} \right]. \tag{24.61}$$

The annihilation and creation operators a_k and $a_k^\dagger$ satisfy the usual commutation relations, with the physical momenta replaced by the conformal momenta. However, we have still to choose the initial vacuum state, which amounts to fixing the coefficients A_k and B_k.

We observe modes which exited the horizon $\Delta N \sim 60$ e-folding before the end of inflation. If there is no special choice of the initial conditions for inflation, we expect that the total number N_{tot} of e-folding is much larger. Thus the physical momentum of these modes at the beginning of inflation, $p(t_i) \sim H e^{N_{\text{tot}} - \Delta N}$, is extremely high. A natural assumption is therefore that these modes were empty at the start of inflation. Thus we should require that for early times, $\eta \to -\infty$, only positive frequencies survive, what implies $B_k = 0$ and $A_k = 1$. This choice is called the *Bunch–Davies vacuum*. Inserting this choice into (24.60), we find for the fluctuations inside the horizon, $k \gg |1/\eta| = aH$,

$$|\delta\phi_k| = \left|\frac{\chi_k}{a}\right| = \frac{H\eta}{\sqrt{2k}}. \tag{24.62}$$

Modes outside the horizon, $k \ll aH$, are frozen-in with amplitude

$$|\delta\phi_k| = \left|\frac{\chi_k}{a}\right| = \frac{H}{\sqrt{2k^3}}. \tag{24.63}$$

Power spectrum of perturbations. The two-point correlation function for the scalar field fluctuation $\delta\phi$ is given by

$$\langle \delta\phi(\boldsymbol{x}', t')\delta\phi(\boldsymbol{x}, t)\rangle = \int \frac{\mathrm{d}^3 k}{(2\pi)^3} |\delta\phi_k|^2 \, \mathrm{e}^{-ik(x'-x)}. \tag{24.64}$$

Introducing spherical coordinates in Fourier space and choosing $x = x'$ results in

$$\langle \delta\phi^2(\boldsymbol{x}, t)\rangle = \int \frac{4\pi k^2 \mathrm{d}k}{(2\pi)^3} |\delta\phi_k|^2 = \int \mathrm{d}k \underbrace{\frac{k^2}{2\pi^2} |\delta\phi_k|^2}_{\equiv P(k)} = \int \frac{\mathrm{d}k}{k} \Delta_\phi^2(k). \tag{24.65}$$

The functions $P(k)$ and $\Delta_\phi^2(k)$ are the linear and logarithmic *power spectra* of the fluctuations, respectively. The spectrum $\Delta_\phi^2(k)$ of fluctuations outside of the horizon is given by

$$\Delta_\phi^2(k) = \frac{k^3}{2\pi^2} |\delta\phi_k|^2 = \frac{H^2}{4\pi^2}. \tag{24.66}$$

Hence, the power spectrum of superhorizon fluctuations is independent of the wavenumber in the approximation that H is constant during inflation. The total area below the function $\Delta_\phi^2(k)$ plotted versus $\ln(k)$ gives $\langle \phi^2(\boldsymbol{x}, t)\rangle$. Therefore a spectrum with $\Delta_\phi^2(k) = $ const. contains the same amount of fluctuations in each decade of k. Such a spectrum of fluctuations is called a *Harisson–Zel'dovich* spectrum, and is produced by inflation in the limit of an infinitely slow-rolling inflaton.

Remark 24.2: Let us compare the power spectrum of fluctuations in inflation to those of normal Minkowski space. Recalling (3.56), the latter are given by

$$\langle 0|\phi^2(x)|0\rangle = \int \frac{\mathrm{d}^3 k}{(2\pi)^3 2\omega_k} = \int_0^\infty \frac{\mathrm{d}k}{k} \frac{k^2}{(2\pi)^2}. \tag{24.67}$$

Thus the amplitude of fluctuations in Minkowski space is given by $\Delta_\phi(k) \equiv (\Delta_\phi^2(k))^{1/2} = k/(2\pi)$, and their relative size $\Delta_\phi(k)/k = 1/(2\pi)$ is constant. Moving to smaller and smaller scales, fluctuations keep their importance—this is another point of view for understanding the problem of UV divergences. Comparing the relative size of fluctuations at the end of inflation and in Minkowski space, we find

$$\frac{\Delta\phi(k)|_{\mathrm{dS}}}{\Delta_\phi(k)|_{\mathrm{M}}} = H/k \gtrsim \mathrm{e}^{N_{\mathrm{f}}} \sim \mathrm{e}^{60}.$$

This explains why quantum fluctuations become macroscopically important via inflation.

The fluctuations in the inflaton field, $\phi = \phi_0 + \delta\phi$, lead to fluctuations in the stress tensor $T^{\mu\nu} = T_0^{\mu\nu} + \delta T^{\mu\nu}$, and thus to metric perturbations $g^{\mu\nu} = g_0^{\mu\nu} + \delta g^{\mu\nu}$. Thus as the next step we have to examine metric perturbations.

24.4.2   Gauge-invariant variables for perturbations

Metric scalar, vector and tensor perturbations.   The spatial uniformity and isotropy of the FLRW background metric suggests that we decompose perturbations of the metric tensor into irreducible components under spatial rotations. Thus we split the full metric tensor $g_{\mu\nu}$ into its background part $g_{\mu\nu}^0$ and scalar, vector and tensor perturbations,

$$g_{\mu\nu} = g_{\mu\nu}^0 + \delta g_{\mu\nu}^{\mathrm{s}} + \delta g_{\mu\nu}^{\mathrm{v}} + \delta g_{\mu\nu}^{\mathrm{t}}. \tag{24.68}$$

This decomposition is useful, since perturbations with different helicities develop independently in the linear approximation (problem 24.10).

The line element of the flat FLRW metric using conformal time is

$$\mathrm{d}s^2 = a^2[\mathrm{d}\eta^2 - \delta_{ij}\mathrm{d}x^i\mathrm{d}x^j] \tag{24.69}$$

and thus

$$g_{\mu\nu} = a^2(\eta_{\mu\nu} + h_{\mu\nu}) \quad \text{and} \quad g^{\mu\nu} = \frac{1}{a^2}(\eta^{\mu\nu} + h^{\mu\nu}). \tag{24.70}$$

We break up the perturbation $h_{\mu\nu}$ in a first step into

$$h_{\mu\nu} = \begin{pmatrix} 2A & B_i \\ B_i & -C_{ij} \end{pmatrix}, \tag{24.71}$$

which gives as line element

$$\mathrm{d}s^2 = a^2\left[(1-2A)\mathrm{d}\eta^2 + 2B_i\mathrm{d}x^i\mathrm{d}\eta - (\delta_{ij} + C_{ij})\mathrm{d}x^i\mathrm{d}x^j\right]. \tag{24.72}$$

The function A is already a scalar, while we have to find the irreducible components of B_i and C_{ij}. Any vector in $\mathbb{R}^3$ can be written as the sum of a divergence-free and a rotation-free vector; the latter is the gradient of a scalar. Thus we can perform the replacements

$$B_i = -\partial_i B + V_i \tag{24.73}$$

and

$$C_{ij} = 2D\delta_{ij} + 2\partial_i\partial_j E + (\partial_i E_j + \partial_j E_i) + h_{ij}. \tag{24.74}$$

The six degrees of freedom of the reducible tensor C_{ij} are decomposed into two scalar (D, E), two vector (E_i with the constraint $\partial_i E_i = 0$) and two tensor (h_{ij}) degrees of freedom. The tensor h_{ij} corresponds to gravitational waves in the TT gauge, with $h_{ii} = 0$ and $\partial_i h_{ij} = 0$. The three degrees of freedom of the reducible vector B_i are decomposed into one scalar (B) and two vector (B_i with the constraint $\partial_i B_i = 0$) degrees of freedom. Thus altogether, $\delta g_{\mu\nu}$ contains four scalar, four vector and two tensor degrees of freedom. Now we can split the line element into scalar, vector and tensor perturbations around the uniform FLRW background,

$$\mathrm{d}s^2 = a^2 \left\{ (1 + 2A)\mathrm{d}\eta^2 + 2\partial_i B \mathrm{d}x^i \mathrm{d}\eta - [(1 - 2D)\delta_{ij} + \partial_i\partial_j E]\,\mathrm{d}x^i \mathrm{d}x^j \right\} \tag{24.75a}$$

$$\mathrm{d}s^2 = a^2 \left[\mathrm{d}\eta^2 + 2V_i \mathrm{d}x^i \mathrm{d}\eta - (\delta_{ij} + (\partial_i E_j + \partial_j E_i))\mathrm{d}x^i \mathrm{d}x^j \right] \tag{24.75b}$$

$$\mathrm{d}s^2 = a^2 \left[\mathrm{d}\eta^2 - (\delta_{ij} + h_{ij})\mathrm{d}x^i \mathrm{d}x^j \right]. \tag{24.75c}$$

As next step, we have to determine the source terms for the different perturbations.

Perturbations of a (non-) ideal fluid. The perturbations of the stress tensor serve as source for the perturbations of the metric. If we model the energy content of the universe as an ideal fluid,

$$T_{\mu\nu} = (\rho + P)u_\mu u_\nu - P g_{\mu\nu} \tag{24.76}$$

with $u^\alpha = (1, 0, 0, 0)$ for a comoving observer, then $\delta T_{\mu\nu}$ is parameterised by the perturbations $\delta\rho$, δP and velocities v^i. Thus the perturbed stress tensor $\delta T_\mu{}^\nu$ of an ideal fluid contains five degrees of freedom,

$$\delta T_0{}^0 = \delta\rho\,, \qquad \delta T_i{}^j = \delta_i^j \delta P \quad \text{and} \quad \delta T_0{}^j = (\rho + P)v^j. \tag{24.77}$$

We split the vector v_i again in its irreducible components, $v_i = \tilde{v}_i + \partial_i v$. Hence the perturbed stress tensor of an ideal fluid contains three scalars ($\delta\rho$, δP, v) and one vector ($\tilde{v}_i$), summing to five degrees of freedom. The remaining five degrees of freedom of a general stress tensor $\delta T_{\mu\nu}$ are contained in the anisotropic pressure tensor Π_{ij}. The presence of this term is characteristic for a non-ideal fluid, that is, a fluid with viscosity. Since we extracted the isotropic pressure $P\delta_{ij}$, the anisotropic pressure tensor is traceless, $\Pi_{ii} = 0$. Looking back at the decomposition of C_{ij}, we see that the anisotropic pressure contains two tensor ($\propto h_{ij}$), one vector ($\propto (\partial_i E_j + \partial_j E_i)$ and one scalar ($\propto \partial_i\partial_j E$) degrees of freedom.

We can now associate the metric perturbations with the various contribution to the perturbation of the stress tensor. Tensor perturbations h_{ij} have as their source only the anisotropic pressure Π_{ij}. This term is generated, for example, by freely streaming neutrinos after weak decoupling, and its effects are always subleading. Therefore, we will set $\Pi_{ij} = 0$ in the following and treat gravitational waves as freely propagating. Vector perturbations correspond to rotational flows of matter—these perturbations are without practical interest for two reasons: first, it is unlikely that they are generated during inflation and, second, they have only decaying solutions. Finally, scalar perturbations are sourced by $\delta\rho$, δP and v_i. We will see that they contain a growing solution and are connected to the inhomogeneities of matter in the universe.

Gauge-invariant variables for scalar perturbations. In a second step, we have to identify how the perturbations are connected to the physical degrees of freedom.

We know already that tensor perturbations, or gravitational waves, contain only two physical degrees of freedom. Similarly, the line element for scalar perturbations agrees with the Newtonian weak-field limit (19.72), if one sets $A = D = \Phi$ and $E_{,ij} = B_{,i} = 0$. Therefore, one may suspect that not all the four scalar variables describing scalar perturbations are physical.

In order to identify the physical degrees of freedom, we examine how the splitting into scalar perturbations changes under a finite gauge transformation,

$$\tilde{h}_{\mu\nu} = h_{\mu\nu} - \nabla_\mu \xi_\nu - \nabla_\nu \xi_\mu, \tag{24.78}$$

of the metric. We decompose the gauge vector ξ_μ as usually in its irreducible components,

$$\xi^\mu = (\xi^0, \xi^i) = (\xi^0, \partial_i \xi + \xi_\perp). \tag{24.79}$$

Consider now, for example, the 00 component of the metric tensor, $\delta g_{00}^{(s)} = h_{00} = 2Aa^2$, from which we can read off the transformation law of A. With $\Gamma^0{}_{00} = \mathcal{H}$, $\nabla_0 \xi_0 = (a^2 \xi_0)' - \Gamma^\mu{}_{00} \xi_\mu$ and $\xi_0 = a^2 \xi^0$, we find

$$\delta g_{00}^{(s)} = 2Aa^2 \to 2\tilde{A}a^2 = 2Aa^2 - 2\nabla_0 \xi_0 = 2Aa^2 - 2(a^2\xi^0)' - 2\mathcal{H}a^2\xi^0. \tag{24.80}$$

Thus the scalar perturbation A changes under a gauge transformation as

$$\tilde{A} = A - \mathcal{H}\xi^0 - \xi^{0\prime}. \tag{24.81}$$

Proceeding in the same way for the other perturbations, we obtain

$$\tilde{B} = B + \xi^0 - \xi', \tag{24.82}$$

$$\tilde{D} = D + \mathcal{H}\xi^0, \tag{24.83}$$

$$\tilde{E} = E - \xi. \tag{24.84}$$

The transformations of the potentials are parameterised by only two arbitrary parameters, ξ^0 and ξ', since $\xi_\perp$ influences only vector perturbations. Moreover, the gauge vector (24.79) contains no tensor component, which expresses the fact that $h_{\mu\nu}$ in the TT gauge contains only physical degrees of freedom. As a result, we can eliminate two variables. For instance, choosing $\xi^0 = -D/\mathcal{H}$ and $\xi = E$ we can set $\tilde{D} = \tilde{E} = 0$.

There are two ways to eliminate the gauge ambiguity in the scalar perturbations. In the first one, we combine the gauge-dependent variables into invariant combinations. Consider, for example, the quantity

$$\Psi = D - \mathcal{H}(B - E') \tag{24.85}$$

which is invariant under gauge transformations,

$$\Psi \to \tilde{\Psi} = \tilde{D} - \tilde{\mathcal{H}}(\tilde{B} - \tilde{E}') = D - \mathcal{H}(B - E'). \tag{24.86}$$

Similarly, the combination

$$\Phi = A + \mathcal{H}(B - E') + (B - E')' \tag{24.87}$$

is shown to be invariant. This method was suggested by Bardeen and therefore Ψ and Φ are called Bardeen potentials. An alternative way is to choose a gauge condition which eliminates the gauge freedom partly or completely, analogously to the Coulomb gauge in electrodynamics.

The conformal Newtonian gauge has the virtue that the metric for scalar perturbations is diagonal, since one sets $B = E = 0$. Then the Bardeen potentials become $\Psi = D$ and $\Phi = A$ and we can write the scalar part of metric as

$$ds^2 = a^2 \left\{ (1 + 2\Phi)d\eta^2 - (1 - 2\Psi)\delta_{ij}dx^i dx^j \right\}. \tag{24.88}$$

We show in the appendix that the combination $\Phi - \Psi$ is only sourced by the anisotropic pressure. In the absence of anisotropic pressure, $\Psi = \Phi$, and thus the line element (except for the conformal factor a^2) coincides with the Schwarzschild metric in the Newtonian limit. This makes an understanding of the perturbation, especially on sub-horizon scales, easier.

Gauge-invariant curvature perturbation. In the full theory, gauge transformations couple matter and curvature fluctuations. Thus we have to find as final step a gauge invariant combination of both fluctuations. We determine first how the matter perturbation $\delta\phi$ transforms under a gauge transformation $\tilde{x}^\mu = x^\mu + \xi^\mu(x^\nu)$,

$$\widetilde{\delta\phi}(\tilde{x}) = \tilde{\phi}(\tilde{x}) - \phi_0(\tilde{x}) = \phi(x) - \phi_0(x + \xi) = \underbrace{\phi(x) - \phi_0(x)}_{\delta\phi(x)} - \xi^0 \partial_0 \phi_0(x). \tag{24.89}$$

Here we used the fact that ϕ is a scalar field, $\tilde{\phi}(\tilde{x}) = \phi(x)$, and that ϕ_0 is uniform. Thus the perturbation transforms as

$$\widetilde{\delta\phi}(\tilde{x}) = \delta\phi(x) - \xi^0 \phi_0' \simeq \delta\phi(x) - \xi^0 \phi', \tag{24.90}$$

where we could also replace $\phi_0' \sim \phi'$ neglecting a quadratic term. We already know that D transforms as $\tilde{D} = D + \mathcal{H}\xi^0$. Thus the quantity

$$\mathcal{R} \equiv D + \mathcal{H}\frac{\delta\phi}{\phi_0'} \to \tilde{\mathcal{R}} = D + \mathcal{H}\xi^0 + \mathcal{H}\frac{\delta\phi - \xi^0 \phi'}{\phi_0'} = \mathcal{R} \tag{24.91}$$

called the curvature perturbation is a gauge-invariant combination of the metric perturbation D and the matter perturbation $\delta\phi$.

24.4.3 Fluctuations in the full linear theory

Up to now we have considered the evolution of the inflaton field ϕ in a fixed FLRW background, neglecting the back-reaction of the fluctuation $\delta\phi$ on the metric $g_{\mu\nu}$. The Einstein equations couple however the inflaton field ϕ and the metric $g_{\mu\nu}$ already at the linear level. Having identified gauge-invariant variables for the perturbations, it remains to perform the straightforward but tedious linearisation of the Einstein equations.

Perturbed stress tensor of the inflaton. Inserting into the stress tensor (24.12) of a scalar field,

$$T_\mu^{\ \nu} = g^{\nu\lambda}\partial_\lambda\phi\partial_\mu\phi - \delta_\mu^\nu \left[\frac{1}{2}g^{\rho\sigma}\partial_\rho\phi\partial_\sigma\phi - V(\phi)\right], \tag{24.92}$$

the decomposition $\phi(x) = \phi_0(t) + \delta\phi(x)$ into a homogeneous background field and fluctuation, we find

$$\delta T_0^0 = \frac{1}{a^2} \left(-\Phi\phi_0'^2 + \phi_0'\delta\phi + V_{,\phi_0}\delta\phi \right) \tag{24.93a}$$

$$= \frac{1}{a^2} \left[-\Phi\phi_0'^2 + \phi_0'\delta\phi - \left(\phi_0'' + 2\frac{a'}{a}\phi_0' \right)\delta\phi \right], \tag{24.93b}$$

$$\delta T_i^0 = \frac{1}{a^2}\phi_0'\partial_i\delta\phi. \tag{24.93c}$$

Here we used the Klein–Gordon equation for ϕ_0 to replace $V_{,\phi_0}$. The only information we will need about the spatial components δT_i^j of the perturbed stress tensor of the inflaton is that the uniformity of the background requires that $\delta T_i^j \propto \delta_i^j$.

Next we observe that $\rho = \dot\phi^2/2 + V(\phi)$ and $P = \dot\phi^2/2 - V(\phi)$ imply for the background field ϕ_0 that

$$\rho + P = \dot\phi_0^2 = \frac{\phi_0'^2}{a^2}. \tag{24.94}$$

Comparing $\delta T_i^0 = a^{-2}\phi_0'\partial_i\delta\phi$ to an ideal fluid, $\delta T_i^0 = -(\rho + P)\partial_i v = -(\phi_0'^2/a^2)\partial_i v$, we have thus

$$v = -\frac{\delta\phi}{\phi_0'}. \tag{24.95}$$

Hence the choice of a comoving gauge, where $v = 0$, leads to $\delta\phi = 0$: the inflaton fluctuations are zero in the frame where the observer is at rest relative to the inflaton. Thus the curvature perturbation $\mathcal{R}$ becomes $\mathcal{R} = D$ on the hypersurfaces defined by $v = 0$. Evaluating then the three-dimensional curvature $R^{(3)}$ using the spatial part of the metric (24.75a), one finds $R^{(3)} = 4/a^2\Delta\mathcal{R}$. This explains the name curvature perturbation for $\mathcal{R}$.

Perturbed Einstein equations. As next step, we need the linearised Einstein tensor for perturbations around the FLRW metric. The result of the straightforward but lengthy calculations are given in the appendix 24.A. We use the conformal Newtonian gauge where the metric contains the two Bardeen potentials Φ and Ψ. Since the anisotropic pressure is zero, it follows $\Phi = \Psi$, and thus the Einstein equations contain only two free variables, which we choose as Φ and $\delta\phi$. Therefore we can select out of three Einstein equations (00, 0i and ij) the two most convenient ones. Choosing the time–time and the time–space equations, we avoid second-order time derivatives and find

$$\Delta\Phi - 3\frac{a'}{a}\Phi' - 3\frac{a'^2}{a^2}\Phi = 4\pi G\left[-\Phi\phi_0'^2 + \phi_0'\delta\phi' - \left(\phi_0'' + 2\frac{a'}{a}\phi_0' \right)\delta\phi \right], \tag{24.96a}$$

$$\Phi' + \frac{a'}{a}\Phi = 4\pi G\phi_0'\delta\phi. \tag{24.96b}$$

Here we could integrate the last equation immediately, since H and ϕ_0 are uniform.

Our aim is to combine these two first-order equations into a single second-order equation for the gauge-invariant variable $\mathcal{R}$. An often-employed strategy is to simplify first Eq. (24.96a), neglecting terms which are small in the slow-roll approximation. We prefer to derive the exact equation for single-field inflation, massaging the complete Eq. (24.96a) into a suitable form. We use the fact that the background field ϕ_0 and the scale factor a are connected by the unperturbed Einstein equations. Changing Eqs. (20.48) and (20.49) to conformal time, we have

$$\frac{2a''}{a^3} + \frac{a'^2}{a^4} = -8\pi GP \quad \text{and} \quad 3\frac{a'^2}{a^4} = 8\pi G\rho. \tag{24.97}$$

Subtracting these two equations, we can write

$$\frac{a''}{a} - 2\frac{a'^2}{a^2} = -4\pi G(\rho + P)a^2 = -4\pi G\phi_0'^2, \tag{24.98}$$

where we used also (24.94). This allows us to eliminate the Φ term on the RHS of (24.96a),

$$\Delta\Phi - 3\frac{a'}{a}\Phi' - \left(\frac{a'^2}{a^2} + \frac{a''}{a}\right)\Phi = 4\pi G\left[\phi_0'\delta\phi' - \left(\phi_0'' + 2\frac{a'}{a}\phi_0'\right)\delta\phi\right]. \tag{24.99}$$

Next we eliminate the Φ term on the LHS. We substitute Φ by $\delta\phi$ and Φ' with the help of (24.96b), and use then again (24.98). The LHS of (24.99) becomes thereby

$$\Delta\Phi - 4\pi G(\phi_0'^2\Phi' + \phi_0'\delta\phi).$$

Then we bring all terms linear in G to the RHS and combine them as

$$\Delta\Phi = 4\pi G\frac{a}{a'}\phi_0'^2\frac{\mathrm{d}}{\mathrm{d}\eta}\left(\Phi + \frac{a'}{a\phi_0'}\delta\phi\right). \tag{24.100}$$

Now we recognise the term in the parenthesis as the gauge-invariant variable $\mathcal{R}$ we were out for. We can combine the dependence of $\mathcal{R}$ on Φ and $\delta\phi$ introducing as new variable

$$\tilde{\phi} = \delta\phi + \frac{a\phi_0'}{a'}\Phi. \tag{24.101}$$

Analogous to the uniform case, we replace next $\tilde{\phi}$ by the auxiliary field $u = a\tilde{\phi}$, which is often called the Mukhanov–Sasaki variable. It is connected to $\mathcal{R}$ by

$$u = -\frac{a^2\phi_0'}{a'}\mathcal{R} \equiv z\mathcal{R}. \tag{24.102}$$

Expressed by the Mukhanov–Sasaki variable, Eq. (24.100) becomes

$$\Delta\Phi = 4\pi G\frac{z}{a}\phi_0'^2\frac{\mathrm{d}}{\mathrm{d}\eta}\left(\frac{u}{z}\right). \tag{24.103}$$

Then we rewrite the not yet used Eq. (24.96b) as function of u,

$$\frac{a'}{a^2}\frac{\mathrm{d}}{\mathrm{d}\eta}\left(\frac{a^3}{a'}\Phi\right) = 4\pi G\phi_0'u. \tag{24.104}$$

The remaining task is to combine these two equations. Applying the Laplace operator Δ on (24.100), replacing $\Delta\Phi$ via (24.103) and inserting $\phi_0' = za'/a^2$, we arrive at

$$\frac{\mathrm{d}}{\mathrm{d}\eta}\left(z^2\left(\frac{\mathrm{d}}{\mathrm{d}\eta}\frac{u}{z}\right)\right) = z\Delta u \tag{24.105}$$

or

$$u'' - \frac{z''}{z}u^2 - \Delta u = 0. \tag{24.106}$$

This equation describes linear perturbations in single-field inflation exactly, that is, without implying the slow-roll approximation or any specific shape of the inflation potential.

For superhorizon modes, $\Delta u \approx 0$, and the growing modes satisfies $u \propto z$. The definition of the Mukhanov–Sasaki variable implies then that $\mathcal{R}$ is conserved during inflation, that is $\mathcal{R}$ does not depend on time for superhorizon modes. More generally, one can show that $\mathcal{R}$ is conserved on superhorizon scales also after inflation, if the perturbations are adiabatic.

In order to quantise u, we need to find its action $S[u]$. Formally, we could derive $S[u]$ by integrating out the the gravitational potential Φ from the combined action $S[g_{\mu\nu}, \phi]$. We use instead that the equation of motion (24.106) fixes the action up to an unknown constant A as

$$S[u] = A\int \mathrm{d}^4x \left[u'^2 - (\partial_i u)^2 + \frac{z''}{z}u^2\right]. \tag{24.107}$$

Then we determine A by requiring that the term χ'^2, which comes from the scalar action, has the correct coefficient $1/2$. With $u = \chi + z\Phi$, we obtain $A = 1/2$. The effective mass of the u field is now

$$m_{\text{eff}}^2 = -\frac{z''}{z} = -\frac{H}{a\dot\phi}\frac{\partial^2}{\partial\eta^2}\left(\frac{a\dot\phi}{H}\right), \tag{24.108}$$

where we used

$$z = -\frac{a^2\phi_0'}{a'} = -\frac{a\dot\phi_0}{H} = -\sqrt{2}a\widetilde{M}_{\text{Pl}}\varepsilon. \tag{24.109}$$

The back-reaction between the scalar and the gravitational field is encoded in the time behaviour of $\dot\phi$ and H. The exact solutions of Eqs. (24.106) and (24.107) have to be found by numerical integration. To proceed analytically, we employ instead the slow-roll approximation. Then the Hubble parameter is close to constant, $H \approx$ const., and the kinetic energy $\dot\phi_0$ should be small for a sufficiently long time, while a is increasing exponentially. Thus we can treat $\dot\phi$ and H as constant, obtaining in the slow-roll approximation

$$m_{\text{eff}}^2 = -\frac{z''}{z} \to -\frac{\eta''}{\eta}. \tag{24.110}$$

Therefore we can identify in the slow-roll approximation the Mukhanov–Sasaki variable u with χ, while $\tilde\phi$ coincides with χ/a. Hence, our results obtained for a fixed FLRW background remain valid in the lowest order of the slow-roll approximation, if we express $\mathcal{R}$ by $\tilde{\phi}$. Inserting (24.109) in $u = a\tilde{\phi} = z\mathcal{R}$, we obtain this relation as

$$\mathcal{R} = \frac{a}{z}\,\tilde{\phi} = \frac{H}{\dot{\phi}_0}\,\tilde{\phi}. \tag{24.111}$$

We can now use the connection between perturbations in $\mathcal{R}$ and $\tilde{\phi}$ using our result in a fixed background,

$$\Delta_{\mathcal{R}}^2(k) = \frac{H^2}{\dot{\phi}^2}\Delta_{\phi}^2(k) = \frac{H^2}{\dot{\phi}^2}\frac{H^2}{4\pi^2} = \left(\frac{H^2}{2\pi\dot{\phi}}\right)^2\bigg|_{k=aH} = \frac{1}{8\pi^2\varepsilon}\left(\frac{H}{\widetilde{M}_{\rm Pl}}\right)^2\bigg|_{k=aH}. \tag{24.112}$$

Here, we used also $\dot{H} = -4\pi G\dot{\phi}^2$ from Eq. (24.27) and $\varepsilon = -\dot{H}/H^2$. Moreover, we accounted for deviations from de Sitter (i.e. the time-dependence of H) by evaluating the power spectrum for wave-vectors at horizon crossing, $k = aH$. Since during slow-roll inflation $H^2 \propto V$, we see that the normalisation of the power spectrum informs us about the ratio V/ε.

Tensor perturbations.   While scalar perturbations are a gauge-dependent mixture of perturbations in $\delta g_{\mu\nu}^{\rm s}$ and $\delta\phi$, tensor perturbations $\delta g_{\mu\nu}^{\rm t}$ are fully fixed by the two physical degrees of freedom in the gravitational wave tensor $h_{\mu\nu}$ present in the TT gauge. Moreover, the source term for gravitational waves corresponds to anisotropic pressure, which is absent in the perturbed stress tensor of the inflaton. We have seen that the action (19.80) for gravitational waves in the TT gauge is identical to the one of a free minimally coupled scalar field, changing its normalisation by $2 \times 32\pi G$. Hence the power spectrum $\Delta_T^2(k)$ of the metric tensor perturbations is connected to the scalar perturbations $\Delta_\phi^2(k)$ by

$$\Delta_{\rm t}^2(k) = 2 \times 32\pi G\Delta_\phi^2(k)\big|_{k=aH} = \frac{2}{\pi^2}\left(\frac{H}{\widetilde{M}_{\rm Pl}}\right)^2\bigg|_{k=aH}. \tag{24.113}$$

Measuring the amplitudes of scalar and tensor perturbations on the scale $k = aH$ determines thus both the slow-roll parameter ε and the Hubble parameter H at horizon crossing of this mode.

Deviations from a scale invariant spectrum.   The spectrum of fluctuations is scale invariant only for a de Sitter universe. Since inflation has to end, we expect deviations from the scale invariant spectrum $P(k) \propto 1/k$ which we parameterise via the scalar spectral index n_s of the scalar fluctuations. Thus $\Delta_{\mathcal{R}}^2(k) \propto k^{-(1-n_s)}$ and

$$n_{\rm s} - 1 \equiv \frac{\ln\Delta_{\mathcal{R}}^2(k)}{\ln(k)} = \frac{{\rm d}\ln\Delta_{\mathcal{R}}^2(k)}{{\rm d}N}\frac{{\rm d}N}{{\rm d}\ln(k)}. \tag{24.114}$$

Since the slow-roll parameters determine the deviations from de Sitter expansion, we should be able to express n_s via ε and η. Using $\Delta_{\mathcal{R}}^2(k) = H^2/(8\pi^2\varepsilon)\big|_{k=aH}$, we find for the first factor

$$\frac{{\rm d}\ln\Delta_{\mathcal{R}}^2(k)}{{\rm d}N} = 2\frac{{\rm d}\ln H}{{\rm d}N} - \frac{{\rm d}\ln\varepsilon}{{\rm d}N} = -2\varepsilon - \eta, \tag{24.115}$$

where we used the definition of the slow-roll parameters. For the evaluation of the second factor, we use that modes crossing the horizon satisfy $k = aH$, or

$$\ln k = \ln a + \ln H. \tag{24.116}$$

Differentiating this expression and recalling $\mathrm{d}N = d\ln(a)$ gives

$$\frac{\mathrm{d}\ln(k)}{\mathrm{d}N} = 1 - \varepsilon. \tag{24.117}$$

Combining the results for the two factors and neglecting second-order terms, we arrive at

$$n_{\mathrm{s}} - 1 \approx (-2\varepsilon - \eta)(1 + \varepsilon) \approx -2\varepsilon - \eta. \tag{24.118}$$

Thus the slope of the scalar fluctuations informs us about deviations from a perfect de Sitter phase, or via ε_V and η_V, on the shape of the inflaton potential.

Next we consider tensor fluctuations, defining their spectral slope by

$$n_{\mathrm{t}} \equiv \frac{\ln \Delta_{\mathrm{r}}^2(k)}{\ln(k)}. \tag{24.119}$$

In contrast to the scalar slope, n_{t} contains no ε term and thus it is given simply by

$$n_{\mathrm{t}} = -2\varepsilon. \tag{24.120}$$

Tensor–scalar ratio. The ratio r of perturbations in tensor and scalar modes is determined in single-field inflation fully by the slow-roll parameter ε,

$$r \equiv \frac{\Delta_{\mathrm{t}}^2}{\Delta_{\mathcal{R}}^2} = \frac{2}{\pi^2}\left(\frac{H}{\widetilde{M_{\mathrm{Pl}}}}\right)^2 \Big/ \frac{H^2}{8\pi^2\varepsilon} = \frac{8\dot{\phi}^2}{\widetilde{M_{\mathrm{Pl}}^2}H^2} = 16\varepsilon. \tag{24.121}$$

This result has two important consequences. First, we can derive an upper limit on $r|_{k=aH}$ as function of the inflaton field value at the time of horizon crossing of the scale k. This limit, often called the Lyth bound, can be derived using $\mathrm{d}\phi/(\mathrm{d}tH) = \mathrm{d}\phi/\mathrm{d}N$. Then the amount $\Delta\phi$ the inflaton evolved between the horizon exit of the CMB modes and the end of inflation is given by

$$\Delta\phi \geq \widetilde{M_{\mathrm{Pl}}} \int_0^{N_{\mathrm{CMB}}} \mathrm{d}N \sqrt{r/8} \approx \text{few} \times \widetilde{M_{\mathrm{Pl}}}\left(\frac{r}{0.01}\right)^{1/2}. \tag{24.122}$$

Here, we used that r during slow-roll should be nearly constant. This relation connects the tensor–scalar ratio on CMB scales and the minimal initial value of the inflaton. Hence large, observable tensor perturbations require trans-Planckian initial values of the inflaton.

As second consequence, we can derive a so-called consistency relation. We can combine $r = 16\varepsilon$ and $n_{\mathrm{t}} = -2\varepsilon$ as follows

$$n_{\mathrm{t}} = -2\varepsilon = -\frac{r}{8}. \tag{24.123}$$

The slope of the tensor perturbations is fixed by the ratio of the amplitudes of scalar and tensor perturbations. Measuring n_{t} independently provides therefore a consistency check of single-field inflation.

Emergence of classical fields. Inflation amplifies the length-scales of perturbations which have been generated as quantum fluctuations. Since field operators as

$$\chi(\eta, \boldsymbol{x}) = \int \frac{\mathrm{d}^3 k}{(2\pi)^{3/2}} \left(\chi_k^{(+)}(\eta) a_{\boldsymbol{k}}^\dagger \mathrm{e}^{-\mathrm{i}\boldsymbol{k}\boldsymbol{x}} + \chi_k^{(-)}(\eta) a_{\boldsymbol{k}} \mathrm{e}^{\mathrm{i}\boldsymbol{k}\boldsymbol{x}} \right) \tag{24.124}$$

are complex, the expectation values of products of fields are complex too. Moreover, they are depending on their order. In contrast, the fluctuations of a fluid are classical and real. In general, we expect that the initial quantum fluctuations are converted into classical fluctuations by the phenomenon of decoherence, that is, by the coupling to a thermal bath. In the case of superhorizon fluctuations, the coupling between the inflaton and the gravitational field is sufficient for this conversion process. For modes after horizon exit, $k_* \ll 1/|\eta|$, which are frozen-in, the field operator simplifies to

$$\chi(\eta, \boldsymbol{x}) = \int_{k_*} \frac{\mathrm{d}^3 k}{\sqrt{(2\pi)^3 2\omega_k}} \left(\frac{-1}{k\eta} \right) \left(a_{\boldsymbol{k}}^\dagger \mathrm{e}^{-\mathrm{i}(\boldsymbol{k}\boldsymbol{x} + \alpha_{\boldsymbol{k}})} + a_{\boldsymbol{k}} \mathrm{e}^{\mathrm{i}(\boldsymbol{k}\boldsymbol{x} - \alpha_{\boldsymbol{k}})} \right), \tag{24.125}$$

where $\alpha_{\boldsymbol{k}}$ are arbitrary phases. Thus the mode functions are real, $\chi_k^{(+)}(\eta) = \chi_k^{(-)}(\eta)$. As a result, also the expectation values of products of fields become real and do not depend on the order. This can be also seen calculating the canonically conjugated field operator on superhorizon scales,

$$\pi(\eta, \boldsymbol{x}) = \frac{\partial \mathscr{L}}{\partial \chi'} = -\frac{1}{\eta}\chi(\eta, \boldsymbol{x}) \tag{24.126}$$

which is proportional to $\chi(\eta, \boldsymbol{x})$. Therefore, the two operators commute and χ behaves as a classical field. Thus we can treat the perturbations as classical random fields and replace quantum averages by statistical averages. Since we included only the quadratic part of the potential, cf. Eq. (24.54), the quantum field χ is a free Gaussian field. As a result, the corresponding classical random field is Gaussian too, implying that all information is contained in the two-point correlation functions. Deviations from Gaussianity should be non-zero but tiny.

24.5 Outlook: Further evolution of fluctuations

We have found that inflation generates a nearly power law-like spectrum of curvature fluctuations and gravitational waves. Superhorizon modes are frozen-in and can be described by classical Gaussian random fields. These fluctuations become observable through the temperature fluctuations of the CMB and the large-scale structure of the universe. The evolution of these fluctuations after inflation follows classical physics, and thus this topic is outside the focus of this book. We therefore offer a sketchy overview only in this section.

After reheating, the matter content of the universe consists of a plasma containing the SM degrees of freedom at least. Additionally, there should be new particles associated to a dark matter sector and to baryogenesis. Initially, all the species[3] are

---

[3] An exception might be e.g. non-thermal DM which presence would lead to isothermal fluctuations.

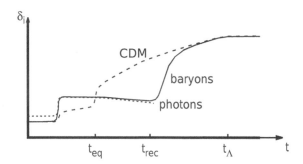

Fig. 24.4 Schematic evolution of perturbations in cold dark matter, photons, and baryons.

tightly coupled and can described therefore as an ideal fluid with a single mode of perturbation $\delta\rho$. This implies that the fluctuations are *adiabatic*, that is, that they satisfy

$$\delta\rho_i = \dot{\rho}_i\delta(x) \quad \text{and} \quad \delta P_i = \dot{P}_i\delta(x) \tag{24.127}$$

with a single function $\delta(x)$ describing the density and pressure fluctuations of all the components i. In this case, the conservation law for $\mathcal{R}$ continues to hold for super-horizon modes. As the universe cools down, some species first go out of chemical and later out of kinetic equilibrium. As a result, the evolution of these species has to be described either by a non-ideal fluid including anisotropic pressure or by a set of coupled Boltzmann equations. Examples where such a treatment is necessary are the free-streaming of neutrinos after weak decoupling and of photons after recombination.

During the evolution of the universe at least two episodes occurred where unknown physics beyond the SM can influence the evolution of fluctuations. These two episodes are the generation of the baryon asymmetry and of the dark matter abundance. If both are generated through the freeze-out mechanism discussed in chapter 21 ! and 22 then the same ratio n_B/s and $n_{\rm DM}/s$ is generated in the whole universe. As a result, the fluctuations remain adiabatic. Another option is that the ratios n_B/s or $n_{\rm DM}/s$ vary in space. Such fluctuations are called isocurvature perturbations. This may happen, for example, in axion models, if the Peceei–Quinn phase transition takes place at a smaller temperature than reheating. Then the observable universe today contains many patches with different values of the misalignment angle ϑ. Since ϑ is space-dependent, the resulting axion dark matter density $n_{\rm DM}/s$ varies in space, too. Observations of the CMB are consistent with a purely adiabatic fluctuation spectrum and can be thus used to limit models predicting isocurvature perturbation.

Matter perturbations. The evolution of the perturbations $\delta\rho_i$ and δP_i can be determined from the Einstein equations, following the same strategy as in the case of the inflaton but using the stress tensor for a sum of the various fluid components. The evolution of the density contrast in different components of the energy density of the universe is schematically shown in Fig. 24.4 assuming adiabatic fluctuations.

An important feature to note is that the density contrast of radiation is approximately constant. This can be understood from the fact that the scale on which radiation is gravitationally bound is comparable to the Hubble radius (problem 24.12). As a result, also the density contrast of baryons is approximately constant as long the photon–baryon fluid is tightly coupled. In contrast, the density contrast of CDM grows as $\delta_{CDM} \propto (1 + z)$ starting from matter–radiation equilibrium z_{eq}. At recombination, baryons decouple from photons and start to fall into the potential walls formed by CDM. Note also that fluctuations of CDM on small scales k crossed earlier the horizon. They started therefore to grow and to form gravitationally bound systems earlier, leading to the hierarchical formation of structures in a CDM universe. Observations strongly favour this picture. Finally, the growth of the density contrast stops outside of already gravitationally bound structures, when the accelerated expansion of universe starts at $z \simeq 0.5$.

Example 24.4: Derive the connection between $\mathcal{R}$ and the gravitational potentials. We have already matched the fluctuations of the inflaton and the plasma, see Eqs. (24.94) and (24.95), finding $v = -\delta\phi/\phi_0'$. Thus we can express the curvature perturbation in the conformal Newtonian gauge as

$$\mathcal{R} = \Psi - \mathcal{H}v. \tag{24.128}$$

Now using the (integrated) $0i$ component of the Einstein equation, $\Psi' + \mathcal{H}\Phi = 4\pi G a^2 (\rho + P)v$ and $\mathcal{H}^2 = 8\pi G a^2 \rho/3$, we obtain

$$\mathcal{R} = \Psi + \frac{2}{3}\frac{\Phi + \Psi'/\mathcal{H}}{1 + w}. \tag{24.129}$$

Here, $w = P/\rho$ is the effective EoS with $P = \sum_i P_i$ and $\rho = \sum_i \rho_i$ for the different components of the plasma. This equation allows us to deduce two important results, cf. problem 24.13. First, differentiating it for adiabatic perturbations shows that $\mathcal{R}$ is also after inflation conserved on superhorizon scales. Second, we can use it to deduce from $\mathcal{R}$ the gravitational potentials, that is, we can connect the predictions of inflationary models with the input for the formation of large-scale structures. In particular, Eq. (24.129) simplifies neglecting anisotropic pressure and assuming $w = $ const. on superhorizon scales to

$$\Phi = \Psi = \frac{3 + 3w}{5 + 3w}\mathcal{R}. \tag{24.130}$$

This gives $\Phi = 2\mathcal{R}/3$ for the radiation and $\Phi = 3\mathcal{R}/5$ for the matter-dominated epoch.

Angular power spectrum. We will next examine in somewhat more detail the cosmic microwave background (CMB). Fluctuations in the CMB temperature are seen on the sphere of last scattering. Thus it is convenient to decompose a map of CMB temperatures $T(\vartheta, \phi)$ into spherical harmonics $Y_{lm}(\vartheta, \phi)$,

$$T = \sum_{l=0}^{\infty} \sum_{m=-l}^{l} a_{lm} Y_{lm}(\vartheta, \phi). \tag{24.131}$$

The first two moments are usually considered separately. The monopole moment $l = 0$ of a CMB temperature map corresponds to the average temperature of the CMB,

$T_0 = 2.725$ K. The relative motion of the Sun with respect to the CMB introduces (mainly) a dipole $l = 1$ anisotropy. More precisely, the temperature transforms as

$$T = \frac{T_0\sqrt{1 - \beta^2}}{1 - \beta\cos\vartheta} = T(1 + \cos\beta + \mathcal{O}(\beta^2)), \tag{24.132}$$

where we can neglect the higher-order terms because the peculiar velocity of the Sun is small, $\beta = v/c \ll 1$: from the size of the dipole one deduces that the Sun moves with 370 km/h relative to the CMB.

If this dipole anisotropy is subtracted, temperature differences of the order $\delta T/T \sim 10^{-5}$ remain between different directions on the sky. However, in each direction the spectrum is the one of a perfect black-body. The moments $l \geq 2$ of the fluctuations $\delta T/T$,

$$\Theta(\vartheta, \phi) \equiv \frac{\delta T}{T} = \sum_{l=2}^{\infty}\sum_{m=-l}^{l} a_{lm}Y_{lm}(\vartheta, \phi) \tag{24.133}$$

are connected with the cosmological parameters and physical processes between recombination and today. For an isotropic universe, the m dependence of the coefficients a_{lm} contains no information and one therefore defines

$$C_l = \langle a_{lm}a_{lm}^* \rangle = \frac{1}{2l+1}\sum_{m=-l}^{l} a_{lm}a_{lm}^*. \tag{24.134}$$

Since a single spherical harmonic Y_{lm} corresponds roughly to angular variations of $\vartheta \sim \pi/l$, the coefficient C_l determines the power of fluctuation with the angular scale π/l. Note that this procedure allows one to replace an average over an ensemble of universes by an average over different patches of the observed universe. For small ℓ, this introduces an irreducible statistical error $\delta C_\ell/C_\ell \sim 1/\sqrt{2\ell + 1}$ which is called *cosmic variance*.

We are not interested in the temperature fluctuations $\Theta(\vartheta, \phi)$ themselves, but only in their statistical properties. For Gaussian initial quantum fluctuations, all the information is contained in the two-point correlation function,

$$C(\vartheta) = \langle \Theta(\vartheta, \phi)\Theta(\vartheta', \phi')\rangle. \tag{24.135}$$

(Since the initial curvature fluctuations are small, the perturbations of all other quantities are linearly related to the curvature fluctuations and therefore Gaussian too.) A Legendre transformation gives then

$$C_l = 2\pi\int_{-1}^{1} d\cos\vartheta\, C(\vartheta)P_l(\cos\vartheta), \tag{24.136}$$

which summarise in an efficient way the experimental data $\delta T(\vartheta, \phi)$ contained in the many pixels of an CMB experiment. In the next step, we have to connect the temperature fluctuations observed today to the fluctuations at recombination.

Temperature fluctuations. Our aim is to connect the frequency of a photon emitted on the last scattering surface (LSS) to its frequency today. This calculation is simplified performing it in the frame which is conformally flat for zero perturbations, $\tilde{g}_{\mu\nu} = g_{\mu\nu}/a^2(\eta)$, using the fact that light-like geodesics coincide in two conformally related frames.

Let us consider the covariant momentum $P^\mu = \mathrm{d}x^\mu/\mathrm{d}\lambda$ of a photon as function of conformal time $\eta = x^0(\lambda)$. Then we can rewrite the geodesic equation $\mathrm{d}P^\mu/\mathrm{d}\lambda + \Gamma^\mu{}_{\rho\sigma} P^\rho P^\sigma = 0$ using

$$\frac{\mathrm{d}P^\mu}{\mathrm{d}\lambda} = \frac{\mathrm{d}\eta}{\mathrm{d}\lambda}\frac{\mathrm{d}P^\mu}{\mathrm{d}\eta} = P^0 \frac{\mathrm{d}P^\mu}{\mathrm{d}\eta} \tag{24.137}$$

as

$$\frac{\mathrm{d}P^\mu}{\mathrm{d}\eta} + \Gamma^\mu{}_{\rho\sigma} \frac{P^\rho}{P^0}\frac{P^\sigma}{P^0}P^0 = 0. \tag{24.138}$$

In the following, we will need only the component of this equation describing the evolution of P^0. Using the conformal Newtonian gauge and curvature perturbations, the Christoffel symbols related to the frame $\tilde{g}_{\mu\nu}$ are given by $\Gamma^0{}_{i0} = \partial_i \Phi$, and $\Gamma^0{}_{ij} = \Psi' \delta_{ij}$. Inserting them into the geodesic equation for P^0, it follows

$$\frac{\mathrm{d}P^0}{\mathrm{d}\eta} + \left(\Phi' + \Psi'\delta_{ij}\frac{P^i}{P^0}\frac{P^j}{P^0} + 2\partial_i\Phi\frac{P^i}{P^0} \right) P^0 = 0. \tag{24.139}$$

Without perturbations, $\mathrm{d}P^\mu/\mathrm{d}\eta = 0$ and thus the vector $n_i = n^i = P^i/P^0$ is a constant tangent vector along the photon trajectory with unit norm. Next we rewrite the parenthesis as

$$\Phi' + \Psi'\delta_{ij}n^i n^j + 2\boldsymbol{n} \cdot \boldsymbol{\nabla}\Phi = -(\Phi' - \Psi') + 2(\phi' + \boldsymbol{n} \cdot \boldsymbol{\nabla}\Phi). \tag{24.140}$$

Here, we combined the terms such that the second parenthesis on the RHS is a total derivative, since

$$\frac{\mathrm{d}\Phi(\eta, \boldsymbol{x}(\eta))}{\mathrm{d}\eta} = \Phi' + \frac{\partial \boldsymbol{x}}{\partial \eta} \cdot \boldsymbol{\nabla}\Phi = \Phi' + \frac{\partial \boldsymbol{x}/\partial\lambda}{\partial\eta/\partial\lambda} \cdot \boldsymbol{\nabla}\Phi = \Phi' + \frac{\boldsymbol{P}}{P^0} \cdot \boldsymbol{\nabla}\Phi = \phi' + \boldsymbol{n} \cdot \boldsymbol{\nabla}\Phi. \tag{24.141}$$

Integrating (24.140) along the photon trajectory and using the fact that P^μ is constant at zero order, we obtain

$$\frac{P^0(\eta_0) - P^0(\eta_1)}{P^0(\eta_1)} = \int_{\eta_1}^{\eta_0} \mathrm{d}\eta(\Phi' - \Psi') - 2[\phi(\eta_0) - \phi(\eta_1)]. \tag{24.142}$$

Now we have to connect this expression to the observed photon frequency today in the frame $g_{\mu\nu}$. Let us denote the frequency in the rest-frame of the plasma of a photon emitted at the point x of the LSS with $\bar{\omega}$. In the frame connected to the $g_{\mu\nu}$ coordinates, the plasma moves with the four-velocity u^α, with the components $u^0 = (1 - \Phi)$ and $u^i = v^i$ (recall that we use conformal quantities). Thus it follows

$$u_0 = (1 + \Phi) \quad \text{and} \quad u^i = u_i = -v^i.$$

We choose in the rest-frame of the plasma at the position where the photon is emitted normal coordinates, $\bar{g}_{\mu\nu}(x) = \eta_{\mu\nu}$. Then $\bar{u}^\alpha = (1, \mathbf{0})$ and $\omega = \bar{u}_\alpha \bar{P}^\alpha = \bar{P}^0$. Since $\omega = u_\alpha P^\alpha = \bar{u}_\alpha \bar{P}^\alpha$, we can use

$$\omega = u_\alpha P^\alpha = (1 + \Phi)P^0 - \mathbf{v} \cdot \mathbf{P} = (1 + \Phi - \mathbf{n} \cdot \mathbf{v})P^0 \tag{24.143}$$

to calculate the frequency $\omega(\eta_1)$ of the photon emitted at LSS. The same formula applies to the frequency $\omega(\eta_0)$ measured by an observer at present time η_0 with velocity v^i in the conformal frame. Thus the relative frequency shift is

$$\frac{\omega(\eta_0) - \omega(\eta_1)}{\omega(\eta_1)} = \int_{\eta_1}^{\eta_0} d\eta (\Phi' - \Psi') - [\phi(\eta_0) - \phi(\eta_1) + \mathbf{n} \cdot \mathbf{v}|_{\eta_0} - \mathbf{n} \cdot \mathbf{v}|_{\eta_1}]. \tag{24.144}$$

The RHS is independent of the photon frequency and thus an initially thermal photon spectrum remains thermal. Consequently, the same formula gives also the relative temperature change $\Theta(-\mathbf{n}) = \delta T/T$ of the CMB in the direction $\mathbf{k} = -k\mathbf{n}$. The monopole contribution $\Phi(\eta_0)$ is not observable, since it can be absorbed into the average CMB temperature. Similarly, the term $\mathbf{n} \cdot \mathbf{v}|_{\eta_0}$ corresponds to the dipole term due to the movement of the observer versus the CMB which we subtracted. Finally, we should add the temperature fluctuations on the LSS which are connected to the energy-density fluctuation on the LSS via $\delta\rho_\gamma/\rho_\gamma = 4\delta T/T$. Combining everything, we obtain

$$\Theta(\mathbf{n}, \eta_0) = \underbrace{\int_{\eta_r}^{\eta_0} d\eta (\Phi' - \Psi')}_{\text{ISW}} + \underbrace{\frac{1}{4}\frac{\delta\rho_\gamma}{\rho_\gamma} + \Phi(\eta_r)}_{\text{SW}} + \underbrace{\mathbf{n} \cdot \mathbf{v}|_{\eta_r}}_{\text{Doppler}}. \tag{24.145}$$

The last term accounts for the Doppler effect induced by the movement of the plasma relative to the coordinate frame. The two terms in the middle describe the Sachs–Wolfe effect, which has two contributions: an over-dense spot emits photons with larger average energies but also has a larger (negative) gravitational potential which in turn leads to a larger redshift. Thus the two contributions partly cancel each other. Finally, the first terms called the integrated Sachs–Wolfe effect take into account the change of the gravitational potentials between LSS and today. Next we should stress the approximations we made. First, we have assumed that recombination happens instantaneously. Second, we have neglected that some of the photons may scatter on free electrons after the universe become reionised by the first stars at $z \lesssim 10$.

We omit from now on the argument η_0 and write simply $\Theta(\mathbf{n}, \eta_0) \equiv \Theta(\mathbf{n})$. Once again, it is useful to move to Fourier space. The integral in $\Theta(\mathbf{n})$ receives contributions along the path $x(\eta) = \eta\mathbf{n}$ with $\eta \in [\eta_r, \eta_0]$. Thus the Fourier transformed $\Theta(\mathbf{k})$ is a function only of $\eta\mathbf{n}$ and the magnitude k, or $\Theta(\mathbf{k}) = \Theta(k\mathbf{n}, k) \equiv \Theta(k\cos\vartheta, k)$.

Performing the Fourier transformation, we can expand thus the phase in Legendre polynomials,

$$\Theta(\boldsymbol{n}) = \frac{\delta T(-\boldsymbol{n})}{T} = \int \mathrm{d}^3 k \, \Theta(\cos\vartheta, k) \mathrm{e}^{\mathrm{i}k\eta\cos\vartheta} \tag{24.146}$$

$$= \sum_l \mathrm{i}^l (2l+1) \int \mathrm{d}^3 k \, \widetilde{\Theta}_l(k) P_l(\cos\vartheta). \tag{24.147}$$

We denoted the Legendre transformed by $\widetilde{\Theta}_l(k)$, because one usually extracts the effect of primordial fluctuations in the gravitational potential setting

$$\widetilde{\Theta}_l(k) \equiv \Phi_i(k)\Theta_l(k). \tag{24.148}$$

Inserting $\Theta(\boldsymbol{n})$ into (24.135) and the result then in (24.136), we arrive at our final formula for the angular power spectrum induced by scalar perturbations,

$$C_l = 4\pi \int \frac{\mathrm{d}k}{k} \Delta_{\Phi_i}(k) \Theta_l^2(k). \tag{24.149}$$

Hence the CMB power spectrum is determined by the product of the power spectrum $\Delta_{\Phi_i}(k)$ of initial fluctuations and the functions $\Theta_l^2(k)$. The latter contain projection effects of the plane waves onto the sphere of last scattering, encode the physics of the cosmological fluid at recombination and the evolution of the gravitational potential between inflation and today.

As simplest application, we consider the Sachs–Wolfe effect which is relevant on large angular scales. From an analysis of the fluid equations one finds that on super-horizon scales $\Theta = \Phi/3 = 3\Phi_i/10 = \mathcal{R}/5$ holds (cf. example 24.4),

$$\Theta(\boldsymbol{n}) = \int \mathrm{d}^3 k \Theta(\boldsymbol{n} \cdot \boldsymbol{n}, k) = \frac{1}{3} \int \mathrm{d}^3 k \Theta(\boldsymbol{k}) \mathrm{e}^{\mathrm{i}\boldsymbol{k}\cdot\boldsymbol{n}(\eta_0-\eta_r)}. \tag{24.150}$$

Expanding the plane wave in Legendre polynomials, we obtain as coefficients spherical Bessel functions,

$$\widetilde{\Theta}_l = \frac{1}{3}\Phi(\boldsymbol{k})j_l(k(\eta_0-\eta_r)) = \frac{3}{10}\Phi_i(\boldsymbol{k})j_l(k(\eta_0-\eta_r)), \tag{24.151}$$

or, if we extract the primordial spectrum and use $\eta_0 \gg \eta_r$,

$$\Theta_l = \frac{3}{10}j_l(k(\eta_0-\eta_r)) \simeq \frac{3}{10}j_l(k\eta_0). \tag{24.152}$$

Inserting Θ_l into Eq. (24.149) results in

$$C_l = \frac{36\pi}{100} \int \frac{\mathrm{d}k}{k} \Delta_\Phi(k) j_l^2(k\eta_0). \tag{24.153}$$

Choosing a power law $\Delta_\Phi(k) \simeq A_\Phi (k/k_0)^{n_s-1}$, we can evaluate the integral (problem 24.15) and obtain finally in the limit of a flat spectrum

$$C_l = \frac{18\pi}{100} \frac{A_\Phi}{l(l+1)}. \tag{24.154}$$

Thus a flat primordial spectrum results in a flat temperature power spectrum (at $l \lesssim 100$). Comparing this result to observations determines the normalisation of the

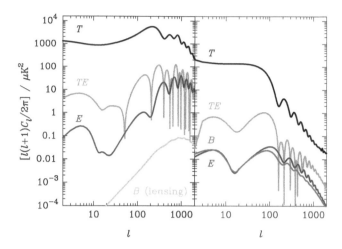

Fig. 24.5 CMB temperature and polarisation power spectra from scalar (left) and tensor perturbations (right) for a (unrealistic) tensor-to-scalar ratio $r = 0.38$, from Challinor (2006).

primordial fluctuations as $A_\Phi \simeq 2.6 \times 10^{-9}$, which in turn fixes $(V/\varepsilon)^{1/4} \sim 6 \times 10^{16}\,\mathrm{GeV}$ using (24.112).

In the left panel of Fig. 24.5, numerical results for the temperature power spectra from scalar perturbations are shown as a black line. After the plateau up to $l \sim 100$, a series of peaks with declining amplitude is visible. These peaks are caused by the coherent oscillations of the baryon–photon fluid, with gravitation as the driving and photon pressure as the restoring force. The fundamental frequency of these sound waves corresponds to the sound horizon, and thus the first peak at $l \sim 200$ indicates the horizon scale at recombination. The fact that these peaks are visible in the power spectrum requires that the oscillations are in phase on the whole LSS—this is a natural prediction of inflation. Since Fourier modes δ_k are frozen-in outside the horizon, their initial condition is $\delta_k = \mathrm{const.}$ and $\dot{\delta}_k = 0$ at horizon crossing. In other words, all modes $\delta_k \sim \cos(\boldsymbol{kx} + \boldsymbol{\phi_k})$ start with $\boldsymbol{\phi_k} = 0$ at horizon crossing. The relative size of these peaks depends on the cosmological parameters; and the precise measurement of the CMB power spectrum has led to the standard model of cosmology.

Metric perturbations and CMB polarisation. Another important prediction of inflation are tensor perturbations which lead to additional temperature fluctuations. Their derivation proceeds analogous to the scalar case, but now the non-vanishing Christoffel symbols in the conformal Newtonian gauge are given by $\Gamma^0{}_{ij} = -h'_{ij}$. Thus the geodesic equation becomes

$$\frac{\mathrm{d}P^0}{\mathrm{d}\eta} - \frac{1}{2}P^0 h'_{ij} n_i n_j = 0. \tag{24.155}$$

In linear order, we can neglect again the time-dependence of P^0 and obtain immediately

$$\frac{P^0(\eta_0) - P^0(\eta_1)}{P^0(\eta_1)} = \frac{1}{2} \int_{\eta_1}^{\eta_0} \mathrm{d}\eta\, h'_{ij} n_i n_j. \tag{24.156}$$

The perturbations of an ideal fluid contain no tensor component and thus the plasma four-velocity is undisturbed by metric perturbations, $u^\alpha = (1,0)$. Therefore it is $\omega = P^0$, and only the ISW effect contributes to temperature fluctuations induced by tensor perturbations,

$$\Theta(\boldsymbol{n}, \eta_0) = \frac{1}{2} \int_{\eta_1}^{\eta_0} \mathrm{d}\eta \, h'_{ij} n_i n_j. \tag{24.157}$$

The connection between the angular power spectrum C_l and the primordial spectrum of tensor fluctuations is derived following the same logic as in the case of scalar perturbation, and we therefore summarise the result only: the temperature fluctuations induced by tensor perturbations have also a plateau up to $\lesssim 100$, and decay then faster than those of scalar perturbations, cf. with the black line in the right panel of Fig. 24.5. Since the relative size of tensor and scalar perturbations is bounded as $r < 0.1$, it seems thus hopeless to try to disentangle the two using only temperature fluctuations. What comes to our rescue is that the CMB is polarised and that one of the two polarisation states can be generated only by gravitational perturbations.

In order to understand how the CMB became polarised, we have to abandon the approximation of instantaneous recombination. Let us model instead the LSS as a layer of thickness $2\Delta\eta_r$: then for $\eta_r - \Delta\eta_r$, photons are tightly coupled to the baryon fluid, while for $\eta_r + \Delta\eta_r$ they are free-streaming. In the intermediate region, they scatter on free electrons. The scattering is described by the non-relativistic Thomson cross-section, $\sigma \propto |\boldsymbol{\varepsilon}' \cdot \boldsymbol{\varepsilon}|^2$, cf. with Eq. (9.69). If we choose linear polarisation vectors with $\boldsymbol{\varepsilon}_\parallel$ contained in the scattering plane spanned by $\boldsymbol{k}$ and $\boldsymbol{k}'$, and $\boldsymbol{\varepsilon}_\perp$ perpendicular to the plane, then $\sigma_\parallel \propto \cos^2\vartheta$ and $\sigma_\perp \propto 1$. Thus Thomson scattering generates linearly polarised photons. However, this is not sufficient for the generation of polarised radiation. If the initial photon intensity is isotropic, the polarisation is averaged out integrating over the directions of the initial photons. In the case of the CMB, the initial photon intensity is, however, anisotropic because the intensity of black-body radiation is a function of T. Thus inhomogeneities $\delta T/T$ lead to fluctuations of the intensity which result in turn in fluctuations of the polarisation. This implies first that the fluctuations in temperatures and the degree of polarisation are correlated and second that the degree of polarisation is bounded by $\delta T/T$. Third, the CMB polarisation disappears in the limit of instantaneous recombination, $\Delta\eta_r \to 0$. More precisely, the polarisation is proportional to the ratio of the mean-free-path l of photons and the thickness $\Delta\eta_r$ of the LSS.

Next we discuss how we can describe the polarisation states of the CMB photons. The intensity I of a photon beam is determined by the square of the electric field strength vector, $I \propto \langle E^2 \rangle$, where the average $\langle \cdots \rangle$ is taken over one oscillation period. If we distinguish the two polarisation states of a photon, that is we set $E \to E_a = \{E_+, E_-\}$, then the intensity becomes a tensor, I_{ab}. This tensor[4] is a Hermitian 2×2 matrix and has thus four real components, while the polarisation of a photon beam is fully described by two real parameters. We are only interested in the polarisation

---

[4]Expanding I_{ab} on the sphere, we have to use spin-2 spherical harmonics, $^2Y_m^l(\vartheta, \phi)$. This indicates already that the (irreducible) spin-2 part of the polarisation signal is sourced by tensor perturbations of the metric.

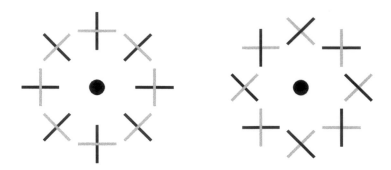

Fig. 24.6 The two E polarisation (left panel) and B polarisation (right panel) modes around a polarisation extremum shown by a circle.

and we introduce therefore instead of I_{ab} the normalised, traceless 2×2 polarisation tensor

$$\mathcal{P}_{ab} = \frac{1}{\langle E_c E^c \rangle} \left(\langle E_a E_b \rangle - \frac{1}{2} \langle E_c E_c \rangle g_{ab} \right). \tag{24.158}$$

Since we consider the polarisation on the LSS, we have to use for g_{ab} the two-dimensional metric tensor on S^2. We break the vector E_a into its rotation and divergence-free part, which we call E and B, respectively,

$$\mathcal{P}_{ab} = \left(\nabla_a \nabla_b - \frac{1}{2} g_{ab} \right) E + \left(\varepsilon_{ac} \nabla_b \nabla_c + \varepsilon_{bc} \nabla_a \nabla_c \right) B. \tag{24.159}$$

From the three quantities T, E, and B, we can form six bilinear observables: the auto-correlations TT, EE, BB and the cross-correlations TE, TB, EB. Since T and E are scalars, but B is a pseudo-scalar, the combinations TB and EB are zero (if parity is conserved). Moreover, the polarisation signal is weak and thus the cross-correlation TE is (after TT) easiest to observe. Figure 24.5 shows additional to the temperature power spectra also the cross-correlation and the E and B power spectrum.

Typical E and B modes are plotted in Fig. 24.6. The expression for $\mathcal{P}_{ab}$ shows that E polarisation transforms as a scalar, while B polarisation is a pseudo-scalar. As a result, E modes are symmetric with respect to reflections at a line through the centre, while B modes change sign. This distinction makes B modes on superhorizon scales[5] to a "smoking gun" for the presence of gravitational waves during recombination: Since scalar perturbations are characterised by scalar functions, they cannot lead to any B polarisation. On the other hand, gravitational waves consists of left- and right-circularly polarised waves, whose amplitudes $h^{\pm}$ are random variables. Thus in some directions left- and in other direction right-circular polarised waves dominate, leading in turn to a (locally) parity-breaking B polarisation.

The detection of tensor perturbations would provide important information on the inflationary models. Such a measurement would inform us that i) B modes are

---

[5]Weak gravitational lensing can transform E into B-modes, as a causal mechanism however only on subhorizon scales: this contribution is shown in the left panel of Fig. 24.5.

correlated with temperature fluctuations. No causal mechanism can generate these correlations on superhorizon scales. ii) The tensor–scalar-ratio r measures the energy scale of inflation, $V^{1/4} \sim (r/0.01)^{1/4} 10^{16}$ GeV. iii) A large value $r > 0.01$ requires large-field models and thus trans-Planckian field values. iv) The observation of tensor perturbations would be a proof that gravity is a usual quantum theory, since they are sourced by quantum fluctuations.

Successes and problems of the inflationary paradigm. Inflation has a set of successful predictions. First, the flatness of the universe, $\Omega \simeq 1$, which seemed in the 1980s and early 1990s at odds with observations. Second, inflation seeds scalar perturbations which are very close to Gaussian, nearly scale invariant with a small red tilt, and lead, at least in the simplest models, to adiabatic temperature fluctuations of the CMB. Third, inflation predicts tensor perturbations which can be detected via B modes. Last but not least, the auto- and cross-correlation of these fluctuations have fixed phase relations on superhorizon scales, which are difficult to understand in any causal mechanism. From these predictions, only the existence of tensor perturbations still awaits confirmation.

Having summarised the successes, we turn now to possible problems of the inflationary paradigm. First, we come back to the question if trans-Planckian field values of the inflaton are problematic. More precisely, we ask if, we can trust our analysis of large field models such as $V = \frac{1}{2} m^2 \phi^2$. From the point of view of general relativity, the answer is yes. If m is sufficiently small, the energy density satisfies $\rho \sim V \ll M_{\rm pl}^4$ and classical relativity holds. Moreover, the virtuality of the quantum fluctuations is close to zero and thus special relativity ensures that the large spatial momenta are not dangerous as long as their four-momentum squared is small. However, if we conceive a specific inflationary model $\mathscr{L}_{\rm infl}$ as part of a complete effective field theory $\mathscr{L}_{\rm BSM}$ beyond the SM, then

$$\mathscr{L} = \mathscr{L}_{\rm BSM} + \mathscr{L}_{\rm infl} + \sum_{n=1}^{\infty} \frac{c_n}{M_{\rm Pl}^n} \mathcal{O}_{4+n}. \tag{24.160}$$

Usually, we can neglect the operators $\mathcal{O}_{4+n}$ of dimension five and higher, which are suppressed by powers of $(E/M_{\rm Pl})^n$. However, in large field models the whole infinite tower of higher-dimensional operators should contribute during inflation with a priori equal weight. Thus there is no reason to trust the analysis of these models restricted to the operators with $d \leq 4$ contained in $\mathscr{L}_{\rm BSM} + \mathscr{L}_{\rm infl}$. If the upper limits on r continue to improve, this problem might become in the future a purely academic one.

Another potential problem becomes clear looking back at Eqs. (24.65) and (24.66). For a constant power spectrum, $\Delta_\phi^2(k) = H^2/(4\pi^2)$, the two-point function $\langle \delta\phi^2 \rangle$ is logarithmically IR- and UV-divergent. The former divergence is caused by exponentially large wavelengths. These modes are homogeneous over the horizon scale both in the present and in the inflationary epoch. Therefore this IR divergence can be absorbed into a rescaling of the classical inflaton field. Considering the UV divergence, we have to recall that for field modes $k \gtrsim aH$ the spectrum is the one of a free field in Minkowski space. Thus the divergence of $\langle \delta\phi^2 \rangle = \int {\rm d}k/k \, \Delta_\phi^2(k)$ is the same as in Minkowski space, and should be cured with our usual renormalisation procedure.

Since the modes ϕ_k of a free field are independent, the subtraction of subhorizon modes should not affect the spectra of superhorizon modes in $\Delta_\phi^2(k)$. While the use of $\Delta_\phi^2(k) = H^2/(4\pi^2)$ seems to be in accord with our standard practice, other subtraction procedures are possible and would lead to different predictions. Moreover, we should keep in mind that our predictions depend on the choice of the vacuum state. Again, our selection of the Bunch–Davies vacuum is well-motivated, but cannot be derived from first principles.

Finally, let us comment on the naturalness of the initial conditions in inflation. We have seen that inflation in the homogeneous limit happens for a large set of initial values, considering, for example, the phase portrait for the classical $V = m^2\phi^2/2$ potential in Fig 24.2. Such large fluctuations should be rare, but the probability per volume to sit inside an inflationary patch should be large precisely because these patches are exponentially inflated. In the "eternal inflation" scenario, this simple classical picture is changed since quantum fluctuations modify the classical inflation trajectories. It is unclear what measure for the initial conditions should be used and thus it is also disputed how natural the initial conditions for successful inflation are in this scenario.

24.A   Appendix: Perturbed Einstein equations

The derivation of the Einstein equations in the conformal Newtonian gauge (24.88) is straight-forward but tedious. One may either calculate directly the linearised Einstein tensor for this metric, or one may use that the $k = 0$ FLRW metric is conformally flat. Therefore one can map the linearised Einstein equations for perturbations around Minkowski space derived in chapter 19.3 on the FLRW metric using the transformation rules (23.2) between conformally related spacetimes, see for example, Gorbunov and Rubakov (2011a). Alternatively, one may use a program like `differentialGeometry.py` to calculate first the Einstein tensor $G_{\mu\nu}$ for the metric (24.88), expanding then $G_{\mu\nu}$ in Φ and Ψ. Either way, the Einstein tensor in the conformal Newtonian gauge at linear order in the perturbations Ψ and Φ follows as

$$\delta G_0^0 = \frac{2}{a^2}\left(\Delta\Psi - 3\frac{a'}{a}\Psi' - 3\frac{a'^2}{a^2}\Phi\right) \tag{24.161a}$$

$$\delta G_i^0 = \frac{2}{a^2}\left(\partial_i\Psi' + \frac{a'}{a}\partial_i\Phi\right) \tag{24.161b}$$

$$\delta G_i^j = \frac{1}{a^2}\partial_i\partial^j(\Phi - \Psi) - \frac{2}{a^2}\delta_i^j\left(\Psi'' + \frac{1}{2}\Delta(\Phi - \Psi) + \frac{a'}{a}(\Phi' + 2\Psi') + 2\frac{a''}{a}\Phi - \frac{a'^2}{a^2}\Phi\right). \tag{24.161c}$$

The spatial components δG_i^j of the Einstein tensor contain the two independent tensor structures $\partial_i\partial^j$ and δ_i^j. In the case of an ideal fluid or a scalar field, anisotropic pressure is absent, $\delta T_i^j = \delta_i^j P$. As a result, the Einstein equation implies $\partial_i\partial^j(\Phi - \Psi) = 0$ for $i \neq j$, but $\Phi - \Psi$ cannot be constant, since the perturbed Einstein tensor should vanish in the limit $\Phi, \Psi \to 0$. Moreover, isotropy and homogeneity forbid a linear dependence of the potentials on the coordinates. Thus we found that

$$\Phi = \Phi \quad \text{and} \quad A = D \tag{24.162}$$

in the absence of anisotropic pressure.

Summary

Inflation denotes a phase of nearly exponential expansion of the early universe, which solves the horizon and flatness problem of the standard big bang model. It generates a nearly scale invariant spectrum of Gaussian density fluctuations which is typically red-tilted, $n_s < 1$, and in the simplest models adiabatic. The fluctuations have fixed phase relations on superhorizon scales what results in characteristic oscillations of the CMB temperature fluctuations. The ratio r of power in gravitational waves and curvature perturbations is determined by the slow-roll parameter $\varepsilon = r/16$, which controls also the slope of the tensor perturbations $n_T = -2\varepsilon = -r/8$.

Further reading. Our derivation of the perturbation spectrum follows the one of Gorbunov and Rubakov (2011*a*). For a somewhat complementary approach see Lyth and Liddle (2009). Both references as well as Mukhanov (2005) discuss also in much more detail the connection to CMB fluctuations. A more detailed review of preheating is given by Allahverdi *et al.* (2010) and Gorbunov and Rubakov (2011*a*).

Problems

24.1 Before t_{PL}. Check how one should change the time interval $t \in [0 : t_{\mathrm{PL}}]$ such that the problem of the standard big bang theory are solved.

24.2 Minimal number of e-foldings. Find the minimal number of e-foldings required that inflation solves the horizon and flatness problem as function of the reheating temperature T_{rh}. Account for the matter-dominated phase.

24.3 Slow-roll condition and parameters. a.) Show that the slow-roll condition $\varepsilon_H \equiv -\dot{H}/H^2 \to 1$ ensures that $w \to -1$. b.) Determine the connection between $\{\varepsilon_H, \eta_H\}$ and $\{\varepsilon_V, \eta_V\}$ and show that $n_s - 1 \approx 2\eta_V - 6\varepsilon_V$. c.) Determine the slow-roll parameters of the $V(\phi) = V_0 - \lambda\phi^4/4$ model.

24.4 Contraction. Show that the horizon and flatness problems are also solved, if the universe went through a contracting phase before "bouncing".

24.5 Pressure of a scalar gas II. The pressure is often defined by $P = T_i^i/3$. Show that then the gradient term $(\nabla\phi)^2$ in Eq. (24.17) has the coefficient $-1/6a^2$.

24.6 Reheating 1. a.) Derive $\langle T \rangle = -n/2\langle V \rangle$ by averaging the KG equation over several oscillations, when the Hubble friction can be neglected. b.) Show that inflaton energy density decreases as $\rho \propto 1/a^3$ as long as $\Gamma_{\mathrm{tot}} \ll H$.

24.7 Reheating 2. Find the maximal decay width of the inflaton.

24.8 Preheating. a.) Find the transmission and reflection coefficient for the Schrödinger equation (24.49). Determine the Bogolyubov coefficients and find thereby the change of the particle number. b.) Solve the problem in the WKB approximation.

24.9 Dependence of Einstein equations. Show that the space–space part of the Einstein equation with a scalar field contains no additional information at linear order.

24.10 Helicity decomposition. Show that perturbations of different helicity h develop independently in the linear approximation.

24.11 Expansion in the slow-roll parameter. Use $z^2 = -2a^2 \widetilde{M}_{Pl}^2 \varepsilon$ for a systematic expansion of $\Delta_{\mathcal{R}}^2(k)$ in powers of the slow-roll parameter.

24.12 Development of matter perturbations. Consider the following toy model. Starting from a homogeneous universe, add matter inside a sphere of radius R, $\bar{\rho} \to \bar{\rho}(1+\delta)$. Show that the radius evolves as $R(t) \propto a(t)[1+\delta]^{-1/3}$ and that for $\delta \ll 1$ the relation $\ddot{\delta} + 2H\dot{\delta} - 4\pi G\rho\delta = 0$ holds. Find the solution $\delta(t)$ for a radiation- and matter-dominated universe.

24.13 Conservation of $\mathcal{R}$. Show that $\mathcal{R}$ is constant for superhorizon modes also after inflation, if the fluctuations are adiabatic. (Hint: differentiate (24.129).) Derive the limiting case Eq. (24.130).

24.14 Jeans length and the sound speed. Consider perturbations of the non-relativistic fluid equation. Find the Jeans length which is defined as the maximal size of a stable perturbation, and the sound speed as function of the EoS.

24.15 Normalisation of $\Delta_{\Phi}(k)$. Choose a power-law in (24.153) and evaluate the integral.

25
Black holes

In 1784 John Michell speculated already that the gravitational attraction of a mass concentrated inside a sufficiently small radius can become so strong that not even light escapes. The advent of general relativity put Michell's premise that energy is subject to gravity on a firm footing, nevertheless the idea of black holes was accepted very slowly. In modern language, we call a black hole a solution of Einstein's equations containing a physical singularity which in turn is covered by an event horizon. Such a horizon acts as a perfect unidirectional membrane which any causal influence can cross only towards the singularity. After this view became accepted in the 1960s, the discovery by Hawking that quantum effects lead to the emission of thermal radiation by a black holes came as big surprise. Before we examine this process of Hawking radiation we discuss the essential features of stationary and rotating black holes.

25.1 Schwarzschild black holes

Definitions. Let us start by introducing few definitions. First, we need to distinguish between *physical* and *coordinate singularities*. The latter arise only for a specific coordinate choice and all measurable quantities remain finite at a coordinate singularity. By contrast, physical singularities cannot be eliminated by a change of coordinates and physical quantities as the curvature or the stress tensor diverge. Next, we recall our definition of an *event horizon* as a three-dimensional hypersurface which limits a region of a spacetime which can never influence an observer. The event horizon is formed by light-rays and is therefore a null surface. Hence we require that at each point of such a surface defined by $f(x^\mu) = 0$ a null tangent vector n^μ exists that is orthogonal to two space-like tangent vectors. The normal n^μ to this surface is parallel to the gradient along the surface, $n^\mu = h\nabla^\mu f = h\partial^\mu f$, where h is an arbitrary non-zero function. From

$$0 = n_\mu n^\mu = g_{\mu\nu} n^\mu n^\nu \tag{25.1}$$

we see that the line element vanishes on the horizon, $\mathrm{d}s = 0$. Hence the (future) light-cones at each point of an event horizon are tangential to the horizon.

We add two additional definitions for spacetimes with special symmetries. A *stationary* spacetime has a time-like Killing vector field. In appropriate coordinates, the metric tensor is independent of the time coordinate,

$$\mathrm{d}s^2 = g_{00}(\boldsymbol{x})\mathrm{d}t^2 + 2g_{0i}(\boldsymbol{x})\mathrm{d}t\mathrm{d}x^i + g_{ij}(\boldsymbol{x})\mathrm{d}x^i\mathrm{d}x^j . \tag{25.2}$$

A stationary spacetime is *static* if it is invariant under time reversal. Thus the off-diagonal terms g_{0i} have to vanish, and the metric simplifies to

Quantum Fields–From the Hubble to the Planck Scale. Michael Kachelriess. © Michael Kachelriess 2018.
Published in 2018 by Oxford University Press. DOI 10.1093/oso/9780198802877.001.0001

$$ds^2 = g_{00}(\boldsymbol{x})dt^2 + g_{ij}(\boldsymbol{x})dx^i dx^j. \tag{25.3}$$

An example of a stationary spacetime is the metric around a spherically symmetric mass distribution which rotates with constant velocity. If the mass distribution is at rest then the spacetime becomes static.

Schwarzschild metric. The Schwarzschild solution describes the static spacetime outside a spherically symmetric mass distribution and is therefore of the form (25.3). Our discussion of symmetric spaces in section 20.1 implies that the spatial part g_{ij} of the metric tensor is given by Eq. (20.4) with $S(t) = \text{const.}$ Moreover, the metric tensor can depend only on the radial distance r to the centre of the mass distribution. Thus the complete line element is

$$ds^2 = A(r)dt^2 - B(r)dr^2 - r^2(d\vartheta^2 + \sin^2\vartheta d\phi^2), \tag{25.4}$$

where $A(r)$ and $B(r)$ are two arbitrary functions. Following the steps from Eq. (20.9a) to (20.13), but using now $T_{\mu\nu} = 0$ appropriate for the vacuum outside a spherical mass distribution, leads[1] to (problem 25.1)

$$ds^2 = \left(1 - \frac{2M}{r}\right)dt^2 - \left(1 - \frac{2M}{r}\right)^{-1}dr^2 - r^2(d\vartheta^2 + \sin^2\vartheta d\phi^2). \tag{25.5}$$

Here we also required that the metric is asymptotically flat, so that we recover Minkowski space for $M/r \to 0$. The weak-field limit $r \gg 2M$ implies that M is the total mass as measured by an observer at infinity. The specific coordinates used in (25.5) which make the static property of the spacetime manifest are called Schwarzschild coordinates. The Schwarzschild solution which is parameterised only by the mass M is the unique spherically symmetric vacuum solution of the Einstein equations. Allowing for time-dependent functions, $A(t,r)$ and $B(t,r)$ would result in the same *static* spacetime, a result known as Birkhoff's theorem.

The main properties of the Schwarzschild solution can be summarised as follows. The time-independence and spherically symmetry of the metric imply the existence of four Killing vectors. If we order coordinates as $\{t, r, \phi, \vartheta\}$, then the two Killing vectors leading to the conservation of energy and z component of the angular momentum are $\xi^\mu \equiv (\xi^t, \xi^r, \xi^\phi, \xi^\vartheta) = (1,0,0,0)$ and $\eta^\mu = (0,0,0,1)$. The Schwarzschild coordinates have two singularities at $r = 2M$ and $r = 0$. The radius $2M$ is called Schwarzschild radius R_s and numerically has the value

$$R_s = 2M = \frac{2G_N M}{c^2} \simeq 3\,\text{km}\,\frac{M}{M_\odot}, \tag{25.6}$$

where $M_\odot$ denotes the mass of the Sun. In a stationary, radial-symmetric spacetime the general equation of a surface, $f(x^\mu) = 0$ simplifies to $f(r) = 0$. Then the condition defining a horizon becomes simply $g^{rr} = 0$ or $g_{rr} = 1/g^{rr} = \infty$. Thus $r = R_s$ is an event horizon. Moreover, at R_s the coordinate t and r switch their character: for $r < R_s$, the "time coordinate" t becomes space-like, while r is time-like. In order

---

[1]We set $G_N = 1$ in this chapter, if not otherwise stated.

to decide if $r = 2M$ and $r = 0$ are coordinate or physical singularities, one can calculate the scalar invariants formed from the Riemann tensor. For instance, one finds $R_{\mu\nu\rho\sigma}R^{\mu\nu\rho\sigma} = 48M^2/r^6$, indicating that $r = 2M$ is a coordinate and $r = 0$ a physical singularity. Approaching $r = 0$, any macroscopic body would be destroyed by tidal forces. However, the question if at $r = 0$ a true singularity exists cannot be addressed within classical gravity which is expected to breakdown for curvatures larger than $R \sim M_{\text{Pl}}^2$.

Gravitational redshift. An observer with four-velocity $\boldsymbol{u}$ measures the frequency

$$\omega = \boldsymbol{k} \cdot \boldsymbol{u} \qquad (25.7)$$

of a photon with four-momentum $\boldsymbol{k}$. For an observer at rest,

$$\boldsymbol{u} \cdot \boldsymbol{u} = 1 = g_{tt}(u^t)^2, \qquad (25.8)$$

and hence we can express $\boldsymbol{u}$ through the Killing vector $\boldsymbol{\xi}$ as

$$\boldsymbol{u} = (1 - 2M/r)^{-1/2}\boldsymbol{\xi}. \qquad (25.9)$$

Inserting this expression into Eq. (25.7), we find for the frequency measured by an observer at the position r

$$\omega(r) = (1 - 2M/r)^{-1/2}\boldsymbol{\xi} \cdot \boldsymbol{k}. \qquad (25.10)$$

Since $\boldsymbol{\xi} \cdot \boldsymbol{k}$ is conserved and $\omega_\infty = \boldsymbol{\xi} \cdot \boldsymbol{k}$, we obtain

$$\omega_\infty = \omega(r)\sqrt{1 - \frac{2M}{r}}. \qquad (25.11)$$

Thus a photon climbing out of the potential wall of the mass M looses energy, in agreement with the principle of equivalence. In the same way, a signal sent towards an observer at infinity by a spaceship falling towards $r = 2M$ will be more and more redshifted, with $\omega_\infty \to 0$ for $r \to 2M$. Thus the event horizon at $r = 2M$ is also an infinite redshift surface.

Radial infall into a black hole. We investigate now the time-dependence of a trajectory describing the infall of an object into a black hole. The geodesics of the Schwarzschild metric are most easily derived combining the normalisation condition of the four-velocity with the conserved quantities defined by the Killing vectors $\boldsymbol{\xi}$ and $\boldsymbol{\eta}$.

A test particle moving in the Schwarzschild metric initially in the radial direction will continue to do so because then $\dot{u}^\phi = \dot{u}^\vartheta = 0$ (problem 25.4). The normalisation condition $\boldsymbol{u} \cdot \boldsymbol{u} = 1$ written out for a radial trajectory simplifies to

$$1 = A\left(\frac{dt}{d\tau}\right)^2 - A^{-1}\left(\frac{dr}{d\tau}\right)^2, \qquad (25.12)$$

where we set also $A \equiv 1 - 2M/r$. Now we replace the velocity u^t by the conserved quantity

$$e \equiv \boldsymbol{\xi} \cdot \boldsymbol{u} = A \frac{\mathrm{d}t}{\mathrm{d}\tau}, \tag{25.13}$$

obtaining

$$1 = -\frac{e^2}{A} + \frac{1}{A}\left(\frac{\mathrm{d}r}{\mathrm{d}\tau}\right)^2. \tag{25.14}$$

We consider the free fall of a particle that was at rest at spatial infinity. Then the proper and coordinate time coincide for $r \to \infty$, $\mathrm{d}t/\mathrm{d}\tau = 1$, and thus $e^2 = 1$. The radial equation (25.14) then simplifies to

$$\frac{1}{2}\left(\frac{\mathrm{d}r}{\mathrm{d}\tau}\right)^2 = -\frac{M}{r} \tag{25.15}$$

and can be integrated by separation of variables,

$$\int_r^0 \mathrm{d}r\, r^{1/2} = \sqrt{2M} \int_{\tau_*}^{\tau} \mathrm{d}\tau, \tag{25.16}$$

with the result

$$\frac{2}{3} r^{3/2} = \sqrt{2M}(\tau_* - \tau). \tag{25.17}$$

Hence a freely falling particle needs only a finite proper time to fall from finite r to $r = 0$. In particular, it passes the Schwarzschild radius $2M$ in finite proper time, and no singular behaviour of the trajectory at $2M$ is apparent.

We can answer the same question using the coordinate time t by combining Eq. (25.13) and (25.15),

$$\frac{\mathrm{d}t}{\mathrm{d}r} = \frac{\mathrm{d}t}{\mathrm{d}\tau}\frac{\mathrm{d}\tau}{\mathrm{d}r} = -\left(\frac{2M}{r}\right)^{-1/2}\left(1 - \frac{2M}{r}\right)^{-1}. \tag{25.18}$$

Integration gives

$$t - t_0 = \int_{r_0}^r \mathrm{d}r'\left(\frac{2M}{r'}\right)^{-1/2}\left(1 - \frac{2M}{r'}\right)^{-1} = \tag{25.19}$$

$$= -2M\left[-\frac{2}{3}\left(\frac{r'}{2M}\right)^{3/2} - 2\left(\frac{r'}{2M}\right)^{1/2} + \ln\left|\frac{\sqrt{r'/2M}+1}{\sqrt{r'/2M}-1}\right|\right]_{r_0}^r \to \infty \quad \text{for} \quad r \to 2M.$$

Since the coordinate time t is the proper time for an observer at infinity, a freely falling particle reaches the Schwarzschild radius $r = 2M$ only for $t \to \infty$ as seen from spatial infinity. The last result can be derived immediately for light-rays. Choosing a light-ray in radial direction with $\mathrm{d}\phi = \mathrm{d}\vartheta = 0$, the metric (25.5) simplifies with $\mathrm{d}s^2 = 0$ to

$$\frac{\mathrm{d}r}{\mathrm{d}t} = 1 - \frac{2M}{r}. \tag{25.20}$$

Thus light travelling towards the star, as seen from the outside, will travel at a slower and slower rate as it comes closer to the Schwarzschild radius $r = 2M$. The coordinate time is $\propto \ln|1 - 2M/r|$ and thus for an observer at infinity the signal will reach $r = 2M$ again only asymptotically for $t \to \infty$.

We noted already that at $R_s = 2M$ the coordinates t and r switch their character, r becoming time-like. As the proper time τ of an observer has to increase continuously, the time-like coordinate has to change continuously too. Because of $dr/dt = 1 - 2M/r < 0$, we anticipate therefore that r has to decrease continuously for any particle that crossed the horizon until it hits the singularity at $r = 0$.

Eddington–Finkelstein coordinates. We next try to find new coordinates which are regular at $r = 2M$ and valid in the whole range $0 < r < \infty$. Such a coordinate transformation has to be singular at $r = 2M$, otherwise we cannot hope to cancel the singularity present in the Schwarzschild coordinates. We can eliminate the troublesome factor $g_{rr} = (1 - \frac{2M}{r})^{-1}$ introducing a new radial coordinate r^* defined by

$$dr^* = \frac{dr}{1 - \frac{2M}{r}}. \tag{25.21}$$

Integrating (25.21) results in

$$r^*(r) = r + 2M \ln\left|\frac{r}{2M} - 1\right| + A, \tag{25.22}$$

with $A \equiv -2Ma$ as integration constant. The coordinate $r^*(r)$ is often called tortoise coordinate, because $r^*(r)$ changes only logarithmically close to the horizon. This coordinate change maps the range $r \in [2M, \infty]$ of the radial coordinate onto $r^* \in [-\infty, \infty]$. A radial null geodesics satisfies $d(t \pm r^*) = 0$, and thus in- and out-going light-rays are given by

$$\tilde{u} \equiv t - r^* = t - r - 2M \ln\left|\frac{r}{2M} - 1\right| - A, \qquad \text{outgoing rays}, \tag{25.23}$$

$$\tilde{v} \equiv t + r^* = t + r + 2M \ln\left|\frac{r}{2M} - 1\right| + A, \qquad \text{ingoing rays}. \tag{25.24}$$

For $r > 2M$, Eq. (25.20) implies that $dr/dt > 0$ so that r increases with t. Therefore (25.23) describes outgoing light-rays, while (25.24) corresponds to ingoing light-rays for $r > 2M$.

We can extend now the Schwarzschild metric using as coordinate the "advanced time parameter $\tilde{v}$" instead of t. Forming the differential,

$$d\tilde{v} = dt + dr + \left(\frac{r}{2M} - 1\right)^{-1} dr = dt + \left(1 - \frac{2M}{r}\right)^{-1} dr, \tag{25.25}$$

we can eliminate dt from the Schwarzschild metric and find

$$ds^2 = \left(1 - \frac{2M}{r}\right) d\tilde{v}^2 - 2d\tilde{v}dr - r^2 d\Omega. \tag{25.26}$$

This metric was found first by Eddington and was later rediscovered by Finkelstein. Although $g_{\tilde{v}\tilde{v}}$ vanishes at $r = 2M$, the determinant $g = r^4 \sin^2 \vartheta$ is non-zero at the

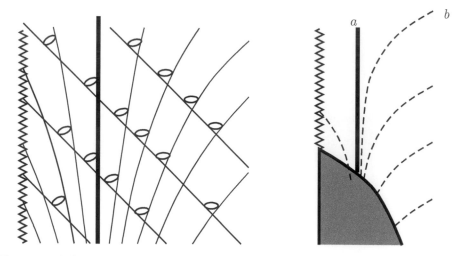

Fig. 25.1 Left: The Schwarzschild spacetime using advanced Eddington–Finkelstein coordinates; the singularity is shown by a zigzag line, the horizon by a thick line and geodesics by thin lines. Right: Collapse of a star modelled by pressureless matter; dashes lines show geodesics, the thin solid line encompasses the collapsing stellar surface.

horizon and thus the metric is invertible. Moreover, r^* was defined by (25.22) initially only for $r > 2M$, but we can use this definition also for $r < 2M$, arriving at the same expression (25.26). Therefore, the metric using the advanced time parameter $\tilde{v}$ is regular at $2M$ and valid for all $r > 0$. We can view this metric hence as an extension of the $r > 2M$ part of the Schwarzschild solution, similar to the process of analytic continuation of complex functions. The price we have to pay for a non-zero determinant at $r = 2M$ are non-diagonal terms in the metric. As a result, the spacetime described by (25.26) is not symmetric under the exchange $t \to -t$. We will see shortly the consequences of this asymmetry.

We now study the behaviour of radial light-rays, which are determined by $\mathrm{d}s^2 = 0$ and $\mathrm{d}\phi = \mathrm{d}\vartheta = 0$. Thus radial light-rays satisfy $A\mathrm{d}\tilde{v}^2 - 2\mathrm{d}\tilde{v}\mathrm{d}r = 0$, which is trivially solved by ingoing light-rays, $\mathrm{d}\tilde{v} = 0$ and thus $\tilde{v} = $ const. The solutions for $\mathrm{d}\tilde{v} \neq 0$ are given by (25.23). Additionally, the horizon $r = 2M$ which is formed by stationary light-rays satisfies $\mathrm{d}s^2 = 0$. In order to draw a spacetime diagram, it is more convenient to replace the light-like coordinate $\tilde{v}$ by a new time-like coordinate. We show in the left panel of Fig. 25.1 geodesics using as new time coordinate $\tilde{t} = \tilde{v} - r$. Then the ingoing light-rays are straight lines at 45° to the r axis. Radial light-rays which are outgoing for $r > 2M$ and ingoing for $r < 2M$ follow Eq. (25.24). A few future light-cones are indicated: they are formed by the intersection of light-rays, and they tilt towards $r = 0$ as they approach the horizon. At $r = 2M$, one light-ray forming the light-cone becomes stationary and part of the horizon, while the remaining part of the cone lies completely inside the horizon.

Let us now discuss how Fig. 25.1 would like using the retarded Eddington–Finkelstein coordinate $\tilde{u}$. Now the outgoing radial null geodesics are straight lines at 45°. They start from the singularity, crossing smoothly $r = 2M$ and continue to spatial infinity. Such a situation, where the singularity is not covered by an event horizon is called a "white hole". The cosmic censorship hypothesis postulates that singularities formed in gravitational collapse are always covered by event horizons. This implies that the time-invariance of the Einstein equations is broken by its solutions. In particular, only the BH solution using the retarded Eddington–Finkelstein coordinates should be realised by nature—otherwise we should expect causality to be violated. This behaviour may be compared to classical electrodynamics, where all solutions are described by the retarded Green function, while the advanced Green function seems to have no relevance.

Collapse to a BH. After a star has consumed its nuclear fuel, gravity can be balanced only by the Fermi degeneracy pressure of its constituents. Increasing the total mass of the star remnant, the stellar EoS is driven towards the relativistic regime until the star becomes unstable. As a result, the collapse of its core to a BH seems to be inevitable for a sufficiently heavy star.

Let us consider a toy model for such a gravitational collapse. We describe the star by a spherically symmetric cloud of pressureless matter. While the assumption of negligible pressure is unrealistic, it implies that particles at the surface of the star follow radial geodesics in the Schwarzschild spacetime. Thus we do not have to bother about the interior solution of the star, where $T_{\mu\nu} \neq 0$ and our vacuum solution does not apply. In advanced Eddington–Finkelstein coordinates, the collapse is schematically shown in the right panel of Fig. 25.1. At the end of the collapse, a stationary Schwarzschild BH has formed. Note that in our toy model the event horizon forms before the singularity, as required by the cosmic censorship hypothesis. The horizon grows from $r = 0$ following the light-like geodesic a shown by the thin black line until it reaches its final size $R_s = 2M$. What happens if we drop a lump of matter δM on a radial geodesics into the BH? Since we do not add angular momentum to the BH, the final stage is, according to the Birkhoff's theorem, still a Schwarzschild BH. All deviations from spherical symmetry corresponding to gradient energy in the intermediate regime are being radiated away as gravitational waves. Thus in the final stage, the only change is an increase of the horizon, size $R_s \to 2(M + \delta M)$. Therefore some light-rays (e.g. b) which we expected to escape to spatial infinity will be trapped. Similarly, light-ray a, which we thought to form the horizon, will be deflected by the increased gravitational attraction towards the singularity. In essence, knowing only the spacetime up to a fixed time t, we are not able to decide which light-rays form the horizon. The event horizon of a black hole is a global property of the spacetime: It is not only independent of the observer but also influenced by the complete spacetime.

How does the stellar collapse looks like for an observer at large distances? Let us assume that the observer uses a neutrino detector and is able to measure the neutrino luminosity $L_\nu(r) = dE_\nu/dt = N_\nu\omega_\nu/dt$ emitted by a shell of stellar material at radius r. In order to determine the luminosity $L_\nu(r)$, we have to connect r and t. Linearising Eq. (25.19) around $r = 2M$ gives

$$\frac{r - 2M}{r_0 - 2M} = e^{-(t-t_0)/2M}. \qquad (25.27)$$

For an observer at large distance r_0, the time difference between two pulses sent by a shell falling into a BH increases thus exponentially for $r \to 2M$. As a result the energy ω_ν of an individual neutrino is also exponentially redshifted

$$\omega_\nu(r) = \omega_\nu(r_0)e^{-(t-t_0)/2M}. \tag{25.28}$$

A more detailed analysis confirms the expectation that then also the luminosity decreases exponentially. Thus an observer at infinity will not see shells which slow down logarithmically as they fall towards $r \to 2M$, as suggested by Eq. (25.19). Instead the signal emitted by the shell will fade away exponentially, with the short characteristic time scale of $M = Mt_{\mathrm{Pl}}/M_{\mathrm{Pl}} \approx 10^{-5}\,\mathrm{s}$ for a stellar-size BH.

Kruskal coordinates. We have been able to extend the Schwarzschild solution into two different branches; a BH solution using the advanced time parameter $\tilde{v}$ and a white hole solution using the retarded time parameter $\tilde{u}$. The analogy with the analytic continuation of complex functions leads naturally to the question of whether we can combine these two branches into one common solution. Moreover, our experience with the Rindler metric suggests that an event horizon where energies are exponentially redshifted implies the emission of a thermal spectrum. If true, our BH would not be black after all. One way to test this suggestion is to relate the vacua as defined by different observers via a Bogolyubov transformation. In order to simplify this process, we would like to find new coordinates for which the Schwarzschild spacetime is conformally flat.

An obvious attempt to proceed is to use both the advanced and the retarded time parameters. For most of our discussion, it is sufficient to concentrate on the t, r coordinates in the line element $\mathrm{d}s^2 = \mathrm{d}\bar{s}^2 + r^2\mathrm{d}\Omega$, and to neglect the angular dependence from the $r^2\mathrm{d}\Omega$ part. We start by eliminating r in favour of r^*,

$$\mathrm{d}\bar{s}^2 = \left(1 - \frac{2M}{r(r^*)}\right)(\mathrm{d}t^2 - \mathrm{d}r^{*2}), \tag{25.29}$$

where r has to be expressed through r^*. This metric is conformally flat but the definition of $r(r^*)$ on the horizon contains the ill-defined factor $\ln(2m/r - 1)$. Clearly, a new set of coordinates where this factor is exponentiated is what we are seeking.

This is achieved introducing both Eddington–Finkelstein parameters,

$$\tilde{u} = t - r^*, \qquad \tilde{v} = t + r^*, \tag{25.30}$$

for which the metric simplifies to

$$\mathrm{d}\bar{s}^2 = \left(1 - \frac{2M}{r(\tilde{u}, \tilde{v})}\right)\mathrm{d}\tilde{u}\mathrm{d}\tilde{v}. \tag{25.31}$$

From (25.22) and (25.30), it follows

$$\frac{\tilde{v} - \tilde{u}}{2} = r^*(r) = r + 2M\ln\left|\frac{r}{2M} - 1\right| - 2Ma, \tag{25.32}$$

or

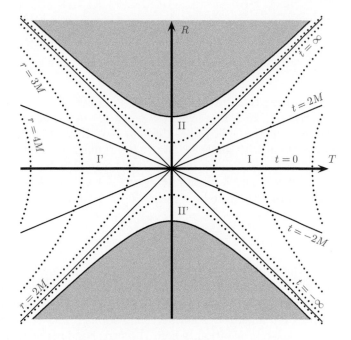

Fig. 25.2 spacetime diagram for the Kruskal coordinates T and R.

$$1 - \frac{2M}{r} = \frac{2M}{r} \exp\left(\frac{\tilde{v} - \tilde{u}}{4M}\right) \exp\left(a - \frac{r}{2M}\right). \tag{25.33}$$

This allows us to eliminate the singular factor $1 - 2M/r$ in (25.31), obtaining

$$\mathrm{d}\bar{s}^2 = \frac{2M}{r} \exp\left(a - \frac{r}{2M}\right) \exp\left(-\frac{\tilde{u}}{4M}\right) \mathrm{d}\tilde{u} \exp\left(\frac{\tilde{v}}{4M}\right) \mathrm{d}\tilde{v}. \tag{25.34}$$

Finally, we change to Kruskal light-cone coordinates u and v defined by

$$u = -4M \exp\left(-\frac{\tilde{u}}{4M}\right) \quad \text{and} \quad v = 4M \exp\left(\frac{\tilde{v}}{4M}\right), \tag{25.35}$$

arriving at

$$\mathrm{d}s^2 = \frac{2M}{r} \exp\left(a - \frac{r}{2M}\right) \mathrm{d}u\mathrm{d}v + r^2 \mathrm{d}\Omega. \tag{25.36}$$

Kruskal diagram. The coordinates $\tilde{u}, \tilde{v}$ cover only the exterior $r > 2M$ of the Schwarzschild spacetime, and thus u, v are initially only defined for $r > 2M$. Since they are regular at the Schwarzschild radius, we can extend these coordinates towards $r = 0$. In order to draw the spacetime diagram of the full Schwarzschild spacetime shown in Fig. 25.2, it is useful to go back to time- and space-like coordinates via

$$u = T - R \quad \text{and} \quad v = T + R. \tag{25.37}$$

Then the connection between the pair of coordinates $\{T, R\}$, $\{u, v\}$ and $\{t, r\}$ is given by

$$uv = T^2 - R^2 = -16M^2 \exp\left(\frac{r^*}{2M}\right) = -16M^2 \left(\frac{r}{2M} - 1\right) \exp\left(\frac{r}{2M} - a\right), \quad (25.38a)$$

$$\frac{u}{v} = \frac{T - R}{T + R} = \exp\left[-t/(2M)\right]. \quad (25.38b)$$

Lines with $r = $ const. are given by $uv = T^2 - R^2 = $ const. They are thus parabola shown as dotted lines in Fig. 25.2. Lines with $t = $ const. are determined by $u/v = $ const. and are thus given by straight (solid) lines through zero. In particular, null geodesics correspond to straight lines with angle $45°$ in the $R - T$ diagram. The horizon $r = 2M$ is given by to $u = 0$ or $v = 0$. Hence two separate horizons exist: a past horizon at $t = -\infty$ (for $v = 0$) and a future horizon at $t = +\infty$ (for $u = 0$). Also, the singularity at $r = 0$ corresponds to two separate lines in the $R - T$ Kruskal diagram[2] and is given by

$$T = \pm\sqrt{16M^2 + R^2}. \quad (25.39)$$

The horizon lines $\{t = -\infty, r = 2M\}$ and $\{t = \infty, r = 2M\}$ divide the spacetime in four parts. The future singularity is unavoidable in part II, while in region II' all trajectories start at the past singularity. Region I corresponds to the original Schwarzschild solution outside the horizon $r > 2M$, while region I and II encompass the advanced Eddington–Finkelstein solution. The regions I' and II' represent the retarded Eddington–Finkelstein solution, where II' corresponds to a white hole. Note that I' represents a new asymptotically flat Schwarzschild exterior solution.

The presence of a past horizon $v = 0$ at $t = -\infty$ makes the complete BH solutions time-symmetric and corresponds to an eternal BH. If we model a realistic BH, that is, one that was created at finite t by a collapsing mass distribution, with Kruskal coordinates, then any effect induced by the past horizon should be considered as unphysical.

25.2 Kerr black holes

The stationary spacetime outside a rotating mass distribution can be derived by symmetry arguments similarly to the case of the Schwarzschild metric, but it was found first accidentally by R. Kerr in 1963. The black hole solution of this spacetime is fully characterised by two quantities, the mass M and the angular momentum L of the Kerr BH. Both parameters can be manipulated, at least in a gedankenexperiment, dropping material into the BH. Examining the response of a Kerr black hole to such changes was crucial for the discovery of "black hole thermodynamics".

In Boyer–Lindquist coordinates, the metric outside of a rotating mass distribution is given by

$$
\begin{aligned}
ds^2 = \left(1 - \frac{2Mr}{\rho^2}\right) dt^2 &+ \frac{4Mar\sin^2\vartheta}{\rho^2}d\phi dt - \frac{\rho^2}{\Delta}dr^2 - \rho^2 d\vartheta^2 \\
&- \left(r^2 + a^2 + \frac{2Mra^2\sin^2\vartheta}{\rho^2}\right)\sin^2\vartheta d\phi^2,
\end{aligned}
\quad (25.40)
$$

---

[2]Recall that we suppress two space dimension: Thus a point in the $R - T$ Kruskal diagram correspond to a sphere S^2, and a line to $\mathbb{R} \times S^2$.

with the abbreviations

$$a = L/M, \qquad \rho^2 = r^2 + a^2 \cos^2 \vartheta, \qquad \Delta = r^2 - 2Mr + a^2. \qquad (25.41)$$

The metric is time-independent and axially symmetric. Hence two obvious Killing vectors are, as in the Schwarzschild case, $\boldsymbol{\xi} = (1, 0, 0, 0)$ and $\boldsymbol{\eta} = (0, 0, 0, 1)$, where we again order coordinates as $\{t, r, \vartheta, \phi\}$. The presence of the mixed term $g_{t\phi}$ means that the metric is stationary, but not static—as one expects for a star or BH rotating with constant rotation velocity. Finally, the metric is asymptotically flat and the weak-field limit shows that L is the angular momentum of the rotating black hole.

Singularity. First we examine the potential singularities at $\rho = 0$ and $\Delta = 0$. The calculation of the scalar invariants formed from the Riemann tensor shows that only $\rho = 0$ is a physical singularity, while $\Delta = 0$ corresponds to a coordinate singularity. The physical singularity at $\rho^2 = 0 = r^2 + a^2 \cos \vartheta^2$ corresponds to $r = 0$ and $\vartheta = \pi/2$. Thus the value $r = 0$ is surprisingly not compatible with all ϑ values. To understand this point, we consider the $M \to 0$ limit of the Kerr metric (25.40) keeping $a = L/M$ fixed,

$$ds^2 = dt^2 - \frac{\rho}{r^2 + a^2} dr^2 - r^2 d\vartheta^2 - (r^2 + a^2) \sin^2 \vartheta d\phi^2. \qquad (25.42)$$

The comparison with the Minkowski metric shows that

$$
\begin{aligned}
x &= \sqrt{r^2 + a^2} \sin \vartheta \cos \phi, \qquad z = r \cos \vartheta, \\
y &= \sqrt{r^2 + a^2} \sin \vartheta \sin \phi,
\end{aligned}
\qquad (25.43)
$$

Hence the singularity at $r = 0$ and $\vartheta = \pi/2$ corresponds to a ring of radius a in the equatorial plane $z = 0$ of the Kerr black hole.

Horizons. We have defined an event horizon as a three-dimensional hypersurface, $f(x^\mu) = 0$, that is null. In a stationary, axisymmetric spacetime the general equation of a surface, $f(x^\mu) = 0$, simplifies to $f(r, \vartheta) = 0$. The condition for a null surface becomes

$$0 = g^{\mu\nu} (\partial_\mu f)(\partial_\nu f) = g^{rr} (\partial_r f)^2 + g^{\vartheta\vartheta} (\partial_\vartheta f)^2. \qquad (25.44)$$

In the case of the surface defined by the coordinate singularity $\Delta = r^2 - 2Mr + a^2 = 0$ that depends only on r,

$$r_\pm = M \pm \sqrt{M^2 - a^2}, \qquad (25.45)$$

the condition defining a horizons becomes simply $g^{rr} = 0$ or $g_{rr} = 1/g^{rr} = \infty$. Hence, r_- and r_+ define an inner and outer horizon around a Kerr black hole.

The surface A of the outer horizon follows from inserting r_+ together with $dr = dt = 0$ into the metric,

$$ds^2 = \rho_+^2 d\vartheta^2 + \left(r_+^2 + a^2 + \frac{2Mr_+ a^2 \sin^2 \vartheta}{\rho_+^2} \right) \sin^2 \vartheta d\phi^2, \qquad (25.46)$$

Using $r_\pm^2 + a^2 = 2Mr_\pm$, we obtain

$$ds^2 = \rho_+^2 d\vartheta^2 + \left(\frac{2Mr_+}{\rho_+}\right)^2 \sin^2 \vartheta d\phi^2. \tag{25.47}$$

Hence the metric determinant g_2 restricted to the angular variables is given by $\sqrt{g_2} = g_{\vartheta\vartheta}g_{\phi\phi} = 2Mr_+ \sin\vartheta$ and integration gives the area A of the horizon as

$$A = \int_0^{2\pi} d\phi \int_0^\pi d\vartheta \sqrt{g_2} = 8\pi M r_+ = 8\pi M (M + \sqrt{M^2 - a^2}). \tag{25.48}$$

Note that the area depends on the angular momentum of the black hole that can in turn be manipulated by dropping material into the hole. The horizon area A for fixed mass M becomes maximal for a non-rotating black hole, $A = 16\pi M^2$, and decreases to $A = 8\pi M^2$ for a maximally rotating one with $a = M$. For $a > M$, the metric component $g^{rr} = \Delta$ has no real zero and thus no event horizon exists.

Ergosphere and dragging of inertial frames. The Kerr metric is a special case of a metric with $g_{t\phi} \neq 0$. As result, both massive and massless particles with zero angular momentum falling into a Kerr black hole will acquire a non-zero angular rotation velocity $\omega = d\phi/dt$ as seen by an observer from infinity.

We consider a light-ray with $d\vartheta = dr = 0$. Then the line element becomes

$$g_{tt}dt^2 + 2g_{t\phi}dtd\phi + g_{\phi\phi}d\phi^2 = 0. \tag{25.49}$$

Dividing by $g_{\phi\phi}dt^2$ we find as two possible solutions for the angular rotation velocity

$$\dot\phi_\pm = -\frac{g_{t\phi}}{g_{\phi\phi}} \pm \sqrt{\left(\frac{g_{t\phi}}{g_{\phi\phi}}\right)^2 - \frac{g_{tt}}{g_{\phi\phi}}}. \tag{25.50}$$

There are two interesting special cases of this equation. First, on the surface $g_{tt} = 0$, the two possible solutions of $\omega = d\phi/dt$ for light-rays satisfy

$$\dot\phi_+ = -2\frac{g_{t\phi}}{g_{\phi\phi}} \qquad \text{and} \qquad \dot\phi_- = 0. \tag{25.51}$$

Hence, the rotating black hole drags spacetime at $g_{tt} = 0$ so strongly that even a photon can only co-rotate. Similarly, this condition specifies a surface inside which no stationary observers are possible. The normalisation condition $\boldsymbol{u} \cdot \boldsymbol{u} = 1$ is inconsistent with $u^a = (1,0,0,0)$ and $g_{tt} < 0$: however strong your rocket engines are, your spaceship will not be able to hover at the same point inside the region with $g_{tt} < 0$. Therefore one calls a surface with $g_{tt} = 0$ a *stationary limit surface*. Solving

$$g_{tt} = 1 - \frac{2Mr}{\rho^2} = 0, \tag{25.52}$$

we find the position of the two stationary limit surfaces at

$$r_{1/2} = M \pm \sqrt{M^2 - a\cos\vartheta}. \tag{25.53}$$

The *ergosphere* is the space bounded by these two surfaces.

The other interesting special case of (25.50) occurs when the allowed range of values, $\phi_- \leq \omega \leq \phi_+$, shrinks to a single value, i.e. when

$$\omega^2 = \frac{g_{tt}}{g_{\phi\phi}} = \left(\frac{g_{t\phi}}{g_{\phi\phi}}\right)^2. \tag{25.54}$$

This happens at the outer horizon r_+ and defines the rotation velocity ω_H of the black hole. In the case of a Kerr black hole, we find

$$\omega_H = \frac{a}{2Mr_+}. \tag{25.55}$$

Thus the rotation velocity of the black hole corresponds to the rotation velocity of the light-rays forming its horizon, as seen by an observer at spatial infinity.

Penrose process and the area theorem. The total energy of a Kerr BH consists of its rest energy and its rotational energy. These two quantities control the size of the event horizon and therefore it is important to understand how they change dropping matter into the BH.

The energy of any particle moving on a geodesics is conserved, $E = -\boldsymbol{p}\cdot\boldsymbol{\xi}$. Inside the ergosphere, the Killing vector $\boldsymbol{\xi}$ is space-like and the quantity E is thus the component of a spatial momentum which can have both signs. This led Penrose to entertain the following gedankenexperiment: Suppose the spacecraft A starts at infinity and falls into the ergosphere. There it splits into two parts: B is dropped into the BH, while C escapes to infinity. In the splitting process, four-momentum has to be conserved, $\boldsymbol{p}_A = \boldsymbol{p}_B + \boldsymbol{p}_C$. We can now choose a time-like geodesics for B falling into the BH such that $E_B < 0$. Then $E_C > E_A$ and the escaping part C of the spacecraft has at infinity a higher energy than initially.

The Penrose process decreases both the mass and the angular momentum of the BH by an amount equal to that of the space craft B falling into the BH. Now we want to show that the changes are correlated in such a way that the area of the BH increases. Let us first define a new Killing vector,

$$\boldsymbol{K} = \boldsymbol{\xi} + \omega_H \boldsymbol{\eta}.$$

This Killing vector is null on the horizon and time-like outside. It corresponds to the four-velocity with the maximal possible rotation velocity. Now we use $E_B = -\boldsymbol{p}_C \cdot \boldsymbol{\xi}$ and $L_B = -\boldsymbol{p}_C \cdot \boldsymbol{\eta}$ and

$$\boldsymbol{p}_B \cdot \boldsymbol{K} = \boldsymbol{p}_B \cdot (\boldsymbol{\xi} + \omega_H \boldsymbol{\eta}) = -(E_B - \omega_H L_B) < 0, \tag{25.56}$$

to obtain the bound $L_B < E_B/\omega_H$. Since $E_B < 0$, the added angular momentum is negative, $L_B < 0$.

The mass and the angular momentum of the BH change by $\delta M = E_B$ and $\delta L = L_B$, when particle B drops into the BH. Thus

$$\delta M > \omega_H \delta L = \frac{a\delta L}{r_+^2 + a^2}. \tag{25.57}$$

Now we define the irreducible mass of BH as the mass of that Schwarzschild BH whose event horizon has the same area,

$$M_{\text{irr}}^2 = \frac{1}{2}(M^2 + \sqrt{M^2 - L^2}) \tag{25.58}$$

or

$$M^2 = M_{\text{irr}}^2 + \left(\frac{L}{2M_{\text{irr}}}\right)^2. \tag{25.59}$$

Thus we can interpret the total mass as the Pythagorean sum of the irreducible mass and a contribution related to the rotational energy. Differentiating the relation (25.58) results in

$$\delta M_{\text{irr}} = \frac{a}{4M_{\text{irr}}\sqrt{M^2 - a^2}}\left(\omega_H^{-1}\delta M - \delta L\right). \tag{25.60}$$

Our bound implies now $\delta M_{\text{irr}} > 0$ or $\delta A > 0$. Thus the surface of a Kerr BH can only increase, even when its mass decreases.

25.3 Black hole thermodynamics and Hawking radiation

Bekenstein entropy. We have shown that classically the horizon of a black hole can only increase with time. The only other quantity in physics with the same property is the entropy, $\mathrm{d}S \geq 0$. This suggests a connection between the horizon area and its entropy. To derive this relation, we apply the first law of thermodynamics $\mathrm{d}U = T\mathrm{d}S - P\mathrm{d}V + \ldots$ to a Kerr black hole. Its internal energy U is given by $U = M$ and thus

$$\mathrm{d}U = \mathrm{d}M = T\mathrm{d}S - \boldsymbol{\omega}\mathrm{d}\boldsymbol{L}, \tag{25.61}$$

where $\boldsymbol{\omega}\mathrm{d}\boldsymbol{L}$ denotes the mechanical work done on a rotating macroscopic body.

Our experience with the thermodynamics of non-gravitating systems suggests that the entropy is an extensive quantity and thus proportional to the volume, $S \propto V$. We now offer an argument that shows that the entropy S of a black hole is proportional to its area A. We introduce the "rationalised area" $\alpha = A/4\pi = 2Mr_+$, cf. (25.48), or

$$\alpha = 2M^2 + 2\sqrt{M^4 - L^2}. \tag{25.62}$$

The parameters describing a Kerr black hole are its mass M and its angular momentum L and thus $\alpha = \alpha(M, L)$. We form the differential $\mathrm{d}\alpha$ and find after some algebra (problem 25.3)

$$\frac{\sqrt{M^2 - a^2}}{2\alpha}\,\mathrm{d}\alpha = \mathrm{d}M + \frac{\boldsymbol{a}}{\alpha}\mathrm{d}\boldsymbol{L}. \tag{25.63}$$

Using now Eq. (25.48) and (25.55), we can rewrite the RHS as

$$\frac{\sqrt{M^2 - a^2}}{2\alpha}\,\mathrm{d}\alpha = \mathrm{d}M + \boldsymbol{\omega}_H\mathrm{d}\boldsymbol{L}. \tag{25.64}$$

Thus the first law of black hole thermodynamics predicts the correct angular velocity ω_H of a Kerr black hole. Including the term $\Phi\mathrm{d}q$ representing the work done by adding the charge $\mathrm{d}q$ to a black hole, the area law of a charged black hole together with the first law of BH thermodynamics reproduces the correct surface potential Φ of a charged black hole.

The factor in front of $d\alpha$ is positive, as its interpretation as temperature requires. We identify

$$T dS = \frac{\sqrt{M^2 - a^2}}{2\alpha} \, d\alpha \tag{25.65}$$

and thus $S = f(A)$. The validity of the area theorem requires that f is a linear function, the proportionality coefficient between S and A can be only determined by calculating the temperature of black hole. Hawking could show 1974 that a black hole in vacuum emits black-body radiation ("Hawking radiation") with temperature

$$T = \frac{2\sqrt{M^2 - a^2}}{A} \tag{25.66}$$

and thus

$$S = \frac{kc^3}{4\hbar G} \, A = \frac{A}{4L_{\text{Pl}}^2}. \tag{25.67}$$

The entropy of a black hole is not extensive but is proportional to its surface. It is large, because its basic unit of entropy, $4L_{\text{Pl}}^2$, is so tiny. The presence of $\hbar$ in the first formula, where we have inserted the natural constants, signals that the black hole entropy is a quantum property.

The heat capacity C_V of a Schwarzschild black hole follows with $U = M = 1/(8\pi T)$ from the definition

$$C_V = \frac{\partial U}{\partial T} = -\frac{1}{8\pi T^2} < 0. \tag{25.68}$$

As it is typical for self-gravitating systems, its heat capacity is negative. Thus a black hole surrounded by a cooler medium emits radiation, heats up the environment and becomes hotter.

Black hole temperature. If the interpretation of the BH area as its entropy is correct then we should be able to treat a BH as a thermal system. In particular, a BH should emit thermal radiation with temperature T. As the entropy and the temperature of a BH are quantum properties, we should examine the evolution of a quantum field in the background of a gravitational field describing a BH solution. We consider the generating functional $Z[J]$ of a real scalar field in a Schwarzschild background

$$Z[J] = \int \mathcal{D}\phi \, e^{i(S[\phi, g^{\mu\nu}] + \langle J\phi\rangle)} \tag{25.69}$$

together with the generating functional $Z[J_{\mu\nu}]$ of the gravitational field coupled to an external stress tensor $J_{\mu\nu}$ as source,

$$Z[J^{\mu\nu}] = \int \mathcal{D}g_{\mu\nu} \, e^{i(S[g_{\mu\nu}] + \langle J^{\mu\nu} g_{\mu\nu}\rangle)}. \tag{25.70}$$

Here $S[g_{\mu\nu}]$ is the Einstein–Hilbert action of gravity coupled to source $J^{\mu\nu}$. The path integral over $S[g_{\mu\nu}]$ is even less well-behaved than those of other quantum fields. In particular, the gravitational action is not bounded from below and the classical limit for a BH contains singularities. Luckily, we can read the quantity of our interest, the BH temperature, from the path integral without the need to evaluate it.

Using Kruskal coordinates, we have eliminated the coordinate singularity at $r = 2M$ but we are left with the physical curvature singularity at $r = 0$. Recalling the connection between the Kruskal coordinates T and R and the Schwarzschild coordinates t and r given by (25.38) and setting $A = 0$ gives

$$T^2 - R^2 = \exp\left(\frac{r}{2M}\right)\left(\frac{r}{2M} - 1\right) \tag{25.71a}$$

$$\frac{T + R}{T - R} = -\exp\left(\frac{-t}{2M}\right). \tag{25.71b}$$

Thus the singularity at $r = 0$ is mapped onto the surface $R^2 - T^2 = 1$. With the change $T \to \tau = iT$, the metric becomes (up to an overall sign) Euclidean

$$-ds^2 = \frac{32M^3}{r}\exp\left(-\frac{r}{2M}\right)(d\tau^2 + dR^2) + r^2 d\Omega. \tag{25.72}$$

Now (25.71a) implies that r is real and larger than $2M$ for real τ and R. Thus we avoid the singularity at $r = 0$ by the change to a Euclidean metric.

If we set $R + i\tau = \rho e^{i\phi}$, then (25.71b) becomes

$$\frac{R + i\tau}{R - i\tau} = e^{2i\phi} = \exp\left(i\frac{t_E}{2M}\right), \tag{25.73}$$

where we also introduced Euclidean Schwarzschild time $it_E = t$. Our coordinates are single-valued functions only if $2\phi = t_E/(2M)$ is a periodic function with period $\beta = 2\pi$. Thus the Euclidean time has the period $\beta = 8\pi M$ and consequently the Euclidean path integral has to be restricted to periodic fields, $\phi(t_E, \boldsymbol{x}) = \phi(t_E + \beta, \boldsymbol{x})$ and $g_{\mu\nu}(t_E, \boldsymbol{x}) = g_{\mu\nu}(t_E + \beta, \boldsymbol{x})$. However, this condition describes the partition function of a thermal system with temperature $T = 1/\beta = 1/(8\pi M)$. In this picture, the BH is in thermal equilibrium with its environment filled by the scalar field ϕ, emitting and absorbing the same amount of radiation.

Hawking radiation in $1 + 1$ dimension.   An alternative, more direct approach to Hawking radiation is to calculate the rate of particle production measured by an observer at $r \to \infty$ using the method of Bogolyubov transformations. As usual, the calculation simplifies if we consider the conformally invariant case of a massless scalar field in $1 + 1$ dimensions. Then we can express the solutions of the action

$$S = \int d^2x \sqrt{|g|}\, \frac{1}{2} g^{\alpha\beta} \nabla_\alpha \phi \nabla_\beta \phi \tag{25.74}$$

through light-cone coordinates as

$$\phi = f(\tilde{u}) + g(\tilde{v}) = f(u) + g(v). \tag{25.75}$$

Here, f and g are arbitrary smooth functions specifying the wave form, and $(\tilde{u}, \tilde{v})$ and (u, v) are the tortoise and Kruskal light-cone coordinates, respectively.

Unruh effect	Hawking effect
acceleration a	surface gravity $\kappa = 1/(4M)$
$u = -\exp(-a\tilde{u})/a$	$u = -\exp(-\kappa\tilde{u})/\kappa$
$v = \exp(a\tilde{v})/a$	$v = \exp(\kappa\tilde{v})/\kappa$
Minkowski vacuum $|0_{\mathrm{M}}\rangle$	Kruskal vacuum $|0_{\mathrm{K}}\rangle$
Rindler vacuum $|0_{\mathrm{R}}\rangle$	Boulware vacuum $|0_{\mathrm{B}}\rangle$

Table 25.1 Comparison of variables used in the Unruh and Hawking effect.

For an observer at spatial infinity, we can use the tortoise light-cone coordinates $\tilde{u}, \tilde{v}$ and the metric (25.31)

$$ds^2 = \left(1 - \frac{2M}{r(\tilde{u}, \tilde{v})}\right) d\tilde{u}d\tilde{v}. \tag{25.76}$$

With $r^* \to r$ for $r \to \infty$, the metric becomes at large distances Minkowskian,

$$ds^2 \to d\tilde{u}d\tilde{v} = dt^2 - dr^2, \tag{25.77}$$

and thus coordinate time t and proper time τ coincide for such an observer. This observer will split the modes using the Schwarzschild coordinate time t,

$$\phi(x) = \sum_i \left[b_i e^{-i\Omega\tilde{u}} + b_i^\dagger e^{i\Omega\tilde{u}}\right] + \text{left-movers}. \tag{25.78}$$

The vacuum defined by $b_i |0_{\mathrm{B}}\rangle = 0$ is called the Boulware vacuum $|0_{\mathrm{B}}\rangle$. The tortoise coordinates cover however only the Schwarzschild spacetime with $r > 2M$. They, and as a result also the Boulware vacuum $|0_{\mathrm{B}}\rangle$, are singular on the horizon: the regularised vacuum $\langle 0_{\mathrm{B}}| T_{\mu\nu}(r = 2M) |0_{\mathrm{B}}\rangle$ energy diverges at $r = 2M$ using these coordinates. On the other hand, we know that the Kruskal coordinates u, v cover the whole Schwarzschild spacetime; in particular they are not singular on the horizon. Close to the horizon,

$$ds^2 \to dudv = dT^2 - dR^2, \tag{25.79}$$

and thus the time T should be used to split the modes into positive and negative frequency modes,

$$\phi(x) = \sum_i \left[a_i e^{-i\Omega u} + a_i^\dagger e^{i\Omega u}\right] + \text{left-movers}. \tag{25.80}$$

Now the Kruskal vacuum is defined by $a_i |0_{\mathrm{a}}\rangle = 0$ for all i. The energy density of $|0_{\mathrm{a}}\rangle$ is finite (after subtracting as usually the zero-point energies) in the whole manifold, and increasing towards the singularity. For a sufficiently large BH, we can treat the metric as a classical static background, neglecting any quantum back-reaction.

We find now the spectrum of particle measured by an observer at $r \to \infty$ connecting the two vacua by a Bogolyubov transformation. Comparing the case at hand with the Unruh effect, we find the following analogy shown in Table 25.1. With the identification $a = \kappa = 1/(4M)$, we can translate the result for the Unruh effect to the spectrum emitted by a BH,

$$\left\langle \tilde{N} \right\rangle = \langle 0_K | \, b^\dagger b \, | 0_K \rangle = \frac{1}{\exp(2\pi\Omega/\kappa) - 1} \delta(0) \tag{25.81}$$

obtaining a thermal spectrum with temperature $T = \kappa/(2\pi) = 1/(8\pi M)$. The quantity κ is the surface gravity of the BH, i.e. it corresponds to the acceleration of a test particle experienced on the horizon, cf. problem 25.6.

The state $|0_K\rangle$ contains not only outgoing right-movers, but also incoming left-movers with the same temperature. Since we are assuming a static system in thermal equilibrium, the amount of energy emitted and absorbed by the BH should be equal. This assumes that the BH is static and eternal, i.e. does not originate from the collapse of star. In the latter case, the past horizon at $v = 0$ does not exist and the spacetime approaches the flat Minkowski metric. Therefore it is more appropriate to choose a_i as annihilation operators only for the out-going right-movers, and b_i for the left-movers. Hence an observer at spatial infinity sees only thermal radiation emitted by the BH.

Comment on $1+3$ dimension.   How are our results modified in $d > 2$ dimensions? We know that the classical Newtonian potential $V(r) = M/r$ acquires an additional, effective term due to the centrifugal barrier. The same happens in the relativistic case. We can separate the Klein–Gordon equation $\Box\phi = 0$ if we insert the ansatz

$$\phi(r_*, t, \vartheta, \phi) = \sum_{l,m} f_l^m(r_*, t) Y_l^m(\vartheta, \phi). \tag{25.82}$$

The resulting equation for $f_l^m(r_*, t)$ is

$$\left[\frac{\partial^2}{\partial t^2} - \frac{\partial^2}{\partial r^2} + \left(1 - \frac{2M}{r} \right) \left(\frac{2M}{r^3} + \frac{l(l+1)}{r^2} \right) \right] f_l^m(t, r) = 0, \tag{25.83}$$

where the two terms in the round brackets define the effective potential $V_{\mathrm{eff}} = dV_l/dr$. A plot of the effective potential V_{eff} describing the centrifugal barrier is shown in Fig. 25.3. Note that even for $l = 0$ a barrier exists and therefore a particle has to tunnel through it. Since the tunnelling probability is energy-dependent, the spectrum emitted by a BH is modified by a grey-factor $\Omega_l(E)$. The condition for thermal radiation, namely an equal emission and absorption rate, is, however, satisfied also for $\Omega_l(E) < 1$.

Information paradox.   The discovery by Hawking that black holes evaporate led to the so-called information paradox. The event horizon formed during the collapse of a massive star hides all information about its initial state in a set of conserved quantum numbers, M, J and Q. If the black hole is stable, the information about the infalling material can be stored in its microphysical states. If the black hole evaporates however, then the question arises where the information goes—or if it is lost.

In order to see why a loss of information during black hole evaporation seems to contradict basic properties of quantum theory, let us recall first the familiar case of scattering in Minkowski space. The S-matrix maps initial states $|\psi_{\mathrm{in}}\rangle$ at $t = -\infty$ on final states $|\psi_{\mathrm{out}}\rangle = S\,|\psi_{\mathrm{in}}\rangle$ at $t = +\infty$. Since the S-matrix is unitary, we can recover

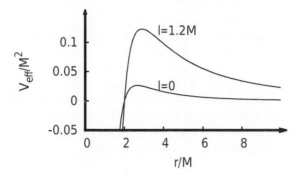

Fig. 25.3 Effective potential d V_{eff} describing the centrifugal barrier in $d = 4$ for two different values of the angular momentum l/M.

the initial state by measuring the final state, $|\psi_{\text{in}}\rangle = S^\dagger |\psi_{\text{out}}\rangle$. In other words, the unitary time evolution of quantum theory guarantees the preservation of information.

Next we consider the case that a BH is formed at finite time t. We first neglect Hawking radiation so that the BH is stable. At $t \to -\infty$, we are in Minkowski space and we can choose the initial state $|\psi_{\text{in}}\rangle$ as a pure state from the elements of the usual Minkowski Hilbert space, $|\psi_{\text{in}}\rangle \in \mathscr{H}_{\text{in}}(M)$. Some of the scattered particles will end in the BH and hit finally the singularity; others will stay outside the horizon at $t \to \infty$. We denote the Hilbert space of states ending in the singularity as $\mathscr{H}_{\text{out}}(BH)$, while the Hilbert space of states escaping is $\mathscr{H}_{\text{out}}(M)$. Now the question arises what the complete Hilbert space $\mathscr{H}_{\text{out}}(M, BH)$ of final states is. Hawking argued that all operators defined on the Hilbert spaces $\mathscr{H}_{\text{out}}(M)$ and $\mathscr{H}_{\text{out}}(BH)$ commute because the BH singularity is at space-like distances to the outside future. Therefore the complete Hilbert space of final states, $\mathscr{H}_{\text{out}}(M, BH)$, is the tensor product of the individual Hilbert spaces, $\mathscr{H}_{\text{out}}(M, BH) = \mathscr{H}_{\text{out}}(M) \otimes \mathscr{H}_{\text{out}}(BH)$. Consequently, the final states $|\psi_{\text{out}}\rangle$ are product states, $|\psi_{\text{out}}(M, BH)\rangle = |\psi_{\text{out}}(M)\rangle \otimes |\psi_{\text{out}}(BH)\rangle$. An observer outside the BH horizon can only perform measurements on $|\psi_{\text{out}}(M)\rangle$. Thus the outcome of its measurements is described by a density matrix,

$$\rho = \text{tr}_{BH}\{|\psi_{\text{out}}(M)\rangle \langle \psi_{\text{out}}(M)|\}, \tag{25.84}$$

where the trace is over a complete set of states in $\mathscr{H}_{\text{out}}(BH)$. Now we add the effect of Hawking radiation. If the black hole evaporates completely, then a pure state has been transformed into a mixed state described by the density matrix (25.84). Since the time evolution in quantum theory is unitary, this cannot happen without assuming that quantum gravity violates the basic principles of quantum theory. Various proposals regarding how this paradox can be avoided have been made, but no clear picture has emerged yet.

Summary

Black holes are solutions of Einstein's equations containing a physical singularity which—according to the cosmic censorship hypothesis—is always covered by an event horizon. This horizon is a global property of the spacetime, being independent of the observer and influenced by the complete spacetime. Within classical physics, the event horizon can only increase and has been therefore associated by Bekenstein with the entropy of a black holes. However, the event horizon is a infinite redshift surface and emits in the semi-classical picture thermal radiation. This Hawking radiation leads in turn to the information paradox.

Further reading.   The Kerr solution is derived using symmetry arguments by Ludvigsen (1999). Additional material on classical BHs can be found, for example, in Hobson *et al.* (2006). Birrell and Davies (1982) treat Hawking radiation in four dimensions. The entropy of a BH is calculated by Bailin and Love (2004). For a description of the information paradox see, for example, Mathur (2009), the more recent discussion can be traced from Almheiri *et al.* (2013) and its descendants.

Problems

25.1 Schwarzschild metric. Derive the Schwarzschild metric (25.5).

25.2 Orbits of massive particles. Derive the orbits of massive particles in the Schwarzschild metric. (Hint: Replace in the normalisation condition $\mathbf{u} \cdot \mathbf{u} = 1$ the velocities u^t and u^r by conserved quantities and rewrite this as $\frac{1}{2}\left(\frac{dr}{d\tau}\right)^2 + V_{\text{eff}} = \text{const.}$). Classify the possible orbits (conditions for stable/bound trajectories, for reaching $r = 0$?).

25.3 Differential $d\alpha$ of the area. Show that the differential of the rationalised horizon area α satisfies (25.63).

25.4 Radial infall. Show that the geodesic equation implies that a particle moving in the Schwarzschild metric initially on radial trajectory will continue to do so.

25.5 Tidal forces. Estimate the minimal mass of a BH such that the tidal forces at the Schwarzschild radius equal those on the Earth surface.

25.6 Surface gravity ♣. Derive the surface gravity of a Schwarzschild BH.

25.7 Thermodynamic stability of a BH. a.) Find the specific heat capacity of a star using the virial theorem and show that it is negative. b.) Consider a BH immersed in a (finite) heat reservoir and determine the condition for its stability.

25.8 Entropy of a stellar BH. Estimate the entropy of the iron core of a star just before a type II supernova explosion ($M = 1.5\mathcal{M}_\odot$, $T = 10^9 \, \text{K}$) and compare it to the corresponding BH entropy.

25.9 Entropy of the universe. Estimate the total entropy of the observable universe, comparing the entropy of the CMBR with $T = 2.7 \, \text{K}$ to the entropy of all supermassive black holes.

26
Cosmological constant

We have encountered three contributions to the cosmological constant. A non-zero Λ term in the Einstein–Hilbert Lagrangian would mean that an empty spacetime has classically a non-zero energy density. Another classical contribution to the cosmological constant arises naturally in theories with SSB, because the minimum of the potential is non-zero either before or after the symmetry breaking. We expect a sequence of broken symmetries during the cosmological evolution, and it looks therefore mysterious why we end up today with $V \simeq 0$. Finally, the vacuum fluctuations of quantum fields contribute to the cosmological constant and it is often only this aspect which is called the cosmological constant problem. We start this chapter reconsidering these quantum fluctuations, studying the influence of the used regularisation scheme. Then we will introduce alternative explanations for the observed accelerated expansion of the universe in the present epoch which either modify gravity or add a new component of matter, dubbed dark energy. This approach assumes that the cosmological constant problem is solved, that is, that $\rho_\Lambda + \langle \rho \rangle = 0$. One may argue that we need a theory of quantum gravity to understand how this works, and we close with some comments on this issue.

26.1 Vacuum energy density

Let us recall that we obtained for the regularised vacuum energy density $\langle \rho \rangle$ two expressions with a very different dependence on the regularisation parameter. Using in the Wilsonian approach an effective action including only modes up to the scale M, we found that $\rho \propto M^4$. In contrast, in DR power-like divergences are absent and thus ρ depends only logarithmically on the renormalisation scale μ. We will now reconsider these calculations, evaluating in addition the pressure P exerted by vacuum fluctuations. This will allow us to check which one of the two results reproduce correctly the equation of state $w = -1$ required for a Lorentz invariant vacuum.

As usually, we look at the simplest case, a real scalar field, where we only have to recollect our previous results from section 3.4 and example 5.1. The contribution of a free scalar field to the expectation value of the vacuum energy density $\langle \rho \rangle$ measured by an observer at rest is given by

$$\langle \rho \rangle = \langle 0 | T^{00} | 0 \rangle = \frac{1}{2} \langle 0 | \dot{\phi}^2 + (\boldsymbol{\nabla} \phi)^2 + m^2 \phi^2 | 0 \rangle = \int \frac{\mathrm{d}^3 k}{(2\pi)^3} \frac{1}{2} \omega_k \qquad (26.1)$$

with $\omega_k = \sqrt{m^2 + \boldsymbol{k}^2}$. The same observer measures an isotropic contribution of the scalar zero-point fluctuations to the vacuum pressure, $P^{ij} = P \delta^{ij}$, given by

Quantum Fields–From the Hubble to the Planck Scale. Michael Kachelriess. © Michael Kachelriess 2018.
Published in 2018 by Oxford University Press. DOI 10.1093/oso/9780198802877.001.0001

$$\langle P \rangle = \frac{1}{3}\langle 0|T^{ii}|0\rangle = \frac{1}{6}\langle 0|(\dot{\phi})^2 + (\boldsymbol{\nabla}\phi)^2 - m^2\phi^2|0\rangle = \frac{1}{3}\int \frac{\mathrm{d}^3k}{(2\pi)^3}\frac{\boldsymbol{k}^2}{2\omega_k}\,. \qquad (26.2)$$

The scalar Feynman propagator $\Delta_F(0)$ at coincident points can be written as the limit of two fields at nearby points,

$$\mathrm{i}\Delta_F(0) = \langle 0|\phi(x')\phi(x)|0\rangle_{x'\searrow x} = \int \frac{\mathrm{d}^3k}{(2\pi)^3 2\omega_k}\mathrm{e}^{-\mathrm{i}k(x'-x)}\Big|_{x'\searrow x} = \int \frac{\mathrm{d}^3k}{(2\pi)^3 2\omega_k}\,. \qquad (26.3)$$

Thus we can connect the Feynman propagator $\mathrm{i}\Delta_F(0)$ with the stress tensor at the one-loop level forming the trace of the stress tensor of an ideal fluid and inserting Eqs. (26.1) and (26.2) for the energy density and pressure, respectively,

$$\langle T^\mu_\mu \rangle = \langle \rho \rangle - 3\langle P \rangle = \frac{m^2}{4\pi^2}\int_0^M \mathrm{d}k\,\frac{k^2}{\sqrt{m^2+\boldsymbol{k}^2}} = m^2\,\mathrm{i}\Delta_F(0)\,. \qquad (26.4)$$

Using again the equation of state $w = \langle P \rangle/\langle \rho \rangle = -1$ valid for a contribution to the cosmological constant, we can rewrite this relation as

$$\langle \rho \rangle = -\langle P \rangle = \frac{m^2}{4}\,\mathrm{i}\Delta_F(0)\,. \qquad (26.5)$$

Thus we should impose two physical conditions on the vacuum fluctuations: First, Lorentz invariance requires that the energy density and the pressure of the vacuum satisfy the EoS $w = -1$. Second, the trace $\langle T^\mu_\mu \rangle$ of the stress tensor and the Feynman propagator $\Delta_F(0)$ are connected by (26.5) at the one-loop level (or at $\mathcal{O}(\lambda^0)$). Since at $\mathcal{O}(\lambda^0)$ the mass m is cut-off independent, this implies that an M^4 term is absent at one-loop in the vacuum energy density.

As a side remark, we recall that classically $T^\mu_\mu \to 0$ for $m \to 0$. The trace anomaly will generate an additional contribution proportional to the beta function $\beta(\lambda) = \mu\partial\lambda/\partial\mu$ of the $\lambda\phi^4$ interaction,

$$\langle T^\mu_\mu \rangle = \frac{1}{2}m^2\phi^2 + \frac{\beta}{2\lambda}\phi^4\,. \qquad (26.6)$$

The last term is only logarithmically sensitive to the UV cut-off M. This suggests that a term $\propto M^4$ is absent in general. Note also that while the trace anomaly gives a contribution to the vacuum energy, its EoS does not qualify it as dark energy.

Cut-off in spatial momenta. We start investigating a sharp cut-off in the spatial momenta, as used in many discussions of the cosmological constant problem. Integrating the energy density (26.7a) up to the maximal scale $|\boldsymbol{k}| \le M$, we obtain

$$\langle \rho \rangle = \frac{1}{4\pi^2}\int_0^M \mathrm{d}|\boldsymbol{k}|\,\boldsymbol{k}^2\sqrt{m^2+\boldsymbol{k}^2} = \frac{1}{4\pi^2}M^4\int_0^1 \mathrm{d}z\,z^2\sqrt{x^2+z^2} \qquad (26.7a)$$

$$= \frac{1}{16\pi^2}\left[M^4\sqrt{1+x^2}\left(1+\frac{1}{2}x^2\right) - \frac{1}{2}m^4\mathrm{arcsinh}\left(\frac{1}{x}\right)\right], \qquad (26.7b)$$

with $z \equiv |\boldsymbol{k}|/M$ and $x \equiv m/M \ll 1$. Similarly, we find for the pressure

$$\langle P \rangle = = \frac{1}{3} \frac{1}{4\pi^2} \int_0^M \mathrm{d}|\boldsymbol{k}| \frac{\boldsymbol{k}^4}{\sqrt{m^2 + \boldsymbol{k}^2}} = \frac{1}{3} \frac{1}{4\pi^2} M^4 \int_0^1 \mathrm{d}z \frac{z^4}{\sqrt{x^2 + z^2}} \tag{26.8a}$$

$$= \frac{1}{3} \frac{1}{16\pi^2} \left[M^4 \sqrt{1 + x^2} \left(1 + \frac{1}{2}x^2 \right) + \frac{3}{2} m^4 \mathrm{arcsinh} \left(\frac{1}{x} \right) \right] \tag{26.8b}$$

and for the Feynman propagator at coincident points

$$\mathrm{i}\Delta_F(0) = \frac{1}{4\pi^2} \int_0^M \mathrm{d}|\boldsymbol{k}| \frac{\boldsymbol{k}^2}{\sqrt{m^2 + \boldsymbol{k}^2}} = \frac{1}{4\pi^2} M^2 \int_0^1 \mathrm{d}z \frac{z^2}{\sqrt{x^2 + z^2}} \tag{26.9a}$$

$$= \frac{1}{8\pi^2} \left[M^2 \sqrt{1 + x^2} - m^2 \mathrm{arcsinh} \left(\frac{1}{x} \right) \right]. \tag{26.9b}$$

All three results show the behaviour expected from naive power-counting, $\rho \propto P \propto M^4$ and $\Delta_F(0) \propto M^2$ in the limit $M \gg m$. We can check now, whether these results fulfil the two constraints. We first neglect the mass of the scalar particle, considering the limit $x \to 0$. Then vacuum fluctuations are predicted to have the EoS of radiation, $w = \langle P \rangle/\langle \rho \rangle = 1/3 + \mathcal{O}(x^4)$ and to break scale-invariance, $\langle T^\mu_\mu \rangle \neq 0$ for $m^2 \to 0$. Moreover, the relation (26.5) between the stress tensor and the Feynman propagator $\Delta_F(0)$ is violated.

The natural explanation for this behaviour is that regularisation schemes that break symmetries lead to spurious terms reflecting this violation. In our case, a momentum cut-off breaks Lorentz invariance and thus we cannot expect that a relation like $w = -1$ for the vacuum fluctuations is correctly reproduced. Using a regularisation scheme which breaks symmetries requires therefore adding symmetry-breaking counter-terms and subtracting the offending terms by hand. This suggests that the subleading m^4 terms do satisfy the two conditions, which is indeed the case.

Dimensional regularisation. Now we replace the three-dimensional integrals in Eqs. (26.1)–(26.3) by $d - 1$ dimensional ones. Using then DR and the definition of Euler's Beta function, we obtain

$$\langle \rho \rangle = \frac{\mu^{4-d}}{(2\pi)^{(d-1)}} \frac{1}{2} \int \mathrm{d}^{d-1}k \sqrt{m^2 + \boldsymbol{k}^2} = \frac{\mu^4}{2 (4\pi)^{(d-1)/2}} \frac{\Gamma(-d/2)}{\Gamma(-1/2)} \left(\frac{m}{\mu} \right)^d. \tag{26.10}$$

Similarly, it follows

$$\langle P \rangle = \frac{\mu^{4-d}}{(2\pi)^{(d-1)}} \frac{1}{2(d-1)} \int \mathrm{d}^{d-1}k \frac{\boldsymbol{k}^2}{\omega_k} = \frac{\mu^4}{4 (4\pi)^{(d-1)/2}} \frac{\Gamma(-d/2)}{\Gamma(1/2)} \left(\frac{m}{\mu} \right)^d. \tag{26.11}$$

and

$$\mathrm{i}\Delta_F(0) = \mu^{4-d} \frac{(m^2)^{\frac{d}{2}-1}}{(4\pi)^{d/2}} \Gamma\left(1 - \frac{d}{2} \right). \tag{26.12}$$

Expanding next in $\varepsilon = 2 - d/2$ results in

$$\langle\rho\rangle = -\langle P\rangle = -\frac{m^4}{64\pi^2}\left[\frac{1}{\varepsilon} + \frac{3}{2} - \gamma + \ln\left(\frac{4\pi\mu^2}{m^2}\right)\right] + \mathcal{O}(\varepsilon). \qquad (26.13)$$

It follows that DR reproduces the correct EoS as well as the relation (26.5). The latter ensures also that $T_\mu^\mu \to 0$ for $m \to 0$, as expected at $O(\lambda^0)$. Note also that the integrands in both Eqs. (26.10) and (26.11) are positive definite. Therefore, any cut-off scheme has to fail in reproducing the correct EoS. In contrast, we know that the integration measure of DR is not positive definite and here we see an example where this property is required to reproduce the correct physics.

In order to obtain the observed value of ρ and P at a certain scale μ, we have to add a corresponding counter-term. After subtraction of the M^4 term and identifying $4\pi\mu^2$ with the cut-off scale M^2, we find in both schemes the same logarithmic dependence

$$\langle\rho\rangle = -\frac{m^4}{32\pi^2}\,\text{arcsinh}\,(M/m) \approx -\frac{m^4}{32\pi^2}\ln\,(M/m), \qquad (26.14)$$

in the limit $M^2 = 4\pi\mu^2 \gg m^2$. We thus conclude that the natural value of the vacuum energy density is $\langle\rho\rangle \sim m^4$. While this behaviour reduces the "numerical size" of the cosmological constant problem, it does not solve it, since the natural value of $\langle\rho\rangle$ in the SM would be determined by the top quark, $m_t^4 \gg \rho_\Lambda$.

26.2 Dark energy

Since the theoretically expected value of the cosmological constant is much larger than the observed one, one may hope that the cosmological constant is set to zero by some principle yet to be discovered. In this case, we require an alternative explanation for the accelerated expansion of the universe in the present epoch. Such explanations can be divided in modifications of gravity and dark energy, depending on if the LHS or RHS of the Einstein equations are changed. Since we can reshuffle matter and gravity by field redefinitions, such a distinction is somewhat ambiguous. As a practical criterion, one can use the presence of additional long-range forces and thus the violation of the strong equivalence principle as the characteristic feature of modified gravity models.

Quintessence.   These models introduce a scalar field ϕ with a canonically normalised kinetic term that evolves in the slow-roll regime. Its potential $V(\phi)$ has to be chosen such that the slow-roll regime sets in only recently, around redshift $z \simeq 0.7$. In general, quintessence models predict time-dependent deviations from $w = -1$, which can be observationally tested. While these models cannot solve the cosmological constant problem on the fundamental level, they may address the coincidence problem. The observed value of ρ_Λ implies that the potential energy of the new scalar field is tiny. For instance, the potential $V(\phi) = \lambda\phi^4$ requires as coupling $\lambda \sim 10^{-122}$, while $V(\phi) = m^2\phi^2/2$ leads to $m \sim 10^{-33}$ eV. Thus quintessence models are extremely fine-tuned. They therefore require some stabilising mechanism, if they are embedded in a more complete theory. We neglect this problem and discuss here only how these models attempt to solve the coincidence problem.

We considered already in Eqs. (24.26a) and (24.27) the inflaton ϕ coupled to the Friedmann equations as a dynamical system. Reinterpreting ϕ as a quintessence field and adding normal matter with density ρ and EoS $w_\mathrm{m} = P/\rho$, we obtain as equations of motion

$$H^2 = \frac{8\pi G}{3}\left(\frac{1}{2}\dot{\phi}^2 + V + \rho\right),\tag{26.15a}$$

$$\dot{H} = -4\pi G\left[\dot{\phi}^2 + (1 + w_\mathrm{m})\rho\right].\tag{26.15b}$$

Additionally, we have the usual Klein–Gordon equation for ϕ. Next we introduce the dimensionless variables

$$x = \frac{\dot{\phi}}{\sqrt{48\pi GH}}\quad\text{and}\quad y = \frac{\sqrt{V}}{\sqrt{24\pi GH}}.\tag{26.16}$$

The energy fraction of the quintessence field, $\Omega_\phi = \rho_\phi/\rho_\mathrm{cr}$, becomes then $\Omega_\phi = x^2 + y^2$, while its EoS is given by

$$w_\phi = \frac{x^2 - y^2}{x^2 + y^2}.\tag{26.17}$$

We obtain dimensionless equations of motion taking derivatives of x and y with respect to $N = \ln(a)$. Setting also $\lambda \equiv -(8\pi G)^{1/2}V_{,\phi}/V$, they are given by

$$\frac{\mathrm{d}x}{\mathrm{d}N} = 3x + \frac{\sqrt{6}}{2}\lambda y^2 + \frac{3}{2}x[(1 - w_\mathrm{m})x^2 + (1 + w_\mathrm{m})(1 - y^2)],\tag{26.18a}$$

$$\frac{\mathrm{d}y}{\mathrm{d}N} = -\frac{\sqrt{6}}{2}\lambda xy + \frac{3}{2}y[(1 - w_\mathrm{m})x^2 + (1 + w_\mathrm{m})(1 - y^2)].\tag{26.18b}$$

For comparisons with observations, it is useful to define the effective equation of state $w_\mathrm{eff} \equiv P_\mathrm{tot}/\rho_\mathrm{tot}$, since this is the EoS deduced from measuring H and $\dot{H}$. Using $\dot{H} = \ddot{a}/a - H^2$, we can express the Einstein equations (20.47) and (20.48) for a flat FLRW metric as

$$3H^2 = \kappa\rho\quad\text{and}\quad 3H^2 + 2\dot{H} = -\kappa P.\tag{26.19}$$

Thus the effective equation of state becomes

$$w_\mathrm{eff} = \frac{P_\mathrm{tot}}{\rho_\mathrm{tot}} = -1 - \frac{2\dot{H}}{3H^2}.\tag{26.20}$$

Applied to quintessence, Eq. (26.15), we find

$$\frac{\dot{H}}{H^2} = -3x^2 - \frac{3}{2}(1 + w_\mathrm{m})(1 - x^2 - y^2)\tag{26.21}$$

and thus

$$w_\mathrm{eff} = w_\mathrm{m} + (1 - w_\mathrm{m})x^2 - (1 + w_\mathrm{m})y^2.\tag{26.22}$$

An interesting simplification appears, if λ is constant. Then the two equations (26.18) are closed and can be analysed as a two-dimensional dynamical system. The derivation of the fixed points of this dynamical system, which can be obtained setting

$dx/dN = dy/dN = 0$, is the topic of problem 26.4. In particular, there exist trajectories connecting the saddle-point $(x, y) = (0, 0)$ with $\Omega_\phi = 0$ and $w = w_m$ to the stable fixed point $(x, y) = (\lambda/\sqrt{6}, \sqrt{1 - \lambda^2/6})$ with $\Omega_\phi = 1$ and $w_{\text{eff}} = w_\phi = -1 + \lambda^2/3$. Thus this case corresponds to a transition from a matter-dominated universe to an accelerated expansion with $w_{\text{eff}} = -1 + \lambda^2/3 > -1$. Integrating the definition of λ, we see that the case $\lambda = \text{const.}$ corresponds to the special case of exponential potentials, $V(\phi) = V_0 \exp(-\lambda\phi/\widetilde{M}_{\text{Pl}})$. For other potentials, Eqs. (26.18) have to analysed combined with the Klein–Gordon equation for the quintessence field.

Tracker solutions.   Another subclass of quintessence models uses the potentials

$$V(\phi) = \frac{M^{4+n}}{n\phi^n} \tag{26.23}$$

with $n > 2$. Since $V(\phi)$ is unbounded from below, it violates our standard stability requirement and should be therefore modified by quantum correction at small ϕ. The Klein–Gordon equation in a FLRW background with $a(t) \propto t^\alpha$ becomes for this potential

$$\ddot{\phi} + \frac{3\alpha}{t}\dot{\phi} - \frac{M^{4+n}}{\phi^{n+1}} = 0. \tag{26.24}$$

A special solution called tracker solution is provided by

$$\phi_*(t) = CM^{1+\alpha}t^\alpha \tag{26.25}$$

with $\alpha = 2/(2 + n)$, cf. problem 26.5. This solutions is an attractor, as we can check considering the driving force

$$\mathcal{F}(\phi) = -V'(\phi) = \frac{M^{4+n}}{\phi^{n+1}} \tag{26.26}$$

acting on ϕ: For $\phi(t) < \phi_*(t)$, the force is larger, $\mathcal{F}(\phi) > \mathcal{F}(\phi_*)$, and the solution ϕ makes up leeway. Similarly, a field with $\phi(t) > \phi_*(t)$ has a smaller driving force and falls back to the tracking solution. At late times, the evolution of the scalar field $\phi(t)$ is thus given by the tracking solution for a wide range of initial conditions.

We consider in the following always the tracking regime and set for simplicity $\phi \simeq \phi_*$. Using (26.25), it follows that the kinetic field energy scales as $\dot{\phi}^2 \propto t^{2\alpha-2}$ and that $\dot{\phi}^2 \simeq V(\phi)$ holds. Thus the tracker solution violates the slow-roll condition. The energy density ρ_ϕ decreases with $a \propto t^\beta$ as

$$\rho_\phi \propto \frac{1}{t^{2-2\alpha}} \propto \frac{1}{a^{(2-2\alpha)/\beta}}. \tag{26.27}$$

Using $\rho \propto a^{-3(1+w)}$, we obtain as EoS for the tracker field

$$w_\phi = -1 + \frac{2}{3}\frac{1-\alpha}{\beta} = \frac{n}{2+n}w_m - \frac{2}{2+n}. \tag{26.28}$$

Thus for large n, it tracks the EoS of the dominant form of usual matter in the universe. In particular, w_ϕ is able to follow the change of w_m during the radiation–matter transition.

Since the field does not evolve in the slow-roll regime, it seems that it cannot lead to the desired accelerated expansion. However, the relative contribution Ω_ϕ of the field to the total energy density increases, $\Omega_\phi \propto t^{2\alpha}$, and at some point Eq. (26.24) assuming an evolution in a fixed background is no longer valid. This happens when $\rho_\phi \sim (\phi/t)^2$. The corresponding matter density follows with $H \sim 1/t$ as $\rho_m \sim 3H^2\widetilde{M}_{\rm Pl}^2 \sim (\widetilde{M}_{\rm Pl}/t)^2$. Thus the contributions of ordinary matter and the tracker field are equal for $\phi \sim M_{\rm Pl}$. This coincides with the generic slow-roll condition[1] for power-law potentials: Thus the tracker potential leads to an accelerated expansion, as soon as it dominates the energy density of the universe. Since Ω_ϕ in this class of models was never extremely small, the coincidence problem is alleviated relative to the ΛCDM model. However, the question why $\phi \sim M_{\rm Pl}$ happens at the present epoch has still to be addressed in a specific dark energy model.

Additionally to reproduce correctly the evolution of $a(t)$, dark energy models have to reproduce the observed large-scale structure. Deviations to the standard ΛCDM model can become important only at late times, and thus these models can be constrained mainly using the evolution of large-scale structures at redshifts $z \lesssim 5$.

26.3 Modified gravity

An important class of modified gravity models are the so-called $f(R)$ gravity models, which generalise the Einstein–Hilbert form replacing R by a general function $f(R)$. Thus the action of $f(R)$ gravity coupled to matter has the form

$$S = \int \mathrm{d}^4x \sqrt{|g|} \left\{ \frac{1}{2\tilde{\kappa}} f(R) + \mathscr{L}_{\rm m} \right\}, \tag{26.29}$$

where $\mathscr{L}_{\rm m}$ may contain both non-relativistic matter and radiation. Note that for $f(R) \neq R$, the gravitational constant $\tilde{\kappa} = 8\pi\widetilde{G}$ deviates from Newton's constant G measured in a Cavendish experiment. The field equations can be derived from the action (26.29) either by a variation w.r.t. the metric or the connection. The dynamics and the number of the resulting degrees of freedom differ in the two treatments. Following the first approach, generalising the derivation in section 19.2 we obtain

$$F(R)R_{\mu\nu} - \frac{1}{2}f(R)g_{\mu\nu} - \nabla_\mu\nabla_\nu F(R) + g_{\mu\nu}\Box F(R) = -\kappa T_{\mu\nu} \tag{26.30}$$

with $F \equiv \mathrm{d}f/\mathrm{d}R$. Taking the trace of this expression, we find

$$F(R)R - 2f(R)g_{\mu\nu} + 3\Box F(R) = -\kappa T. \tag{26.31}$$

The term $\Box F(R)$ acts as a kinetic term so that these models contain an additional propagating scalar degree of freedom, $\phi = F(R)$. Applied to a flat FLRW metric, one obtains from the 00 and ii part of the field equation (26.30) the modified Friedmann equations as

---

[1]The model belongs to the class of large field models, with its associated problems.

$$3FH^2 = \tilde{\kappa} \left(\rho_{\mathrm{m}} + \rho_{\mathrm{rad}} \right) + \frac{1}{2}(FR - f) - 3H\dot{F}, \tag{26.32a}$$

$$-2F\dot{H} = \tilde{\kappa} \left[\rho_{\mathrm{m}} + \frac{4}{3}\rho_{\mathrm{rad}} \right] + \ddot{F} - H\dot{F}. \tag{26.32b}$$

Dividing (26.32a) by $3FH^2$, we can express the RHS through dimensionless variables x_i which correspond at the present epoch to the density parameters $\Omega_i = \tilde{\kappa}\rho_i/(3F_0 H_0^2)$,

$$x_1 = -\frac{\dot{F}}{HF}, \qquad x_2 = -\frac{f}{6FH^2}, \qquad x_3 = \frac{R}{6H^2} \quad \text{and} \quad x_4 = \frac{\tilde{\kappa}^2 \rho_{\mathrm{rad}}}{3FH^2}. \tag{26.33}$$

For a flat universe, the matter component then has to satisfy

$$\frac{\tilde{\kappa}\rho_{\mathrm{m}}}{3FH^2} = 1 - x_1 - x_2 - x_3 - x_4. \tag{26.34}$$

We are interested in the "late" universe, when the effect of radiation is negligible, and set therefore $x_4 = 0$. Following the example of quintessence, we obtain dimensionless equations of motion taking derivatives of x_i with respect to $N = \ln(a)$,

$$\frac{\mathrm{d}x_1}{\mathrm{d}N} = -1 - x_3 - 3x_2 + x_1^2 - x_1 x_3, \tag{26.35a}$$

$$\frac{\mathrm{d}x_2}{\mathrm{d}N} = \frac{x_1 x_3}{m} - x_2(2x_3 - x_1 - 4), \tag{26.35b}$$

$$\frac{\mathrm{d}x_3}{\mathrm{d}N} = -\frac{x_1 x_3}{m} - 2x_3(x_3 - 2). \tag{26.35c}$$

Here, $m \equiv \mathrm{d}\ln F/\mathrm{d}\ln R$ characterises the deviation from a ΛCDM model which has $m = 0$, while the second parameter r is defined by

$$r \equiv -\frac{\mathrm{d}\ln f}{\mathrm{d}\ln R} = \frac{x_3}{x_2}. \tag{26.36}$$

In order to simplify the comparison with observations which assume typically standard Einstein gravity, we keep the definition (26.20) of the effective equation of state w_{eff}. Then we recall from Eq. (20.46) the curvature of a flat FLRW metric,

$$R = g^{\mu\nu} R_{\mu\nu} = 6\frac{a\ddot{a} + \dot{a}^2}{a^2} = 6(\dot{H} + 2H^2). \tag{26.37}$$

Combined with the definition of x_3, we can express the effective EoS as a function of only x_3,

$$w_{\mathrm{eff}} = -\frac{2}{3}\left(x_3 - \frac{1}{2} \right). \tag{26.38}$$

The cosmological evolution can now be studied for a chosen $f(R)$ solving numerically the system (26.35). Alternatively, one can determine analytically the fixed points of this dynamical system, study their stability and the behaviour of w_{eff}. As result, one finds that models with an accelerated expansion as attractor solution either contradict observations (behaving as $a(t) \propto t^{1/2}$ before acceleration), violate stability constraints (having $w_{\text{eff}} < -1$) or are hardly distinguishable from quintessence models in standard gravity (Amendola *et al.*, 2007). In addition to reproducing the observed large-scale structure, modified gravity models have to pass solar system tests. Since they contain typically a light scalar (as $\phi = F(R)$ in $f(R)$ gravity) which mediates a 5th force, they require a screening mechanism to evade constraints of local gravity tests.

26.4 Comments on quantising gravity

Gravity as an effective theory. The Einstein–Hilbert action contains with the Planck scale $\widetilde{M}_{\text{Pl}}$ a dimensionfull coupling, and gravity is therefore a non-renormalisable theory. As any other non-renormalisable theory, we can treat Einstein gravity as an effective field theory, at least in the limit that horizons play no role. Including higher-order operators, we should be able to calculate quantum corrections to observables as, for example, the Newtonian potential following the scheme outlined in section 12.5. If we recall that $[R] = m^2$ and thus also $[R_{\mu\nu}] = [R_{\mu\nu\rho\sigma}] = m^2$, we can order possible higher-order terms according their dimension $d = 6, 8, \ldots$ as

$$S_{\text{eff}} = \kappa \int \mathrm{d}^4 x \sqrt{|g|} \left\{ (R + 2\Lambda) + L^2 \left[A_1 R^2 + A_2 R_{\mu\nu} R^{\mu\nu} + A_3 R_{\mu\nu\rho\sigma} R^{\mu\nu\rho\sigma} \right] + \mathcal{O}(L^4) \right\}.$$
$$(26.39)$$

Here, the length scale $L \sim 1/M$ indicates when the higher-order operators become important. Suppressing Lorentz indices, we can schematically write the gravitational wave equation (with $A_i \sim 1$) as

$$\Box h + L^2 \Box \Box h \sim \kappa T.$$
$$(26.40)$$

Solving for the propagator, we obtain in momentum space

$$D(k^2) \sim \frac{1}{k^2 + L^2 k^4} = \frac{1}{k^2} - \frac{1}{k^2 + M^2}.$$
$$(26.41)$$

In the static limit, the first term corresponds to the usual Newtonian $1/r$ potential, while the second one is of the Yukawa type,

$$\Phi \sim -GM \left[\frac{1}{r} - \frac{\exp(-r/L)}{r} \right].$$
$$(26.42)$$

Thus the higher-order operators lead in tree-level amplitudes to short-range interactions, which are exponentially suppressed on scales $r \gg L$. Newton's law is not tested on sub-mm scales and thus L is only bounded as $L \lesssim 1\,\mathrm{mm}$. If L would be close to this limit, as e.g. in scenarios with large extra dimensions, new gravitational effects may be detectable both in Cavendish-like experiments and at accelerators. However, one usually associates L with the Planck length L_{Pl}, and then the action becomes

$$S_{\text{eff}} = \int \mathrm{d}^4 x \sqrt{|g|} \left\{ \kappa (R - 2\Lambda) - a\kappa^2 R^2 + b\kappa^2 R_{\mu\nu} R^{\mu\nu} + \ldots \right\}.$$
$$(26.43)$$

Performing perturbation theory, we expand the metric $g_{\mu\nu}$ around the classical Minkowski background. Applied to the inverse $g^{\mu\nu}$ and $\sqrt{|g|} = \sqrt{\det(-g_{\mu\nu})}$, this will generate an infinite tower of gravitational self-interactions and couplings to matter. Suppressing again Lorentz indices, we can write schematically

$$S_{\text{eff}} = \frac{\widetilde{M}_{\text{Pl}}^2}{4} \int \mathrm{d}^4x \left[\left(\frac{1}{2} h \Box h + h \Box h^2 + \ldots \right) + \left(-a h \Box^2 h + b h \Box^2 h^2 + \ldots \right) + \right]. \quad (26.44)$$

Since we are interested in quantum effects, it is convenient to rescale the gravitational field as $\mathfrak{h}_{\mu\nu} = \frac{1}{2} \widetilde{M}_{\text{Pl}} h_{\mu\nu}$ such that it is canonically normalised and has mass dimension one,

$$S_{\text{eff}} = \int \mathrm{d}^4x \left[\left(\frac{1}{2} \mathfrak{h} \Box \mathfrak{h} + \frac{1}{\widetilde{M}_{\text{Pl}}} \mathfrak{h} \Box \mathfrak{h}^2 + \ldots \right) + \left(\frac{a}{\widetilde{M}_{\text{Pl}}^2} \mathfrak{h} \Box^2 \mathfrak{h} + \frac{b}{\widetilde{M}_{\text{Pl}}^3} \mathfrak{h} \Box^2 \mathfrak{h}^2 + \ldots \right) + \right].$$
$$(26.45)$$

The terms in the first round bracket correspond to an expansion of the Einstein–Hilbert action. Their tree-level contributions to the Newtonian potential of a static source with mass M have long-range. The leading term corresponding to single graviton exchange, cf. with Eq. (7.46), is given by

$$\text{} \quad = \Phi_2 \sim \frac{M}{\widetilde{M}_{\text{Pl}}} \frac{1}{r} \frac{1}{\widetilde{M}_{\text{Pl}}} \sim \frac{M}{\widetilde{M}_{\text{Pl}}^2} \frac{1}{r}. \quad (26.46)$$

Here, the heavy source contributes the factor $\sqrt{G}M$, while the coupling to the test particle adds only $\sqrt{G}$. We can estimate the contribution of the next, trilinear term as follows

$$\text{} \quad = \Phi_3 \sim \left(\frac{M}{\widetilde{M}_{\text{Pl}}} \right)^2 \frac{1}{\widetilde{M}_{\text{Pl}}} \frac{1}{r^2} \frac{1}{\widetilde{M}_{\text{Pl}}} \sim \frac{1}{\widetilde{M}_{\text{Pl}}^2} \left(\frac{M}{\widetilde{M}_{\text{Pl}}} \right)^2 \frac{1}{r^2}. \quad (26.47)$$

The two heavy sources add each a factor $M/\widetilde{M}_{\text{Pl}}$, while the trilinear coupling contributes the factor $1/\widetilde{M}_{\text{Pl}}$. The dependence $1/r^2$ follows then by dimensional analysis. At first sight, one might guess that the effect of the higher-order term Φ_3 is unobservable: However, classical sources of gravitational fields can be strong, $M/\widetilde{M}_{\text{Pl}} \gg 1$, and compact, $r \sim$ few $\times R_s \sim M/\widetilde{M}_{\text{Pl}}^2$. If the latter condition is satisfied, then $\Phi_2 \sim \Phi_3 \sim \Phi_n \sim 1$ and all higher-order terms in the post-Newtonian expansion become important. Ideal objects to test this expansion are therefore close binary systems of neutron stars or black holes. Using the virial theorem, we can express Φ_3 as $\Phi_3 \sim GM/r \times (v/c)^2$, which shows that Φ_3 is classical, $\Phi_3 \propto \hbar^0$.

Next we consider how loop correction to the Einstein–Hilbert action affect observables like the Newtonian potential. As we know from our general discussion in section 11.4.1, the divergences have the structure of local operators. Since the one-loop corrections come with a factor κ, the one-loop divergences of the classical Einstein–Hilbert Lagrangian are connected to κ^2 terms in the effective Lagrangian. Thus the sole effect of these terms is to renormalise the coefficients a and b. Moreover, we know that the non-analytic terms induced by the loop corrections are finite and computable.

In particular, the one-loop correction to the graviton propagator in the harmonic gauge is given by

$$\frac{k^4}{M_{\text{Pl}}^2}\left[\frac{21}{120}(\eta_{\mu\rho}\eta_{\nu\sigma}+\eta_{\mu\sigma}\eta_{\nu\rho})+\frac{1}{120}\eta_{\mu\nu}\eta_{\rho\sigma}\right]\left[\frac{1}{\varepsilon}-\ln(k^2/\mu^2)+\text{const.}\right]. \qquad (26.48)$$

As in the case of QCD, one has to add the effect of Faddeev–Popov ghosts to obtain a consistent results using a covariant gauge. Performing then e.g. on-shell renormalisation, we are left with the logarithm which turns in the static limit after Fourier transformation into a $1/r^3$ term. Combining these results, the Newtonian potential is

$$\Phi=-\frac{GM}{r}\left[1-\frac{3M}{c^2r^2}+\frac{41}{10\pi}\frac{\hbar G}{c^2r^3}-\frac{2}{3}\exp\left(-\frac{1}{a}\sqrt{\frac{\hbar G}{96\pi c^3}}r\right)\right], \qquad (26.49)$$

where we set for simplicity $b=0$, and added c, $\hbar$ and the numerical prefactors (Stelle, 1978; Bjerrum-Bohr *et al.*, 2003). This example shows that loop corrections to observables in classical gravity can be calculated following the usual effective theory approach. These corrections are—as expected—tiny for $E\ll M_{\text{Pl}}$. Additional loop corrections due to higher-order operators will be even more suppressed. Therefore any consistent theory of gravity will lead to experimentally indistinguishable predictions for the loop corrections in the limit $E\ll M_{\text{Pl}}$. However, deviations from Einstein gravity are testable via gravitational wave emission of compact objects and observations on astronomical and cosmological scales.

Approaches beyond Einstein gravity. In addition to being a non-renormalisable theory, gravity poses specific technical problems. For instance, canonical quantisation relies on the light-cone structure of spacetime as input, while the metric should be the output of a quantum theory of gravity. Using the path integral avoids this problem and allows even the summation over spacetimes with different topologies. However, the path-integral formulation of quantum gravity is plagued with problems too. The justification for a Wick rotation from a metric with Lorentzian to Euclidean signature is obscure—but even if we formally perform the Wick rotation, the Euclidean gravitational action remains unbounded from below.

In the traditional view, one neglects these principal problems and stays within the framework of QFT in $d=4$. The question arises then what UV completion the Einstein–Hilbert action has. One possibility is that the graviton is a composite particle, similar to the pion. However, the constraint that the agent of gravity couples in $d=4$ QFT to the conserved stress tensor seems to forbid this option. Another possibility is that the graviton is a fundamental particle, and additional particles and symmetries lead to an unitarisation of gravity. In particular, adding supersymmetries improves the UV behaviour of gravity and makes it renormalisable up to the two-loop level, since all possible one- and two-loop divergences are forbidden by supersymmetric Ward identities. While no explicit three-loop calculations have been performed to date, power counting arguments suggest however that divergences exist at the three-loop level. An even more minimalistic ansatz is the idea of asymptotic safety which assumes that Einstein gravity, defined non-perturbatively, for example, on a lattice, is a consistent theory. This requires that the non-perturbatively calculated running gravitational coupling constant $G_N(q^2)$ has a UV fixed point. In this case, a consistent continuum limit of gravity could be defined. However, even the concept of a universal running coupling constant is, at least in perturbative calculations, ill-defined (Donoghue, 2012). While it is not excluded that this is an artifact of perturbation theory, little evidence exists to support asymptotic safety.

We give next a simple argument suggesting that quantum gravity does not admit local observables. As a starter, let us consider the case of gauge theories. Here, measurable quantities are associated to n-point functions $\langle O(x_1) \cdots O(x_n) \rangle$ of gauge-invariant operators $O(x)$, or to S-matrix elements. While the one-particle states used in the initial and final states of the S-matrix are not gauge-invariant, the possible gauge transformations are reduced in the limit $t \to \pm\infty$ to global transformations which map one physical state onto a different physical state. Tying the states to $t = \pm\infty$ thus eliminates the redundancy of local gauge transformations.

Now we consider the case of gravity where the gauge group is the group of all invertible coordinate transformations. This means that no local operators $O(x)$ exist which are gauge-invariant, leaving the S-matrix as observable. More explicitly, we can show this as follows. Let us consider the expectation value of observables which are scalar functions of the metric and a scalar field, $O(x) \equiv O(g_{\mu\nu}(x), \phi(x))$. We write $\langle O(x_1) \cdots O(x_n) \rangle$ as a path-integral average, and use the fact that a scalar transforms as $O(x) = \tilde{O}(\tilde{x})$ under a general coordinate transformation,

$$\langle O(x_1) \cdots O(x_n) \rangle = \int \mathcal{D}\phi \mathcal{D}g_{\mu\nu}\, O(x_1) \cdots O(x_n)\, \mathrm{e}^{\mathrm{i}S[\phi, g_{\mu\nu}]} \tag{26.50a}$$

$$= \int \mathcal{D}\phi \mathcal{D}g_{\mu\nu}\, \tilde{O}(\tilde{x}_1) \cdots \tilde{O}(\tilde{x}_n)\, \mathrm{e}^{\mathrm{i}S[\phi, g_{\mu\nu}]}. \tag{26.50b}$$

Next we apply the invariance of the action, $S[\tilde{\phi}, \tilde{g}_{\mu\nu}] = S[\phi, g_{\mu\nu}]$, and the measure $\mathcal{D}\tilde{\phi}\mathcal{D}\tilde{g}_{\mu\nu} = \mathcal{D}\phi\mathcal{D}g_{\mu\nu}$, and then relabel the integration variables as ϕ and $g_{\mu\nu}$,

$$\langle O(x_1) \cdots O(x_n) \rangle = \int \mathcal{D}\tilde{\phi}\mathcal{D}\tilde{g}_{\mu\nu}\, \tilde{O}(\tilde{x}_1) \cdots \tilde{O}(\tilde{x}_n)\, \mathrm{e}^{\mathrm{i}S[\tilde{\phi}, \tilde{g}_{\mu\nu}]} \tag{26.51a}$$

$$= \int \mathcal{D}\phi \mathcal{D}g_{\mu\nu}\, O(\tilde{x}_1) \cdots O(\tilde{x}_n)\, \mathrm{e}^{\mathrm{i}S[\phi, g_{\mu\nu}]} = \langle O(\tilde{x}_1) \cdots O(\tilde{x}_n) \rangle. \tag{26.51b}$$

Thus $\langle O(x_1) \cdots O(x_n) \rangle$ cannot depend on the spacetime points x_i and has to be constant.

We can compare this behaviour to the case of spacetime symmetries in Minkowski space. Applying as a symmetry transformation translations, for example, results in the constraint that an observable can depend only on the differences $\tilde{x}_i - x_i$, that is, it should be translation invariant. Increasing the symmetry group to general coordinate transformations, the restriction on observables becomes so severe that no non-trivial solution is possible. This argument suggests that a quantum theory of gravity is not simply a version of Einstein–Hilbert gravity with improved UV properties but should include some fundamentally new features compared to the local quantum field theories of point particles we have considered. Examples for such non-local theories may contain a minimal length scale like loop gravity and string theory, or may be based on the non-commutativity of spacetime.

Finally, we comment on the idea of gravity as an emergent phenomenon. We have usually assumed that the universe at higher energies becomes more and more symmetric. However, several condensed matter systems show the opposite behaviour, where new symmetries emerge at low energies. Applying this picture to gravity, the gravitational field and thus spacetime would be no fundamental degrees of freedom, but would appear at $E \ll M_{\rm Pl}$ as an approximate symmetry of the low-energy world. Support for this idea comes from a few directions. First, the cosmological constant problem is solved, if the coupling of the gravitational field to the stress tensor is shift-invariant, $T_{\mu\nu} \to T_{\mu\nu} + cg_{\mu\nu}$. However, this simple solution excludes that the gravitational action is a functional of the metric $g_{\mu\nu}$. Second, one can derive the Einstein equations using a thermodynamical language, an attempt which is suggested by the thermodynamical description of black holes and horizons in general. More importantly, it is possible to reinterpret the gravitational field equations as equations describing the heating and cooling of null surfaces within this thermodynamical picture. This thermodynamic description poses the question of whether the gravitational field is not merely a macroscopic description for the unknown fundamental degrees of freedom. A concrete example of a theory where spacetime is emergent is string theory since there the metric is not a fundamental degree of freedom. Last but not least, black hole formation provides a fundamental limitation to the measurement of spacetime: probing spacetime distances below the Planck scale is impossible, because otherwise the measuring device would collapse to a black hole (problem 26.7). This implies that the uncertainty relations derived from the usual commutation relations of quantum fields break down on scales $\lesssim 1/M_{\rm Pl}$, or in other words that local quantum field theory should be replaced by a new theoretical setting. Matvei Bronstein (1936) used first this argument to conclude:

"The elimination of the logical inconsistencies connected with this requires a radical reconstruction of the theory, and ... perhaps also the rejection of our ordinary concepts of space and time, modifying them by some much deeper and nonevident concepts. Wer's nicht glaubt, bezahlt einen Thaler."

Summary

Alternatives to the standard ΛCDM model involve either new scalar fields (dark energy) or modify gravity. While dark energy models are generically plagued by fine-tuning problems, modifications of gravity contain new additional degrees of freedom which pose both theoretical and observational challenges. The cosmological constant problem raises the question, if our (effective) field theory approach has a restricted validity. Few hints suggest that locality has to be abandoned in a quantum theory of gravity and that spacetime may be an emergent phenomenon.

Further reading. Martin (2012) discusses the cosmological constant problem in great detail. Dark energy and modified gravity models and their observational consequences are reviewed by Joyce *et al.* (2016). The effective field-theory approach to gravity is discussed by Donoghue (2012), while Padmanabhan (2016) reviews links between gravity and thermodynamics.

Problems

26.1 Smooth cut-off. Recalculate ρ and P using a smooth exponential cut-off instead of a sharp cut-off (for simplicity, consider only the case $m = 0$).

26.2 Vacuum energy in QED. Consider the vacuum energy in QED at two loop. Argue that one of the two diagrams vanishes for a charge-neutral vacuum. Connect the other one to the one-loop photon polarisation $\Pi_{\mu\nu}$ and use our old results to calculate it.

26.3 Fixed points of a $\Omega_\Lambda + \Omega_m = 1$ universe. a.) Derive the relation $q = \frac{1}{2}(\Omega_m + 2\Omega_r - \Omega_\Lambda)$. b.) Show that $\dot{\Omega}_i = \Omega_i H(\Omega_m + 2\Omega_r - 2\Omega_\Lambda - 1 - 3w_i)$. c.) Apply this relation to an universe with $\Omega_\Lambda + \Omega_m = 1$ considering the dynamical system $d\Omega_\Lambda/d\Omega_m$. Draw typical trajectories in the Ω_Λ-Ω_m plane and find the fixed points.

26.4 Fixed points of (26.18). Find the fixed points of the system (26.18) for constant λ.

26.5 Tracker solution. Show that $\phi(t) = CM^{1+\nu}t^\nu$ solves (26.24) and determine $C(\alpha, n)$.

26.6 Lorentz-invariance violation. Assume that Lorentz-invariance violation modifies the dispersion relation of photons as $v = 1 - \sum_n a_n (E/M_{\rm Pl})^n$. Estimate limits on a_1 and a_2 from the observation of photons with energies between $E_{\max} = 30\,{\rm GeV}$ and $E_{\min} = 3\,{\rm GeV}$ within the time-interval $\Delta t = 0.3\,{\rm s}$ from a gamma-ray burst at redshift $z = 0.9$.

26.7 Minimal distance measurement. Imagine to measure the position $x(t)$ of a (freely floating) object with mass M and length L at time $t = 0$ and t. Try to minimise $s = x(t) - x(0)$, accounting for causality and black hole formation.

Appendices

Appendix A
Units, conventions and useful formulae

A.1 Units and conventions

Physical constants and measurements

Gravitational constant	$G_{\mathrm{N}} = 6.674 \times 10^{-8} \mathrm{cm}^3 \, \mathrm{g}^{-1} \, \mathrm{s}^{-2}$
Planck's constant	$\hbar = h/(2\pi) = 1.055 \times 10^{-27} \mathrm{erg\,s}$
velocity of light	$c = 2.998 \times 10^{10} \mathrm{cm/s}$
Boltzmann constant	$k = 1.38 \times 10^{-16} \mathrm{erg/K}$
electron mass	$m_e = 9.109 \times 10^{-28} \mathrm{g} = 0.5110 \, \mathrm{MeV}/c^2$
proton mass	$m_p = 1.673 \times 10^{-24} \mathrm{g} = 938.3 \, \mathrm{MeV}/c^2$
Fine-structure constant	$\alpha = e^2/(4\pi\hbar c) = 1/137.03$
Fermi's constant	$G_{\mathrm{F}}/(\hbar c)^3 = 1.166 \times 10^{-5} \ \mathrm{GeV}^{-2}$
Weak mixing angle	$\sin^2 \vartheta(m_Z) = 0.2312$ $(\overline{\mathrm{MS}})$
W-boson mass	$m_W = 9.109 \times 10^{-28} \mathrm{g} = 80.38 \, \mathrm{GeV}/c^2$
Z-boson mass	$m_Z = 9.109 \times 10^{-28} \mathrm{g} = 91.19 \, \mathrm{GeV}/c^2$
Strong coupling	$\alpha_s(m_Z) = 0.1184$ $(\overline{\mathrm{MS}})$

Astronomical constants and measurements

Astronomical Unit	$\mathrm{AU} = 1.496 \times 10^{13} \mathrm{cm}$
Parsec	$\mathrm{pc} - 3.086 \times 10^{18} \mathrm{cm} = 3.261 \, \mathrm{ly}$
Tropical year	$\mathrm{yr} = 31\,556\,925.2 \, \mathrm{s} \approx \pi \times 10^7 \, \mathrm{s}$
Solar radius	$R_\odot = 6.960 \times 10^{10} \mathrm{cm}$
Solar mass	$M_\odot = 1.998 \times 10^{33} \mathrm{g}$
Solar luminosity	$L_\odot = 3.828 \times 10^{33} \ \mathrm{erg/s}$
Age of the universe	$t_0 = (13.75 \pm 0.13) \ \mathrm{Gyr}$
present Hubble parameter	$H_0 = 100 \, h \, \mathrm{km/(s\,Mpc)}$ with $h = 0.673 \pm 0.012$
critical density	$\rho_{\mathrm{cr}} = 1.878 \times 10^{-29} h^2 \, \mathrm{g/cm}^3 = 1.053 \times 10^{-5} h^2 \, \mathrm{GeV/cm}^3$
	$\quad = 2.775 \times 10^{11} h^2 \, M_\odot/\mathrm{Mpc}^3$
present CMB	$T = 2.725 \, \mathrm{K}, \ s_\gamma/k = 2891/\mathrm{cm}^3, \ n_\gamma = 410.7/\mathrm{cm}^3$
present baryon density	$n_{\mathrm{b}} = (2.5 \pm 0.1) \times 10^{-7} \, \mathrm{cm}^3$
	$\Omega_b = \rho_b/\rho_{\mathrm{cr}} = 0.0223/h^2 \approx 0.0425$
dark matter abundance	$\Omega_{\mathrm{DM}} = \Omega_m - \Omega_b = 0.105/h^2 \approx 0.20$
cosmological constant	$\Omega_\Lambda = 0.685 \pm 0.017$
neutrino abundance	$\Omega_\nu \lesssim 0.0025 h^2$
inflationary spectrum	$n_s = 0.963 \pm 0.014, \ \mathrm{d}n_s/\mathrm{d}\ln(k)|_{0.002\mathrm{Mpc}^{-1}} = 0.03 \pm 0.03, \ r \leq 0.1$

Other useful quantities

cross-section	$\mathrm{mbarn} = 10^{-27} \ \mathrm{cm}^2$
reduced Planck mass	$\widetilde{M}_{\mathrm{Pl}} = 1/\sqrt{8\pi G_{\mathrm{N}}} = 2.435 \times 10^{18} \ \mathrm{GeV}.$

A short review of conventions used in this book can be found on page viii. Some useful conversion factors from GeV to cgs units are

	mass	energy	1/length	1/time	temperature
GeV	1.78×10^{-24} g	1.60×10^{-3} erg	5.06×10^{13} cm^{-1}	1.52×10^{24} s^{-1}	1.16×10^{13}K

For additional constants and particle properties, updates as well as useful mini reviews see the website of the Particle Data Group, `http://pdg.lbl.gov/`.

Minkowski space.   We use a metric with the signature $+, -, -, -$. In the special case of a Cartesian inertial frame, the metric tensor **g** is diagonal with elements

$$g_{\mu\nu} = \begin{pmatrix} 1 & 0 & 0 & 0 \\ 0 & -1 & 0 & 0 \\ 0 & 0 & -1 & 0 \\ 0 & 0 & 0 & -1 \end{pmatrix} \equiv \eta_{\mu\nu}. \tag{A.1}$$

Other invariant tensors are the Kronecker delta, $\delta_\mu^\nu = \eta_\mu^\nu$ with $\delta_\mu^\nu = 1$ for $\mu = \nu$ and 0 otherwise, and the Levi–Civita tensor $\varepsilon^{\mu\nu\rho\sigma}$. The latter tensor is completely antisymmetric and has in four dimensions the elements $+1$ for an even permutation of 0123, -1 for odd permutations and zero otherwise. It satisfies $\varepsilon^{\mu\nu\rho\sigma} = -\varepsilon_{\mu\nu\rho\sigma}$ and the identities

$$\varepsilon^{\mu\nu\rho\sigma}\varepsilon_{\mu\nu\bar\rho\bar\sigma} = -2!(\delta_{\bar\rho}^\rho\delta_{\bar\sigma}^\sigma - \delta_{\bar\sigma}^\rho\delta_{\bar\rho}^\sigma), \qquad \varepsilon^{\mu\nu\rho\sigma}\varepsilon_{\mu\nu\rho\bar\sigma} = -3!\delta_{\bar\sigma}^\sigma \quad \text{and} \quad \varepsilon^{\mu\nu\rho\sigma}\varepsilon_{\mu\nu\rho\sigma} = -4!.$$

In three dimensions, we define the Levi–Civita tensor by $\varepsilon^{123} = \varepsilon_{123} = 1$.

The four-dimensional nabla operator ∇_μ has in a Cartesian inertial frame the components

$$\nabla_\mu = \partial_\mu \equiv \frac{\partial}{\partial x^\mu} = \left(\frac{\partial}{\partial t}, \frac{\partial}{\partial x}, \frac{\partial}{\partial y}, \frac{\partial}{\partial z} \right). \tag{A.2}$$

Then the four-momentum operator becomes $p^\mu = i\partial^\mu$ with spatial components $p^i = i\partial^i = -i\partial_i$ or $\boldsymbol{p} = -i\boldsymbol{\nabla}$. The d'Alembert or wave operator follows as

$$\Box \equiv \eta_{\mu\nu}\partial^\mu\partial^\nu = \frac{\partial^2}{\partial t^2} - \Delta = \frac{\partial^2}{\partial t^2} - \frac{\partial^2}{\partial x^2} - \frac{\partial^2}{\partial y^2} - \frac{\partial^2}{\partial z^2}. \tag{A.3}$$

The components of the electromagnetic field-strength tensor $F^{\mu\nu}$ and its dual $\tilde{F}_{\alpha\beta} = \frac{1}{2}\varepsilon_{\alpha\beta\mu\nu}F^{\mu\nu}$ are given by

$$F^{\mu\nu} = \begin{pmatrix} 0 & -E_x & -E_y & -E_z \\ E_x & 0 & -B_z & B_y \\ E_y & B_z & 0 & -B_x \\ E_z & -B_y & B_x & 0 \end{pmatrix} \quad \text{and} \quad \tilde{F}^{\mu\nu} = \begin{pmatrix} 0 & -B_x & -B_y & -B_z \\ B_x & 0 & E_z & -E_y \\ B_y & -E_z & 0 & E_x \\ B_z & E_y & -E_x & 0 \end{pmatrix}.$$

They are connected to the electric and magnetic fields measured by an observer with four-velocity u_α as $E_\alpha = F_{\alpha\beta}u^\beta$ and $B_\alpha = \tilde{F}_{\alpha\beta}u^\beta$.

General relativity.   The signs of the metric tensor, Riemann's curvature tensor and the Einstein tensor can be fixed arbitrarily,

$$\eta^{\alpha\beta} = S_1 \times [-1, +1, +1, +1], \tag{A.4a}$$

$$R^\alpha{}_{\beta\rho\sigma} = S_2 \times [\partial_\rho\Gamma^\alpha{}_{\beta\sigma} - \partial_\sigma\Gamma^\alpha{}_{\beta\rho} + \Gamma^\alpha{}_{\kappa\rho}\Gamma^\kappa{}_{\beta\sigma} - \Gamma^\alpha{}_{\kappa\sigma}\Gamma^\kappa{}_{\beta\rho}], \tag{A.4b}$$

$$S_3 \times G_{\alpha\beta} = 8\pi G\, T_{\alpha\beta}, \tag{A.4c}$$

$$R_{\alpha\beta} = S_2 S_3 \times R^\rho{}_{\alpha\rho\beta}. \tag{A.4d}$$

The curvature of spherical spaces has the signature $S_1 S_3$, while the free Lagrangian for bosonic fields becomes

$$\mathscr{L} = -\frac{s_1}{2}(\partial_\mu \phi)^2 - \frac{1}{4} F_{\mu\nu} F^{\mu\nu} + \frac{s_1 s_3}{2\kappa} R. \tag{A.5}$$

We choose these three signs as $S_i = \{-,+,+\}$, as for example Gorbunov and Rubakov (2011*b*, 2011*a*). Hobson *et al.* (2006) use $S_i = \{-,+,-\}$, while Carroll (2003) uses $S_i = \{+,+,+\}$.

A.2   Diracology

Pauli matrices.   The Pauli matrices are

$$\sigma^1 = \begin{pmatrix} 0 & 1 \\ 1 & 0 \end{pmatrix}, \qquad \sigma^2 = \begin{pmatrix} 0 & -i \\ i & 0 \end{pmatrix}, \qquad \sigma^3 = \begin{pmatrix} 1 & 0 \\ 0 & -1 \end{pmatrix}. \tag{A.6}$$

They satisfy $\sigma^i \sigma^j = \delta^{ij} + i\varepsilon^{ijk}\sigma^k$. Combining the Pauli matrices with the unit matrix, we can construct the two 4-vectors $\sigma^\mu \equiv (1, \boldsymbol{\sigma})$ and $\bar{\sigma}^\mu \equiv (1, -\boldsymbol{\sigma})$.

Gamma matrices.   They satisfy the Clifford algebra $\{\gamma^\mu, \gamma^\nu\} = 2\eta^{\mu\nu}$, with $\gamma^0 = \beta$ being Hermitian and the $\gamma^i = \beta\alpha^i$ anti-Hermitian. Additionally, one defines

$$\gamma^5 \equiv \gamma_5 \equiv i\gamma^0\gamma^1\gamma^2\gamma^3 = \frac{i}{4!}\,\varepsilon_{\alpha\beta\gamma\delta}\,\gamma^\alpha\gamma^\beta\gamma^\gamma\gamma^\delta = -i\gamma_0\gamma_1\gamma_2\gamma_3,$$

which satisfies $(\gamma^5)^\dagger = \gamma^5$, $(\gamma^5)^2 = 1$, and $\{\gamma^\mu, \gamma^5\} = 0$. From the commutator of gamma matrices one defines the Hermitian matrix

$$\sigma^{\mu\nu} \equiv \frac{i}{2}[\gamma^\mu, \gamma^\nu],$$

which satisfies $[\gamma^5, \sigma^{\mu\nu}] = 0$. The spin matrix $\boldsymbol{\Sigma} = \gamma^5\gamma_0\boldsymbol{\gamma}$ is connected to $\sigma^{\mu\nu}$ by $\Sigma^i = \frac{1}{2}\varepsilon^{ijk}\sigma^{jk}$. The 16 independent combinations of γ-matrices

$$\Gamma_i = \{1, \gamma^\mu, \gamma^\mu\gamma^5, \sigma^{\mu\nu}, \gamma^5\}$$

form a basis of the 4×4 matrices.

The three most useful representations of the Clifford algebra are the Dirac (γ^0 diagonal), the Weyl (γ^5 diagonal) and the real Majorana representation. Denoting by $\otimes$ the tensor product of the two sets of Pauli matrices σ^i and τ^i, we can write the Dirac representation as

$$\gamma^0 = 1 \otimes \tau^3 = \begin{pmatrix} 1 & 0 \\ 0 & -1 \end{pmatrix}, \qquad \gamma^i = \sigma^i \otimes i\tau^2 = \begin{pmatrix} 0 & \sigma^i \\ -\sigma^i & 0 \end{pmatrix}, \tag{A.7}$$

$$\gamma^5 = 1 \otimes \tau^1 = \begin{pmatrix} 0 & 1 \\ 1 & 0 \end{pmatrix}, \qquad C = i\gamma^2\gamma^0 = -i\sigma^1 \otimes \tau^2 = \begin{pmatrix} 0 & -i\sigma^2 \\ -i\sigma^2 & 0 \end{pmatrix}. \tag{A.8}$$

The Weyl or chiral representation

$$\gamma^0 = 1 \otimes \tau^1 = \begin{pmatrix} 0 & 1 \\ 1 & 0 \end{pmatrix}, \qquad \gamma^i = \sigma^i \otimes i\tau^3 = \begin{pmatrix} 0 & \sigma^i \\ -\sigma^i & 0 \end{pmatrix}, \tag{A.9}$$

$$\gamma^5 = 1 \otimes \tau^3 = \begin{pmatrix} -1 & 0 \\ 0 & 1 \end{pmatrix} \qquad C = -\gamma^2\gamma^0 = -i\sigma^3 \otimes \tau^2 = \begin{pmatrix} -i\sigma^2 & 0 \\ 0 & i\sigma^2 \end{pmatrix} \tag{A.10}$$

is connected to the Dirac representation by $\gamma^\mu_{\text{Weyl}} = U\gamma^\mu_{\text{Dirac}}U^\dagger$ and $U = \left(\begin{smallmatrix} 1 & -1 \\ 1 & -1 \end{smallmatrix}\right)/\sqrt{2}$.

The Majorana representation

$$\gamma^0 = \sigma^1 \otimes \tau^2 = \begin{pmatrix} 0 & \sigma^2 \\ \sigma^2 & 0 \end{pmatrix}, \qquad \gamma^1 = i\,1 \otimes i\tau^3 = \begin{pmatrix} i\sigma^3 & 0 \\ 0 & i\sigma^3 \end{pmatrix}, \qquad (A.11)$$

$$\gamma^2 = -i\sigma^2 \otimes i\tau^2 = \begin{pmatrix} 0 & -\sigma^2 \\ \sigma^2 & 0 \end{pmatrix}, \qquad \gamma^3 = -i\,1 \otimes i\tau^1 = \begin{pmatrix} -i\sigma^1 & 0 \\ 0 & -i\sigma^1 \end{pmatrix}, \qquad (A.12)$$

$$\gamma^5 = \sigma^3 \otimes \tau^2 = \begin{pmatrix} \sigma^2 & 0 \\ 0 & -\sigma^2 \end{pmatrix}, \qquad C = -i\sigma^1 \otimes \tau^2 = \begin{pmatrix} 0 & -i\sigma^2 \\ -i\sigma^2 & 0 \end{pmatrix} \qquad (A.13)$$

is connected to the Dirac representation by $\gamma^\mu_{\text{Majorana}} = U\gamma^\mu_{\text{Dirac}}U^\dagger$ and $U = \begin{pmatrix} 1 & \sigma^2 \\ \sigma^2 & -1 \end{pmatrix}/\sqrt{2}$. Majorana spinors have the following flip properties

$$\bar{\xi}\eta = \bar{\eta}\xi, \qquad\qquad \bar{\xi}\gamma^\mu\eta = -\bar{\eta}\gamma^\mu\xi, \qquad (A.14a)$$

$$\bar{\xi}\gamma^5\eta = \bar{\eta}\gamma^5\xi, \qquad\qquad \bar{\xi}\sigma^{\mu\nu}\eta = -\bar{\eta}\sigma^{\mu\nu}\xi, \qquad (A.14b)$$

$$\bar{\xi}\gamma^\mu\gamma^5\eta = \bar{\eta}\gamma^\mu\gamma^5\xi. \qquad (A.14c)$$

The charge conjugation matrix C satisfies

$$C^\dagger = C^{-1}, \qquad C^T = -C, \quad \text{and} \quad C^{-1}\Gamma_i C = \eta_i\Gamma_i^T \qquad (A.15)$$

with $\eta_i = +1$ for $\{1, i\gamma^5, \gamma^\mu\gamma^5\}$ and $\eta_i = -1$ for $\{\gamma^\mu, \sigma^{\mu\nu}\}$.

Helicity spinors. We denote the unit vector in the direction of the three-momentum as $\hat{p} = (\sin\vartheta\cos\phi, \sin\vartheta\sin\phi, \cos\vartheta)$. Defining then $\mathcal{N} = |\boldsymbol{p}|/(E+m)$ and the two-spinors

$$w = \begin{pmatrix} \cos(\vartheta/2) \\ e^{i\phi}\sin(\vartheta/2) \end{pmatrix} \quad \text{and} \quad v = \begin{pmatrix} -\sin(\vartheta/2) \\ e^{i\phi}\cos(\vartheta/2) \end{pmatrix}, \qquad (A.16)$$

the helicity eigenstates of the Dirac equation follow as

$$u_\uparrow = \begin{pmatrix} w \\ \mathcal{N}w \end{pmatrix}, \qquad u_\downarrow = \begin{pmatrix} v \\ -\mathcal{N}v \end{pmatrix}, \qquad v_\uparrow = \begin{pmatrix} -\mathcal{N}v \\ v \end{pmatrix}, \qquad v_\downarrow = \begin{pmatrix} \mathcal{N}w \\ w \end{pmatrix}. \qquad (A.17)$$

Traces and contractions for $d = 4$. The trace of an odd number of γ^μ matrices vanishes, as well as

$$\text{tr}[\gamma^5] = \text{tr}[\gamma^\mu\gamma^5] = \text{tr}[\gamma^\mu\gamma^\nu\gamma^5] = 0. \qquad (A.18)$$

Non-zero traces are

$$\text{tr}[\gamma^\mu\gamma^\nu] = 4\eta^{\mu\nu} \quad \text{and} \quad \text{tr}[\displaystyle{\not}a\displaystyle{\not}b] = 4a \cdot b \qquad (A.19)$$

$$\text{tr}[\displaystyle{\not}a\displaystyle{\not}b\displaystyle{\not}c\displaystyle{\not}d] = 4[(a \cdot b)(c \cdot d) - (a \cdot c)(b \cdot d) + (a \cdot d)(b \cdot c)] \qquad (A.20)$$

$$\text{tr}\left[\gamma^5\gamma_\mu\gamma_\nu\gamma_\alpha\gamma_\beta\right] = 4i\varepsilon_{\mu\nu\alpha\beta} \qquad (A.21)$$

Useful are also $\displaystyle{\not}a\displaystyle{\not}b + \displaystyle{\not}b\displaystyle{\not}a = 2a \cdot b$, $\displaystyle{\not}a\displaystyle{\not}a = a^2$ and the following contractions,

$$\gamma^\mu\gamma_\mu = 4, \qquad \gamma^\mu\displaystyle{\not}a\gamma_\mu = -2\displaystyle{\not}a, \qquad \gamma^\mu\displaystyle{\not}a\displaystyle{\not}b\gamma_\mu = 4a \cdot b, \qquad \gamma^\mu\displaystyle{\not}a\displaystyle{\not}b\displaystyle{\not}c\gamma_\mu = -2\displaystyle{\not}c\displaystyle{\not}b\displaystyle{\not}a. \quad (A.22)$$

Traces and contractions for arbitrary d. In d spacetime dimensions, the Clifford algebra becomes $\{\gamma^\mu, \gamma^\nu\} = 2\eta^{\mu\nu}I_d$ with I_d as the d-dimensional unit matrix. Thus the contractions (A.22) change to

$$\gamma^\mu\gamma_\mu = dI_d, \qquad \gamma^\mu\displaystyle{\not}a\gamma_\mu = (2-d)\displaystyle{\not}a, \qquad \gamma^\mu\displaystyle{\not}a\displaystyle{\not}b\gamma_\mu = 4a \cdot bI_d - (d-4)\displaystyle{\not}a\displaystyle{\not}b. \quad (A.23)$$

It is standard to define $\text{tr}(I_d) = 4$, which has the advantage that trace relations like $\text{tr}[\displaystyle{\not}a\displaystyle{\not}b] = 4a \cdot b$ are unchanged.

A.3 Special functions

A.3.1 Gamma function

Definition. The Gamma function $\Gamma(z)$ is defined as a generalisation of the factorial $n!$ which satisfies $\Gamma(n) = (n-1)!$. Thus $\Gamma(1) = 1$ and $z\Gamma(z) = \Gamma(z+1)$. As the first step of this extension of the factorial into the complex plane, we show that

$$\Gamma(z) = \int_0^\infty dt\, e^{-t} t^{z-1} \tag{A.24}$$

is an integral representation of the Gamma function valid in the positive half-plane $\Re\, z > 0$. First, the RHS equals $\Gamma(1) = 1$. Second, we can rewrite this integral representation as

$$\Gamma(z+1) = -\int_0^\infty dt\, \left(\frac{d}{dt}e^{-t}\right) t^z = \int_0^\infty dt\, z e^{-t} t^{z-1} + e^{-t} t^z \Big|_0^\infty. \tag{A.25}$$

The boundary term vanishes and thus the integral representation also satisfies the recurrence relation $\Gamma(z+1) = z\Gamma(z)$. Next we extend $\Gamma(z)$ into the half-plane $\Re\, z > -1$ by setting $\Gamma_1(z) = \Gamma(z+1)/z$. The function $\Gamma_1(z)$ has a simple pole at $z = 0$. In the second step, we set $\Gamma_2(z) = \Gamma(z+2)/[z(z+1)]$, defining thereby the function $\Gamma_2(z)$ valid in the half-plane $\Re\, z > -2$ with simple poles at $z = 0$ and -1. By induction, we can extend thus the Gamma function as an analytic function in the whole complex plane except for simple poles at $z = 0, -1, -2, \ldots$. Since the requirement $\Gamma(n) = (n-1)!$ fixes the Gamma function only at isolated points, the extension (A.24) is not unique and other generalisations of the factorial $n!$ are possible. The choice (A.24) seems to be, however, the only one of relevance for nature. From the mathematical point of view, it is uniquely characterised by the property of being log-convex, that is, that the second derivative of $\ln\Gamma(x)$ is positive for all $x \in \mathbb{R}^+$.

Often we need to know the values of the Gamma function for half-integer arguments. Inserting $z = 1/2$ into the definition (A.24) and substituting $t = s^2$ gives

$$\Gamma(1/2) = \int_0^\infty dt\, e^{-t} t^{-1/2} = 2\int_0^\infty ds\, e^{-s^2} = \sqrt{\pi}. \tag{A.26}$$

Using then the recurrence relation, we can generate from $\Gamma(1) = 1$ and $\Gamma(1/2) = \sqrt{\pi}$ the values for all (positive) integer and half-integer arguments: $\Gamma(3/2) = \sqrt{\pi}/2$, $\Gamma(5/2) = 3\sqrt{\pi}/4$, etc. The Gamma function satisfies the reflection formula (cf. problem 23.6)

$$\Gamma(z)\Gamma(1-z) = \frac{\pi}{\sin(\pi z)}, \tag{A.27}$$

while Stirling's formula becomes

$$\Gamma(x) = \sqrt{2\pi}\, x^{x-\frac{1}{2}}\, e^{-x}[1 + 1/(12x) + \mathcal{O}(x^{-2})]. \tag{A.28}$$

Euler's Beta function is defined for $\Re(a) > 0, \Re(b) > 0$ by

$$B(a,b) \equiv \int_0^1 dt\, t^{a-1}(1-t)^{b-1} = \int_0^\infty dt\, \frac{t^{a-1}}{(1+t)^{a+b}}. \tag{A.29}$$

The equivalence of the two representations is proven by substituting first $x = \sqrt{t}$ and then $t = x^2/(1-x^2)$. In order to connect the Beta and Gamma functions, we transform the first integral representation substituting $t = \sin^2\vartheta$ into

$$\frac{\Gamma(a)\Gamma(b)}{\Gamma(a+b)} = 2\int_0^{\pi/2} d\vartheta\, (\sin\vartheta)^{2a-1}(\cos\vartheta)^{2b-1}. \tag{A.30}$$

This trigonometric form of Euler's Beta function can be evaluated expressing $m!\,n!$ by the Gamma function using Eq. (A.24), substituting $s = x^2$, $t = y^2$ and changing then to polar coordinates,

$$m!\,n! = \int_0^\infty ds\, e^{-s} s^{m-1} \int_0^\infty dt\, e^{-t} t^{n-1} = 4 \int_0^\infty dx\, e^{-x^2} x^{2m+1} \int_0^\infty dy\, e^{-y^2} y^{2n+1} \quad \text{(A.31a)}$$

$$= \int_0^{2\pi} d\vartheta \int_0^\infty dr\, re^{-r^2} |r\cos\vartheta|^{2m+1} |r\sin\vartheta|^{2n+1} \quad \text{(A.31b)}$$

$$= 4 \int_0^\infty dr\, r^{2m+2n+3}\, e^{-r^2} \int_0^{\pi/2} d\vartheta\, (\cos\vartheta)^{2m+1} (\sin\vartheta)^{2n+1} \quad \text{(A.31c)}$$

$$= 2(m+n+1)! \int_0^{\pi/2} d\vartheta\, (\cos\vartheta)^{2m+1} (\sin\vartheta)^{2n+1}. \quad \text{(A.31d)}$$

The Beta function follows then for integers as

$$B(m+1, n+1) = 2 \int_0^{\pi/2} d\vartheta\, (\cos\vartheta)^{2m+1} (\sin\vartheta)^{2n+1} = \frac{m!\,n!}{(m+n+1)!}. \quad \text{(A.32)}$$

Expressing the factorials through the Gamma function, we arrive at

$$B(a,b) = \frac{\Gamma(a)\Gamma(b)}{\Gamma(a+b)}. \quad \text{(A.33)}$$

Logarithmic derivative. The (first) logarithmic derivative of the Gamma function is called the digamma function,

$$\psi(z) = \frac{d\ln\Gamma(z)}{dz} = \frac{\Gamma'(z)}{\Gamma(z)}. \quad \text{(A.34)}$$

We define moreover the special value

$$\psi(1) = \Gamma'(1) \equiv -\gamma \quad \text{(A.35)}$$

as the Euler–Mascheroni constant γ. Differentiating $z\Gamma(z) = \Gamma(z+1)$ we obtain $\Gamma(z) + z\Gamma'(z) = \Gamma'(z+1)$ or

$$1 + \frac{z\Gamma'(z)}{\Gamma(z)} = \frac{\Gamma'(z+1)}{\Gamma(z)} = \frac{z\Gamma'(z+1)}{\Gamma(z+1)}. \quad \text{(A.36)}$$

This gives us a recurrence relation for the digamma function,

$$\psi(z+1) = \frac{1}{z} + \psi(z), \quad \text{(A.37)}$$

which becomes for the special case of non-negative integers $n = 0, 1, 2, \ldots$

$$\psi(n+1) = \frac{1}{n} + \psi(n) = \frac{1}{n} + \frac{1}{n-1} + \ldots + \psi(1). \quad \text{(A.38)}$$

Thus we can express the digamma function for integer values through the Euler–Mascheroni constant and a finite harmonic series,

$$\psi(n+1) = -\gamma + \sum_{k=1}^n \frac{1}{k}. \quad \text{(A.39)}$$

Considering this relation in the limit $n \to \infty$, we can determine the value of γ. First we obtain the asymptotic behaviour of $\psi(x)$ using Stirling's formula,

$$\ln \Gamma(x+1) = \left(x + \frac{1}{2}\right) \ln x - x + \frac{1}{2} \ln 2\pi + \mathcal{O}(x^{-1}). \tag{A.40}$$

Differentiating then this expression results in

$$\psi(x+1) = \ln x + \frac{1}{2x} + \mathcal{O}(x^{-2}). \tag{A.41}$$

In the limit $x \to \infty$, we find thus $\psi(x+1) = \ln x$. Inserting this equality into (A.39), we obtain an expression for the Euler–Mascheroni constant,

$$\gamma = \lim_{n \to \infty} \left(\sum_{k=1}^{n} \frac{1}{k} - \ln(n) \right) = 0.5772\ldots . \tag{A.42}$$

Expansion near a pole.   We are now in the position to derive formulae like (4.45) for the expansion of the Gamma function around its poles. We start expanding $\Gamma(1+\varepsilon)$ around one,

$$\Gamma(1+\varepsilon) = \Gamma(1) + \varepsilon\Gamma'(1) + \mathcal{O}(\varepsilon^2) = 1 + \varepsilon\Gamma(1)\psi(1) + \mathcal{O}(\varepsilon^2) = 1 - \varepsilon\gamma + \mathcal{O}(\varepsilon^2). \tag{A.43}$$

From $z\Gamma(z) = \Gamma(z+1)$ it follows

$$\Gamma(\varepsilon) = \frac{\Gamma(1+\varepsilon)}{\varepsilon} = \frac{1}{\varepsilon} - \gamma + \mathcal{O}(\varepsilon). \tag{A.44}$$

Applying $z\Gamma(z) = \Gamma(z+1)$ again, we obtain

$$\Gamma(-1+\varepsilon) = \frac{-\Gamma(\varepsilon)}{1-\varepsilon} = -\left[1 + \varepsilon + \mathcal{O}(\varepsilon^2)\right]\left[\frac{1}{\varepsilon} - \gamma + \mathcal{O}(\varepsilon)\right] = -\frac{1}{\varepsilon} - 1 + \gamma + \mathcal{O}(\varepsilon). \tag{A.45}$$

For general n, the formula

$$\Gamma(-n+\varepsilon) = \frac{(-1)^n}{n!}\left[\frac{1}{\varepsilon} + \psi(n+1) + \mathcal{O}(\varepsilon)\right], \tag{A.46}$$

follows by induction for the expansion of the Gamma function near a pole.

A.3.2   Spherical functions

Bessel functions.   We define the Bessel functions $J_\nu(z)$ as

$$J_\nu(z) = \left(\frac{z}{2}\right)^\nu \sum_{k=0}^{\infty} \frac{(-1)^k \left(\frac{z}{2}\right)^{2k}}{k!\Gamma(\nu+k+1)} \tag{A.47}$$

for $\nu \in \mathbb{C}$. This definition implies the recurrence relation

$$J_{\nu\pm1}(z) = \frac{\nu}{z} J_\nu(z) \pm J'_\nu(z) \tag{A.48}$$

between adjunct Bessel functions. For half-integral values of the index, we can express Bessel functions through trigonometric functions. From the series solutions, we can obtain first

$$J_{1/2}(x) = \sqrt{\frac{2}{\pi x}} \sin(x) \quad \text{and} \quad J_{-1/2}(x) = \sqrt{\frac{2}{\pi x}} \cos(x). \tag{A.49}$$

Using then the recurrence relation, we can progressively generate Bessel functions with increasing index $|n|$,

$$J_{3/2}(x) = \sqrt{\frac{2}{\pi x}} \left(\frac{\sin(x)}{x} - \cos(x) \right) \quad \text{and} \quad J_{-3/2}(x) = \sqrt{\frac{2}{\pi x}} \left(-\frac{\cos(x)}{x} - \sin(x) \right), \dots. \tag{A.50}$$

For large $|x|$, this implies the asymptotic expansion

$$J_\nu(x) = \sqrt{\frac{2}{\pi x}} \cos\left(x - \nu\frac{\pi}{2} - \frac{\pi}{4} \right) + \mathcal{O}(x^{-3/2}). \tag{A.51}$$

(While we derived and will need this expansion only for half-integer values of ν, it holds for all $\nu \in \mathbb{C}$.) Finally, we define the spherical Bessel functions $j_n(r)$ for $n \in \mathbb{N}$ as

$$j_n(r) \equiv \sqrt{\frac{\pi}{2r}} J_{n+1/2}(r) = \frac{1}{r} \cos\left(r - (l+1)\frac{\pi}{2} \right) + \mathcal{O}(1/r^2), \tag{A.52}$$

where we noted in the second step its corresponding asymptotic expansion.

Legendre polynomials.   The Legendre polynomials $P_l(x)$ are a complete, orthogonal basis on the interval $[-1:1]$, normalised as

$$\int_{-1}^1 \mathrm{d}(\cos\vartheta) P_l(\cos\vartheta) P_{l'}(\cos\vartheta) = \frac{2}{(2l+1)} \delta_{ll'}. \tag{A.53}$$

We can define them by the recurrence relation

$$(l+1)P_{l+1}(x) = (2l+1)P_l(x) - lP_{l-1}(x), \tag{A.54}$$

specifying the first three polynomials as $P_0(x) = 1$, $P_1(x) = x$ and $P_2(x) = (3x^2 - 1)/2$. They satisfy $P_l(1) = 1$ and $P_l(-x) = (-1)^l P_l(x)$.

We want to decompose a plane wave into Legendre polynomials and try as ansatz

$$\mathrm{e}^{ikr\cos\vartheta} = \sum_{m=0}^\infty c_m j_m(kr) P_m(\cos\vartheta). \tag{A.55}$$

Setting $x = \cos\vartheta$, multiplying with $P_l(x)$ and integrating from $x = -1$ to 1 we obtain

$$\int_{-1}^1 \mathrm{d}x\, \mathrm{e}^{ikrx} P_l(x) = \frac{2c_l}{(2l+1)} j_l(kr) - \frac{2c_l}{(2l+1)} \left[\frac{1}{kr} \cos\left(kr - (l+1)\frac{\pi}{2} \right) + \mathcal{O}(1/(kr)^2) \right], \tag{A.56}$$

where we used the asymptotic expansion (A.51) in the last step. Next we integrate the LHS twice by parts,

$$\int_{-1}^1 \mathrm{d}x\, \mathrm{e}^{ikrx} P_l(x) = \left[\frac{\mathrm{e}^{ikrx}}{ikr} P_l(x) \right]_{-1}^1 - \left[\frac{\mathrm{e}^{ikrx}}{(ikr)^2} P_l'(x) \right]_{-1}^1 + \int_{-1}^1 \mathrm{d}x\, \frac{\mathrm{e}^{ikrx}}{(ikr)^2} P_l''(x) \tag{A.57}$$

$$= \left[\frac{\mathrm{e}^{ikrx}}{ikr} P_l(x) \right]_{-1}^1 + \mathcal{O}(1/(kr)^2). \tag{A.58}$$

Now we compare the $1/kr$ terms. Using $P_l(1) = 1$ and $P_l(-1) = (-1)^l$, the LHS is periodic in l with period two, while the RHS has period four. Thus it is sufficient to consider, for example, $l = \{0, 1, 2, 3\}$ to determine the coefficients c_l as $c_l = (2l + 1)\mathrm{i}^l$.

Spherical harmonics. The spherical harmonics $Y_{l,m}(\vartheta, \phi)$ form a complete, orthogonal basis on the sphere S^2, which is normalised as

$$\int \mathrm{d}\Omega\, Y_{l,m}^*(\vartheta, \phi) Y_{l',m'}(\vartheta, \phi) = \delta_{ll'} \delta_{mm'}. \tag{A.59}$$

They are defined as

$$Y_{l,m}(\vartheta, \phi) = \left[\frac{2l + 1}{4\pi} \frac{(l - m)!}{(l + m)!} \right]^{1/2} P_l^m(\cos \vartheta) \mathrm{e}^{\mathrm{i}m\phi} \tag{A.60}$$

with the associated Legendre polynomials

$$P_l^m(x) = (-1)^m (1 - x^2)^{m/2} \frac{\mathrm{d}^m}{\mathrm{d}x^m} P_l(x) = (-1)^m \frac{(1 - x^2)^{m/2}}{2!\, l!} \frac{\mathrm{d}^{m+l}}{\mathrm{d}x^{m+l}} (x^2 - 1)^l. \tag{A.61}$$

The spherical harmonics satisfy $Y_{l,m}^* = (-1)^m Y_{l,-m}$ and transform under parity as

$$Y_{l,m}(\vartheta, \phi) \to Y_{l,m}(\pi - \vartheta, -\phi) = (-1)^l Y_{l,-m}(\vartheta, \phi). \tag{A.62}$$

A rotation $(\vartheta, \phi) \to (\vartheta', \phi')$ mixes spherical harmonics with fixed l by a unitary transformation,

$$Y_{l,m}^*(\vartheta, \phi) \to Y_{l,m}(\vartheta', \phi') = \sum_{m'} U_{m'm}(l) Y_{l,m}(\vartheta', \phi'), \tag{A.63}$$

and thus

$$\sum_m Y_{l,m}^*(\vartheta_1, \phi_1) Y_{l,m}(\vartheta_2, \phi_2) = \sum_m Y_{l,m}^*(\vartheta_1', \phi_1') Y_{l,m}(\vartheta_2', \phi_2'). \tag{A.64}$$

Now we set $\vartheta_1' = \phi_1' = 0$. Then only the $m = 0$ term contributes to the sum on the RHS, because of $Y_{l,m}(\vartheta, 0) = -Y_{l,-m}(\vartheta, 0)$. Using also

$$Y_{l,0} = \sqrt{\frac{2l + 1}{4\pi}} P_l(\cos \vartheta), \tag{A.65}$$

we find

$$\sum_m Y_{l,m}^*(\vartheta_1, \phi_1) Y_{l,m}(\vartheta_2, \phi_2) = \frac{2l + 1}{4\pi} P_l(\cos \vartheta_{12}), \tag{A.66}$$

where ϑ_{12} denotes the angle between (ϑ_1, ϕ_1) and (ϑ_2, ϕ_2). Inserted into (A.55), we arrive at

$$\mathrm{e}^{\mathrm{i}\boldsymbol{k}\cdot\boldsymbol{x}} = \sum_{l=0}^{\infty} (2l + 1)\mathrm{i}^l j_l(kr) P_l(\cos \vartheta) = 4\pi \sum_{l=0}^{\infty} \sum_{m=-l}^{l} \mathrm{i}^l j_l(kr) Y_{l,m}^*(\hat{\boldsymbol{k}}) Y_{l,m}(\hat{\boldsymbol{x}}). \tag{A.67}$$

Appendix B
Minimal group theory

B.1 Lie groups and algebras

A set of elements $\{a, b, c, \ldots\}$ is called a group $\mathcal{G}$, if the following four properties are satisfied:

- For every pair a, b, the product $ab = c \in \mathcal{G}$ is defined ("closure property").

- A unit element e exists in $\mathcal{G}$ such that for every a, $ae = ea = a$.

- The associative law holds: $(ab)c = a(bc)$.

- Each group element has an inverse, $aa^{-1} = a^{-1}a = e$.

A map from the group $\mathcal{G}$ into the group $\mathcal{G}'$ is called homomorph if it respects the group operation, that is, when $(ab)' = a'b'$ for all $a, b \in \mathcal{G}$ and $a', b' \in \mathcal{G}'$. The map is called isomorphic if there exists additionally a one-to-one relation between the elements of the two groups. A group $T(\mathcal{G})$ of square matrices which is homomorph to the group $\mathcal{G}$ is called a matrix representation of this group. Using matrix multiplication as group operation, the associative law holds automatically.

The invertible $n \times n$ matrices satisfy all four conditions and therefore form a group called $\mathrm{GL}(n, \mathbb{R})$ or $\mathrm{GL}(n, \mathbb{C})$ depending on whether the matrix elements are real or complex. More specific examples are orthogonal ($O^T = O^{-1}$) and unitary ($U^\dagger = U^{-1}$) matrices. Here we have to check the closure property. If O_1 and O_2 are orthogonal then their product $O = O_1 O_2$ is orthogonal too, since

$$O^T O = (O_1 O_2)^T O_1 O_2 = O_2^T O_1^T O_1 O_2 = \mathbf{1}. \tag{B.1}$$

The corresponding group of n-dimensional orthogonal matrices is called $\mathrm{O}(n)$. In the same way, one shows that the product of unitary matrices is again a unitary matrix. The corresponding group of n-dimensional unitary matrices is called $\mathrm{U}(n)$. Adding the restriction that the determinant of the matrices is 1, one obtains the special orthogonal groups $\mathrm{SO}(n)$ and the special unitary groups $\mathrm{SU}(n)$.

Orthogonal and unitary matrices can be defined as linear transformations that keep a real or complex scalar product invariant. By analogy, one defines symplectic matrices S as those set of linear transformations that preserve the natural scalar product on phase space, $S^T \Omega S = \Omega$, with $\Omega = \left(\begin{smallmatrix} 0 & I \\ -I & 0 \end{smallmatrix} \right)$. The determinant of symplectic matrices equals 1 and they form the group called $\mathrm{Sp}(2n)$.

Symmetry transformations in physics satisfy the group axioms. Therefore symmetries can be described by groups, which may be discrete or continuous. An example of a discrete, finite group is the Z_2 symmetry $\phi \to -\phi$ which has only two elements. Analogously, the Z_N symmetry has N elements and connects $\phi \to \exp(2n\pi i/N)\phi$ with $n = 1, \ldots, N-1$. An example of a continuous symmetry is the phase transformation $\phi \to \exp(i\vartheta)\phi$ with the symmetry group $\mathrm{U}(1)$.

Lie groups. A Lie group $\mathcal{G}$ is a continuous group which depends analytically on a finite number n of real parameters ϑ^a. The neighbourhood of any group element $g \in \mathcal{G}$ can be parameterised using ϑ^a as coordinates, and a Lie group is therefore at the same time an n-dimensional manifold. A group element g can be expanded as a power series,

$$g(\vartheta) = 1 + \sum_{a=1}^{n} i\vartheta^a T^a + \mathcal{O}(\vartheta^2) \equiv 1 + i\vartheta^a T^a + \mathcal{O}(\vartheta^2). \tag{B.2}$$

The linear transformation in the arbitrary direction ϑ^a is called an infinitesimal transformation, the T^a the (infinitesimal) generators of the transformation. The generators T^a can be obtained by differentiation, $T^a = -\mathrm{i}\, dg(\vartheta)/d\vartheta^a|_{\vartheta=0}$. Conversely, analyticity implies that the group element $g(\vartheta)$ can be obtained by exponentiation,

$$g(\vartheta) = \lim_{n\to\infty} [1 + i\vartheta^a T^a/n]^n = \exp(i\vartheta^a T^a). \tag{B.3}$$

Note that a Lie group can consist of disconnected pieces. In this case, we can generate via Eq. (B.3) only the group elements in the piece containing the unit element.

The generators T^a form an algebra $\mathfrak{g}$ called the Lie algebra. Three operations are defined in this algebra: addition, multiplication by real numbers, and the Lie bracket $[A, B] = AB - BA$. The latter is true, since the commutator $[A, B] = AB - BA$ can be obtained by expanding

$$e^{sA}\, e^{tB}\, e^{-sA}\, e^{-tB} = 1 + st[A, B] + \ldots$$

and neglecting higher-order terms in st. The LHS is a group element, $e^{uC} \simeq 1 + uC$, and therefore the commutator belongs to the Lie algebra too. Hence one can express the commutator as a linear combination of generators,

$$[T^a, T^b] = i f^{abc} T^c, \tag{B.4}$$

where the real numbers f^{abc} are called structure constants. The Jacobi identity for double commutators,

$$[A, [B, C]] + [B, [C, A]] + [C, [A, B]] = 0, \tag{B.5}$$

implies that the structure constants f^{abc} seen as $a = 1, \ldots, n$ matrices satisfy the same Lie algebra. Hence the matrices

$$(T_{\text{adj}}^a)_{bc} = -i f^{abc} \tag{B.6}$$

also form a representation of the Lie group $\mathcal{G}$ which is called the adjoint representation. Another important representation is the smallest non-trivial matrix representation of the Lie group which is called the fundamental representation.

Lie groups that have the same Lie algebra are locally isomorphic, but they may differ globally. An example of such a global difference are the two groups SU(2) and SO(3). They share the same Lie algebra, but two elements of SU(2) correspond to one element of SO(3). The number of generators that can be simultaneously diagonalised defines the rank of a Lie group. An operator which commutes with all generators is called a Casimir operator. Thus SO(3) and SU(2) have rank one (only J_3 or σ_3 are diagonal) and one Casimir operator ($\boldsymbol{J}^2$ or $\boldsymbol{\sigma}^2$).

The gauge group of the standard model is the direct product SU(3)⊗SU(2)⊗U(1). Groups that cannot be factored further, $\mathcal{G} \neq \mathcal{G}_1 \otimes \mathcal{G}_2$, are called simple. The Lie algebra of a semi-simple Lie group is the direct sum of simple Lie algebras. Thus the gauge group of the standard model is semi-simple. Practically all groups relevant in physics except the Lorentz group (and thus the Poincaré and the conformal group) have a compact parameter space. Cartan classified all simple compact groups. They compromise additionally to the three infinite sets of the orthogonal, unitary and symplectic groups only the five exceptional groups, G_2, F_4, E_6, E_7 and E_8.

B.2   Special unitary groups SU(n)

The unitary group $U(n)$ consists of the $n \times n$ complex matrices satisfying $U^\dagger U = 1$ or $|\det(U)| = 1$. In general, an $n \times n$ complex matrix has $2n^2$ real parameters. Accounting for the n^2 unitarity conditions, an element of $U(n)$ is parameterised by n^2 real numbers. Extracting a phase factor $\exp(i\vartheta_0)$, we can choose $\det(U) = 1$. This extra condition defines the special unitary groups $SU(n)$, which has thus $n^2 - 1$ real parameters. The unitary group $U(n)$ is the direct product $U(n) = U(1) \otimes SU(n)$,

$$U = \exp(i\vartheta_0 + i\vartheta^a T^a) \tag{B.7}$$

where T^a are the generators of $SU(n)$. From

$$1 = \det(U) = \exp[\,i\,\mathrm{tr}(\vartheta^a T^a)] \tag{B.8}$$

we see that the generators of $SU(n)$ are traceless, $\mathrm{tr}(T^a) = 0$, while

$$1 = UU^\dagger = 1 + \vartheta^a(T^a - T^{a\dagger}) + \mathcal{O}(\vartheta^2) \tag{B.9}$$

implies that they are Hermitian, $T^a = T^{a\dagger}$.

Matrix representations.   If T_R is a matrix representation of the Lie group $SU(n)$ and U an unitary matrix of the same dimension, then $UT_R U^\dagger$ provides an equivalent representation. We can fix this arbitrariness by demanding that $D^{ab}(R) \equiv \mathrm{tr}(T_R^a T_R^b)$ is diagonal, $D^{ab}(R) = C(R)\delta^{ab}$. The normalisation factor $C(R)$ depends still on the representation and physicists choose $C_F = 1/2$ for the fundamental representation. This implies $C_A = N$ for the adjoint representation. Calculating then

$$\mathrm{tr}([T_R^a, T_R^b]T_R^c) = i f^{abd}\mathrm{tr}(T_R^d T_R^c) = i f^{abd} C(R)\delta^{cd}, \tag{B.10}$$

we find

$$f^{abc} = -\frac{i}{C(R)}\mathrm{tr}([T_R^a, T_R^b]T_R^b) = -\frac{i}{C(R)}\mathrm{tr}(T_R^a T_R^b T_R^c - T_R^b T_R^a T_R^c), \tag{B.11}$$

which shows that structure constants are completely antisymmetric. This result holds more general for all groups $\mathcal{G}$ that are compact and simple. Note also that we can use this result to simplify colour sums in Feynman amplitudes. Choosing the fundamental representation in Eq. (B.11) we can eliminate all f^{abd} terms coming from triple and quartic gauge boson couplings, obtaining instead an expression containing only the generators T_F.

Since the generators are Hermitian, $T_{\bar R}^a = -T_R^{a*}$ forms also a representation of the group: Using $T^{a*} = T^{aT}$ and thus $T_{\bar R}^a = -T_R^{aT}$ one confirms that they satisfy the Lie algebra

$$[T_{\bar R}^a, T_{\bar R}^b] = [T_R^b, T_R^a]^T = i f^{abc} T_R^{cT} = i f^{abc} T_{\bar R}^c. \tag{B.12}$$

Because of the minus sign in the definition of $T_{\bar R}^a$ states in the complex conjugated representation $\bar R$ have eigenvalues with the opposite sign to the states in R. Hence we can use the representation R and $\bar R$ for particles and antiparticles, respectively. If these two representations are identical, possibly after applying a unitary transformation to both, then the representation is real. If they are not identical but are unitary equivalent, that is, a non-singular transformation $ST_{\bar R}^a S^{-1} = T_R^a$ relates them, one calls the representation pseudo-real, otherwise complex. An example of a group having pseudo-real representations is $SU(2)$, because of $\sigma^a \neq -\sigma^{a*}$ but $S\sigma^a S = \sigma^{a*}$ with $S = \sigma^2$. By contrast, $SU(3)$ and more generally the groups $SU(n)$ with $n \geq 3$ have complex representations.

The fundamental representation as the smallest representation of $SU(n)$ has dimension n, while the adjoint representation has dimension $n^2 - 1$. In the standard model, fermions live in the fundamental representation of the gauge groups: left-chiral fermions transform as doublets under $SU_L(2)$, quarks as triplets under $SU(3)$. Gauge fields and ghosts transform under the adjoint representation, which is real.

The generators of the fundamental representation of $SU(2)$ can be chosen as the Pauli matrices, $T^a = \sigma^a/2$, while the eight generators of $SU(3)$ can be chosen as the Gell-Mann matrices, $T^a = \lambda^a/2$, with

$$\lambda_1 = \begin{pmatrix} 0 & 1 & 0 \\ 1 & 0 & 0 \\ 0 & 0 & 0 \end{pmatrix}, \qquad \lambda_2 = \begin{pmatrix} 0 & -i & 0 \\ i & 0 & 0 \\ 0 & 0 & 0 \end{pmatrix}, \qquad \lambda_3 = \begin{pmatrix} 1 & 0 & 0 \\ 0 & -1 & 0 \\ 0 & 0 & 0 \end{pmatrix},$$

$$\lambda_4 = \begin{pmatrix} 0 & 0 & 1 \\ 0 & 0 & 0 \\ 1 & 0 & 0 \end{pmatrix}, \qquad \lambda_5 = \begin{pmatrix} 0 & 0 & -i \\ 0 & 0 & 0 \\ i & 0 & 0 \end{pmatrix}, \qquad \lambda_8 = \frac{1}{\sqrt{3}} \begin{pmatrix} 1 & 0 & 0 \\ 0 & 1 & 0 \\ 0 & 0 & -2 \end{pmatrix},$$

$$\lambda_6 = \begin{pmatrix} 0 & 0 & 0 \\ 0 & 0 & 1 \\ 0 & 1 & 0 \end{pmatrix}, \qquad \lambda_7 = \begin{pmatrix} 0 & 0 & 0 \\ 0 & 0 & -i \\ 0 & i & 0 \end{pmatrix}.$$

Note that the λ_i with $i = 1, 4, 6$ and $i = 2, 5, 7$ can be generated from the Pauli matrices σ_1 and σ_2 adding a raw and a column with zeros. This reflects the fact that $SU(3)$ contains as subgroups three $SU(2)$ factors (of which only two are independent).

Finally, we summarise some useful relations for $SU(n)$. The normalisation $\text{tr}(T^a T^b) = T_{ji}^a T_{ij}^b = \frac{1}{2}\delta^{ab}$ for the fundamental representation implies

$$(T^a T^a)_{ij} = \frac{N^2 - 1}{2N}, \tag{B.13}$$

and

$$f^{acd} f^{bcd} = N\delta^{ab}. \tag{B.14}$$

Using the relation (B.11) one can replace f^{abd} terms by expressions containing only the generators T_F. The Fierz relation of $SU(N)$ reads

$$T_{ij}^a T_{kl}^b = \frac{1}{2} \left(\delta_{il}\delta_{jk} - \frac{1}{N}\delta_{ij}\delta_{kl} \right). \tag{B.15}$$

Thus products of traces can be simplified using

$$\text{tr}(T^a A)\text{tr}(T^b B) = \frac{1}{2}\text{tr}(AB) - \frac{1}{2N}\text{tr}(A)\text{tr}(B) \tag{B.16}$$

$$\text{tr}(T^a A T^b B) = \frac{1}{2}\text{tr}(A)\text{tr}(B) - \frac{1}{2N}\text{tr}(AB) \tag{B.17}$$

for any A and B.

B.3 Lorentz group

Transformation of spacetime. Lorentz transformations $\Lambda^\mu{}_\nu$ are all those coordinate transformations $x^\mu \to \tilde{x}^\mu = \Lambda^\mu{}_\nu x^\nu$ that keep the distance

$$\Delta s^2 \equiv \Delta t^2 - \Delta x^2 = \Delta \tilde{t}^2 - \Delta \tilde{x}^2 \tag{B.18}$$

between two spacetime points in Minkowski space invariant. If we replace t by $-it$ in Δs^2, the difference between two spacetime events becomes (minus) the normal Euclidean distance.

Similarly, the identity $\cos^2 \alpha + \sin^2 \alpha = 1$ becomes $\cosh^2 \eta - \sinh^2 \eta = 1$ for an imaginary angle $\eta = i\alpha$. Thus a close correspondence exists between rotations R_{ij} in Euclidean space that leave $\Delta \boldsymbol{x}^2$ invariant and Lorentz transformations $\Lambda^\mu{}_\nu$ that leave Δs^2 invariant. We try therefore, as a guess, for a boost along the x direction

$$\tilde{t} = t \cosh \eta + x \sinh \eta, \tag{B.19}$$
$$\tilde{x} = t \sinh \eta + x \cosh \eta, \tag{B.20}$$

with $\tilde{y} = y$ and $\tilde{z} = z$. Direct calculation shows that Δs^2 is invariant as desired. Consider now in the system $\tilde{K}$ the origin of the system K. Then $x = 0$ and

$$\tilde{x} = t \sinh \eta \quad \text{and} \quad \tilde{t} = t \cosh \eta. \tag{B.21}$$

Dividing the two equations gives $\tilde{x}/\tilde{t} = \tanh \eta$. Since $\beta = \tilde{x}/\tilde{t}$ is the relative velocity v of the two systems, the imaginary "rotation angle η" equals the rapidity

$$\eta = \operatorname{arctanh}(\beta). \tag{B.22}$$

Inserting the identities $\cosh \eta = \gamma$ and $\sinh \eta = \gamma \beta$ into Eq. (B.19), we find the standard form of the Lorentz transformations,

$$\tilde{x} = \frac{x + vt}{\sqrt{1 - \beta^2}} = \gamma(x + \beta t) \quad \text{and} \quad \tilde{t} = \frac{t + vx}{\sqrt{1 - \beta^2}} = \gamma(t + \beta x). \tag{B.23}$$

The inverse transformation is obtained by replacing $v \to -v$ and exchanging quantities with and without tilde.

O(1,3) and its subgroups. The group O(1,3) of Lorentz transformations consists of four disconnected pieces. In order to see this, we rewrite first the fact that the metric tensor $\eta_{\mu\nu}$ is invariant under Lorentz transformations, $\tilde{\eta}_{\mu\nu} \Lambda^\mu{}_\rho \Lambda^\nu{}_\sigma = \eta_{\rho\sigma}$, in matrix form, $\eta = \Lambda^T \eta \Lambda$. Taking then the determinant, we obtain

$$(\det \Lambda)^2 = 1 \qquad \text{or} \qquad \det \Lambda = \pm 1. \tag{B.24}$$

Thus O(1,3) consists of at least two disconnected pieces. The part with $\det \Lambda = 1$ contains the proper Lorentz transformations and is called SO(1,3). The second part contains the improper Lorentz transformations that can be written as the product of a proper transformation and a discrete transformation changing the sign of an odd number of coordinates.

Next we consider the 00 component of the equation $\eta = \Lambda^T \eta \Lambda$. With $\eta_{00} = \eta_{\mu\nu} \Lambda^\mu{}_0 \Lambda^\nu{}_0$, it is

$$1 = (\Lambda^0{}_0)^2 - \sum_{i=1}^3 (\Lambda^i{}_0)^2. \tag{B.25}$$

Thus $(\Lambda^0{}_0)^2 \geq 1$ and both the proper and the improper Lorentz transformations consist of two disconnected pieces. The transformations with $\Lambda^0{}_0 \geq 1$ are called orthochronous since they do not change the direction of time. In contrast, transformations with $\Lambda^0{}_0 \leq -1$ transform a future-directed time-like vector into a past-directed one and vice versa, and are therefore called antichronous. These properties are Lorentz invariant, and thus we can split Lorentz transformations into four categories:

- proper, orthochronous (= restricted) transformations $L \in \mathcal{L}^\uparrow_+$ with $\det \Lambda = 1$ and $\Lambda^0{}_0 \geq 1$,

- proper, antichronous transformations $TPL \in \mathcal{L}^\downarrow_+$ with $\det \Lambda = 1$ and $\Lambda^0{}_0 \leq -1$,

- improper, orthochronous transformation $PL \in \mathcal{L}_-^\uparrow$ with $\det \Lambda = -1$ and $\Lambda^0_{\ 0} \geq 1$,

- improper, antichronous transformations $TL \in \mathcal{L}_-^\downarrow$ with $\det \Lambda = -1$ and $\Lambda^0_{\ 0} \leq -1$.

Only those Lorentz transformations that are elements of the restricted Lorentz group, $L \in \mathcal{L}_+^\uparrow$, are connected smoothly to the identity and can thus be built up from infinitesimal transformations. The other three disconnected pieces of the Lorentz group can be obtained as the product of an element $L \in \mathcal{L}_+^\uparrow$ and additional discrete P and T transformations with $P(x^0, \boldsymbol{x}) = (x^0, -\boldsymbol{x})$ and $T(x^0, \boldsymbol{x}) = (-x^0, \boldsymbol{x})$.

Lie algebra.   Let us consider infinitesimal Lorentz transformations. Then $\Lambda^\mu_{\ \nu} = \delta^\mu_\nu + \omega^\mu_{\ \nu} + \mathcal{O}(\omega^2)$ and

$$\Lambda^\mu_{\ \nu}\Lambda_\mu^{\ \sigma} = \delta^\sigma_\nu = (\delta^\mu_\nu + \omega^\mu_{\ \nu})(\delta^\sigma_\mu + \omega_\mu^{\ \sigma}) = \delta^\sigma_\nu + \omega^\sigma_{\ \nu} + \omega_\nu^{\ \sigma} + \mathcal{O}(\omega^2). \tag{B.26}$$

Thus the matrix of infinitesimal generators of Lorentz transformations is antisymmetric, $\omega^{\mu\nu} = -\omega^{\nu\mu}$, and has six independent elements. Under a Lorentz transformation, a set ϕ_a of fields with definite transformations properties is changed into $S_{ba}\phi_a$. For an infinitesimal transformation, the transformed $\tilde{\phi}_a$ depend linearly on $\omega^{\mu\nu}$ and ϕ_a,

$$\tilde{\phi}_a = S_{ba}\phi_a = \exp\left(-\frac{i}{2}\omega_{\mu\nu}(J^{\mu\nu})_{ab}\right)\phi_a \simeq \phi_a - \frac{i}{2}\omega_{\mu\nu}(J^{\mu\nu})_{ab}\phi_b. \tag{B.27}$$

The symmetric part of $(J^{\mu\nu})_{ab}$ does not contribute, because of the antisymmetry of $\omega^{\mu\nu}$. Hence we can choose also the $(J^{\mu\nu})_{ab}$ as antisymmetric and thus there exists six generators $(J^{\mu\nu})_{ab}$ corresponding to the three boosts and three rotations. The explicit form of the generators J depends on the spin of ϕ_a. In the case $s = 1$, we can use $\delta x^\rho = \omega_\nu^{\ \sigma} x^\sigma = -(i/2)\omega_{\mu\nu}(J^{\mu\nu})^\rho_{\ \sigma}x^\sigma$ together with $J^{\mu\nu} = -J^{\nu\mu}$ to find

$$(J^{\mu\nu})^\rho_{\ \sigma} = i(\eta^{\mu\rho}\delta^\nu_\sigma - \eta^{\nu\rho}\delta^\mu_\sigma). \tag{B.28}$$

Let us compare the resulting Lorentz transformations to the "standard" ones. We split $\omega^{\mu\nu}$ into rotations and boosts,

$$\frac{1}{2}\omega_{\mu\nu}J^{\mu\nu} = \omega_{i0}J^{i0} + \omega_{12}J^{12} + \omega_{13}J^{13} + \omega_{23}J^{23}, \tag{B.29}$$

and parameterise the rotation by $\alpha^i = (1/2)\varepsilon^{ijk}\omega^{jk}$ and the boost by $\eta^i = \omega^{i0}$. Choosing $\alpha^3 = \omega^{12} = -\omega^{21} > 0$ gives

$$\delta x^\mu = -i\alpha(J^{12})^\mu_{\ \nu}x^\nu = \alpha(\eta^{1\mu}\delta^2_\nu - \eta^{2\mu}\delta^1_\nu)x^\nu \tag{B.30}$$

or $\delta x = -\alpha y$ and $\delta y = +\alpha x$. This corresponds to a rotation counter-clockwise in the xy plane. Next we evaluate a boost with $\eta^1 = \omega^{10}$, obtaining

$$\delta x^\mu = i\eta(J^{10})^\mu_{\ \nu}x^\nu = -\eta(\eta^{1\mu}\delta^0_\nu - \eta^{0\mu}_{\ \nu}\delta^1_\nu)x^\nu \tag{B.31}$$

or $\delta t = \eta x$ and $\delta x = +\eta t$. Thus an infinitesimal Lorentz transformation with η^i applied to a particle at rest boosts its velocity to $v^i = \eta > 0$. Introducing finally

$$J^i = \frac{1}{2}\varepsilon^{ijk}J^{jk} \quad \text{and} \quad K^i = J^{i0} \tag{B.32}$$

we find

$$\frac{1}{2}\omega_{\mu\nu}J^{\mu\nu} = \boldsymbol{J}\boldsymbol{\alpha} - \boldsymbol{K}\boldsymbol{\eta}, \tag{B.33}$$

which are the generators we used in section 8.1.

We have several ways at our disposal to construct the Lie algebra of the Lorentz group. First, we can use that the generators T^a of any Lie group can be obtained by differentiating the finite transformations, $T^a = -\mathrm{i}\, \mathrm{d}g(\vartheta)/\mathrm{d}\vartheta^a|_{\vartheta=0}$, in an arbitrary representation. Applied to the finite boost $B_x(\eta)$ along the x direction given in (B.19) we find as generator K_x

$$
B_x(\eta) = \begin{pmatrix} \cosh\eta & \sinh\eta & 0 & 0 \\ \sinh\eta & \cosh\eta & 0 & 0 \\ 0 & 0 & 1 & 0 \\ 0 & 0 & 0 & 1 \end{pmatrix}, \qquad K_x = \frac{1}{\mathrm{i}} \left.\frac{\partial B_x(\eta)}{\partial\eta}\right|_{\eta=0} = -\mathrm{i} \begin{pmatrix} 0 & 1 & 0 & 0 \\ 1 & 0 & 0 & 0 \\ 0 & 0 & 0 & 0 \\ 0 & 0 & 0 & 0 \end{pmatrix} \qquad \text{(B.34)}
$$

and similarly for the other two boosts and the three rotations. Second, we can read off the generators directly from the infinitesimal transformations. For instance, rewriting (B.31) in matrix form as $\tilde{x}^\mu = x^\mu + \mathrm{i}\eta (K_x)^\mu_{\ \nu} x^\nu$ reproduces again the boost generator. Third, we can view the generators as differential operators, as we are used from quantum mechanics. Consider, for example, the effect of an infinitesimal rotation around the z axis on a function f,

$$
f(x^\mu) \to f(\tilde{x}^\mu) = f(t, x + \alpha y, y - \alpha x, z) = f + \alpha \left(x\frac{\partial}{\partial y} - y\frac{\partial}{\partial x} \right) f = f + \alpha \mathbf{J}_z f. \qquad \text{(B.35)}
$$

Thus we can view the generator $\mathbf{J}_z = x\partial_y - y\partial_x$ both as a differential operator or as a vector field. Finally, we can use that the Killing vector fields $\mathbf{V}$ of Minkowski space generate its spacetime symmetries. Therefore the generators of the Poincaré group agree with these Killing vector fields which were derived in example 6.2. Having obtained the generators either in matrix form, as differential operators or vector fields, we can calculate their Lie brackets finding the relations given in Eq. (8.1). Expressed by the four-dimensional generators $J^{\mu\nu}$, these commutation relations are

$$
[J^{\mu\nu}, J^{\rho\sigma}] = \mathrm{i} \left(\eta^{\nu\rho} J^{\mu\sigma} - \eta^{\mu\rho} J^{\nu\sigma} - \eta^{\nu\sigma} J^{\mu\rho} + \eta^{\mu\sigma} J^{\nu\rho} \right). \qquad \text{(B.36)}
$$

Remark B.1: Let us compare the Lie groups SO(2) and SO(1,1). Their generators and finite group elements are given by

$$
J = -\mathrm{i} \begin{pmatrix} 0 & 1 \\ -1 & 0 \end{pmatrix} \qquad \text{and} \qquad R(\alpha) = \mathrm{e}^{\mathrm{i}\alpha J} = \begin{pmatrix} \cos\alpha & \sin\alpha \\ -\sin\alpha & \cos\alpha \end{pmatrix},
$$

$$
K = -\mathrm{i} \begin{pmatrix} 0 & 1 \\ 1 & 0 \end{pmatrix} \qquad \text{and} \qquad B(\eta) = \mathrm{e}^{\mathrm{i}\eta K} = \begin{pmatrix} \cosh\eta & \sinh\eta \\ \sinh\eta & \cosh\eta \end{pmatrix},
$$

respectively. First, we note that the parameter α is periodic, $-\pi \le \alpha \le \pi$, and can be identified with S^1, while the boost parameter η is mapped into $\eta \in \mathbb{R}$. Therefore SO(2) and SO(1,1) are examples for compact and non-compact Lie groups, respectively. Next, we calculate $\mathrm{tr}(JJ) = 2$ and $\mathrm{tr}(KK) = -2$. The result that $\mathrm{tr}(T^a T^a)$ is positive definite only for compact Lie groups holds in general. Therefore we have to use compact Lie groups in $\mathscr{L}_{YM} = -F^a_{\mu\nu} F^{a\mu\nu}/4$ such that all $a = 1, \ldots, n$ terms contribute with the same sign and thus the energy is bounded from below.

Representations for fields.   We consider now the transformation properties of fields, that is, functions defined on Minkowski space. Adding to Eq. (B.27) the x dependence gives

$$
\tilde{\phi}_a(\tilde{x}) = S_{ba}\phi_a(\tilde{x}) = \exp\left(-\frac{\mathrm{i}}{2}\omega_{\mu\nu}(J^{\mu\nu})_{ab} \right) \phi_a(x) \simeq \phi_a(x) - \frac{\mathrm{i}}{2}\omega_{\mu\nu}(J^{\mu\nu})_{ab}\phi_b(x), \qquad \text{(B.37)}
$$

where a represents a set of tensor and spinor indices. The infinitesimal change $\delta\phi_a$ can be split into two terms,

$$\delta\phi_a = \tilde{\phi}_a(\tilde{x}) - \phi_a(x) = \tilde{\phi}_a(x) - \phi_a(x) + \partial_\mu\tilde{\phi}_a(x)\delta x^\mu \equiv \delta_0\phi_a(x) + \partial_\mu\tilde{\phi}_a(x)\delta x^\mu, \qquad \text{(B.38)}$$

where we have introduced the total variation $\delta_0\phi_a$ and the transport term $\partial_\mu\tilde{\phi}_a(x)\delta x^\mu$. Note that the local variation $\delta\phi_a(x)$ compares fields at the same points, which are labelled differently in the two set of coordinates, while the total variation $\delta_0\phi_a$ compares fields at different base points. In the first case, the variation $\delta\phi_a$ with $a = 1,\ldots,n$ has n degrees of freedom and we obtain n-dimensional non-unitary representations of the Lorentz group. This is the case we considered in the main text. In contrast, the total variation $\delta_0\phi_a$ compares fields at different base points $x \in \mathbb{R}(1,3)$. Since the number of base points is infinite, we have to study the transformation properties of an infinite dimensional space of functions, as, for example, the space of square-integrable functions $L^2(\mathbb{R}(1,3))$ on Minkowski space. This allows us to obtain unitary representations of the Lorentz group. For instance, for a scalar field ϕ the local variation $\delta\phi(x)$ vanishes and thus the total variation is

$$\delta_0\phi(x) = -\partial_\mu\tilde{\phi}(x)\delta x^\mu. \qquad \text{(B.39)}$$

Inserting $\delta x^\rho = \omega_\nu{}^\sigma x^\sigma = -(\mathrm{i}/2)\omega_{\mu\nu}(J^{\mu\nu})^\rho{}_\sigma x^\sigma$ we find

$$\delta_0\phi(x) = \frac{\mathrm{i}}{2}\omega_{\mu\nu}(J^{\mu\nu})^\rho{}_\sigma x^\sigma\partial_\rho\phi(x) = -\frac{\mathrm{i}}{2}\omega_{\mu\nu}L^{\mu\nu}\phi(x), \qquad \text{(B.40)}$$

where we have introduced also the orbital angular momentum operator

$$L^{\mu\nu} = -(J^{\mu\nu})^\rho{}_\sigma x^\sigma\partial_\rho = \mathrm{i}(x^\mu\partial^\nu - x^\nu\partial^\mu). \qquad \text{(B.41)}$$

As expected for a scalar field, the total angular momentum $J^{\mu\nu}$ is identical with the orbital angular momentum $L^{\mu\nu}$. Using as scalar product Eq. (9.25), the operator $\exp(-\mathrm{i}\omega_{\mu\nu}L^{\mu\nu}/2)$ is unitary.

We consider next the case of a left-chiral Weyl field ϕ_L, where the variation is given by

$$\delta_0\phi_L(x) = \delta\phi_L(x) - \partial_\mu\tilde{\phi}_a(x)\delta x^\mu = (S_L - 1)\phi_L(x) - \partial_\mu\phi_a(x)\delta x^\mu. \qquad \text{(B.42)}$$

The second term gives again $\partial_\mu\phi_L\delta x^\mu = \frac{1}{2}\omega_{\mu\nu}L^{\mu\nu}\phi_L$. Since the angular momentum $L^{\mu\nu}$ part is connected to the transport term $\partial_\mu\tilde{\phi}_a(x)\delta x^\mu$, it is universal. Thus only the spin part acting on the tensor or spinor indices of a given field varies. Setting $S_L = \exp(-\mathrm{i}\omega_{\mu\nu}S^{\mu\nu}/2)$, we can write the variation as

$$\delta_0\phi_L(x) = -\frac{\mathrm{i}}{2}\omega_{\mu\nu}(S^{\mu\nu} + L^{\mu\nu})\phi_L(x) = -\frac{\mathrm{i}}{2}\omega_{\mu\nu}J^{\mu\nu}\phi_L(x). \qquad \text{(B.43)}$$

Comparing with Eq. (8.5a), we can identify

$$J^i = S^i = \frac{1}{2}\varepsilon^{ijk}S^{jk} = \sigma^i/2 \quad\text{and}\quad K^i = S^{i0} = \mathrm{i}\sigma^i/2. \qquad \text{(B.44)}$$

For a right-chiral field $\phi_R(x)$, we find in the same way $J^i = S^i = \frac{1}{2}\varepsilon^{ijk}S^{jk} = \sigma^i/2$ and $K^i = S^{i0} = \mathrm{i}\sigma^i/2$.

B.4   Remarks on the Poincaré and conformal group

Poincaré group.   An isolated system is invariant under uniform translations in spacetime, $x^\mu \to \tilde{x}^\mu = x^\mu + a^\mu$. Adding translation to the Lorentz group SO(1,3), we obtain the complete symmetry group of massive particles in Minkowski space, the Poincaré (or inhomogeneous Lorentz) group ISO(1,3) with group elements (Λ, a). Performing two successive transformations results in

$$x^{\mu} \to \tilde{\tilde{x}} = \Lambda_2 \tilde{x} + a_2 = \Lambda_2 \Lambda_1 x + \Lambda_2 a_1 + a_1. \tag{B.45}$$

The translation a_1 is rotated into Λa_1, since it is a four-vector. Thus group multiplication $(\Lambda_2, a_2) \circ (\Lambda_1, a_1) = (\Lambda_2 \Lambda_1, \Lambda_2 a_1 + a_1)$ corresponds to a semi-direct product. Considering an infinitesimal translation, $\delta x^{\mu} = \varepsilon \xi^{\mu} = \varepsilon \xi^{\nu} \partial_{\nu} x^{\mu}$, we see that translations are generated by the momentum operator $P_{\mu} = i\partial_{\mu}$. Hence finite translations are generated by $\exp(a^{\mu} P_{\mu})$.

The determination of the Lie algebra is straightforward. The additional commutators

$$[P_{\mu}, P_{\nu}] = 0, \tag{B.46a}$$

$$[J_i, P_j] = i\varepsilon_{ijk} P_k \quad \text{and} \quad [J_i, P_0] = 0, \tag{B.46b}$$

show that translations commute, that the energy P_0 transforms as a scalar and the three-momentum P_i as a vector under rotations. Expressed by four-dimensional quantities, the commutators are given by

$$[P^{\mu}, J^{\rho\sigma}] = i\left(\eta^{\mu\rho} P^{\sigma} - \eta^{\mu\sigma} P^{\rho}\right). \tag{B.47}$$

Next we have to find the Casimir operators whose eigenvalues we can use to label its representations. Recalling that such operators have to commute with all generators of the group, we know that they have to be Lorentz scalars in the case of the Poincaré group. The first, rather obvious, Casimir invariant is the momentum squared, or

$$C_1 = P_{\mu} P^{\mu} = m^2. \tag{B.48}$$

A second Casimir invariant should be connected to the spin of the particle, and we try therefore as ansatz $C_2 = W_{\mu} W^{\mu}$ with the Pauli–Lubanski spin vector W_{μ} defined in Eq. (5.27). In the rest-frame of a massive particle, $W^i = mJ^i$, and

$$C_2 = W_{\mu} W^{\mu} = -m^2 s(s+1) \text{ for } m > 0. \tag{B.49}$$

Direct calculation shows that P^2 and W^2 commute with all generators as required. Thus the different representations of the Poincaré group for massive particles can be labelled by m and s. The states of free, massive particles within a single representation can be characterised by their momentum p^{μ} and a component of the spin vector W_i.

Finally, we comment briefly on the case of massless particles. Then the two Casimir operators P^2 and W^2 have zero eigenvalues and thus both P^{μ} and W^{μ} are light-like vectors. Moreover, the definition of W^{μ} implies that they are orthogonal, $W_{\mu} P^{\mu} = 0$. Two light-like orthogonal vectors have to be parallel,

$$W^{\mu} = \lambda P^{\mu}. \tag{B.50}$$

Now we compare the $\mu = 0$ component of this equation, $W^0 = \lambda P^0$, with $W^0 = \boldsymbol{P} \cdot \boldsymbol{J}$. Using then also the definition of helicity, $h = \hat{P} \cdot \boldsymbol{J} = \boldsymbol{P} \cdot \boldsymbol{J}/P^0$, we see that the proportionality factor λ corresponds to the helicity of the massless particle,

$$\lambda = h = \frac{W^0}{P^0}. \tag{B.51}$$

Hence, independently of the value of s, there is only one value of h per massless particle type allowed. If we consider additionally the effect of a parity transformation then we obtain a second state with $h = -\lambda = -W^0/P^0$. For particles whose interactions respect parity like the photon and graviton, it is natural to identify the two states as the two helicity states of a single particle. The proof that the allowed helicity values are only $h = \pm 1/2, \pm 1, \dots$ is based on the topological properties of the Poincaré group. It is technically more involved and can be found, for example, in Weinberg (2005).

Conformal group.   We obtain the conformal group C(1,3) as the maximal symmetry group of Minkowski space, if we admit coordinate transformations which keep only the light-cone structure but not distances invariant. A straightforward way to obtain the Lie algebra of $C(1, d - 1)$ is to determine the conformal Killing vector fields of Minkowski space, cf. problem 6.4. These fields are the solution of the conformal Killing equation which simplifies in Minkowski space to

$$\partial_\mu \xi_\nu + \partial_\nu \xi_\mu = \kappa \eta_{\mu\nu}. \tag{B.52}$$

Taking the trace $\kappa = 2\partial_\mu \xi^\mu / d$ follows. Inserted into the Killing equation, we obtain

$$\partial_\mu \xi_\nu + \partial_\nu \xi_\mu = \frac{2}{d} \partial_\rho \xi^\rho \, \eta_{\mu\nu} = \kappa \, \eta_{\mu\nu}. \tag{B.53}$$

Next we take one more derivative ∂_ρ, exchange then indices $\rho \leftrightarrow \mu$ and subtract the two equations, arriving at

$$\partial_\rho(\partial_\mu \xi_\nu - \partial_\nu \xi_\mu) = \partial_\mu \kappa \eta_{\nu\rho} - \partial_\nu \kappa \eta_{\mu\rho}. \tag{B.54}$$

Integrating this equation, we find

$$\partial_\mu \xi_\nu - \partial_\nu \xi_\mu = \int (\partial_\mu \kappa \mathrm{d}x_\nu - \partial_\nu \kappa \mathrm{d}x_\mu) + 2\omega_{\mu\nu}, \tag{B.55}$$

where $\omega_{\mu\nu}$ is an antiymmetric tensor containing the integration constant. Now we add to this the Killing eq. (B.52). Integrating then again, we obtain

$$\xi^\mu = a^\mu + \omega_\nu{}^\mu x^\nu + \alpha x^\mu \kappa + \frac{1}{2} \int \mathrm{d}x^\nu \int (\partial_\nu \kappa \mathrm{d}x^\mu - \partial_\mu \kappa \mathrm{d}x^\nu). \tag{B.56}$$

Finally, we have to determine the function κ. Acting with ∂^μ on Eq. (B.53) we find

$$d \, \Box \xi_\nu = (2 - d) \partial_\nu \partial_\mu \xi^\mu. \tag{B.57}$$

Hence in $d = 2$ any harmonic function (i.e. a function satisfying $\Box \xi_\nu = 0$) determines a conformal Killing vector field and the conformal group $C(1, 1)$ is thus infinite dimensional. In contrast, for $d > 2$ the condition $\partial_\nu \partial_\mu \xi^\mu = 0$ follows. Hence the function κ can be at most linear in the coordinates, and we choose it as

$$\kappa = -2\alpha + 4\beta_\mu x^\mu. \tag{B.58}$$

Inserting κ into Eq. (B.55), we find the conformal Killing vector fields as

$$\xi^\mu = a^\mu + \omega_\nu{}^\mu x^\nu + \alpha x^\mu + \beta_\nu(2x^\mu x^\nu - \eta^{\mu\nu} x^2). \tag{B.59}$$

They depend on $(d+1)(d+2)/2$ parameters: d translations, $d(d-1)/2$ Lorentz transformations, one dilatation and d special conformal transformations. The Lie algebra of the conformal group can be derived now in a straightforward way and turns out to be the same as the one of $SO(d - 2)$, for details cf. Zee (2013).

References

Abbott, B.P. et al. (2016). Observation of Gravitational Waves from a Binary Black Hole Merger. *Phys. Rev. Lett.*, **116**, 061102. arXiv:1602.03837.

Allahverdi, R., Brandenberger, R., Cyr-Racine, F.-Y., and Mazumdar, A. (2010). Reheating in Inflationary Cosmology: Theory and Applications. *Ann. Rev. Nucl. Part. Sci.*, **60**, 27–51. arXiv:1001.2600.

Almheiri, A., Marolf, D., Polchinski, J., and Sully, J. (2013). Black Holes: Complementarity or Firewalls? *JHEP*, **02**, 062. arXiv:1207.3123.

Amaldi, U., de Boer, W., and Fürstenau, H. (1991). Comparison of Grand Unified Theories with Electroweak and Strong Coupling Constants Measured at LEP. *Phys. Lett.*, **B260**, 447–455.

Amendola, L., Gannouji, R., Polarski, D., and Tsujikawa, S. (2007). Conditions for the Cosmological Viability of $f(R)$ Dark Energy Models. *Phys. Rev.*, **D75**, 083504. arXiv:gr-qc/0612180.

Anisimov, A., Buchmüller, W., Drewes, M., and Mendizabal, S. (2011). Quantum Leptogenesis I. *Ann. Phys.*, **326**, 1998–2038. [Erratum: Ann. Phys. 338, 376 (2011)]. arXiv:1012.5821.

Appelquist, T. and Carazzone, J. (1975). Infrared Singularities and Massive Fields. *Phys. Rev.*, **D11**, 2856.

Ask, S. et al. (2012). From Lagrangians to Events: Computer Tutorial at the MC4BSM-2012 Workshop. arXiv:1209.0297.

Bailin, D. and Love, A. (1993). *Introduction to Gauge Field Theory (Graduate Student Series in Physics)* (Revised edn). Institute of Physics Publishing, Bristol.

Bailin, D. and Love, A. (2004). *Cosmology in Gauge Field Theory and String Theory (Graduate Student Series in Physics)*. Institute of Physics Publishing, Bristol.

Birrell, N.D. and Davies, P.C.W. (1982). *Quantum Fields in Curved Space*. Cambridge University Press, Cambridge.

Bjerrum-Bohr, N.E.J, Donoghue, J.F., and Holstein, B.R. (2003). Quantum Gravitational Corrections to the Nonrelativistic Scattering Potential of two Masses. *Phys. Rev.*, **D67**, 084033. [Erratum: Phys. Rev. D71, 069903 (2005)]. arXiv:hep-th/0211072.

Blaizot, J.-P. (2011). Quantum Fields at Finite Temperature from Tera to nano Kelvin. arXiv:1108.3482.

Bronstein, M. (1936). Kvantovanie Gravitatsionnykh Voln. *Zh. Eksp. Tear. Fiz.*, **6**, 195–236.

Buchmüller, W., Di Bari, P., and Plümacher, M. (2005). Leptogenesis for Pedestrians. *Annals Phys.*, **315**, 305–351. arXiv:hep-ph/0401240.

Callan, C.G., Jr. and Coleman, S.R. (1977). The Fate of the False Vacuum. 2. First Quantum Corrections. *Phys.Rev.*, **D16**, 1762–1768.

Carroll, S. (2003). *Spacetime and Geometry: An Introduction to General Relativity.* Addison-Wesley. A draft version is available as arXiv:gr-qc/9712019.

Challinor, A. (2006). Constraining fundamental physics with the cosmic microwave background. In *International Scientific Workshop on Cosmology and Gravitational Physics Thessaloniki, Greece, December 15-16, 2005.* arXiv:astro-ph/0606548.

Cheng, T.-P. and Li, L.-F. (1988). *Gauge Theory of Elementary Particle Physics.* Oxford University Press, New York, NY.

Coleman, S. (1988). *Aspects of Symmetry: Selected Erice Lectures.* Cambridge University Press, Cambridge.

Coleman Miller, M. (2016). Implications of the Gravitational Wave Event GW150914. *Gen. Rel. Grav.*, **48**, 95. arXiv:1606.06526.

Coles, P. and Lucchin, F. (2002). *Cosmology* (2nd edn). Wiley, Chichester.

Das, A. (2006). *Field Theory: A Path Integral Approach (World Scientific Lecture Notes in Physics)* (2nd edn). World Scientific, Singapore.

Degrassi, G., Di Vita, S., Elias-Miro, J., Espinosa, J.R., Giudice, G.F., Isidori, G., and Strumia, A. (2012). Higgs Mass and Vacuum Stability in the Standard Model at NNLO. *JHEP*, **08**, 098. arXiv:1205.6497.

Delamotte, B. (2004). A Hint of renormalization. *Am. J. Phys.*, **72**, 170–184. arXiv:hep-th/0212049.

Dine, M. (2016). *Supersymmetry and String Theory: Beyond the Standard Model* (2nd edn). Cambridge University Press.

Dissertori, G., Knowles, I.G., and Schmelling, M. (2009). *Quantum Chromodynamics: High Energy Experiments and Theory.* Oxford University Press, Oxford.

Dodelson, S. (2003). *Modern Cosmology.* Academic Press, San Diego.

Donoghue, J.F. (2012). The Effective Field Theory Treatment of Quantum Gravity. *AIP Conf. Proc.*, **1483**, 73–94. arXiv:1209.3511.

Donoghue, J.F., Golowich, E., and Holstein, Barry R. (2014). *Dynamics of the Standard Model* (2 edn). Cambridge University Press, Cambridge.

Ellis, R.K., Stirling, W.J., and Webber, B.R. (2003). *QCD and Collider Physics.* Cambridge University Press, Cambridge.

Espinosa, J.R. (2014). Vacuum Stability and the Higgs Boson. *PoS*, **LATTICE2013**, 010. arXiv:1311.1970.

Flory, M., Helling, R.C., and Sluka, C. (2012). How I Learned to Stop Worrying and Love QFT. arXiv:1201.2714.

Gorbunov, D.S. and Rubakov, V.A. (2011*a*). *Introduction to the Theory of the Early Universe: Cosmological Perturbations and Inflationary Theory.* World Scientific, Singapore.

Gorbunov, D.S. and Rubakov, V.A. (2011*b*). *Introduction to the Theory of the Early Universe: Hot Big Bang Theory.* World Scientific, Singapore.

Greiner, W. (2000). *Relativistic Quantum Mechanics. Wave Equations* (3rd edn). Springer, Berlin.

Greiner, W. and Reinhardt, J. (2008). *Field Quantization.* Springer, Berlin.

Griest, K. and Kamionkowski, M. (1990). Unitarity Limits on the Mass and Radius of Dark Matter Particles. *Phys. Rev. Lett.*, **64**, 615.

Haber, H.E. (1994). Spin Formalism and Applications to New Physics Searches.

arXiv:hep-ph/9405376.

Hill, E.L. (1951). Hamilton's Principle and the Conservation Theorems of Mathematical Physics. *Rev. Mod. Phys.*, **23**, 253–260.

Hisano, J., Nagai, R., and Nagata, N. (2015). Effective Theories for Dark Matter Nucleon Scattering. *JHEP*, **05**, 037. arXiv:1502.02244.

Hobson, M.P., Efstathiou, G.P., and Lasenby, A.N. (2006). *General Relativity: An Introduction for Physicists*. Cambridge University Press, Cambridge.

Hui, L. (2001). Unitarity Bounds and the Cuspy Halo Problem. *Phys. Rev. Lett.*, **86**, 3467–3470. arXiv:astro-ph/0102349.

Jackson, J.D. and Okun, L.B. (2001). Historical Roots of Gauge Invariance. *Rev. Mod. Phys.*, **73**, 663–680. arXiv:hep-ph/0012061.

Jegerlehner, F. (2007). Essentials of the Muon g-2. *Acta Phys. Polon.*, **B38**, 3021. arXiv:hep-ph/0703125.

Joyce, A., Lombriser, L., and Schmidt, F. (2016). Dark Energy vs. Modified Gravity. *Ann. Rev. Nucl. Part. Sci.*, **66**, 95–122. arXiv:1601.06133.

Kapusta, J.I. and Gale, Ch. (2011). *Finite-Temperature Field Theory: Principles and Applications* (2 edn). Cambridge University Press, Cambridge.

Kawasaki, M. and Nakayama, K. (2013). Axions: Theory and Cosmological Role. *Ann. Rev. Nucl. Part. Sci.*, **63**, 69–95. arXiv:1301.1123.

Kazakov, D.I. (2000). Beyond the Standard Model: In Search of Supersymmetry. In *2000 European School of High-Energy Physics, Caramulo, Portugal, 20 Aug-2 Sep 2000: Proceedings*, pp. 125–199. arXiv:hep-ph/0012288.

Kolb, E.W. and Turner, M.S. (1994). *The Early Universe (Frontiers in Physics 69)*. Addison-Wesley, Redwood City.

Kragh, H. (2013). *Conceptions of Cosmos: From Myths to the Accelerating Universe* (Reprint edn). Oxford University Press, Oxford.

Laine, M. and Vuorinen, A. (2016). *Basics of Thermal Field Theory (Lecture Notes in Physics 925)*. Springer, Heidelberg. arXiv:1701.01554.

Landau, L.D. and Lifshitz, E.M. (1976). *Mechanics (Course of Theoretical Physics, Volume 1)* (3 edn). Butterworth-Heinemann, Oxford.

Landau, L.D. and Lifshitz, E.M. (1980). *The Classical Theory of Fields (Course of Theoretical Physics Series, Volume 2)* (4 edn). Butterworth-Heinemann, Oxford.

Landau, L.D. and Lifshitz, L.M. (1981). *Quantum Mechanics (Course of Theoretical Physics Series, Volume 3)* (3 edn). Butterworth-Heinemann, Oxford.

Le Bellac, M. (1992). *Quantum and Statistical Field Theory (Oxford Science Publications)*. Clarendon Press, Oxford.

Leader, E. and Predazzi, E. (2013). A Note on the Implications of Gauge Invariance in QCD. *J. Phys.*, **G40**, 075001. arXiv:1101.3425.

Lisanti, M. (2016). Lectures on Dark Matter Physics. In *Theoretical Advanced Study Institute in Elementary Particle Physics: New Frontiers in Fields and Strings (TASI 2015) Boulder, CO, USA, June 1-26, 2015*. arXiv:1603.03797.

Ludvigsen, M. (1999). *General Relativity: A Geometric Approach*. Cambridge University Press, Cambridge.

Lyth, D.H. and Liddle, R. (2009). *The Primordial Density Perturbation: Cosmology, Inflation and the Origin of Structure*. Cambridge University Press, Cambridge.

MacKenzie, R. (2000). Path Integral Methods and Applications – Lectures given at Rencontres du Vietnam: VIth Vietnam School of Physics. arXiv:quant-ph/0004090.

Maggiore, M. (2007). *Gravitational Waves: Volume 1: Theory and Experiments*. Oxford University Press, USA.

Martin, J. (2012). Everything You always Wanted to Know about the Cosmological Constant Problem (But Were Afraid to Ask). *Comptes Rendus Physique*, **13**, 566–665. arXiv:1205.3365.

Mathur, S.D. (2009). What Exactly is the Information Paradox? *Lect. Notes Phys.*, **769**, 3–48. arXiv:0803.2030.

Mukhanov, V. (2005). *Physical Foundations of Cosmology*. Cambridge University Press, Cambridge.

Mukhanov, V. and Winitzki, S. (2007). *Introduction to Quantum Effects in Gravity*. Cambridge University Press, Cambridge.

Nachtmann, O. (1990). *Elementary Particle Physics: Concepts and Phenomena*. Springer, Heidelberg.

Olive, K.A. et al. (2014). Review of Particle Physics. *Chin. Phys.*, **C38**, 090001. For updates see the online version at `http://pdg.lbl.gov`.

O'Raifeartaigh, L. (1997). *The Dawning of Gauge Theory*. Princeton University Press, Princeton.

Padmanabhan, T. (2016). Exploring the Nature of Gravity. In *The Planck Scale II Wroclaw, Poland, September 7-12, 2015*. arXiv:1602.01474.

Parker, L. and Toms, D. (2009). *Quantum Field Theory in Curved Spacetime: Quantized Fields and Gravity*. Cambridge University Press, Cambridge.

Peebles, P.J.E., Page, L.A., and Partridge, R.B. (ed.) (2009). *Finding the Big Bang*. Cambridge University Press, Cambridge.

Peskin, M.E. (2011). Simplifying Multi-Jet QCD Computation. arXiv:1101.2414.

Poisson, E. (2007). *A Relativist's Toolkit: The Mathematics of Black-Hole Mechanics*. Cambridge University Press, Cambridge.

Pokorski, S. (1987). *Gauge Field Theories*. Cambridge University Press, Cambridge.

Quigg, C. (2015). Electroweak Symmetry Breaking in Historical Perspective. *Ann. Rev. Nucl. Part. Sci.*, **65**, 25–42. arXiv:1503.01756.

Ramond, P. (1994). *Field Theory: A Modern Primer (Frontiers in Physics)* (2 Revised edn). Addison-Wesley, Redwood City.

Romao, J.C. and Silva, J.P. (2012). A Resource for Signs and Feynman Diagrams of the Standard Model. *Int. J. Mod. Phys.*, **A27**, 1230025. arXiv:1209.6213.

Rubakov, V.A. (2002). *Classical Theory of Gauge Fields*. Princeton University Press.

Schutz, B. (2009). *A First Course in General Relativity* (2nd edn). Cambridge University Press, Cambridge.

Schwartz, M.D. (2013). *Quantum Field Theory and the Standard Model*. Cambridge University Press, Cambridge.

Schweber, S.S. (2005). The Sources of Schwinger's Green's Functions. *Proc. Nat. Acad. Sci.*, **102**, 7783–7788.

Schweber, S. S. (1994). *QED and the Men Who Made It*. Princeton University Press. Princeton.

Srednicki, M. (2007). *Quantum Field Theory*. Cambridge University Press,

Cambridge. A draft version is available at `http://web.physics.ucsb.edu/~mark/qft.html`.

Stelle, K. S. (1978). Classical Gravity with Higher Derivatives. *Gen. Rel. Grav.*, **9**, 353–371.

Sterman, G. (1993). *An Introduction to Quantum Field Theory.* Cambridge University Press, Cambridge.

Valle, J.W.F. and Romao, J. (2015). *Neutrinos in High Energy and Astroparticle Physics.* Wiley VCH, Weinheim.

Weinberg, S. (1965). Infrared Photons and Gravitons. *Phys.Rev.*, **B140**, 516–524.

Weinberg, S. (2005). *The Quantum Theory of Fields, Volume 1: Foundations.* Cambridge University Press, Cambridge.

Weinzierl, S. (2016). Tales of 1001 Gluons. arXiv:1610.05318.

White, C.D. (2015). An Introduction to Webs. arXiv:1507.02167.

Zee, A. (2013). *Einstein Gravity in a Nutshell.* Princeton University Press, Princeton, NJ.

Index